NIVERSITY OF
ESTER

D0302893

WITHDRAWN FROM
THE LIBRARY
UNIVERSITY OF
WINCHESTER

KA 0443027 1

Modern Statistics for the Social and Behavioral Sciences

A Practical Introduction

SECOND EDITION

Modern Statistics for the Social and Behavioral Sciences

A Practical Introduction

SECOND EDITION

Rand Wilcox
University of Southern California
Los Angeles, USA

CRC Press is an imprint of the
Taylor & Francis Group, an **informa** business
A CHAPMAN & HALL BOOK

CRC Press
Taylor & Francis Group
6000 Broken Sound Parkway NW, Suite 300
Boca Raton, FL 33487-2742

Printed on acid-free paper

International Standard Book Number-13: 978-1-4987-9678-1 (Hardback)

Visit the Taylor & Francis Web site at
http://www.taylorandfrancis.com

and the CRC Press Web site at
http://www.crcpress.com

CONTENTS

PREFACE to the SECOND EDITION

This second edition contains major changes and additions. All of the chapters have been updated; for some there are only minor changes, but for others there are major changes. Many new R functions have been added that reflect recent advances. Briefly, they deal with new methods for comparing groups and studying associations. For example, it is now possible to compare a collection of quantiles in a manner that controls the probability of one or more Type I errors, even when there are tied values. This new approach can provide a deeper understanding of how groups compare in contrast to methods that focus on a single measure of location. Some new multiple comparison procedures are covered as well as some new techniques related to testing global hypotheses. New methods for comparing regression lines are covered and several new and improved methods for dealing with ANCOVA are described. There is even a new method for determining which independent variables are most important in a regression model. There are many ways of estimating which variables are most important. What is new here is the ability to determine the strength of the empirical evidence that the most important variables are correctly identified.

Details about basic bootstrap methods now appear when sampling distributions are introduced. This is in contrast to the first edition where bootstrap methods were introduced in a separate chapter. Introducing bootstrap methods in conjunction with sampling distributions seems to help some students understand sampling distributions.

A solution manual, which contains a detailed solution to all of the exercises, is now available on the author's web site:

Dornsife.usc.edu/cf/labs/wilcox/wilcox-faculty-display.cfm.

Perhaps an easier way of reaching the author's web page is to search for Software Rand Wilcox. Clicking on this link will take you to the author's web page. Go to the section labeled books. The file CRC_answers.pdf the contains solution manual. Most of the data used in this book are available on the author's web page and are stored in the section labeled datasets.

PREFACE

There are two general goals in this book. Assuming the reader has had no prior training in statistics, the first is to provide a graduate-level introduction to basic, routinely used statistical techniques relevant to the social and behavioral sciences. As can be seen from the table of contents, a wide range of topics is covered. Included is a nontechnical description of three major insights regarding when routinely used methods perform well and when they are highly unsatisfactory. One of these insights has to do with situations where data are sampled from what is called a heavy-tailed distribution. Such distributions are routinely encountered. During the last half century, hundreds of published papers describe practical concerns and how they might be addressed. The second general insight has to do with the effects of skewness. A deeper understanding of how to deal with skewed distributions and why skewed distributions can wreak havoc on any inferential method based in part on a least squares estimator, have been developed. The third insight has to with heteroscedasticity. In practical terms, routinely taught and used methods can result in relatively low power and highly misleading results, contrary to what was once believed. Explanations are provided about why it was once thought that classic methods are robust to violations of assumptions and why it is now known that under general conditions, this is not the case.

The second general goal is to describe and illustrate methods developed during the last half century that deal with known problems associated with classic techniques. There is increasing awareness among nonstatisticians that these newer methods can make a considerable difference in the conclusions reached when analyzing data. And numerous illustrations are provided that support this view. An important point is that no single method is always best. So what is needed is a general understanding of the relative merits of various techniques so that the choice of method can be made in an informed manner.

As for software, this book focuses on R, which is introduced in Chapter 1. In terms of taking advantage of modern statistical techniques, R clearly dominates. When analyzing data, it is undoubtedly the most important software development during the last quarter of a century. And it is free. Although classic methods have fundamental flaws, it is not suggested that they be completely abandoned. And it is important to know how to apply and interpret these methods simply because they are still routinely used. Consequently, illustrations are provided on how to apply standard methods with R. Of particular importance here is that a vast array of modern methods can be applied via an R package that contains over 1300 R functions, many of which are illustrated in this book.

Rand R. Wilcox
Los Angeles, CA

CHAPTER 1

INTRODUCTION

CONTENTS

Statistical methods that are used by a wide range of disciplines consist of at least three basic components:

- Experimental design, meaning the planning and carrying out of a study.
- Summarizing data, using what are called descriptive statistics.
- Inferential techniques, which roughly are methods aimed at making predictions or generalizations about a population of individuals or things when not all individuals or things can be measured.

The fundamental goal in this book is to summarize the basic statistical techniques associated with these three components, with an emphasis on the latter two components, in a manner that makes them accessible to students not majoring in statistics. Of particular importance is fostering the ability of the reader to think critically about how data are summarized and analyzed.

The mathematical foundation of the statistical tools routinely used today was developed about two centuries ago by Pierre-Simon Laplace and Carl Friedrich Gauss in a series of remarkable advances. About a century ago, important refinements and extensions were made by Karl Pearson, Jerzy Neyman, Egon Pearson, William Gosset, and Sir Ronald Fisher. The strategies and methods that they developed are routinely used today.

During the last half century, however, literally hundreds of journal articles have made it abundantly clear that there are three basic concerns associated with these routinely used techniques that are of fundamental importance. This is not to say that they should be abandoned, but it is important to understand their limitations as well as how these limitations might be addressed with methods developed during the last half century. It is evident that any routinely used statistical method that addresses basic issues needs to be covered in any introductory statistics book aimed at students and researchers trying to understand their data. Simultaneously, it seems equally evident that when relevant insights are made regarding the proper use and interpretation of these methods, they should be included in an introductory book as well. Omission of some modern insights might be acceptable if the results were at some level controversial among statisticians familiar with the underlying principles. But when there are hundreds of papers acknowledging a problem with a routinely used method, with no counterarguments being offered in a reputable statistics journal, surely it is important to discuss the practical implications of the insight in a book aimed at non-statisticians. This is the point of view adopted here.

1.1 SAMPLES VERSUS POPULATIONS

Assuming the reader has no prior training in statistics, we begin by making a distinction between a population of individuals of interest and a sample of individuals. A *population* of participants or objects consists of all those participants or objects that are relevant in a particular study.

> *Definition*: A *sample* is any subset of the population of individuals or things under study.

EXAMPLE

Imagine a study dealing with the quality of education among high-school students. One aspect of this issue might be the number of hours students spend on homework. Imagine that 100 students are interviewed at a particular school and 40 say they spend less than 1 hour on homework. The 100 students represent a sample; they are a subset of the population of interest, which is all high-school students.

EXAMPLE

Imagine a developmental psychologist studying the ways children interact. One aspect of interest might be the difference between males and females in terms of how they handle certain situations. For example, are boys more aggressive than girls in certain play situations? Imagine that the psychologist videotapes 4-year-old children playing, and then raters rate each child on a 10-point scale in terms of the amount of aggressive behavior they display. Further imagine that 30 boys get an average rating of 5, while 25 girls get an average rating of 4. The 30 boys represent a sample from the entire population of 4-year-old boys and the 25 girls represent a sample from the population of all 4-year-old girls.

Inferential methods are broadly aimed at assessing the implications of a sample regarding the characteristics of a population. In the last example, the 30 boys have a higher average rating than the 25 girls. Based on this result, is it reasonable to conclude that if all 4-year-old boys could be measured, as well as all 4-year-old girls, it would again be the case that the average rating for boys would be higher than the average rating for girls? This is one of the many issues addressed in this book.

Notice that in this last example, an average was used with the goal of characterizing the typical boy and girl. Are there other ways of summarizing the data that have practical importance? The answer is an unequivocal yes, as will be made evident in subsequent chapters. Indeed, one of the major advances during the last half century is a better understanding of the relative merits associated with various ways of summarizing data.

1.2 SOFTWARE

There are a number of software packages aimed at providing easy access to routinely used statistical techniques. The focus in this book is on the free (open source) software package R, which within statistics is arguably the most important software development during the last quarter of a century. It is powerful, flexible, and it provides a relatively simple way of applying cutting-edge techniques. R packages are routinely written by statisticians for applying recently developed methods, which are easily installed by anyone using R. This is extremely important because relying on commercial software has proven to be highly unsatisfactory when it comes to applying the many important methods developed during the last half century. R can be installed by going to the website

www.R-project.org.

S-plus is very similar to R with one important difference: it can be expensive. All of the methods in this book can be applied with R, but this is not the case when using S-plus.

SPSS is popular among academics, but gaining access to modern methods is difficult and typically impossible. Generally, using SPSS requires ignoring the three major problems associated with classic methods that are described in this book. Moreover, compared to R, SPSS is highly inflexible when it comes to applying the many new and improved methods that have appeared during the last half century. Because SPSS does not update its software in an adequate manner, it was excluded from consideration when choosing software for this book.

SAS is another well-known software package that offers high power and a fair degree of flexibility. SAS software could be written for applying the modern methods mentioned in this book, but for many of the techniques to be described, this has not been done.

Another well-known software package is EXCEL. It was excluded from consideration based on reviews by McCullough and Wilson (2005) as well as Heiser (2006), which conclude that this package is not maintained in an adequate manner. Indeed, even certain basic features have fundamental flaws.

1.3 R BASICS

As previously noted, R can be downloaded from www.R-project.org. A free and very useful interface for R is R Studio (www.rstudio.com). Many books are available that are focused on the basics of R (e.g., Crawley, 2007; Venables and Smith, 2002; Verzani, 2004; Zuur et al., 2009). The book by Verzani (2004) is available on the web at

http://cran.r-project.org/doc/contrib/Verzani-SimpleR.pdf.

R comes with a vast array of methods for handling and analyzing data. Explaining all of its features and built-in functions is impossible here and to a large extent not necessary for present purposes. Here, the focus is on the more basic features needed to apply standard statistical methods, as well as the more modern methods covered in this book.

Once you start R you will see this prompt:

```
>
```

This prompt is not typed, it merely means that R is waiting for a command. To quit R, use the command

```
> q()
```

1.3.1 Entering Data

To begin with the simplest case, imagine you want to store the value 5 in an R variable called `dat`. This can be done with the command

```
dat=5.
```

Typing dat and hitting Enter will produce the value 5 on the computer screen.

To store the values 2, 4, 6, 8, 12 in the R variable dat, use the c command which stands for "combine." That is, the command

```
dat=c(2,4,6,8,12))
```

will store these values in the R variable dat.

To read data stored in a file into an R variable, use the scan command or the read.table command. Both of these commands assume the data are stored in the directory where R expects to find the file. A simple way of finding a particular file is with the the file.choose, which will be illustrated. The scan command is convenient when working with a single variable; read.table is convenient when working with two or more variables. In the simplest case, it is assumed that values are separated by one or more spaces. *Missing values* are assumed to be recorded as NA for "not available." For example, imagine that a file called *ice.dat* contains

6 3 12 8 9

Then the command

```
dat=scan(file="ice.dat")
```

will read these values from the file and store them in the R variable dat having what is called a vector mode. Roughly, a vector is just a collection of values associated with a single variable. When using the scan command, the file name must be in quotes.

If instead a file called *dis.data* that contains

12	6	4
7	NA	8
1	18	2

then the command

```
dat2=scan(file="dis.data")
```

will store the data in the R variable dat2. Typing dat2 and hitting Enter returns

```
12  6  4  7 NA  8  1 18  2
```

When you quit R with the command q(), R will ask whether you want to save the workspace. If you enter y, values stored in R variables stay there until they are removed. So in this last example, if you turn off your computer, and then turn it back on, typing dat2 will again return the values just displayed. To remove data, use the rm command. For example,

```
rm(dat)
```

would remove the R variable dat.

R variables are case sensitive. So, for example, the command

```
Dat2=5
```

would store the value 5 in Dat2, but the R variable dat2 would still contain the nine values listed previously, unless, of course, they had been removed.

The R command read.table is another commonly used method for reading data into an R. It is convenient when dealing with two or more variables, with the values for each variable stored in columns. It has the general form

```
read.table(file, header = FALSE, sep = "",na.strings = "NA",skip=0)
```

where the argument file indicates the name of the file where the data are stored and the other arguments are explained momentarily. Notice that the first argument, file, does not contain an equal sign. This means that a value for this argument must be specified. With an equal sign, the argument defaults to the value shown, but it can be altered when desired, as will be illustrated. If missing values are not stored as NA, but rather as *, then use the command

```
dat=read.table(file="dis.data",na.strings = "*").
```

EXAMPLE

Suppose a file called quake.dat contains three measures related to earthquakes:

magnitude	length	duration
7.8	360	130
7.7	400	110
7.5	75	27
7.3	70	24
7.0	40	7
6.9	50	15
6.7	16	8
6.7	14	7
6.6	23	15
6.5	25	6
6.4	30	13
6.4	15	5
6.1	15	5
5.9	20	4
5.9	6	3
5.8	5	2

Then the R command

```
dat=read.table("quake.dat",skip=1)
```

will read the data into the R variable `dat`, where the argument `skip=1` indicates that the first line of the data file is to be ignored, which in this case contains the three labels magnitude, length, and duration. As previously noted, this command assumes that the file quake.dat is stored in the directory where R expects to find it. A simple way of finding the file is to use the command

```
dat=read.table(file.choose(),skip=1).
```

This will open a window where all the files are stored. Click on the appropriate file, click open, and hit Return. Typing `dat` and hitting Enter produces

```
   V1  V2  V3
1 7.8 360 130
2 7.7 400 110
3 7.5  75  27
4 7.3  70  24
5 7.0  40   7
```

```
 6 6.9  50  15
 7 6.7  16   8
 8 6.7  14   7
 9 6.6  23  15
10 6.5  25   6
11 6.4  30  13
12 6.4  15   5
13 6.1  15   5
14 5.9  20   4
15 5.9   6   3
16 5.8   5   2
```

So the columns of the data are given the names V1, V2, and V3, as indicated. Typing the R command

```
dat$V1
```

and hitting enter would return the first column of data, namely the magnitude of the earthquakes. The command

```
dat[,1]
```

would accomplish the same goal.

Now consider the command

```
dat=read.table("quake.dat",header=TRUE).
```

This command says that the first row of the data file contains labels for the columns of data, which for the earthquake data are magnitude, length, and duration. Now the command

```
dat$magnitude
```

would print the first column of data on the computer screen.

Situations arise where it is convenient to store data in what is called a CSV file, where CSV stands for Comma Separated Values. Storing data in a CSV file helps when trying to move data from one software package to another. Imagine, for example, that data are stored in an EXCEL file. EXCEL has an option for storing the data in a CSV file. Once this is done, the data can be read into R using the `read.csv` command, which is used in the same way as the `read.table` command.

By default, R assumes that values stored in a data set are separated by one or more spaces. But suppose the values are separated by some symbol, say &. This can be handled via the argument `sep`. Now the command would look something like this:

```
dat=read.table("quake.dat",sep="&").
```

R has different ways of storing data called modes. The three modes that are relevant here are vector, matrix, and data frame. A vector simply refers to a string of numbers associated with a single variable. When reading data with the `scan` command, data are stored in a vector. Matrix mode refers to an R variable that contains multiple columns of data. A data frame is explained momentarily You can determine whether the R variable `dat` has matrix mode with the command `is.matrix`. For the earthquake data stored in `dat`, the command is

```
is.matrix(dat)
```

which will return true or false.

When using the R command `read.table`, data are not stored in a matrix, but rather in what is called a data frame. Data frames are useful when some columns of data have different types. For instance, one column might contain numeric data, while other columns have character data such as M for male and F for female. With a data frame, data can be stored in a single R variable, with some of the columns containing numeric data and others containing character values. But if the data were stored in an R variable that has matrix mode, having both numeric and character strings is not allowed. All of the values must be character data or all of them must be numeric.

Once stored in a matrix, it is a simple matter to access a subset of the data. For example, if `m` is a matrix, `m[1,1]` contains the value in the first row and first column, `m[1,3]` contains the value in the first row and third column, and `m[2,4]` contains the value in row 2 and column 4. The symbol [,1] refers to the first column and [2,] is the second row. So here, typing m[,2] and hitting Enter returns

```
[1] 120 100 120  85  90
```

which is the data in the second column.

Imagine that `m` is a matrix with three columns. The columns of this matrix are taken to be a vector, not a matrix. So the command `is.matrix(m[,1])` would return FALSE and `is.vector(m[,1])` would return TRUE. Any individual row is viewed as a vector as well.

Another method for storing data is called list mode. Imagine that data on three groups of individuals are under study. For example, the three groups might correspond to three methods for treating migraines. List mode is a convenient way of storing all of the data in one R variable, even when the number of observations differs among the groups. If the data are stored in the R variable `mig`, then `mig[[1]]` would contain the data for group 1, `mig[[2]]` would contain the data for group 2, and so on. The practical utility of list mode, in terms of the methods covered in this book, will become clear in subsequent chapters.

In this book, typically data are stored in list mode via the R functions `fac2list` and `split`, which are described in Chapter 9. In case the need arises, another way is to proceed as follows. First create a variable having list mode. If you want the variable to be called `gdat`, use the command

```
gdat=list().
```

Then the data for group 1 can be stored via the command

```
gdat[[1]]=c(36, 24, 82, 12, 90, 33, 14, 19)
```

the group 2 data would be stored via the command

```
gdat[[2]]=c(9, 17, 8, 22, 15)
```

and the group 3 data would be stored by using the command

```
gdat[[3]]=c(43, 56, 23, 10).
```

Typing the command `gdat` and hitting Enter returns

```
[[1]]:
[1] 36 24 82 12 90 33 14 19

[[2]]:
[1]  9 17  8 22 15

[[3]]:
[1] 43 56 23 10
```

That is, gdat contains three vectors of numbers corresponding to the three groups under study.

1.3.2 R Functions and Packages

R has many built-in functions; the ones relevant to this book will be introduced as needed. For illustrative purposes, here it is noted that one of these functions is called mean, which computes the average of the values stored in some R variable. For example, the command

```
mean(x)
```

would compute the average of the values stored in the R variable x and print it on the screen.

R Packages for this Book

Of particular importance is that many statistical techniques that are not built into R are easily accessed using what are called R packages. Perhaps the easiest way of gaining access to the functions written for this book is to download the file Rallfun located at www-rcf.usc.edu/~rwilcox/. (Or on the web, search for software Rand Wilcox.) Save the most current version of the file Rallfun, which at present is Rallfun-v32. Next, use the R command

```
source(file.choose())
```

and then click on the file Rallfun.

A second way of gaining access is via the R package WRS (Wilcox and Schönbrodt, 2016). The R commands for installing WRS are located at

https://github.com/nicebread/WRS

Copy and paste the R commands into R. Then use the R command

```
library(WRS)
```

to gain access to the functions. This web site also contains commands for installing the R package WRScpp. Most of the R functions in this book have very low execution time, but situations are encountered where this is not the case. The R functions in WRScpp address this issue by using C++ versions of the functions.

An advantage of downloading Rallfun from the author's web page is that it is updated more frequently and it is more easily installed. Help files for the functions in Rallfun are not available. But these R functions contain information at the top of the code that summarizes the information provided in this book. For example, there is a function called yuen, which is described in Chapter 7. Typing yuen and hitting Enter, you will see

```
#
#  Perform Yuen's test for trimmed means on the data in x and y.
#  The default amount of trimming is 20%
#  Missing values (values stored as NA) are automatically removed.
#
#  A confidence interval for the trimmed mean of x minus the
#  the trimmed mean of y is computed and returned in yuen$ci.
#  The p-value is returned in yuen$p.value
#
```

followed by some R code. These first few lines provide a quick summary of what the function does and how it is used. To see the various arguments used by the function, use the `args` command. For example,

```
args(yuen)
```

returns

```
function (x, y, tr = 0.2, alpha = 0.05).
```

So the function expects to find data stored in two variables, labeled here as `x` and `y`. And there are two optional arguments, the first (labeled `tr`), which defaults to 0.2, and the second (`alpha`), which defaults to 0.05.

A subset of the R functions used in this book is available in the R package WRS2, which is stored on CRANS. A possible appeal of this package is that it contains help files and it is easily installed. To install it, use the R command

```
install.packages("WRS2").
```

The R command

```
library(WRS2)
```

provides access to the functions and it lists the functions that are currently available. Currently, a negative feature is that WRS2 does not contain all of the functions described and illustrated in this book. See Mair and Wilcox (2016) for a description of the functions currently available. (A copy of this paper is stored on the author's web in the directory software.)

Many of the R functions written for this book are based in part on other R packages available at CRANS. They include the following packages:

- akima
- cobs
- MASS
- mgcv
- multicore
- plotrix
- pwr
- quantreg
- robust

- rrcov
- scatterplot3d
- stats

All of these packages can be installed with the `install.packages` command (assuming you are connected to the web). For example, the R command

```
install.packages('akima')
```

will install the R package akima, which is used when creating three-dimensional plots.

1.3.3 Data Sets

Numerous data sets are used to illustrate the methods in this book, some of which are built into R. A list of all of the data sets that come with R can be seen with the R command

```
?datasets
```

or the command

```
data().
```

Other data sets used in this book are stored on the web page

http:college.usc.edu/labs/rwilcox.

One of these data sets, which is used in subsequent chapters, deals with plasma retinol and was downloaded from a site maintained by Carnegie Mellon University. Retinol is vitamin A, and plasma retinol appears to be related to the utilization of vitamin A in rats.

EXAMPLE

Imagine that the plasma retinol data are stored in the file plasma.dat2. Then the R command

```
plasma=read.table('plasma.dat2')
```

will store the data in the R variable `plasma` as a data frame. (Again, if the data file is not stored where R expects to find it, use the command `source(file.choose())`.) The R command

```
plasma=as.matrix(plasma)
```

would change the storage mode to a matrix, as previously noted. This last command is of no particular consequence in this case because all of the data are numeric. But if one of the columns had contained nonnumeric data, the result would be that all of the data would be converted to characters. So, for example, the value 2 would now be stored as `'2'` and any attempt to perform an arithmetic operation would result in an error. The error could be avoided by converting `'2'` back to a numeric value using the R command `as.numeric`. More generally, if the entries in column two were stored as `'1'` (male) or `'2'` (female), we could sum the values with the command

```
sum(as.numeric(plasma[,2])).
```

But with a data frame, it is possible to have the column 2 entries stored as numeric values with other columns stored as characters, as previously indicated.

For future reference, the variable names in the plasma retinol data set are:

```
1   AGE: Age (years)
2   SEX: Sex (1=Male, 2=Female).
3   SMOKSTAT: Smoking status (1=Never,2=Former,3=Current Smoker)
4   QUETELET: Quetelet (weight/(height^2))
5   VITUSE: Vitamin Use (1=Yes,fairly often,2=Yes,not often,3=No)
6   CALORIES: Number of calories consumed per day
7   FAT: Grams of fat consumed per day
8   FIBER: Grams of fiber consumed per day
9   ALCOHOL: Number of alcoholic drinks consumed per week
10  CHOLESTEROL: Cholesterol consumed (mg per day)
11  BETADIET: Dietary beta-carotene consumed (mcg per day)
12  RETDIET: Dietary retinol consumed (mcg per day)
13  BETAPLASMA: Plasma beta-carotene (ng/ml)
14  RETPLASMA: Plasma Retinol (ng/ml)
```

The first few lines of the data set look like this:

```
64  2  2 21.48380  1 1298.8 57.0  6.3  0.0 170.3 1945 890 200 915
76  2  1 23.87631  1 1032.5 50.1 15.8  0.0  75.8 2653 451 124 727
38  2  2 20.01080  2 2372.3 83.6 19.1 14.1 257.9 6321 660 328 721
40  2  2 25.14062  3 2449.5 97.5 26.5  0.5 332.6 1061 864 153 615
```

1.3.4 Arithmetic Operations

In the simplest case, arithmetic operations can be performed on numbers using the operators +, −, * (multiplication), / (division), and ^ (exponentiation). For example, to compute 1 plus 5 squared, use the command

1+5^2,

which returns

[1] 26.

To store the answer in an R variable, say `ans`, use the command

ans=1+5^2.

If a vector of observations is stored in an R variable, arithmetic operations applied to the variable name will be performed on all the values. For example, if the values 2, 5, 8, 12, and 25 are stored in the R variable `vdat`, then the command

```
vinv=1/vdat
```

will compute 1/2, 1/5, 1/8, 1/12, and 1/25 and store the results in the R variable `vinv`.

Most R commands consist of a name of some function followed by one or more arguments enclosed in parentheses. There are hundreds of functions that come with R. Some of the more basic functions are listed here:

Function	Description
exp	exponential
log	natural logarithm
sqrt	square root
cor	correlation
mean	arithmetic mean (with a trimming option)
median	median
min	smallest value
max	largest value
quantile	quantiles
range	max value minus the min value
sd	standard deviation
sum	arithmetic sum
var	variance and covariance

EXAMPLE

If the values 2, 7, 9, and 14 are stored in the R variable `x`, the command

```
min(x)
```

returns 2, the smallest of the four values stored in x. The average of the numbers is computed with the command `mean(x)` and is 8. The command range(x) returns the largest and smallest values stored in `x`, and `sum(x)` returns the value $2 + 7 + 9 + 14 = 32$.

Suppose you want to subtract the average from each value stored in the R variable blob. The command

```
blob-mean(blob)
```

accomplishes this goal. If, in addition, you want to square each of these differences and then sum the results, use the command

`sum(blob-mean(blob))`^{2}.

You can apply arithmetic operations to specific rows or columns of a matrix. For example, to compute the average of all values in column 1 of the matrix m, use the command

```
mean(m[,1]).
```

The command

```
mean(m[2,])
```

will compute the average of all values in row 2. An easy way to compute the average for each row is with the command

```
apply(m,1,mean).
```

To compute the average for each column, use

```
apply(m,2,mean).
```

In contrast, the command `mean(m)` will average all of the values in `m`.

In a similar manner, if `x` has list mode, then

```
mean(x[[2]])
```

will average the values in x[[2]]. To compute the average for each of the groups, use the command

```
lapply(m,mean).
```

This chapter has summarized the basic features of R needed to analyze data with the methods covered in this book. R has many other features, but a complete description of all of them is not needed for present purposes. Included are a variety of ways of managing and manipulating data. Some special data management issues are relevant in subsequent chapters, and special R functions needed to deal with these issues will be described when the need arises. In case it helps, a few of these functions are listed here, but explanations regarding how they are used are postponed for now. The functions are:

- `fac2list(x,g)`. This function divides data stored in `x` based on the corresponding values stored in `g`. The arguments `x` and `g` are vectors having the same length. For example, the argument `g` might indicate whether the corresponding `x` value corresponds to a participant from a control group or an experimental group; see Chapter 9
- `mat2grp`. This function divides data stored in a matrix into groups based on values in a specified column; see Chapter 13
- `matsplit`. This function splits a matrix into two matrices based on the binary values in a specified column; see Chapter 13
- `bw2list`. This function is useful when dealing with data management issues described in Chapter 11
- `bbw2list`. This function is useful when dealing with data management issues described in Chapter 11
- `binmat(m, col, lower, upper)`. This function pulls out the rows of the matrix m based on the values in the column indicated by the argument `col` that are between values indicated by the arguments `lower` and `upper`, inclusive. Example: Imagine that age is stored in column 3 of the matrix `m` and the goal is to focus on participants between the ages of 10 and 15. The command `binmat(m,3,10,15)` will return the rows of `m` such that the values in column 3 are between 10 and 15, inclusive.

CHAPTER 2

NUMERICAL AND GRAPHICAL SUMMARIES OF DATA

CONTENTS

A fundamental goal is summarizing data in a manner that helps convey some of its important properties in a relatively simple manner. Consider, for example, a study dealing with a new medication aimed at lowering blood pressure and imagine that 200 participants take the new medication for a month. If we simply record the changes in blood pressure, we have information relevant to the effectiveness of the drug. Further imagine that another group of 250 individuals takes a placebo and the changes in their blood pressure are recorded as well. How can we summarize the data in a manner that provides a useful and informative sense of how these two groups of participants differ? In what sense, if any, does the new drug improve upon the placebo or some alternative medication currently in use? The topics covered in this chapter play a major role when addressing these questions.

This chapter includes some of the more basic plots that can provide a useful summary of data. It is noted that many additional methods are available via the R package ggplot2 (e.g., Chang, 2013; Field et al., 2012; Wickham, 2009).

Before continuing, some comments about the style of this chapter might help. In the age of the computer, there is in some sense little need to spend time illustrating basic computations. The bulk of this book takes this view, particularly when dealing with computations that are quite involved. Rather, the focus is on making decisions and inferences based on the results of an analysis. This includes making inferences that are reasonable based on the analysis done as well as understanding what inferences are unreasonable. Another general goal is to provide perspective on the relative merits of competing methods for addressing a particular issue of interest. For example, when comparing two groups of individuals, a topic discussed in Chapter 8, there are now many methods that might be used. To get the most out of a data analysis, the relative merits of each method need to be understood. Accomplishing this goal is aided by developing an understanding and awareness of certain properties of the methods summarized in this chapter. For this reason, portions of this chapter illustrate basic computations with an eye toward building a foundation for understanding when and why commonly used methods might be unsatisfactory, as well as why more recently developed methods might have considerable practical value.

2.1 BASIC SUMMATION NOTATION

This section introduces some basic notation that will be needed. To be concrete, imagine that 15 college students are asked to rate their feelings of optimism based on a 10-point scale. If the first student gives a rating of 6, this result is typically written $X_1 = 6$, where the subscript 1 indicates that this is the first student interviewed. If you sample a second student who gets a score of 4, you write this as $X_2 = 4$, where now the subscript 2 indicates that this is the second student you measure. Here we assume 15 students are interviewed in which case their ratings are represented by $X_1, \ldots, X_{15}$. The notation X_i is used to represent the ith participant. In the example with a total of 15 participants, the possible values for i are the integers 1, 2, …, 15. Typically, the sample size (the number of observations) is represented by n. Here there are 15 participants and this is written as $n = 15$. Table 2.1 illustrates the

Table 2.1 Hypothetical Data Illustrating a Commonly Used Notation

i	X_i
1	3
2	7
3	6
4	4
5	8
6	9
7	10
8	4
9	5
10	4
11	5
12	6
13	5
14	7
15	6

notation along with the ratings you might get. The first participant ($i = 1$) got a score of 3, so $X_1 = 3$. The next participant ($i = 2$) got a score of 7, so $X_2 = 7$.

Summation notation is just a shorthand way of indicating that the observations $X_1, \ldots, X_n$ are to be summed. This is denoted by

$$\sum_{i=1}^{n} X_i = X_1 + X_2 + \cdots + X_n. \tag{2.1}$$

The subscript i is the *index of summation* and the 1 and n that appear, respectively, below and above the symbol $\sum$ designate the range of the summation. For the data in Table 2.1, the number of observations is $n = 15$ and

$$\sum_{i=1}^{n} X_i = 3 + 7 + 6 + \cdots + 7 + 6 = 89,$$

while

$$\sum_{i=12}^{15} X_i = 6 + 5 + 7 + 6 = 24.$$

In most situations, the sum extends over all n observations, in which case it is customary to omit the index of summation. That is, simply use the notation

$$\sum X_i = X_1 + X_2 + \cdots + X_n.$$

2.2 MEASURES OF LOCATION

One of the most common approaches to summarizing a sample of individuals, or a batch of numbers, is to use a so-called measure of location. A common description of a *measure of location* is that it is a number intended to reflect the typical individual or thing under study. But from a technical point of view, this description is inaccurate and misleading as will become evident. (A more precise definition is given in Section 2.2.8.) For now, we simply describe several measures of location that play an important role when trying to summarize data in a meaningful manner.

2.2.1 The Sample Mean

Chapter 1 made a distinction between a population of individuals or things versus a sample. Here, to be concrete, imagine a study aimed at determining the average cholesterol level of all adults living in Hungary. That is, in this study, the population consists of all adults living in Hungary. If the cholesterol level of all of these adults could be measured and averaged, we would have what is called the population mean. (A more formal definition is given in Chapter 3.) Typically the population mean is represented by μ, a lower case Greek mu.

Typically, not everyone in the population can be measured. Rather, only a subset of the population is available, which is called a sample, as noted in Chapter 1. In symbols, the n observed values from a sample are denoted by $X_1, \ldots, X_n$, as previously explained, and the average of these n values is

$$\bar{X} = \frac{1}{n}\sum X_i, \tag{2.2}$$

where the notation $\bar{X}$ is read X bar. (The R function mean, described in Section 2.2.5, computes the mean.) In statistics, $\bar{X}$ is called the *sample mean.* The sample mean is intended as an estimate of the population mean. But typically they differ, which raises the issue of how well the sample mean estimates the population mean, a topic that we will begin to discuss in Chapter 4.

EXAMPLE

You sample 10 married couples and determine the number of children they have. The results are 0, 4, 3, 2, 2, 3, 2, 1, 0, 8, and the sample mean is $\bar{X} = 2.5$. Based on this result, it is estimated that if we could measure all married couples, the average number of children would be 2.5. In more formal terms, $\bar{X} = 2.5$ is an estimate of μ. In all likelihood the population mean is not 2.5, so there is the issue of how close the sample mean is likely to be to the population mean. Again, we will get to this topic in due course.

There are two properties of the sample mean worth stressing. The first is that if a constant, say c, is added to every possible value, the sample mean is increased by c as well. The second is that if every value is multiplied by b, the mean is multiplied by b. In symbols,

$$\frac{1}{n}\sum(bX_i + c) = b\bar{X} + c.$$

(The technical term for this property is that the sample mean is location and scale *equivariant.*)

EXAMPLE

Consider a sample of 1000 six-year-old boys and imagine that their average height is 4 feet. To convert to inches, we could multiply each value by 12 and recompute the average, but, of course, it is easier to simply multiply the sample mean by 12 yielding 48 inches. If every child grows exactly 2 inches, their average height is now 50.

Computing the Sample Mean Based on Relative Frequencies

There is another way of describing how to compute the sample mean that helps convey a conceptual issue discussed in Chapter 3. The notation f_x indicates the number of times

Table 2.2 Desired Number of Sexual Partners for 105 Males

x:	0	1	2	3	4	5	6	7	8	9
f_x:	5	49	4	5	9	4	4	1	1	2
x:	10	11	12	13	15	18	19	30	40	45
f_x:	3	2	3	1	2	1	2	2	1	1
x:	150	6000								
f_x:	2	1								

the value x occurred among a sample of size n; it is called the *frequency* of x. The *relative frequency* is f_x/n, the proportion of values equal to x. The sample mean is

$$\bar{X} = \sum_x x\frac{f_x}{n},$$

where now the notation $\sum_x$ indicates that the sum is over all possible values of x. In words, multiply every possible value of x by its relative frequency and add the results.

EXAMPLE

In a sexual attitude study conducted by Pedersen et al. (2002), 105 young males were asked how many sexual partners they desire over the next 30 years. The observed responses and the corresponding frequencies are shown in Table 2.2. For instance, $f_0 = 5$ indicates that 5 said they want 0 sexual partners and $f_1 = 49$ means that 49 said 1. Because the sample size is $n = 105$, the relative frequency corresponding to 0 is $f_0/n = 5/105$ and the relative frequency corresponding to 1 is 49/105. The sample mean is

$$\bar{X} = 0\frac{5}{105} + 1\frac{49}{105} + \cdots = 64.9.$$

This is, however, a dubious indication of the typical response because 97% of the responses are less than the sample mean. This occurs because of a single extreme response: 6000. If this extreme response is removed, now the mean is 7.9. But even 7.9 is rather misleading because over 77% of the remaining observations are less than 7.9.

Unusually large or small values are called *outliers*. One way of quantifying the sensitivity of the sample mean to outliers is with the so-called *finite sample breakdown point*. The finite sample breakdown point of the sample mean is the smallest proportion of observations that can make it arbitrarily large or small. Said another way, the finite sample breakdown point of the sample mean is the smallest proportion of n observations that can render it meaningless. A single observation can make the sample mean arbitrarily large or small, regardless of what the other values might be, so its finite sample breakdown point is $1/n$.

2.2.2 R Function Mean

R has a built-in function for computing a mean:

```
mean(x),
```

where x is any R variable containing data as described in Chapter 1.

2.2.3 The Sample Median

As previously illustrated, a single extreme value can result in a sample mean that poorly reflects the typical response. If this is viewed as undesirable, there are two general strategies for dealing with this problem. The first is to simply trim a specified portion of the smallest and largest values and average the values that remain. Certainly the best-known example of this strategy is the sample median, which is described here. The other general strategy is to use some empirical rule for identifying unusually small or large values and then eliminate these values when computing a measure of location. Two examples of this type are described in Section 2.4.8.

Generally, trimmed means are computed by first removing a specified proportion of the smallest and largest values and averaging what remains. The sample median represents the most extreme amount of trimming where all but one or two values are removed. Simply put, if the sample size is odd, the sample median is the middle value after putting the observations in ascending order. If the sample size is even, the sample median is the average of the two middle values.

It helps to describe the sample median in a more formal manner in order to illustrate a commonly used notation. For the observations $X_1, \ldots, X_n$, let $X_{(1)}$ represent the smallest number, $X_{(2)}$ is the next smallest, and $X_{(n)}$ is the largest. More generally,

$$X_{(1)} \leq X_{(2)} \leq X_{(3)} \leq \cdots \leq X_{(n)}$$

is the notation used to indicate that n values are to be put in ascending order. The sample median is computed as follows:

1. If the number of observations, n, is odd, compute $m = (n+1)/2$. Then the sample median is
$$M = X_{(m)},$$
the mth value after the observations are put in order.
2. If the number of observations, n, is even, compute $m = n/2$. Then the sample median is
$$M = \frac{X_{(m)} + X_{(m+1)}}{2},$$
the average of the mth and $(m+1)$th observations after putting the observed values in ascending order.

EXAMPLE

Consider the values 1.1, 2.3, 1.7, 0.9, and 3.1. Then the smallest of the five observations is 0.9, so $X_{(1)} = 0.9$. The smallest of the remaining four observations is 1.1, and this is written as $X_{(2)} = 1.1$. The smallest of the remaining three observations is 1.7, so $X_{(3)} = 1.7$; the largest of the five values is 3.1, and this is written as $X_{(5)} = 3.1$.

EXAMPLE

Seven participants are given a test that measures feelings of optimism. The observed scores are

$$34, 29, 55, 45, 21, 32, 39.$$

Because the number of observations is $n = 7$, which is odd, $m = (7 + 1)/2 = 4$. Putting the observations in order yields

$$21, 29, 32, \mathbf{34}, 39, 45, 55.$$

The fourth observation is $X_{(4)} = 34$, so the sample median is $M = 34$.

EXAMPLE

We repeat the last example, only with six participants having test scores

$$29, 55, 45, 21, 32, 39.$$

Because the number of observations is $n = 6$, which is even, $m = 6/2 = 3$. Putting the observations in order yields

$$21, 29, \mathbf{32}, \mathbf{39}, 45, 55.$$

The third and fourth observations are $X_{(3)} = 32$ and $X_{(4)} = 39$, so the sample median is $M = (32 + 39)/2 = 35.5$.

Notice that nearly half of any n values can be made arbitrarily large without making the value of the sample median arbitrarily large as well. Consequently, the finite sample breakdown point is approximately 0.5, the highest possible value. So the mean and median lie at two extremes in terms of their sensitivity to outliers. The sample mean can be affected by a single outlier, but about half of the observations must be altered to make the median arbitrarily large or small. That is, the sample median is resistant to outliers. For the sexual attitude data in Table 2.2, the sample median is $M = 1$ which gives a decidedly different picture of what is typical versus the mean, which is 64.9.

Based on the single criterion of having a high breakdown point, the median beats the mean. But it is not being suggested that the mean is always inappropriate when extreme values occur.

EXAMPLE

An individual invests $200,000 and reports that the median amount earned per year, over a 10-year period, is $100,000. This sounds good, but now imagine that the earnings for each year are: $100,000, $200,000, $200,000, $200,000, $200,000, $200,000, $200,000, $300,000, $300,000, $−1,800,000. So at the end of 10 years, this individual has earned nothing and in fact lost the $200,000 initial investment. Certainly the long-term total amount earned is relevant in which case the sample mean provides a useful summary of the investment strategy that was followed.

EXAMPLE

Consider the daily rainfall in Boston, Massachusetts during the year 2005. The median is zero. No rain is typical, but, of course, it does rain in Boston.

Chapter 4 describes other criteria for judging measures of location, and situations will be described where both the median and mean are unsatisfactory. That is, an alternative measure of location has practical value, one of which is described in Section 2.2.5.

2.2.4 R Function for the Median

R has a built-in function for computing the median:

```
median(x).
```

2.3 A CRITICISM OF THE MEDIAN: IT MIGHT TRIM TOO MANY VALUES

The sample median represents the most extreme amount of trimming. Despite this, there are situations where the median has advantages over other measures of location, but the reality is that under general conditions there can be practical advantages to trimming less, for reasons best postponed for now. For the moment, attention is focused on how, in general, trimmed means are computed. As previously indicated, a *trimmed mean* removes a specified proportion of both the lowest and highest values and then the average of the remaining values is computed. Removing the lowest 10% of the values, as well as the highest 10%, and then averaging the values that remain is called a 10% trimmed mean. Trimmed means contain as special cases the sample mean where no observations are trimmed, and the median where the maximum possible amount of trimming is used.

EXAMPLE

Consider the values

37, 14, 26, 17, 21, 43, 25, 6, 9, 11.

Putting the observations in ascending order yields

6, 9, 11, 14, 17, 21, 25, 26, 37, 43.

There are 10 observations, so the 10% trimmed mean removes the smallest and largest observations and averages the remaining 8, which yields

$$\bar{X}_t = \frac{9+11+14+17+21+25+26+37}{8} = 20.$$

The 20% trimmed mean removes the two largest and two smallest values, and the remaining 6 observations are averaged, which gives

$$\bar{X}_t = \frac{11+14+17+21+25+26}{6} = 19.$$

Here is a more formal and more precise description of how to compute a 20% trimmed mean. Set $g = [0.2n]$, where the notation $[0.2n]$ means that $.2n$ is rounded down to the nearest integer. For example, if the sample size is $n = 99$, then $g = [0.2n] = [19.8] = 19$. The 20% trimmed mean is just the average of the values after the g smallest and g largest observations are removed. In symbols, the 20% trimmed mean is

$$\bar{X}_t = \frac{1}{n-2g}(X_{(g+1)} + X_{(g+2)} + \cdots + X_{(n-g)}). \tag{2.3}$$

A 10% trimmed mean is obtained by taking $g = [0.1n]$.

The finite sample breakdown point of a trimmed mean is approximately equal to the proportion of points trimmed. For the 10% trimmed mean it is approximately 0.1, and for the

20% trimmed mean it is 0.2. This says that when using the 20% trimmed mean, for example, about 20% of the values must be altered to make the 20% trimmed mean arbitrarily large or small.

A fundamental issue is deciding how much to trim. No amount is always optimal. But a good choice for routine use is 20% for reasons that are described in subsequent chapters. This is not to say that alternative amounts of trimming should be excluded, but it is too soon to discuss this issue in any detail. For now it is merely noted that when addressing a variety of practical goals, often 20% trimming offers a considerable advantage over no trimming and the median with the understanding that exceptions will be described and illustrated. Huber (1993) has argued that any estimator with a breakdown point less than or equal to 0.1 is dangerous and should be avoided. But classic methods based on means, which are routinely used, do have some redeeming features.

2.3.1 R Function for the Trimmed Mean

The R function mean, introduced in Section 2.2.2, can be used to compute a trimmed mean via the argument tr. That is, a more general form of this R function is

```
mean(x,tr=0).
```

Following the standard conventions used by R functions, the notation tr=0 indicates that the amount of trimming defaults to 0. That is, the sample mean will be computed if a value for tr is not specified. For example, if the values 2, 6, 8, 12, 23, 45, 56, 65, 72 are stored in the R variable x, then `mean(x)` will return the value 32.111, which is the sample mean. The command `mean(x,.2)` returns the 20% trimmed mean, which is 30.71.

Because 20% trimming is often used, for convenience, the R function

```
tmean(x,tr=0.2)
```

is included in the library of functions written for this book, where again the argument `tr` indicates the amount of trimming, only now it defaults to 20%. (Chapter 1 explained how to download the R functions used in this book.)

2.3.2 A Winsorized Mean

In order to deal with some technical issues described in Chapter 4, we will need the so-called Winsorized mean. The first step in the computations is to Winsorize the observed values. Recall that when computing the 10% trimmed mean, you remove the smallest 10% of the observations. Winsorizing the observations by 10% simply means that rather than remove the smallest 10%, their values are set equal to the smallest value not trimmed when computing the 10% trimmed mean. Simultaneously, the largest 10% are reset to the largest value not trimmed. In a similar manner, 20% Winsorizing means that the smallest 20% of the observations are pulled up to the smallest value not trimmed when computing the 20% trimmed mean, and the largest 20% are pulled down to the largest value not trimmed. The *Winsorized mean* is just the average of the Winsorized values, which is labeled $\bar{X}_w$. In symbols, the Winsorized mean is

$$\bar{X}_w = \frac{1}{n}\{(g+1)X_{(g+1)} + X_{(g+2)} + \cdots + X_{(n-g-1)} + (g+1)X_{(n-g)}\}. \tag{2.4}$$

The finite sample breakdown point of a Winsorized mean is approximately equal to the proportion of points Winsorized. For example, the finite sample breakdown point of the 20% Winsorized mean is 0.2, and the 10% Winsorized mean has a breakdown point of 0.1.

EXAMPLE

Consider again the ten values

$$37, 14, 26, 17, 21, 43, 25, 6, 9, 11.$$

Because $n = 10$, with 10% trimming $g = [0.1(10)] = 1$, meaning that the smallest and largest observations are removed when computing the 10% trimmed mean. The smallest value is 6, and the smallest value not removed when computing the 10% trimmed mean is 9. So 10% Winsorization of these values means that the value 6 is reset to the value 9. In a similar manner, the largest observation, 43, is pulled down to the next largest value, 37. So 10% Winsorization of the values yields

$$37, 14, 26, 17, 21, 37, 25, 9, 9, 11.$$

The 10% Winsorized mean is just the average of the Winsorized values:

$$\bar{X}_w = \frac{37 + 14 + 26 + 17 + 21 + 37 + 25 + 9 + 9 + 11}{10} = 20.6.$$

EXAMPLE

To compute a 20% Winsorized mean using the values in the previous example, first note that when computing the 20% trimmed mean, $g = [0.2(10)] = 2$, so the two smallest values, 6 and 9, would be removed and the smallest value not trimmed is 11. Winsorizing the values means that the values 6 and 9 become 11. Similarly, when computing the 20% trimmed mean, the largest value not trimmed is 26, so Winsorizing means that the two largest values become 26. Consequently, 20% Winsorization of the data yields

$$26, 14, 26, 17, 21, 26, 25, 11, 11, 11.$$

The 20% Winsorized mean is $\bar{X}_w = 18.8$, the average of the Winsorized values.

2.3.3 R Function winmean

R does not have a built-in function for computing the Winsorized mean, so one has been provided in the library of functions written especially for this book. (Again, these functions can be obtained as described in Chapter 1.) The function has the form

```
winmean(x,tr=0.2),
```

where now `tr` indicates the amount of Winsorizing, which defaults to 0.2. So if the values 2, 6, 8, 12, 23, 45, 56, 65, 72 are stored in the R variable x, the command `winmean(x)` returns 31.78, which is the 20% Winsorized mean. The command `winmean(x,0)` returns the sample mean, 32.11.

2.3.4 What is a Measure of Location?

Notice that the mean, median, and trimmed mean all satisfy three basic properties:

- Their value always lies between the largest and smallest value, inclusive. When working with the sample mean, for example, $X_{(1)} \le \bar{X} \le X_{(n)}$.
- If all of the values are multiplied by the constant b, the values of the mean, median, and trimmed mean are multiplied by b as well.
- If the constant c is added to every value, the mean, median, and trimmed mean are increased by c as well.

Any summary of the data that has these three properties is called a *measure of location.*

EXAMPLE

The largest value, $X_{(n)}$, is a measure of location.

2.4 MEASURES OF VARIATION OR SCALE

As is evident, not all people are alike; they respond differently to conditions. If we ask 15 people to rate how well a particular political leader is performing on a 10-point scale, some might give a rating of 1 or 2, and others might give a rating of 9 or 10. There is *variation* among their ratings. In many ways, it is variation that provides the impetus for sophisticated statistical techniques. To illustrate why, imagine that all adults would give the same rating for the performance of some political leader, say 7, if they were polled. In statistical terms, the population mean is $\mu = 7$ because, as is evident, if all ratings are equal to 7, the average rating will be 7. This implies that we need sample only one adult in order to determine that the population mean is 7. That is, with $n = 1$ adult, the sample mean, $\bar{X}$, will be exactly equal to the population mean. Because in reality there is variation, in general the sample mean will not be equal to the population mean. If the population mean is $\mu = 7$ and we poll 15 adults, we might get a sample mean of $\bar{X} = 6.2$. If we poll another 15 adults, we might get $\bar{X} = 7.3$. We get different sample means because of the variation among the population of adults we want to study. One common goal in statistics is finding ways of taking into account how variation affects our ability to estimate the population mean with the sample mean. In a similar manner, there is the issue of how well the sample trimmed mean estimates the population trimmed mean.

To make progress, we need appropriate measures of variation, which are also called *measures of scale.* Like measures of location, many measures of scale have been proposed and studied.

2.4.1 Sample Variance and Standard Deviation

Imagine you sample ten adults ($n = 10$), ask each to rate a newly released movie, and the responses are:

3, 9, 10, 4, 7, 8, 9, 5, 7, 8.

The sample mean is $\bar{X} = 7$ and this is your estimate of the population mean, μ. The *sample variance* is

$$s^2 = \frac{1}{n-1} \sum (X_i - \bar{X})^2. \tag{2.5}$$

In other words, the sample variance is computed by subtracting the sample mean from each observation and squaring. Then you add the results and divide by $n-1$, the number of observations minus one. For the data at hand, the calculations can be summarized as follows:

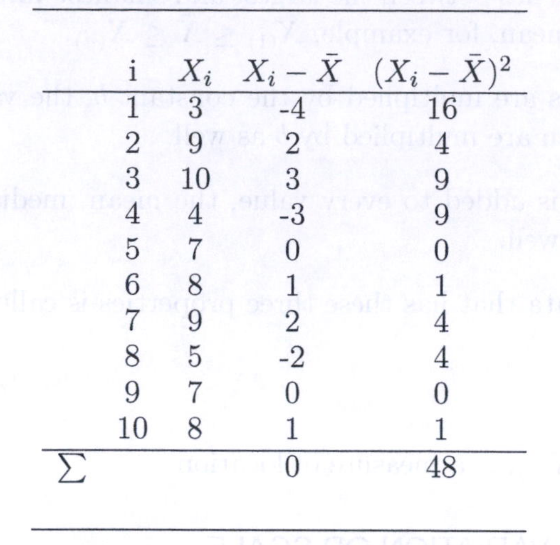

i	X_i	$X_i - \bar{X}$	$(X_i - \bar{X})^2$
1	3	-4	16
2	9	2	4
3	10	3	9
4	4	-3	9
5	7	0	0
6	8	1	1
7	9	2	4
8	5	-2	4
9	7	0	0
10	8	1	1
$\sum$		0	48

The sum of the observations in the last column is $\sum(X_i - \bar{X})^2 = 48$. Then, because there are ten participants ($n = 10$), the sample variance is

$$s^2 = \frac{48}{10-1} = 5.33.$$

The positive square root of the sample variance, s, is called the sample *standard deviation.*

It is important to realize that a single unusual value can dominate the sample variance. This is one of several facts that wreaks havoc with classic statistical techniques based on the mean. To provide a glimpse of problems to come, consider the values

$$8, 8, 8, 8, 8, 8, 8, 8, 8, 8, 8.$$

The sample variance is $s^2 = 0$, meaning there is no variation. If we increase the last value to 10, the sample variance is $s^2 = 0.36$. Increasing the last observation to 12, $s^2 = 1.45$, and increasing it to 14, $s^2 = 3.3$. The point is, even though there is no variation among the bulk of the observations, a single value can make the sample variance arbitrarily large. In modern terminology, the sample variance is not ***resistant*** to outliers, meaning that a single unusual value can inflate the sample variance and give a misleading indication of how much variation there is among the bulk of the observations. Said more formally, the sample variance has a finite sample breakdown point of only $1/n$. In some cases, this sensitivity to extreme values is desirable, but for many applied problems it is not, as will be seen.

2.4.2 R Functions var and sd

R has a built-in function for computing the sample variance:

```
var(x).
```

The function

```
sd(x)
```

computes the standard deviation.

2.4.3 The Interquartile Range

Another measure of scale or dispersion that is frequently used in applied work, particularly when the goal is to detect outliers, is called the interquartile range. To describe the interquartile range in rough terms, imagine we remove the smallest 25% of the data as well as the largest 25%. Then one way of computing the interquartile range is to take the difference between the largest and smallest values remaining. That is, it reflects the range of values among the middle 50% of the data. There are, however, many alternative proposals regarding how the interquartile range should be computed that are based on criteria not discussed here (e.g., Harrell and Davis, 1982; Dielman, Lowry, and Pfaffenberger, 1994; and Parrish, 1990). The important point here is that the choice of method can depend in part on the how the interquartile range is to be used. In this book its main use is to detect outliers, in which case results in Cleveland (1985), Hoaglin and Iglewicz (1987), Hyndman and Fan (1996), Frigge et al. (1989) as well as Carling (2000) are relevant.

As usual, let $X_{(1)} \leq \cdots \leq X_{(n)}$ be the observations written in ascending order. The so-called lower ideal fourth is

$$q_1 = (1-h)X_{(j)} + hX_{(j+1)}, \tag{2.6}$$

where j is the integer portion of $(n/4) + (5/12)$, meaning that j is $(n/4) + (5/12)$ rounded down to the nearest integer, and

$$h = \frac{n}{4} + \frac{5}{12} - j.$$

The upper ideal fourth is

$$q_2 = (1-h)X_{(k)} + hX_{(k-1)}, \tag{2.7}$$

where $k = n - j + 1$, in which case the interquartile range is

$$\text{IQR} = q_2 - q_1. \tag{2.8}$$

It should be noted that q_1 and q_2 are attempting to estimate what are called the lower and upper *quartiles*, respectively. To illustrate roughly what this means, imagine that if all participants could be measured, 25% of the values would be less than or equal to 36 and that 75% would be less than or equal to 81. Then 36 and 81 are the 0.25 and 0.75 quartiles, respectively.[1] In practice, typically all participants cannot be measured, in which case the quartiles are estimated with q_1 and q_2.

EXAMPLE

Consider the values

$$-29.6, -20.9, -19.7, -15.4, -12.3, -8.0, -4.3, 0.8, 2.0, 6.2, 11.2, 25.0.$$

There are twelve observations ($n = 12$), so

$$\frac{n}{4} + \frac{5}{12} = 3.41667.$$

Rounding this last quantity down to the nearest integer gives $j = 3$, so $h = 3.416667 - 3 = 0.41667$. Because $X_{(3)} = -19.7$ and $X_{(4)} = 15.4$,

$$q_1 = (1 - 0.41667)(-19.7) + 0.41667(-15.4) = -17.9.$$

[1]This crude description of quartiles is unsatisfactory when dealing with certain technical issues, but the details are not important in this book.

In a similar manner,

$$q_2 = (1 - 0.41667)(6.2) + 0.41667(2) = 4.45,$$

so the interquartile range, based on the ideal fourths, is

$$\text{IQR} = 4.45 - (-17.9) = 22.35.$$

2.4.4 R Functions idealf and ideafIQR

The R function

```
idealf(x),
```

written for this book, computes the ideal fourths for the data stored in x, where x can be any R variable. The R function

```
idealfIQR(x)
```

computes the interquartile range based on the ideal fourths.

2.4.5 Winsorized Variance

When working with the trimmed mean, we will see that the so-called Winsorized variance plays an important role. To compute the Winsorized variance, simply Winsorize the observations as was done when computing the Winsorized mean in Section 2.2.7. The Winsorized variance, s_w^2, is just the sample variance of the Winsorized values. Its finite sample breakdown point is equal to the amount Winsorized. So, for example, when computing a 20% Winsorized sample variance, more than 20% of the observations must be changed in order to make the sample Winsorized variance arbitrarily large.

2.4.6 R Function winvar

The R functions written for this book include the function

```
winvar(x,tr=0.2),
```

which computes the Winsorized variance, where again tr represents the amount of Winsorizing and defaults to 0.2. So if the values

$$12, 45, 23, 79, 19, 92, 30, 58, 132$$

are stored in the R variable x, winvar(x) returns the value 937.9, which is the 20% Winsorized variance. The command winvar(x,0) returns the sample variance, s^2, which is 1596.8. Typically the Winsorized variance will be smaller than the sample variance s^2 because Winsorizing pulls in extreme values.

2.4.7 Median Absolute Deviation

Another measure of dispersion, which plays an important role when trying to detect outliers, is the *median absolute deviation* (MAD) statistic. To compute it, first compute the sample median, M, subtract it from every observed value, and then take absolute values. In symbols, compute

$$|X_1 - M|, \ldots, |X_n - M|.$$

The median of the n values just computed is MAD. Its finite sample breakdown point is 0.5.

EXAMPLE

Again using the values

12, 45, 23, 79, 19, 92, 30, 58, 132,

the median is $M = 45$, so $|X_1 - M| = |12 - 45| = 33$ and $|X_2 - M| = 0$. Continuing in this manner for all nine values yields

33, 0, 22, 34, 26, 47, 15, 13, 87.

Putting these values in ascending order yields

0 13 15 22 **26** 33 34 47 87.

MAD is the median of the nine values: 26.

Typically, MAD is divided by 0.6745 (for reasons related to the normal distribution, introduced in Chapter 3). For convenience, we set

$$\text{MADN} = \frac{\text{MAD}}{0.6745}. \tag{2.9}$$

2.4.8 R Function mad

Statisticians define MAD in the manner just described. R has a built-in function called

```
mad(x),
```

but it computes MADN, not MAD, again because in most situations, MADN is used.

2.4.9 Average Absolute Distance from the Median

There is a measure of dispersion closely related to MAD that should be mentioned. Rather than take the median of the values $|X_1 - M|, \ldots, |X_n - M|$, take the average of these values instead. That is, use

$$D = \frac{1}{n}\sum |X_i - M|. \tag{2.10}$$

Despite using the median, D has a finite sample breakdown point of only $1/n$. If, for example, we increase the largest of the X_i values ($X_{(n)}$), M does not change, but the difference between the largest value and the median becomes increasingly large, which, in turn, can make D arbitrarily large as well.

2.4.10 Other Robust Measures of Variation

There are many methods for measuring the variation among a batch of numbers, over 150 of which were compared by Lax (1985). Two that performed well, based on the criteria used by Lax, are the *biweight midvariance* and *percentage bend midvariance*, what Lax calls *A-estimators.* (Also see Randal, 2008.) And more recently some new measures of scatter have been proposed, two of which should be mentioned: Rocke's (1996) TBS (translated biweight S) estimator, and the tau measure of scale introduced by Yohai and Zamar (1988), which can be estimated as described by Maronna and Zamar (2002). Basically, these robust measures of variation offer protection against outliers and they might have certain advantages over

MAD or the Winsorized variance when dealing with issues described in later chapters. Recall that the Winsorized variance and MAD measure the variation of the middle portion of your data. In effect, they ignore a fixed percentage of the more extreme values. In contrast, both the biweight and percentage bend midvariances, as well as the tau measure of variation and TBS, make adjustments according to whether a value is flagged as being unusually large or small. The biweight midvariance, for example, empirically determines whether a value is unusually large or small using a slight modification of the outlier detection method that will be described in Section 2.4.2. These values are discarded and the variation among the remaining values is computed. Computational details are described in Box 2.1. Presumably readers will use R to perform the computations, so detailed illustrations are omitted. A method for computing the tau scale is described by Maronna and Zamar (2002, p. 310) and is used by the R function described in the next section. The TBS estimator is computationally complex, but an R function for computing it is described here as well. (Precise computational details for all of these measures of scatter are summarized in Wilcox, 2017.)

It is noted that the most common strategy for comparing two groups of individuals is in terms of some measure of location, as indicated in Chapter 1. However, comparing measures of variation can be of interest in which case the robust measures of variation introduced here might have a practical advantage for reasons best postponed until Chapter 7.

2.4.11 R Functions bivar, pbvar, tauvar, and tbs

The R functions

`pbvar(x)` and `bivar(x)`

(written for this book) compute the percentage bend midvariance and the biweight midvariance, respectively. Storing the values

12, 45, 23, 79, 19, 92, 30, 58, 132

in the R variable `x`, `pbvar(x)` returns the value 1527.75. If the largest value, 132, is increased to 1000, pbvar still returns the value 1527.75. If the two largest values (92 and 132) are increased to 1000, again `pbvar` returns the value 1527.75. Essentially, the two largest and two smallest values are ignored.

R functions

`tauvar(x)` and `tbsvar(x)`

compute the tau measure of variation and the tbs measure of variation, respectively. (There is also a tau measure of location, which is computed by the R function `tauloc`, but it does not play a role in this book.)

For the original values used to illustrate `pbvar`, the R function `bivar` returns 1489.4, a value very close to the percentage bend midvariance. But increasing the largest value (132) to 1000, bivar now returns the value 904.7. Its value decreases because it did not consider the value 132 to be an outlier, but increasing it to 1000, `pbvar` considers 1000 to be an outlier and subsequently ignores it. Increasing the value 92 to 1000, `bivar` returns 739. Now `bivar` ignores the two largest values because it flags both as outliers.

Box 2.1: How to compute the percentage bend midvariance and the biweight midvariance.

The percentage bend midvariance:

As usual, let $X_1, \ldots, X_n$ represent the observed values. Choose a value for the finite sample breakdown point and call it β. A good choice for general use is 0.2. Set $m = [(1-\beta)n + 0.5]$, the value of $(1-\beta)n + 0.5$ rounded down to the nearest integer. Let $W_i = |X_i - M|$, $i = 1, \ldots, n$, and let $W_{(1)} \leq \ldots \leq W_{(n)}$ be the W_i values written in ascending order. Set

$$\hat{\omega}_\beta = W_{(m)},$$

$$Y_i = \frac{X_i - M}{\hat{\omega}_\beta},$$

$$a_i = \begin{cases} 1, & \text{if } |Y_i| < 1 \\ 0, & \text{if } |Y_i| \geq 1, \end{cases}$$

in which case the estimated percentage bend midvariance is

$$\hat{\zeta}^2_{pb} = \frac{n\hat{\omega}^2_\beta \sum\{\Psi(Y_i)\}^2}{(\sum a_i)^2}, \tag{2.11}$$

where

$$\Psi(x) = \max[-1, \min(1, x)].$$

How to compute the biweight midvariance:

Set

$$Y_i = \frac{X_i - M}{9 \times MAD},$$

$$a_i = \begin{cases} 1, & \text{if } |Y_i| < 1 \\ 0, & \text{if } |Y_i| \geq 1, \end{cases}$$

$$\hat{\zeta}_{\text{bimid}} = \frac{\sqrt{n}\sqrt{\sum a_i(X_i - M)^2(1 - Y_i^2)^4}}{|\sum a_i(1 - Y_i^2)(1 - 5Y_i^2)|}. \tag{2.12}$$

The biweight midvariance is $\hat{\zeta}^2_{\text{bimid}}$.

It is not the magnitude of the biweight midvariance that will interest us in future chapters. Rather, the biweight midvariance plays a role when comparing groups of individuals, and it plays a role when studying associations among variables.

Some Comments on Measures of Variation

Several ways of measuring the variation among a batch of numbers have been described. So there is the practical issue of which one should be used. The answer depends in part on the type of problem that is of interest. For example, MADN and the interquartile range are important when checking for outliers. For some purposes yet to be described, MADN is preferable, and for others it is not. As for the other measures of variation described here, comments on which one to use will be made in subsequent chapters.

2.5 DETECTING OUTLIERS

For various reasons that will become clear, detecting outliers, unusually large or small values among a batch of numbers, can be very important. This section describes several strategies for accomplishing this goal. The first, which is based on the sample mean and variance, is motivated by certain results covered in Chapter 3. Unfortunately it can be highly unsatisfactory for reasons that will be illustrated, but it is important to know because currently it is commonly used.

2.5.1 A Method Based on the Mean and Variance

One strategy for detecting outliers is to check whether any values are more than two standard deviations away from the mean. Any such value is declared an outlier. In symbols, declare X an outlier if

$$\frac{|X - \bar{X}|}{s} > 2. \tag{2.13}$$

(The motivation for using 2 in this last equation stems from results covered in Chapter 3.)

EXAMPLE

Consider the values

2, 2, 2, 2, 2, 3, 3, 3, 3, 3, 4, 4, 4, 4, 4, 1000.

The sample mean is $\bar{X} = 65.94$; the sample standard deviation is $s = 249.1$;

$$\frac{|1000 - 65.94|}{249.1} = 3.75.$$

3.75 is greater than 2, so the value 1000 is declared an outlier. As is evident, the value 1000 is certainly unusual and, in this case, this outlier detection rule gives a reasonable result.

A concern with this outlier detection method is *masking*, which refers to the fact that the very presence of outliers can cause them to be missed. The reason is that in a very real sense, outliers have more of an influence on the sample standard deviation than the mean. The following illustrations demonstrate the problem.

EXAMPLE

Consider

2, 2, 2, 2, 2, 3, 3, 3, 3, 3, 4, 4, 4, 4, 4, 1000, 10000.

These are the same values as in the last example, but with another outlier added. The value 10,000 is declared an outlier using Equation (2.13). Surely 1000 is unusual compared to the bulk of the observations, but it is not declared an outlier. The reason is that the two outliers inflate the sample mean and especially the sample standard deviation. Moreover, the influence of the outliers on s is so large, the value 1000 is not declared an outlier. In particular, $\bar{X} = 650.3$, $s = 2421.4$, so

$$\frac{|1000 - \bar{X}|}{s} = 0.14.$$

EXAMPLE

Now consider the values

$$2, 2, 3, 3, 3, 4, 4, 4, 100000, 100000.$$

It is left as an exercise to verify that the value 100,000 is not declared an outlier using Equation (2.13), yet surely it is unusually large.

To avoid the problem of masking, we need a method for detecting outliers that is not itself affected by outliers. One way of accomplishing this goal is to switch to measures of location and variation that are relatively insensitive to outliers, meaning that they have a reasonably high breakdown point.

2.5.2 A Better Outlier Detection Rule: The MAD-Median Rule

Here is a simple outlier detection rule that has received a great deal of attention. Declare X to be an outlier if

$$\frac{|X - M|}{\text{MADN}} > 2.24, \tag{2.14}$$

where the value 2.24 in Equation (2.14) stems from Rousseeuw and van Zomeren (1990).[2] An important advantage of this outlier detection rule is that it addresses the problem of masking by employing measures of location and variation that have a breakdown point of 0.5. That is, the method can handle a large number of outliers without the problem of masking becoming an issue.

EXAMPLE

Consider again the values

$$2, 2, 3, 3, 3, 4, 4, 4, 100000, 100000.$$

Using our rule based on the sample mean and variance, the two values equal to 100,000 are not declared outliers. It can be seen that $M = 3.5$,

$$\text{MADN} = \text{MAD}/0.6745 = 0.7413,$$

and

$$\frac{100000 - 3.5}{0.7413} = 134893.4.$$

So in contrast to the rule based on the mean and variance, 100,000 would now be declared an outlier.

2.5.3 R Function out

The R function

```
out(x)
```

[2]Equation (2.14) is known as the *Hampel identifier*, but Hampel used the value 3.5 rather than 2.24. For some refinements, see Davies and Gather (1993).

has been supplied, which detects outliers using Equation (2.14). If the values from the last example are stored in the R variable `data`, then part of the output from the command `out(data)` is

```
$out.val:
[1] 100000 100000

$out.id:
[1]  9 10
```

That is, there are two values declared outliers, both equal to 100,000, and they are the ninth and tenth observations stored in the R variable data. That is, the outliers are stored in `data[9]` and `data[10]`.

2.5.4 The Boxplot

Proposed by Tukey (1977), a boxplot is a commonly used graphical summary of data that provides yet another method for detecting outliers. An example of a boxplot is shown in Figure 2.1. As indicated, the ends of the rectangular box mark the lower and upper quartiles. That is, the box indicates where the middle half of the data lie. The horizontal line inside the box indicates the position of the median. The lines extending out from the box are called whiskers.

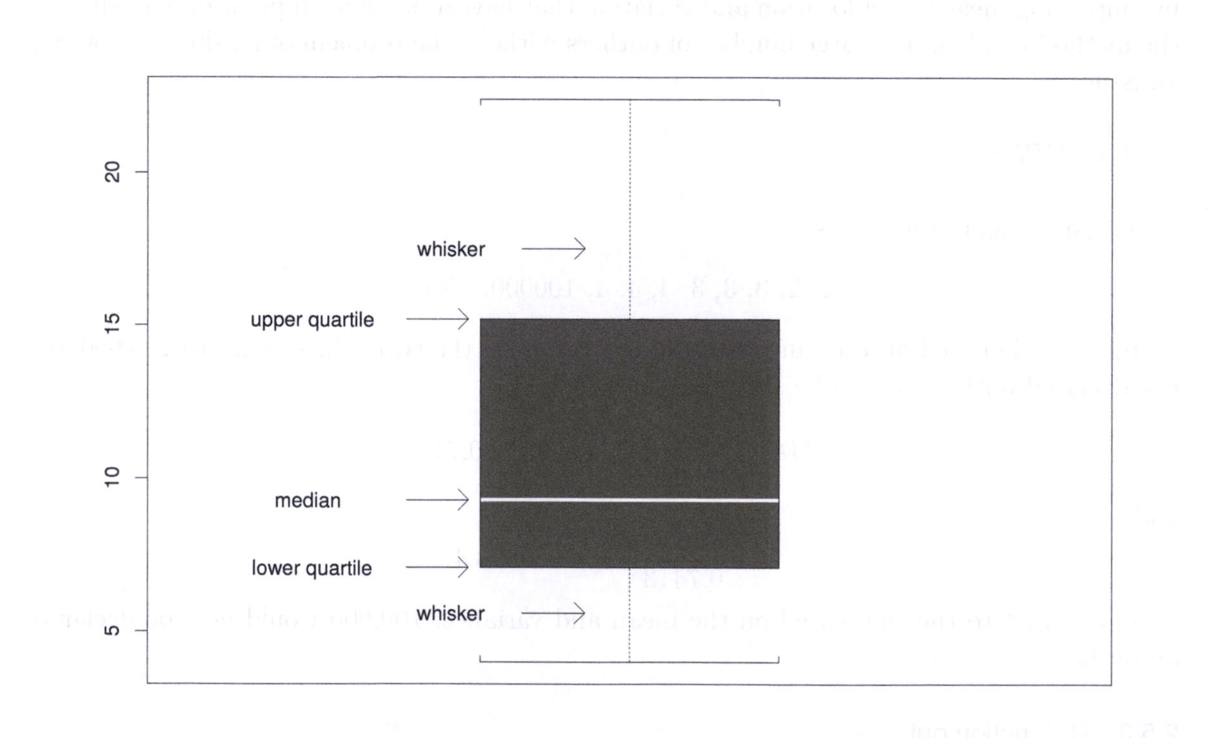

Figure 2.1 Example of a boxplot with no outliers.

The *boxplot rule* declares the value X to be an outlier if

$$X < q_1 - 1.5(\text{IQR}), \tag{2.15}$$

or

$$X > q_2 + 1.5(\text{IQR}), \tag{2.16}$$

where IQR is the interquartile range defined in Section 2.3.3, and q_1 and q_2 are again the ideal fourths given by Equations (2.6) and (2.7).

EXAMPLE

Figure 2.2 shows a boxplot with two outliers. The ends of the whiskers are called *adjacent values.* They are the smallest and largest values not declared outliers. Because the interquartile range has a finite sample breakdown point of 0.25, more than 25% of the data must be outliers before the problem of masking occurs. A breakdown point of 0.25 seems to suffice in most situations, but exceptions occur.

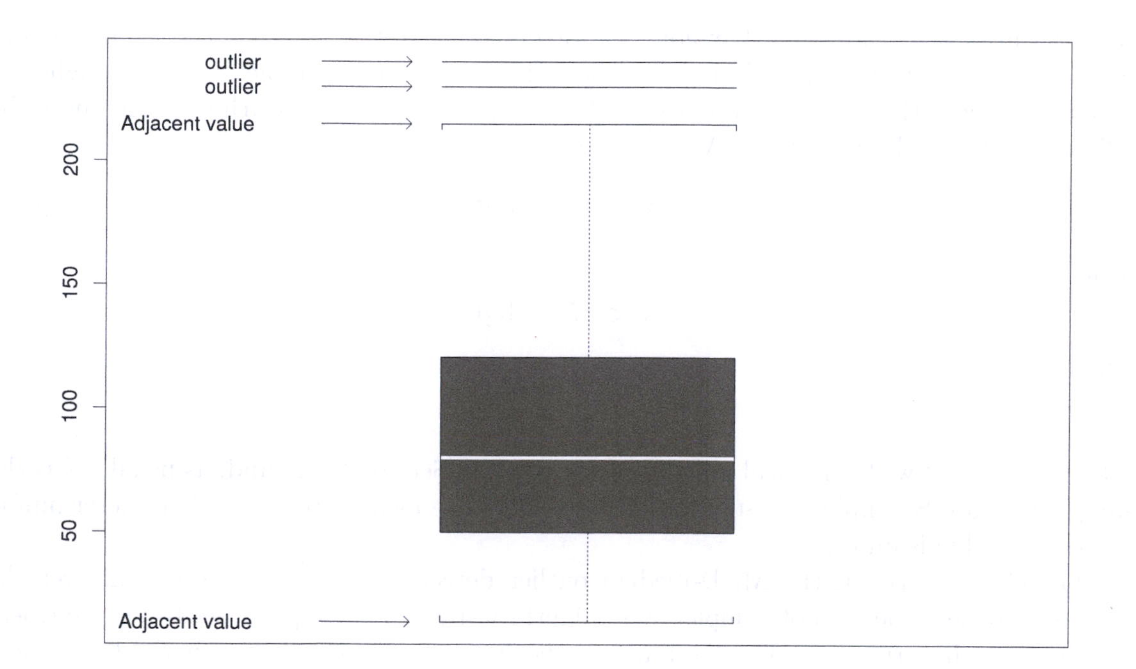

Figure 2.2 Example of a boxplot with two outliers.

2.5.5 R Function boxplot

The built-in R function

```
boxplot(x)
```

creates a boxplot. A numerical summary of which points are declared outliers with the R function boxplot can be determined with the command

```
print(boxplot(x,plot=F)$out).
```

If no outliers are found, the function returns `NULL` on the computer screen.

The built-in R function **boxplot** does not use the ideal fourths when estimating the quartiles. To detect outliers using a boxplot rule that does use the ideal fourths, see Section 2.4.7.

EXAMPLE

For the data in Table 2.2, a boxplot declares all values greater than 13.5 to be outliers. This represents 11.4% of the data. In contrast, the MAD-median rule in Section 2.4.2 declares all values greater than or equal to 6 to be outliers, which is 27.6% of the data. This suggests that masking might be a problem for the boxplot because the proportion of outliers using the rule in Section 2.4.2 exceeds 0.25, the breakdown point of the boxplot. If we use the sample mean and standard deviation to detect outliers, only the value 6000 is declared an outlier.

2.5.6 Modifications of the Boxplot Rule for Detecting Outliers

A criticism of the traditional boxplot rule for detecting outliers is that the expected proportion of numbers that are declared outliers depends on the sample size; it is higher in situations where the sample size is small. To correct this problem, Carling (2000) uses the following rule: Declare the value X an outlier if

$$X > M + k\text{IQR} \tag{2.17}$$

or if

$$X < M - k\text{IQR} \tag{2.18}$$

where

$$k = \frac{17.63n - 23.64}{7.74n - 3.71},$$

IQR is estimated with the ideal fourths as described in Section 2.3.2, and, as usual, M is the sample median. So unlike the standard boxplot rule, the median plays a role in determining whether a value is an outlier.

The choice between the MAD-median outlier detection rule in Section 2.4.2 over the method just described is not completely straightforward. For some purposes to be described, the MAD-median rule is preferable, but for other purposes it is not. The method used here has a breakdown point of 0.25 in contrast to a breakdown point of 0.5 using the MAD-median rule. As previously noted, generally a breakdown point of 0.25 suffices, but exceptions can occur. However, based on the breakdown point alone, the MAD-median rule is superior to the boxplot rule or Carling's modification. Furthermore, based on other criteria and goals described in later chapters, for some purposes the boxplot rule is more satisfactory.

Finally, it is noted that there is a vast literature on detecting outliers. A few additional issues will be covered in subsequent chapters. Readers interested in a book-length treatment of this topic are referred to Barnett and Lewis (1994).

2.5.7 R Function outbox

The R function

```
outbox(x,mbox=F),
```

written for this book, checks for outliers using one of two methods. With mbox=F, it uses the rule given by Equations (2.15) and (2.16), but unlike R, the quartiles are estimated with the ideal fourths. Setting mbox=TRUE, outliers are detected using Carling's method given by Equations (2.17) and (2.18).

EXAMPLE

Here is an example of how the R function outbox reports results, using the boxplot rule (mbox=FALSE), again using the data in Table 2.2.

```
$out.val:
 [1]   45  150   19   18  150   40   30   19   15   30 6000   15

$out.id:
 [1]  30  35  37  40  62  64  66  79  89  98 100 105

$keep:
  1 2 3 4 5 6 7 8 9 10  11  12  13  14  15  16  17  18  19 20
  21 22 23 24 25 26 27 28 29 31 32 33  34 36 38 39 41 42 43
 44 45 46 47 48 49 50 51 52 53 54 55 56 57 58 59 60 61  63
 65 67 68 69 70 71 72 73 74 75 76 77 78 80 81 82 83 84  85
 86 87 88 90 91 92 93  94 95 96 97 99 101 102 103 104

$cl:
[1] -6.5

$cu:
[1] 13.5
```

The first line lists the values declared outliers. The next line indicates where in the R variable, containing the data, the values are located. For instance, if the data are stored in the variable `z`, the first outlier is stored in `z[30]` and contains the value 45. The locations of the values not declared outliers are indicated by the output called keep. This is useful if you want to eliminate outliers and store the remaining data in another R variable. For example, the R command

```
w=z[outbox(z)$keep]
```

will store all values not declared outliers in the R variable `w`. The values denoted by `cl` and `cu` indicate, respectively, how small and how large a value must be to be declared an outlier.

To use Carling's method, set the argument mbox=TRUE. For the data in Table 2.2, now values greater than or equal to 13 are declared outliers. But as previously indicated, with mbox=FALSE, 13 is not declared an outlier. Situations also arise where a value is declared an outlier with mbox=FALSE, but not when mbox=TRUE.

2.5.8 Other Measures of Location

Section 2.2 mentioned that, when considering measures of location, there are two general strategies for dealing with outliers. The first is to trim a specified portion of the largest and smallest values, with the median being the best-known example. The second is to use a measure of location that somehow identifies and eliminates any outliers. So if no outliers are found, no values are trimmed. There are many variations of this latter approach, two examples of which are given here.

The first is called a modified one-step M-estimator. It simply checks for any outliers using the MAD-median rule in Section 2.4.2, eliminates any outliers that are found, and then averages the values that remain. In symbols, if ℓ (for lower) values are declared unusually small, and u (for upper) are declared unusually large, the modified one-step M-estimator is

$$\bar{X}_{\text{mom}} = \frac{1}{n-\ell-u} \sum_{i=\ell+1}^{n-u} X_{(i)}. \tag{2.19}$$

The finite sample breakdown point is 0.5. This modified one-step M-estimator, *MOM*, is an example of what are known as skipped estimators of location, which were studied by Andrews et al. (1972). That is, check for outliers, remove any that are found and average the remaining values. (Early versions of skipped estimators used a boxplot rule to detect outliers rather than the MAD-median rule.)

EXAMPLE

Computing a modified one-step M-estimator is illustrated with the following observations:

77, 87, 88, 114, 151, 210, 219, 246, 253, 262, 296, 299, 306, 376, 428, 515, 666, 1310, 2611.

It can be seen that $n = 19$, $M = 262$, MADN=MAD/0.6745=169, and the three largest values are declared outliers using the MAD-median rule. That is, $u = 3$. None of the smallest values are declared outliers, so $\ell = 0$. Consequently,

$$\bar{X}_{\text{mom}} = \frac{1}{16}(77 + 87 + \cdots + 515) = 245.4.$$

The measure of location just illustrated is a modification of what is called a one-step M-estimator. The most commonly used version of the one-step M-estimator, which has received considerable attention in the statistics literature, uses a slight variation of the MAD-median mean rule. In particular, now the value X is declared an outlier if

$$\frac{|X - M|}{\text{MADN}} > 1.28. \tag{2.20}$$

There is one other fundamental difference. If the number of small outliers (ℓ) is not equal to the number of large outliers (u), the one-step M-estimator is given by

$$\bar{X}_{\text{os}} = \frac{1.26(\text{MADN})(u-\ell)}{n-u-\ell} + \frac{1}{n-u-\ell} \sum_{i=\ell+1}^{n-u} X_{(i)}. \tag{2.21}$$

Roughly, remove any outliers, average the values that remain, but if $u \neq \ell$ (the number of large outliers is not equal to the number of small outliers), adjust the measure location by adding

$$\frac{1.26(\text{MADN})(u-\ell)}{n-u-\ell}.$$

When first encountered, this one-step M-estimator looks peculiar. Why does MADN, a measure of variation, play a role when computing a measure of location? An adequate explanation goes beyond the scope of this book. (For a relatively nontechnical explanation, see Wilcox, 2017.) It is, in part, a consequence of how a one-step M-estimator is formally defined, which is not made clear here. The important point in this book is that it has a 0.5 breakdown point and that based on criteria described in subsequent chapters, there are conditions where it has a practical advantage over the modified one-step M-estimator, $\bar{X}_{\text{mom}}$.

EXAMPLE

The previous example is repeated, only now a one-step M-estimator is computed. Based on Equation (2.20), the four largest values are outliers ($u = 4$). There are no small values declared outliers, so $\ell = 0$. Removing the four outliers and averaging the values left yields 227.47. Because $u = 4$ and $\ell = 0$, $1.26(MADN)(4 - 0)/(19 - 4) = 56.79$. So,

$$\bar{X}_{\text{os}} = 56.79 + 227.47 = 284.26.$$

Many alternative measures of location have been proposed (Andrews et al., 1972) that are not described here. But comments on one collection of estimators should be made. They are called *R-estimators*, that include the Hodges–Lehmann estimator as a special case. To compute the Hodges–Lehmann estimator, first average every pair of observations. The median of all such averages is the Hodges–Lehmann estimator. There are conditions where this measure of location, as well as R-estimators in general, have good properties. But there are general conditions under which they perform poorly (e.g., Bickel and Lehmann, 1975; Huber, 1981, p. 65; Morgenthaler and Tukey, 1991, p. 15.). Consequently, further details are omitted.

A point that cannot be stressed too strongly is that when outliers are discarded, this is not to say that they are uninteresting or uninformative. Outliers can be very interesting, but for some goals they do more harm than good for reasons made clear in subsequent chapters.

2.5.9 R Functions mom and onestep

The R function

```
onestep(x)
```

computes the one-step M-estimator just illustrated. The function

```
mom(x)
```

computes the modified one-step M-estimator.

2.6 HISTOGRAMS

This section introduces a classic graphical tool for summarizing data called a histogram. The basic idea is illustrated with data from a heart transplant study conducted at Stanford University between October 1, 1967 and April 1, 1974. Of primary concern is whether a transplanted heart will be rejected by the recipient. With the goal of trying to address this issue, a so-called T5 mismatch score was developed by Dr. C. Bieber. It measures the degree of dissimilarity between the donor and the recipient tissue with respect to HL-A antigens. Scores less than 1 represent a good match and scores greater than 1 represent a poor match. Of course, of particular interest is how well a T5 score predicts rejection, but this must wait for now. The T5 scores, written in ascending order, are shown in Table 2.3 and are taken from Miller (1976).

A histogram simply groups the data into categories and plots the corresponding frequencies. A simple and commonly used variation is to plot the relative frequencies instead. To illustrate the basic idea, we group the T5 values into seven categories: (1) values greater than 0.0 but less than or equal to 0.5, (2) values greater than 0.5 but less than or equal to 1.0, and

Table 2.3 T5 Mismatch Scores from a Heart Transplant Study

0.00	0.12	0.16	0.19	0.33	0.36	0.38	0.46	0.47	0.60	0.61	0.61	0.66	0.67	0.68
0.69	0.75	0.77	0.81	0.81	0.82	0.87	0.87	0.87	0.91	0.96	0.97	0.98	0.98	1.02
1.06	1.08	1.08	1.11	1.12	1.12	1.13	1.20	1.20	1.32	1.33	1.35	1.38	1.38	1.41
1.44	1.46	1.51	1.58	1.62	1.66	1.68	1.68	1.70	1.78	1.82	1.89	1.93	1.94	2.05
2.09	2.16	2.25	2.76	3.05										

Table 2.4 Frequencies and Relative Frequencies for Grouped T5 Scores, $n = 65$

Test Score (x)	Frequency	Relative Frequency
0.0 - 0.5	8	$9/65 = 0.138$
0.5 - 1.0	20	$20/65 = 0.308$
1.0 - 1.5	18	$18/65 = 0.277$
1.5 - 2.0	12	$12/65 = 0.138$
2.0 - 2.5	4	$4/65 = 0.062$
2.5 - 3.0	1	$1/65 = 0.015$
3.0 - 3.5	1	$1/65 = 0.015$

so on. The frequency and relative frequency associated with each of these intervals is shown in Table 2.4. For example, there are 9 T5 mismatch scores in the interval extending from 0.0 to 0.5 and the proportion of all scores belonging to this interval is 0.138.

Figure 2.3 shows the histogram for the T5 scores that was created with the built-in R function `hist`. Notice that the base of the left most rectangle extends from 0 to 0.5 and has a height of 9. This means that there are nine cases where a T5 score has a value between 0 and 0.5. The base of the next rectangle extends from 0.5 to 1 and has a height of 20. This means that there are 20 T5 scores having a value between 0.5 and 1. The base of the next rectangle extends from 1 to 1.5 and has a height of 18, so there are 18 T5 scores between 1 and 1.5.

Histograms attempt, among other things, to tell us something about the overall shape of the data. In particular, they provide some indication of whether the data are reasonably symmetric about some central value, which is an important issue for reasons described later in this book. Sometimes it is suggested that histograms be used to check for outliers. But in general, histograms are not good outlier detection methods compared to the boxplot rule or the MAD-median rule.

EXAMPLE

Data were collected on the net worth of all billionaires, a histogram of which is shown in Figure 2.4. The histogram might suggest that among billionaires, outliers correspond to 20 billion or higher. But the boxplot rule indicates that values greater than or equal to 5.5 billion are outliers. And the MAD-median rule indicates that values greater than or equal to 3.8 billion are outliers, the only point being that the boxplot rule and the MAD-median rule can result in a substantially different sense about which values are outliers. One fundamental problem is that, when using a histogram, no precise rule is available for telling us when a value should be declared an outlier. Without a precise rule there can be no agreement

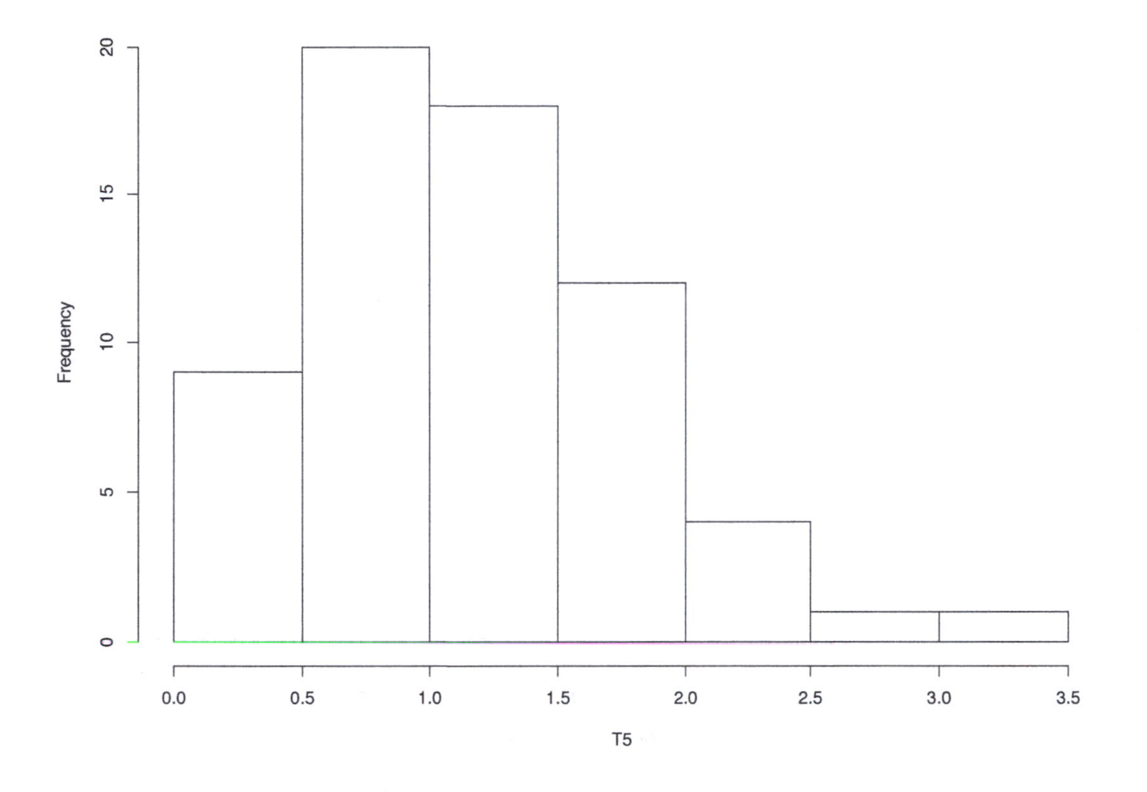

Figure 2.3 Example of a histogram based on the T5 mismatch data.

regarding the extent to which masking is avoided. Simultaneously, it is difficult to judge the relative merits of the histogram in terms of its ability to detect truly unusual values.

There is another aspect of histograms that should be discussed. A fundamental goal is understanding the shape of the histogram if all participants could be measured. In particular, if data on everyone in the population of interest were available, would the plot of the data be reasonably symmetric or would it be noticeably skewed? In some situations, histograms often accomplish this goal reasonably well with a sample size of 100. But in other situations, this is not the case; a larger sample size is needed.

To elaborate, imagine that the goal is to study liver damage caused by a drug. To be concrete, imagine that we are interested in all adults over the age of 60, and to keep the illustration simple, further imagine that there are exactly 1 million such individuals. Further suppose that if we could measure all 1 million individuals, we would get a histogram as shown in Figure 2.5 (Here, the y-axis indicates the relative frequencies.)

Now imagine that 100 individuals are selected from the 1 million individuals, with every individual having the same probability of being chosen. Mimicking this process on a computer resulted in the histogram shown in Figure 2.6. So in this particular case, the histogram provides a reasonable reflection of the population histogram, roughly capturing its bell shape. A criticism of this illustration is that maybe we just got lucky. That is, in general, perhaps with only 100 individuals, the histogram will not accurately reflect the population. This might indeed happen, but for the special case under consideration, typically it gives a reasonable sense of the shape of the population histogram.

Now consider the population histogram in Figure 2.7. This histogram has the same bell

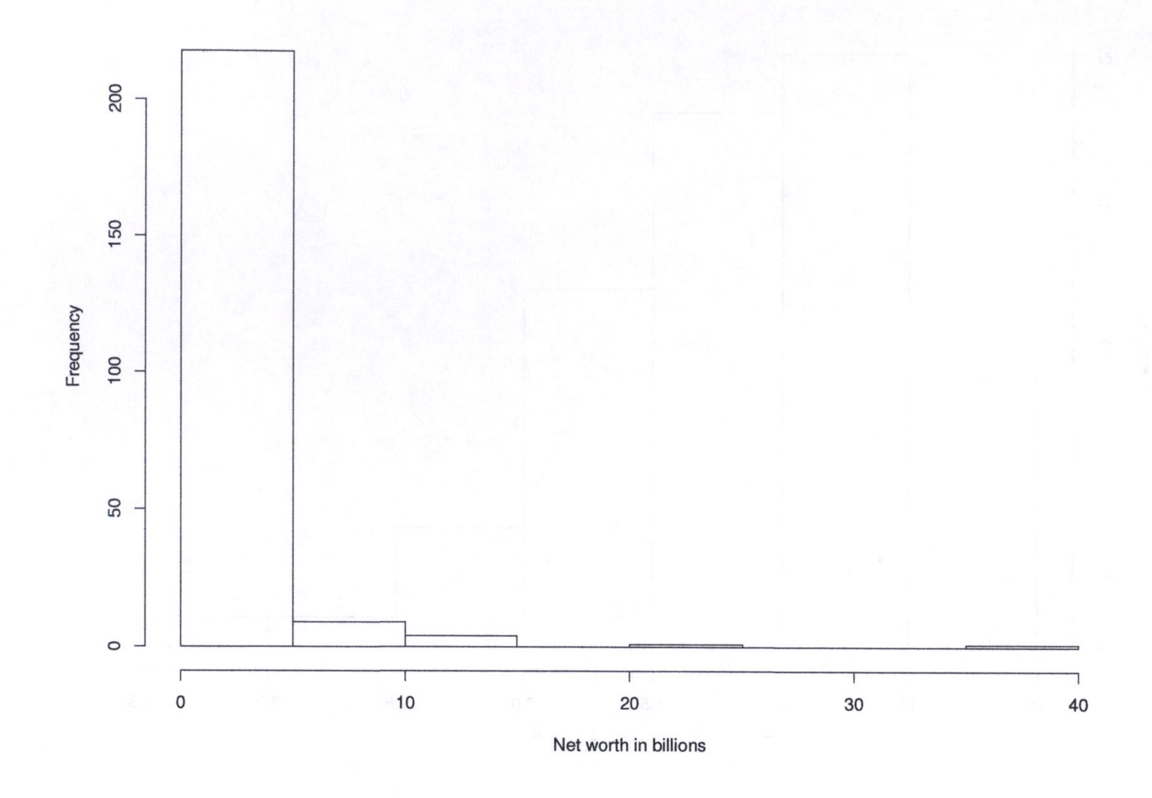

Figure 2.4 This histogram might suggest that any value greater than 20 is an outlier. But a boxplot rule indicates that values greater than or equal to 5.5 are outliers.

shape as in Figure 2.5, but the tails extend out farther. This reflects the fact that for this particular population, there are more outliers or extreme values. Now look at Figure 2.8, which is based on 100 individuals sampled from the population histogram in Figure 2.7. As is evident, it provides a poor indication of what the population histogram looks like. Figure 2.8 also provides another illustration that the histogram can perform rather poorly as an outlier detection rule. It suggests that values greater than 10 are highly unusual, which turns out to be true based on how the data were generated. But values less than −5 are also highly unusual, which is less evident here. The fact that the histogram can miss outliers limits its ability to deal with problems yet to be described.

A rough characterization of the examples just given is that when the population histogram is symmetric and bell-shaped, and outliers tend to be rare, it performs tolerably well with 100 observations, in terms of indicating the shape of the population histogram. But when outliers are relatively common, the reverse is true.

2.6.1 R Functions hist and splot

The R function

```
hist(x)
```

creates a histogram. The x-axis will be labeled with the name of the variable containing the data. For example, `hist(zz)` will result in the x-axis being labeled zz. To use another label,

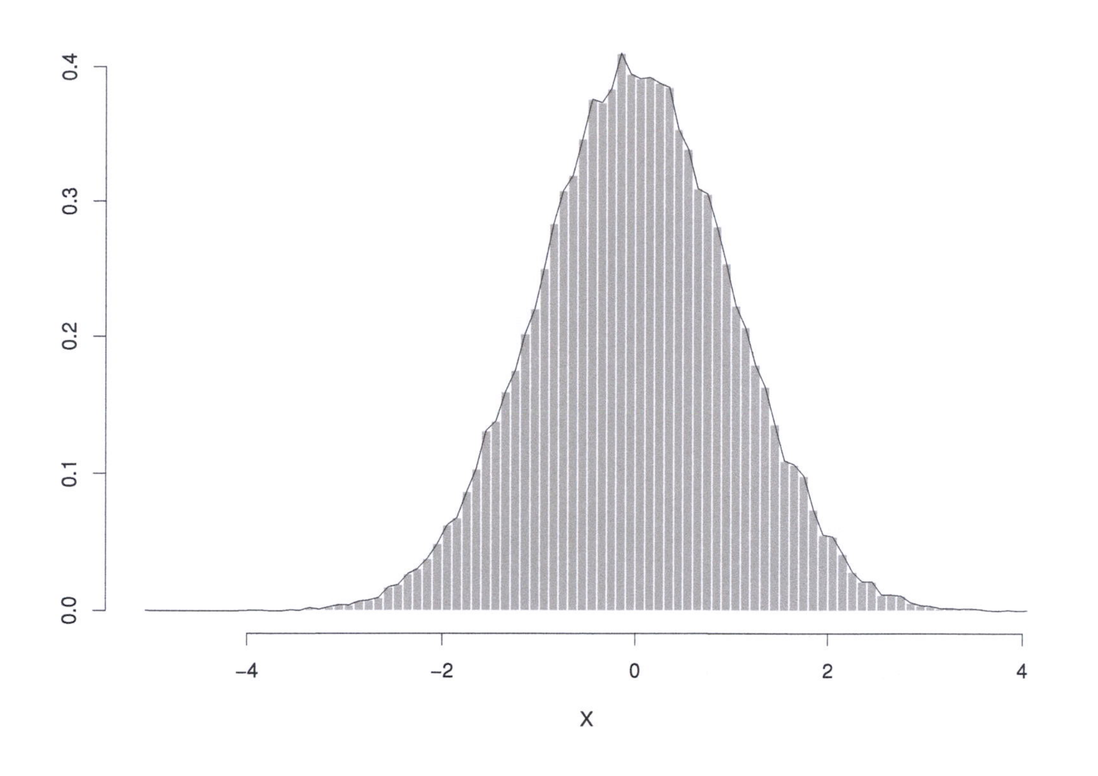

Figure 2.5 Histogram of hypothetical liver-damage data based on 1 million individuals.

use the argument `xlab`. For example, `hist(zz,xlab="protein")` results in the x-axis being labeled protein.

When the number of observed values among a batch of numbers is relatively small, rather than group the values into bins as done by a histogram, it might be preferable to use the height of a spike to indicate the relative frequencies. For example, if the possible responses are limited to the values 1, 2, 3, 4, and 5, and if the frequencies are $f_1 = 12$, $f_2 = 22$, $f_3 = 32$, $f_4 = 19$, and $f_5 = 8$, then a simple graphical summary of the data consists of an x-axis that indicates the observed values and a y-axis that indicates the relative frequencies by the height of a spike. The R function

```
splot(x,xlab="X",ylab = "Rel. Freq."),
```

written for this book, creates such a plot. The arguments `xlab` and `ylab` can be used to label the x-axis and y-axis, respectively.

2.7 KERNEL DENSITY ESTIMATORS

Kernel density estimators (e.g., Silverman, 1986) represent a large class of methods aimed at improving on the histogram. Some of these estimators appear to perform fairly well over a broad range of situations. From an applied point of view, kernel density estimators can be very useful when dealing with certain issues described in Chapter 7. But, for now, the only goal is to introduce two versions and illustrate that they can be more satisfactory than a histogram when trying to get a sense of the overall shape of the data.

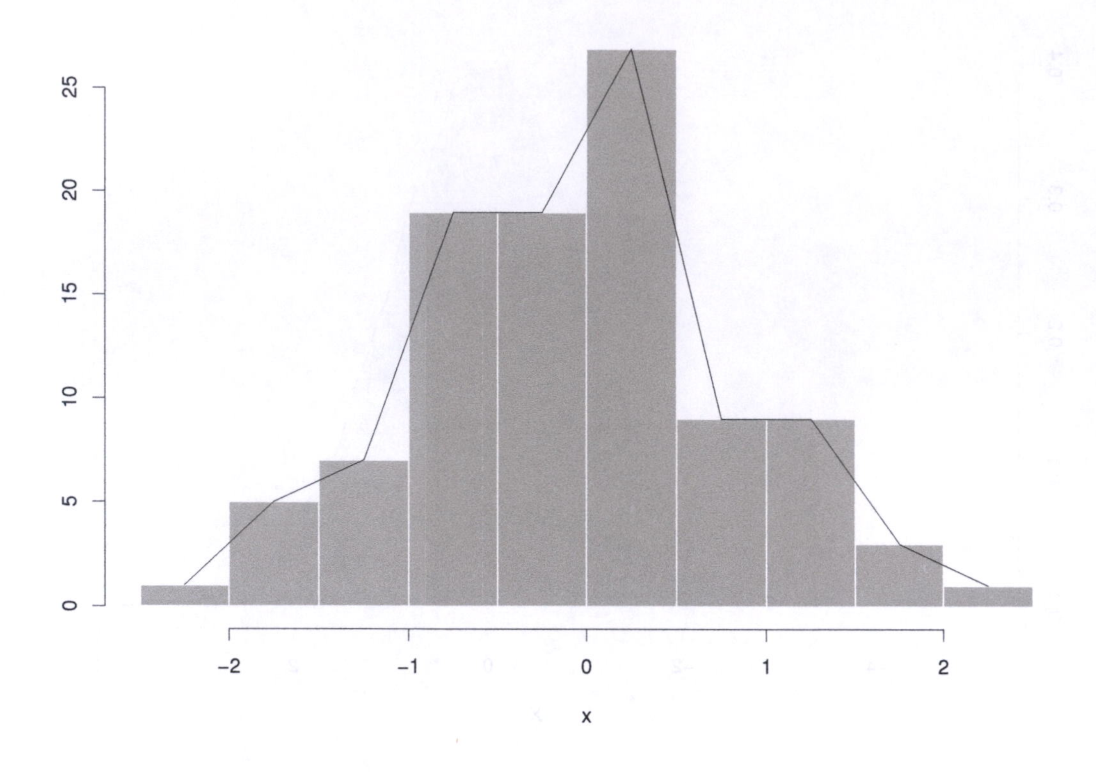

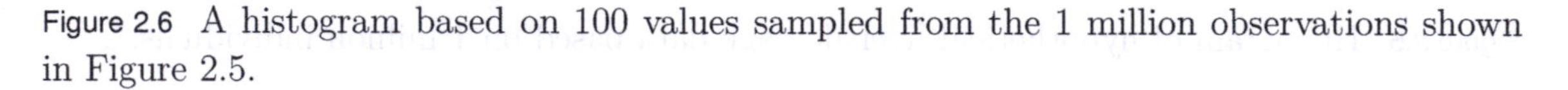

Figure 2.6 A histogram based on 100 values sampled from the 1 million observations shown in Figure 2.5.

In the previous illustration involving liver damage, imagine that you want to determine the proportion of individuals who would get a value close to 2. A very rough characterization of kernel density estimators is that they empirically determine which observed values are close to 2, or whatever value happens to be of interest, based on some measure of variation, such as the interquartile range or the sample variance. Then the proportion of such points is computed. (This greatly oversimplifies the mathematical issues, and it ignores some key tools used by kernel density estimators, but this description suffices for present purposes.)

Rosenblatt's shifted histogram provides a relatively simple example. It uses results derived by Scott (1979) as well as Freedman and Diaconis (1981). In particular, let

$$h = \frac{1.2(\text{IQR})}{n^{1/5}},$$

where IQR is the interquartile range. Let A be the number of observations less than or equal to $x + h$. Let B be the number of observations strictly less than $x - h$. Then the proportion of values close to x, say $f(x)$, among all individuals we might measure, is estimated with

$$\hat{f}(x) = \frac{A - B}{2nh}.$$

By computing $\hat{f}(x)$ for a range of x values and plotting the results, we get an estimate of what a plot of all of the data would look like if all participants could be measured.

There are numerous suggestions regarding how we might improve on Rosenblatt's shifted

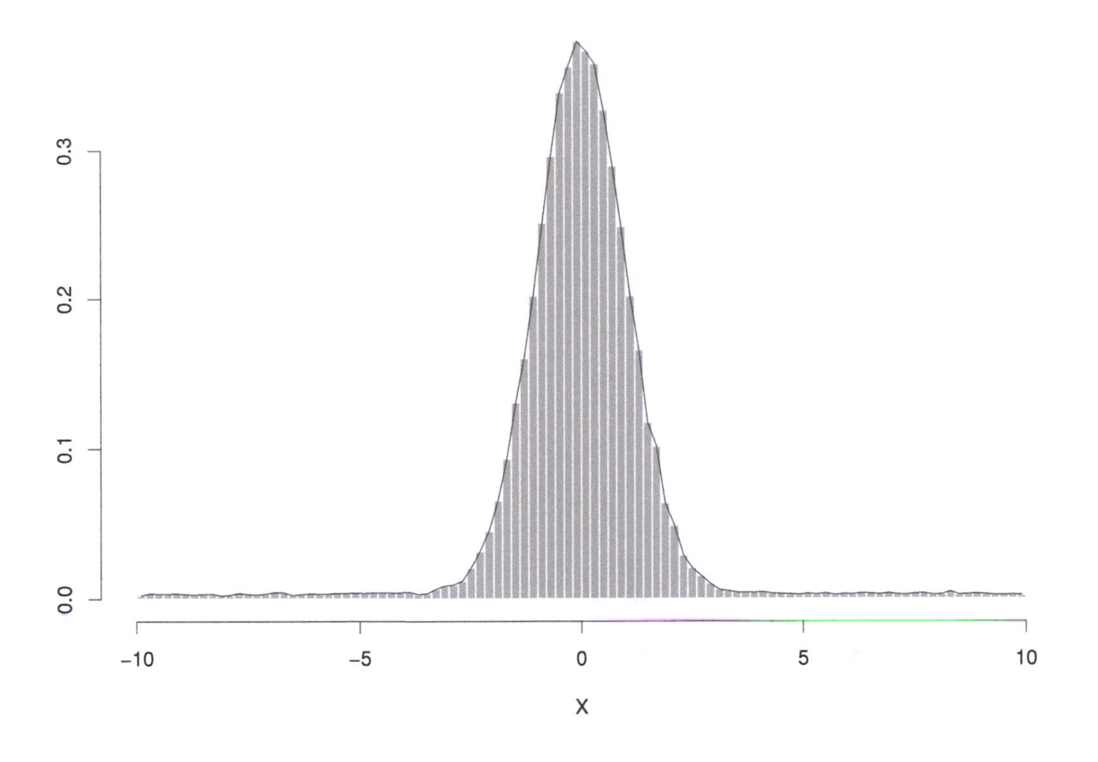

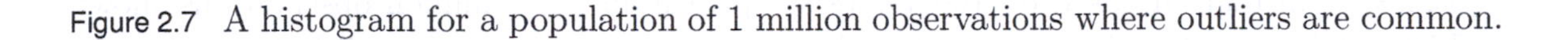

Figure 2.7 A histogram for a population of 1 million observations where outliers are common.

histogram, but the details go beyond the scope of this book. Here it is merely remarked that a certain variation of what is called an *adaptive kernel density estimator* appears to perform relatively well, but this is not to suggest that it is always best. (Computational details can be found in Wilcox, 2017. For recent results related to kernel density estimators, see Harpole et al., 2014. Also see Bourel et al., 2014, for an alternative approach.)

2.7.1 R Functions kdplot and akerd

The R function

```
kdplot(x,xlab = "X", ylab = "Y")
```

computes Rosenblatt's shifted histogram using the values stored in x. The function

```
akerd(x,xlab = "X", ylab = "Y")
```

uses the adaptive kernel density estimator. In general, it seems best to use `akerd`. But with extremely large sample sizes, execution time on a computer might become an issue, in which case `kdplot` can be used.

EXAMPLE

Consider again the histogram in Figure 2.8, which was based on data generated from the histogram in Figure 2.7. If instead we use the R function `akerd` to plot the data, which uses

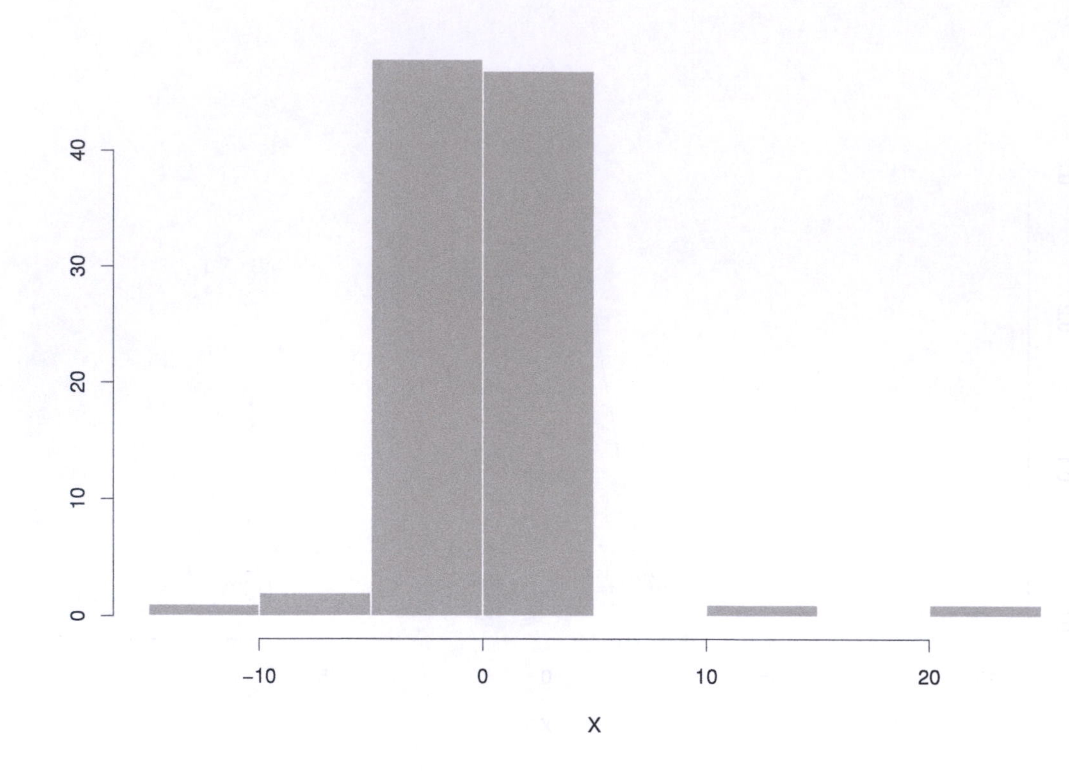

Figure 2.8 A histogram based on 100 values sampled from the 1 million observations shown in Figure 2.7. Note how it suggests that there are outliers in the right tail but not the left. There are, in fact, outliers in both tails.

an adaptive kernel density estimate, we get the plot in Figure 2.9. Note that it does much better in terms of providing a sense of what the population histogram is like.

2.8 STEM-AND-LEAF DISPLAYS

A stem-and-leaf display is another method of gaining some overall sense of what data are like. The method is illustrated with measures taken from a study aimed at understanding how children acquire reading skills. A portion of the study was based on a measure that reflects the ability of children to identify words.[3] Table 2.5 lists the observed scores in ascending order.

The construction of a stem-and-leaf display begins by separating each value into two components. The first is the *leaf*, which, in this example, is the number in the ones position (the single digit just to the left of the decimal place). For example, the leaf corresponding to the value 58 is 8. The leaf for the value 64 is 4 and the leaf for 125 is 5. The digits to the left of the leaf are called the *stem*. Here the stem of 58 is 5, the number to the left of 8. Similarly, 64 has a stem of 6 and 125 has a stem of 12. We can display the results for all 81 children as follows:

[3]These data were supplied by L. Doi.

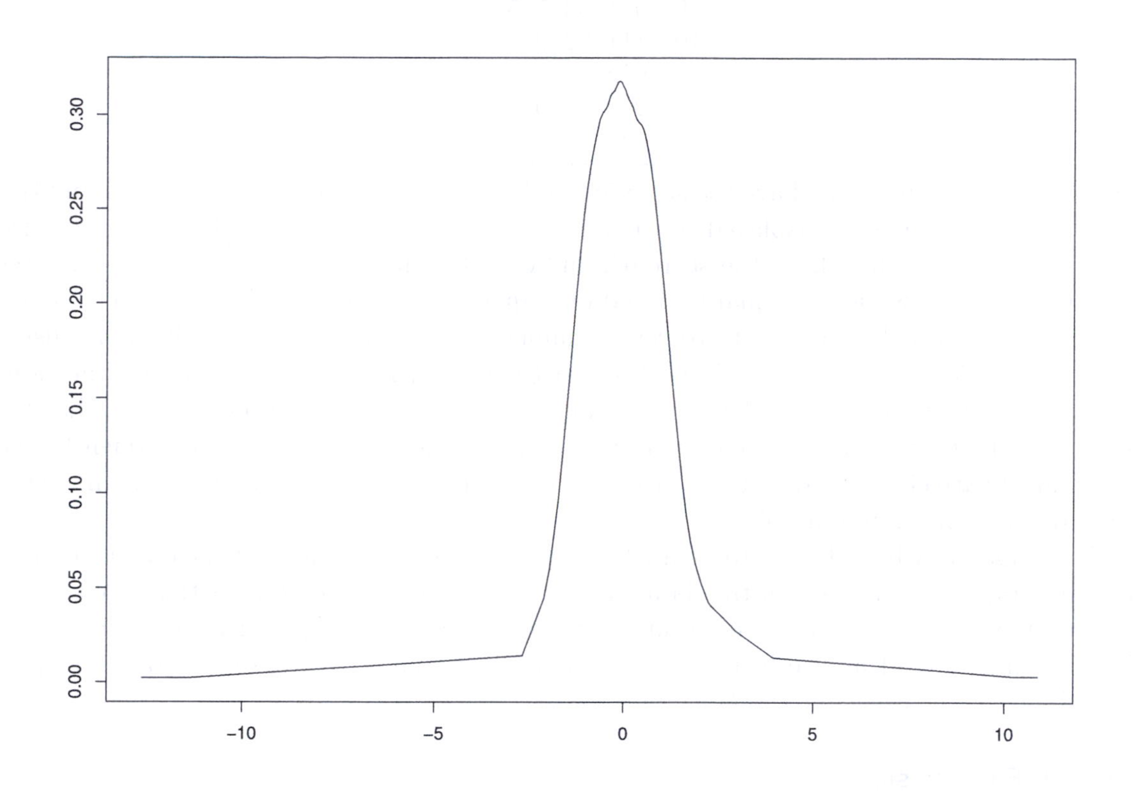

Figure 2.9 A kernel density plot based on the same 100 values used in Figure 2.8. Note how the shape of this plot does a better job, compared to Figure 2.8, of capturing the shape of the 1 million values shown in Figure 2.7.

Table 2.5 Word Identification Scores

58 58 58 58 58 64 64 68 72 72 72 75 75 77 77 79 80 82 82
82 82 82 84 84 85 85 90 91 91 92 93 93 93 95 95 95 95 95
95 95 95 98 98 99 101 101 101 102 102 102 102 102 103 104 104 104 104
104 105 105 105 105 105 107 108 108 110 111 112 114 119 122 122 125 125 125
127 129 129 132 134

Stems	Leaves
5	88888
6	448
7	22255779
8	0222224455
9	011233355555555889
10	1112222234444455555788
11	01249
12	22555799
13	24

There are five children who have the score 58, so there are five scores with a leaf of 8, and this is reflected by the five 8s displayed to the right of the stem 5 and under the column headed by Leaves. Two children have the score 64, and one child has the score 68. That is, for the stem 6, there are two leaves equal to 4 and one equal to 8, as indicated by the list of leaves in the display. Now look at the third row of numbers where the stem is 7. The leaves listed are 2, 2, 2, 5, 5, 7, 7, and 9. This indicates that the value 72 occurred three times, the value 75 occurred two times, as did the value 77, and the value 79 occurred once. Notice that the display of the leaves gives us some indication of the values that occur most frequently and which are relatively rare. Like the histogram, the stem-and-leaf display gives us an overall sense of what the values are like.

The choice of which digit is to be used as the leaf depends in part on which digit provides a useful graphical summary of the data. But details about how to address this problem are not covered here. Suffice it to say that algorithms have been proposed for deciding which digit should be used as the leaf and determining how many lines a stem-and-leaf display should have (e.g., Emerson and Hoaglin, 1983).

2.8.1 R Function stem

The R built-in function

```
stem(x)
```

creates a stem-and-leaf display. For the T5 mismatch scores, the stem-and-leaf display created by R is as follows.

EXAMPLE

```
Decimal point is at the colon

   0 : z122344
   0 : 556667777788889999
   1 : 00000111111223334444
   1 : 5566777788999
   2 : 0122
   2 : 8
   3 : 0
```

The z in the first row stands for zero. This function also reports the median and the quartiles.

2.9 SKEWNESS

When applying classic, routinely used inferential techniques described later in this book, an issue of considerable practical importance is the extent to which a plot of the data is symmetric around some central value. If it is symmetric about some value, various technical issues, to be described, are avoided. But when the plots are asymmetric, serious concerns arise that require more modern techniques, as will become evident. In Figures 2.3 and 2.4, the histograms are not symmetric. Both are skewed to the right, roughly meaning that the longer tail occurs on the right side of the plot. This plot suggests that if the entire population could be measured, a plot of the data might be skewed as well. In practice, one might simply hope that skewness is not a practical issue, but a seemingly better strategy is to consider methods aimed at dealing with known problems, some of which are described in subsequent chapters. The only goal for the moment is to draw attention to the issue of skewness and to make a few comments about the strategy of transforming the data.

2.9.1 Transforming Data

One suggestion for dealing with skewness is to transform the data. A well-known approach is to take logarithms, and more complex transformations have been suggested. But there are two concerns about this strategy. The first is that after applying commonly suggested transformations, plots of the data can remain noticeably skewed. Second, after transforming the data, outliers and their deleterious effects can remain (e.g., Rasmussen, 1989). In some cases, after transforming data, a plot of the data will be more symmetric, but Doksum and Wong (1983) concluded that when dealing with measures of location, trimming can remain beneficial (based on criteria yet to be described). Subsequent chapters will describe the relative merits of alternative techniques for dealing with skewness.

EXAMPLE

Consider again the data in Figure 2.4. As is evident, the histogram indicates that the data are highly skewed to the right. If we take logarithms, a (kernel density) plot of the results is shown in Figure 2.10, which is again highly skewed to the right. A boxplot rule indicates that for the data in Figure 2.4, the largest 13 values are declared outliers. For the data in Figure 2.10 the largest 6 values are declared outliers. So, in this case, the number of outliers is reduced, but situations are encountered where the number stays the same or actually increases. Perhaps the more salient point is that outliers remain. We will see that even a single outlier can create practical problems when applying routinely used techniques.

2.10 CHOOSING A MEASURE OF LOCATION

When analyzing data, it is common to encounter situations where different measures of location result in substantially different conclusions and perspectives about the population under study. How does one decide which measure of location to use? An informed and comprehensive discussion of this topic is impossible at this stage; results in subsequent chapters are required in order to answer the question in an effective manner. For the moment, it is merely remarked that the optimal measure of location depends in part on the nature of the histogram for the entire *population* of individuals. If the population histogram is reasonably symmetric and light-tailed, roughly meaning outliers are relatively rare, the sample mean performs well compared to other location estimators that might be used (based on criteria yet to be described). An example of such a histogram is shown in Figure 2.5. But this is only partially helpful because in practice the histogram for the entire population is not known.

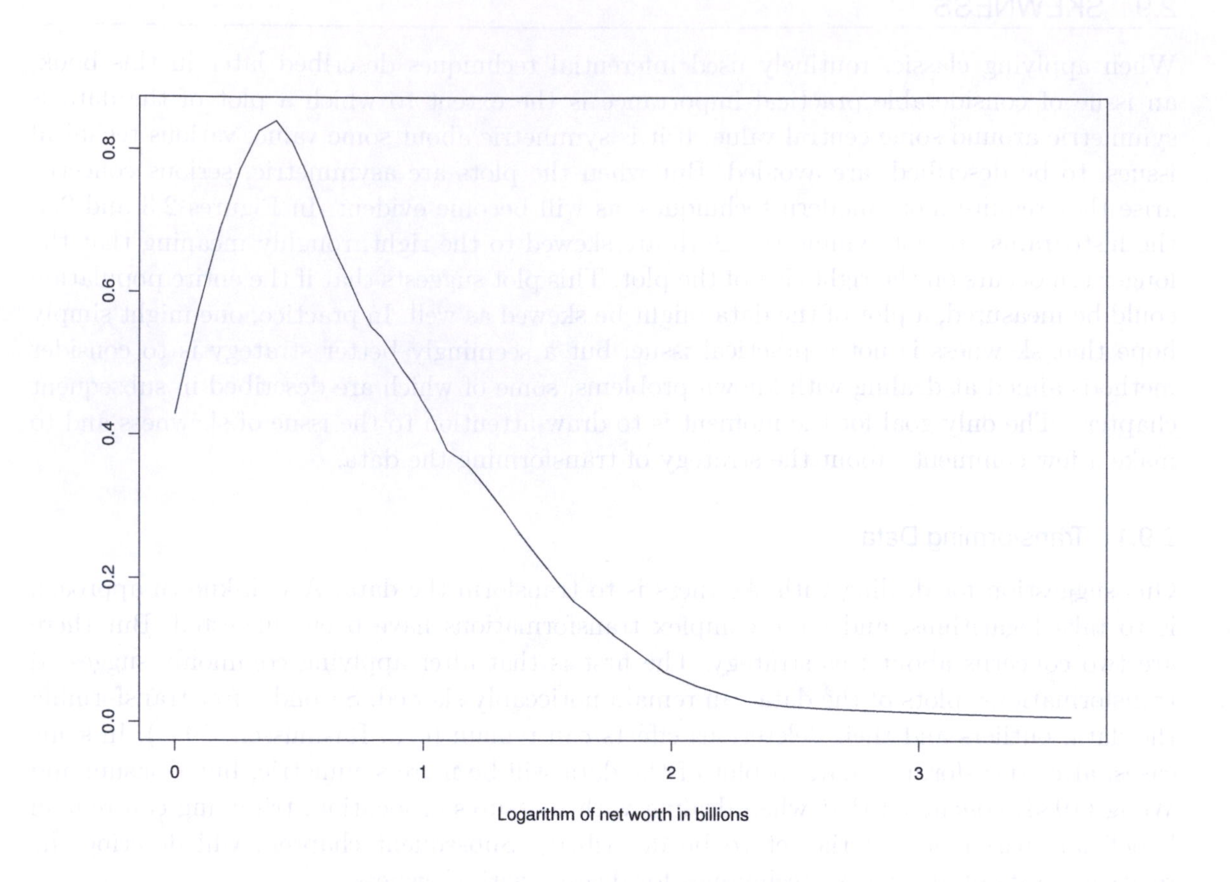

Figure 2.10 An illustration that, even after taking logarithms, a plot of the data can remain highly skewed.

We can estimate the population histogram using data available to us, but as illustrated, the histogram based on a sample might poorly reflect the histogram we would get if all individuals could be measured. This is particularly true when dealing with sample sizes less than 100. A better estimate of the population histogram might be obtained with a kernel density estimator, but, again, when is it safe to assume that the population histogram is both symmetric and light-tailed? A completely satisfactory answer to this question appears to be unavailable. (The final sections of Chapters 5 and 8 comment further on this issue.)

Another concern is that when dealing with population histograms that are skewed, the mean can poorly reflect the typical response. In some circumstances, surely this is undesirable; the median or 20% trimmed mean might be a better choice. But as illustrated by the last two examples in Section 2.2.1, there are exceptions. So when analyzing data, some thought is required about whether the mean remains satisfactory when a population histogram is skewed.

When dealing with situations where outliers are relatively common, there are certain practical advantages to using a median or 20% trimmed mean, *even when the population histogram is symmetric.* Even with no outliers, there are advantages to using a 20% trimmed mean and median when a population histogram is skewed for reasons explained in Chapters 4 and 5. For now, the following point is stressed. A reasonable suggestion, given the goal of justifying the use of the sample mean, is to check for skewness and outliers using the methods in this chapter. Unfortunately, this can be unsatisfactory roughly because these diagnostic tools can fail to detect a situation where the mean has undesirable properties (discussed in detail later in this book).

Finally, the practical advantages of a 20% trimmed mean are impossible to appreciate at this stage. While no measure of location is always optimal, we will see that often a 20% trimmed mean is an excellent choice. Some researchers use a 20% trimmed mean, but it should be noted that it is less well known compared to the mean and median.

2.11 EXERCISES

Some of the exercises are designed to provide practice using R to perform basic computations. Others illustrate some basic principles that will play a major role in subsequent chapters. Brief answers to most of the exercises are given in Appendix A. More detailed answers for all of the exercises can be downloaded from http://dornsife.usc.edu/labs/rwilcox/books. (Or google the author's name and follow the links to his personal web page.) The answers are stored in the file CRC_answers.pdf.

1. For the observations

$$21, 36, 42, 24, 25, 36, 35, 49, 32$$

verify that the sample mean, 20% trimmed mean, and median are $\bar{X} = 33.33$, $\bar{X}_t = 32.9$, and $M = 35$.

2. The largest observation in the first problem is 49. If 49 is replaced by the value 200, verify that the sample mean is now $\bar{X} = 50.1$ but the trimmed mean and median are not changed. What does this illustrate about the resistance of the sample mean?

3. For the data in Exercise 1, what is the minimum number of observations that must be altered so that the 20% trimmed mean is greater than 1000?

4. Repeat the previous problem but use the median instead. What does this illustrate about the resistance of the mean, median, and trimmed mean?

5. For the observations

$$6, 3, 2, 7, 6, 5, 8, 9, 8, 11$$

verify that the sample mean, 20% trimmed mean, and median are $\bar{X} = 6.5$, $\bar{X}_t = 6.7$, and $M = 6.5$.

6. A class of fourth graders was asked to bring a pumpkin to school. Each of the 29 students counted the number of seeds in their pumpkin and the results were

$$250, 220, 281, 247, 230, 209, 240, 160, 370, 274, 210, 204, 243,$$
$$251, 190, 200, 130, 150, 177, 475, 221, 350, 224, 163, 272, 236,$$
$$200, 171, 98.$$

(These data were supplied by Mrs. Capps at the La Cañada elementary school.) Verify that the sample mean, 20% trimmed mean, and median are $\bar{X} = 229.2$, $\bar{X}_t = 220.8$, and $M = 221$.

7. Suppose health inspectors rate sanitation conditions at restaurants on a five-point scale where a 1 indicates poor conditions and a 5 is excellent. Based on a sample of restaurants in a large city, the frequencies are found to be $f_1 = 5$, $f_2 = 8$, $f_3 = 20$, $f_4 = 32$, and $f_5 = 23$. What is the sample size, n? Verify that the sample mean is $\bar{X} = 3.7$.

8. For the frequencies $f_1 = 12$, $f_2 = 18$, $f_3 = 15$, $f_4 = 10$, $f_5 = 8$, and $f_6 = 5$, verify that the sample mean is 3.
9. For the observations

$$21, 36, 42, 24, 25, 36, 35, 49, 32$$

verify that the sample variance and the sample Winsorized variance are $s^2 = 81$, $s_w^2 = 51.4$.
10. In the previous problem, what is the Winsorized variance if the value 49 is increased to 102?
11. In general, will the Winsorized sample variance, s_w^2, be less than the sample variance, s^2?
12. Among a sample of 25 participants, what is the minimum number of participants that must be altered to make the sample variance arbitrarily large?
13. Repeat the previous problem but for s_w^2 instead. Assume 20% Winsorization.
14. For the observations

$$6, 3, 2, 7, 6, 5, 8, 9, 8, 11$$

verify that the sample variance and Winsorized variance are 7.4 and 1.8, respectively.
15. For the data in Exercise 6, verify that the sample variance is 5584.9 and the Winsorized sample variance is 1375.6.
16. For the data in Exercise 14, determine the ideal fourths and the corresponding interquartile range.
17. For the data used in Exercise 6, which values would be declared outliers using the rule based on MAD?
18. Devise a method for computing the sample variance based on the frequencies, f_x.
19. One million couples are asked how many children they have. Suppose the relative frequencies are $f_0/n = 0.10$, $f_1/n = 0.20$, $f_2/n = 0.25$, $f_3/n = 0.29$, $f_4/n = 0.12$, and $f_5/n = 0.04$. Compute the sample variance.
20. For a sample of n participants, the relative frequencies are $f_0/n = 0.2$, $f_1/n = 0.4$, $f_2/n = 0.2$, $f_3/n = 0.15$, and $f_4/n = 0.05$. The sample mean is 1.45. Using your answer to Exercise 18, verify that the sample variance is 1.25.
21. Snedecor and Cochran (1967) report results from an experiment on weight gain in rats as a function of source of protein and levels of protein. One of the groups was fed beef with a low amount of protein. The weight gains were

$$90, 76, 90, 64, 86, 51, 72, 90, 95, 78.$$

Verify that there are no outliers among these values when using a boxplot, but there is an outlier using the MAD-median rule.
22. For the values 1, 2, 3, 4, 5, 6, 7, 8, 9, and 10 there are no outliers. If you increase 10 to 20, then 20 would be declared an outlier by a boxplot. (a) If the value 9 is also increased to 20, would the boxplot find two outliers? (b) If the value 8 is also increased to 20, would the boxplot find all three outliers?
23. Use the results of the last problem to come up with a general rule about how many outliers a boxplot can detect.

24. For the data in Table 2.3, use R to verify that the value 3.05 is an outlier based on the boxplot rule.
25. Looking at Figure 2.2, what approximately is the interquartile range? How large or small must a value be to be declared an outlier?
26. In a study conducted by M. Earleywine, one goal was to measure hangover symptoms after consuming a specific amount of alcohol in a laboratory setting. The resulting measures, written in ascending order, were

    0, 0, 0, 0, 0, 0, 0, 0, 0, 0, 0, 0, 0, 0, 0, 0, 0, 0, 0, 0, 0, 0, 1, 2, 2, 2, 3, 3, 3, 6, 8, 9, 11, 11, 11, 12, 18, 32, 32, 41.

    Use R to construct a histogram. Then use the functions `outbox` and `out` to determine which values are outliers. Comment on using the histogram as an outlier detection method.

CHAPTER 3

PROBABILITY AND RELATED CONCEPTS

CONTENTS

There are two general goals in this chapter. The first is to summarize the basic principles of probability that are relevant to the methods and techniques covered in this book. This includes formal definitions of quantities called parameters, which characterize the population of individuals or things under study. The second general goal is to explain some fundamental principles that are not typically covered in an introductory statistics book but which have become highly relevant to methods developed during the last half century. The motivation for the first goal seems rather evident. But some general introductory comments about the second goal seem warranted.

During the last half century, there has been an increasing awareness that commonly used methods for analyzing data, based on the sample mean, have serious fundamental shortcomings, which will be described and illustrated at numerous points in this book. One of these concerns has to do with outliers. It might seem that a reasonable way of dealing with outliers is to discard unusually large and small values and apply methods based on means using the remaining data. This strategy is often used in the context of what is called *data cleaning*, but it can result in serious technical difficulties that can have a substantially negative impact on the veracity of the results, as will be illustrated in subsequent chapters. (See, for instance, the last example in Section 4.7.1.) Consequently, it is important to develop some understanding of why this strategy is unsatisfactory, and some of the results in this chapter will help achieve this goal. A closely related issue is understanding that even with an extremely large sample size, practical problems associated with outliers can persist. Technically sound methods for dealing with known problems due to outliers are not evident based on standard training. So one goal is to attempt to develop some understanding about which strategies are theoretically sound and which are not. There are other serious concerns associated with methods based on means, beyond outliers, and some of the results in this chapter will help lay the foundation for dealing with them and simultaneously help underscore the practical importance of more modern methods.

3.1 BASIC PROBABILITY

Defining the term *probability* turns out to be a non-trivial matter. A simple way of viewing probability is in terms of proportions. For example, if the proportion of adults with high blood pressure is 25%, a seemingly natural claim is that the probability of sampling an adult with high blood pressure is 0.25. But from a technical point of view, defining probability in terms of proportions is unsatisfactory. To provide a rough explanation, imagine a bowl containing 100 marbles, half of which are green and the other half are blue. Thinking of probability in terms of proportions, a natural reaction is that if someone picks a marble, the probability that it will be blue is 0.5. But suppose all of the blue marbles are on top of the green marbles. If someone has a tendency to pick a marble from the top, with near certainty they will pick a blue marble. That is, the probability of a blue marble is not 0.5. We could mix the marbles, but how do we characterize the extent to which they are mixed to a sufficient degree? A response might be that each marble has the same probability of being chosen. But this is unsatisfactory because we cannot use the term probability when trying to define what probability means.

For present purposes, defining probability is not important. What is important is having some understanding of the basic properties of what are called probability functions, which is one of the topics covered in this chapter.

As indicated in Chapter 2, an uppercase Roman letter is typically used to represent whatever measure happens to be of interest, the most common letter being X. To be concrete, imagine a study aimed at measuring the optimism of college students. Further imagine that the possible values based on some measure of optimism are the integers 1 through 5. In symbols, there are five possible events: $X = 1, X = 2, \ldots, X = 5$. The notation $X = 1$ refers to the event that a college student receives a score of 1 for optimism, $X = 2$ means a student got a score of 2, and so on.

Unless stated otherwise, it is assumed that the possible responses we might observe are mutually exclusive and exhaustive. In the illustration, describing the five possible ratings of optimism as being *mutually exclusive* means that a student can get one and only one rating. By assumption, it is impossible, for example, to have a rating of both 2 and 3. *Exhaustive* means that a complete list of the possible values we might observe has been specified. If we consider only those students who get a rating between 1 and 5, meaning, for example, that we exclude the possibility of no response, then the ratings 1–5 are exhaustive. If instead we let 0 represent no response, then an exhaustive list of the possible responses would be 0, 1, 2, 3, 4, and 5.

The set of all possible responses is called the *sample space*. If in our ratings illustration the only possible responses are the integers 1–5, then the sample space consists of the numbers 1, 2, 3, 4, and 5. If instead we let 0 represent no response, then the sample space is 0, 1, 2, 3, 4, and 5. If our goal is to study birth weight among humans, the sample space can be viewed as all numbers greater than or equal to zero. Obviously some birth weights are impossible — there seems to be no record of someone weighing 100 pounds at birth — but for convenience the sample space might contain outcomes that have zero probability of occurring.

A probability function is a function

$$p(x) \tag{3.1}$$

that has the following properties:

- $p(x) \geq 0$ for any x.
- For any two mutually exclusive outcomes, say x and y, $p(x \text{ or } y) = p(x) + p(y)$.
- $\sum p(x) = 1$, where the notation $\sum p(x)$ means that $p(x)$ is evaluated for all possible values of x and the results are summed. In the optimism example where the sample space is x: 1, 2, 3, 4, 5, $\sum p(x) = p(1) + p(2) + p(3) + p(4) + p(5) = 1$.

In other words, the first criterion is that any probability must be greater than or equal to zero. The second criterion says, for example, that if the responses 1 and 2 are mutually exclusive, then the probability of observing the value 1 or 2 is equal to the probability of a 1 plus the probability of a 2. Notice that this criterion makes perfect sense when probabilities are viewed as relative frequencies. If, for example, 15% of students have an optimism rating of 1, and 20% have a rating of 2, then the probability of a rating of 1 or 2 is just the sum of the proportions: $0.15 + 0.20 = 0.35$. The third criterion is that if we sum the probability of all possible events that are mutually exclusive, we get 1. (In more formal terms, the probability that an observation belongs to the sample space is 1.)

3.2 EXPECTED VALUES

A fundamental tool in statistics is the notion of *expected values*. To convey the basic principle, it helps to start with a simple but unrealistic situation. Still using our optimism illustration, imagine that the entire population of college students consists of 10 people. That is, we are interested in these 10 individuals only. So, in particular, we have no desire to generalize to a larger group of college students. Further assume that two students have an optimism rating of 1, three a rating of 2, two a rating of 3, one a rating of 4, and two a rating of 5. As is evident, the average of these 10 ratings is

$$\frac{1+1+2+2+2+3+3+4+5+5}{10} = 2.8.$$

Now suppose the probability function for this population of individuals corresponds to the proportion of students giving a particular rating. For example, the proportion of students with the rating 1 is 2/10 and we are assuming that $p(1) = 2/10$. Notice that the left side of this last equation can be written as

$$\frac{1(2)+2(3)+3(2)+4(1)+5(2)}{10} = 1\frac{2}{10} + 2\frac{3}{10} + 3\frac{2}{10} + 4\frac{1}{10} + 5\frac{2}{10}.$$

The fractions in this last equation are just the probabilities associated with the possible outcomes. That is, the average rating for all college students, which is given by the right side of this last equation, can be written as

$$1p(1) + 2p(2) + 3p(3) + 4p(4) + 5p(5).$$

EXAMPLE

If there are 1 million college students, and the proportion of students associated with the five possible ratings 1, 2, 3, 4, and 5 are 0.1, 0.15, 0.25, 0.3, and 0.2, respectively, then the average rating for all 1 million students is

$$1(0.1) + 2(0.15) + 3(0.25) + 4(0.3) + 5(0.2) = 3.35.$$

EXAMPLE

If there are 1 billion college students, and the probabilities associated with the five possible ratings are 0.15, 0.2, 0.25, 0.3, and 0.1, respectively, then the average rating of all 1 billion students is

$$1(0.15) + 2(0.2) + 3(0.25) + 4(0.3) + 5(0.1) = 3.$$

Next we introduce some general notation for computing an average based on the view just illustrated. Again let a lowercase x represent a particular value associated with the variable X. The *expected value* of X corresponding to some probability function $p(x)$ is

$$E(X) = \sum xp(x), \tag{3.2}$$

where the notation $\sum xp(x)$ means that you compute $xp(x)$ for every possible value of x and sum the results. So if, for example, the possible values for X are the integers 0, 1, 2, 3, 4, 5, and 6, then

$$\sum xp(x) = 0p(0) + 1p(1) + 2p(2) + 3p(3) + 4p(4) + 5p(5).$$

The expected value of X is so fundamental it has been given a special name: the *population mean*, which was informally introduced in Chapter 2. Typically the population mean is represented by μ, a lowercase Greek mu. So,

$$\mu = E(X)$$

is the average value for all individuals in the population of interest.

EXAMPLE

Imagine that an auto manufacturer wants to evaluate how potential customers will rate handling for a new car being considered for production. So here, X represents ratings of how well the car handles, and the population of individuals who are of interest consists of all individuals who might purchase it. If all potential customers were to rate handling on a four-point scale, 1 being poor and 4 being excellent, and if the corresponding probabilities associated with these ratings are $p(1) = 0.2$, $p(2) = 0.4$, $p(3) = 0.3$, and $p(4) = 0.1$, then the population mean is

$$\mu = E(X) = 1(0.2) + 2(0.4) + 3(0.3) + 4(0.1) = 2.3.$$

That is, the average rating is 2.3.

3.3 CONDITIONAL PROBABILITY AND INDEPENDENCE

Conditional probability refers to the probability of some event given that some other event has occurred; it plays a fundamental role in statistics. The notion of conditional probability is illustrated in two ways. The first is based on what is called a *contingency table*, an example

Table 3.1 Hypothetical Probabilities for Sex and Political Affiliation

Sex	Democrat (D)	Republican (R)	
M	0.25	0.20	0.45
F	0.28	0.27	0.55
	0.53	0.47	1.00

of which is shown in Table 3.1. In the contingency table are the probabilities associated with four mutually exclusive groups: individuals who are (1) both male and belong to the Republican party, (2) male and belong to the Democratic party, (3) female and belong to the Republican party, and (4) female and belong to the Democratic party. So according to Table 3.1, the proportion of people who are both female and Republican is 0.27. The last column shows what are called the *marginal probabilities.* For example, the probability of being male is $0.20 + 0.25 = 0.45$, which is just the proportion of males who are Democrats plus the proportion who are Republicans. The last line of Table 3.1 shows the marginal probabilities associated with party affiliation. For example, the probability of being a Democrat is $0.25 + 0.28 = 0.53$.

Now consider the probability of being a Democrat given that the individual is male. According to Table 3.1, the proportion of people who are male is 0.45. So among the people who are male, the proportion who belong to the Democratic party is $0.25/0.45 = 0.56$. Put another way, the probability of being a Democrat, given that the individual is male, is 0.56.

Notice that a conditional probability is determined by altering the sample space. In the illustration, the proportion of all people who belong to the Democratic party is 0.53. But restricting attention to males, meaning that the sample space has been altered to include males only, the proportion is $0.25/0.45 = 0.56$. In a more general notation, if A and B are any two events, and if we let $P(A)$ represent the probability of event A and $P(A \text{ and } B)$ represent the probability that events A and B occur simultaneously, then the conditional probability of A, given that B has occurred, is

$$P(A|B) = \frac{P(A \text{ and } B)}{P(B)}. \quad (3.3)$$

In the illustration, A is the event of being a Democrat, B is the event that a person is male. So according to Table 3.1, $P(A \text{ and } B) = 0.25$, and $P(B) = 0.45$, so $P(A|B) = 0.25/0.45$, as previously indicated.

EXAMPLE

From Table 3.1, the probability that someone is a female, given that she is Republican, is

$$0.27/0.47 = 0.5745.$$

Roughly, two events are *independent* if the probability associated with the first event is not altered when the second event is known. If the probability is altered, the events are *dependent.*

EXAMPLE

According to Table 3.1, the probability that someone is a Democrat is 0.53. The event

Table 3.2 Hypothetical Probabilities for Presidential Effectiveness

	Husband (X)			
Wife (Y)	1	2	3	
1	0.02	0.10	0.08	0.2
2	0.07	0.35	0.28	0.7
3	0.01	0.05	0.04	0.1
	0.1	0.5	0.4	

that someone is a Democrat is independent of the event someone is male if when we are told that someone is male, the probability of being a Democrat remains 0.53. We have seen, however, that the probability of being a Democrat, given that the person is male, is 0.56, so these two events are dependent.

Consider any two variables, say X and Y, and let x and y be any two possible values corresponding to these variables. We say that the variables X and Y are independent if for any x and y we might pick

$$P(Y = y|X = x) = P(Y = y). \tag{3.4}$$

Otherwise they are said to be dependent.

EXAMPLE

Imagine that married couples are asked to rate the effectiveness of the President of the United States. To keep things simple, assume that both husbands and wives rate effectiveness with the values 1, 2, and 3, where the values stand for fair, good, and excellent, respectively. Further assume that the probabilities associated with the possible outcomes are as shown in Table 3.2. We see that the probability a wife (Y) gives a rating of 1 is 0.2. In symbols, $P(Y = 1) = 0.2$. Furthermore, $P(Y = 1|X = 1) = 0.02/0.1 = 0.2$, where $X = 1$ indicates that the wife's husband gave a rating of 1. So the event $Y = 1$ is independent of the event $X = 1$. If the probability had changed, we could stop and say that X and Y are dependent. But to say that they are independent requires that we check all possible outcomes. For example, another possible outcome is $Y = 1$ and $X = 2$. We see that $P(Y = 1|X = 2) = 0.1/0.5 = 0.2$, which again is equal to $P(Y = 1)$. Continuing in this manner, it can be seen that for any possible values for Y and X, the corresponding events are independent, so we say that X and Y are independent. That is, they are independent regardless of what their respective values might be.

Now, the notion of dependence is described and illustrated in another manner. A common and fundamental question in applied research is whether information about one variable influences the probabilities associated with another variable. For example, in a study dealing with diabetes in children, one issue of interest was the association between a child's age and the level of serum C-peptide at diagnosis. For convenience, let X represent age and Y represent C-peptide concentration. For any child we might observe, there is some probability that her C-peptide concentration is less than 3, or less than 4, or less than c, where c is any constant we might pick. The issue at hand is whether information about X (a child's age) alters the probabilities associated with Y (a child's C-peptide level). That is, does the conditional probability of Y, given X, differ from the probabilities associated with Y when X is not known or ignored? If knowing X does not alter the probabilities associated with

Y, we say that X and Y are independent. Equation (3.4) is one way of providing a formal definition. An alternative way is to say that X and Y are independent if

$$P(Y \leq y|X = x) = P(Y \leq y) \tag{3.5}$$

for any x and y values we might pick. Equation (3.5) implies Equation (3.4). Yet another way of describing independence is that for any x and y values we might pick

$$\frac{P(Y = y \text{ and } X = x)}{P(X = x)} = P(Y = y), \tag{3.6}$$

which follows from Equation (3.4). From this last equation it can be seen that if X and Y are independent, then

$$P(X = x \, \text{and} \, Y = y) = P(X = x)P(Y = y). \tag{3.7}$$

Equation (3.7) is called the *product rule* and says that if two events are independent, the probability that they occur simultaneously is equal to the product of their individual probabilities.

EXAMPLE

If two wives rate presidential effectiveness according to the probabilities in Table 3.2, and if their responses are independent, then the probability that both give a response of 2 is $0.7 \times 0.7 = 0.49$.

EXAMPLE

Suppose that for all children we might measure, the probability of having a C-peptide concentration less than or equal to 3 is $P(Y \leq 3) = 0.4$. Now consider only children who are 7 years old and imagine that for this subpopulation of children, the probability of having a C-peptide concentration less than 3 is 0.2. In symbols, $P(Y \leq 3|X = 7) = 0.2$. Then C-peptide concentrations and age are said to be dependent because knowing that the child's age is 7 alters the probability that the child's C-peptide concentration is less than 3. If instead $P(Y \leq 3|X = 7) = 0.4$, the events $Y \leq 3$ and $X = 7$ are independent. More generally, if, for any x and y we pick, $P(Y \leq y|X = x) = P(Y = y)$, then C-peptide concentrations and age are independent.

Attaining a graphical intuition of independence will be helpful. To be concrete, imagine a study where the goal is to study the association between a person's general feeling of well-being (Y) and the amount of chocolate they consume (X). Assume that an appropriate measure for these two variables has been devised and that the two variables are independent. If we were to measure these two variables for a very large sample of individuals, what would a plot of the results look like? Figure 3.1 shows a scatterplot of observations where values were generated on a computer with X and Y independent. As is evident, there is no visible pattern.

If X and Y are dependent, generally — but not always — there is some discernible pattern. But it is important to keep in mind that there are many types of patterns that can and do arise. Figure 3.2 shows four types of patterns where feelings of well-being and chocolate consumption are dependent. The two upper scatterplots show some rather obvious types of dependence that might arise. The upper left scatterplot, for example, shows a linear association where feelings of well-being increase with chocolate consumption. The upper right scatterplot shows a curved, nonlinear association. The type of dependence shown in the lower

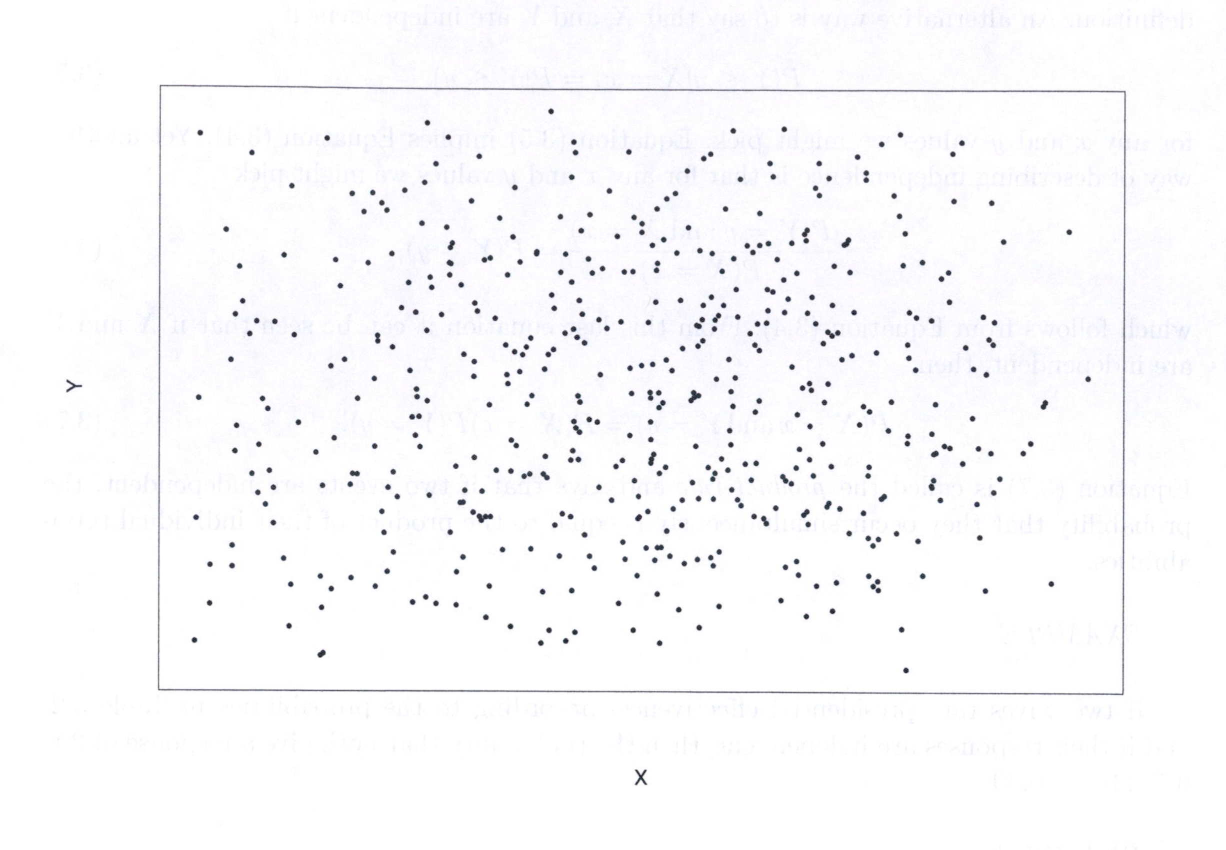

Figure 3.1 A scatterplot of points, corresponding to two independent variables, exhibits no particular pattern, as shown here.

two scatterplots are, perhaps, less commonly considered when describing dependence, but in recent years both have been found to be relevant and very important in applied work, as we shall see. In the lower left scatterplot we see that the variation in feelings of well-being differs depending on how much chocolate is consumed. The points in the left portion of this scatterplot are more tightly clustered together. For the left portion of this scatterplot there is, for example, virtually no possibility that someone's feeling of well-being exceeds 1. But for the right portion of this scatterplot, the data were generated so that among individuals with a chocolate consumption of 3, there is a 0.2 probability that the corresponding value of well-being exceeds 1. That is, $P(Y \leq 1|X)$ increases as X gets large, so X and Y are dependent. Generally, any situation where the variation among the Y values changes with X implies that X and Y are dependent. Finally, the lower right scatterplot shows a situation where feelings of well being tend to increase for consumption less than 3, but for $X > 3$ this is no longer the case. Considered as a whole, X and Y are dependent, but in this case, if attention is restricted to $X > 3$, X and Y are independent.

The lower left scatterplot of Figure 3.2 illustrates a general principle that is worth stressing: If knowing the value of X alters the range of possible values for Y, then X and Y are dependent. Also, if the variance of Y depends on the value of X, there is dependence. In the illustration, the variance of the well-being measures depends on the amount of chocolate consumed; it increases, which means these two measures are dependent. Said another way, even if the mean of Y does not depend on the value of X, if the variance of Y depends on the

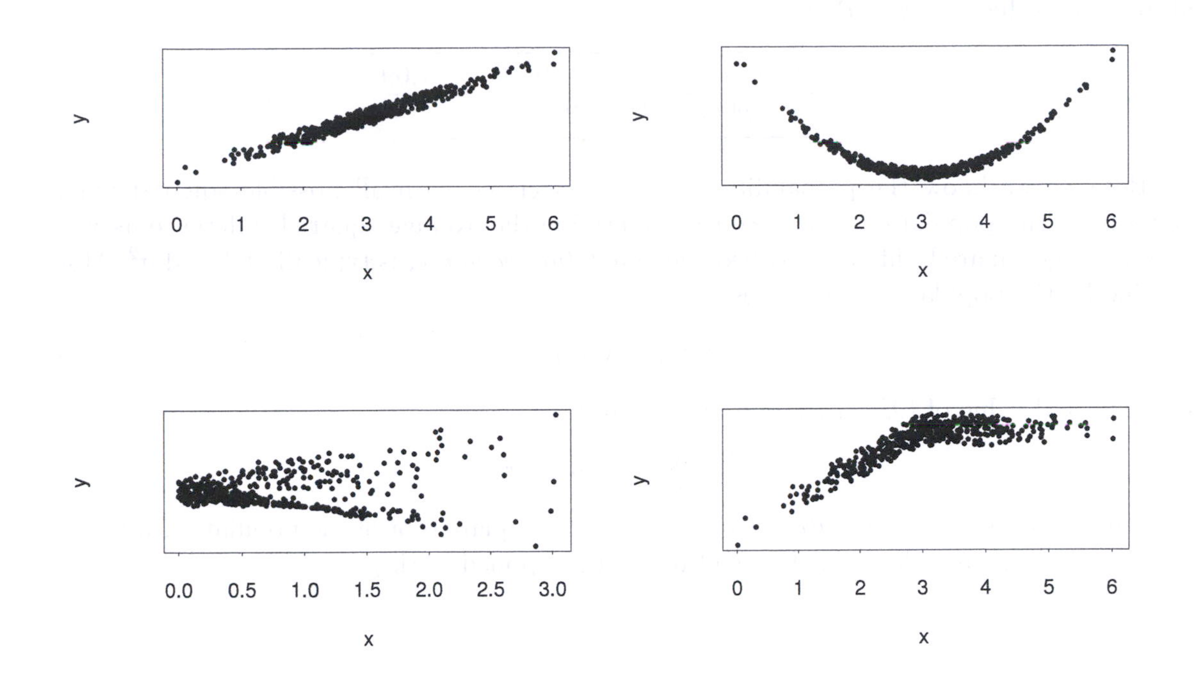

Figure 3.2 The scatterplots illustrate different types of associations that play an important role when trying to understand the relative merits of competing methods for analyzing data.

value of X, X and Y are dependent. (Chapter 6 elaborates on this issue; see the discussion of homoscedasticity.)

3.4 POPULATION VARIANCE

Associated with every probability function is a quantity called the *population variance*. The population variance reflects the average squared difference between the population mean and an observation you might make.

Consider, for example, the following probability function:

x:	0	1	2	3
$p(x)$:	0.1	0.3	0.4	0.2

The population mean is $\mu = 1.7$. If, for instance, we observe the value 0, its squared distance from the population mean is $(0-1.7)^2 = 2.89$ and reflects how far away the value 0 is from the population mean. Moreover, the probability associated with this squared difference is .1, the probability of observing the value 0. In a similar manner, the squared difference between 1 and the population mean is 0.49, and the probability associated with this squared difference is 0.3, the same probability associated with the value 1. More generally, for any value x, it has some squared difference between it and the population mean, namely, $(x-\mu)^2$, and the probability associated with this squared difference is $p(x)$. So if we know the probability function, we know the probabilities associated with all squared differences from the population mean. For the probability function considered here, we see that the probability function associated with all possible values of $(x-\mu)^2$ is

$(x-\mu)^2$:	2.89	0.49	0.09	1.69
$p(x)$:	0.1	0.3	0.4	0.2

Because we know the probability function associated with all possible squared differences from the population mean, we can determine the average squared difference as well. This average squared difference, called the *population variance*, is typically labeled σ^2. More succinctly, the population variance is

$$\sigma^2 = E[(X-\mu)^2], \tag{3.8}$$

the expected value of $(X-\mu)^2$. Said another way,

$$\sigma^2 = \sum (x-\mu)^2 p(x).$$

The *population standard deviation* is σ, the (positive) square root of the population variance. (Often it is σ, rather than σ^2, that is of interest in applied work.)

EXAMPLE

Suppose that for a five-point scale of anxiety, the probability function for all adults living in New York City is

x:	1	2	3	4	5
$p(x)$:	0.05	0.1	0.7	0.1	0.05

The population mean is

$$\mu = 1(0.05) + 2(0.1) + 3(0.7) + 4(0.1) + 5(0.05) = 3,$$

so the population variance is

$$\sigma^2 = (1-3)^2(0.05) + (2-3)^2(0.1) + (3-3)^2(0.7) + (4-3)^2(0.1) + (5-3)^2(0.05) = 0.6,$$

and the population standard deviation is $\sigma = \sqrt{0.6} = 0.775$.

Understanding the practical implications associated with the magnitude of the population variance is a complex task that is addressed at various points in this book. There are circumstances where knowing σ is very useful, but there are common situations where it can mislead and give a highly distorted view of what a variable is like. So a general goal is to develop a sense of when the population variance is useful and when it can be highly unsatisfactory. For the moment, complete details must be postponed. But to begin to provide some sense of what σ tells us, consider the following probability function:

x:	1	2	3	4	5
$p(x)$:	0.2	0.2	0.2	0.2	0.2

It can be seen that $\mu = 3$, the same population mean associated with the probability function in the last example, but the population variance is

$$\sigma^2 = (1-3)^2(0.2) + (2-3)^2(0.2) + (3-3)^2(0.2) + (4-3)^2)(0.2) + (5-3)^2(0.2) = 2.$$

Notice that this variance is larger than the variance in the previous example, where $\sigma^2 = 0.6$. The reason is that in the former example, it is much less likely for a value to be far from the mean than is the case for the probability function considered here. Here, for example, there is a 0.4 probability of getting the value 1 or 5. In the previous example, this probability is only 0.1. Here the probability that an observation differing from the population mean is 0.8, but in the previous example it was only 0.3. This illustrates the crude rule of thumb that larger values for the population variance reflect situations where observed values are likely to be far from the mean, and small population variances indicate the opposite.

For discrete data, it is common to represent probabilities graphically with the height of spikes. Figure 3.3 illustrates this approach with the last two probability functions used to illustrate the variance. The left panel shows the probability function

x:	1	2	3	4	5
$p(x)$:	0.05	0.1	0.7	0.1	0.05

The right panel graphically shows the probability function

x:	1	2	3	4	5
$p(x)$:	0.2	0.2	0.2	0.2	0.2

Look at the graphed probabilities in Figure 3.3 and notice that the graphed probabilities in the left panel indicate that an observed value is more likely to be close to the mean; in the right panel they are more likely to be further from the mean. That is, the graphs suggest that the variance is smaller in the left panel because, probabilistically, observations are more tightly clustered around the mean.

3.5 THE BINOMIAL PROBABILITY FUNCTION

The most important discrete distribution is the binomial, which was first derived by James Bernoulli in 1713. It arises in situations where only two possible outcomes are possible when making a single observation. The outcomes might be yes and no, success and failure, agree and disagree. Such random variables are called *binary*. Typically the number 1 is used to represent a success and a failure is represented by 0. A common convention is to let p represent the probability of success and to let $q = 1 - p$ be the probability of a failure.

Before continuing, a comment about notation might help. Consistent with Section 2.1, we follow the common convention of letting X be a variable that represents the number of successes among n observations. The notation $X = 2$, for example, means we observed two successes; more generally, $X = x$ means we observed x successes, where the possible values for x are $0, 1, \ldots, n$.

In applied work, often the goal is to estimate p, the probability of success, given some data. But before taking up this problem, we must first consider how to compute probabilities given p. For example, suppose you ask five people whether they approve of a certain political leader. If these five people are allowed to respond only yes or no, and if the probability of a yes response is $p = 0.6$, what is the probability that exactly 3 of 5 randomly sampled people will say yes? There is a convenient formula for solving this problem based on the *binomial probability function*. It says that among n observations, the probability of exactly x successes, $P(X = x)$, is given by

$$p(x) = \binom{n}{x} p^x q^{n-x}. \tag{3.9}$$

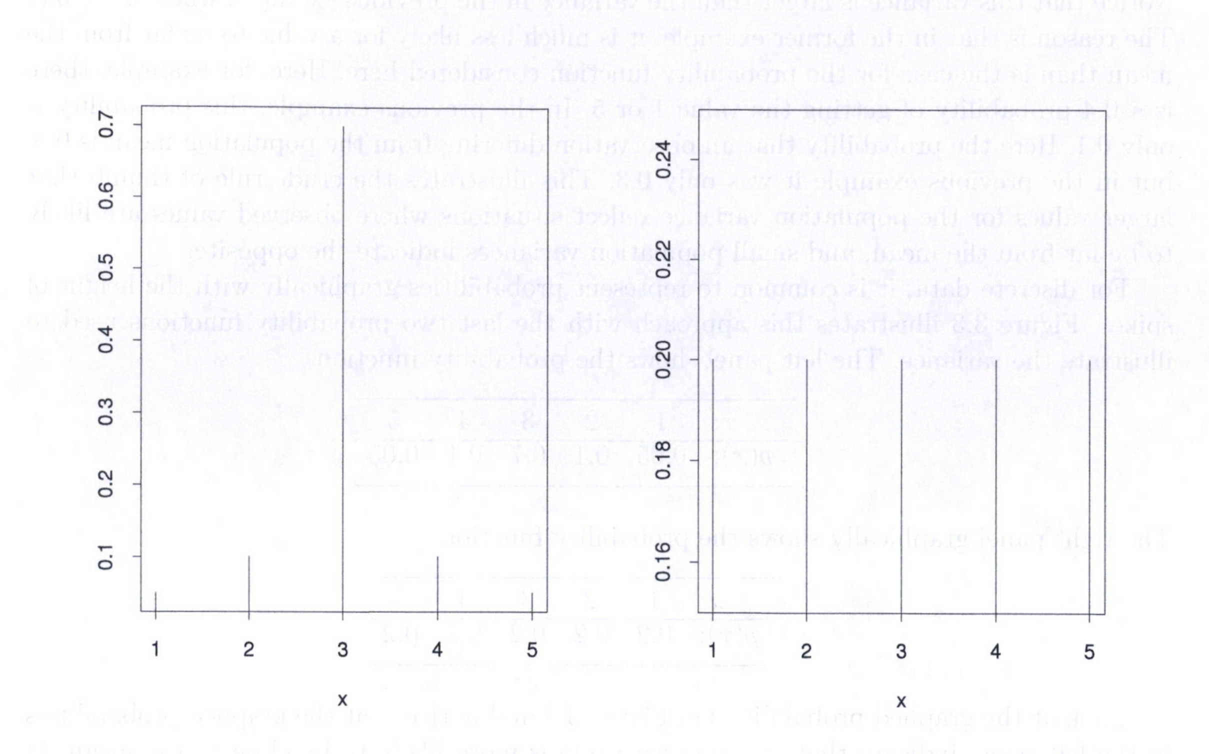

Figure 3.3 For both probability functions, the population mean is 3. The variance is larger in the right panel, roughly because there is a higher probability of getting a value that is relatively far from the population mean.

The first term on the right side of this equation, called the binomial coefficient, is defined to be

$$\binom{n}{x} = \frac{n!}{x!(n-x)!},$$

where $n!$ represents n factorial. That is,

$$n! = 1 \times 2 \times 3 \times \cdots \times (n-1) \times n.$$

For example, $1! = 1, 2! = 2$, and $3! = 6$. By convention, $0! = 1$.

In the illustration, there are $n = 5$ randomly sampled people and the goal is to determine the probability that exactly $x = 3$ people will respond yes when the probability of a yes is $p = 0.6$. To solve this problem, compute

$$\begin{aligned} n! &= 1 \times 2 \times 3 \times 4 \times 5 = 120, \\ x! &= 1 \times 2 \times 3 = 6, \\ (n-x)! &= 2! = 2, \end{aligned}$$

in which case

$$p(3) = \frac{120}{6 \times 2}(0.6^3)(0.4^2) = 0.3456.$$

As another illustration, suppose you randomly sample 10 couples who recently got married, and your experiment consists of assessing whether they are happily married at the end of one year. If the probability of success is $p = 0.3$, the probability that exactly $x = 4$ couples will report that they are happily married is

$$p(4) = \frac{10!}{4! \times 6!}(0.3^4)(0.7^6) = 0.2001.$$

Often attention is focused on the probability of *at least* x successes in n trials or *at most* x successes, rather than the probability of getting *exactly* x successes. In the last illustration, you might want to know the probability that four couples or fewer are happily married as opposed to exactly four. The former probability consists of five mutually exclusive events, namely, $x = 0, x = 1, x = 2, x = 3$, and $x = 4$. Thus, the probability that four couples or fewer are happily married is

$$P(X \le 4) = p(0) + p(1) + p(2) + p(3) + p(4).$$

In summation notation,

$$P(X \le 4) = \sum_{x=0}^{4} p(x).$$

More generally, the probability of k successes or less in n trials is

$$\begin{aligned} P(X \le k) &= \sum_{x=0}^{k} p(x) \\ &= \sum_{x=0}^{k} \binom{n}{x} p^x q^{n-x}. \end{aligned}$$

Table 2 in Appendix B gives the values of $P(X \le k)$ for various values of n and p. Returning to the illustration where $p = 0.3$ and $n = 10$, Table 2 reports that the probability of four successes or less is 0.85. Notice that the probability of five successes or more is just the complement of getting four successes or less, so

$$\begin{aligned} P(X \ge 5) &= 1 - P(X \le 4) = 1 - 0.85 \\ &= 0.15. \end{aligned}$$

In general,

$$P(X \ge k) = 1 - P(X \le k - 1),$$

so $P(X \ge k)$ is easily evaluated with Table 2.

Expressions like

$$P(2 \le x \le 8)$$

meaning you want to know the probability that the number of successes is between 2 and 8, inclusive, can also be evaluated with Table 2 by noting that

$$P(2 \le x \le 8) = P(x \le 8) - P(x \le 1).$$

In other words, the event of eight successes or less can be broken down into the sum of two mutually exclusive events: the event that the number of successes is less than or equal to 1 and the event that the number of successes is between 2 and 8, inclusive. Rearranging terms yields the last equation. The point is that $P(2 \le x \le 8)$ can be written in terms of two expressions that are easily evaluated with Table 2 in Appendix B.

EXAMPLE

Assume $n = 10$ and $p = 0.5$. From Table 2 in Appendix B, $P(X \le 1) = 0.011$ and

$P(X \leq 8) = 0.989$, so

$$P(2 \leq X \leq 8) = 0.989 - 0.011 = 0.978.$$

A related problem is determining the probability of one success or less or nine successes or more. The first part is simply read from Table 2 and can be seen to be 0.011. The probability of nine successes or more is the complement of eight successes or less, so $P(X \geq 9) = 1 - P(X \leq 8) = 1 - 0.989 = 0.011$, again assuming that $n = 10$ and $p = 0.5$. Thus, the probability of one success or less or nine successes or more is $0.011 + 0.011 = 0.022$. In symbols,

$$P(X \leq 1 \text{ or } X \geq 9) = 0.022.$$

There are times when you will need to compute the mean and variance of a binomial probability function once you are given n and p. It can be shown that the mean and variance are given by

$$\begin{aligned} \mu &= E(X) \\ &= np, \end{aligned}$$

and

$$\sigma^2 = npq.$$

For example, if $n = 16$ and $p = 0.5$, the mean of the binomial probability function is $\mu = np = 16(0.5) = 8$. That is, on average, 8 of the 16 observations in a random sample will be a success, while the other 8 will not. The variance is $\sigma^2 = npq = 16(0.5)(0.5) = 4$, so the standard deviation is $\sigma = \sqrt{4} = 2$. If instead, $p = 0.3$, $\mu = 16(0.3) = 4.8$. That is, the average number of successes is 4.8.

In most situations, p, the probability of a success, is not known and must be estimated based on x, the observed number of successes. The result $E(X) = np$ suggests that x/n be used as an estimator of p; and indeed this is the estimator that is typically used. Often this estimator is written as

$$\hat{p} = \frac{x}{n}.$$

Note that $\hat{p}$ is just the proportion of successes in n trials. It can be shown (using the rules of expected values covered in Section 3.9) that

$$E(\hat{p}) = p.$$

That is, if you were to repeat an experiment infinitely many times, each time randomly sampling n observations, the average of these infinitely many $\hat{p}$ values is p. It can also be shown that the variance of $\hat{p}$ is

$$\sigma^2_{\hat{p}} = \frac{pq}{n}.$$

EXAMPLE

If you sample 25 people and the probability of success is 0.4, the variance of $\hat{p}$ is

$$\sigma^2_{\hat{p}} = \frac{0.4 \times 0.6}{25} = 0.098.$$

The characteristics and properties of the binomial probability function can be summarized as follows:

- The experiment consists of exactly n independent trials.
- Only two possible outcomes are possible on each trial, usually called *success* and *failure*.
- Each trial has the same probability of success, p.
- $q = 1 - p$ is the probability of a failure.
- There are x successes among the n trials.
- $p(x) = \binom{n}{x} p^x q^{n-x}$ is the probability of x successes in n trials, $x = 0, 1, \ldots, n$.
- $\binom{n}{x} = \frac{n!}{x!(n-x)!}$.
- The usual estimate p is $\hat{p} = \frac{x}{n}$, where x is the total number of successes.
- $E(\hat{p}) = p$.
- The variance of $\hat{p}$ is $\sigma^2 = \frac{pq}{n}$.
- The average or expected number of successes in n trials is $\mu = E(X) = np$.
- The variance of X is $\sigma^2 = npq$.

3.5.1 R Functions dbinom and pbinom

The R function

```
dbinom(x, size, prob)
```

computes the probability of x successes, given by Equation (3.9), where the argument size corresponds to n, the number of trials, and prob indicates the probability of success, p. The R function

```
pbinom(q, size, prob)
```

computes the probability of q successes or less. For example, if the probability of success is 0.4 and the number of trials is $n = 10$, the probability of six successes or less is returned by the command `pbinom(6,10,0.4)` yielding 0.945.

3.6 CONTINUOUS VARIABLES AND THE NORMAL CURVE

For various reasons (described in subsequent chapters), continuous variables, meaning that the variables can have any value over some range of values, play a fundamental and useful role in statistics. In contrast to discrete variables, probabilities associated with continuous variables are given by the area under a curve. The equation for this curve is called a *probability density function.* If, for instance, we wanted to know the probability that a variable has a value between 2 and 5, say, this is represented by the area under the curve and between 2 and 5.

EXAMPLE

Suppose X represents the proportion of time someone spends on pleasant tasks at their job. So, of course, for any individual we observe, X has some value between zero and one. Assume that for the population of all working adults, the probability density function is as shown in Figure 3.4. Further assume that we want to know the probability that the proportion of time spent working on pleasant tasks is less than or equal to 0.4. In symbols, we want to know $P(X \leq 0.4)$. This probability is given by the area under the curve and to the left of 0.4 in Figure 3.4, which is 0.096. In symbols, $P(X \leq 0.4) = 0.096$.

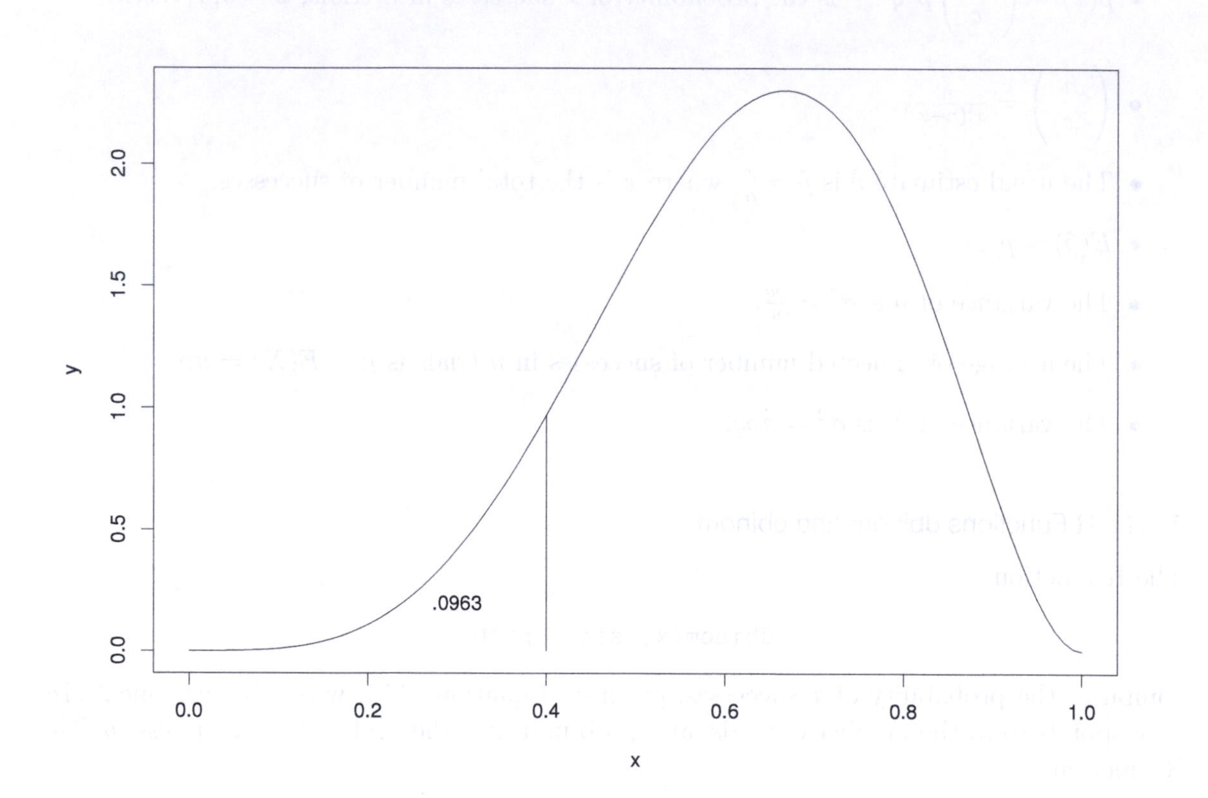

Figure 3.4 In contrast to discrete distributions, such as shown in Figure 3.3, probabilities are not given by the height of the curve given by the probability density function. Rather, probabilities are given by the area under the curve. Here, the area under the curve for values less than or equal to 0.4 is 0.096. That is, $P(X \leq 0.4) = 0.096$.

If $P(X \leq 5) = 0.8$ and X is a continuous variable, then the value 5 is called the 0.8 quantile. If $P(X \leq 3) = 0.4$, then 3 is the 0.4 quantile. In general, if $P(X \leq c) = q$, then c is called the qth *quantile*. In Figure 3.4, for example, 0.4 is the 0.096 quantile. *Percentiles* are just quantiles multiplied by 100. So in Figure 3.4, 0.4 is the 9.6 percentile. There are some mathematical difficulties when defining quantiles for discrete data. There is a standard method for dealing with this issue (e.g., Serfling, 1980, p. 3), but the details are not important here.

The 0.5 quantile is called the *population median*. If $P(X \leq 6) = 0.5$, then 6 is the population median. The median is centrally located in a probabilistic sense because there is a 0.5 probability that a value is less than the median, and there is a 0.5 probability that a value is greater than the median instead.

The Normal Distribution

The best known and most important probability density function is the normal distribution, an example of which is shown in Figure 3.5. Normal distributions have the following important properties:

1. The total area under the curve is 1. (This is a requirement of any probability density function.)
2. All normal distributions are bell shaped and symmetric about their mean, μ. It follows that the population mean and median are identical.
3. Although not indicated in Figure 3.5, all normal distributions extend from $-\infty$ to ∞ along the x-axis.
4. If the variable X has a normal distribution, the probability that X has a value within one standard deviation of the mean is 0.68 as indicated in Figure 3.5. In symbols, if X has a normal distribution, $P(\mu - \sigma < X < \mu + \sigma) = 0.68$ regardless of what the population mean and variance happen to be. The probability of being within two standard deviations is approximately 0.954. In symbols, $P(\mu - 2\sigma < X < \mu + 2\sigma) = 0.954$. The probability of being within three standard deviations is $P(\mu - 3\sigma < X < \mu + 3\sigma) = 0.9975$.
5. The probability density function of a normal distribution is

$$f(x) = \frac{1}{\sigma\sqrt{2\pi}} e^{-(x-\mu)^2/(2\sigma^2)}, \tag{3.10}$$

   where e is Euler's number (not Euler's constant), which is also called Napier's constant, and is approximately equal to 2.718, and as usual μ and σ^2 are the mean and variance. This rather complicated looking equation does not play a direct role in applied work, so no illustrations are given on how it is evaluated. Be sure to notice, however, that the probability density function is determined by the mean and variance. If for example we want to determine the probability that a variable is less than 25, this probability is completely determined by the mean and variance *if* we assume normality.

Figure 3.6 shows three normal distributions, two of which have equal means of 0 but standard deviations $\sigma = 1$ and $\sigma = 1.5$, and the third again has standard deviation $\sigma = 1$, but now the mean is $\mu = 2$. There are two things to notice. First, if two normal distributions have equal variances but unequal means, the two probability curves are centered around different values but otherwise they are identical. Second, for *normal* distributions, there is a distinct and rather noticeable difference between the two curves when the standard deviation increases from 1 to 1.5.

3.6.1 Computing Probabilities Associated with Normal Curves

Assume that human infants have birth weights that are normally distributed with a mean of 3700 g and a standard deviation of 200 g. What is the probability that a baby's birth weight will be less than or equal to 3000 g? As previously explained, this probability is given by the area under the normal curve, but simple methods for computing this area are required. Today the answer is easily obtained on a computer.

3.6.2 R Function pnorm

The R function

```
pnorm(q, mean = 0, sd = 1)
```

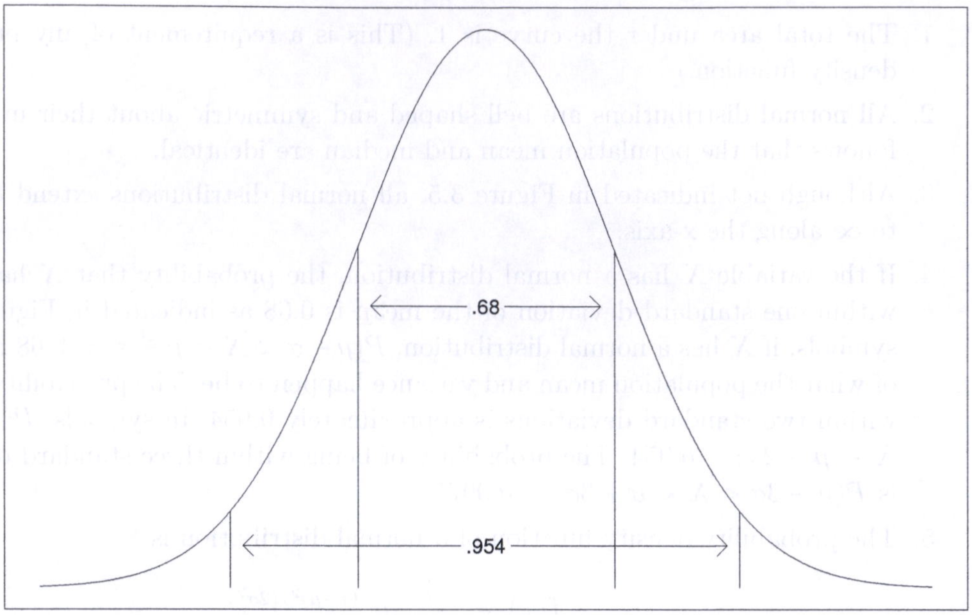

Figure 3.5 For any normal curve, the probability that an observation is within one standard deviation of the mean is 0.68. The probability of being within two standard deviations is always 0.954.

returns the probability that a normal random variable has a value less than or equal to q given a value for the mean and standard deviation. By default, the mean and standard deviation are taken to be 0 and 1, but any mean and standard deviation can be used via the arguments `mean` and `sd`.

EXAMPLE

If a normal distribution has mean 6 and standard deviation 2, what is the probability that an observation will be less than or equal to 7? The R command

```
pnorm(7, mean = 6, sd = 2)
```

returns the value 0.691.

For pedagogical reasons, a more traditional method is covered here for determining the probabilities associated with a normal distribution. We begin by considering the special case where the mean is zero and the standard deviation is one ($\mu = 0$, $\sigma = 1$), after which we illustrate how to compute probabilities for any mean and standard deviation.

Standard Normal

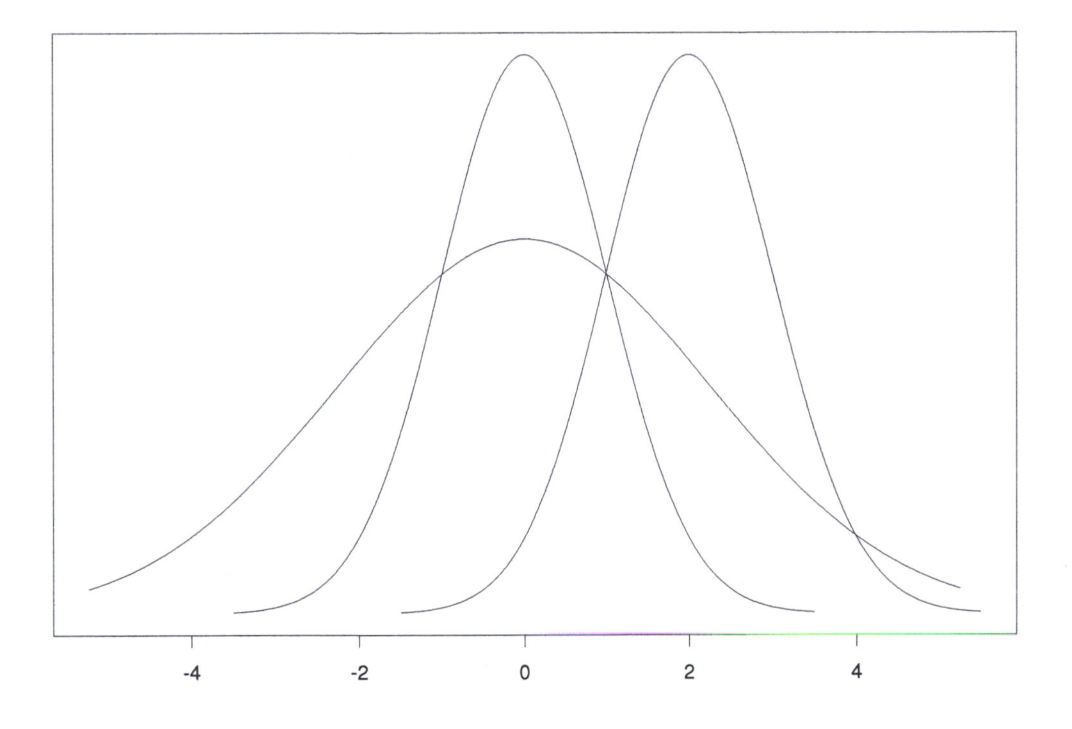

Figure 3.6 Two of the normal distributions have a population mean equal to 0, but they have unequal variances. One has a standard deviation of 1; the other has a standard deviation of 1.5. One of the distributions has mean 2 and variance 1.

The *standard normal distribution* is a normal distribution with mean $\mu = 0$ and standard deviation $\sigma = 1$; it plays a central role in many areas of statistics. As is typically done, Z is used to represent a variable that has a standard normal distribution. Our immediate goal is to describe how to determine the probability that an observation randomly sampled from a standard normal distribution is less than any constant c we might choose.

These probabilities are easily determined using Table 1 in Appendix B, which reports the probability that a standard normal random variable has probability less than or equal to c for $c = -3.00, -2.99, -2.98, \ldots, -0.01, 0, 0.01, \ldots 3.00$. The first entry in the first column shows -3. The column next to it gives the corresponding probability, 0.0013. That is, the probability that a standard normal random variables is less than or equal to -3 is $P(Z \leq -3) = 0.0013$. Put another way, -3 is the 0.0013 quantile of the standard normal distribution. Going down the first column we see the entry -2.08, and the column next to it indicates that the probability of a standard normal variable being less than or equal to -2.08 is 0.0188. This says that -2.08 is the 0.0188 quantile. Looking at the last entry in the third column, we see -1.55, the entry just to the right, in the fourth column, is 0.0606, so $P(Z \leq -1.55) = 0.0606$. This probability corresponds to the area in the left portion of Figure 3.7. Because the standard normal curve is symmetric about zero, the probability that X is greater than 1.55 is also 0.0606, which is shown in the right portion of Figure 3.7. Again looking at the first column of Table 1 in Appendix B, we see the value $c = 1.53$, and next to it is the value 0.9370 meaning that $P(Z \leq 1.53) = 0.9370$.

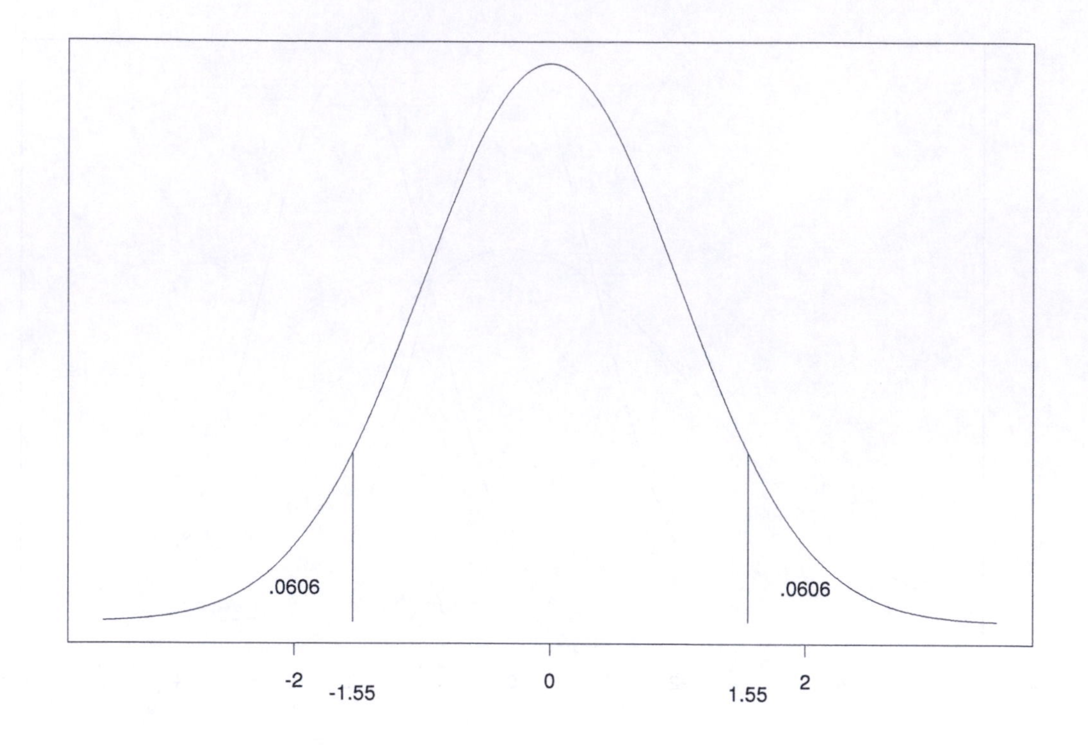

Figure 3.7 The standard normal probability curve. The probability that an observation is less than or equal to -1.55 is 0.0606. The probability that an observation is greater than or equal to -1.55 is 0.0606 as well.

In applied work, there are three types of probabilities that need to be determined:

1. $P(Z \leq c)$, the probability that a standard normal random variable is less than or equal to c,
2. $P(Z \geq c)$, the probability that a standard normal random variable is greater than or equal to c, and
3. $P(a \leq Z \leq b)$, the probability that a standard normal random variable is between the values a and b.

The first of these is determined from Table 1 in Appendix B, as already indicated. Because the area under the curve is 1, the second is given by

$$P(Z \geq c) = 1 - P(Z \leq c).$$

The third is given by

$$P(a \leq Z \leq b) = P(Z \leq b) - P(Z \leq a).$$

EXAMPLE

Determine $P(Z \geq 1.5)$, the probability that a standard normal random variable is greater than 1.5. From Table 1 in Appendix B, $P(Z \leq 1.5) = 0.9332$. Therefore, $P(Z \geq 1.5) = 1 - 0.9332 = 0.0668$.

EXAMPLE

Next we determine $P(-1.96 \leq Z \leq 1.96)$, the probability that a standard normal random variable is between -1.96 and 1.96. From Table 1 in Appendix B, $P(Z \leq 1.96) = 0.975$. Also, $P(Z \leq -1.96) = 0.025$, so $P(-1.96 \leq Z \leq 1.96) = 0.975 - 0.025 = 0.95$.

In some situations it is necessary to use Table 1 (in Appendix B) backwards. That is, we are given a probability and the goal is to determine c. For example, if we are told that $P(Z \leq c) = 0.99$, what is c? We simply find where 0.99 happens to be in Table 1 under the columns headed by $P(Z \leq c)$, and then read the number to the left, under the column headed by c. The answer is 2.33.

Two related problems also arise. The first is determining c given the value of

$$P(Z \geq c).$$

A solution is obtained by noting that the area under the curve is 1, so $P(Z \geq c) = 1 - P(Z \leq c)$, which involves a quantity we can determine from Table 1. That is, you compute $d = 1 - P(Z \geq c)$ and then determine c such that

$$P(Z \leq c) = d.$$

EXAMPLE

To determine c if $P(Z \geq c) = 0.9$, first compute $d = 1 - P(Z \leq c) = 1 - 0.9 = 0.1$. Then c is given by $P(Z \leq c) = 0.1$. Referring to Table 1 in Appendix B, $c = -1.28$.

The other type of problem is determining c given

$$P(-c \leq Z \leq c).$$

Letting $d = P(-c \leq Z \leq c)$, the answer is given by

$$P(Z \leq c) = \frac{1+d}{2}.$$

EXAMPLE

To determine c if $P(-c \leq Z \leq c) = 0.9$, let $d = P(-c \leq Z \leq c) = 0.9$ and then compute $(1+d)/2 = (1+0.9)/2 = 0.95$. Then c is given by $P(Z \leq c) = 0.95$. Referring to Table 1 in Appendix B, $c = 1.645$.

Solution for Any Normal Distribution

Now consider any normal random variable having mean μ and standard deviation σ. The next goal is to describe how to determine the probability of an observation being less than c, where, as usual, c is any constant that might be of interest. The solution is based on *standardizing* a normal random variable, which means that we subtract the population mean μ and divide by the standard deviation, σ. In symbols, we standardize a normal random variable X by transforming it to

$$Z = \frac{X - \mu}{\sigma}. \tag{3.11}$$

It can be shown that if X has a normal distribution, then the distribution of Z is standard normal. In particular, the probability that a normal random variable X is less than or equal to c is

$$P(X \le c) = P\left(Z \le \frac{c - \mu}{\sigma}\right). \tag{3.12}$$

EXAMPLE

Suppose it is claimed that the cholesterol levels in adults have a normal distribution with mean $\mu = 230$ and standard deviation $\sigma = 20$. If this is true, what is the probability that an adult will have a cholesterol level less than or equal to $c = 200$? Referring to Equation (2.12), the answer is

$$P(X \le 200) = P\left(Z \le \frac{200 - 230}{20}\right) = P(Z < -1.5) = 0.0668,$$

where 0.0668 is read from Table 1 in Appendix B. This means that the probability of an adult having a cholesterol level less than 200 is 0.0668.

In a similar manner, we can determine the probability that an observation is greater than or equal to 240 or between 210 and 250. More generally, for any constant c that is of interest, we can determine the probability that an observation is greater than c with the equation

$$P(X \ge c) = 1 - P(X \le c),$$

the point being that the right side of this equation can be determined with Equation (3.12). In a similar manner, for any two constants a and b,

$$P(a \le X \le b) = P(X \le b) - P(X \le a).$$

EXAMPLE

Continuing the last example, determine the probability of observing an adult with a cholesterol level greater than or equal to 240. We have that

$$P(X \ge 240) = 1 - P(X \le 240).$$

Referring to Equation (3.12),

$$P(X \le 240) = P\left(Z \le \frac{240 - 230}{20}\right) = P(Z < 0.5) = 0.6915,$$

so

$$P(X \ge 240) = 1 - 0.6915 = 0.3085.$$

In other words, the probability of an adult having a cholesterol level greater than or equal to 240 is 0.3085.

EXAMPLE

Continuing the cholesterol example, we determine

$$P(210 \le X \le 250).$$

We have that

$$P(210 \le X \le 250) = P(X \le 250) - P(X \le 210).$$

Now

$$P(X \le 250) = P\left(Z < \frac{250 - 230}{20}\right) = P(Z \le 1) = 0.8413$$

and

$$P(X \le 210) = P\left(Z < \frac{210 - 230}{20}\right) = P(Z \le -1) = 0.1587,$$

so

$$P(210 \le X \le 250) = 0.8413 - 0.1587 = 0.6826,$$

meaning that the probability of observing a cholesterol level between 210 and 250 is 0.6826.

3.6.3 R Function qnorm

The R function

```
qnorm(p, mean = 0, sd = 1)
```

computes the pth quantile of a normal distribution.

3.7 UNDERSTANDING THE EFFECTS OF NONNORMALITY

Classic, routinely used statistical methods are based on the assumption that observations follow a normal curve. It was once thought that violating the normality assumption rarely had a detrimental impact on these methods, but theoretical and empirical advances have made it clear that two general types of non-normality cause serious practical problems in a wide range of commonly occurring situations. Indeed, even very slight departures from normality can be a source for concern. To appreciate the practical utility of modern statistical techniques, as well as when and why conventional methods can be unsatisfactory, it helps to build an intuitive sense of how non-normality influences the population mean and variance and how this effect is related to determining probabilities.

The so-called contaminated or mixed normal distribution is a classic way of illustrating some of the more important effects of non-normality. Consider a situation where we have two subpopulations of individuals or things. Assume each subpopulation has a normal distribution, but they differ in terms of their means, or variances, or both. When we mix the two populations together we get what is called a *mixed* or *contaminated normal.* Generally, mixed normals fall outside the class of normal distributions. That is, for a distribution to qualify as normal, the equation for its curve must have the form given by Equation (3.10), and the mixed normal does not satisfy this requirement. When the two normals mixed together have a common mean, but unequal variances, the resulting probability curve is again symmetric about the mean, but even now the mixed normal is not a normal curve.

To provide a more concrete description of the mixed normal, consider the entire population of adults living around the world and let X represent the amount of weight they have gained or lost during the last year. Let's divide the population of adults into two groups: those who have tried some form of dieting to lose weight and those who have not. For illustrative purposes, assume that for adults who have not tried to lose weight, the distribution of their weight loss is standard normal (so $\mu = 0$ and $\sigma = 1$). As for adults who have dieted to lose weight, assume that their weight loss is normally distributed again with mean $\mu = 0$ but with standard deviation $\sigma = 10$. Finally, suppose there is a 10% chance of selecting someone who has dieted. That is, there is a 10% chance of selecting an observation from a normal distribution having standard deviation 10, in which case there is a 90% chance of selecting an observation from a normal curve having a standard deviation of 1.

Now, if we mix these two populations of adults together, the exact distribution of X (the weight loss for a randomly sampled adult) can be derived and is shown in Figure 3.8. Also shown is the standard normal distribution, and as is evident there is little separating the two curves. Let $P(X \leq c)$ be the probability that an observation is less than c when sampling from the mixed normal, and let $P(Z \leq c)$ be the probability when sampling from the standard normal instead. For any constant c we might pick, it can be shown that $P(X \leq c)$ does not differ from $P(Z \leq c)$ by more than 0.04. For example, for a standard normal curve, we see from Table 1 in Appendix B that $P(Z \leq 1) = 0.8413$. If X has the mixed normal distribution considered here, then the probability that X has a value less than or equal to 1 will not differ from 0.8413 by more than 0.04; it will be between 0.8013 and 0.8813. The exact value happens to be 0.81.

Here is the point: Very small departures from normality can greatly influence the value of the population variance. For the standard normal in Figure 3.8 the variance is 1, but for the mixed normal it is 10.9. The full implications of this result are impossible to appreciate at this point, but they will become clear in subsequent chapters. The main goal now is to lay the foundation for understanding some of the problems associated with conventional methods to be described.

To illustrate one of the many implications associated with the mixed normal, consider the following problem: Given the population mean and variance, how can we determine the probability that an observation is less than c? If, for example, $\mu = 0$ and $\sigma^2 = 10.9$, and if we want to know the probability that an observation is less than 1, we get an answer if we assume normality and use the method described in Section 2.5.1. The answer is 0.619. But for the mixed normal having the same mean and variance, the answer is 0.81 as previously indicated. So determining probabilities assuming normality, when in fact a distribution is slightly non-normal, can lead to a fairly inaccurate result. Figure 3.9 graphically illustrates the problem. Both curves have equal means and variances, yet there is a very distinct difference.

Figure 3.9 illustrates another closely related point. As previously pointed out, normal curves are completely determined by their mean and variance, and Figure 3.6 illustrated that, under normality, increasing the variance from 1 to 1.5 results in a very noticeable difference in the graphs of the probability curves. If we assume that curves are normal, or at least approximately normal, this might suggest that, in general, if two distributions have equal variances, surely they will appear very similar in shape. But this is not necessarily true even when the two curves are symmetric about the population mean and are bell-shaped. Again, knowing σ is useful in some situations to be covered, but there are many situations where it can mislead.

Figure 3.10 provides another illustration that two curves can have equal means and variances yet differ substantially.

Here is another way σ might mislead. We saw that for a normal distribution, there is a 0.68 probability that an observation is within one standard deviation of the mean. It is incorrect to conclude, however, that for non-normal distributions, this rule always applies.

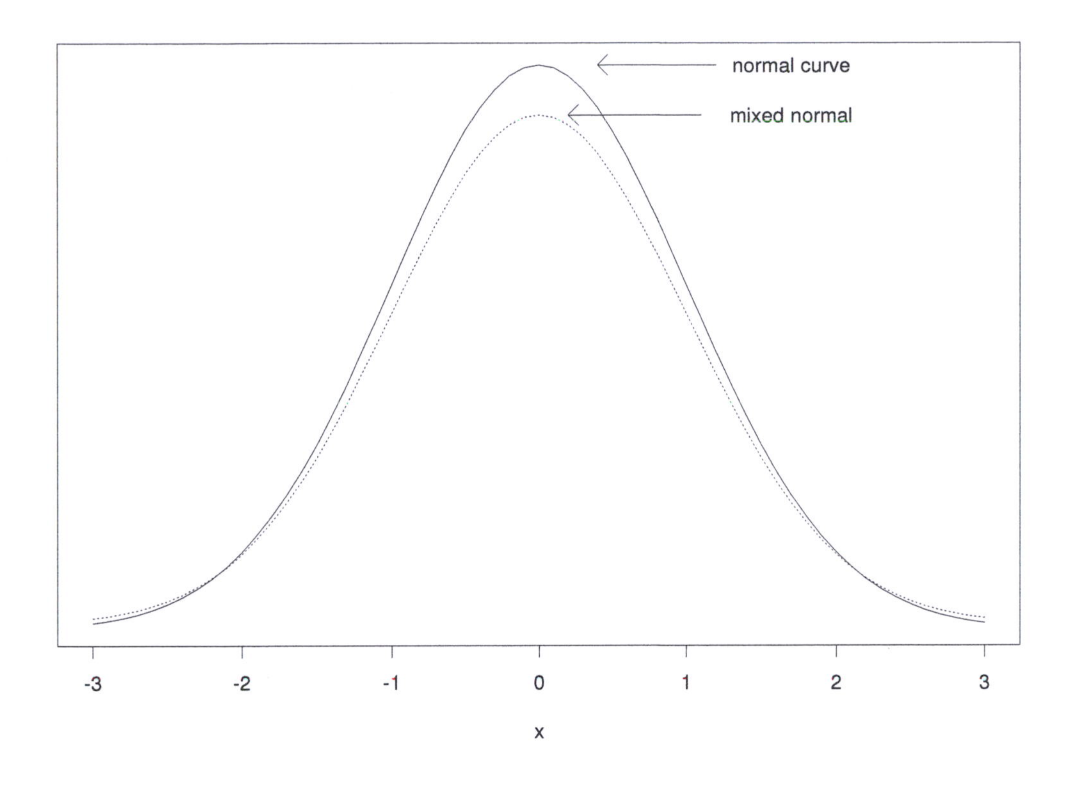

Figure 3.8 For normal distributions, increasing the standard deviation from 1 to 1.5 results in a substantial change in the distribution, as illustrated in Figure 3.6. But when considering non-normal distributions, seemingly large differences in the variances does not necessarily mean that there is a large difference in the graphs of the distributions. The two curves shown here have an obvious similarity, yet the variances are 1 and 10.9.

The mixed normal is approximately normal in a sense already described, yet the probability of being within one standard deviation of the mean now exceeds 0.925.

One reason this last point is important in applied work is related to the notion of outliers. *Outliers* are values that are unusually large or small. For a variety of reasons to be described in subsequent chapters, detecting outliers is important. Assuming normality, a common rule is to declare a value an outlier if it is more than two standard deviations from the mean. In symbols, declare X an outlier if

$$|X - \mu| > 2\sigma. \tag{3.13}$$

So, for example, if $\mu = 4$ and $\sigma = 3$, the value $X = 5$ would not be declared an outlier because $|5-4|$ is less than $2 \times 3 = 6$. In contrast, the value 12 would be labeled an outlier. The idea is that if a value lies more than two standard deviations from the mean, then probabilistically it is unusual. For normal curves, the probability that an observation is more than two standard deviations from the mean is 0.046.

To illustrate a concern about this rule, consider what happens when the probability density function is the mixed normal in Figure 3.8. Because the variance is 10.9, we would declare X an outlier if

$$|X - \mu| > 2\sqrt{10.9} = 6.6.$$

But $\mu = 0$, so we declare X to be an outlier if $|X| > 6.6$. It can be seen that now the

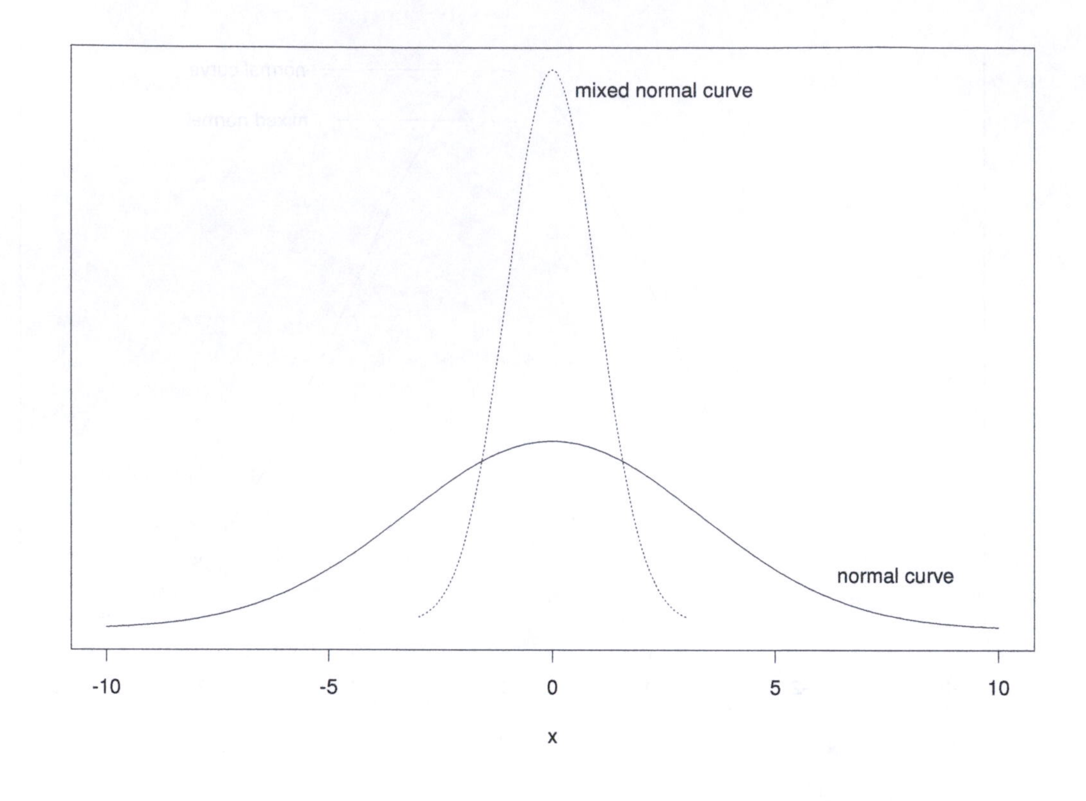

Figure 3.9 Two probability curves having equal means and variances.

probability of declaring a value an outlier is 4×10^{-11}— it is virtually impossible. (The method used to derive this probability is not important here.) The value 6, for example, would not be declared an outlier even though the probability of getting a value greater than or equal to 6 is 9.87×10^{-10}. That is, from a probabilistic point of view, 6 is unusually large because the probability of getting this value or larger is less than one in a billion, yet Equation (3.13) does not flag it as being unusual.

Note that in Figure 3.8, the tails of the mixed normal lie above the tails of the normal. For this reason, the mixed normal is often described as being *heavy-tailed.* Because the area under the extreme portions of a heavy-tailed distribution is larger than the area under a normal curve, extreme values or outliers are more likely when sampling from the mixed normal. Generally, very slight changes in the tail of any probability density function can inflate the variance tremendously, which in turn can make it difficult and even virtually impossible to detect outliers using the rule given by Equation (3.13) even though outliers are relatively common. The boxplot and MAD-median rule for detecting outliers, described in Chapter 2, are aimed at dealing with this problem.

3.7.1 Skewness

Heavy-tailed distributions are one source of concern when employing conventional statistical techniques based on means. Another is skewness, which generally refers to distributions that are not exactly symmetric. It is too soon to discuss all the practical problems associated with skewed distributions, but one of the more fundamental issues can be described here.

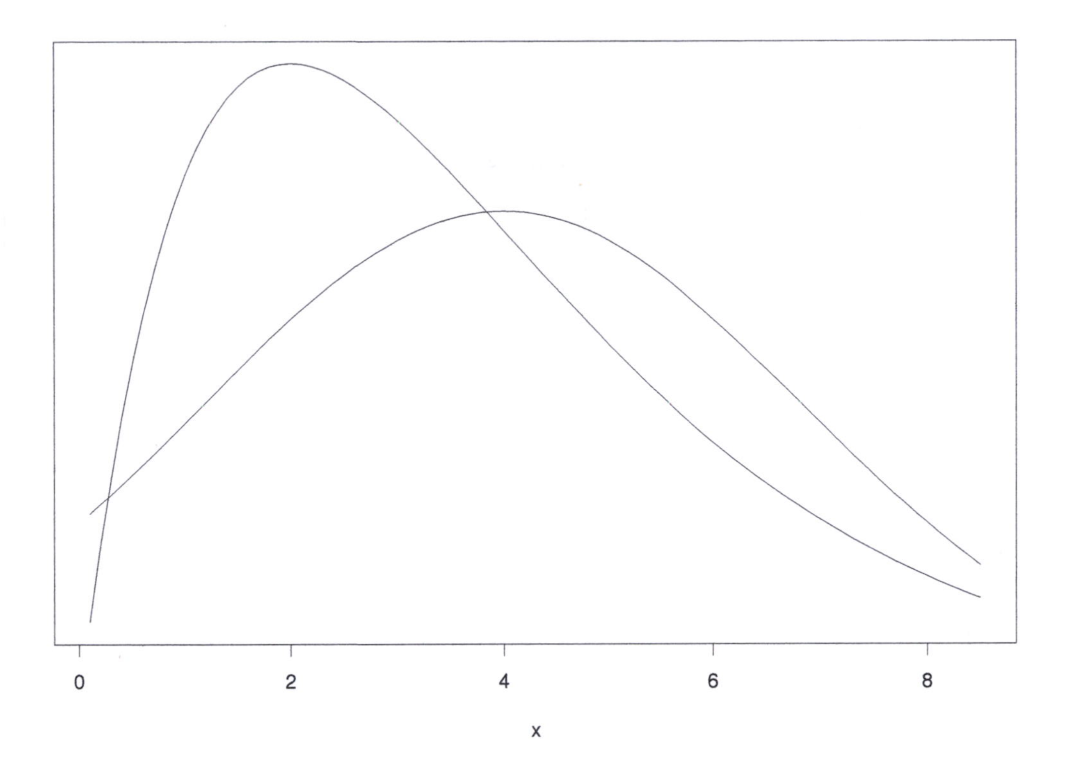

Figure 3.10 Two probability curves that have identical means and variances.

Consider how we might choose a single number to represent the typical individual or thing under study. A seemingly natural approach is to use the population mean. If a distribution is symmetric about its mean, as is the case when a distribution is normal, there is general agreement that the population mean is indeed a reasonable reflection of what is typical. But when distributions are skewed, at some point doubt begins to arise as to whether the mean is a good choice. Consider, for example, the distribution shown in Figure 3.11 which is skewed to the right. In this particular case, the population mean is located in the extreme right portion of the curve. In fact, the probability that an observation is less than the population mean is 0.74. So from a probabilistic point of view, the population mean is rather atypical. In contrast, the median is located near the more likely outcomes and would seem to better reflect what is typical.

One strategy is to routinely use means and hope that situations are never encountered where extreme skewness occurs. We will see empirical evidence, however, that such situations arise in practice. Another strategy is to use the median rather than the mean. If a distribution is symmetric, the population mean and median are identical, but if a distribution is skewed, the median can be argued to be a better indication of what is typical. In some applied settings, the median is a good choice, but unfortunately the routine use of the median can be rather unsatisfactory as well for reasons that will be made clear in subsequent chapters. For the moment it is merely remarked that dealing with skewness is a complex issue that has received a great deal of attention. In addition to the concern illustrated by Figure 3.11, there are a variety of other problems that will become evident in Chapter 5. Yet another strategy

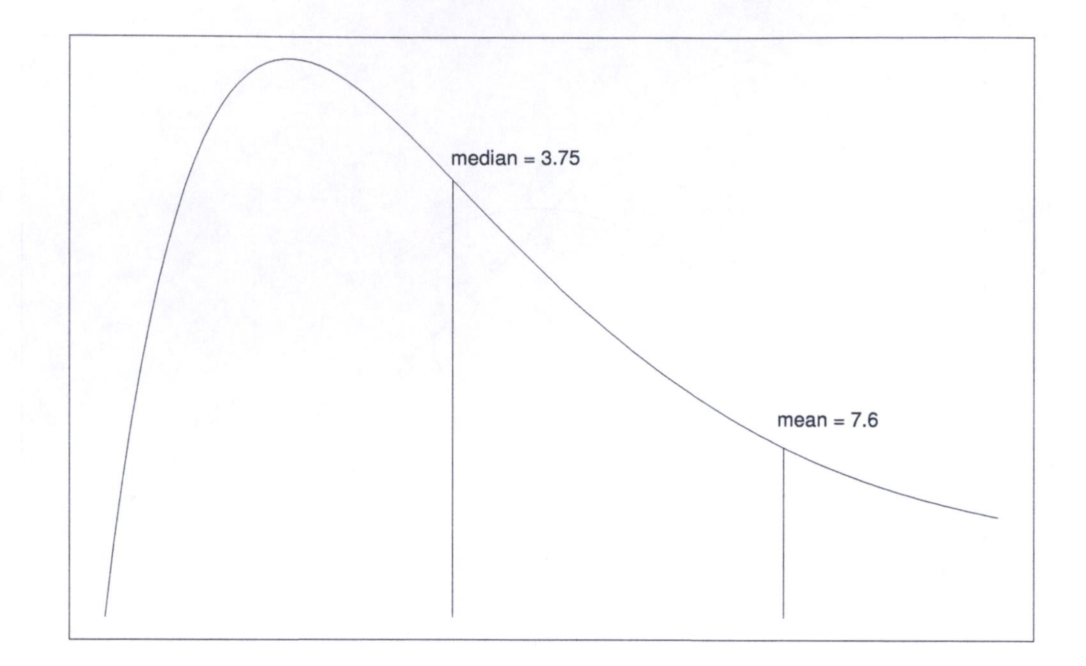

Figure 3.11 The population mean can be located in the extreme portion of the tail of a distribution. That is, the mean can represent a highly atypical response.

is to use some simple transformation of the data in an attempt to deal with skewness. A common method is to take logarithms, but this can fail as well for reasons to be described.

3.8 PEARSON'S CORRELATION AND THE POPULATION COVARIANCE (OPTIONAL)

This section is optional. It is included for those readers interested in the details associated with an important technical issue regarding outliers. Subsequent chapters describe classic, routinely used methods for analyzing data based on the sample mean. A common strategy is to delete any outliers that are found and apply these techniques to the remaining data. However, this strategy is technically unsound and can yield highly erroneous conclusions, a fact that cannot be stressed too strongly. There are well developed methods for dealing with outliers in a theoretically correct manner, as will be seen. This section provides some technical details that help explain why discarding outliers and applying standard methods based on means violates certain basic principles that are described in Section 4.2.1. The technical issue described here is based in part on what is called Pearson's correlation. Of particular importance here is how Pearson's correlation is related to the variance of the sum of two variables, which is given below by Equation (3.15).

Imagine any situation where we have two measures. For example, for each individual among a population of individuals, the two measures might be height and weight, or measures of gregariousness and severity of heart disease. Or another situation might be where we sample married couples and the two measures are the cholesterol levels of the wife and husband. For convenience, we label the two measures X and Y.

The *population covariance* between X and Y is

$$\sigma_{xy} = E[(X - \mu_x)(Y - \mu_y)].$$

In other words, if for some population of individuals we subtract the mean of X from every possible value for X, and do the same for Y, then the covariance between X and Y is defined to be the average of the products of these differences. It might help to note that the covariance of X with itself is just its (population) variance, and the same is true for Y. That is, the idea of covariance generalizes the notion of variance to two variables. *Pearson's correlation* is the covariance divided by the product of the standard deviations and is typically labeled ρ (a lower case Greek rho). That is,

$$\rho = \frac{\sigma_{xy}}{\sigma_x \sigma_y}. \tag{3.14}$$

Here are the properties that will be important in some of the chapters to follow:

- $-1 \le \rho \le 1$. (Pearson's correlation always has a value between -1 and 1.)
- If X and Y are independent, then $\rho = \sigma_{xy} = 0$.
- If $\rho \neq 0$, then X and Y are dependent.
- For any two variables, the variance of their sum is

$$\text{VAR}(X + Y) = \sigma_x^2 + \sigma_y^2 + 2\rho\sigma_x\sigma_y. \tag{3.15}$$

- For any two variables, the variance of their difference is

$$\text{VAR}(X - Y) = \sigma_x^2 + \sigma_y^2 - 2\rho\sigma_x\sigma_y. \tag{3.16}$$

Note that when we add any two measures together, their sum will have some average value. Using the rules of expected values (covered in Section 3.9), the average of this sum is simply $\mu_x + \mu_y$, the sum of the means. Equation (3.15) says that if we add two measures together, the average squared difference between any sum we might observe and the mean of the sum is completely determined by the individual variances plus the correlation. The practical importance of Equation (3.15) is explained in Sections 4.2.1 and 4.7.1.

EXAMPLE

There are two variables in Table 3.2: A wife's rating and a husband's rating. The variance associated with the wives can be seen to be $\sigma_y^2 = 0.29$. As for the husband's ratings, $\sigma_x^2 = 0.41$. It was already pointed out that the ratings for husbands and wives are independent, so $\rho = 0$. Consequently, the variance of the sum of the ratings is $0.29 + 0.41 = 0.7$. That is, without even determining the probability function associated with this sum, its variance can be determined.

A cautionary note should be added. Although independence implies $\rho = 0$, it is *not* necessarily the case that $\rho = 0$ implies independence. In the lower left panel of Figure 3.2, for example, points were generated on a computer with $\rho = 0$, yet there is dependence because, as already indicated, the variation in the Y values increases with X. (This issue will be discussed in more detail in Chapter 6.)

Table 3.3 Hypothetical Probabilities for Pearson's Correlation

		X			
		1	2	3	
	1	0.13	0.15	0.06	0.34
Y	2	0.04	0.08	0.18	0.30
	3	0.10	0.15	0.11	0.36
		0.27	0.38	0.35	

Table 3.4 How to Compute the Covariance for the Probabilities in Table 3.3

x	y	$x-\mu_x$	$y-\mu_y$	$p(x,y)$	$(x-\mu_x)(y-\mu_y)p(x,y)$
1	1	−1.08	−1.02	0.13	0.143208
1	2	−1.08	−0.02	0.04	0.000864
1	3	−1.08	0.98	0.10	−0.063504
2	1	−0.08	−1.02	0.15	0.003264
2	2	−0.08	−0.02	0.08	0.000128
2	3	−0.08	0.98	0.15	−0.014112
3	1	0.92	−1.02	0.06	−0.093840
3	2	0.92	−0.02	0.18	−0.002760
3	3	0.92	0.98	.11	0.099176
					0.0748

3.8.1 Computing the Population Covariance and Pearson's Correlation

In case it helps, this section illustrates how to compute the population covariance and Pearson's correlation when the probability function is known.

Consider the probabilities shown in Table 3.3. It can be seen that the expected value of X is $\mu_x = 2.08$ and the expected value of Y is $\mu_y = 2.02$. Let $p(x,y)$ be the probability of observing the values $X = x$ and $Y = y$ simultaneously. So according to Table 3.3, the probability that $Y = 1$ and $X = 1$ is $p(1,1) = 0.13$, and $p(2,3) = 0.18$. To compute the population covariance, you simply perform the calculations shown in Table 3.4. That is, for every combination of values for X and Y, you subtract the corresponding means yielding the values in columns three and four of Table 3.4. The probabilities associated with all possible pairs of values are shown in column 5. Column 6 shows the product of the values in columns 3, 4, and 5. The population covariance is the sum of the values in column 6, which is 0.0748. Under independence, X and Y must have a covariance of zero, so we have established that the variables considered here are dependent because the covariance differs from zero.

EXAMPLE

To compute the population correlation for the values and probabilities shown in Table 3.3, first compute the covariance, which we just saw is $\sigma_{xy} = 0.0748$. It is left as an exercise to show that the variances are $\sigma_x^2 = 0.6136$ and $\sigma_y^2 = 0.6996$. Consequently,

$$\rho = \frac{0.0748}{\sqrt{0.6136} \times \sqrt{0.6996}} = 0.11.$$

So according to Equation (3.15), the variance of the sum, $X + Y$, is

$$0.6136 + 0.6996 + 2 \times 0.0748 = 1.46.$$

3.9 SOME RULES ABOUT EXPECTED VALUES

This section summarizes some basic rules about expected values that will be useful in subsequent chapters. The first rule is that if we multiply a variable by some constant c, its expected value is multiplied by c as well. In symbols,

$$E(cX) = cE(X). \tag{3.17}$$

This is just a fancy way of saying, for example, that if the average height of some population of children is five feet ($\mu = 5$), then the average in inches is 60 (the average in feet multiplied by 12). More formally, if $E(X) = \mu = 5$, then $E(12X) = 12 \times 5 = 60$.

The second rule is that if we add c to every possible value, the expected value increases by c as well. That is,

$$E(X + c) = E(X) + c. \tag{3.18}$$

So if $\mu = 6$ and 4 is added to every possible value for X, the average becomes 10. Or in terms of Equation (3.18), $E(X + 4) = E(X) + 4 = 6 + 4 = 10$.

EXAMPLE

Because μ is a constant, $E(X - \mu) = E(X) - \mu = \mu - \mu = 0$. That is, if we subtract the population mean from every possible value we might observe, the average value of this difference is always zero. If, for instance, the average height of all adult men is 5.9 feet, and we subtract 5.9 from everyone's height, the average will be zero.

To provide some intuition about the next rule, imagine that for the population of all married women, if they were to rate their marital satisfaction, the probability function would be

x:	1	2	3	4	5
$p(x)$:	0.2	0.1	0.4	0.2	0.1

Now consider two individuals, Mary and Jane, and suppose they are asked about how they would rate their level of satisfaction regarding their married life. For convenience, label Mary's response X_1 and Jane's response X_2. So before Mary rates her marriage, there is a 0.2 probability that she will rate her marriage satisfaction as 1, a 0.1 probability she will rate it as a 2, and so on. The same is assumed to be true for Jane. Now consider the sum of their two ratings, which we label $X = X_1 + X_2$. What is the expected value of this sum? That is, on average, what is the value of X?

One way of solving this problem is to attempt to derive the probability function of the sum, X. The possible values for X are $1 + 1 = 2$, $1 + 2 = 3, \ldots, 6 + 6 = 12$, and if we could derive the probabilities associated with these values, we could determine the expected value of X. But there is a much simpler method because it can be shown that the expected value of a sum is just the sum of the expected values. That is,

$$E(X_1 + X_2) = E(X_1) + E(X_2), \tag{3.19}$$

so the expected value of the sum can be determined if we know the probability function associated with each of the observations we make. But given the probability function, we can do just that. We see that $E(X_1) = 2.9$; in a similar manner $E(X_2) = 2.9$, so the expected value of their sum is 5.8. That is, if two women are asked to rate their marital satisfaction, the average sum of their ratings, over all pairs of women we might interview, is 5.8.

This last illustration demonstrates a more general principle that will be helpful. If X_1 and X_2 have identical probability functions, so in particular the variables have a common mean, μ, then the expected value of their sum is 2μ. So using our rule for constants, we see that the average of these two ratings is μ. That is,

$$E\left[\frac{1}{2}(X_1 + X_2)\right] = \frac{1}{2}(\mu + \mu) = \mu.$$

Here is a summary of the rules for expected values where c is any constant:

- $E(cX) = cE(X) = c\mu$.
- $E(X + c) = E(X) + c = \mu + c$.
- $E(X_1 + X_2) = E(X_1) + E(X_2)$. For the special case where X_1 and X_2 have a common mean μ, which occurs when they have identical probability functions, $E(X_1+X_2) = 2\mu$.

3.10 CHI-SQUARED DISTRIBUTIONS

A number of inferential methods in statistics are based in part on what are called chi-squared distributions, including some classic chi-squared tests that are described in Chapter 15. When first encountered, it might seem that these methods have nothing to do with the normal distribution introduced in this chapter. However, this is not the case. For any method based on a chi-squared distribution, the family of normal distributions plays a role. To stress this point, this section provides a brief description of how chi-squared distributions are defined.

Suppose Z has a standard normal distribution and let $Y = Z^2$. The distribution of Y is so important, it has been given a special name: A chi-squared distribution with one degree of freedom. Next, suppose two independent observations are made, Z_1 and Z_2, both of which have standard normal distributions. Then the distribution of

$$Y = Z_1^2 + Z_2^2$$

is called chi-square distribution with two degrees of freedom. More generally, for n independent standard normal variables, $Z_1, \ldots, Z_n$,

$$Y = Z_1^2 + \cdots + Z_n^2$$

is said to have a chi-squared distribution with n degrees of freedom.

3.11 EXERCISES

Brief answers to most of the exercises are given in Appendix A. More detailed answers for all of the exercises can be downloaded from http://dornsife.usc.edu/labs/rwilcox/books. (Or google the author's name and follow the links to his personal web page.) The answers are stored in the file CRC_answers.pdf.

1. For the probability function

x: 0, 1

p(x): 0.7, 0.3

verify that the mean and variance are 0.3 and 0.21. What is the probability of getting a value less than the mean?

2. Standardizing the possible values in the previous exercise means that we transform the possible values (0 and 1) by subtracting the population mean and dividing by the population standard deviation. Here this yields $(1-0.3)/\sqrt{0.21} = 0.7/\sqrt{0.21}$ and $(0-0.3)/\sqrt{0.21} = -0.3/\sqrt{0.21}$, respectively. The probabilities associated with these two values are 0.7 and 0.3. Verify that the expected value of the standardized values is 0 and the variance is 1.
3. For the probability function

   x: 1, 2, 3, 4, 5
   p(x): 0.15, 0.2, 0.3, 0.2, 0.15

   determine the mean, the variance, and $P(X \leq \mu)$.
4. For the probability function

   x: 1, 2, 3, 4, 5
   p(x): 0.1, 0.25, 0.3, 0.25, 0.1

   would you expect the variance to be larger or smaller than the variance associated with the probability function used in the previous exercise? Verify your answer by computing the variance for the probability function given here.
5. For the probability function

   x: 1, 2, 3, 4, 5
   p(x): 0.2, 0.2, 0.2, 0.2, 0.2

   would you expect the variance to be larger or smaller than the variance associated with the probability function used in the previous exercise. Verify your answer by computing the variance.
6. Verify that if we standardize the possible values in the previous exercise, the resulting mean is zero and the variance is one.
7. For the following probabilities

	Income		
Age	High	Medium	Low
< 30	0.030	0.180	0.090
30-50	0.052	0.312	0.156
> 50	0.018	0.108	0.054

   determine (a) the probability someone is under 30, (b) the probability that someone has a high income given that they are under 30, (c) the probability of someone having a low income given that they are under 30, and (d) the probability of a medium income given that they are over 50.
8. For the previous exercise, are income and age independent?
9. Coleman (1964) interviewed 3398 schoolboys and asked them about their self-perceived membership in the "leading crowd." Their response was either yes, they were a member, or no, they were not. The same boys were also asked about their attitude concerning the leading crowd. In particular, they were asked whether membership meant that it does not require going against one's principles sometimes or whether they think it does. Here, the first response will be indicated by a 1, while the second will be indicated by a 0. The results were as follows:

Member	Attitude	
	1	0
Yes	757	496
No	1071	1074

These values, divided by the sample size, 3398, are called relative frequencies, as noted in Chapter 2. For example, the relative frequency of the event (yes, 1) is 757/3398. Treat the relative frequencies as probabilities and determine (a) the probability that an arbitrarily chosen boy responds yes, (b) $P(\text{yes}|1)$, (c) $P(1|\text{yes})$, (d) whether the response yes is independent of the attitude 0, (e) the probability of a (yes and 1) or a (no and 0) response, (f) the probability of not responding (yes and 1), (g) the probability of responding yes or 1.

10. The probability density function associated with a so-called *uniform distribution* is given by $f(x) = 1/(b - a)$, where a and b are given constants and $a \le x \le b$. That is, the possible values you might observe range between the constants a and b with every value between a and b equally likely. If for example $a = 0$ and $b = 1$, the possible values you might observe lie between 0 and 1. For a uniform distribution over the interval $a = 1$ and $b = 4$, draw the probability density function and determine the median, the 0.1, and the 0.9 quantiles.
11. For the uniform distribution over the interval -3 to 2, determine (a) $P(X \le 1)$, (b) $P(X < -1.5)$, (c) $P(X > 0)$, (d) $P(-1.2 \le X \le 1.)$, (e) $P(X = 1)$.
12. For the uniform distribution in the previous problem, determine the median, the 0.25, and the 0.9 quantiles.
13. For the uniform distribution with $a = -1$ and $b = 1$, determine c such that (a) $P(X \le c) = 0.9$, (b) $P(X \le c) = 0.95$, (c) $P(X > c) = 0.99$.
14. For the uniform distribution with $a = -1$ and $b = 1$, determine c such that (a) $P(-c \le X \le c) = 0.9$, (b) $P(-c \le X \le c) = 0.95$, (c) $P(-c \le X \le c) = 0.99$.
15. Suppose the waiting time at a traffic light has a uniform distribution from 0 to 20 seconds. Determine (a) the probability of waiting exactly 12 seconds, (b) less than 5 seconds, (c) more than 10 seconds.
16. When you look at a clock, the number of minutes past the hour, say X, is some number between 0 and 60. Assume the number of minutes past the hour has a uniform distribution. Determine (a) $P(X = 30)$, (b) $P(X \le 10)$, (c) $P(X \ge 20)$, (d) $P(10 \le X < 20)$.
17. For the previous problem, determine the 0.8 quantile.
18. Given that Z has a standard normal distribution, use Table 1 in Appendix B to determine (a) $P(Z \ge 1.5)$, (b) $P(Z \le -2.5)$, (c) $P(Z < -2.5)$, (d) $P(-1 \le Z \le 1)$.
19. If Z has a standard normal distribution, determine (a) $P(Z \le 0.5)$, (b) $P(Z > -1.25)$, (c) $P(-1.2 < Z < 1.2)$, (d) $P(-1.8 \le Z < 1.8)$.
20. If Z has a standard normal distribution, determine (a) $P(Z < -0.5)$, (b) $P(Z < 1.2)$, (c) $P(Z > 2.1)$, (d) $P(-0.28 < Z < 0.28)$.
21. If Z has a standard normal distribution, find c such that (a) $P(Z \le c) = 0.0099$, (b) $P(Z < c) = .9732$, (c) $P(Z > c) = 0.5691$, (d) $P(-c \le Z \le c) = 0.2358$.
22. If Z has a standard normal distribution, find c such that (a) $P(Z > c) = 0.0764$, (b) $P(Z > c) = 0.5040$, (c) $P(-c \le Z < c) = 0.9108$, (d) $P(-c \le Z \le c) = 0.8$.

23. If X has a normal distribution with mean $\mu = 50$ and standard deviation $\sigma = 9$, determine (a) $P(X \leq 40)$, (b) $P(X < 55)$, (c) $P(X > 60)$, (d) $P(40 \leq X \leq 60)$.
24. If X has a normal distribution with mean $\mu = 20$ and standard deviation $\sigma = 9$, determine (a) $P(X < 22)$, (b) $P(X > 17)$, (c) $P(X > 15)$, (d) $P(2 < X < 38)$.
25. If X has a normal distribution with mean $\mu = 0.75$ and standard deviation $\sigma = 0.5$, determine (a) $P(X < 0.25)$, (b) $P(X > 0.9)$, (c) $P(0.5 < X < 1)$, (d) $P(0.25 < X < 1.25)$.
26. If X has a normal normal distribution, determine c such that

$$P(\mu - c\sigma < X < \mu + c\sigma) = 0.95.$$

27. If X has a normal distribution, determine c such that

$$P(\mu - c\sigma < X < \mu + c\sigma) = 0.8.$$

28. Assuming that the scores on a math achievement test are normally distributed with mean $\mu = 68$ and standard deviation $\sigma = 10$, what is the probability of getting a score greater than 78?
29. In the previous problem, how high must someone score to be in the top 5%? That is, determine c such that $P(X > c) = 0.05$.
30. A manufacturer of car batteries claims that the life of its batteries is normally distributed with mean $\mu = 58$ months and standard deviation $\sigma = 3$. Determine the probability that a randomly selected battery will last at least 62 months.
31. Assume that the income of pediatricians is normally distributed with mean $\mu = 100000$ and standard deviation $\sigma = 10000$. Determine the probability of observing an income between \$85,000 and \$115,000.
32. Suppose the winnings of gamblers at Las Vegas are normally distributed with mean $\mu = -300$ (the typical person loses \$300), and standard deviation $\sigma = 100$. Determine the probability that a gambler does not lose any money.
33. A large computer company claims that its salaries are normally distributed with mean \$50,000 and standard deviation 10,000. What is the probability of observing an income between \$40,000 and \$60,000?
34. Suppose the daily amount of solar radiation in Los Angeles has a normal distribution with mean 450 calories and standard deviation 50. Determine the probability that for a randomly chosen day, the amount of solar radiation is between 350 and 550.
35. If the cholesterol levels of adults are normally distributed with mean 230 and standard deviation 25, what is the probability that a randomly sampled adult has a cholesterol level greater than 260?
36. If after one year, the annual mileage of privately owned cars is normally distributed with mean 14,000 miles and standard deviation 3500, what is the probability that a car has mileage greater than 20,000 miles?
37. Can small changes in the tails of a distribution result in large changes in the population mean, μ, relative to changes in the median?
38. Explain in what sense the population variance is sensitive to small changes in a distribution.

39. For normal random variables, the probability of being within one standard deviation of the mean is 0.68. That is, $P(\mu - \sigma \leq X \leq \mu + \sigma) = 0.68$ if X has a normal distribution. For non-normal distributions, is it safe to assume that this probability is again 0.68? Explain your answer.
40. If a distribution appears to be bell-shaped and symmetric about its mean, can we assume that the probability of being within one standard deviation of the mean is 0.68?
41. Can two distributions differ by a large amount yet have equal means and variances?
42. If a distribution is skewed, is it possible that the mean exceeds the 0.85 quantile?
43. Determine $P(\mu - \sigma \leq X \leq \mu + \sigma)$ for the probability function

x: 1, 2, 3, 4
p(x): 0.2, 0.4, 0.3, 0.1

44. The Department of Agriculture of the United States reports that 75% of all people who invest in the futures market lose money. Based on the binomial probability function, with $n = 5$, determine (a) the probability that all 5 lose money, (b) the probability that all 5 make money, (c) the probability that at least 2 lose money.
45. If for a binomial, $p = 0.4$ and $n = 25$, determine (a) $P(X < 11)$, (b) $P(X \leq 11)$, (c) $P(X > 9)$, and (d) $P(X \geq 9)$
46. In the previous problem, determine $E(X)$, the variance of X, $E(\hat{p})$, and the variance of $\hat{p}$.
47. Bayes' theorem states the following: $P(A|B) = P(B|A)P(A)/P(B)$. Imagine that with probability 0.9, a diagnostic test correctly identifies those individuals who have an illness, and correctly rules out a disease with probability 0.95 among individuals who do not have it. Further imagine that the probability of having this illness is 0.02. If a diagnostic test indicates that someone has this illness, what is the probability that the test is correct?

CHAPTER 4

SAMPLING DISTRIBUTIONS AND CONFIDENCE INTERVALS

CONTENTS

As indicated in Chapter 3, when dealing with the binomial, the proportion of observed successes, $\hat{p}$, estimates p, the true probability of success. But, in general, this estimate will be incorrect. That is, typically, $\hat{p} \neq p$, and this raises a fundamental question: How well does $\hat{p}$ estimate p? If, for example, $\hat{p} = 0.45$ based on a sample of 50 individuals, is it reasonable to conclude that the actual value of p is between 0.3 and 0.5? More generally, what range of values for p are reasonable based on the available data? Can values less than 0.1 and greater than 0.85 be ruled out with a high degree of certainty?

In a similar manner, the sample mean, $\bar{X}$, estimates the population mean, μ, but typically $\bar{X} \neq \mu$. If, for example, the sample mean is $\bar{X} = 10$, this means that the population mean is estimated to be 10, but in all likelihood this estimate is wrong. What is needed is a method for judging the precision of this estimate. That is, what range of values for μ are likely to be correct based on the available data? Which values for μ can be ruled out with a reasonable degree of certainty? This chapter describes classic and modern methods for addressing this issue stemming from the notion of a sampling distribution, which is introduced in Section 4.2.

The notion of a sampling distribution has other important roles. First, it provides a useful perspective on the relative merits of the measures of location introduced in Chapter 2. Of particular importance is understanding when the sample mean performs relatively well and when it can be highly unsatisfactory. The notion of a sampling distribution also provides a basis for understanding and appreciating a modern technique called the bootstrap, which is introduced in Section 5.9. Finally, sampling distributions help foster and understanding of the deleterious effects of nonnormality and the practical benefits of numerous methods developed during the last half century.

4.1 RANDOM SAMPLING

To begin, it helps to be a bit more formal about the notion of a random sample. The variables $X_1, \ldots, X_n$ are said to be a random sample if any two have the same distribution and are independent.

EXAMPLE

Consider a sample $X_1, \ldots, X_n$ where every outcome is either a success, denoted by the value 1, or a failure, denoted by 0. That is, $X_1 = 1$ if the first observation is a success, otherwise, $X_1 = 0$. Similarly, $X_2 = 1$ if the second observation is a success, and so on. Saying that $X_1, \ldots X_n$ have the same distribution simply means that every observation has the same probability function, which for the situation at hand is the same as saying that each observation has the same probability of success, which we label p. In symbols,

$$P(X_1 = 1) = P(X_2 = 1) = \cdots = P(X_n = 1) = p.$$

If in addition any two observations are independent, then $X_1, \ldots X_n$ is a random sample. Under random sampling, the total number of successes in n trials is

$$X = \sum X_i,$$

which has the binomial probability function described in Chapter 3.

4.2 SAMPLING DISTRIBUTIONS

Generally, a *sampling distribution* refers to the probabilities associated with some estimator of interest. The estimator might be the sample mean, the median, the sample variance, or the proportion of successes, $\hat{p}$, when dealing with the binomial.

To elaborate, we first focus on the binomial probability function where the goal is to estimate the probability of success, p, with the proportion of successes, $\hat{p}$. To begin with a relatively simple case, suppose a random sample of size five is obtained and we compute $\hat{p}$. Note that if we were to get a new sample, we might get a different value for $\hat{p}$. The first sample might result in $\hat{p} = 3/5$ and the second might result in $\hat{p} = 2/5$. The sampling distribution of $\hat{p}$ refers to the probabilities associated with the possible values of $\hat{p}$.

If we are told the probability of success p, the sampling distribution of $\hat{p}$ is readily determined. Momentarily assume that, unknown to us, the probability of success is $p = 0.5$. Then from Chapter 3, the probability function for the total number of successes is given in Table 2 of Appendix B and can be summarized as follows:

x:	0	1	2	3	4	5
$P(X = x)$:	0.03125	0.15625	0.31250	0.31250	0.15625	0.03125

For example, the probability of exactly one success is 0.15625. Also note that as a consequence, the probabilities associated with $\hat{p}$ are determined as well.

$\hat{p}$:	0/5	1/5	2/5	3/5	4/5	5/5
$P(\hat{p})$:	0.03125	0.15625	0.31250	0.31250	0.15625	0.03125

For instance, the probability of exactly one success ($X = 1$) is 0.15625, so the probability that $\hat{p} = 1/5$ is also 0.15625. That is, given the probability of success p, the sampling distribution of $\hat{p}$ is readily determined. Moreover, the mean of this sampling distribution can be seen to be p, the probability of success, and the variance is $p(1-p)/n$. We say that $\hat{p}$ is an *unbiased* estimator of p because on average it is equal to p. More succinctly, $E(\hat{p}) = p$.

One way of conceptualizing the probabilities associated with $\hat{p}$ is in terms of the proportion of times a value for $\hat{p}$ occurs if a study were repeated (infinitely) many times. Imagine we repeat a study (infinitely) many times with each study based on a random sample of n observations resulting in $\hat{p}$ values that are labeled

$$\hat{p}_1, \hat{p}_2, \ldots .$$

Then, for instance, the probability that $\hat{p} = 1/5$ can be viewed as the proportion of times the value 1/5 occurs. Moreover, the average value of $\hat{p}$ over these many studies is equal to the probability of success, p. In addition, the variance of the $\hat{p}$ values over many studies is $p(1-p)/n$.

EXAMPLE

Imagine that, unknown to us, a new drug has probability 0.35 of being successful when taken by a randomly sampled individual. Further imagine that 1 million random samples are generated, each of size 15, and that for each sample we compute the proportion of successes, $\hat{p}$. If these $\hat{p}$ values were averaged, the result would be approximately 0.35. If the variance of these one million $\hat{p}$ values were computed, the result would be, to a close approximation, $0.35(1-0.35)/15 = 0.0152$.

4.2.1 Sampling Distribution of the Sample Mean

Similar to $\hat{p}$, the sampling distribution of the sample mean, $\bar{X}$, refers to the probabilities associated with all possible values for $\bar{X}$. Consider, for example, a study where the outcome of interest is the amount of weight lost after a year of following a particular diet plan. Imagine that 20 participants take part and their average weight loss is $\bar{X} = 12$ lb. Now imagine that the study is replicated, again with $n = 20$, and that this time $\bar{X} = 19$. If the study were replicated (infinitely) many times, the results could be summarized as follows:

Study:	1	2	3	...
Means:	$\bar{X}_1$	$\bar{X}_2$	$\bar{X}_3$	...

The sampling distribution of the sample mean refers to the associated cumulative probabilities. For example, if the proportion of sample means less than or equal to 6 is 0.3 under random sampling, we write this as $P(\bar{X} \leq 6) = 0.3$. Roughly, over millions of studies, 30% will result in a sample mean that is less than or equal to 6. Moreover, it can be shown that the average of all sample means, under random sampling, is μ, the population mean of the distribution from which the values were sampled. In symbols, $E(\bar{X}) = \mu$, meaning that the sample mean is an *unbiased* estimator of the population mean. It can be shown that the variance of the sample mean is

$$\text{VAR}(\bar{X}) = \frac{\sigma^2}{n},$$

which is sometimes written as

$$\sigma^2_{\bar{X}} = \frac{\sigma^2}{n}.$$

This result follows from Equation (3.15) by noting that under random sampling, Pearson's correlation for any two observations is zero. That is, independence among the n observations plays a crucial role in the derivation of $\text{VAR}(\bar{X})$.

The variance of an estimator over many studies plays such an important role, that it has been given a special name: *squared standard error*. The standard deviation of an estimator is called its standard error, which is just the (positive) square root of the squared standard error. As just indicated, the squared standard error of the sample mean, under random sampling, is σ^2/n and so its standard error is $\sigma/\sqrt{n}$. For the binomial, the squared standard error of $\hat{p}$ is $p(1-p)/n$.

With the binomial, we are able to determine the sampling distribution of $\hat{p}$ once we know its mean, which is p. As just indicated, the mean of $\bar{X}$ is μ, but in general knowing the population mean does not determine the sampling distribution of $\bar{X}$. If, for example, we are told that $\mu = 32$, in most cases this is not enough information to determine the probability that the sample mean is less than 35, or less than 30, or less than c for any constant c we might pick. Under random sampling, we know that the variance of the sample mean is σ^2/n, but for most situations, even if we are told the value of the population variance σ^2 and the sample size n, this is not sufficient for determining $P(\bar{X} \leq c)$. A common strategy for dealing with this problem is to impose an assumption: sampling is from a normal distribution. Then under random sampling, it can be shown that the distribution of the sample mean also has a normal distribution. In summary, then, we have three key results when randomly sampling from a normal distribution:

- $E(\bar{X}) = \mu$
- $\sigma^2_{\bar{X}} = \frac{\sigma^2}{n}$
- $\bar{X}$ has a normal distribution

with the first two results requiring random sampling only.

Notice that when we have a random sample of n observations, we can view this sample in one of two ways: (1) a single study consisting of n values or (2) n studies, each based on a sample size of one. For the latter view, if the sample size n is sufficiently large, a plot of the values will give us a fairly accurate indication of the distribution from which the values were randomly sampled. If, for example, we randomly sample many observations from a (standard) normal distribution with mean $\mu = 0$ and variance $\sigma^2 = 1$, then a plot of the data (using, for example, a kernel density estimator described in Chapter 2) will result in a normal distribution having mean 0 and variance 1. The same plot is obtained if the number of studies is large with each study having a single observation. If instead each study results in a sample mean based on 16 observations, and if sampling is from a normal distribution, again a plot of the sample means will have a normal distribution with mean $\mu = 0$, but now the variance is 1/16 as shown in Figure 4.1.

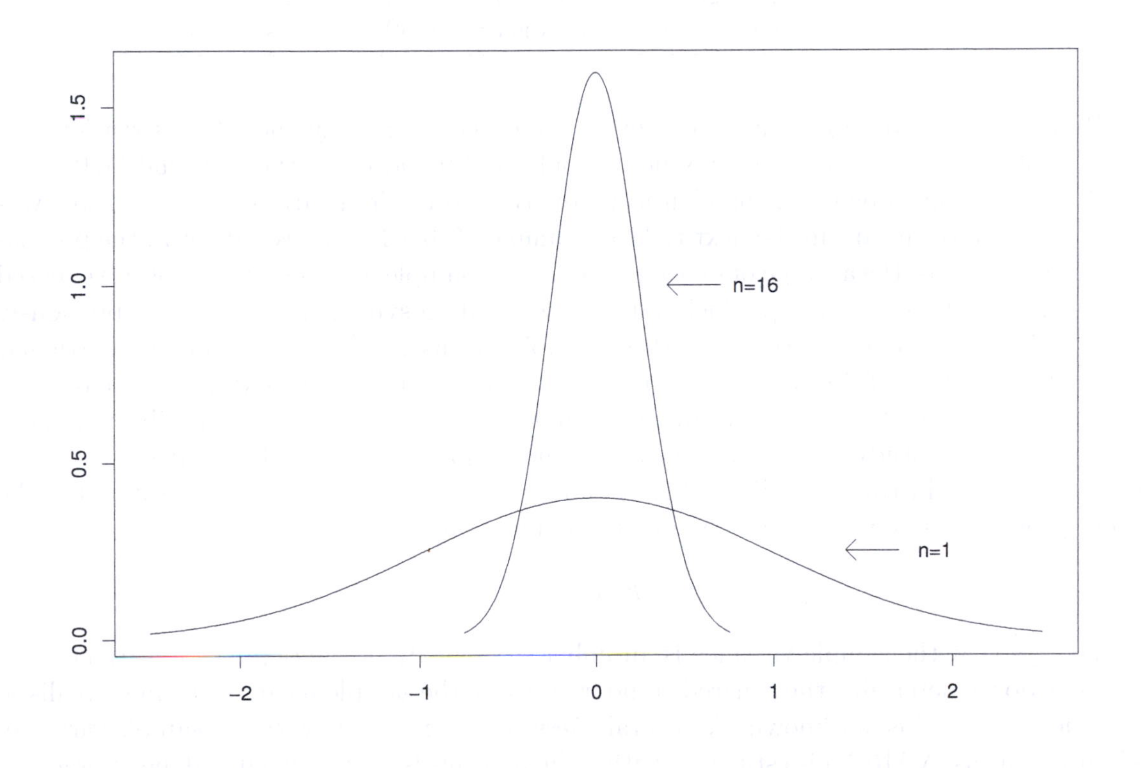

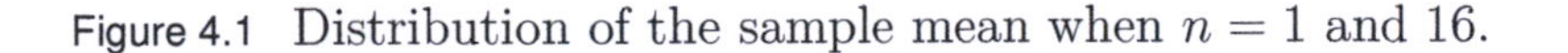

Figure 4.1 Distribution of the sample mean when $n = 1$ and 16.

Figure 4.1 graphically demonstrates the extent to which sample means based on sample sizes of 16 improve upon sample means based on a sample size of 1. With $n = 16$, the sample means are more tightly clustered around the population mean, indicating that typically a more accurate estimate of the population mean will be obtained.

A slightly closer look at the notion of a random sample might help when dealing with certain conceptual issues covered later in this chapter. Suppose $X_1, \ldots, X_n$ is a random sample from a normal distribution having mean $\mu = 1.9$ and variance $\sigma^2 = 0.49$. Then, in particular, X_1 has a normal distribution with mean $\mu = 1.9$ and variance $\sigma^2 = 0.49$, X_2 has a mean of 1.9 and variance 0.49, and so on.

Here is way of conceptualizing what this means. Imagine we have a random sample of size 100, which is depicted in Table 4.1. Focus on the column headed by X_1. When we say that X_1 has a normal distribution with mean 1.9 and variance 0.49, this means that if we were to plot all of the X_1 values over many studies, we would get a normal distribution with mean 1.9 and variance 0.49. This is written in a more formal manner as $E(X_1) = 1.9$ and

Table 4.1 A Way of Conceptualizing a Sampling Distribution When Each Sample Has $n = 100$

Sample	X_1	X_2	X_3	...	X_{100}	$\bar{X}$	s^2
1	2	3	2	...	2	2.4	0.43
2	1	3	2	...	1	2.2	0.63
3	1	2	1	...	3	2.1	0.57
4	3	1	3	...	1	2.3	0.62
⋮	⋮	⋮	⋮	⋮	⋮	⋮	⋮

Average of $\bar{X}$ values is $\mu = 1.9$, i.e., $E(\bar{X}) = \mu$
Average of s^2 values is $\sigma^2 = 0.49$, i.e., $E(s^2) = \sigma^2$

$\text{VAR}(X_1) = 0.49$. More generally, for any of the observations, say the ith observation, the average of the X_i values over many studies would be 1.9 and the variance would be 0.49.

Now, an implication of having identically distributed random variables is that if we average all the sample means in the next to last column of Table 4.1, we would get μ, the population mean. That is, the average or expected value of all sample means we might observe, based on n participants, is equal to μ which in this case is 1.9. In symbols, $E(\bar{X}) = \mu$ as previously indicated. Moreover, the variance of these sample means is σ^2/n, which in this particular case is $0.49/100 = 0.0049$, so the standard error of the sample mean is $\sqrt{0.0049} = 0.07$.

Notice that like the sample mean, the sample variance s^2 also has a distribution. That is, if we repeat a study many times, typically the sample variance will vary from one study to the next as illustrated in Table 4.1. It can be shown that under random sampling, the average of these sample variances will be the population variance, σ^2. That is,

$$E(s^2) = \sigma^2,$$

which says that the sample variance is an unbiased estimate of the population variance.

How do we determine the squared standard error of the sample mean in the more realistic situation where σ^2 is not known? A natural guess is to estimate σ^2 with the sample variance, s^2, in which case $\text{VAR}(\bar{X})$ is estimated with s^2/n, and this is exactly what is done in practice.

EXAMPLE

Imagine you are a health professional interested in the effects of medication on the diastolic blood pressure of adult women. For a particular drug being investigated, you find that for nine women ($n = 9$), the sample mean is $\bar{X} = 85$ and the sample variance is $s^2 = 160.78$. An estimate of the squared standard error of the sample mean, assuming random sampling, is $s^2/n = 160.78/9 = 17.9$. An estimate of the standard error of the sample mean is $\sqrt{17.9} = 4.2$, the square root of s^2/n.

4.2.2 Computing Probabilities Associated with the Sample Mean

This section illustrates how to compute probabilities associated with the sample mean when sampling is from a normal distribution and the standard deviation, σ, is known. (The more realistic situation where σ is not known, plus issues and methods when dealing with nonnormality, are described and discussed later in this chapter.) To be concrete, consider an experiment aimed at understanding the effect of ozone on weight gain in rats. Suppose a claim is made that if all rats could be measured, we would find that weight gain is normally distributed with $\mu = 14$ and a standard deviation of $\sigma = 6$. Under random sampling, this claim implies that the distribution of the sample mean has a mean of 14 as well. If, for example, 22 rats are randomly sampled ($n = 22$), the variance of the sample mean is $6^2/22$. In symbols, $E(\bar{X}) = 14$ and $\text{VAR}(\bar{X}) = \sigma^2/22 = 6^2/22 = 1.636$. Further imagine that based on the 22 rats in the experiment, the sample mean is $\bar{X} = 11$. Is it reasonable to expect a sample mean less than or equal to 11 if the claimed values for the population mean and standard deviation are correct? If $\bar{X} \leq 11$ is small, this suggests that there might be something wrong with the claims made about the distribution of how much weight is gained. That is, if the claims are correct, this has implications about which values for the sample mean are likely and which are not. Under normality, the sample mean has a normal distribution as well, so the R function `pnorm` described in Chapter 3 can be used to determine $P(\bar{X} \leq 11)$:

```
pnorm(11,14,sqrt(1.636)),
```

which returns the value 0.0095 suggesting that perhaps the claims made are incorrect.

A more classic method for solving the problem just illustrated is based on standardizing the sample mean. Standardizing a random variable is a recurrent theme in statistics. It plays an explicit role in the methods covered in Chapter 5, so the method is illustrated here.

Recall from Chapter 3 that if X has a normal distribution, $P(X \leq c)$ can be determined by standardizing X. That is, subtract the mean and divide by the standard deviation yielding

$$Z = \frac{X - \mu}{\sigma}.$$

If X is normal, Z has a standard normal distribution. That is, we solve the problem of determining $P(X \leq c)$ by transforming it into a problem involving the standard normal distribution. A similar strategy is useful when determining $P(\bar{X} \leq c)$ for any constant c.

Returning to the ozone experiment, the goal is to determine $P(\bar{X} \leq 11)$. When sampling from a normal distribution, $\bar{X}$ also has a normal distribution, so the strategy of standardizing the sample mean can be used to determine $P(\bar{X} \leq 11)$. That is, we convert the problem into one involving a standard normal distribution. This means that we *standardize the sample mean* by subtracting its (population) mean and dividing by its standard deviation which is $\sigma/\sqrt{n}$. That is, we compute

$$Z = \frac{\bar{X} - \mu}{\sigma/\sqrt{n}}.$$

When sampling from a normal distribution, it can be mathematically verified that Z has a standard normal distribution. In formal terms,

$$P(\bar{X} \leq 11) = P\left(Z \leq \frac{11 - \mu}{\sigma/\sqrt{n}}\right).$$

For the problem at hand,

$$\frac{11 - \mu}{\sigma/\sqrt{n}} = \frac{11 - 14}{6/\sqrt{22}} = -2.35,$$

and Table 1 in Appendix B tells us that the probability that a standard normal variable is less than -2.35 is 0.0095. More succinctly, $P(Z \leq -2.35) = 0.0095$, which means that the probability of getting a sample mean less than or equal to 11 is 0.0095. That is, if simultaneously the assumption of random sampling from a normal distribution is true, the population mean is $\mu = 14$, and the population standard deviation is $\sigma = 6$, then the probability of getting a sample mean less than or equal to 11 is 0.0095.

More generally, given c, some constant of interest, the probability that the sample mean is less than or equal to c is

$$\begin{aligned} P(\bar{X} \leq c) &= P\left(\frac{\bar{X}-\mu}{\sigma/\sqrt{n}} \leq \frac{c-\mu}{\sigma/\sqrt{n}}\right) \\ &= P\left(Z < \frac{c-\mu}{\sigma/\sqrt{n}}\right). \end{aligned} \tag{4.1}$$

In other words, $P(\bar{X} \leq c)$ is equal to the probability that a standard normal random variable is less than or equal to $(c-\mu)/(\sigma/\sqrt{n})$.

EXAMPLE

If 25 observations are randomly sampled from a normal distribution with mean 50 and standard deviation 10, what is the probability that the sample mean will be less than or equal to 45? We have that $n = 25$, $\mu = 50$, $\sigma = 10$, and $c = 45$, so

$$\begin{aligned} P(\bar{X} \leq 45) &= P\left(Z \leq \frac{45-50}{10/\sqrt{25}}\right) \\ &= P(Z \leq -2.5) \\ &= 0.0062. \end{aligned}$$

EXAMPLE

A company claims that after years of experience, students who take their training program typically increase their SAT scores by an average of 30 points. They further claim that the increase in scores has a normal distribution with standard deviation 12. As a check on their claim, you randomly sample 16 students and find that the average increase is 21 points. The company argues that this does not refute their claim because getting a sample mean of 21 or less is not that unlikely. To determine whether their claim has merit, you compute

$$\begin{aligned} P(\bar{X} \leq 21) &= P\left(Z \leq \frac{21-30}{12/\sqrt{16}}\right) \\ &= P(Z \leq -3) \\ &= 0.0013, \end{aligned}$$

which indicates that getting a sample mean as small or smaller than 21 is a relatively unlikely event. That is, there is empirical evidence that the claim made by the company is probably incorrect.

Analogous to results in Chapter 3, we can determine the probability that the sample mean is greater than some constant c, and we can determine the probability that the sample mean is between two numbers, say a and b. The probability of getting a sample mean greater than c is

$$P(\bar{X} > c) = 1 - P\left(Z < \frac{c-\mu}{\sigma/\sqrt{n}}\right), \tag{4.2}$$

and the probability of the sample mean being between the numbers a and b is

$$P(a < \bar{X} < b) = P\left(Z < \frac{b-\mu}{\sigma/\sqrt{n}}\right) - P\left(Z < \frac{a-\mu}{\sigma/\sqrt{n}}\right). \tag{4.3}$$

EXAMPLE

A researcher claims that for college students taking a particular test of spatial ability, the scores have a normal distribution with mean 27 and variance 49. If this claim is correct, and you randomly sample 36 participants, what is the probability that the sample mean will be greater than $c = 28$? First compute

$$\frac{c-\mu}{\sigma/\sqrt{n}} = \frac{28-27}{\sqrt{49/36}} = 0.857.$$

Because $P(Z \leq .857) = 0.80$, $P(\bar{X} > 28) = 1 - P(Z \leq 0.857) = 1 - 0.80 = 0.20$. This says that if we randomly sample $n = 25$ participants, and the claims of the researcher are true, the probability of getting a sample mean greater than 28 is 0.2.

EXAMPLE

Suppose observations are randomly sampled from a normal distribution with $\mu = 5$ and $\sigma = 3$. If $n = 36$, what is the probability that the sample mean is between 4 and 6? To find out, compute

$$\frac{b-\mu}{\sigma/\sqrt{n}} = \frac{6-5}{3/\sqrt{36}} = 2.$$

Referring to Table 1 in Appendix B, $P(\bar{X} < 4) = P(Z < 2) = 0.9772$. Similarly,

$$\frac{a-\mu}{\sigma/\sqrt{n}} = \frac{4-5}{3/\sqrt{36}} = -2,$$

and $P(Z < -2) = 0.0228$. So according to Equation (4.3),

$$P(4 < \bar{X} < 6) = 0.9772 - 0.0228 = 0.9544.$$

This means that if $n = 36$ observations are randomly sampled from a normal distribution with mean $\mu = 5$ and standard deviation $\sigma = 3$, there is a 0.9544 probability of getting a sample mean between 4 and 6.

4.3 A CONFIDENCE INTERVAL FOR THE POPULATION MEAN

An important practical consequence of knowing the sampling distribution of the sample mean is that it is possible to specify a range of values, called a *confidence interval*, which is likely to contain the unknown population mean, and simultaneously indicate which values for the population mean appear to be unlikely. The probability that this interval contains the unknown population mean is called the *probability coverage*. For the moment, normality is assumed. nonnormality is a serious practical concern that will be discussed at length later in this chapter, but for now the focus is on basic principles.

4.3.1 Known Variance

We begin with the case where the population variance, σ^2, is known. In most cases it is not known, but this simplifies the description of the underlying strategy. The case where σ^2 is not known is discussed in the next section of this chapter.

As noted in the previous section, under normality, random sampling, and when the variance σ^2 is known,

$$Z = \frac{\bar{X} - \mu}{\sigma/\sqrt{n}}$$

has a standard normal distribution. This means, for example, that

$$P\left(-1.96 \leq \frac{\bar{X} - \mu}{\sigma/\sqrt{n}} \leq 1.96\right) = 0.95.$$

Rearranging terms in this last equation, we see that

$$P\left(\bar{X} - 1.96\frac{\sigma}{\sqrt{n}} \leq \mu \leq \bar{X} + 1.96\frac{\sigma}{\sqrt{n}}\right) = 0.95.$$

This says that although the population mean, μ, is not known, there is a 0.95 probability that its value is between

$$\bar{X} - 1.96\frac{\sigma}{\sqrt{n}}$$

and

$$\bar{X} + 1.96\frac{\sigma}{\sqrt{n}}.$$

The interval

$$\left(\bar{X} - 1.96\frac{\sigma}{\sqrt{n}}, \bar{X} + 1.96\frac{\sigma}{\sqrt{n}}\right) \tag{4.4}$$

is called a 0.95 confidence interval for μ. This means that if the experiment were repeated billions of times (and in theory infinitely many times), and each time a confidence interval is computed as just described, 95% of the resulting confidence intervals will contain μ if observations are randomly sampled from a normal distribution. Moreover, the probability that the confidence interval does not contain μ is 0.05. That is, there is a 5% chance that the confidence interval is in error.

There are several ways in which a confidence interval can be interpreted incorrectly. (Also see Morey et al., 2016.) To be concrete, imagine a study aimed at computing a confidence interval for the mean of IQ scores at some university. Further imagine that the 0.95 confidence was found to be (103.9, 116.1). Here are three misinterpretations of the confidence interval:

- 95% of all students have an IQ between 103.9 and 116.1. The error here is interpreting the ends of the confidence intervals as quantiles. That is, if among all students, the 0.025 quantile is 100.5 and the 0.975 quantile is 120.2, this means that 95% of all students have an IQ between 100.5 and 120.2. But confidence intervals for the population mean tell us nothing about the quantiles associated with all students at this university.
- There is a 0.95 probability that a randomly sampled student will have an IQ between 100.5 and 120.2. This erroneous interpretation is similar to the one just described. Again, confidence intervals do not indicate the likelihood of observing a particular IQ. Rather, if the study were replicated (infinitely) many times, 95% of the resulting confidence intervals would contain the population mean. So here, there is a 95% chance that (103.9, 116.1) contains μ.

Table 4.2 Common Choices for $1-\alpha$ and c

$1-\alpha$	c
0.90	1.645
0.95	1.96
0.99	2.58

- All sample means among future studies will have a value between 103.9 and 116.1 with probability 0.95. This statement is incorrect because it is about the sample mean, not the population mean. (For more details about this particular misinterpretation, see Cumming and Maillardet, 2006.)

Generalizing, imagine the goal is to compute a confidence interval that has probability $1-\alpha$ of containing the population mean μ. Let c be the $1-\alpha/2$ quantiles of a standard normal distribution. Then from Chapter 3,

$$P(-c \leq Z \leq c) = 1-\alpha.$$

A little algebra shows that as a result

$$P\left(\bar{X} - c\frac{\sigma}{\sqrt{n}} \leq \mu \leq \bar{X} + c\frac{\sigma}{\sqrt{n}}\right) = 1-\alpha.$$

The interval

$$\left(\bar{X} - c\frac{\sigma}{\sqrt{n}}, \bar{X} + c\frac{\sigma}{\sqrt{n}}\right) \tag{4.5}$$

is a $1-\alpha$ confidence interval for μ. For convenience, the values of c for $1-\alpha = 0.9, 0.95$, and 0.99 are listed in Table 4.2.

EXAMPLE

For 16 observations randomly sampled from a normal distribution, imagine that $\bar{X} = 32$ and $\sigma = 4$. To compute a 0.9 confidence interval (meaning that $1-\alpha = 0.9$), first note from Table 4.2, $c = 1.645$. So a 0.9 confidence interval for μ is

$$\left(32 - 1.645\frac{4}{\sqrt{16}}, 32 + 1.645\frac{4}{\sqrt{16}}\right) = (30.355, 33.645).$$

Note that although $\bar{X}$ is not, in general, equal to μ, the length of the confidence interval provides some sense of how well $\bar{X}$ estimates the population mean.

EXAMPLE

A psychologist is interested in a new method for treating depression and how it compares to a standard, commonly used technique. Assume that based on a widely used measure of effectiveness, the standard method has been applied thousands of times and found to have a mean effectiveness of $\mu = 48$. That is, the standard method has been used so many times, for all practical purposes we know that the population mean is 48. Suppose we estimate the effectiveness of the new method by trying it on $n = 25$ participants resulting in $\bar{X} = 54$. That is, the new method resulted in a sample mean larger than the mean of the current treatment, but is it reasonable to conclude that the population mean associated with the new method is larger than 48? For illustrative purposes assume that the standard deviation is $\sigma = 9$ and sampling is from a normal distribution. Then a 0.95 confidence interval for μ is

$$\left(54 - 1.96\frac{9}{\sqrt{25}}, 54 + 1.96\frac{9}{\sqrt{25}}\right) = (50.5, 57.5).$$

This interval does not contain 48 suggesting that the new method is more effective than the standard treatment on average.

EXAMPLE

A college president claims that IQ scores at her institution are normally distributed with a mean of $\mu = 123$ and a standard deviation of $\sigma = 14$. Suppose you randomly sample $n = 20$ students and find that $\bar{X} = 110$. Does the $1 - \alpha = 0.95$ confidence interval for the mean support the claim that the average of all IQ scores at the college is $\mu = 123$? Because $1 - \alpha = 0.95$, $c = 1.96$, so the 0.95 confidence interval is

$$\left(110 - 1.96\frac{14}{\sqrt{20}}, 110 + 1.96\frac{14}{\sqrt{20}}\right) = (103.9, 116.1).$$

The interval (103.9, 116.1) does not contain the value 123 suggesting that the president's claim might be false. Note that there is a 0.05 probability that the confidence interval will not contain the true population mean, so there is some possibility that the president's claim is correct.

AN IMPORTANT PRACTICAL POINT

Notice that the length of a 0.95 confidence interval (the difference between the upper and lower ends of the confidence interval) is

$$(\bar{X} + 1.96\frac{\sigma}{\sqrt{n}}) - (\bar{X} - 1.96\frac{\sigma}{\sqrt{n}}) = 2(1.96)\frac{\sigma}{\sqrt{n}}.$$

As is evident, the larger the standard deviation σ, the longer the confidence interval. As indicated in Chapter 3, small changes in a distribution might result in large changes in the standard deviation. In practical terms, when analyzing data, any situation where outliers tend to occur can result in a confidence interval that is relatively long.

EXAMPLE

For illustrative purposes, imagine that $\bar{X} = 11$ based on a random sample of 22 observations from a standard normal distribution, in which case $\sigma^2 = 1$. Then the 0.95 confidence interval for the mean has length $2(1.96)(1)/\sqrt{22} = 0.836$. But suppose that sampling is from a mixed normal instead, which was discussed in Chapter 3. Although the mixed normal differs only slightly from the standard normal, the mixed normal has variance 10.9. Consequently, the 0.95 confidence interval for μ now has length 2.76, which is more than three times longer than the situation where we sampled from a standard normal distribution. This illustrates that in particular, *small shifts away from normality toward any heavy-tailed distribution can substantially increase the length of a confidence interval.* Modern methods have been found for getting much shorter confidence intervals in situations where the length of the confidence interval, based on the sample mean, is relatively long, some of which will be described.

4.3.2 Confidence Intervals When σ Is Not Known

The previous section described how to compute a confidence interval for μ when the standard deviation, σ, is known. But typically σ is not known, so a practical concern is finding a reasonably satisfactory method for dealing with this issue. This section describes a classic method for addressing this problem that was derived by William Gosset about a century ago. The method described here is routinely used today and is based on the assumption that observations are randomly sampled from a normal distribution. It is now known, however, that sampling from a nonnormal distribution can be a serious practical concern. Details about this issue and how it might be addressed are covered later in this chapter and in Chapter 5.

Consider again a situation where the goal is to determine the average diastolic blood pressure of adult women taking a certain drug. Based on $n = 9$ women, we have that $\bar{X} = 85$, and the unknown variance is estimated to be $s^2 = 160.78$. So although the standard error of the sample mean, $\sigma/\sqrt{n}$, is not known, it can be estimated with $s/\sqrt{n} = 4.2$. If we assume that 4.2 is indeed an accurate estimate of $\sigma/\sqrt{n}$, then a reasonable suggestion is to assume that $\sigma/\sqrt{n} = 4.2$ when computing a confidence interval. In particular, a 0.95 confidence interval for the mean would be

$$(85 - 1.96(4.2), 85 + 1.96(4.2)) = (76.8, 93.2),$$

which is a special case of a strategy developed by Laplace about two centuries ago. However, even when sampling from a normal distribution, a concern is that when the sample size is small, the population standard deviation, σ, might differ enough from its estimated value, s, to cause practical problems. Gosset realized that problems can arise and derived a solution assuming random sampling from a normal distribution. Gosset worked for a brewery and was not immediately allowed to publish his results, but eventually he was allowed to publish under the pseudonym Student.

Let

$$T = \frac{\bar{X} - \mu}{s/\sqrt{n}}. \tag{4.6}$$

The random variable T is the same as Z used in Section 4.3.1, only σ has been replaced by s. Note that like $\bar{X}$ and Z, T has a distribution. That is, if we repeat a study many times, then for any constant c we might pick, there is a certain probability that $T < c$ based on a random sample of n participants. If the distribution of T can be determined, then a confidence interval for μ can be computed without knowing σ.

If we assume that observations are randomly sampled from a normal distribution, then the distribution of T, called *Student's T Distribution*, can be determined exactly. It turns out that the distribution depends on the sample size, n. By convention, the quantiles of the distribution are reported in terms of *degrees of freedom*: $\nu = n - 1$, where ν is a lower case Greek nu. Figure 4.2 shows Student's T distribution with $\nu = 4$ degrees of freedom. Note that the distribution is similar to a standard normal. In particular, it is symmetric about zero, so its mean (its average value over many studies) is $E(T) = 0$. With infinite degrees of freedom, Student's t distribution and the standard normal are identical.

Table 4 in Appendix B reports some quantiles of Student's T distribution. The first column gives the degrees of freedom. The next column, headed by $t_{0.9}$, reports the 0.9 quantiles. For example, with $\nu = 1$, we see 3.078 under the column $t_{0.9}$. This means that $P(T < 3.078) = 0.9$. That is, if we randomly sample two observations from a normal distribution, in which case $\nu = n - 1 = 1$, there is a 0.9 probability that the resulting value for T is less than 3.078. Similarly, if $\nu = 24$, $P(T < 1.318) = 0.9$. The column headed by $t_{0.99}$ lists the 0.99 quantiles. For example, if $\nu = 3$, we see 4.541 under the column headed $t_{0.99}$, so the probability that T is less than 4.541 is 0.99. If $\nu = 40$, Table 4 indicates that $P(T < 2.423) = 0.99$.

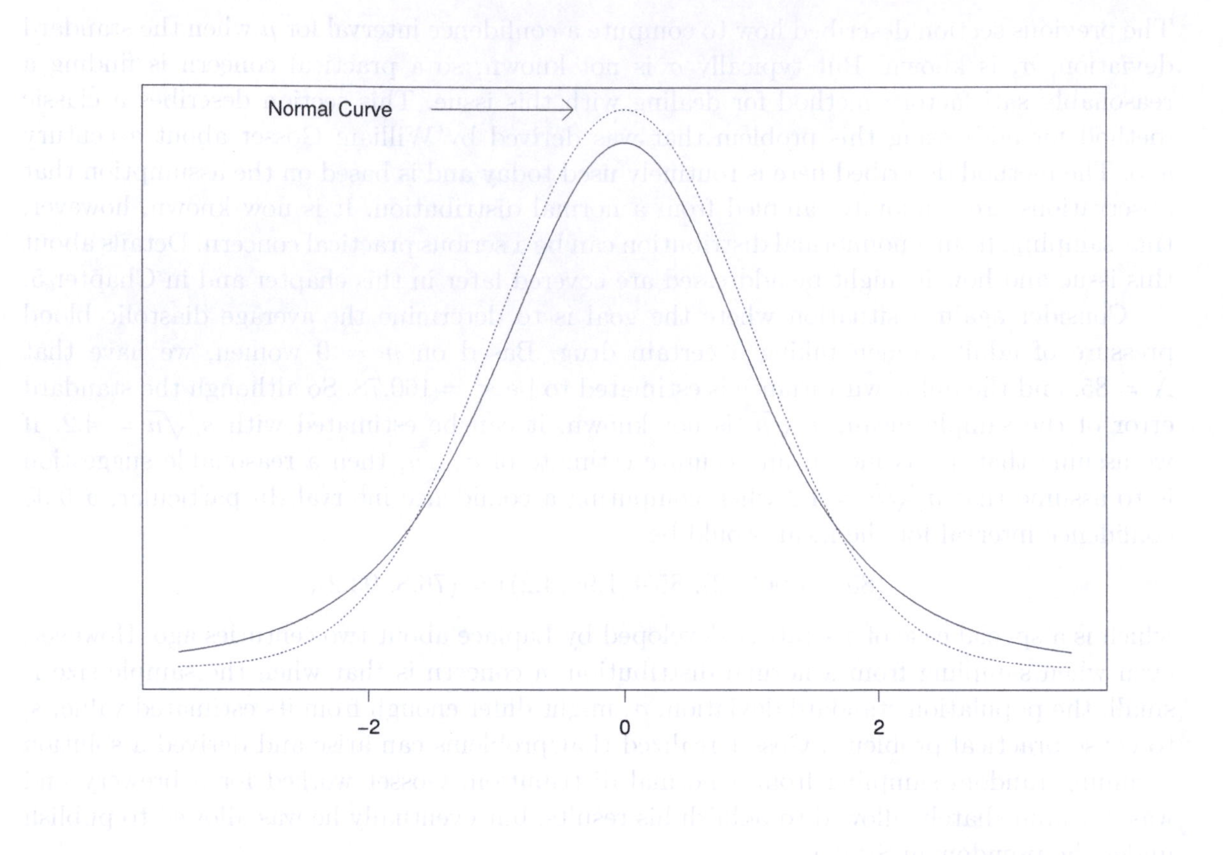

Figure 4.2 Distribution of Student's T with 4 degrees of freedom and a standard normal.

4.3.3 R Functions pt and qt

Many statistical software packages contain functions that compute Student's T distribution for any $\nu \geq 1$. The R function

```
pt(q,df)
```

computes $P(T \leq q)$, the probability that T is less than or equal to q with the degrees of freedom indicated by the argument `df`. So in a very real sense, Table 4 is not required. (It is included in case some instructors feel that it is useful from a pedagogical point of view.) For example, pt(1,5) will return the probability that T is less than or equal to 1 with $\nu = 5$ degrees of freedom.

There is also an R function for determining quantiles:

```
qt(p,df).
```

For example, to determine the 0.975 quantile with 10 degrees of freedom, use the R command `qt(0.975,10)`, which returns the value 2.228139. That is, $P(T \leq 2.228139) = 0.975$.

Similar to the situation when working with normal distributions,

$$P(T \geq c) = 1 - P(T \leq c), \tag{4.7}$$

where c is any constant that might be of interest. For example, with $\nu = 4$, $P(T \leq 2.132) = 0.95$, as previously indicated, so $P(T \geq 2.132) = 1 - P(T \leq 2.132) = 0.05$.

Note that like Z, the distribution of T does not depend on the mean μ when normality is assumed. For example, with $\nu = 4$, $P(T \leq 2.132) = 0.95$ regardless of what the value of μ happens to be. This property makes it possible to compute confidence intervals for the mean, as is demonstrated momentarily.

EXAMPLE

Imagine a study on the effects of alcohol on reaction times. Assuming normality, you randomly sample 13 observations ($n = 13$) and compute the sample mean and variance. To determine the probability that $T = (\bar{X} - \mu)/(s/\sqrt{n})$ is less than 2.179, first note that the degrees of freedom are $\nu = n - 1 = 13 - 1 = 12$. From Table 4 in Appendix B, looking at the row with $\nu = 12$, we see 2.179 in the column headed by $t_{0.975}$, so $P(T < 2.179) = 0.975$. Alternatively, using R, this probability is given by the R command `pt(2.179,12)`.

EXAMPLE

If $\nu = 30$ and $P(T > c) = 0.005$, what is c? Based on Equation (4.7), if $P(T > c) = 0.005$, then $P(T \leq c) = 1 - P(T > c) = 1 - 0.005 = 0.995$. Looking at the column headed by $t_{0.995}$ in Table 4, we see that with $\nu = 30$, $P(T < 2.75) = 0.995$, so $c = 2.75$.

4.3.4 Confidence Interval for the Population Mean Using Student's T

With Student's T distribution, we can compute a confidence interval for μ when σ is not known, assuming that observations are randomly sampled from a normal distribution:

$$\left(\bar{X} - c\frac{s}{\sqrt{n}}, \bar{X} + c\frac{s}{\sqrt{n}}\right), \tag{4.8}$$

where now c is the $1-\alpha/2$ quantile of Student's T distribution with $n-1$ degrees of freedom and read from Table 4 of Appendix B. If observations are randomly sampled from a normal distribution, then the probability coverage is exactly $1 - \alpha$.

4.3.5 R Function t.test

The R built-in function

```
t.test(x,conf.level=0.95)
```

computes a confidence interval in the manner just described. By default, a 0.95 confidence interval is reported. (This function reports additional results described in Chapter 5.)

EXAMPLE

Returning to the ozone experiment, we compute a $1 - \alpha = 0.95$ confidence interval for μ. Because there are 22 rats ($n = 22$), the degrees of freedom are $n - 1 = 22 - 1 = 21$. Because $1 - \alpha = 0.95$, $\alpha = 0.05$, so $\alpha/2 = 0.025$, and $1 - \alpha/2 = 0.975$. Referring to Table 4 in Appendix B, we see that the 0.975 quantile of Student's T distribution with 21 degrees of freedom is approximately $c = 2.08$. Because $\bar{X} = 11$ and $s = 19$, a 0.95 confidence interval is

$$11 \pm 2.08\frac{19}{\sqrt{22}} = (2.6, 19.4).$$

That is, although both the population mean and variance are not known, there is a 0.95 probability that a confidence interval has been computed that contains the population mean μ if the assumption of random sampling from a normal distribution is true.

EXAMPLE

Imagine that you are interested in the reading abilities of fourth graders. A new method for enhancing reading is being considered. You try the new method on 11 students and then administer a reading test yielding the scores

$$12, 20, 34, 45, 34, 36, 37, 50, 11, 32, 29.$$

For illustrative purposes, imagine that after years of using a standard method for teaching reading, the average scores on the reading test have been found to be $\mu = 25$. Someone claims that if the new teaching method is used, the population mean will remain 25. Assuming normality, we determine whether this claim is consistent with the 0.99 confidence interval for μ. That is, does the 0.99 confidence interval contain the value 25? It can be seen that the sample mean is $\bar{X} = 30.9$ and $s/\sqrt{11} = 3.7$. Because $n = 11$, the degrees of freedom are $\nu = 11 - 1 = 10$. Because $1 - \alpha = 0.99$, it can be seen that $1 - \alpha/2 = 0.995$, so from Table 4 in Appendix B, $c = 3.169$. Consequently, the 0.99 confidence interval is

$$30.9 \pm 3.169(3.7) = (19.2, 42.6).$$

This interval contains the value 25, so the claim that $\mu = 25$ cannot be refuted based on the available data. Note, however, that the confidence interval also contains 35 and even 40. Although we cannot rule out the possibility that the mean is 25, there is some possibility that the new teaching method enhances reading by a substantial amount, but with only 11 participants, the confidence interval is too long to resolve how effective the new method happens to be.

4.4 JUDGING LOCATION ESTIMATORS BASED ON THEIR SAMPLING DISTRIBUTION

Another useful feature of sampling distributions is that they provide a perspective on the relative merits of location estimators. Note that the notion of a sampling distribution generalizes to any of the location estimators described in Chapter 2. For example, if we conduct a study and compute the median based on 20 observations, and if we repeat the study billions of times (and in theory infinitely many times), each time computing a median based on 20 observations, for all practical purposes we would know the sampling distribution of the median. That is, we would know the probability that the median is less than 4, less than 10, or less than any constant c that might be of interest. Moreover, the squared standard error of the sample median would be known as well; it is simply the variance of the sample medians. Temporarily assume sampling is from a normal distribution, and for convenience only, suppose the population mean is zero and the variance is one. Because sampling is from a symmetric distribution, the sample median, M, is a reasonable estimate of the population mean, μ. But, of course, the sample median will be in error as was the case when using the sample mean.

Now, when sampling from a normal distribution, theory tells us that the sample mean has a smaller standard error than the median or any other location estimator we might use. Roughly, on average, the mean is a more accurate estimate of the population mean. To illustrate the extent to which this is true, 20 observations were generated on a computer from a standard normal distribution, the sample mean, 20% trimmed, MOM and median were computed, and this process was repeated 5000 times. Figure 4.3 shows boxplots of the results. Note that the means are more tightly centered around the value zero, the value all four estimators are trying to estimate. In more formal terms, the sample mean has a smaller standard error.

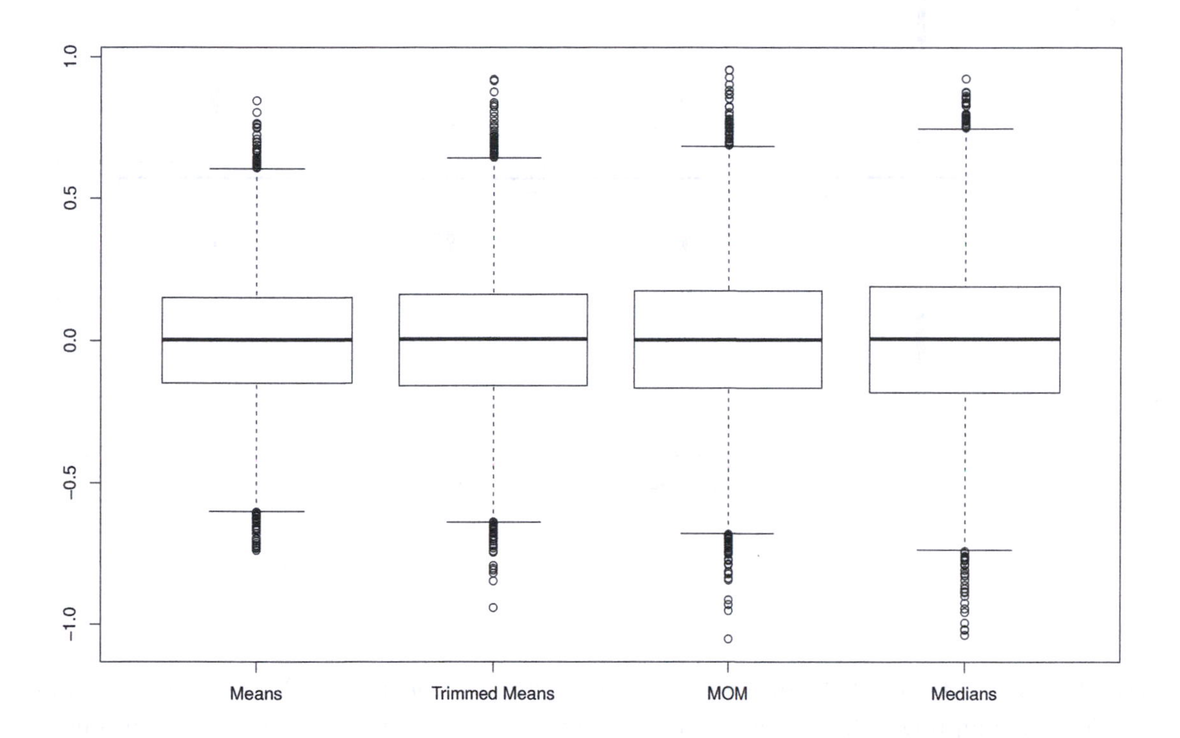

Figure 4.3 Boxplots of 5000 sample means, 20% trimmed means, MOMs and medians when $n = 20$ and sampling is from a normal distribution. Note that the sample means tend to be a more accurate estimate of the population mean, 0.

But what about nonnormal distributions? To illustrate an important point, we repeat the computer experiment used to create Figure 4.3, only now sampling is from the mixed (contaminated) normal introduced in Chapter 3. Recall that the mixed normal appears to be approximately normal, but it represents a heavy-tailed distribution, roughly meaning that outliers are relatively common.

Figure 4.4 shows the results of our computer experiment, and in contrast to Figure 4.3, now the sample mean is substantially less accurate, on average, relative to the other measures of location. This illustrates a general result of considerable importance. Even for an extremely small departure from normality, the mean can have a substantially larger standard error compared to the 20% trimmed mean, MOM and median.

But a more important issue is whether the standard error of the mean can be relatively large when dealing with data from actual studies. We repeat the process used to create Figures 4.3 and 4.4, only now data are randomly sampled with replacement from measures of the cortisol awakening response (CAR) stemming from the Well Elderly 2 study (Clark et al., 2011). The CAR is the change in cortisol upon awakening and measured again 30–60

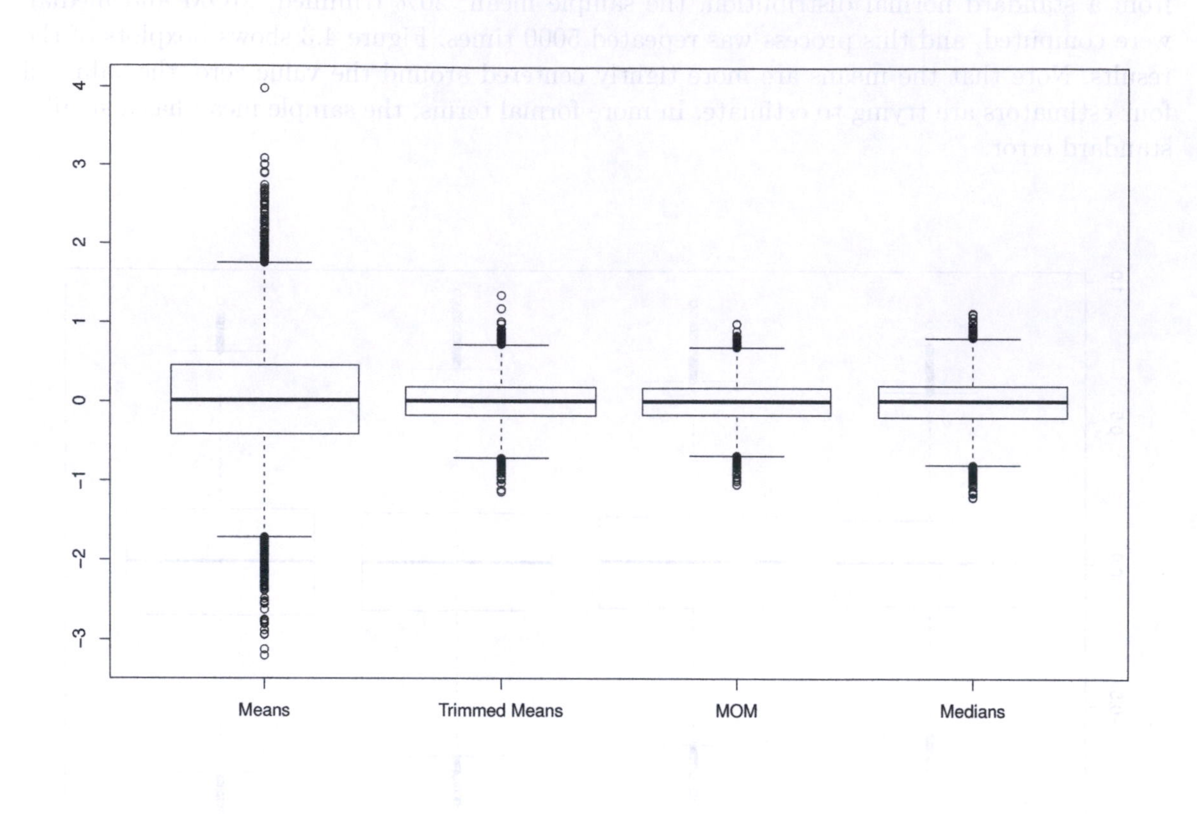

Figure 4.4 Boxplots of 5000 sample means, 20% trimmed means, MOM, and medians when $n = 20$ and sampling is from the contaminated normal distribution. Now the mean tends to be a less accurate estimate of the population mean.

minutes later, which has been found to be associated with various measures of stress. The boxplots demonstrate that the standard error of the mean is relatively large.

As another example, data are randomly sampled with replacement from measures of hangover symptoms. (The data were supplied by M. Earlywine.) The resulting boxplots are shown in Figure 4.6. As is evident, there is substantially less variability among the medians relative to the other estimators that were used. Note that the values based on MOM are the most variable. The 20% trimmed mean is less variable than the mean.

The results illustrated in Figure 4.3 are not surprising in the sense that the 20% trimmed mean, MOM, and M-estimator were designed to give nearly the same accuracy as the mean when sampling from a normal distribution. That is, theoretical results tell us how much we can trim without sacrificing too much accuracy (as measured by the squared standard error) under normality, and theory also suggests how to design the modified one-step M-estimator (MOM) so that again relatively good accuracy is obtained. Simultaneously, theory tells us that these estimators will continue to perform relatively well when sampling from a heavy-tailed distribution such as the mixed normal, in contrast to the mean. However, no single estimator dominates. The optimal estimator depends on the nature of the population distribution, which is unknown. The only certainty is that the choice of estimator can make a substantial difference.

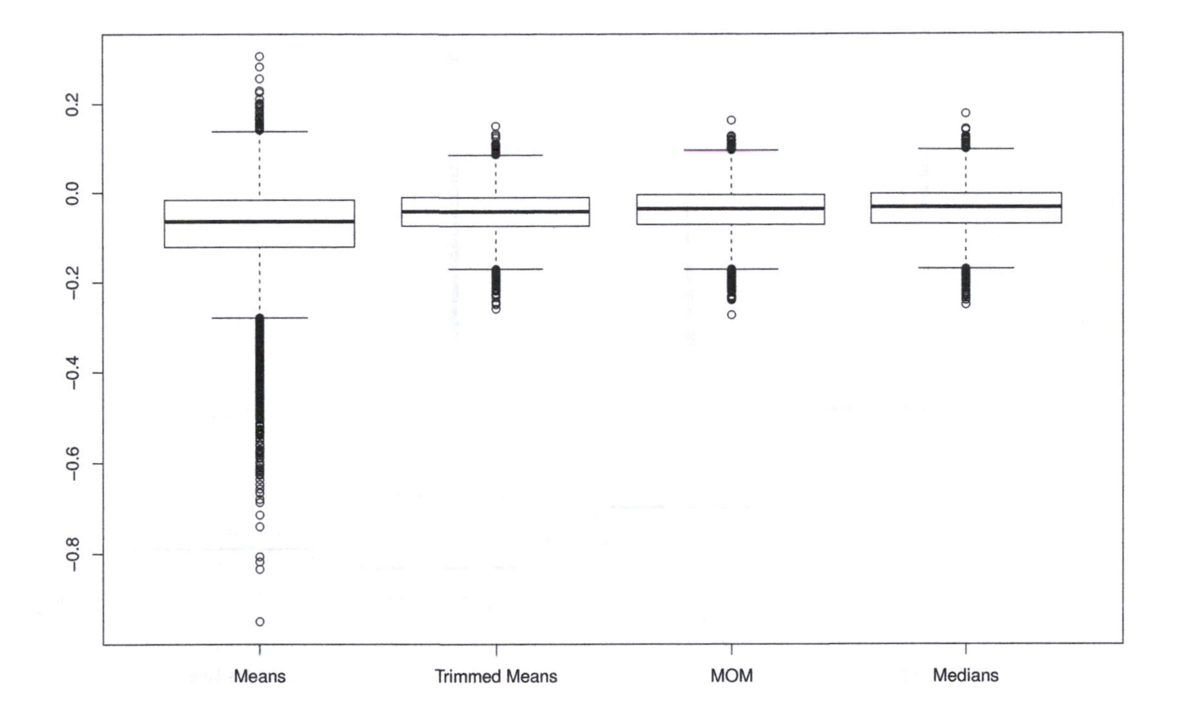

Figure 4.5 Boxplots of 5000 sample means, 20% trimmed means, MOM, and medians, $n = 20$, when sampling with replacement from measures of the cortisol awakening response.

4.4.1 Trimming and Accuracy: Another Perspective

Despite the illustrations just given, it might seem surprising that trimming values can result in a lower standard error versus no trimming, and so some elaboration might help. From the perspective about to be described, it is not surprising that a trimmed mean beats the mean, but it is rather amazing that the sample mean is optimal in any situation at all. To explain, suppose we sample 20 observations from a standard normal distribution. So the population mean and population 20% trimmed mean are zero. Now consider the smallest of the 20 values. It can be shown that with probability 0.983, this value will be less than −0.9, and with probability 0.25 it will be less than −1.5. That is, with fairly high certainty, it will not be close to the population mean, zero, the value we are trying to estimate. In a similar manner, the largest of the 20 observations has probability 0.983 of being greater than 0.9 and probability 0.25 of being greater than 1.5. Simultaneously, if we put the observations in ascending order, the probability that the two middle values do not differ by more than 0.5 from the population mean (zero) is 0.95. So a natural reaction is that extreme values should be given less weight in comparison to the observations in the middle. A criticism of this simple argument is that the smallest value will tend to be less than zero, the largest will be greater than zero, and their average value will be exactly zero, so it might seem that there is no harm in using them to estimate the population mean. However, the issue is how much these extreme values contribute to the variance of our estimator. When sampling from a normal distribution, we are better off on average using the sample mean despite the fact that the extreme values are highly likely to be inaccurate. But as we move away from a

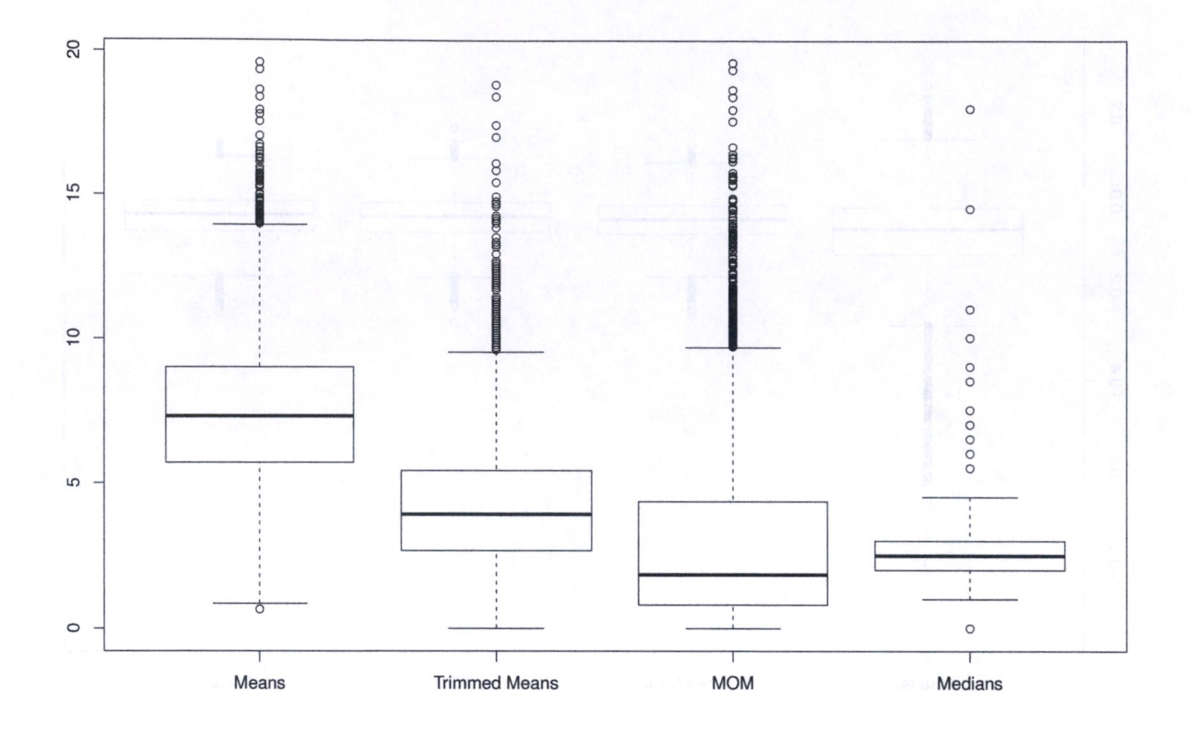

Figure 4.6 Boxplots of 5000 sample means, 20% trimmed means, MOM, and medians, $n = 20$, when sampling with replacement from measures of hangover symptoms

normal distribution toward a heavier tailed distribution, roughly meaning that outliers are more likely to occur, the sample mean becomes extremely inaccurate relative to the 20% trimmed mean.

Although the sample mean can have a much larger standard error than the 20% trimmed mean, median, and M-estimator, there remains the issue of whether they ever make a practical difference in applied work. The answer is an unequivocal yes, as will be seen at various points in this book.

4.5 AN APPROACH TO NONNORMALITY: THE CENTRAL LIMIT THEOREM

About two centuries ago, it was realized that assuming random sampling from a normal distribution greatly simplifies theoretical and computational issues. But how might the assumption of normality be justified? Gauss worked on this problem for a number of years resulting in a famous theorem, but it is a result by Laplace, publicly announced in the year 1810, that is routinely used today to justify the assumption of normality. Laplace dubbed his result the *central limit theorem*, where the term "central" is intended to mean fundamental. Roughly, this theorem says that under very general conditions, if observations are randomly sampled from a nonnormal distribution, the sampling distribution of the sample mean will approach a normal distribution as the sample size gets large. In more practical terms, if n is sufficiently large, we can pretend that $\bar{X}$ has a normal distribution with mean μ and variance σ^2/n in which case an accurate confidence interval for the population mean can be computed.

Said another way, with a sufficiently large sample size, it can be assumed that

$$Z = \frac{\bar{X} - \mu}{\sigma/\sqrt{n}}$$

has a standard normal distribution.

Of course, the statement that the sample size must be sufficiently large is rather vague. How large must n be to assume normality? There is no theorem or other mathematical result that answers this question in a satisfactory manner. The answer depends in part on the skewness of the distribution from which observations are sampled. (Boos and Hughes-Oliver, 2000, summarize relevant details.) Generally, we must rely on empirical investigations, at least to some extent, in our quest to address this problem, such as those described momentarily. Many books claim that $n = 25$ suffices and others claim that $n = 40$ is sufficient. These are not wild speculations, and in terms of the sample mean, these claims are generally reasonable although some exceptions will be described. For many years it was thought that if the sample mean has, approximately, a normal distribution, then Student's T distribution will provide reasonably accurate confidence intervals and good control over the probability of a Type I error. But as will be illustrated, this is not necessarily true. Indeed, using Student's T distribution can be highly unsatisfactory even with a fairly large sample size.

To elaborate, we begin by describing early studies regarding how large the sample size must be in order for the sample mean to have a distribution that is approximately normal. Classic illustrations focused on two particular nonnormal distributions. The first is the uniform distribution shown in the left panel of Figure 4.7. This distribution says that all values between 0 and 1 are equally likely. The second is called an exponential distribution, which is shown in the right panel of Figure 4.7.

Using the software R, $n = 20$ observations were generated from the uniform distribution, the sample mean was computed, and this process was repeated 5000 times. The left panel of Figure 4.8 shows the distribution of the resulting means (which was estimated with the R function `akerd` described in Chapter 2). Also shown is the normal distribution based on the central limit theorem. As we see, the two curves are very similar indicating that the central limit theorem is performing rather well in this particular case. That is, if we sample 20 observations from a uniform distribution, for all practical purposes we can assume the sample mean has a normal distribution.

This same process was repeated, only now observations are sampled from the exponential distribution shown in Figure 4.7. The right panel of Figure 4.8 shows the distribution of the means and the normal distribution based on the central limit theorem. Again, with only $n = 20$, a normal distribution provides a fairly good approximation of the actual distribution of $\bar{X}$.

We repeat our computer experiment one more time, only now we randomly sample (with replacement) $n = 40$ values from the sexual attitude data reported in Chapter 2. The left panel shows the distribution of the means. As is evident, now the distribution is not well approximated by a normal distribution. A criticism might be that the data include an extreme outlier. So the right panel is the same as the left, only the extreme value was removed. The distribution of the means looks more like a normal distribution, but assuming normality remains unsatisfactory.

4.6 STUDENT'S T AND NONNORMALITY

It is particularly important to understand the effects of nonnormality when using Student's T distribution for the simple reason that it is routinely used. If, for example, the goal is to compute a 0.95 confidence interval for the population mean, in reality the actual probability

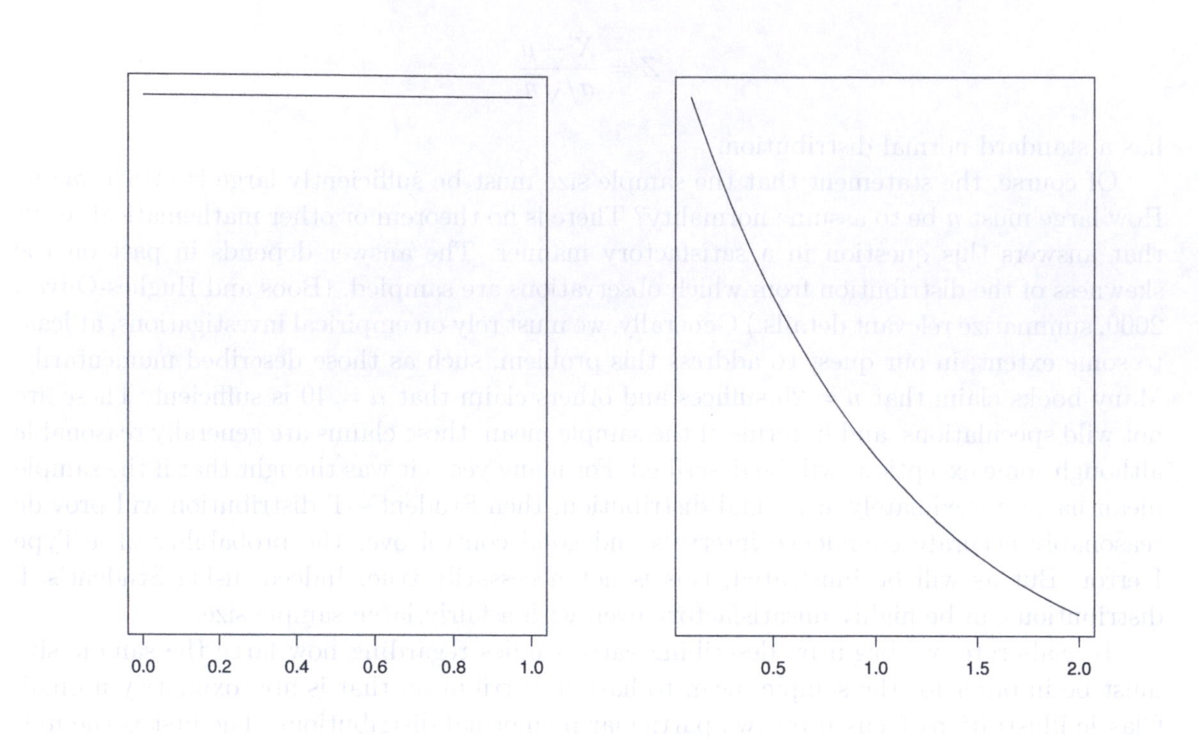

Figure 4.7 The left panel shows a uniform distribution. The right panel shows an exponential distribution. Both of these distributions were used in classic studies regarding the central limit theorem.

coverage will differ from 0.95 under nonnormality. A basic issue is whether the actual probability coverage can differ from the nominal level by a substantial amount, and we will see that the answer is an unequivocal yes. A related issue is understanding the circumstances under which accurate results are obtained and when the actual probability coverage will be unsatisfactory.

To be a bit more concrete, consider the example dealing with weight gain in an ozone environment. It was shown that the 0.95 confidence interval for the mean is $(2.6, 19.4)$ *if* observations are randomly sampled from a normal distribution. But can we be reasonably certain that it contains μ if sampling is from a nonnormal distribution instead? There are situations where the answer is yes, but under general conditions the answer is no. If, for instance, observations are randomly sampled from a skewed distribution, the actual probability coverage can be less than 0.7. That is, the confidence interval is too short and there is at least a $1 - 0.7 = 0.3$ probability that it does not contain μ. To get a confidence interval with 0.95 probability coverage, we need a longer interval. If we increase the sample size, the actual probability coverage will be closer to 0.95, as desired, but probability coverage might be unsatisfactory even with $n = 160$ (e.g., Westfall and Young, 1993, p. 40). In addition, as indicated here, even $n = 300$ might not suffice

One of the main points in this section is that even when the sample mean has, approximately, a normal distribution, Student's T distribution might result in a relatively inaccurate confidence interval for the mean, as will be illustrated. To provide a framework for under-

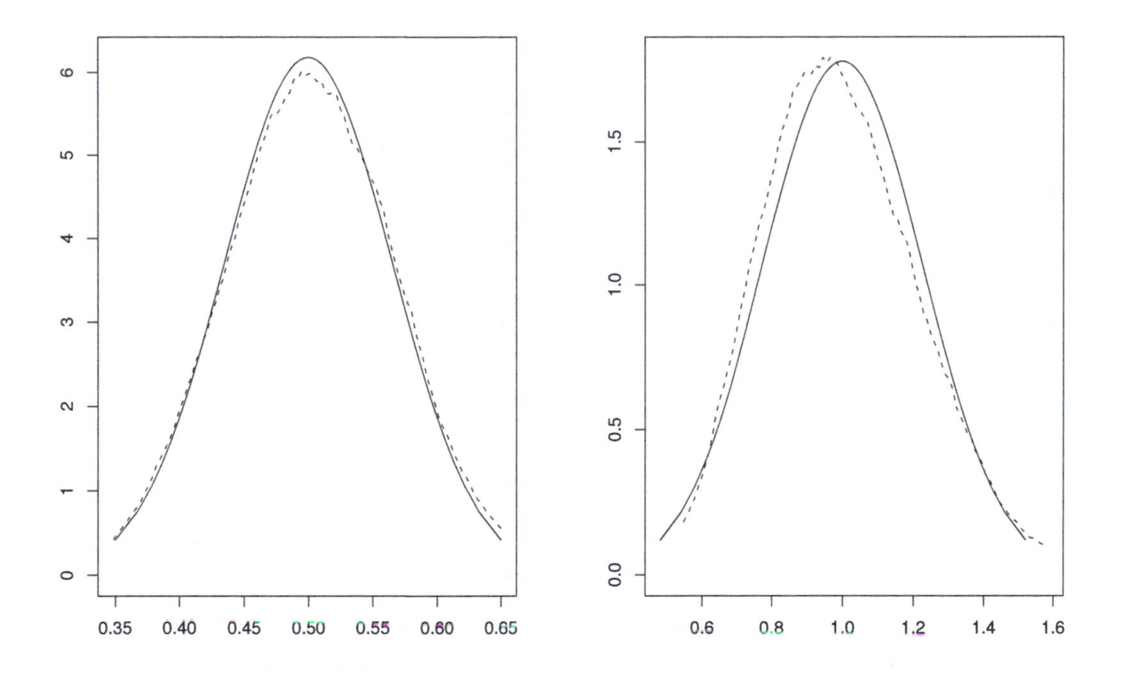

Figure 4.8 The left panel shows an approximation of the distribution of the sample mean, indicated by the dashed line, based on 5000 sample means. The sample size for each mean is 20 and observations were generated from the uniform distribution in Figure 4.7. The smooth symmetric curve is the distribution of the sample means based on the central limit theorem. The right panel is the same as the left, only values were generated from the exponential distribution.

standing when Student's t provides accurate confidence intervals, and when it fails to do so, it helps to consider four types of distributions.

Symmetric, Light-Tailed Distributions

The first of the four types of distributions consists of symmetric, light-tailed distributions, roughly meaning that outliers are relatively rare. This family of distributions includes the normal distribution as a special case. In this case, Student's T distribution provides relatively accurate confidence intervals.

Symmetric, Heavy-Tailed Distributions

Symmetric, heavy-tailed distributions are distributions where outliers tend to be common. An example is the mixed (or contaminated) normal distribution introduced in Chapter 3. Another example is a Student's t distribution with 2 degrees of freedom. One important feature of such distributions is that we tend to get relatively large sample variances, which in turn means that we get long confidence intervals relative to other methods we might use (such

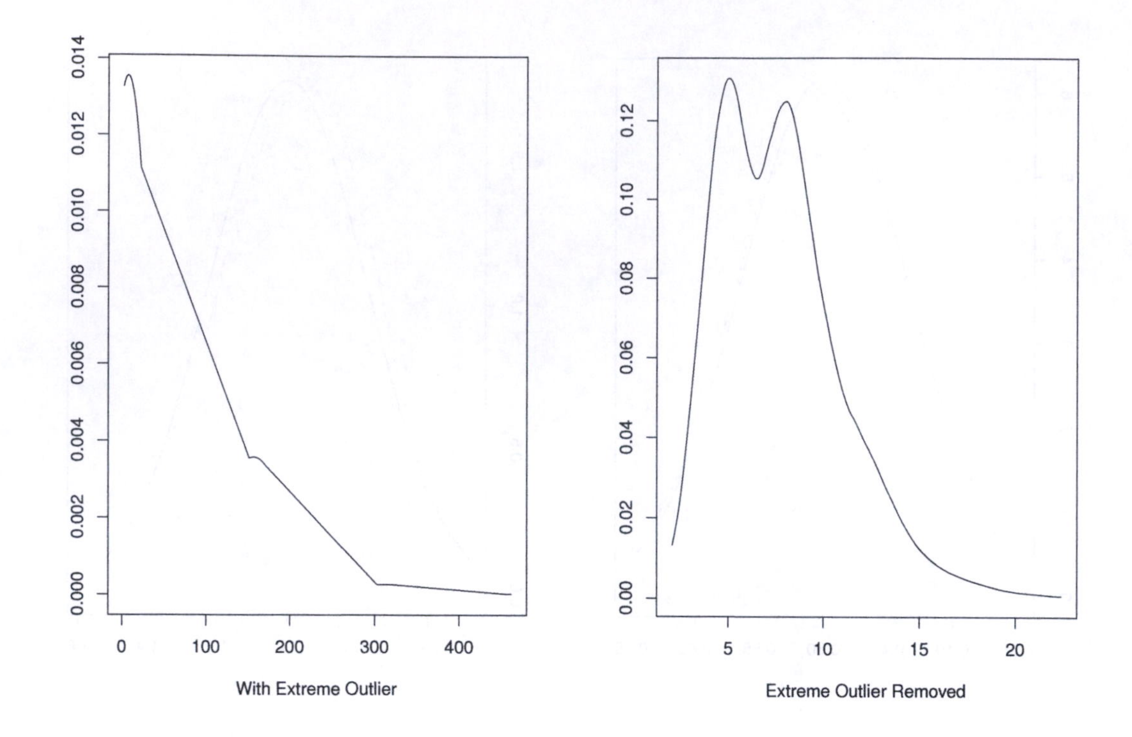

Figure 4.9 The left panel shows the distribution of 5000 sample means, each based on 40 observations generated from the sexual attitude data in Table 2.2. The right panel is the same as the left, only the extreme outlier was removed before sampling values.

as the method for trimmed means described later in this chapter). Put another way, the actual probability coverage tends to be larger than stated. Although the goal might be to compute a 0.95 confidence interval, in reality we might actually be computing a 0.99 confidence interval. In some sense this is a positive feature because we have a higher probability of computing a confidence interval that contains the true population mean. But simultaneously we are less certain about which values for the population mean are unlikely, which turns out to be an important issue, as will become clear.

Skewed, Light-Tailed Distributions

In contrast to symmetric distributions, sampling from a skewed distribution can result in confidence intervals that have probability coverage substantially smaller than what was intended. To elaborate a little, imagine that, unknown to us, observations have the distribution shown in Figure 4.10. This is an example of a *lognormal distribution*, which is skewed to the right and has relatively light tails, meaning that outliers can occur but the number of outliers is relatively low on average. (With $n = 20$, the expected number of outlier is about 1.4 based on the boxplot rule. The median number of outliers is approximately one.) Gleason (1993), also argues that the lognormal distribution is light-tailed. The symmetric smooth curve in the left panel of Figure 4.11 shows Student's T distribution when $n = 20$ and sampling is from a normal distribution. The other curve shows a close approximation of the actual distribution of T when sampling from the distribution in Figure 4.10. (The approximation is based on 5000 T values generated on a computer.) Note that the actual distribution is skewed, not symmetric. Moreover, its mean is not zero but −0.5, approximately. The right panel of Figure 4.11 shows the distribution of T when $n = 100$. There is closer agreement between the two distributions, but the tails of the distributions differ enough that practical problems arise. To get a reasonably accurate 0.95 confidence interval, a sample size of about 200 is required.

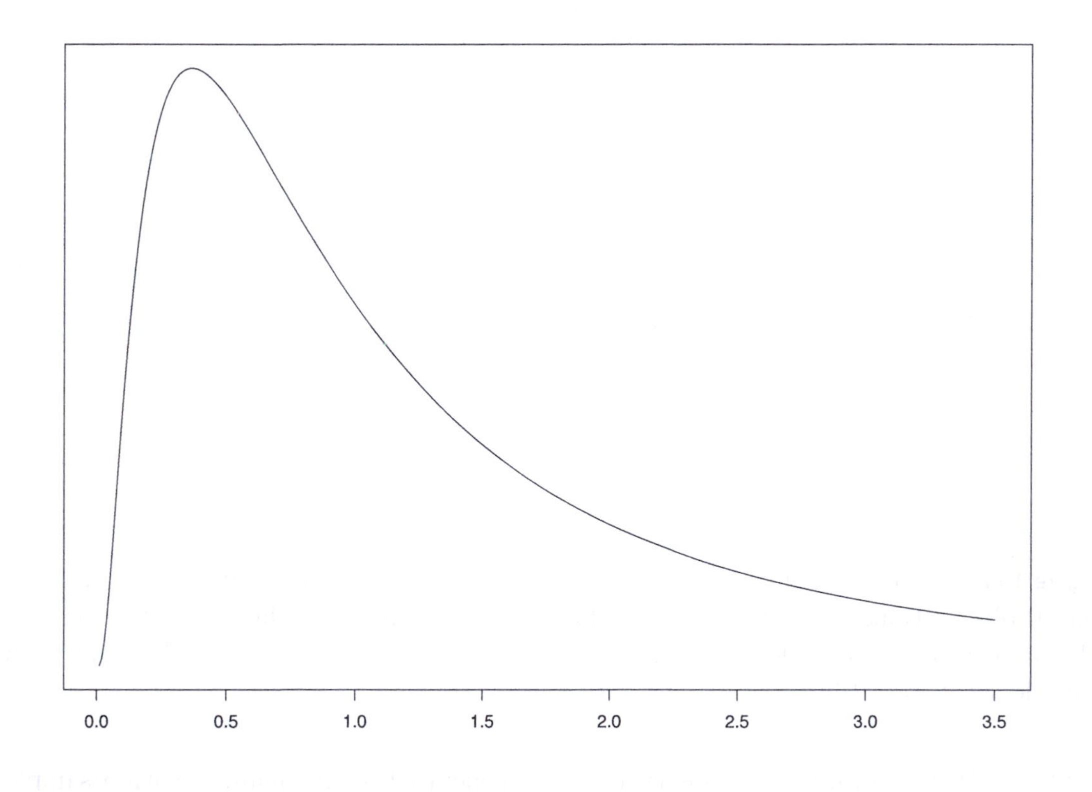

Figure 4.10 Shown is a lognormal distribution, which is skewed and relatively light-tailed, roughly meaning that the proportion of outliers found under random sampling is relatively small.

It was noted that in the left panel of Figure 4.11, T has a mean of −0.5. This might seem to be impossible because the numerator of T is $\bar{X} - \mu$, which has a mean of zero. Under normality, T does indeed have a mean of zero, the proof of which is based on the result that under normality, $\bar{X}$ and s are independent. But for nonnormal distributions, $\bar{X}$ and s are dependent and this makes it possible for T to have a mean that differs from zero. (Gosset was aware of this problem but did not have the tools and technology to study it to the degree he desired.)

Skewed, Heavy-Tailed Distributions

When sampling from a skewed distribution where outliers are rather common, the validity of Student's T deteriorates even further. As an illustration, consider again the lognormal distribution, only now, with probability 0.1, a value is multiplied by 10. Figure 4.12 shows

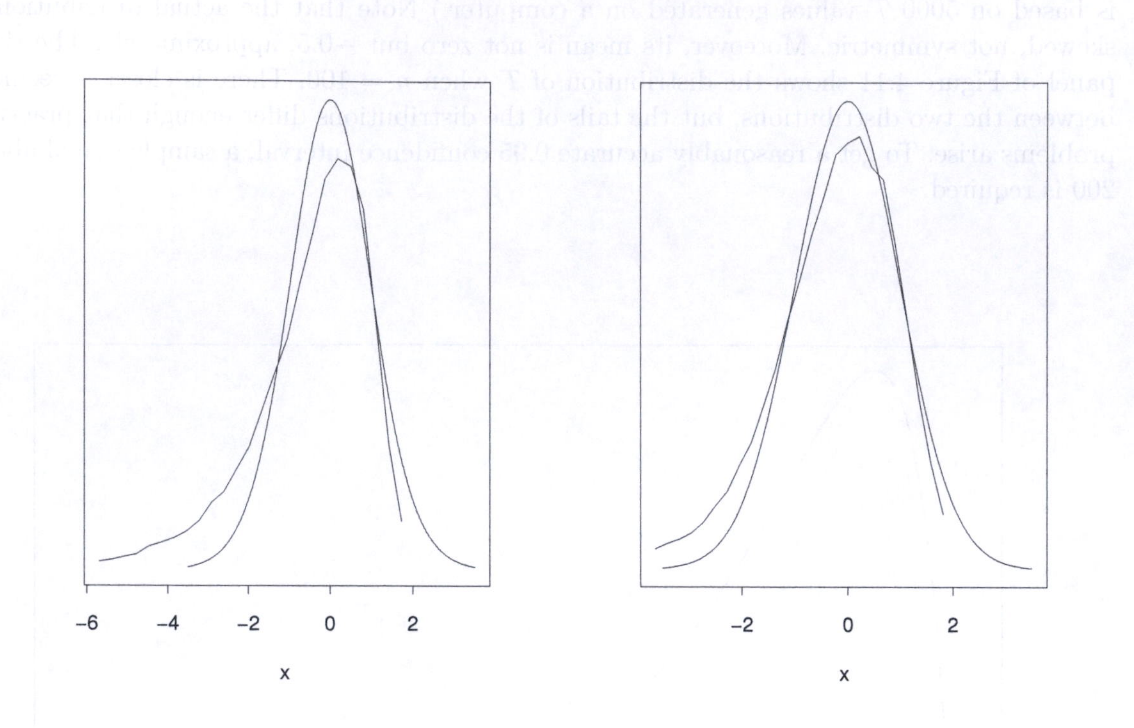

Figure 4.11 The left panel shows the distribution of 5000 T values, with each T value based on 20 observations generated from the lognormal distribution. The symmetric solid line is the distribution of T under normality. The right panel is the same as the left, only now the sample size is $n = 100$.

a plot of 5000 T values when observations are generated in this manner with a sample size of $n = 20$. The actual distribution of T is highly skewed and differs substantially from the distribution under normality, as shown in Figure 4.12.[1] Now over 300 observations are needed to get an accurate confidence interval.

The illustrations thus far are based on hypothetical distributions. Experience with actual data suggests that the problems just illustrated are real and, in at least some situations, these theoretical illustrations appear to underestimate problems with Student's T distribution. To describe one reason for this remark, we consider data from a study on hangover symptoms reported by the sons of alcoholics. The 20 observed values were

$$1\ 0\ 3\ 0\ 3\ 0\ 15\ 0\ 6\ 10\ 1\ 1\ 0\ 2\ 24\ 42\ 0\ 0\ 0\ 2.$$

(These data were supplied by M. Earleywine.) Figure 4.13 shows an approximation of the sampling distribution of T (which was obtained using methods covered in Chapter 5). Also shown is the distribution of T when sampling from a normal distribution instead. As is evident, there is a considerable discrepancy between the two distributions. The practical implication is that when using T, the actual probability coverage might differ substantially from the nominal level.

[1]When a distribution is skewed to the right, it can be shown that the distribution of T is skewed to the left.

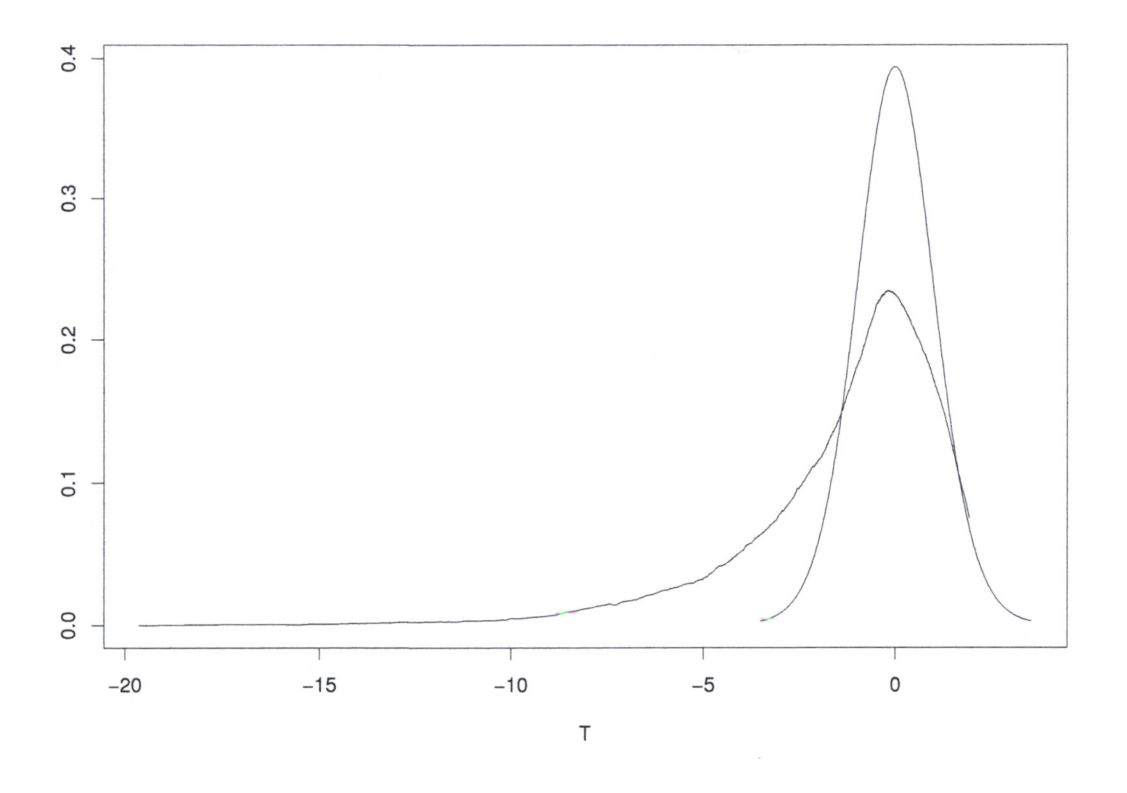

Figure 4.12 The asymmetric curve is the distribution of T when sampling 20 observations from a contaminated lognormal distribution, which is heavy-tailed. The symmetric curve is the distribution of T under normality.

It might be argued that this last example is somehow unusual or that with a slightly larger sample size, satisfactory probability coverage will be obtained. The previous illustration is repeated, only now the data in Table 2.2 are used where $n = 105$. Figure 4.14 shows the result. As is evident, there is extreme skewness indicating that any confidence interval based on Student's T distribution will be highly inaccurate.

An objection to this last illustration might be that there is an extreme outlier among the data. Although the data are from an actual study, it might be argued that having such an extreme outlier is a highly rare event. So we repeat the last illustration with this extreme outlier removed. Figure 4.15 shows an approximation of the distribution of T. Again we see that there is a substantial difference compared to the distribution implied by the central limit theorem.

Yet another possible objection to illustrations based on data from actual studies is that although we are trying to empirically determine the correct distribution for T, nevertheless we are approximating the correct distribution, and perhaps the method used to approximate the distribution is itself in error. That is, if we were to take millions of samples from the population under study, each time computing T, and if we were to then plot the results, perhaps this plot would better resemble a Student's T distribution. There is in fact reason to suspect that the approximation of the distribution of T used here is in error, but all indications are that problems with T are being *underestimated.* For example, in Figure 4.15,

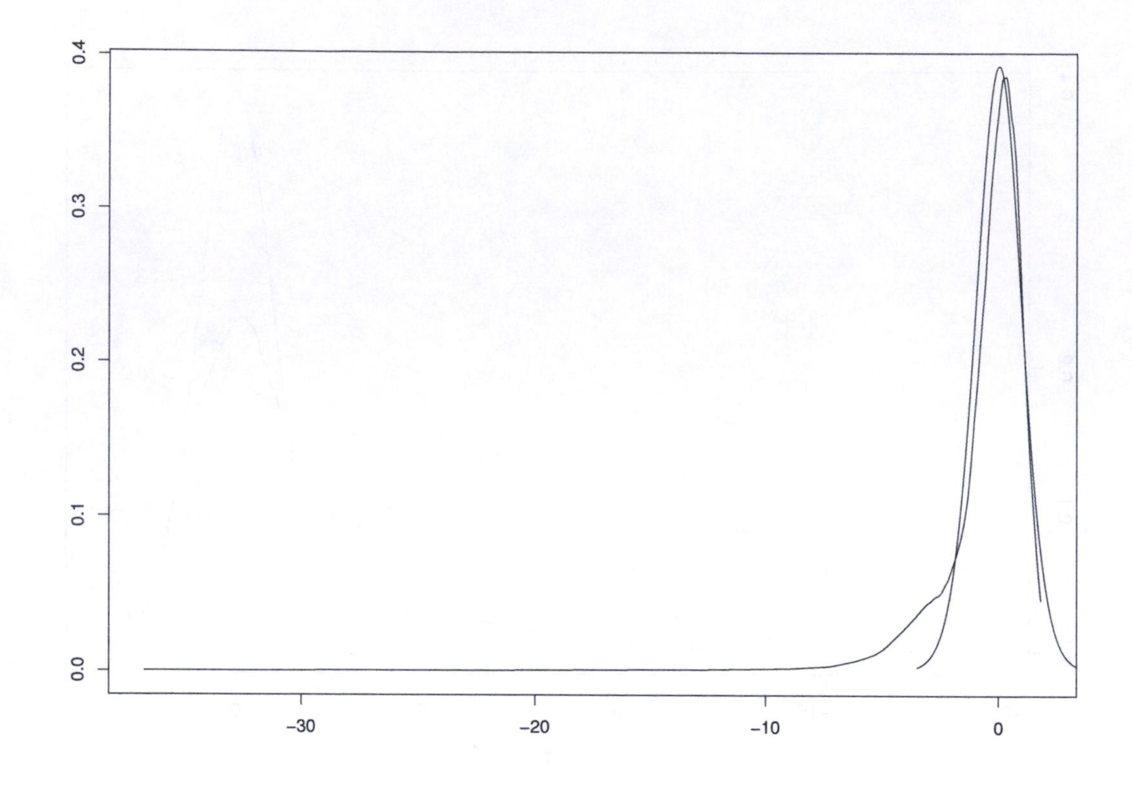

Figure 4.13 The distribution of T when randomly sampling from the hangover data. The symmetric curve is the distribution of T under normality.

it is highly likely that the actual distribution of T is more skewed and that the left tail extends even further to the left than is indicated.

4.7 CONFIDENCE INTERVALS FOR THE TRIMMED MEAN

There are at least three practical concerns with computing confidence intervals for the mean with Student's T distribution.

- The probability coverage can be unsatisfactory for reasons explained in the last section.
- When distributions are skewed, the population mean might provide a poor reflection of the typical participant under study.
- Slight departures from normality can greatly inflate the length of the confidence interval, regardless of whether sampling is from a skewed or symmetric distribution.

Theoretical results (e.g., Huber, 1981; Staudte and Sheather, 1990; Wilcox, 1993a) suggest a strategy that addresses all of these concerns: Switch to the Tukey-McLaughlin confidence interval for the population trimmed mean, μ_t, keeping in mind that for skewed distributions, computing a confidence interval for the population trimmed mean is not the same as computing a confidence interval for the population mean. But before describing this method, we first consider the more fundamental problem of how the standard error of the trimmed mean might be estimated.

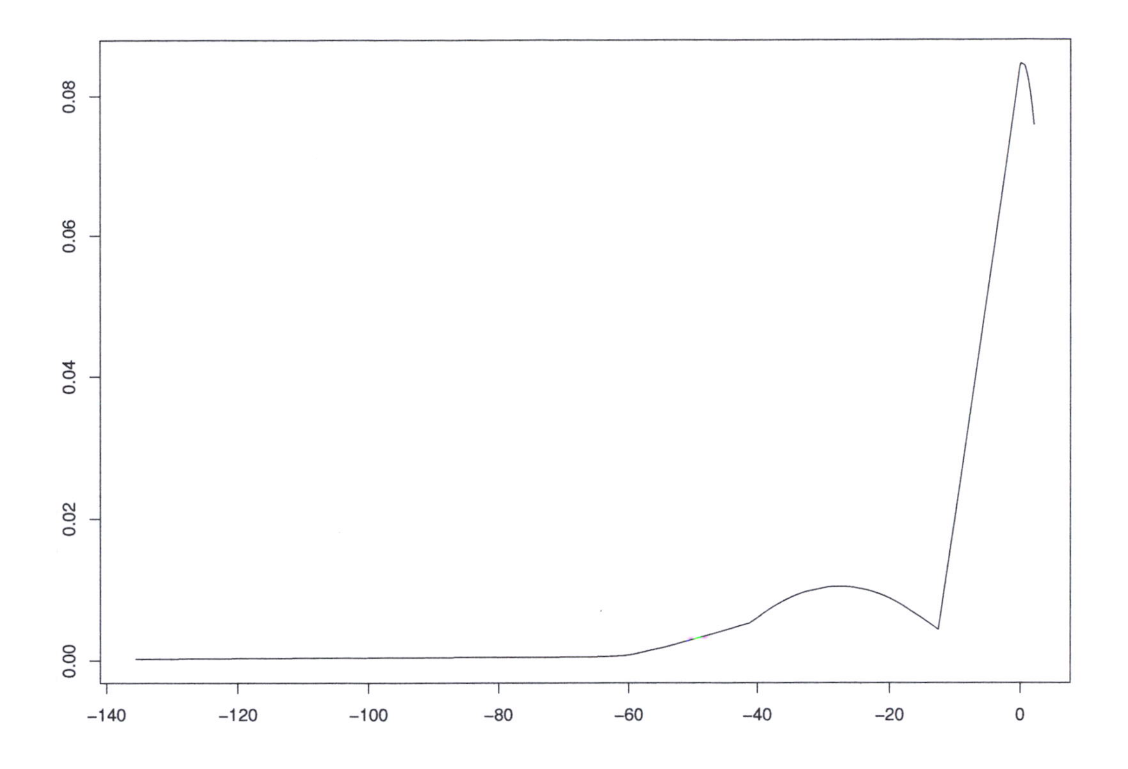

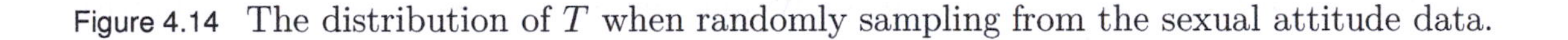
Figure 4.14 The distribution of T when randomly sampling from the sexual attitude data.

4.7.1 Estimating the Standard Error of a Trimmed Mean

This section takes up the important practical problem of estimating the standard error of the sample trimmed mean. Two approaches are described. The first might appear to be the more natural approach, but it is important to be aware that it is theoretically unsound and can yield highly inaccurate results. The second approach is theoretically sound, and as will be illustrated, it can result in a substantially different result than the incorrect method about to be described. (See the last example in this section.)

A Seemingly Natural But Incorrect Method

Recall that the squared standard error of the sample mean is estimated by s^2/n, the sample variance divided by the sample size. The justification of this method depends in a crucial way on the independence among the observations. It might seem that a natural strategy for estimating the squared standard error of a trimmed mean is to compute the sample variance of the values not trimmed and divide by the number of observations not trimmed. But this strategy is unsatisfactory because after any extreme values are trimmed, the remaining observations are no longer identically distributed and they are not independent, which results in using a theoretically unsound estimate of the standard error. In practical terms, we run the risk of getting an estimate that can differ substantially from a theoretically correct method.

When first encountered, the statement that values left after trimming are dependent

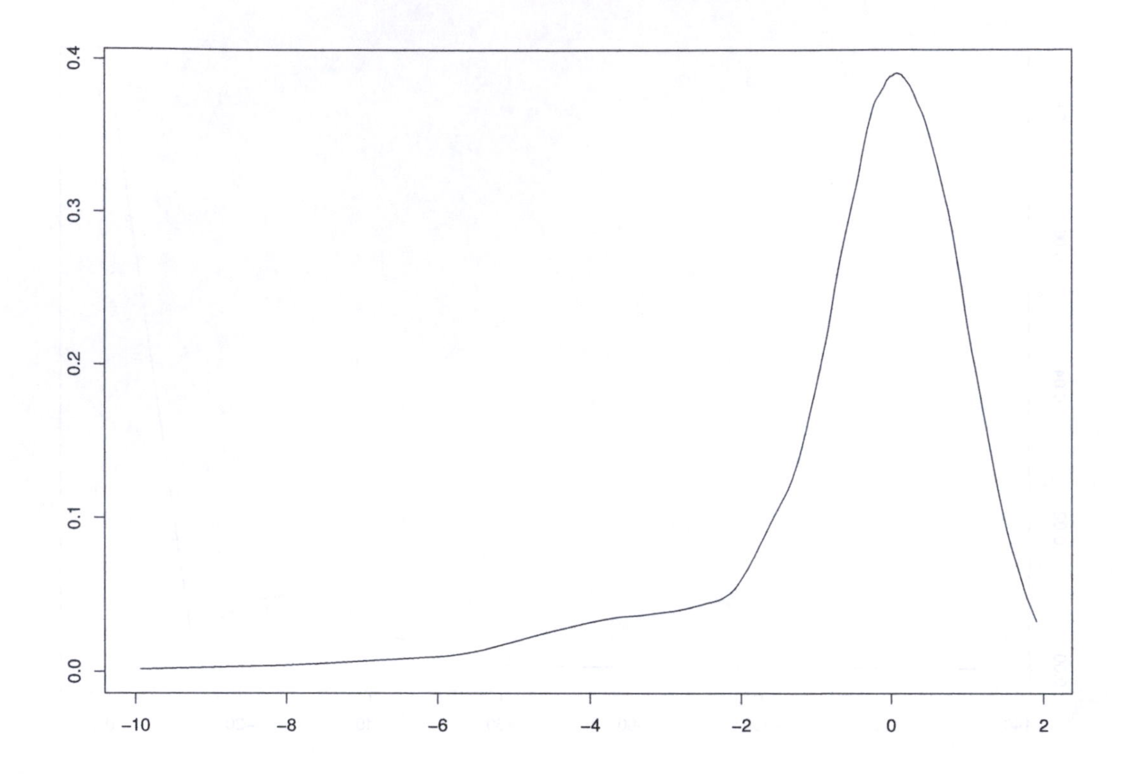

Figure 4.15 The distribution of T when randomly sampling from the sexual attitude data with the extreme outlier removed.

might seem counterintuitive. To provide some indication of why this occurs, suppose $n = 5$ observations are randomly sampled from a standard normal distribution. We might get the values $X_1 = 1.5$, $X_2 = -1.2$, $X_3 = 3.89$, $X_4 = 0.4$, and $X_5 = -0.6$. Random sampling means that the observations are independent as previously explained. Now suppose we repeat this process 500 times, each time generating five observations but only recording the fourth and fifth values we observe. Figure 4.16 shows a scatterplot of the 500 pairs of points, and this is the type of scatterplot we should observe if the observations are independent.

Next, suppose we generate five values from a normal distribution as was done before, but now we put the values in ascending order. As was done in Chapter 2, we label the ordered values $X_{(1)} \leq X_{(2)} \leq X_{(3)} \leq X_{(4)} \leq X_{(5)}$. If we observe $X_1 = 1.5$, $X_2 = -1.2$, $X_3 = 0.89$, $X_4 = 0.4$, and $X_5 = -0.6$, then $X_{(1)} = -1.2$ is the smallest of the five values, $X_{(2)} = -0.6$ is the second smallest, and so on. Now we repeat this process 500 times, each time randomly sampling five observations, but this time we record the two largest values. Figure 4.17 shows the resulting scatterplot for $X_{(4)}$ (the x-axis) versus $X_{(5)}$. There is a discernible pattern because they are dependent.

An important point is that we get dependence when sampling from any distribution, including normal distributions as a special case. If, for example, you are told that $X_{(4)}$ is 0.89, it follows that the largest value ($X_{(5)}$, still assuming $n = 5$) cannot be 0.8, 0.2, 0, or −1; it must be as large or larger than 0.89. Put another way, if we focus attention on the largest value, there is some probability—greater than zero—that its value is less than 0.89. In symbols, $P(X_{(5)} < 0.89) > 0$. But this probability is altered if you are told that $X_{(4)} = 0.89$;

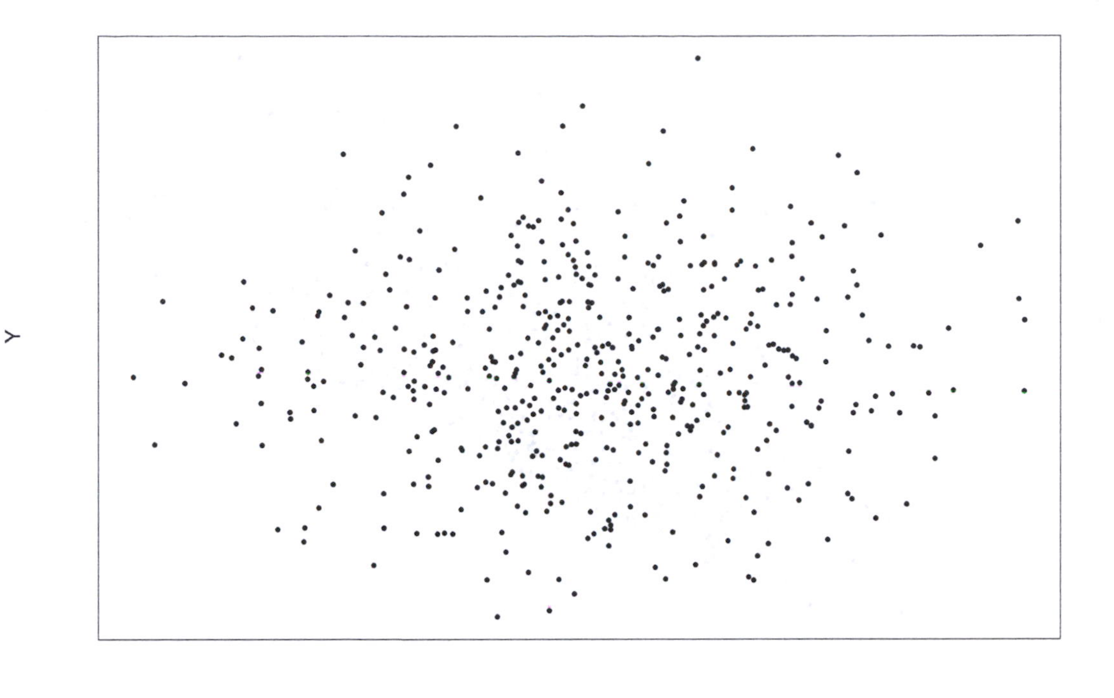

Figure 4.16 A scatterplot of 500 pairs of independent observations.

now it is exactly equal to zero. That is, it is impossible for the largest value to be less than 0.89 if the second largest value is equal to 0.89. Said another way, if knowing the value of $X_{(4)}$ alters the range of possible values for $X_{(5)}$, then $X_{(4)}$ and $X_{(5)}$ are dependent. This argument generalizes: Any two ordered values, say $X_{(i)}$ and $X_{(j)}$, $i \neq j$, are dependent. Moreover, if we discard even one unusually large or small observation, the remaining observations are dependent.

The point of all this is that the method for determining the variance of the sample mean cannot be used to determine the variance of the trimmed mean. The derivation of the variance of the sample mean makes use of the fact that under random sampling, all pairs of observations have zero correlation. But when we discard extreme observations, the remaining observations are dependent and have correlations that differ from zero, and this needs to be taken into account when trying to derive an expression for the variance of the trimmed mean as indicated by Equation (3.15). A simple strategy is to ignore this technical issue and hope that it does not have any practical consequences. But it will be illustrated that this can by highly unsatisfactory.

On a related matter, consider the strategy of discarding any outliers and using the sample mean based on the data that remain. For convenience, assume m observations are left. A tempting strategy is to estimate the squared standard error of this sample mean using the sample variance of these m values divided by m. But it is theoretically unsound because as just indicated, the observations not discarded are dependent, they have correlations that differ from zero, in which case the approach just described is invalid. More formally, after removing extreme values, Pearson's correlation among the remaining values is not equal to

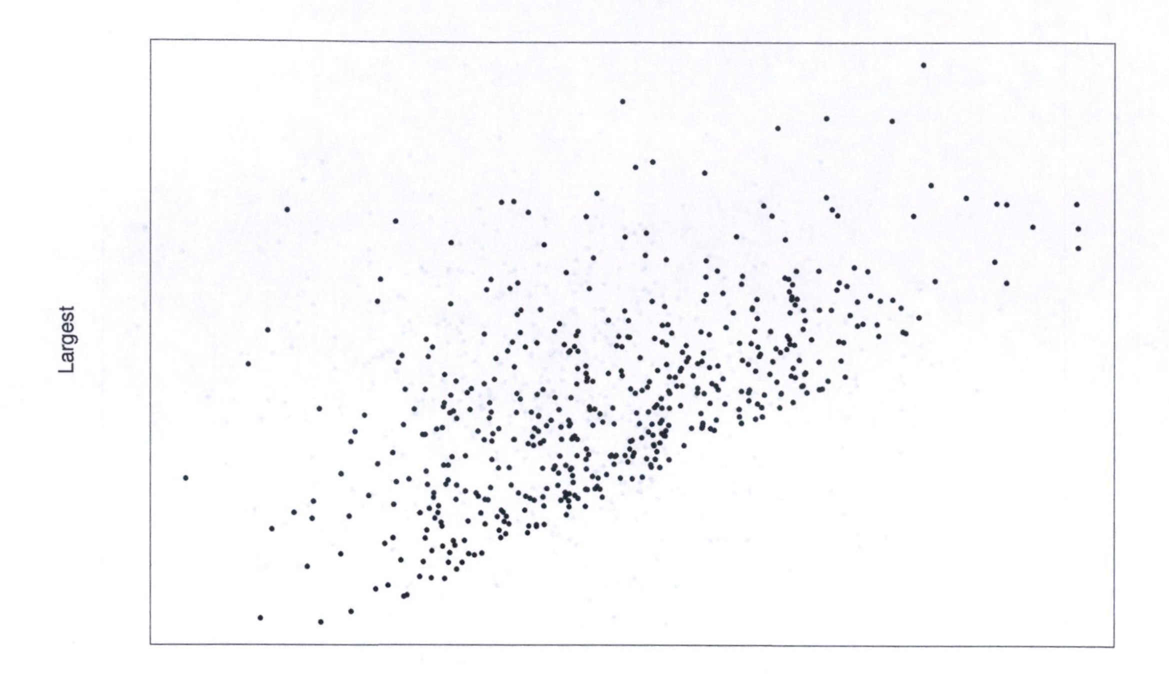

Figure 4.17 Five observations were generated from a normal distribution. The two largest values were recorded and this process was repeated 500 times. Shown is the plot of the resulting 500 pairs of values. This illustrates that the two largest values are dependent.

zero. And from Equation (3.15), it is now necessary to take into account this fact when deriving an expression for the squared standard error.

Data cleaning generally refers to the process of detecting erroneous data. Of course, removing or correcting erroneous data is important and valid. One way of checking for erroneous data is to look for outliers and then verify that they are indeed correct. But a point worth stressing is that the mere fact that a value is an outlier does not necessarily mean that it is erroneous.

EXAMPLE

Imagine that 20 observations are sampled from a normal distribution and outliers are removed using the MAD-median rule. For the mean of the remaining observations, it can be shown that the squared standard error is approximately 0.064. But if the squared standard error is estimated using the variance of the remaining data, divided by the number of observations not declared outliers, the average estimate of the squared standard error is 0.041. Moreover, there is a 89% chance that the estimate will be smaller than the true value, 0.064. If sampling is from the (lognormal) distribution in Figure 4.10, the squared standard error is approximately 0.094. But if the data cleaning strategy is used, the average estimate of the squared standard error is only 0.033 and the median estimate is 0.026. That is, data cleaning results in using an estimate of the standard error that is typically inaccurate by a large amount.

A Theoretically Correct Approach

During the 1960s, some mathematical techniques were developed that provide a convenient and useful method for estimating the squared standard error of a trimmed mean. The theoretical details are described in Huber (1981) and deal with the fact that after trimming, the remaining variables are dependent and have correlations that differ from zero. Here we merely describe the resulting method. In particular, the variance of the 20% trimmed mean over many studies (its squared standard error) can be estimated with

$$\frac{s_w^2}{0.6^2 n}, \tag{4.9}$$

where s_w^2 is the 20% Winsorized sample variance introduced in Chapter 2. The 0.6 in the denominator is related to the amount of trimming, which is assumed to be 20%. More generally, if the proportion of observations trimmed is G, the squared standard error is estimated with

$$\frac{s_w^2}{(1-2G)^2 n}, \tag{4.10}$$

where now s_w^2 is the G-Winsorized variance. For example, if 10% trimming is used, meaning that $G = 0.1$, the 0.6 in Equation (4.9) is replaced by $1-2(0.1) = 0.8$ and a 10% Winsorized variance is used. The standard error of the trimmed mean is estimated with the square root of this last equation.

EXAMPLE

Again consider the weight-gain data for rats and assume 20% trimming is to be used. The sample Winsorized standard deviation can be seen to be $s_w = 3.927$. The trimmed mean is $\bar{X}_t = 23.27$. Because there are $n = 23$ rats, the estimated standard error of the sample trimmed is

$$\frac{3.927}{0.6\sqrt{23}} = 1.4.$$

In contrast, the estimated standard error of the sample mean is $s/\sqrt{n} = 2.25$. Note that the ratio of these two values is $1.4/2.25 = 0.62$, indicating that the 20% trimmed mean has a substantially smaller standard error. Put another way, it might seem that the trimmed mean would have a larger standard error because only 6 of the 10 observations are used to compute the trimmed mean. In fact the exact opposite is true and this turns out to have considerable practical importance.

EXAMPLE

Consider the data in Table 4.3, which are from a study on self-awareness conducted by E. Dana. The estimated standard error of the 20% trimmed mean is 56.1. In contrast, the standard error of the sample mean is $s/\sqrt{n} = 136$, a value approximately 2.4 times larger than the sample standard error of the trimmed mean. This difference will be seen to be substantial.

An Important Point Regarding Standard Errors

Table 4.3 Self-awareness Data

77	87	88	114	151	210	219	246	253	262
296	299	306	376	428	515	666	1310	2611	

The next example illustrates that the theoretically unsound method for estimating the standard error of the 20% trimmed mean, described at the beginning of this section, can give a result that differs substantially from the theoretically correct method just described.

EXAMPLE

For the data in Table 2.2, the estimated standard error of the 20% trimmed mean is 0.532 using Equation (4.9). Now imagine that we again trim 20% and simply use the method for the sample mean on the remaining 63 values. That is, we compute s using these 63 values only and then compute $s/\sqrt{63}$. This yields 0.28, which is about half of the value based on Equation (4.9). So we see that using a theoretically motivated estimate of the standard error of the trimmed mean, rather than using methods for the sample mean based on data not trimmed, is not an academic matter. The incorrect estimate can differ substantially from the estimate based on theory.

4.7.2 R Function trimse

The R function

```
trimse(x,tr=0.2),
```

written for this book, estimates the standard error of a trimmed mean.

4.7.3 A Confidence Interval for the Population Trimmed Mean

As was the case when working with the population mean, we want to know how well the sample trimmed mean, $\bar{X}_t$, estimates the population trimmed mean, μ_t. What is needed is a method for computing a $1-\alpha$ confidence interval for μ_t. A solution was derived by Tukey and McLaughlin (1963) and is computed as follows. Let h be the number of observations left after trimming. Let c be the $1-\alpha/2$ quantile of Student's T distribution with $h-1$ degrees of freedom and let s_w be the Winsorized sample standard deviation, which is described in Section 2.3.5. A confidence interval for a trimmed mean is

$$\left(\bar{X}_t - c\frac{s_w}{(1-2G)\sqrt{n}}, \bar{X}_t + c\frac{s_w}{(1-2G)\sqrt{n}}\right), \tag{4.11}$$

where G is the proportion of values trimmed. So for the special case of 20% trimming, a $1-\alpha$ confidence interval is given by

$$\left(\bar{X}_t - c\frac{s_w}{0.6\sqrt{n}}, \bar{X}_t + c\frac{s_w}{0.6\sqrt{n}}\right). \tag{4.12}$$

In terms of probability coverage, we get reasonably accurate confidence intervals for a much broader range of nonnormal distributions compared to confidence intervals for μ based on Student's T.

Section 4.4 provided one reason why a trimmed mean can be a more accurate estimate of the population mean when sampling from a symmetric distribution. The method for computing a confidence interval just described provides another perspective and explanation.

We saw in Chapter 2 that generally the Winsorized standard deviation, s_w, can be substantially smaller than the standard deviation s. Consequently, a confidence interval based on a trimmed mean can be substantially shorter. However, when computing a confidence interval based on 20% trimming, for example, the estimate of the standard error of the trimmed mean is $s_w/(0.6\sqrt{n})$. Because $s_w/0.6$ can be greater than s, such as when sampling from a normal distribution, it is possible to get a shorter confidence interval using means. Generally, however, any improvement achieved with the mean is small, but substantial improvements based on a trimmed mean are often possible.

EXAMPLE

Suppose a test of open mindedness is administered to 10 participants yielding the observations

$$5, 60, 43, 56, 32, 43, 47, 79, 39, 41.$$

We compute a 0.95 confidence interval for the 20% trimmed mean and compare the results to the confidence interval for the mean. With $n = 10$, the number of trimmed observations is four, as explained in Chapter 2. That is, the two largest and two smallest observations are removed, leaving $h = 6$ observations, and the average of the remaining observations is the trimmed mean, $\bar{X}_t = 44.8$. The mean using all 10 observations is $\bar{X} = 44.5$. This suggests that there might be little difference between the population mean, μ, and population trimmed mean, μ_t. With $\nu = 6 - 1 = 5$ degrees of freedom, Table 4 in Appendix B indicates that the 0.975 quantile of Student's T distribution is $c = 2.57$. It can be seen that the Winsorized sample variance is $s_w^2 = 54.54$, so $s_w = \sqrt{54.54} = 7.385$, and the resulting confidence interval for the trimmed mean is

$$44.8 \pm 2.57\frac{7.385}{0.6\sqrt{10}} = (34.8,\ 54.8).$$

In contrast, the 0.95 confidence interval for the mean is (30.7, 58.3). The ratio of the lengths of the confidence intervals is $(54.8 - 34.8)/(58.3 - 30.7) = 0.72$. That is, the length of the confidence interval based on the trimmed mean is substantially shorter.

In the previous example, a boxplot of the data reveals that there is an outlier. This explains why the confidence interval for the mean is longer than the confidence interval for the trimmed mean: The outlier inflates the sample variance, s^2, but has no effect on the Winsorized sample variance, s_w^2. Yet another method for trying to salvage means is to check for outliers, and if none are found, compute a confidence interval for the mean. Recall, however, that even when sampling from a skewed light-tailed distribution, the distribution of T can differ substantially from the case where observations are normal. This means that even though no outliers are detected, when computing a 0.95 confidence interval for μ, the actual probability coverage could be substantially smaller than intended unless the sample size is reasonably large. When attention is turned to comparing multiple groups of participants, this problem becomes exacerbated, as will be seen. Modern theoretical results tell us that trimmed means reduce this problem substantially.

EXAMPLE

Table 4.4 shows the average LSAT scores for the 1973 entering classes of 15 American law schools. (LSAT is a national test for prospective lawyers.) The sample mean is $\bar{X} = 600.3$ with an estimated standard error of 10.8. The 20% trimmed mean is $\bar{X}_t = 596.2$ with an estimated standard error of 14.92. The 0.95 confidence interval for μ_t is $(561.8, 630.6)$.

Table 4.4 Average LSAT Scores for 15 Law Schools

545	555	558	572	575	576	578	580
594	605	635	651	653	661	666	

In contrast, the 0.95 confidence interval for μ is $(577.1, 623.4)$, assuming T does indeed have a Student's t distribution. Note that the length of the confidence interval for μ is smaller, and in fact is a subset of the confidence interval for μ_t. This might suggest that the sample mean is preferable to the trimmed mean for this particular set of data, but closer examination suggests that this might not be true. The concern here is the claim that the confidence interval for the mean has probability coverage 0.95. If sampling is from a light-tailed, skewed distribution, the actual probability coverage for the sample mean can be substantially smaller than the nominal level. Figure 4.18 shows a boxplot of the data indicating that the central portion of the data is skewed to the right. Moreover, there are no outliers suggesting the possibility that sampling is from a relatively light-tailed distribution. Thus, the actual probability coverage of the confidence interval for the mean might be too low—a longer confidence interval might be needed to achieve 0.95 probability coverage. That is, an unfair comparison of the two confidence intervals has probably been made because they do not have the same probability coverage. If we were able to compute a 0.95 confidence interval for the mean, there is some possibility that it would be longer than the confidence interval for the trimmed mean. When sampling from a skewed, heavy-tailed distribution, problems with the mean can be exacerbated (as illustrated in Chapter 5).

The confidence interval for the 20% trimmed mean assumes that

$$T_t = \frac{0.6(\bar{X}_t - \mu_t)}{s_w/\sqrt{n}}$$

has a Student's T distribution with $h-1 = n-2g-1$ degrees of freedom, where $g = [0.2n]$ and $[0.2n]$ is $0.2n$ rounded down to the nearest integer. To graphically illustrate how nonnormality affects this assumption, we repeat the method used to create Figure 4.11. That is, we sample $n = 20$ observations from the (lognormal) distribution in Figure 4.10, compute T_t, and repeat this 5000 times yielding 5000 T_t values. The left panel of Figure 4.19 shows a plot of these T_t values versus Student's T distribution with 11 degrees of freedom. To facilitate comparisons, the right panel shows a plot of 5000 T values (based on the mean and variance) versus Student's T distribution with 19 degrees of freedom. (The right panel in Figure 4.19 is just a reproduction of the left panel in Figure 4.16.) Student's T distribution gives a better approximation of the actual distribution of T_t versus T. Of particular importance is that the tails of the distribution are better approximated when using a trimmed mean. This indicates that more accurate probability coverage will be achieved using a 20% trimmed mean compared to using the mean with no trimming. Generally, as the sample size increases, problems with nonnormality diminish more rapidly when using a trimmed mean versus using a mean. But switching to a trimmed mean does not eliminate all practical problems when sample sizes are small. Fortunately there are methods for getting even more accurate results, as we shall see.

4.7.4 R Function trimci

The R function

```
trimci(x,tr=0.2,alpha=0.05)
```

Figure 4.18 A boxplot of the law data suggesting that observations were sampled from a skewed distribution.

computes a $1 - \alpha$ confidence interval for μ_t. The amount of trimming, indicated by the argument `tr`, defaults to 20%. If the argument alpha is unspecified, $\alpha = 0.05$ is used.

EXAMPLE

If the data in Table 4.4 are stored in the R variable `blob`, the command `trimci(blob)` returns a 0.95 confidence interval for the 20% trimmed: (561.8, 630.6). The command `trimci(blob,tr=0,alpha=0.01)` returns a 0.99 confidence interval for the population mean using Student's T.

EXAMPLE

Table 2.2 reports the responses of 105 males regarding the number of desired sexual partners over the next 30 years. In another study, 2282 women were asked the same question. (The data were generously supplied by Lynn Miller.) Storing the data in the R variable sexf, the command

```
trimci(sexf)
```

returns a 0.95 confidence for the 20% trimmed mean: (1.082, 1.197). The command

```
trimci(sexf,tr=0)
```

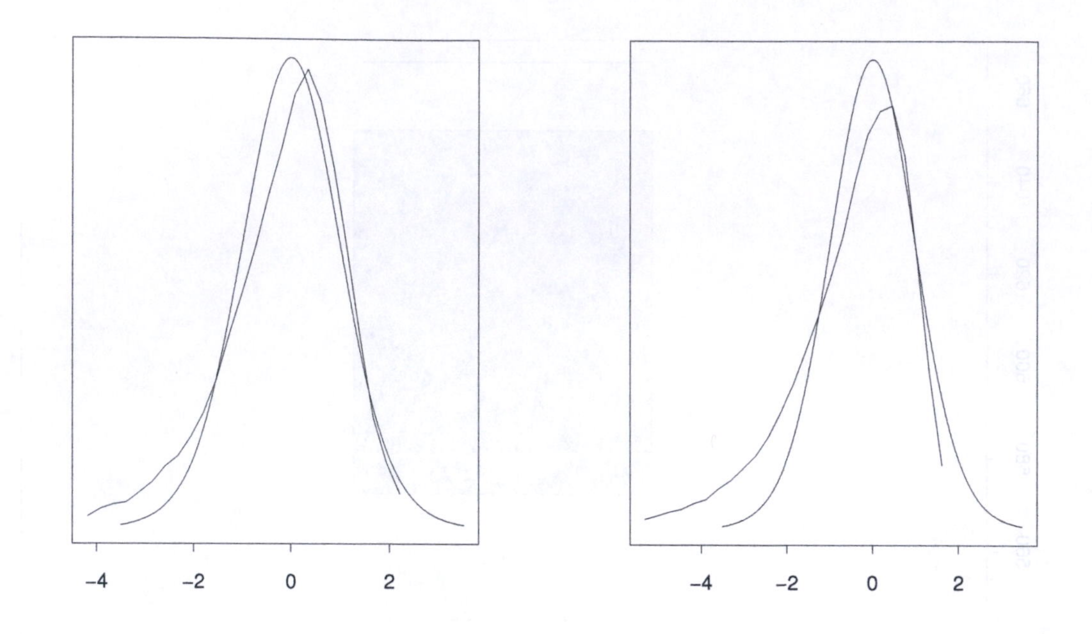

Figure 4.19 The left panel shows actual distribution of T_t and the approximation based on Student's T distribution when sampling from the lognormal distribution in Figure 4.10. The right panel shows the actual distribution of T. In practical terms, using Student's T to compute confidence intervals for the 20% trimmed mean tends to be more accurate than using Student's T to compute a confidence interval for the mean.

returns a 0.95 confidence for the mean: (0.773, 6.174). Notice that the length of the confidence interval for the mean is substantially higher than the length of the confidence interval for the 20% trimmed mean despite the relatively large sample size.

4.8 TRANSFORMING DATA

This section expands a bit on comments made in Chapter 2 regarding the transformation of data. As previously noted, simple transformations of data are often recommended for dealing with problems due to nonnormality, a common recommendation being to replace each value with its logarithm and then use methods based on means. Another common strategy when all observations are positive is to replace each value with its square root. These simple transformations can be useful for certain purposes, but for a wide range of situations they can be relatively unsatisfactory. One problem is that they do not eliminate the deleterious effect of outliers (e.g., Rasmussen, 1989). Transformations can make a distribution appear to be reasonably normal, at least in some cases, but when using simple transformations it has been found that trimming is still beneficial (Doksum and Wong, 1983). Also, as noted in Chapter 2, even after transforming data, a plot of the data can remain skewed as illustrated in Figure 2.10. For these reasons, simple transformations of data are not discussed further.

4.9 CONFIDENCE INTERVAL FOR THE POPULATION MEDIAN

Although the median is a member of the class of trimmed means, the method for computing a confidence interval for a trimmed mean gives absurd results for the extreme case of the median where all but one or two values are trimmed. There is, however, a relatively simple method for dealing with this problem that is *distribution free*. That is, assuming random sampling only, the exact probability coverage can be determined. To provide at least a glimpse of why this is true, note that a randomly sampled observation has probability 0.5 of being less than the population median. Imagine that we randomly sample n observations. Then the probability that y of these n values are less than the population median is given by the binomial probability function with probability of success $p = 0.5$. In general terms, the probability that y observations are less than or equal to the median is equal to the probability of getting y success or less when working with the binomial probability function with probability of success $p = 0.5$. This is true, even though the population median is not known and this fact can be exploited in a manner that yields a confidence interval for the median.

EXAMPLE

If we randomly sample 10 observations, the probability that 3 observations will be less than or equal to the median is equal to the probability of 3 successes or less in 10 trials when dealing with the binomial probability function with probability of success $p = 0.5$. From Table 2 of Appendix B, this probability is 0.1.

Now, consider any integer k between 1 and $n/2$, and for the binomial probability function, let P be the probability of $k-1$ successes or less when the probability of success is $p = 0.5$. As usual, we denote the ordered observations by $X_{(1)} \leq \cdots \leq X_{(n)}$. Then

$$(X_{(k)}, X_{(n-k+1)}) \tag{4.13}$$

is a confidence interval for the median of X that has probability coverage exactly equal to $1-2P$.

Said another way, imagine that we want the probability coverage to be approximately $1-\alpha$. Then from Table 2 (or using appropriate software) we can accomplish this goal by choosing k so that it is approximately true that the number of successes less than k, when $p = 0.5$, is $\alpha/2$, in which case the probability coverage based on Equation (4.13) is approximately equal to α. (Recall from Chapter 3 that for the binomial, the probability of having less than k successes is the same as the probability of having the number of successes less than or equal to $k-1$.)

EXAMPLE

Imagine that for a sample size of 15, the goal is to compute a 0.9 confidence interval for the median, in which case $\alpha/2 = 0.05$. From Table 2 in Appendix B we see that the probability of four successes or less is 0.059. This means that if we use $k-1 = 4$, meaning, of course, that $k = 5$, the probability coverage for the median will be exactly equal to $1 - 2(0.059) = 0.892$.

EXAMPLE

Staudte and Sheather (1990) report data from a study on the lifetimes of EMT6 cells. The results were

10.4, 10.9, 8.8, 7.8, 9.5, 10.4, 8.4, 9.0, 22.2, 8.5,
9.1, 8.9, 10.5, 8.7, 10.4, 9.8, 7.7, 8.2, 10.3, 9.1.

The sample size is $n = 20$. Putting the observations in ascending order yields

7.7, 7.8, 8.2, 8.4, 8.5, 8.7, 8.8, 8.9, 9.0, 9.1, 9.1,
9.5, 9.8, 10.3, 10.4, 10.4, 10.4, 10.5, 10.9, 22.2.

From Table 2 in Appendix B, we see that the probability of 5 successes or less among 20 trials is 0.0207. So if we set $k - 1 = 5$, in which case $k = 6$ and $n - k + 1 = 20 - 6 + 1 = 15$, the probability that the confidence interval

$$(X_{(6)}, X_{(15)}) = (8.7, 10.4)$$

contains the population median is exactly equal to $1 - 2(0.0207) = 0.9586$.

Note that because the binomial distribution is discrete, it is not possible, in general, to choose k so that the probability coverage is exactly equal to $1 - \alpha$. There is a slight generalization of the method just described aimed at dealing with this issue that was derived by Hettmansperger and Sheather (1986). Results supporting the use of this method are reported by Sheather and McKean (1987) as well as Hall and Sheather (1988). No details are given here, but an R function that performs the calculations is provided.

4.9.1 R Function sint

The R function

```
sint(x,alpha=0.05)
```

computes a confidence interval for the median using the Hettmansperger and Sheather method just mentioned.

EXAMPLE

The last example dealing with the lifetimes of EMT6 cells is repeated, only now we use the R function `sint`. Storing the values in the R variable `blob`, the 0.95 confidence interval returned by the command `sint(blob)` is (8.72, 10.38), which is not much different from the confidence interval given in the last example, namely (8.7, 10.4). The 0.99 confidence interval returned by command `sint(blob,0.01)` is (8.5, 10.4).

4.9.2 Estimating the Standard Error of the Sample Median

The method just described for computing a confidence interval for the median does not require an estimate of the standard error of M, the sample median. However, for situations to be covered, an explicit estimate will be needed. Many estimators have been proposed, comparisons of which can be found in Price and Bonett (2001). Here, the method derived by McKean and Schrader (1984) is described because it is very simple and currently appears to have practical value for problems addressed in subsequent chapters.

Compute

$$k = \frac{n+1}{2} - z_{0.995}\sqrt{\frac{n}{4}},$$

round k to the nearest integer, and let $z_{0.995}$ be the 0.995 quantile of a standard normal distribution, which is approximately equal to 2.576. Put the observed values in ascending order yielding $X_{(1)} \leq \cdots \leq X_{(n)}$. Then the McKean–Schrader estimate of the squared standard error of M is

$$\left(\frac{X_{(n-k+1)} - X_{(k)}}{2z_{0.995}}\right)^2.$$

There is, however, a limitation of this estimate of the squared standard error that should be stressed. Tied values mean that one or more values occur more than once. For example, for the values 2, 23, 43, 45, 7, 8, 33 there are no tied values; each value occurs only once. But for the values 3, 5, 6, 34, 3, 7, 11, 3, 7 there are tied values; the value 3 occurs three times and the value 7 occurs twice. *If there are tied values, the McKean–Schrader estimate of the standard error can be highly inaccurate.* Moreover, all indications are that all alternative estimates of the standard error can be highly unsatisfactory as well. In practical terms, when dealing with medians, special methods for dealing with tied values will be needed. Currently, the most effective techniques are based on what is called a percentile bootstrap method, which is introduced in Section 5.11.

4.9.3 R Function msmedse

The R function

```
msmedse(x)
```

computes an estimate of the standard error of sample median, M, using the McKean–Schrader estimator just described.

4.9.4 More Concerns About Tied Values

To underscore why tied values can cause problems when working with the median, and to illustrate a limitation of the central limit theorem, imagine a situation where the possible observed values are the integers $0, 1, \ldots, 15$, with the corresponding probabilities given by the binomial probability function with probability of success $p = 0.7$. So, for example, the probability that a randomly sampled participant responds with the value 13 is 0.09156. Further imagine that 20 individuals are randomly sampled, in which case tied values are guaranteed since there are only 16 possible responses. Next, compute the median and repeat this process 5000 times. The left panel of Figure 4.20 shows a plot of the relative frequencies of the resulting medians. The plot somewhat resembles a normal curve, but note that only five values for the sample median occur. Now look at the right panel, which was created in the same manner as the left panel, only with a sample size of $n = 100$ for each median. Blind reliance on the central limit theorem would suggest that the plot will look more like a normal distribution than the left panel, but clearly this is not the case. Now only three values for the sample median are observed. In practical terms, methods for making inferences about the median, which assume the sample median has a normal distribution, can be disastrous when tied values can occur.

4.10 A REMARK ABOUT MOM AND M-ESTIMATORS

How do we compute a confidence interval based on an M-estimator of location or the MOM estimator introduced in Chapter 3? An expression for the standard error of an M-estimator has been derived and can be used to derive a confidence interval using methods similar to those described in this chapter. It generally yields a reasonably accurate confidence interval when sampling from a perfectly symmetric distribution. But for skewed distributions it can be highly inaccurate. With n sufficiently large, an accurate confidence interval can be computed when sampling from a skewed distribution, but it remains unclear just how large n must be.

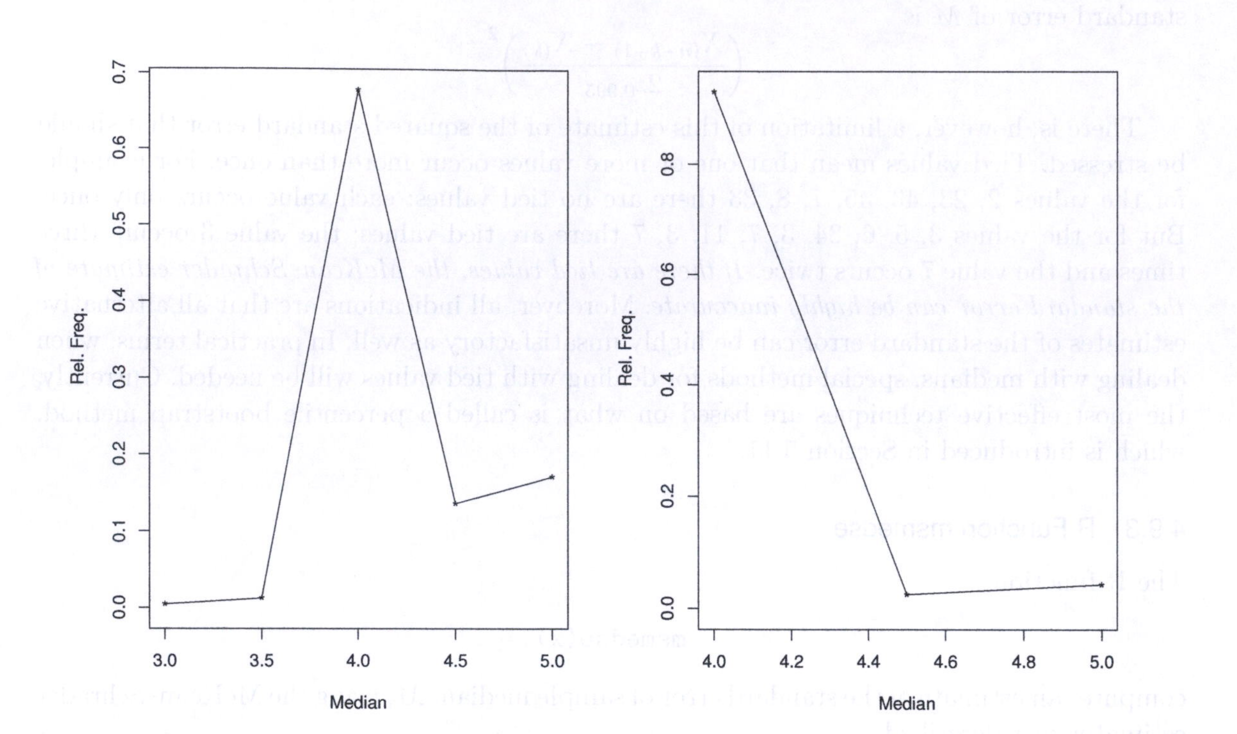

Figure 4.20 The left panel shows the distribution of 5000 sample medians, each based on a sample size of $n = 20$. Note that only five values for sample median occur. The right panel shows the distribution of the medians when the sample size is increased to $n = 100$. The plot is even less similar to a normal curve, contrary to what might be thought based on the central limit theorem, which illustrates why tied values can cause problems when making inferences based on the median.

As for MOM, no expression for the standard error has been derived. The percentile bootstrap method, which is introduced in Section 5.11, provides an effective method for dealing with these problems.

4.11 CONFIDENCE INTERVALS FOR THE PROBABILITY OF SUCCESS

This section considers the problem of computing a confidence interval for p, the probability of success, when dealing with the binomial probability function. As previously noted, if X represents the number of successes among n randomly sampled observations, the usual estimate of p is simply

$$\hat{p} = \frac{X}{n},$$

the proportion of successes among the n observations. Moreover, the squared standard error of $\hat{p}$ is $p(1-p)/n$, and the central limit theorem indicates that if the sample size is sufficiently large, $\hat{p}$ will have, approximately, a normal distribution. This suggests a simple method for computing a $1-\alpha$ confidence interval for p:

$$\hat{p} \pm c\sqrt{\frac{p(1-p)}{n}},$$

where c is the $1-\alpha/2$ quantile of a standard normal distribution. We do not know the value of the quantity under the radical, but it can be estimated with $\hat{p}$, in which case a simple $1-\alpha$ confidence interval for p is

$$\hat{p} \pm c\sqrt{\frac{\hat{p}(1-\hat{p})}{n}}. \tag{4.14}$$

The resulting probability coverage will be reasonably close to $1-\alpha$ if n is not too small and p is not too close to zero or one. Just how large n must be depends on how close p is to zero or one. An obvious concern is that we do not know p, so there is some difficulty in deciding whether n is sufficiently large. Moreover, Brown, Cai, and DasGupta (2002) argue that when using Equation 4.14, the sample size must be much larger than is commonly thought.

Numerous methods have been proposed for dealing with this issue. For the special cases $x = 0$, 1, $n-1$, and n, Blyth (1986) suggests proceeding as follows:

- If $x = 0$, use

$$(0,\, 1-\alpha^{1/n}).$$

- If $x = 1$, use

$$\left(1-(1-\frac{\alpha}{2})^{1/n},\, 1-(\frac{\alpha}{2})^{1/n}\right).$$

- If $x = n-1$, use

$$\left((\frac{\alpha}{2})^{1/n},\, (1-\frac{\alpha}{2})^{1/n}\right).$$

- If $x = n$, use

$$(\alpha^{1/n},\, 1).$$

For all other situations, Blyth's comparisons of various methods suggest using Pratt's (1968) approximate confidence interval, which is computed as folllows.

Let c be the $1-\alpha/2$ quantile of a standard normal distribution read from Table 1 in Appendix B. That is, if Z is a standard normal random variable, $P(Z \le c) = 1-\alpha/2$. To determine c_U, the upper end of the confidence interval, compute

$$A = \left(\frac{x+1}{n-x}\right)^2$$

$$B = 81(x+1)(n-x) - 9n - 8$$

$$C = -3c\sqrt{9(x+1)(n-x)(9n+5-c^2)+n+1}$$

$$D = 81(x+1)^2 - 9(x+1)(2+c^2) + 1$$

$$E = 1 + A\left(\frac{B+C}{D}\right)^3$$

$$c_U = \frac{1}{E}.$$

To get the lower end of the confidence interval, c_L, compute

$$A = \left(\frac{x}{n-x-1}\right)^2$$

$$B = 81(x)(n - x - 1) - 9n - 8$$

$$C = 3c\sqrt{9x(n - x - 1)(9n + 5 - c^2) + n + 1}$$

$$D = 81x^2 - 9x(2 + c^2) + 1$$

$$E = 1 + A\left(\frac{B + C}{D}\right)^3$$

$$c_L = \frac{1}{E}.$$

An approximate $1 - \alpha$ confidence interval for p is

$$(c_L, c_U).$$

For completeness, it is noted that Brown et al. (2002) recommend an alternative method for computing a confidence interval for p. They call it the *Agresti–Coull method* because it generalizes a method that was studied by Agresti and Coull (1998). Again let c be the $1 - \alpha/2$ quantile of a standard normal distribution and let X denote the total number of successes. Let $\tilde{X} = X + c^2/2$, $\tilde{n} = n + c^2$, $\tilde{p} = \tilde{X}/\tilde{n}$, $\tilde{q} = 1 - \tilde{p}$, in which case the $1 - \alpha$ confidence interval for p is

$$\tilde{p} \pm c\sqrt{\frac{\tilde{p}\tilde{q}}{n}}.$$

Brown et al. (2002) compared several methods, but they did not consider Pratt's method, which was recommended by Blythe. Moreover, Pratt's method was not considered by Agresti and Coull (1998), and it seems that a comparison of Pratt's method and the Agresti–Coull method has not been made.

Yet another method was derived by Schilling and Doi (2014). When computing a $1 - \alpha$ confidence interval, their method yields the shortest possible confidence interval such that the probability coverage is at least $1 - \alpha$. For example, if the goal is to compute a 0.95 confidence interval, the actual probability that the confidence interval contains the true probability of success, p, is at least 0.95. The complex computational details are not described but an R function for applying their method is provided.

4.11.1 R Functions binomci, acbinomci and and binomLCO

The R function

```
binomci(x = sum(y), nn = length(y), y = NA, n = NA, alpha = 0.05)
```

has been supplied to compute Pratt's approximate confidence interval for p. In the event $x = 0$, 1, $n - 1$, or n, Blythe's method is used instead. The first argument, x, is the number of successes and the second argument, nn, indicates the value of n, the number of observations. If the data are stored as a vector of 1s and 0s in some R variable, where a 1 indicates a success and 0 a failure, use the third argument y. (Generally, the fourth argument can be ignored.)

The function

```
acbinomci(x = sum(y), nn = length(y), y = NA, n = NA, alpha = 0.05)
```

is used exactly like the function binomci, only it applies the Agresti–Coull method.[2]

The R function

[2]R contains yet another built-in method, accessed via the function binom.test, which stems from Clopper and Pearson (1934). It guarantees that the probability coverage is greater than or equal to the specified level, but in general it does not give the shortest-length confidence interval.

```
binomLCO(x = sum(y), nn = length(y), y = NULL, alpha = 0.05)
```

applies the Schilling–Doi method and is used in the same manner as the R function `binomci`. A possible concern is that with a large sample size, execution time using the Schilling–Doi method can be prohibitive.

EXAMPLE

The command

```
binomci(5,25)
```

returns (0.07, 0.41) as a 0.95 confidence interval for p based on five successes among 25 observations. If the values 1, 1, 1, 0, 0, 1, 1, 0, 0, 0, 0, 1 are stored in the R variable `obs`, the command

```
binomci(y=obs)
```

returns (0.25, 0.79) as a 0.95 confidence interval for p.

EXAMPLE

The 0.95 confidence interval returned by the command

```
binomci(0,10)
```

is (0.0, 0.259). In contrast, the 0.95 confidence interval returned by `binomLCO` is (0.0, 0.291).

4.12 BAYESIAN METHODS

There is another approach to making inferences about the population mean μ and the probability of success, p, that should be mentioned. The first version of this approach was derived by Thomas Bayes with the goal of making inferences about p. His results were published posthumously in the years 1764 and 1765. His strategy was to assume that p has a uniform distribution, which was introduced in Exercise 10 in Section 3.11. Bayes showed that if the number of successes, X, given n and p, has a binomial distribution, then the (posterior) distribution of p, given X and n, is a beta distribution. The point is that given X and n, one can then compute what is called a posterior interval for p, simply meaning an interval that contains p with some probability that can be computed. (Details for making inferences based on Bayesian methods can be found in Gelman et al., 2013.) In the year 1774, Laplace developed the same method, apparently independently of Bayes (Hald, 1988, p. 134). Laplace made major contributions that included extensions to making inferences about the population mean of a normal distribution. The basic strategy, in the simplest case, is based in part on two assumptions: the distribution of X, given μ and σ^2, is a normal distribution, and μ has a normal distribution with mean μ' and standard deviation σ'. The distributions for unknown parameters, such as μ and p, are called prior distributions that are specified by the researcher. That is, they do not stem from the data being analyzed, but rather from judgments made about what seems like a reasonable distribution. So for the situation considered here, a researcher would choose values for μ' and σ'. (For a discussion of how this process might be addressed, see Gelman and Shalizi, 2010; cf. Efron and Hastie, 2016.)

Bayesian inferential methods dominated for about 50 years simply because they were the only inferential methods available. Then, in the year 1811, Laplace developed the frequentist approach, which refers to methods based on the notion of a sampling distribution as described in Section 4.1. In the year 1814, Laplace derived a strategy for computing confidence intervals that contains as a special case the method described in Section 4.3.1. Rather than view parameters as random variables, as done by Bayes, Laplace viewed parameters as fixed, unknown constants. Moreover, inferences are made based solely on the available data. In effect, Laplace created one of the most controversial issues in statistics: when and how should a Bayesian method be used? Gauss, for example, was highly critical of the Bayesian approach and preferred the frequentist approach (Hald, 1998, section 22.2). Today the controversy remains. For example, Gelman (2008, p. 445), who is an advocate of the Bayesian approach, states: "Bayesian inference is one of the more controversial approaches to statistics." His paper is devoted to summarizing why. Certainly one concern is justifying any choice for the prior distribution. More broadly, there is the issue of deciding which assumptions are reasonable in any given situation. Bayesian methods make assumptions that some researchers are willing to adopt while others are not. For additional criticisms of Bayesian methods, see Norton (2011).

In fairness, there are limitations to non-Bayesian (frequentist) methods that Bayesian methods attempt to address. Morey et al. (2016) provide more details and they describe how Bayesian methods attempt to deal with the limitations associated with the frequentist methods. Simultaneously, there are limitations associated with Bayesian methods that can be addressed using frequentist (non-Bayesian) techniques. For example, there are no Bayesian analogs for many of the methods described in subsequent chapters. It seems that all inferential methods are limited in what they tell us, or they are highly controversial, or both.

The main point here is that even if one is willing to adopt a Bayesian approach, there is another issue that needs to be taken into account: robustness. Holmes (2014) examines the inherent concerns about the robustness of Bayesian methods and how they might be addressed. One strategy is to model the prior with some heavy-tailed distribution (e.g., Fúquene et al., 2009). But this does not deal with the fundamental issues summarized by Hampel et al. (1986), Huber and Ronchetti (2009), and Staudte and Sheather (1990). One issue is that the population mean and variance are not robust based on three criteria not described here. This is in contrast to the population median, trimmed mean, and the population analog of an M-estimator. And there is the related issue that the sample mean and variance are not robust as well. Generally, robust (non-Bayesian) methods are designed to have desirable properties without specifying a particular family of distributions. This is in contrast to the typical Bayesian approach where a researcher uses some parametric family of distributions. From a robustness point of view, this raises concerns. It is suggested that if a Bayesian approach is used, these issues need to be taken into account.

4.13 EXERCISES

1. Explain the meaning of a 0.95 confidence interval.
2. If you want to compute a 0.80, 0.92, or a 0.98 confidence interval for μ when σ is known, and sampling is from a normal distribution, what values for c should you use in Equation (4.8)?
3. Assuming random sampling is from a normal distribution with standard deviation $\sigma = 5$, if you get a sample mean of $\bar{X} = 45$ based on $n = 25$ participants, what is the 0.95 confidence interval for μ?
4. Repeat the previous example, only compute a 0.99 confidence interval instead.
5. A manufacturer claims that its light bulbs have an average life span of $\mu = 1200$ h with a standard deviation of $\sigma = 25$. If you randomly test 36 light bulbs and

find that their average life span is $\bar{X} = 1150$, does a 0.95 confidence interval for μ suggest that the claim $\mu = 1200$ is unreasonable?

6. For the following situations, (a) $n = 12$, $\sigma = 22$, $\bar{X} = 65$, (b) $n = 22$, $\sigma = 10$, $\bar{X} = 185$, (c) $n = 50$, $\sigma = 30$, $\bar{X} = 19$, compute a 0.95 confidence interval for the mean.
7. Describe the two components of a random sample.
8. If $n = 10$ observations are randomly sampled from a distribution with mean $\mu = 9$ and variance $\sigma^2 = 8$, what is the mean and variance of the sample mean?
9. Suppose you randomly sample $n = 12$ observations from a discrete distribution with the following probability function:

x:	1	2	3	4
p(x):	0.2	0.1	0.5	0.2

   Determine $E(\bar{X})$ and $\sigma^2_{\bar{X}}$.

10. In the previous problem, again suppose you sample $n = 12$ participants and compute the sample mean. If you repeat this process 1000 times, each time using $n = 12$ participants, and if you averaged the resulting 1000 sample means, approximately what would be the result? That is, approximate the average of the 1000 sample means.
11. Answer the same question posed in the previous problem, only replace means with sample variances.
12. Suppose you randomly sample $n = 8$ participants and get

$$2, 6, 10, 1, 15, 22, 11, 29.$$

   Estimate the variance and standard error of the sample mean.

13. If you randomly sample a single observation and get 32, what is the estimate of the population mean, μ? Can you get an estimate of the squared standard error? Explain, in terms of the squared standard error, why only a single observation is likely to be a less accurate estimate of μ versus a sample mean based on $n = 15$ participants.
14. As part of a health study, a researcher wants to know the average daily intake of vitamin E for the typical adult. Suppose that for $n = 12$ adults, the intake is found to be

$$450, 12, 52, 80, 600, 93, 43, 59, 1000, 102, 98, 43.$$

   Estimate the squared standard error of the sample mean.

15. In Exercise 14, verify that there are outliers. What are the effects of these outliers on the estimated squared standard error?
16. For the values

$$6, 3, 34, 21, 34, 65, 23, 54, 23,$$

   estimate the squared standard error of the sample mean.

17. In Exercise 16, if the observations are dependent, can you still estimate the standard error of the sample mean?

18. Chapter 3 described a mixed normal distribution that differs only slightly from a standard normal. Suppose we randomly sample $n = 25$ observations from a standard normal distribution. Then the squared standard error of the sample mean is 1/25. What is the squared standard error if sampling is from the mixed normal instead? What does this indicate about what might happen under slight departures from normality?
19. Explain why knowing the mean and squared standard error is not enough to determine the distribution of the sample mean. Relate your answer to results on nonnormality described in Chapter 3.
20. Suppose $n = 16$, $\sigma = 2$, and $\mu = 30$. Assume normality and determine (a) $P(\bar{X} < 29)$, (b) $P(\bar{X} > 30.5)$, (c) $P(29 < \bar{X} < 31)$.
21. Suppose $n = 25$, $\sigma = 5$, and $\mu = 5$. Assume normality and determine (a) $P(\bar{X} < 4)$, (b) $P(\bar{X} > 7)$, (c) $P(3 < \bar{X} < 7)$.
22. Someone claims that within a certain neighborhood, the average cost of a house is $\mu =$\$100,000 with a standard deviation of $\sigma =$ \$10,000. Suppose that based on $n = 16$ homes, you find that the average cost of a house is $\bar{X} =$ \$95,000. Assuming normality, what is the probability of getting a sample mean this low or lower if the claims about the mean and standard deviation are true?
23. In the previous problem, what is the probability of getting a sample mean between \$97,500 and \$102,500?
24. A company claims that the premiums paid by its clients for auto insurance has a normal distribution with mean $\mu =$ \$750 dollars and standard deviation $\sigma = 100$. Assuming normality, what is the probability that for $n = 9$ randomly sampled clients, the sample mean will have a value between \$700 and \$800?
25. You sample 16 observations from a discrete distribution with mean $\mu = 36$ and variance $\sigma^2 = 25$. Use the central limit theorem to determine (a) $P(\bar{X} < 34)$, (b) $P(\bar{X} < 37)$, (c) $P(\bar{X} > 33)$, (d) $P(34 < \bar{X} < 37)$.
26. You sample 25 observations from a nonnormal distribution with mean $\mu = 25$ and variance $\sigma^2 = 9$. Use the central limit theorem to determine (a) $P(\bar{X} < 24)$, (b) $P(\bar{X} < 26)$, (c) $P(\bar{X} > 24)$, (d) $P(24 < \bar{X} < 26)$.
27. Describe a situation where Equation (4.11), used in conjunction with the central limit theorem, might yield a relatively long confidence interval.
28. Describe a type of continuous distribution where the central limit theorem gives good results with small sample sizes.
29. Compute a 0.95 confidence interval if (a) $n = 10$, $\bar{X} = 26$, $s = 9$, (b) $n = 18$, $\bar{X} = 132$, $s = 20$, (c) $n = 25$, $\bar{X} = 52$, $s = 12$.
30. Repeat the previous exercise, but compute a 0.99 confidence interval instead.
31. Table 4.3 reports data from a study on self-awareness. Compute a 0.95 confidence interval for the mean.
32. Rats are subjected to a drug that might affect aggression. Suppose that for a random sample of rats, measures of aggression are found to be

$$5, 12, 23, 24, 18, 9, 18, 11, 36, 15.$$

Compute a .95 confidence for the mean assuming the scores are from a normal distribution.

33. Describe in general terms how nonnormality can affect Student's T distribution.

34. When sampling from a light-tailed, skewed distribution, where outliers are rare, a small sample size is needed to get good probability coverage, via the central limit theorem, when the variance is known. How does this contrast with the situation where the variance is not known and confidence intervals are computed using Student's T distribution?

35. Compute a 0.95 confidence interval for the 20% trimmed mean if (a) $n = 24$, $s_w^2 = 12$, $\bar{X}_t = 52$, (b) $n = 36$, $s_w^2 = 30$, $\bar{X}_t = 10$, (c) $n = 12$, $s_w^2 = 9$, $\bar{X}_t = 16$.

36. Repeat the previous exercise, but compute a 0.99 confidence interval instead.

37. Compute a 0.95 confidence interval for the 20% trimmed mean using the data in Table 4.3.

38. Compare the length of the confidence interval in the previous problem to the length of the confidence interval for the mean you obtained in Exercise 31. Comment on why they differ.

39. In a portion of a study of self-awareness, Dana (1990) observed the values

$$59, 106, 174, 207, 219, 237, 313, 365, 458, 497, 515,$$

$$529, 557, 615, 625, 645, 973, 1065, 3215.$$

Compare the lengths of the confidence intervals based on the mean and 20% trimmed mean. Why is the latter confidence interval shorter?

40. The ideal estimator of location would have a smaller standard error than any other estimator we might use. Explain why such an estimator does not exist.

41. Under normality, the sample mean has a smaller standard error than the trimmed mean or median. If observations are sampled from a distribution that appears to be normal, does this suggest that the mean should be preferred over the trimmed mean and median?

42. Chapter 2 reported data on the number of seeds in 29 pumpkins. The results were

250, 220, 281, 247, 230, 209, 240, 160, 370, 274, 210, 204, 243, 251, 190, 200, 130, 150, 177, 475, 221, 350, 224, 163, 272, 236, 200, 171, 98.

The 20% trimmed mean is $\bar{X}_t = 220.8$ and the mean is $\bar{X} = 229.2$. Verify that the 0.95 confidence interval for μ is $(200.7, 257.6)$, and that for the 20% trimmed mean it is $(196.7, 244.9)$.

43. In the previous problem, the length of the confidence interval for μ is $257.6 - 200.7 = 56.9$ and the length based on the trimmed mean is $244.9 - 196.7 = 48.2$. Comment on why the length of the confidence interval for the trimmed mean is shorter.

44. If the mean and trimmed mean are nearly identical, it might be thought that it makes little difference which measure of location is used. Based on your answer to the previous problem, why might it make a difference?

45. For 16 presidential elections in the United States, the incumbent party won or lost the election depending on whether the American football team, the Washington Redskins, won their last game just prior to the election. Verify that according to Blythe's method, a 0.99 confidence for the probability of agreement is (0.75, 1).

46. An ABC news program reported that a standard method for rendering patients unconscious resulted in patients waking up during surgery. These individuals were not only aware of their plight, they suffered from nightmares later. Some physicians tried monitoring brain function during surgery to avoid this problem, the strategy being to give the patient more medication if they showed signs of regaining consciousness, and they found that among 200,000 trials, zero patients woke during surgery. However, administrators concerned about cost argued that with only 200,000 trials, the probability of waking up using the new method could not be accurately estimated. Verify that a 0.95 confidence interval for p, the probability of waking up, is (0, 0.000015).

47. Using R, set the R variable val=0. Next, generate 25 values from a binomial distribution where the possible number of successes is 0, 1, . . ., 6 and the probability of success is 0.9. This can be accomplished with the command `rbinom(20,6, 0.9)`. The resulting observations will have values between 0 and 6 inclusive. Compute the median and store it in the R variable `val`. Repeat this process 5000 times. You can use the R command `for(i in 1:5000) val[i]=median(rbinom(20,6, 0.9))`. Speculate about what the plot of the resulting values will look like and check your answer using the R command `splot(val)`.

CHAPTER 5

HYPOTHESIS TESTING

CONTENTS

The approach to computing a confidence interval for the mean, as described in Section 4.3, is a special case of a more general method developed by Laplace about two centuries ago. Two key components are the notion of a sampling distribution and the standard error of an estimator. About a century ago, Jerzy Neyman and Egon Pearson developed a new approach that plays a major role when analyzing data. They used Laplace's notion of a sampling distribution and standard error, but they proposed an alternative framework for making inferences about the population mean (and other parameters) based on what is called hypothesis testing. Roughly, in the simplest case, some speculation is made about some unknown parameter, such as the population mean μ, and the goal is to determine whether the speculation is unreasonable based on the data available.

EXAMPLE

A researcher claims that on a test of open mindedness, the population mean (μ) for adult men is at least 50. As a check on this claim, imagine that 10 adult males ($n = 10$) are randomly sampled and their scores on a measure of open mindedness are

$$25, 60, 43, 56, 32, 43, 47, 59, 39, 41.$$

The sample mean is $\bar{X} = 44.5$. Does this make the claim $\mu \geq 50$ unreasonable? If, for example, it happens to be that $\mu = 50$, by chance we might get a sample mean less than 50, as happened here. How small must the sample mean be in order to conclude that the claim $\mu \geq 50$ is unreasonable?

Chapter 4 touched on how the problem just described might be addressed. If the 0.95 confidence interval for μ happens to be $(40, 48)$, this interval does not contain any value greater than or equal to 50 suggesting that the claim $\mu \geq 50$ might not be true. If the 0.95 confidence interval is $(46, 52)$, this interval contains values greater than 50, suggesting that the claim $\mu \geq 50$ should not be ruled out. The purpose of this chapter is to expand on the topic of making decisions about whether some claim about the population mean, or some other parameter of interest, is consistent with data.

5.1 THE BASICS OF HYPOTHESIS TESTING

For the example just described, a typical way of writing the claim $\mu \geq 50$ more succinctly is

$$H_0 : \mu \geq 50,$$

where the notation H_0 is read H naught. This is an example of a *null hypothesis*, which is just a statement—some speculation—about some characteristic of a distribution. In the example, the null hypothesis is a speculation about the population mean, but it could just as easily be some speculation about the population median, or trimmed mean, or even the population variance, σ^2. If someone claims that the mean is greater than or equal to 60, then our null hypothesis would be written as H_0: $\mu \geq 60$. If there is some reason to speculate that $\mu \leq 20$, and the goal is to see whether this speculation is consistent with observations we make, then the null hypothesis is H_0: $\mu \leq 20$.

The goal is to find a decision rule about whether the null hypothesis should be ruled out based on the available data. When the null hypothesis is rejected, this means you decide that the corresponding alternative hypothesis is accepted. For example, if the null hypothesis is H_0: $\mu \geq 50$, the *alternative hypothesis* is typically written as

$$H_1 : \mu < 50,$$

and if you reject H_0, you in effect accept H_1. That is, you conclude that the mean is less than 50 based on the data available in your study.

In the illustration, a simple rule is to reject the null hypothesis if the sample mean is less than the hypothesized value, 50. If, for example, $\bar{X} = 44.5$, μ is estimated to be less than 50, which suggests that the null hypothesis is false and should be rejected. But if unknown to us $\mu = 50$, there is some possibility of observing a sample mean less than or equal to 44.5. That is, if we reject the null hypothesis and conclude that μ is less than 50 based on this observed sample mean, there is some possibility that our decision is in error.

> *Definition*: A *Type I error* refers to a particular type of mistake, namely, rejecting the null hypothesis when in fact it is correct. A common notation for the probability of a Type I error is α, which is often referred to as the *level of significance*.

We can avoid a Type I error by never rejecting the null hypothesis. In this case, $\alpha = 0$ meaning that the probability of erroneously rejecting the null hypothesis is zero. But a problem with this rule is that it is impossible to discover situations where indeed the null hypothesis is false. If in our illustration $\mu = 46$, the null hypothesis is false and we want a method that will detect this. That is, we need a rule that allows the possibility of rejecting, but simultaneously we want to control the probability of a Type I error.

A natural strategy is to try to determine how small the sample mean must be in order to reject the hypothesis that μ is greater than or equal to 50. But rather than work with $\bar{X}$, it is more convenient to work with

$$Z = \frac{\bar{X} - 50}{\sigma/\sqrt{n}},$$

where for the moment we assume the population standard deviation (σ) is known. Notice that as the sample mean gets smaller, Z gets smaller as well. Moreover, if the sample mean is less than the hypothesized value 50, Z is negative. So the issue of whether the sample mean is sufficiently small to reject the null hypothesis is the same as whether Z is sufficiently small. Using Z is convenient because when sampling from a normal distribution, it provides a simple method for controlling the probability of a Type I error.

More generally, let μ_0 be some specified value and consider the goal of testing

$$H_0 : \mu \geq \mu_0$$

such that the Type I error probability is α. Then

$$Z = \frac{\bar{X} - \mu_0}{\sigma/\sqrt{n}}.$$

Decision Rule: Reject the null hypothesis H_0: $\mu \geq \mu_0$ if

$$Z \leq c,$$

where c is the α quantile of a standard normal distribution, which can be read from Table 1 in Appendix B.

EXAMPLE

For the hypothesis H_0: $\mu \geq 80$, imagine that based on 20 participants ($n = 20$), with $\sigma = 16$, the sample mean is $\bar{X} = 78$ and the Type I error probability is to be 0.025. Then $\mu_0 = 80$, $Z = -0.559$, and as indicated in Chapter 3 the 0.025 quantile of a standard normal distribution is $c = -1.96$, which is read from Table 1 in Appendix B. Because -0.559 is greater than -1.96, fail to reject. That is, the sample mean is less than 80, but you do not have convincing evidence for ruling out the possibility that the population mean is greater than or equal to 80. This does *not* mean, however, that it is reasonable to accept H_0 and conclude that $\mu \geq 50$. (This issue is elaborated upon in Section 5.2.)

Figure 5.1 illustrates the decision rule when $\alpha = 0.05$. If the null hypothesis H_0: $\mu \geq \mu_0$ is true, and in particular $\mu = \mu_0$, then Z has a standard normal distribution as shown in Figure 5.1, in which case the probability that Z is less than -1.645 is the area of the shaded region, which is 0.05.

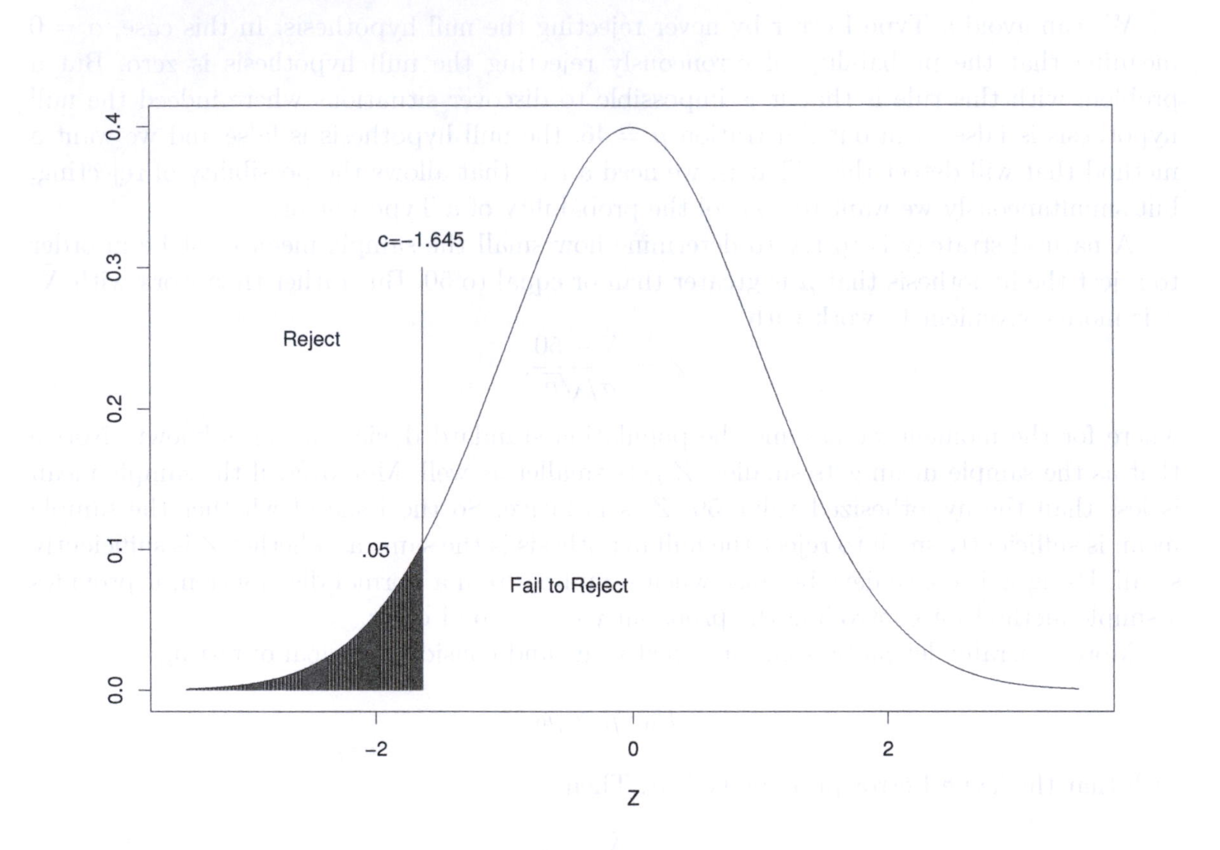

Figure 5.1 If observations are sampled from a normal distribution and the null hypothesis is true, then Z has a standard normal distribution as shown. If the null hypothesis is rejected when $Z \leq -1.645$, the Type I error probability is 0.05.

Definition: A *critical value* is the value used to determine whether the null hypothesis should be rejected.

If it is desired to have a Type I error probability of 0.05 when testing H_0: $\mu \geq \mu_0$, the critical value is $c = -1.645$, meaning that you reject if $Z \leq -1.645$ (assuming normality and that σ is known). The set of all Z values such that $Z \leq -1.645$ is called the *critical region*; it corresponds to the shaded region in Figure 5.1. If you want the probability of a Type I error to be $\alpha = 0.025$, the critical value is -1.96, and the critical region consists of all Z values less than or equal to -1.96. If $\alpha = 0.005$, then the critical value is -2.58 and the critical region is the set of all Z values less than -2.58.

There are two variations of the hypothesis testing method just described. The first is to test

$$H_0 : \mu \leq \mu_0.$$

Decision Rule: Reject if

$$Z \geq c,$$

where now c is the $1 - \alpha$ quantile of a standard normal distribution.

The final variation is to test for exact equality. That is,

$$H_0 : \mu = \mu_0.$$

Now we reject if Z is sufficiently small or sufficiently large. More precisely, let c be the $1-\alpha/2$ quantile of a standard normal distribution.

Decision Rule: Reject if $Z \leq -c$ or if $Z \geq c$, which is the same as rejecting if $|Z| \geq c$.

EXAMPLE

Imagine that you work in the research and development department of a company that helps students train for the SAT examination. After years of experience, it is found that the typical student attending the training course gets an SAT mathematics score of $\mu = 580$ and the standard deviation is $\sigma = 50$. You suspect that the training course could be improved and you want to empirically determine whether this is true. You try the new method on 20 students ($n = 20$) and get a sample mean of $\bar{X} = 610$. For illustrative purposes, assume that the standard deviation is again 50. You have evidence that the new training method is better for the typical student because the estimate of the population mean is 610, which is greater than 580. But you need to convince management, so you assume that the new method is actually worse with the goal of determining whether this assumption can be ruled out based on $\bar{X} = 610$. That is, you decide to test the hypothesis H_0: $\mu \leq 580$; so if you reject, there is empirical evidence suggesting that it is unreasonable to believe that the new method is not beneficial for the typical student. If the Type I error is to be 0.01, then the critical value is $c = 2.23$. Then

$$Z = \frac{\bar{X} - \mu_0}{\sigma/\sqrt{n}} = \frac{610 - 580}{50/\sqrt{20}} = 2.68.$$

Because 2.68 is greater than 2.23, reject and conclude that the mean is greater than 580. That is, you have empirical evidence to present to management that the new training method offers an advantage over the conventional approach, and there is a 0.01 probability that you made a mistake.

EXAMPLE

We repeat the illustration just given, only now imagine you want the probability of a Type I error to be 0.005 instead. From Table 1 in Appendix B, we see that $P(Z \geq 2.58) = 0.005$. This means that if you reject H_0: $\mu \leq 580$ when Z is greater than or equal to 2.58, the probability of a Type I error is $\alpha = 0.005$. As already indicated, $Z = 2.68$, and because this exceeds 2.58, you again reject and conclude that the mean is greater than 580.

EXAMPLE

If null hypothesis H_0: $\mu = \mu_0$ is true and the Type I error is to be 0.05, then the critical value $c = 1.96$ is the .975 quantile of a standard normal distribution. The Type I error probability is

$$P(Z \leq -1.96) + P(Z \geq 1.96) = 0.025 + 0.025 = 0.05,$$

which is the total area of the two shaded regions in Figure 5.2.

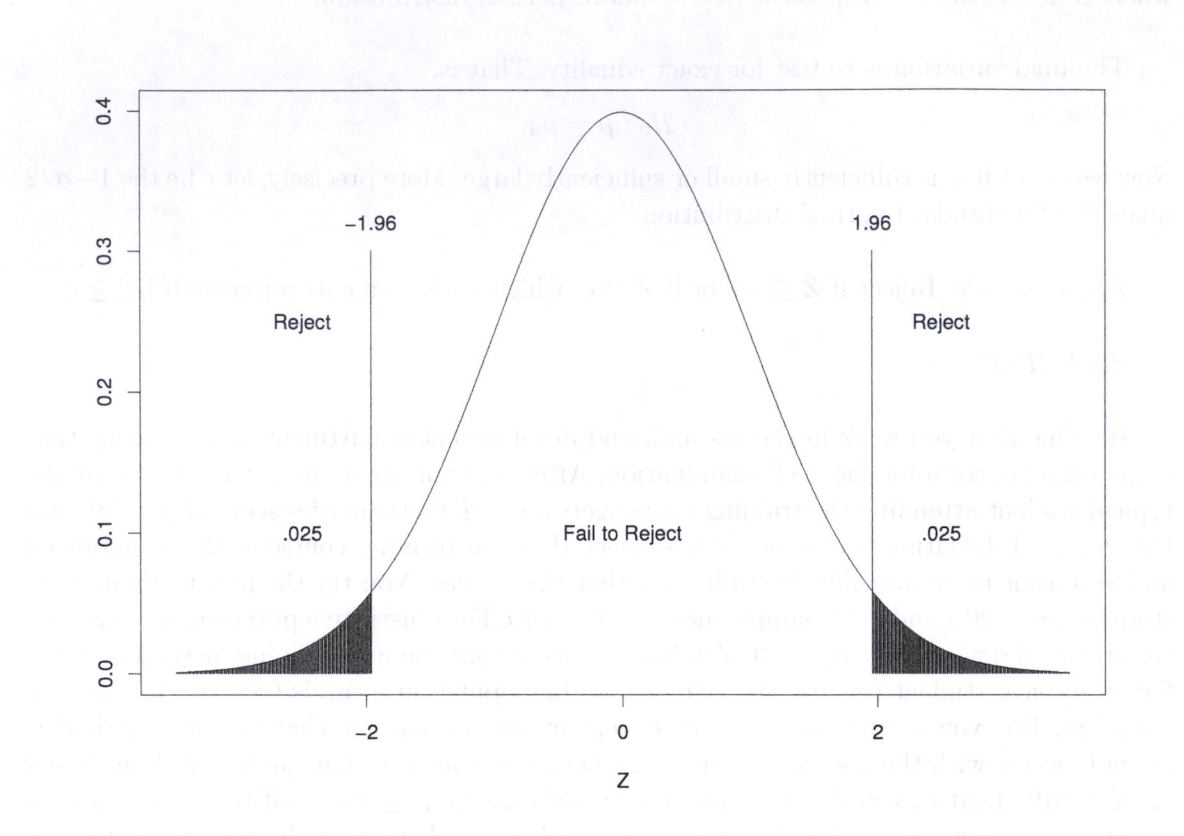

Figure 5.2 For a two-sided test, meaning a test for exact equality, there are two critical regions, one in each tail of the standard normal distribution. Shown are the critical regions when $\alpha = 0.05$.

EXAMPLE

Imagine that for a list of 55 minor malformations babies might have at birth, on average the number of malformations is 15 ($\mu = 15$) and the population standard deviation is $\sigma = 6$. The goal is to determine whether babies born to diabetic women have an average number different from 15. That is, can you reject the hypothesis H_0: $\mu = 15$? To find out, you sample $n = 16$ babies having diabetic mothers, count the number of malformations for each, and find that the average number of malformations is $\bar{X} = 19$. Still assuming that $\sigma = 6$,

$$Z = \frac{19 - 15}{6/\sqrt{16}} = 2.67.$$

If the goal is to have the probability of a Type I error equal to 0.05, then the critical values are -1.96 and 1.96. Because 2.67 is greater than 1.96, reject the null hypothesis and conclude that the average number of malformations is greater than 15.

5.1.1 p-Value or Significance Level

There is an alternative way of describing hypothesis testing that is frequently employed and therefore important to understand. It is based on what is called a *p-value*, sometimes called a *significance level*, an idea that appears to date back to at least the year 1823 in a paper by Laplace (Stigler, 1986, p. 152). The p-value is just the smallest α value for which the null hypothesis is rejected. Alternatively, the p-value is the probability of a Type I error if the observed value of Z is used as a critical value.[1] If you reject when the p-value is less than or equal to 0.05, then the probability of a Type I error will be 0.05, assuming normality. If you reject when the p-value is less than or equal to 0.01, then the probability of a Type I error is 0.01.

EXAMPLE

Again consider the open mindedness example where $\sigma = 12$ and the goal is to test H_0: $\mu \geq 50$. Imagine that you randomly sample $n = 10$ subjects and compute the sample mean, $\bar{X}$. If, for example, you get $\bar{X} = 48$, then

$$Z = \frac{48 - 50}{12/\sqrt{10}} = -0.53.$$

Then the smallest α value for which you would reject is just the probability of a Type I error if you reject when Z is less than or equal to -0.53, which is

$$P(Z \leq -0.53) = 0.298.$$

If you want the probability of a Type I error to be no greater than 0.05, then you would not reject because 0.298 is greater than 0.05. Put another way, if you reject when your test statistic, Z, is less than or equal to -0.53, the probability of a Type I error is 0.298.

One motivation for using the p-value is that it provides more information compared to specifying a Type I error probability (choosing α) and simply reporting whether the null hypothesis is rejected. If you are told that you reject with $\alpha = 0.05$, and nothing else, this leaves open the issue of whether you would also reject with $\alpha = 0.01$. If you are told that the p-value is 0.024, say, then you know that you would reject with $\alpha = 0.05$, or 0.03, or even 0.024, but not $\alpha = 0.01$. If the p-value is 0.003, then in particular you would reject with $\alpha = 0.05$, $\alpha = 0.01$, and even $\alpha = 0.003$.

Computing the p-value When Testing H_0: $\mu \geq \mu_0$

When testing H_0: $\mu \geq \mu_0$, given $\bar{X}$, n, and σ, the p-value is

$$p = P\left(Z \leq \frac{\bar{X} - \mu_0}{\sigma/\sqrt{n}}\right). \tag{5.1}$$

[1] Level of significance refers to α, the Type I error probability specified by the investigator. Consequently, some authorities prefer the term p-value over the expression significance level.

EXAMPLE

If $\bar{X} = 48$, $n = 12$, $\sigma = 12$, and the null hypothesis is H_0: $\mu \geq 50$, then

$$\begin{aligned} p &= P\left(Z \leq \frac{48-50}{12/\sqrt{10}}\right) \\ &= P(Z \leq -0.53) \\ &= 0.298, \end{aligned}$$

which agrees with the previous example.

Computing a p-value When Testing H_0: $\mu \leq \mu_0$

If the null hypothesis is that the mean is less than or equal to some specified value, the p-value is

$$p = P\left(Z \geq \frac{\bar{X}-\mu_0}{\sigma/\sqrt{n}}\right).$$

EXAMPLE

If $\bar{X} = 590$, $\sigma = 60$, $\mu_0 = 580$, and $n = 20$, then the p-value is

$$\begin{aligned} p &= P\left(Z \geq \frac{590-580}{60/\sqrt{20}}\right) \\ &= P(Z \geq 0.745) \\ &= 0.228. \end{aligned}$$

This means that when $\bar{X} = 590$, $Z = 0.745$, and if you reject when $Z \geq 0.745$, the probability of a Type I error is 0.228.

Computing the p-value When Testing H_0: $\mu = \mu_0$

A p-value can be determined when testing H_0: $\mu = \mu_0$, the hypothesis of exact equality, but you must take into account that the critical region consists of both tails of the standard normal distribution. The p-value is

$$p = 2\left(1 - P\left(Z \leq \frac{|\bar{X}-\mu_0|}{\sigma/\sqrt{n}}\right)\right).$$

EXAMPLE

Continuing the last example, where $Z = 2.67$, if you had decided to reject the null hypothesis if $Z \leq -2.67$ or if $Z \geq 2.67$, then the probability of a Type I error is

$$P(Z \leq -2.67) + P(Z \geq 2.67) = 0.0038 + 0.0038 = 0.0076.$$

This means that the p-value is 0.0076. Alternatively,

$$p = 2(1 - P(Z \leq |2.67|)) = 2 \times 0.0038 = 0.0076.$$

5.1.2 Criticisms of Two-Sided Hypothesis Testing and p-Values

Testing for exact equality has been criticized on the grounds that exact equality is impossible. If, for example, one tests H_0: $\mu = 50$, the argument is that surely μ differs from 50 at some decimal place, meaning that the null hypothesis is false and will be rejected with a sufficiently large sample size (e.g., Tukey, 1991). A related criticism is that because H_0 is surely false, the p-value is meaningless.

Assuming that these criticisms have merit, there are at least two ways one might address them. One is to reformulate the goal based on what is called *Tukey's three-decision rule* (cf. Benjamini et al., 1998; Shaffer, 1974). Rather than test H_0: $\mu = \mu_0$, the goal is to determine whether μ is less than or greater than μ_0. Two errors can occur: deciding that $\mu > \mu_0$ when in fact $\mu < \mu_0$, or deciding that $\mu < \mu_0$ when the reverse is true. Suppose we want the probability of making an erroneous decision to be at most α. Again let c be the $1 - \alpha/2$ quantile of a standard normal distribution. Then proceed as follows (assuming sampling from a normal distribution and that σ is known):

1. Conclude that $\mu < \mu_0$ if $Z \leq -c$.
2. If $Z \geq c$, conclude that $\mu > \mu_0$.
3. Otherwise, make no decision.

In this context, the p-value tells us something about the strength of the decision being made about whether μ is less than or greater than the value μ_0. The closer p is to zero, the stronger the empirical evidence is that the decision about μ, based on the sample mean $\bar{X}$, is correct.

A second approach is to rely exclusively on a confidence interval for μ. A confidence interval not only tells us whether we should reject H_0 and conclude that μ is less than or greater than some hypothesized value, it also provides information about the degree to which the population mean differs from the hypothesized value. For example, if the 0.95 confidence interval is (12, 19), we would reject H_0: $\mu = 10$ because this interval does not contain 10. In general, if we compute a $1 - \alpha$ confidence interval for μ, and if we reject whenever this confidence interval does not contain the hypothesized value for μ, the probability of a Type I error is α. Note that a confidence interval provides information not reflected by a p-value. In addition to rejecting $H_0 : \mu = 10$, the 0.95 confidence interval (12, 19) also indicates that we can be reasonably certain that $H_0 : \mu = 11$ would be rejected as well. Or in terms of Tukey's three decision rule, decide that μ is greater than 11. But a p-value provides a succinct way of summarizing the strength of the empirical evidence that the population mean is less than or greater than some specified value. If, for example, the 0.95 confidence interval for the mean is (11,16), we would reject H_0: $\mu = 10$. But if in addition we are told that the p-value is 0.001, we have more information regarding the strength of the empirical evidence that the population mean is greater than 10.

Care must be taken to not read more into a p-value than is warranted. Notice that the expressions for computing a p-value make it clear that it depends on three quantities:

- The difference between the sample mean and the hypothesized value, $\bar{X} - \mu_0$
- the magnitude of the standard deviation, σ
- the value of the sample size, n

Said another way, a p-value is determined by the difference between the sample mean and the hypothesized population mean, $\bar{X} - \mu_0$, as well as the standard error of the sample mean, $\sigma/\sqrt{n}$. The point is that if a p-value is close to zero, and we are told nothing else, it is not necessarily the case that the estimate of the population mean, $\bar{X}$, differs substantially from the hypothesized value. The reason could be that the standard deviation is small relative to the sample size. In particular, a small p-value does *not* necessarily mean that the difference between the hypothesized value of the mean and its actual value, namely $\mu - \mu_0$, is large. (This last issue is discussed in more detail in Chapter 7.) Rather, it quantifies the extent to which empirical evidence supports any conclusions made about whether the population mean is less than or greater than the hypothesized value. If the goal is to test H_0: $\mu = 6$, for example, and if the sample mean is $\bar{X} > 6$, the data suggest that $\mu > 6$. The closer the p-value is to zero, the stronger the evidence that this conclusion is correct, as previously indicated.

A common convention is to describe a test of some hypothesis as *significant* if the p-value is small, say less than or equal to 0.05. But the term significant is a statistical term meaning that there is strong evidence that the null hypothesis should be rejected. Alternatively, in terms of Tukey's three–decision rule when testing H_0: $\mu = \mu_0$, it means that we can be reasonably certain about whether the population mean is less than or greater than the hypothesized value. But as just indicated, it does not necessarily mean that there is an important or large difference between the hypothesized value of the mean and its actual value.

It is noted that if the null hypothesis is true and the probability of a Type I error is controlled exactly, the average p-value over (infinitely) many studies is 0.5. In fact, the p-value has the uniform distribution mentioned in Chapter 4. That is, all p-values are equally likely and are centered around 0.5 (e.g., Sackrowitz and Samuel-Cahn, 1999). In particular, the probability of getting a p-value less than or equal to 0.025 or greater than or equal to 0.975 is exactly 0.05. More generally, when sampling from a non-normal distribution, if a method for testing some hypothesis controls the probability of a Type I error for a sufficiently large sample size (meaning that the central limit theorem applies), then it is approximately the case that the distribution of p has a uniform distribution.

5.1.3 Summary and Generalization

The basics of hypothesis testing, assuming normality and that σ is known, can be summarized in the following manner. Let μ_0 be some specified constant. The goal is to make some inference about how the unknown population mean, μ, compares to μ_0.

Case 1. H_0: $\mu \geq \mu_0$. Reject H_0 if $Z \leq c$, the α quantile of a standard normal distribution. The p-value is

$$p = P\left(Z \leq \frac{\bar{X} - \mu_0}{\sigma/\sqrt{n}}\right), \tag{5.2}$$

which can be determined using Table 1 in Appendix B or the R function `pnorm`.

Case 2. H_0: $\mu \leq \mu_0$. Reject H_0 if $Z \geq c$, the $1 - \alpha$ quantile of a standard normal distribution. The p-value is

$$p = P\left(Z \geq \frac{\bar{X} - \mu_0}{\sigma/\sqrt{n}}\right). \tag{5.3}$$

Case 3. H_0: $\mu = \mu_0$. Reject H_0 if $Z \geq c$ or if $Z \leq -c$, where now c is the $1 - \frac{\alpha}{2}$ quantile of a standard normal distribution. Equivalently, reject if $|Z| \geq c$. The p-value is

$$p = 2\left(1 - P\left(Z \leq \frac{|\bar{X} - \mu_0|}{\sigma/\sqrt{n}}\right)\right). \tag{5.4}$$

In terms of Tukey's three–decision rule, if we fail to reject, make no decision about whether the mean is less than or greater than the hypothesized value.

The hypotheses H_0: $\mu \geq \mu_0$ and H_0: $\mu \leq \mu_0$ are called *one-sided hypotheses* and H_0: $\mu = \mu_0$ is called a *two-sided hypothesis*.

Table 5.1 Four Possible Outcomes When Testing Hypotheses.

	Reality	
Decision	H_0 true	H_0 false
H_0 true	Correct decision	Type II error (probability β)
H_0 false	Type I error (probability α)	Correct decision (power)

5.2 POWER AND TYPE II ERRORS

After years of production, a manufacturer of batteries for automobiles finds that, on average, its batteries last 42.3 months with a standard deviation of $\sigma = 4$. A new manufacturing process is being contemplated and one goal is to determine whether the batteries have a longer life on average. Ten batteries are produced by the new method and their average life is found to be 43.4 months. For illustrative purposes, assume that the standard deviation is again $\sigma = 4$. Based on these $n = 10$ test batteries, it is estimated that the average life of all the batteries produced using the new manufacturing method is greater than 42.3 (the average associated with the standard manufacturing method), in which case the new manufacturing process has practical value. To add support to this speculation, it is decided to test H_0: $\mu \leq 42.3$ versus H_1: $\mu > 42.3$, where μ is the population mean using the new method.

The idea is to determine whether $\bar{X}$ is sufficiently larger than 42.3 to rule out the possibility that $\mu \leq 42.3$. That is, the goal is to determine whether empirical evidence can be found that rules out, to a reasonable degree of certainty, the possibility that the new method is no better and possibly worse on average. If H_0 is rejected, there is empirical evidence that the new method should be adopted. As explained in the previous section, you test this hypothesis by computing $Z = (43.4 - 42.3)/(4/\sqrt{10}) = 0.87$. If you want the probability of a Type I error to be $\alpha = 0.01$, the critical value is 2.33 because $P(Z \leq 2.33) = 0.99$. In the present context, a Type I error is concluding that the new method is better on average when in reality it is not. Because 0.87 is less than 2.33, you fail to reject. Does this imply that you should accept the alternative hypothesis that μ is less than 42.3? In other words, should you conclude that the average battery lasts less than 42.3 months under the new manufacturing method?

Suppose that if the null hypothesis is not rejected, you conclude that the null hypothesis is true and that the population mean is less than 42.3. Then there are four possible outcomes, which are summarized in Table 5.1. The first possible outcome is that the null hypothesis is true and you correctly decide not to reject. The second possible outcome is that the null hypothesis is false, but you fail to reject and therefore make a mistake. That is, your decision that $\mu \leq 42.3$ is incorrect—in reality the mean is greater than 42.3. The third possible outcome is that the null hypothesis is true, but you make a mistake and reject. This is a Type I error, already discussed in Section 5.1. The fourth possible outcome is that in reality $\mu > 42.3$ and you correctly detect this by rejecting H_0.

This section is concerned with the error depicted by the upper right portion of Table 5.1. That is, the null hypothesis is false, but you failed to reject. If, for example, the actual average life of a battery under the new manufacturing method is $\mu = 44$, the correct conclusion is that $\mu > 42.3$. The practical problem is that even if in reality $\mu = 44$, by chance you might get $\bar{X} = 41$ suggesting that the hypothesis H_0: $\mu \leq 42.3$ should be accepted. And even if $\bar{X} > 42.3$, it might be that the sample mean is not large enough to reject even though in reality H_0 is false. Failing to reject when you should reject is called a *Type II error.*

Definition: A *Type II error* is failing to reject a null hypothesis when it should be rejected. The probability of a Type II error is often labeled β.

Definition: *Power* is the probability of rejecting H_0 when in fact it is false. In symbols, power is $1 - \beta$, which is one minus the probability of a Type II error. In the illustration, if the new manufacturing method is actually better, meaning that μ is greater than 42.3, and the probability of rejecting H_0: $\mu \leq 42.3$ is 0.8, say, this means that power is $1 - \beta = 0.8$, and the probability of a Type II error is $\beta = 0.2$.

Power and the probability of making a Type II error are of great practical concern. In the illustration, if $\mu = 44$, the manufacturer has found a better manufacturing method, and clearly it is in its interest to discover this. What is needed is a method for ascertaining power, meaning the probability of correctly determining that the new method is better when in fact $\mu > 42.3$. If power is high, but the company fails to detect an improvement over the standard method of production, the new method can be discarded. That is, there is empirical evidence that H_0 is true and the new method has no practical value. However, if power is low, meaning that there is a low probability of discovering that the new method produces longer-lasting batteries, even when the new method is in fact better, then simply failing to reject does not provide convincing empirical evidence that H_0 is true.

In the present context, power depends on four quantities: σ, α, n, and the value of $\mu - \mu_0$, where μ is the unknown mean of the new manufacturing method. Although μ is not known, you can address power by considering values of μ that are judged to be interesting and important in a given situation. In the illustration, suppose you want to adopt the new manufacturing method if $\mu = 44$. That is, the average life of a battery using the standard method is 42.3 and you want to be reasonably certain of adopting the new method if the average life is now 44. In the more formal terminology of hypothesis testing, you want to test H_0: $\mu \leq 42.3$, and if $\mu = 44$, you want power to be reasonably close to one. What is needed is a convenient way of assessing power given α, σ, μ, μ_0, and the sample size, n, you plan to use.

Summary of How to Compute Power, σ known

Goal: Assuming normality, compute power when testing H_0: $\mu < \mu_0$, or H_0: $\mu > \mu_0$, or H_0: $\mu = \mu_0$ given:

1. n, the sample size,
2. σ, the standard deviation,
3. α, the probability of a Type I error,
4. some specified value for μ, and
5. μ_0, the hypothesized value.

1. *Case 1.* H_0: $\mu < \mu_0$. Determine the critical value c as described in Section 5.1. (The critical value is the $1 - \alpha$ quantile of a standard normal distribution.) Then power, the probability of rejecting the null hypothesis, is

$$1 - \beta = P\left(Z \geq c - \frac{\sqrt{n}(\mu - \mu_0)}{\sigma}\right).$$

In other words, power is equal to the probability that a standard normal random variable is greater than or equal to

$$c - \frac{\sqrt{n}(\mu - \mu_0)}{\sigma}.$$

2. *Case 2.* H_0: $\mu > \mu_0$. Determine the critical value c, which is now the α quantile of a standard normal distribution. Then power is

$$1 - \beta = P\left(Z \leq c - \frac{\sqrt{n}(\mu - \mu_0)}{\sigma}\right).$$

3. *Case 3.* H_0: $\mu = \mu_0$. Now c is the $1 - \frac{\alpha}{2}$ quantile of a standard normal distribution. Power is

$$1 - \beta = P\left(Z \leq -c - \frac{\sqrt{n}(\mu - \mu_0)}{\sigma}\right) + P\left(Z \geq c - \frac{\sqrt{n}(\mu - \mu_0)}{\sigma}\right).$$

EXAMPLE

Continuing the illustration where H_0: $\mu \leq 42.3$, suppose $\alpha = 0.05$, $n = 10$, and the goal is to determine how much power there is when $\mu = 44$. Because $\alpha = 0.05$, the critical value is $c = 1.645$, so

$$\begin{aligned} 1 - \beta &= P\left(Z \geq c - \frac{\sqrt{n}(\mu - \mu_0)}{\sigma}\right) \\ &= P\left(Z \geq 1.645 - \frac{\sqrt{10}(44 - 42.3)}{4}\right) \\ &= P(Z \geq 0.30) \\ &= 0.38. \end{aligned}$$

This says that if battery life has a normal distribution, and, unknown to us, the actual average life of a battery under the new manufacturing method is $\mu = 44$, then the probability of rejecting the hypothesis that the mean is less than 42.3 is 0.38. That is, for this situation where we should reject and conclude that the new manufacturing method is better on average, there is a 0.38 probability of making the correct decision that the null hypothesis is false. Consequently, the probability of committing a Type II error and failing to reject, even though the null hypothesis is false, is $1 - 0.38 = 0.62$.

Figure 5.3 graphically illustrates power when testing H_0: $\mu \leq 42.3$ with $\alpha = 0.05$ and $\mu = 46$. It can be seen that power is $1 - \beta = 0.9$, so the probability of a Type II error is $\beta = 0.1$. The left normal distribution is the distribution of Z when the null hypothesis is true; it is standard normal and you reject if $Z \geq 1.645$, as already discussed. When the null hypothesis is false, and in fact $\mu = 46$, Z still has a normal distribution, but its mean is no longer zero; it is larger as indicated by Figure 5.3. That is, the right distribution reflects the actual distribution of Z when $\mu = 46$. Power is the area under the right (non-null) curve and to the right of the critical value ($c = 1.645$). The area of the shaded region represents the probability of a Type II error, which is 0.1.

Notice that we do not know the actual value of μ, the average life of batteries manufactured with the new method. To deal with this issue, we must ask ourselves a series of questions: what if $\mu = 44$, or 45, or 46, and so on. By computing power for each of these situations, we get some idea about the probability of rejecting when in fact the null hypothesis is false. Figure 5.4 graphs power as μ increases. Notice that the more the mean μ exceeds the hypothesized value of 42.3, the higher the power. This is, of course, a property we want. The larger the difference between the hypothesized value and the actual value of μ, the more likely we are to reject and correctly conclude that μ is greater than 42.3.

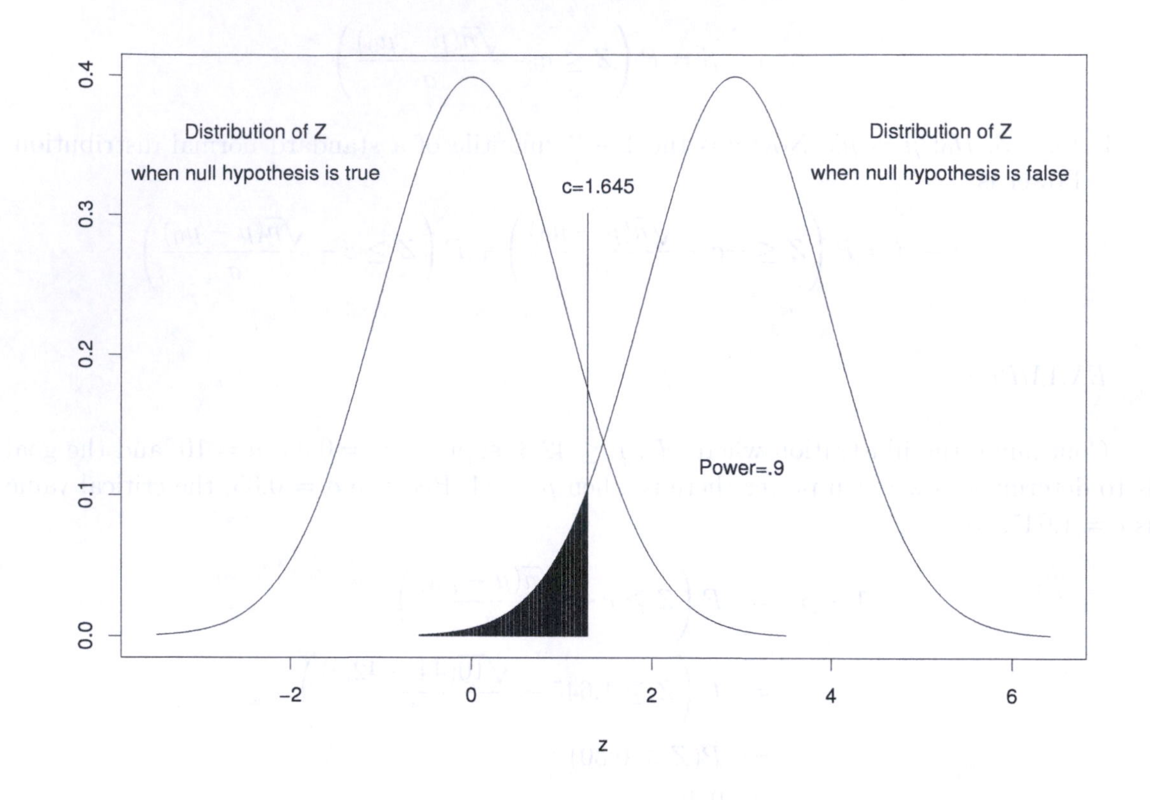

Figure 5.3 An illustration of power. The left distribution is the distribution of Z when the null hypothesis is true. The right distribution is the distribution of Z when the null hypothesis is false. The area of the shaded region, which is the area under the right (non-null) distribution and to the left of 1.645, is 0.1; it is equal to the probability of a Type II error. So power is the area under the non-null distribution and to the right of 1.645.

5.2.1 Understanding How n, α, and σ Are Related to Power

Power is a function of three fundamental components: the sample size, n, the Type I error probability you pick, α, and the population standard deviation, σ. As already explained, power plays a crucial role in applied work, so it is important to understand how each of these quantities is related to power.

Power and the Sample Size

If the null hypothesis is false, we want the probability of rejecting to go up as the sample size increases. That is, as the sample size n gets large, we should have an increasingly higher probability of making a correct decision about H_0 when it is false. Examining the expressions for power reveals that is exactly what happens.

EXAMPLE

Consider the battery example once again where we want to test H_0: $\mu \leq 42.3$ and suppose we want to know how much the power is when $\mu = 44$, but this time we consider three sample

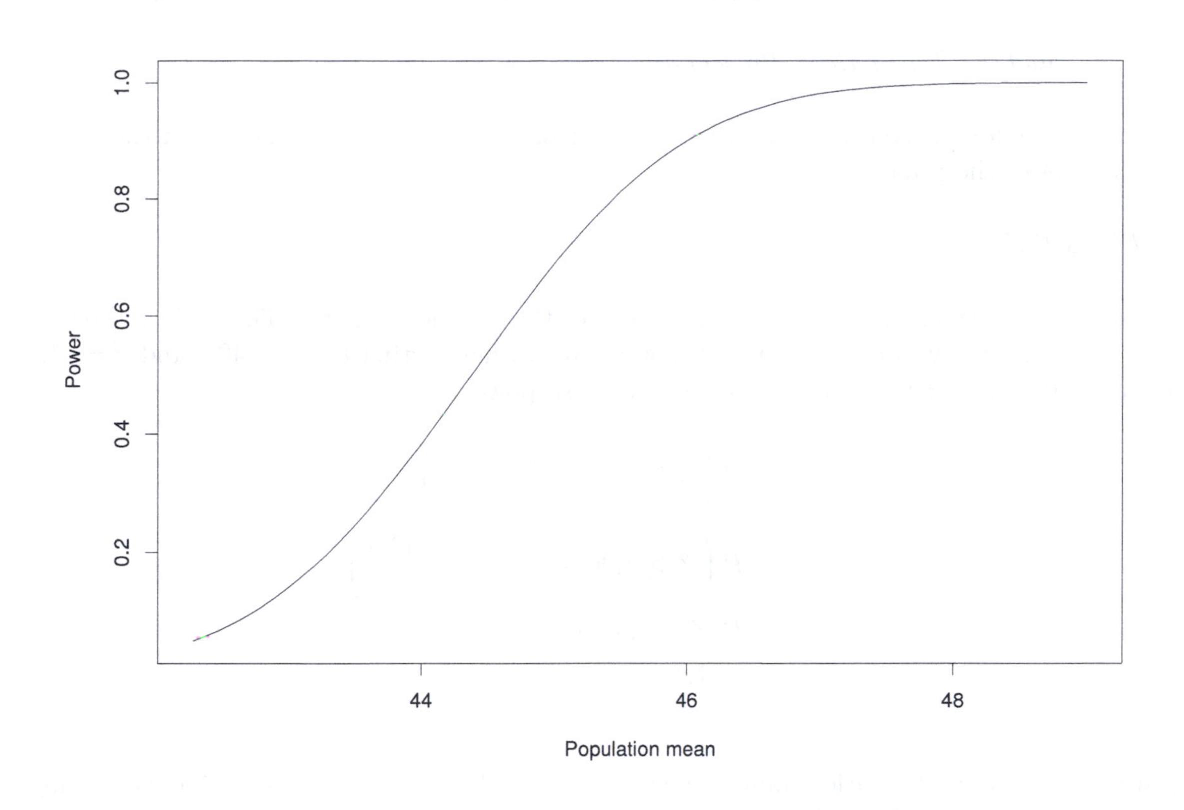

Figure 5.4 Power curve for Z when testing H_0: $\mu \leq 42.3$. As μ increases, meaning that we are moving away from the hypothesized value for μ, power increases.

sizes: 10, 20, and 30. For $n = 10$ we have already seen that power is 0.38. If $n = 20$, then power is now

$$\begin{aligned} 1-\beta &= P\left(Z \geq c - \frac{\sqrt{n}(\mu-\mu_0)}{\sigma}\right) \\ &= P\left(Z \geq 1.645 - \frac{\sqrt{20}(44-42.3)}{4}\right) \\ &= P(Z \geq -0.256) \\ &= 0.60. \end{aligned}$$

Increasing n to 30, it can be seen that power is now $1 - \beta = 0.75$, meaning that your probability of making a correct decision and rejecting H_0 when it is false is now 0.75.

This example illustrates how the sample size might be determined in applied work. First determine the difference between μ and μ_0 that is important in a given situation. In the battery illustration, it might be decided that if $\mu - \mu_0 = 44 - 42.3 = 1.7$, we want to be reasonably certain of rejecting the null hypothesis and deciding that the new manufacturing method is better on average. Next, compute power for some value of n. If the power is judged to be sufficiently large, use this sample size in your study. If not, increase the sample size.

Power and the Type I Error Probability

The choice for α, the probability of a Type I error, also affects power. The lower the value of α, the lower the power.

EXAMPLE

For the battery example with $n = 20$, consider three choices for α: 0.05, 0.025, and 0.01. For $\alpha = 0.05$, we have already seen that power is 0.75 when testing H_0: $\mu < 42.3$ and $\mu = 44$. For $\alpha = 0.025$, the critical value is now $c = 1.96$, so power is

$$\begin{aligned} 1-\beta &= P\left(Z \geq c - \frac{\sqrt{n}(\mu-\mu_0)}{\sigma}\right) \\ &= P\left(Z \geq 1.96 - \frac{\sqrt{20}(44-42.3}{4}\right) \\ &= P(Z \geq 0.059) \\ &= 0.47. \end{aligned}$$

If instead $\alpha = 0.01$, the critical value is now $c = 2.33$ and power can be seen to be 0.33. This illustrates that if we adjust the critical value so that the probability of a Type I error goes down, power goes down as well. Put another way, the more careful you are not to commit a Type I error by choosing α close to zero, the more likely you are to commit a Type II error if the null hypothesis happens to be false.

Power and the Standard Deviation

The higher the standard deviation σ happens to be, the lower the power will be given α, n, and a value for $\mu - \mu_0$.

EXAMPLE

For the battery example, consider $\alpha = 0.05$, $\mu - \mu_0 = 1.7$ and $n = 30$ with $\sigma = 4$. Then, power is 0.75 as previously explained. But if $\sigma = 8$, power is now 0.31. If $\sigma = 12$, power is only 0.19.

In summary:

- As the sample size, n, gets large, power goes up, so the probability of a Type II error goes down.
- As α goes down, in which case the probability of a Type I error goes down, power goes down and the probability of a Type II error goes up.
- As the standard deviation, σ, goes up, with n, α and $\mu - \mu_0$ fixed, power goes down.

Notice that once you have chosen an outcome variable of interest (X) and the population of individuals you want to study, there are two types of factors that affect power. The first type consists of factors that are under your control: n, the sample size, and α, the probability of a Type I error you are willing to allow. By increasing n or α, you increase power. The population standard deviation also affects power, but it is not under your control; it merely reflects a state of nature. (In some situations, the variance of X can be influenced based on how an outcome variable is designed or constructed.) However, understanding how σ affects power is important in applied work because it plays a role in choosing a hypothesis testing method, as will be seen.

Power, p-values and the Probability of Replicating a Significant Result

Another important point is that p-values do not tell us the likelihood of replicating a decision about the null hypothesis. For example, if we get a p-value of 0.001, what is the probability of rejecting at the 0.05 level if we were to conduct the study again? This is a power issue. If, unknown to us, power is 0.2 when testing with $\alpha = 0.05$, then the probability of rejecting again, if the study is replicated, is 0.2. If the null hypothesis happens to be true, and by chance we got a p-value of 0.001, the probability of rejecting again is 0.05 or whatever α value we happen to use. It is a simple matter to derive an estimate of power. But typically these estimates can be highly inaccurate.

5.3 TESTING HYPOTHESES ABOUT THE MEAN WHEN σ IS NOT KNOWN

Next we describe the classic, routinely used method for testing hypotheses about the population mean when the population standard deviation is not known: Student's T test. Then we describe recent insights into when and why this technique has several practical problems followed by methods for dealing with known concerns.

When σ is known, we can test hypotheses about the population mean if we can determine the distribution of

$$Z = \frac{\bar{X} - \mu_0}{\sigma/\sqrt{n}}.$$

When σ is not known, we estimate σ with s, the sample standard deviation, and we can test hypotheses if the distribution of

$$T = \frac{\bar{X} - \mu_0}{s/\sqrt{n}} \tag{5.5}$$

can be determined. As indicated in Chapter 4, the distribution of T can be determined when sampling from a normal distribution. This means that hypotheses can be tested by reading critical values from Table 4 in Appendix B with the degrees of freedom set to $\nu = n - 1$. The details can be summarized as follows.

Goal: Test hypotheses regarding how the population mean, μ, compares to a specified constant, μ_0. The probability of a Type I error is to be α.

Assumptions:

- Random sampling and normality

Decision Rules:

- For H_0: $\mu \geq \mu_0$, reject if $T \leq c$, where c is the α quantile of Student's T distribution with $\nu = n - 1$ degrees of freedom and T is given by Equation (5.5).

- For H_0: $\mu \leq \mu_0$, reject if $T \geq c$, where now c is the $1-\alpha$ quantile of Student's T distribution with $\nu = n-1$ degrees of freedom.
- For H_0: $\mu = \mu_0$, reject if $T \geq c$ or $T \leq -c$, where now c is the $1-\frac{\alpha}{2}$ quantile of Student's T distribution with $\nu = n-1$ degrees of freedom. Equivalently, reject if $|T| \geq c$.

EXAMPLE

For the measures of open mindedness given at the beginning of this chapter, test the hypothesis H_0: $\mu \geq 50$ with $\alpha = 0.05$. The sample standard deviation is $s = 11.4$ and the sample mean is $\bar{X} = 44.5$. Because $n = 10$, the degrees of freedom are $\nu = n - 1 = 9$ and

$$T = \frac{\bar{X} - \mu_0}{s/\sqrt{n}} = \frac{44.5 - 50}{11.4/\sqrt{10}} = -1.5.$$

Referring to Table 4 in Appendix B, $P(T \leq -1.83) = 0.05$, so the critical value is -1.83. This means that if we reject when T is less than or equal to -1.83, the probability of a Type I error will be 0.05, assuming normality. Because the observed value of T is -1.5, which is greater than the critical value, you fail to reject. In other words, the sample mean is not sufficiently smaller than 50 to be reasonably certain that the speculation $\mu \geq 50$ is false. As you can see, the steps you follow when σ is not known mirror the steps you use to test hypotheses when σ is known.

EXAMPLE

Suppose you observe the values

$$12, 20, 34, 45, 34, 36, 37, 50, 11, 32, 29$$

and the goal is to test H_0: $\mu = 25$ such that the probability of a Type I error is $\alpha = 0.05$. Here, $n = 11$, $\mu_0 = 25$, and it can be seen that $\bar{X} = 30.9$, $T = 1.588$, and the critical value is the $1 - \frac{\alpha}{2} = 0.975$ quantile of Student's T distribution with degrees of freedom $\nu = 11 - 1 = 10$. Table 4 in Appendix B indicates that $P(T \leq 2.228) = 0.975$, so our decision rule is to reject H_0 if the value of T is greater than or equal to 2.228 or less than or equal to -2.228. Because the absolute value of T is less than 2.228, you fail to reject. In terms of Tukey's three–decision rule, we fail to make a decision about whether the population mean is greater than or less than 25.

5.3.1 R Function t.test

The R built-in function

```
t.test(x,mu=0,conf.level=0.95),
```

introduced in Chapter 4, tests H_0: $\mu = 0$ using Student's T test. The function reports the value of T as well as the p-value. The argument `mu` indicates the hypothesized value.

EXAMPLE

Chapter 4 mentioned a study dealing with weight gain among rats exposed to an ozone environment. Here we compute a $1 - \alpha = 0.95$ confidence interval for μ. Because there are 22 rats ($n = 22$), the degrees of freedom are $n - 1 = 22 - 1 = 21$. Because $1 - \alpha = 0.95$,

$\alpha = 0.05$, so $\alpha/2 = 0.025$, and $1 - \alpha/2 = 0.975$. Referring to Table 4 in Appendix B, we see that the 0.975 quantile of Student's T distribution with 21 degrees of freedom is approximately $c = 2.08$. The R function `t.test` returns a p-value of 0.013. To test H_0: $\mu = 3$, use the command

```
t.test(x-3).
```

Now the p-value is 0.062.

5.4 CONTROLLING POWER AND DETERMINING THE SAMPLE SIZE

Problems of fundamental importance are determining what sample size to use and finding methods that ensure power will be reasonably close to one. Two approaches are described in this section, both of which assume random sampling from a normal distribution. The first is based on choosing n prior to collecting any data. The second is used after data are available and is aimed at determining whether n was sufficiently large to ensure that power is reasonably high. One fundamental difference between the two methods is how they measure the extent to which the null hypothesis is false.

5.4.1 Choosing n Prior to Collecting Data

First consider how one might choose n prior to collecting data so that power is reasonably close to one. To begin, we need a measure of the difference between the hypothesized value for the mean (μ_0) and its true value (μ). One possibility is

$$\Delta = \mu - \mu_0, \tag{5.6}$$

which is consistent with how we discussed power in Section 5.2 when σ is known. However, when using T, it is impossible to control power given some value for Δ without first obtaining data because when using T, power depends on the unknown variance (Dantzig, 1940). The standard method for dealing with this problem is to replace Δ with

$$\delta = \frac{\mu - \mu_0}{\sigma}. \tag{5.7}$$

So if $\delta = 1$ for example, the difference between the mean and its hypothesized value is one standard deviation. That is, $\mu - \mu_0 = \sigma$. If $\delta = 0.5$, the difference between the mean is a half standard deviation. (That is, $\mu - \mu_0 = 0.5\sigma$.) We saw in Chapter 3 that for normal distributions, σ has a convenient probabilistic interpretation, but even for a small departure from normality, this interpretation breaks down. This creates practical problems when using δ, but we temporarily ignore this issue. (These practical problems are discussed in detail in Chapter 7.) Here it is merely remarked that under normality, power can be determined for any choice of n, δ, and α. Rather than describe the details, we merely describe an R function that performs the computations.

5.4.2 R Function power.t.test

The R function

```
power.t.test(n = NULL, delta = NULL, sd = 1, sig.level = 0.05, power = NULL,
type = c("two.sample", "one.sample", "paired"), alternative = c("two.sided",
                                 "one.sided"))
```

can be used to determine sample sizes and power when using Student's T test. Note that there are three arguments that default to NULL. If two are specified, the third is determined by this function. The argument `type` indicates the version of Student's T to be used. Here, `type="one.sample"` should be used. (The other two choices for `type` are explained in Chapters 7 and 8.)

EXAMPLE

If the goal is to determine how much power when testing H_0: $\mu \leq 15$ with $n = 10$, $\delta = 0.3$, and $\alpha = 0.05$, the R command

```
power.t.test(10,delta=0.3,type="one.sample",
             alternative="one.sided")
```

returns the value 0.221. That is, the power is 0.221. Increasing n to 30, power is now 0.484. With $n = 100$, power is 0.909. So in this particular case, $n = 10$ is inadequate if $\delta = 0.3$ is judged to be a difference that is important to detect. To ensure high power requires a sample size around 100.

5.4.3 Stein's Method: Judging the Sample Size When Data Are Available

Now consider the case where n randomly sampled observations are available. Assuming normality, Stein (1945) derived a (two-stage) method that indicates whether the sample size n is sufficiently large to achieve some specified amount of power. In contrast to the method in Section 5.4.1, it is used after data are collected and it is based on $\Delta = \mu - \mu_0$ rather than δ. Said another way, if you fail to reject some null hypothesis, Stein's method helps you decide whether this is because power is low due to n being too small. Stein's method does even more; it indicates how many additional observations are needed to achieve some specified amount of power.

For convenience, assume that a one-sided test is to be performed. Also assume that n observations have been randomly sampled from some normal distribution yielding a sample variance, s^2. If the goal is to ensure that power is at least $1 - \beta$, then compute

$$d = \left(\frac{\Delta}{t_{1-\beta} - t_\alpha}\right)^2,$$

where $t_{1-\beta}$ and t_α are the $1 - \beta$ and α quantiles of Student's T distribution with $\nu = n - 1$ degrees of freedom. It might help to note that when using Table 4 in Appendix B, $t_\alpha = -t_{1-\alpha}$. That is, typically α is less than 0.5, but Table 4 does not contain the lower quantiles of Student's T distribution. However, they can be determined from Table 4 as just indicated. The required number of observations is

$$N = \max\left(n, \left[\frac{s^2}{d}\right] + 1\right), \tag{5.8}$$

where the notation $[s^2/d]$ means you compute s^2/d and round down to the nearest integer, and max refers to the larger of the two numbers inside the parentheses.

EXAMPLE

Imagine that the Type I error rate is set at $\alpha = 0.01$, $n = 10$ and you want power to be $1 - \beta = 0.9$. The degrees of freedom are $\nu = 9$ and $t_{1-\beta} = t_{0.9} = 1.383$, which can be read from Table 4 in Appendix B. Also $t_\alpha = t_{0.01} = -t_{0.99}$, which is given in Table 4 and is equal to -2.82. If $s = 21.4$ and $\Delta = 20$, then $\nu = 9$,

$$d = \left(\frac{20}{1.383 - (-2.82)}\right)^2 = 22.6,$$

and

$$N = \max(10,\ [(21.4)^2/22.6] + 1) = \max(10, 21) = 21.$$

If $N = n$, the sample size is adequate. Here, $N - n = 21 - 10 = 11$. That is, 11 additional observations are needed to achieve the desired amount of power.

A two-sided test (H_0: $\mu = \mu_0$) is handled in a similar manner. The only difference is that α is replaced by $\alpha/2$. So if in the last example we wanted the Type I error probability to be 0.02 when testing a two-sided test, then again $N = 21$. If we want the Type I error probability to be 0.05, then $t_{\alpha/2} = t_{0.025} = -2.26$, so if again we want power to be 0.9 when $\Delta = \mu - \mu_0 = 20$,

$$d = \left(\frac{20}{1.383 - (-2.26)}\right)^2 = 30.14,$$

so we need a total of

$$N = \max(10,\ [(21.4)^2/30.14] + 1) = 16$$

observations.

Stein also indicated how to test the null hypothesis if the additional $(N-n)$ observations can be obtained. But rather than simply perform Student's T test on all N values, Stein used instead

$$T_s = \frac{\sqrt{N}(\hat{\mu} - \mu_0)}{s}, \tag{5.9}$$

where $\hat{\mu}$ is the mean of all N observations. You test hypotheses by treating T_s as having a Student's T distribution with $\nu = n - 1$ degrees of freedom. That is, you test hypotheses as described in Section 5.3 but with T replaced by T_s. For example, you reject H_0: $\mu \le \mu_0$ if $T_s \ge c$, where c is the $1 - \alpha$ quantile of Student's T distribution with $\nu = n - 1$ degrees of freedom. A two-sided confidence interval for the population mean is given by

$$\hat{\mu} \pm c\frac{s}{\sqrt{N}}, \tag{5.10}$$

where now c is the $1-\alpha/2$ quantile of Student's T distribution with $n-1$ degrees of freedom. For the special case where $N = n$ (meaning that the original sample size was sufficient for your power needs), $T_s = T$ and you are simply using Student's T test, but with the added knowledge that the sample size meets your power requirements. What is unusual about Stein's method is that if $N > n$, it uses the sample variance of the original n observations—not the sample variance of all N observations. Also, the degrees of freedom remain $n-1$ rather than the seemingly more natural $N - 1$. By proceeding in this manner, Stein showed that power will be at least $1 - \beta$ for whatever value of $1 - \beta$ you pick. (Simply performing Student's T test using all N values, when $N > n$, results in certain technical problems that are described by Stein, 1945. For a survey of related methods, see Hewett and Spurrier, 1983.)

An alternative to Stein's method, when trying to assess power when a nonsignificant result is obtained, is based on what is called *observed power*. The approach assumes that the observed difference between the sample mean and its hypothesized value is indeed equal to the true difference $(\mu - \mu_0)$, it assumes that $s^2 = \sigma^2$, and then based on these assumptions one computes power. Hoenig and Heisey (2001) illustrate that this approach is generally unsatisfactory.

5.4.4 R Functions stein1 and stein2

The R function

```
stein1(x,del,alpha=0.05,pow=0.8,oneside=FALSE)
```

returns N, the sample size needed to achieve the power indicated by the argument `pow` (which defaults to 0.8), given some value for Δ (which is the argument `del`) and α. The function assumes that a two-sided test is to be performed. For a one-sided test, set the argument `oneside` equal TRUE.

The R function

```
stein2(x1,x2,mu0=0,alpha=0.05)
```

tests the hypothesis H_0: $\mu = \mu_0$ using Stein's method assuming that the initial n observations are stored in x1, and that the additional $N - n_1$ observations are stored in x2. The argument `mu0` is the hypothesized value, μ_0, which defaults to 0.

EXAMPLE

The example at the end of Section 5.3 used Student's T to test H_0: $\mu = 25$ with $\alpha = 0.05$, based on the data

12, 20, 34, 45, 34, 36, 37, 50, 11, 32, 29.

A nonsignificant result was obtained. If we want power to be 0.9 when $\mu = 28$, in which case $\Delta = 28 - 25 = 3$, was the sample size sufficiently large? Storing these data in the R variable y, the command `stein1(y,3,pow=0.9)` returns the value 220. That is, we need $N = 220$ observations to achieve this much power, but we only have 11 observations, so $220 - 11 = 209$ additional observations are needed. For a one-sided test, $N = 94$.

5.5 PRACTICAL PROBLEMS WITH STUDENT'S T TEST

Student's T test deals with the common situation where σ is not known, but it assumes observations are randomly sampled from a normal distribution. Because distributions are never exactly normal, it is important to understand how nonnormality affects conclusions based on T. An extension of the central limit theorem (Slutsky's theorem) tells us that as n gets large, the distribution of T becomes more like Student's T distribution, and in fact its distribution approaches a standard normal. That is, if the sample size is large enough, and observations are randomly sampled, then violating the normality assumption is not a serious concern. Conventional wisdom is that assuming T has a Student's T distribution with $\nu = n - 1$ degrees of freedom provides reasonably accurate results with n fairly small and surely accurate results are obtained with $n = 100$. However, in recent years, much more sophisticated methods have been derived for understanding how nonnormality affects T, and serious concerns have been discovered, two of which are described here.

The first is that *very* small departures from normality can drastically reduce power. The main reason is that even small departures from normality can inflate the population variance. This in turn inflates the standard error of the sample mean, so power can be relatively low. As indicated in Section 5.2.1, as σ gets large, power goes down when using Z to test hypotheses, and the same is true when using T. (One of the earliest results indicating theoretical concerns about this problem can be found in Bahadur and Savage, 1956.) More generally, any situation where outliers are likely to occur has the potential of low power when using Student's T test, relative to other methods that might be used. (This last point is illustrated later in this chapter.)

EXAMPLE

As an illustration, first consider using Student's T to test H_0: $\mu = 0$ with $\alpha = 0.05$ when sampling from the normal distribution shown in Figure 5.5, which has mean 0.8 and standard deviation $\sigma = 1$. It can be shown that with $n = 20$, power is 0.93. That is, there is a 93% chance of correctly rejecting the null hypothesis that the mean is $\mu = 0$. Now suppose that sampling is from the other distribution shown in Figure 5.5. This is the mixed normal distribution described in Chapter 2 but with mean 0.8. This distribution is very similar to the normal distribution, in the sense described in Chapter 2, yet power is now 0.39. This demonstrates that if you test hypotheses with Student's T or any method based on the sample mean, small departures from normality can result in a substantial decrease in your ability to detect situations where the null hypothesis is false.

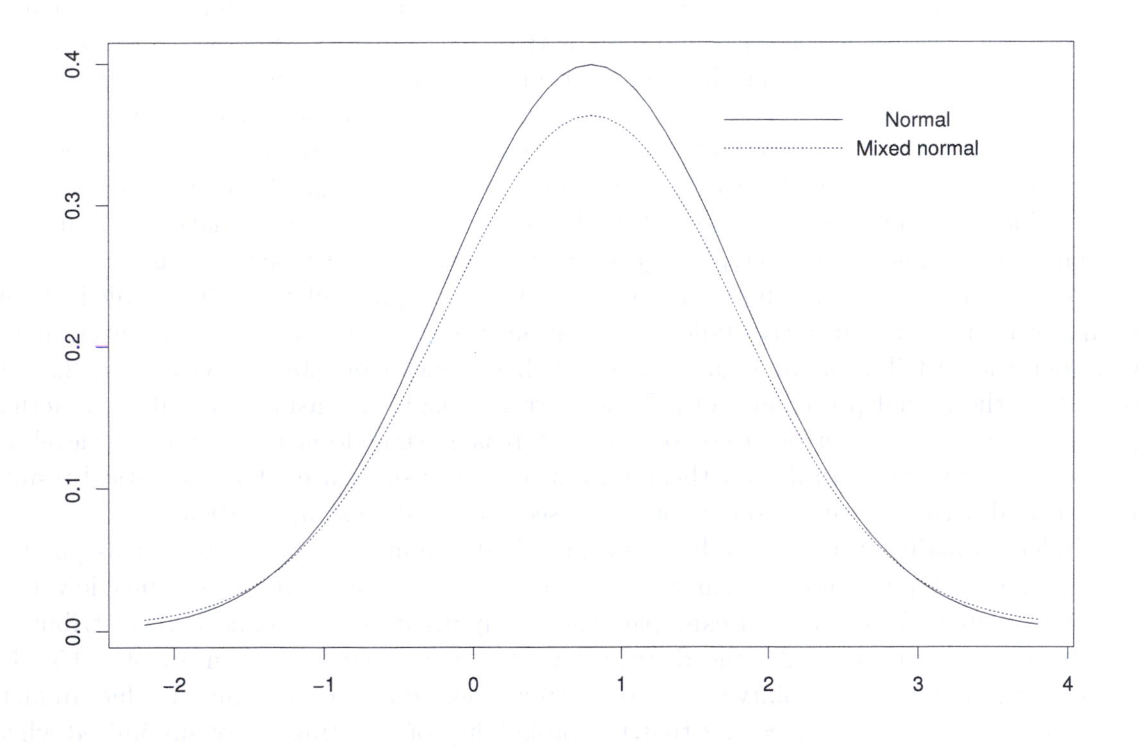

Figure 5.5 Small departures from normality can greatly reduce power when using Student's T. If we sample 20 observations from the normal distribution shown here, which has mean 0.8, and we test H_0: $\mu = 0$ at the 0.05 level, power is 0.93. But if we sample from the mixed normal instead, power is only 0.39.

The second problem is that nonnormality can affect your ability to control the probability of a Type I error or control the probability coverage when computing a confidence interval. First consider situations where sampling is from a perfectly symmetric distribution and imagine you want the probability of a Type I error to be 0.05. When sampling is from a normal distribution, you can accomplish your goal with Student's T test, as already demonstrated. However, if you happen to be sampling observations from a mixed normal instead, the actual probability of a Type I error is only 0.022 with $n = 20$. This might seem desirable because the probability of incorrectly rejecting when the null hypothesis is true is less than the nominal level of 0.05. However, recall that the smaller α happens to be, the lower your power. This means that the probability of correctly rejecting the null hypothesis when it is false might be low, contrary to what you want, because you are inadvertently testing at the 0.022 level. Increasing the sample size to 100, now the actual probability of a Type I error is 0.042, but low power remains a possible concern because sampling is from a heavy-tailed distribution.

If observations are sampled from a skewed distribution, the actual probability of a Type I error can be substantially higher or lower than 0.05. With $n = 12$, there are situations where the actual probability of a Type I error is 0.42 when testing at the 0.05 level and it can be as low as 0.001 when testing a one-sided test at the 0.025 level (e.g., Wilcox, 2017).

As another illustration, imagine you are interested in how response times are affected by alcohol and that, *unknown to you*, response times have the skewed (lognormal) distribution shown in Figure 4.10 of Chapter 4, which has mean $\mu = 1.65$. Further imagine that you want to test the hypothesis H_0: $\mu \geq 1.65$ with $\alpha = 0.05$. That is, unknown to you, the null hypothesis happens to be true, so you should not reject. With $n = 20$ observations, the actual probability of a Type I error is 0.14. That is, your intention was to have a 5% chance of rejecting in the event the null hypothesis is true, but in reality there is a 14% chance of rejecting by mistake. Increasing the sample size to $n = 160$, the actual probability of Type I error is now 0.11. That is, control over the probability of a Type I error improves as the sample size gets large, in accordance with the central limit theorem, but at a rather slow rate. Even with $n = 160$, the actual probability of rejecting might be more than twice as large as intended. The seriousness of a Type I error will depend on the situation, but at least in some circumstances the discrepancy just described would be deemed unsatisfactory.

The lognormal distribution used in the previous paragraph is relatively light-tailed. When sampling from a skewed, heavy-tailed distribution, Student's T can deteriorate even more. Consider the distribution in Figure 5.6 which has a mean of 0.0833. With $n = 20$ and $\alpha = 0.05$, the actual probability of a Type I error is 0.20. Increasing n to 100, the actual probability of a Type I error drops to only 0.19. It is getting closer to the nominal level, in accordance with the central limit theorem, but at a very slow rate. (For theoretical results indicating drastic sensitivity to nonnormality, see Basu and DasGupta, 1995.)

Under normality, Student's T has a symmetric distribution about zero. But as pointed out in Chapter 4, under nonnormality, the actual distribution of T can be asymmetric with a mean that differs from zero. For example, when sampling from the (lognormal) distribution in Figure 4.14, with $n = 20$, the distribution of T is skewed with a mean of -0.5, the result being that the probability of a Type I error is not equal to the nominal value. In fact, Student's T test is *biased* meaning that the probability of rejecting is not minimized when the null hypothesis is true. That is, situations arise where there is a higher probability of rejecting when the null hypothesis is true than in a situation where the null hypothesis is false. (Generally, any hypothesis testing method is said to be unbiased if the probability of rejecting is minimized when the null hypothesis is true. Otherwise it is biased.)

To illustrate the possible effect of skewness on the power curve of Student's T, suppose we sample 20 observations from the (lognormal) distribution shown in Figure 4.14 which has a population mean approximately equal to 1.649. So if we test H_0: $\mu = 1.649$ with $\alpha = 0.05$, the intention is to have the probability of a Type I error equal to 0.05, but in reality it is approximately 0.14. Now suppose we add Δ to every observation. In effect, we shift the distribution in Figure 4.14 so that now its mean is $1.649 + \Delta$. If, for example, we use $\Delta = 0.3$,

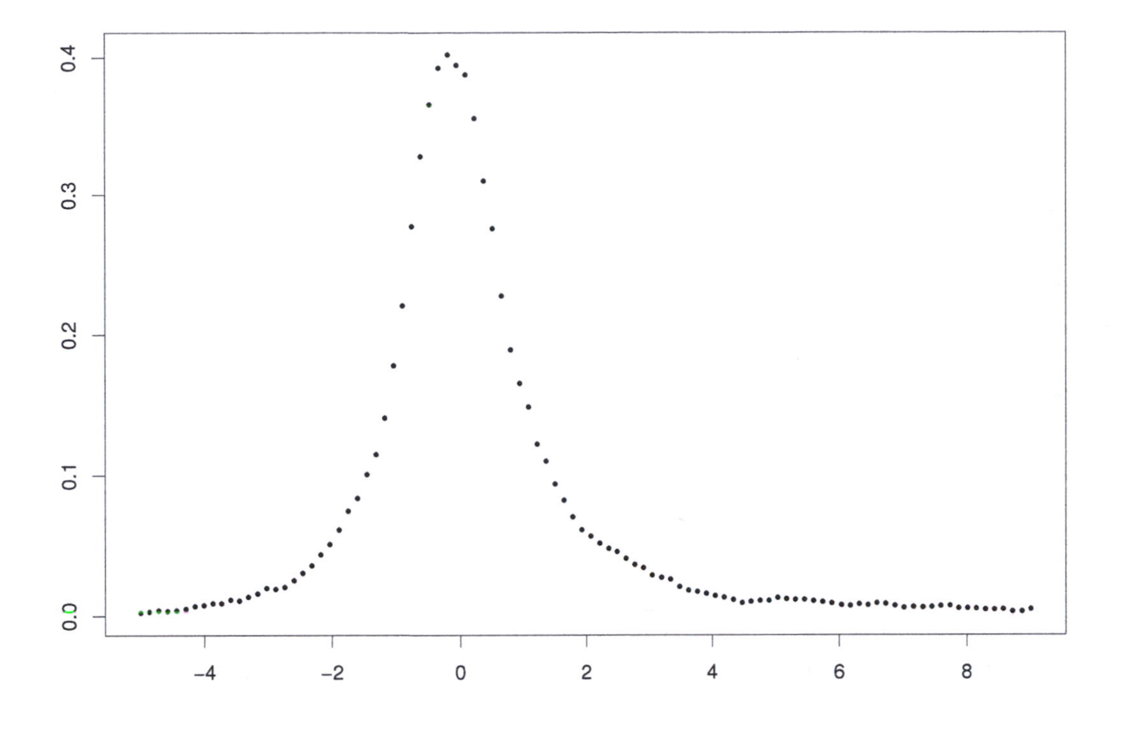

Figure 5.6 An example of a skewed, heavy-tailed distribution used to illustrate the effects of nonnormality on Student's T. This distribution has a mean of 0.8033.

we have in effect increased the population mean from 1.649 to 1.949. So now we should reject H_0: $\mu = 1.649$, but the actual probability of rejecting it is 0.049. That is, we have described a situation where the null hypothesis is false, yet we are less likely to reject than in the situation where the null hypothesis is true. If we set $\Delta = 0.6$, so now $\mu = 2.249$, and we again test H_0: $\mu = 1.649$, the probability of rejecting is 0.131, approximately the same probability of rejecting when the null hypothesis is true. As we increase Δ even more, power continues to increase as well. Figure 5.7 shows the power curve of Student's T for Δ ranging between 0 and 1.

Section 5.2.1 summarized factors that are related to power when using Z. Now we summarize factors that influence power for the more common situation where T is used to make inferences about the population mean. These features include the same features listed in Section 5.2.1 plus some additional features related to nonnormality.

- As the sample size, n, gets large, power goes up, so the probability of a Type II error goes down.
- As α goes down, in which case the probability of a Type I error goes down, power goes down and the probability of a Type II error goes up.
- As the standard deviation, σ, goes up, with n, α, and $\mu - \mu_0$ fixed, power goes down.
- Small departures from normality can inflate the standard error of the sample mean $(\sigma/\sqrt{n})$, which in turn can substantially reduce power.

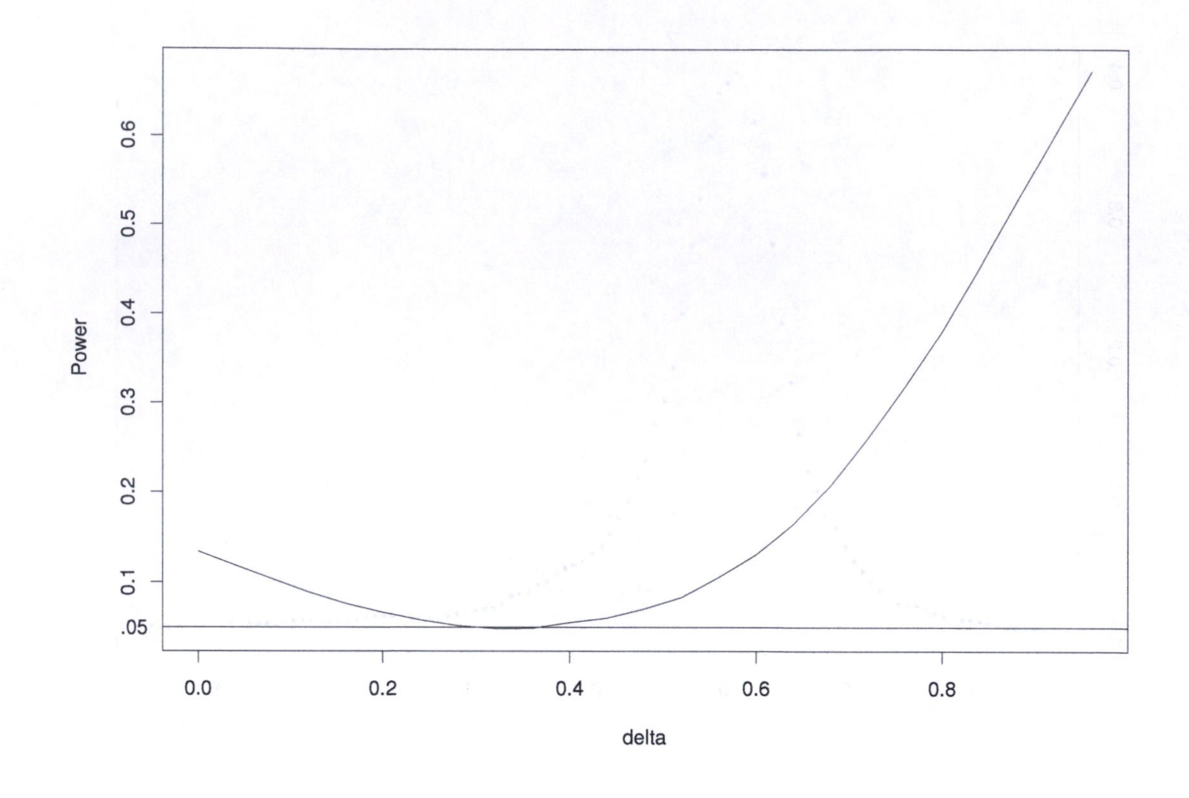

Figure 5.7 Power curve of Student's T when sampling from a lognormal distribution, $n = 20$, $\alpha = 0.05$. The null hypothesis corresponds to delta=0. Ideally the power curve should be strictly increasing as delta gets large.

- Student's T can be biased due to skewness. That is, power might be low (relative to other inferential methods you might use) because as μ moves away from the hypothesized value, the probability of rejecting can actually decrease. Practical problems arise even when sampling from a distribution where outliers are rare.

5.6 HYPOTHESIS TESTING BASED ON A TRIMMED MEAN

An argument for testing hypotheses based on the mean is that under normality, the sample mean has a smaller standard error than any other measure of location we might use. This means that no other hypothesis testing method will have more power than the method based on Student's T. However, this argument is not very compelling because arbitrarily small departures from normality can result in extremely low power relative to other methods you might use, and, of course, there is the concern about getting accurate confidence intervals and good control over the probability of a Type I error. Currently, there seem to be two general strategies for dealing with these problems that are relatively effective. The first is to switch to a robust measure of location, and the trimmed mean is particularly appealing based on published studies. The second is to switch to a rank-based method, some of which are described in subsequent chapters.

Here, attention is focused on the 20% trimmed mean. The method for computing a confidence interval for the trimmed mean described in Chapter 4 is easily extended to the problem of testing some hypothesis about μ_t, the population trimmed mean. The process is the same as Student's T test, only adjust the degrees of freedom, replace the sample mean with the trimmed mean, replace the sample variance with the Winsorized variance, and for technical reasons, multiply by 0.6. In symbols, the test statistic is now

$$T_t = \frac{0.6(\bar{X}_t - \mu_0)}{s_w/\sqrt{n}}, \tag{5.11}$$

where again μ_0 is some specified value of interest and s_w is the 20% Winsorized standard deviation. Then reject H_0: $\mu_t = \mu_0$ if $T_t \leq -c$ or $T_t \geq c$, where now c is the $1 - \frac{\alpha}{2}$ quantile of Student's T distribution with $n - 2g - 1$ degrees of freedom, and g is the number of observations trimmed from each tail, as described and illustrated in Chapter 3. (The total number of trimmed observations is $2g$, so $n - 2g$ is the number of observations left after trimming.)

More generally, when using a γ-trimmed mean,

$$T_t = \frac{(1 - 2\gamma)(\bar{X}_t - \mu_0)}{s_w/\sqrt{n}}, \tag{5.12}$$

where now s_w is the γ-Winsorized standard deviation. (In the previous paragraph, $\gamma = 0.2$.) The degrees of freedom are $\nu = n - 2g - 1$ where $g = [\gamma n]$ and $[\gamma n]$ is γn rounded down to the nearest integer.

As for the one-sided hypothesis H_0: $\mu \geq \mu_0$, reject if $T_t \leq c$, where now c is the α quantile of Student's T distribution with $n - 2g - 1$ degrees of freedom. The hypothesis H_0: $\mu \leq \mu_0$ is rejected if $T_t \geq c$, where now c is the $1 - \alpha$ quantile of Student's T distribution with $n - 2g - 1$ degrees of freedom.

5.6.1 R Function trimci

The R function

```
trimci(x,tr=0.2,alpha=0.05,nv=0),
```

introduced in Chapter 4, also tests hypotheses about the population trimmed mean. By default it tests H_0: $\mu_t = 0$. To test H_0: $\mu_t = 2$, set the argument `nv` equal to 2. In addition to a confidence interval, the function returns a p-value.

EXAMPLE

Doksum and Sievers (1976) report data on weight gain among rats. One group was the control and the other was exposed to an ozone environment. (The data are given in Table 7.6.) For illustrative purposes, attention is focused on the control group and we consider the claim that the typical weight gain is 26.4. If we test the hypothesis H_0: $\mu = 26.4$ with Student's T, we get

$$T = \frac{\bar{X} - \mu_0}{s/\sqrt{n}} = \frac{22.4 - 26.4}{10.77/\sqrt{23}} = -1.8.$$

With $\nu = 23 - 1 = 22$ degrees of freedom, and $\alpha = 0.05$, the critical value is $c = 2.07$. Because $|T| = 1.8$ is less than 2.07, we fail to reject. In contrast, the 20% trimmed mean is $\bar{X}_t = 23.3$, $s_w = 3.9$, and for H_0: $\mu_t = 26.4$ we see that

$$T_t = \frac{0.6(\bar{X}_t - \mu_0)}{s_w/\sqrt{n}} = \frac{0.6(23.3 - 26.4)}{3.9/\sqrt{23}} = -2.3.$$

Because there are 23 rats, $g = 4$, so the number of trimmed observations is $2g = 8$, the degrees of freedom are $\nu = 23 - 8 - 1 = 14$, and the critical value is $c = 2.14$. Because $|T_t| = |-2.3| = 2.3$ is greater than the critical value, reject the hypothesis that the trimmed mean is 26.4. In terms of Tukey's three decision rule, decide that the population 20% trimmed mean is less than 26.4.

In the last example, the sample mean exceeds the hypothesized value by more than the sample 20% trimmed mean. The difference between the hypothesized value of 26.4 and the mean is $22.4 - 26.4 = -4$. The difference between the hypothesized value and the trimmed mean is $23.3 - 26.4 = -3.1$, yet you reject with the trimmed mean but not with the mean. The reason is that the standard error of the trimmed mean is smaller than the standard error of the mean. This illustrates one of the practical advantages of using a trimmed mean. Situations often arise where the trimmed mean has a substantially smaller standard error, and this can translate into a substantial gain in power.

Once again it is stressed that for skewed distributions, population means and trimmed means are generally not equal. So, for example, the null hypothesis H_0: $\mu = 26.4$ is not necessarily the same as H_0: $\mu_t = 26.4$. In the context of hypothesis testing, an argument for the trimmed mean is that good control over the probability of a Type I error can be achieved in situations where Student's T gives poor results. Trimmed means often have a smaller standard error than the mean which can result in substantially higher power. If there is some reason for preferring the mean to the trimmed mean in a particular study, Student's T might be unsatisfactory unless the sample size is very large. Just how large n must be depends on the *unknown* distribution from which observations were sampled. In some cases, even a sample size of 300 is unsatisfactory. A small sample size will suffice in some instances, but in general it is difficult whether this is the case simply by examining the data.

In this book, only two-sided trimming is considered. If a distribution is skewed to the right, for example, a natural reaction is to trim large observations but not small ones. An explanation can now be given as to why one-sided trimming is not recommended. In terms of Type I errors and probability coverage, you get more accurate results if two-sided trimming is used. There is nothing obvious or intuitive about this result, but all of the studies cited by Wilcox (2017) support this view.

Thanks to a generalization of the central limit theorem, we know that when working with means, problems with Student's T diminish as the sample size increases. Theory and simulations indicate that when using a 20% trimmed mean instead, problems diminish much more rapidly. That is, smaller sample sizes are required to get good control over the probability of a Type I error, but with very small sample sizes, practical problems persist. (Bootstrap methods covered later in this chapter provide a basis for dealing with very small sample sizes in an effective manner.)

Finally, no mention has been made about how to determine whether power is adequate based on the sample size used when making inferences about a trimmed mean. One approach is to use an extension and slight variation of Stein methods (Wilcox, 2004b; 2017, section 8.1.6). The computational details are omitted but an R function for applying the method is provided. An alternative approach is described in Section 5.13.

5.6.2 R Functions stein1.tr and stein2.tr

The R function

```
stein1.tr(x, del, alpha = 0.05, pow = 0.8, tr = 0.2)
```

determines the sample size needed to achieve the power indicated by the argument `pow`, when testing the hypothesis H_0: $\mu_t = 0$, when in reality the difference between μ_t and its hypothesized value is equal to the value indicated by the argument `del`. If the additional observations needed to achieve the desired amount of power can be obtained, and are stored in the R variable y, then

```
stein2.tr(x, y, alpha = 0.05, tr = 0.2)
```

will test H_0: $\mu_t = 0$.

5.7 TESTING HYPOTHESES ABOUT THE POPULATION MEDIAN

For reasons outlined in Chapter 4, although the median belongs to the class of trimmed means, special methods for testing hypotheses about the population median are required. To test

$$H_0 : \theta = \theta_0,$$

where θ is the population median and θ_0 is some specified value, use the R function `sint` described in Section 4.9.1. If the confidence interval returned by this function does not contain θ_0, reject H_0. To get a p-value, use the R function described next.

5.7.1 R Function sintv2

The R function

```
sintv2(x,alpha=0.05,nullval=0)
```

returns a p-value when testing H_0: $\theta = \theta_0$, where the argument `nullval` indicates the value of θ_0, which defaults to 0.

5.8 MAKING DECISIONS ABOUT WHICH MEASURE OF LOCATION TO USE

When considering which measure of location to use, one suggestion is to first consider the extent to which the mean remains satisfactory when dealing with skewed distributions. Despite any negative features associated with the mean, even when a distribution is skewed, situations are encountered where the mean is more appropriate than the median or some lower amount of trimming, as indicated in Chapter 2.

Student's T avoids Type I errors well above the nominal level if sampling is from a symmetric distribution. A seemingly natural suggestion is to test the hypothesis that a distribution is symmetric or even normal. Many such methods have been proposed. But this strategy is satisfactory only if such tests have a reasonably high probability of detecting situations where Student's T results in poor control over the Type I error probability. That is, the power of these methods must be sufficiently high to detect a violation of the normality (or symmetry) assumption that can result in invalid conclusions. It is unclear how to establish that tests of normality (or symmetry) have sufficiently high power. A broader approach is to choose a parametric family of distributions to be used based on the available data, which includes the possibility of choosing the family of normal distributions. But results reported by Keselman et al. (2016) do not support this approach. (Section 5.10 describes some possible ways of making more valid inferences about the population mean when dealing with skewed distributions, but as will be seen, practical concerns remain.)

An argument for a 20% trimmed mean is that under normality, little power is lost compared to using Student's T, power can be much higher when dealing with heavy-tailed distributions, and it performs better than Student's T, in terms of Type I errors, when sampling from a skewed distribution. A criticism is that when outliers are sufficiently common, using a median might have more power, keeping in mind that for skewed distributions, the population median and 20% trimmed mean, for example, will differ. Of course, if there is explicit interest in the median, use the R function `sintv2`.

5.9 BOOTSTRAP METHODS

Bootstrap methods provide a way of testing hypotheses and computing confidence intervals without assuming normality that provide major advantages over many competing techniques. For some situations, a bootstrap method is the only known technique that performs reasonably well in terms of Type I errors and achieving accurate confidence intervals. There are, in fact, many bootstrap techniques, but only the most basic methods are described here. It is noted, however, that bootstrap methods are not a panacea for dealing with the many practical problems encountered when trying to understand data. In particular, they are not distribution free, where a distribution free method means that the probability of a Type I error can be controlled assuming random sampling only. When the goal is to test hypotheses about the mean, bootstrap methods can reduce practical concerns associated with Student's T, but not all concerns are eliminated as will be illustrated. However, when combined with robust measures of location, such as a trimmed mean or median, bootstrap methods can offer substantial practical advantages over non-bootstrap methods as will be demonstrated. When studying the association between two or more variables, again bootstrap methods play an invaluable role as will be seen in Chapters 6 and 14. For book-length descriptions of bootstrap methods, see Efron and Tibshirani (1993), Chernick (1999), Davison and Hinkley (1997), Hall and Hall (1995), Lunneborg (2000), Mooney and Duval (1993), and Shao and Tu (1995). Empirical likelihood methods (e.g., Owen, 2001; cf. Vexler et al., 2009) are a competitor of bootstrap methods, but currently it seems that they offer little or no advantage and in some cases their performance is relatively poor when dealing non-normal distributions.

5.10 BOOTSTRAP-T METHOD

One of the most basic bootstrap methods is the bootstrap-t, sometimes called the percentile-t method. Consider the goal of testing H_0: $\mu = \mu_0$ when sampling from some unknown, non-normal distribution. (Again, μ_0 is some specific value of interest chosen by the investigator.) To convey the fundamental idea behind the bootstrap-t method, it helps to digress momentarily and describe how simulation studies can be used to determine an appropriate critical value for some specified non-normal distribution.

Determining Critical Values via Simulation Studies

As previously explained, if the goal is to have a Type I error probability of 0.05, this can be accomplished if the distribution of

$$T = \frac{\bar{X} - \mu_0}{s/\sqrt{n}} \tag{5.13}$$

can be determined. In particular, if the null hypothesis H_0: $\mu = \mu_0$ is true and t can be determined such that

$$P(-t \leq T \leq t) = 0.95,$$

then the probability of a Type I error is exactly 0.05 if the null hypothesis is rejected when $T \leq -t$ or $T \geq t$. Under normality, T has a symmetric distribution, in which case t is the 0.975 quantile. That is, another way of writing this last equation is

$$P(T \leq t) = 0.975.$$

From Chapter 4, a 0.95 confidence interval for μ is given by

$$\left(\bar{X} - t\frac{s}{\sqrt{n}}, \bar{X} + t\frac{s}{\sqrt{n}}\right).$$

Conceptually, when the null hypothesis is true, the critical value t can be determined to a high degree of accuracy if a study could be repeated millions of times. If, for example, it were repeated 1000 times, put the resulting T values in ascending order and label the results $T_{(1)} \leq \cdots \leq T_{(1000)}$. Then one way of estimating t is with $T_{(975)}$, simply because 97.5% of the T values are less than or equal to $T_{(975)}$.

But, of course, repeating a study millions of times, or even a thousand times, is impractical. However, we can use a computer to mimic the process of repeating a study using what is called a simulation study. Here is how simulations can be used to determine t for some specified distribution when the null hypothesis H_0: $\mu = \mu_0$ is true and the goal is to have a Type I error probability equal to 0.05.

1. Generate n observations from the distribution of interest. (For a normal distribution the R function `rnorm` accomplishes this goal.)
2. Compute the sample mean $\bar{X}$, the sample standard deviation s, and then T given by Eq. (5.13) based on the hypothesized value μ_0. For example, if the null hypothesis is H_0: $\mu = 6$, data would be generated from a distribution having mean $\mu = 6$.
3. Repeat steps 1 and 2 B times where B is reasonably large. For illustrative purposes, assume they are repeated $B = 1000$ times yielding 1000 T values: $T_1, \ldots, T_{1000}$.
4. Put these values in ascending order yielding $T_{(1)} \leq \cdots \leq T_{(1000)}$.
5. An estimate of the 0.975 quantile of the distribution of T is $T_{(975)}$ because the proportion of observed T values less than or equal to $T_{(975)}$ is 0.975. If, for example, $T_{(975)} = 2.1$, this suggests using $t = 2.1$ when computing a 0.95 confidence interval.

Chapter 4 illustrated that when dealing with skewed distributions, the distribution of T can be skewed as well. One way of computing a 0.95 confidence interval for this particular situation is to proceed as follows. Let t_ℓ and t_u be the 0.025 and 0.975 quantiles of the distribution of T. In symbols,

$$P(T \leq t_\ell) = 0.025$$

and

$$P(T \leq t_u) = 0.975,$$

in which case

$$P(t_\ell \leq T \leq t_u) = 0.95.$$

A little algebra shows that a 0.95 confidence interval for μ is

$$\left(\bar{X} - t_u\frac{s}{\sqrt{n}}, \bar{X} - t_\ell\frac{s}{\sqrt{n}}\right).$$

Observe that the lower end of the confidence interval is based on t_u, not t_ℓ, contrary to what might be thought. (Also, in this last equation, t_ℓ is always negative, and this helps explain why $t_\ell s/\sqrt{n}$ is subtracted from the sample mean. With 1000 T values for example, estimate t_u with $T_{(975)}$ as previously explained, and estimate t_ℓ with $T_{(25)}$.

EXAMPLE

Consider the (lognormal) distribution in Figure 4.10, which is skewed to the right and has relatively light tails. (Outliers are relatively rare when sampling from this distribution.) The population mean is approximately $\mu = 1.649$. Here is the R code for performing the simulation method just outlined, based on $n = 25$ observations generated from this distribution:

```
set.seed(45) # set the seed of the random number generator
#so that the results can be duplicated by the reader.
t.values=NA # create an R variable where the T values will be stored.
for (i in 1:1000){ # create a loop for generating data, computing T and
repeating this 1000 times.
x=rlnorm(25) # generate 25 values from the distribution in Figure 4.10.
t.values[i]=sqrt(25)*(mean(x)-1.649)/sd(x) # compute T and store the
value in t.values[i]
} # end of loop
```

Notice that this R code mimics how sampling distributions are viewed: a study is repeated many times, each study yielding a value for T. Based on these R commands, it was found that $T_{(25)} = -6$. That is, 2.5% of the T values were less than or equal to -6 indicating that $P(T \leq -6)$ is estimated to be 0.025. In a similar manner, $T_{(975)} = 1.44$ estimates the 0.975 quantile of the distribution of T. This means that if the hypothesis is true, meaning that data are randomly sampled from the distribution in Figure 4.10 having mean $\mu = 1.649$, there is, approximately, a 0.95 probability that T will be between -6 and 1.44. That is, reject the null hypothesis if $T \leq -6$ or if $T \geq 1.44$. An approximate 0.95 confidence interval for the mean is

$$\left(\bar{X} - 1.44\frac{s}{\sqrt{n}}, \bar{X} + 6.0\frac{s}{\sqrt{n}}\right).$$

In particular, the null hypothesis would be rejected if this interval does not contain the value 1.649. So if $\bar{X} = 2.6$ and $s = 3$, the 0.95 confidence interval is

$$\left(2 - 1.44\frac{3}{\sqrt{25}}, 2 + 6.0\frac{3}{\sqrt{25}}\right) = (1.736, 6.2)$$

and the null hypothesis would be rejected. In contrast, under normality, the confidence interval would be (1.36, 3.84) and the null hypothesis would not be rejected.

But in practice, observations are sampled from a distribution that is not known. The bootstrap-t addresses this problem by performing a simulation study using the observed values, rather than performing a simulation study using a hypothetical distribution such as the normal or lognormal distribution, as was done in the last example. This means that data are resampled with replacement from the observed values. In effect, a simulation is performed using an approximation of the distribution that generated the data. Of course, the observed data only provide an approximation of the true distribution, but it can provide a more accurate approximation of the distribution of T compared to assuming normality.

To describe the computational steps in a more formal manner, let $X_1, \ldots, X_n$ represent the observed values and imagine that the goal is to test H_0: $\mu = \mu_0$, where as usual μ_0 is some value of interest specified by the researcher. A bootstrap sample of size n is obtained by randomly sampling, with replacement, n values from $X_1, \ldots, X_n$. As is usually done, this bootstrap sample is labeled $X_1^*, \ldots, X_n^*$. Let

$$\bar{X}^* = \frac{1}{n}\sum X_i^*$$

denote the sample mean based on this bootstrap sample and let s^* denote the standard deviation. Notice that observations are sampled such that every value has probability $1/n$ of being chosen. From Section 3.2, this means that, in the bootstrap world, the population mean is $\bar{X}$, the sample mean associated with the observed data. Consequently, T as given by Equation (5.13) becomes

$$T^* = \frac{\bar{X}^* - \bar{X}}{s^*/\sqrt{n}}.$$

The goal is to estimate the distribution of T and this is accomplished by generating many T^* values.

Here is a more succinct description of the bootsrap-t method.

1. Generate a bootstrap sample $X_1^*, \ldots, X_n^*$ by resampling with replacement n values from the observed data $X_1, \ldots, X_n$.

2. Based on this bootstrap sample, compute the mean $\bar{X}^*$ and the standard deviation s^* followed by

$$T^* = \frac{\bar{X}^* - \bar{X}}{s^*/\sqrt{n}}.$$

3. Repeat steps 1 and 2 B times yielding $T_1^*, \ldots, T_B^*$.

4. Put the T^* values in ascending order yielding $T_{(1)}^* \le T_{(2)}^* \le \cdots \le T_{(B)}^*$.

Let $\ell = \alpha B/2$, rounded to the nearest integer, and $u = B - \ell$. Then an approximate $1-\alpha$ confidence interval for μ is

$$\left(\bar{X} - T_{(u)}^* \frac{s}{\sqrt{n}}, \bar{X} - T_{(\ell)}^* \frac{s}{\sqrt{n}}\right). \tag{5.14}$$

EXAMPLE

Forty observations were generated from a standard normal distribution and then the bootstrap-t method was used to approximate the distribution of T based on $B = 1000$ bootstrap samples. A plot of the 1000 bootstrap T^* values is shown in Figure 5.8. The smooth symmetric curve is the correct distribution (a Student's T distribution with $\nu = 39$). In this particular case, the bootstrap estimate of the distribution of T is fairly accurate. The bootstrap estimates of the 0.025 and 0.975 quantiles are $T_{(\ell)}^* = -1.94$ and $-T_{(u)}^* = -2.01$. The correct answers are -2.022 and 2.022, respectively.

Of course, this last example does not provide compelling evidence that the bootstrap-t method performs reasonably well under normality. But simulation studies suggest that when sampling from a normal distribution, little accuracy is lost. When dealing with non-normal distributions, substantial improvements can be realized (e.g., Westfall and Young, 1993; Wilcox, 2017). But despite the theoretical appeal of the bootstrap-t method, and even though it improves upon Student's T in certain situations, the bootstrap-t can be unsatisfactory when making inferences about the mean, as will be illustrated momentarily (by Table 5.1). This reflects the general fact that accurate inferences about the mean, for a reasonably wide range of non-normal distributions, is difficult at best. But as will be seen, bootstrap methods do offer accurate results for a large range of other situations, even when the sample size is small. Included are situations where the goal is to make inferences based on a 20% trimmed mean or median.

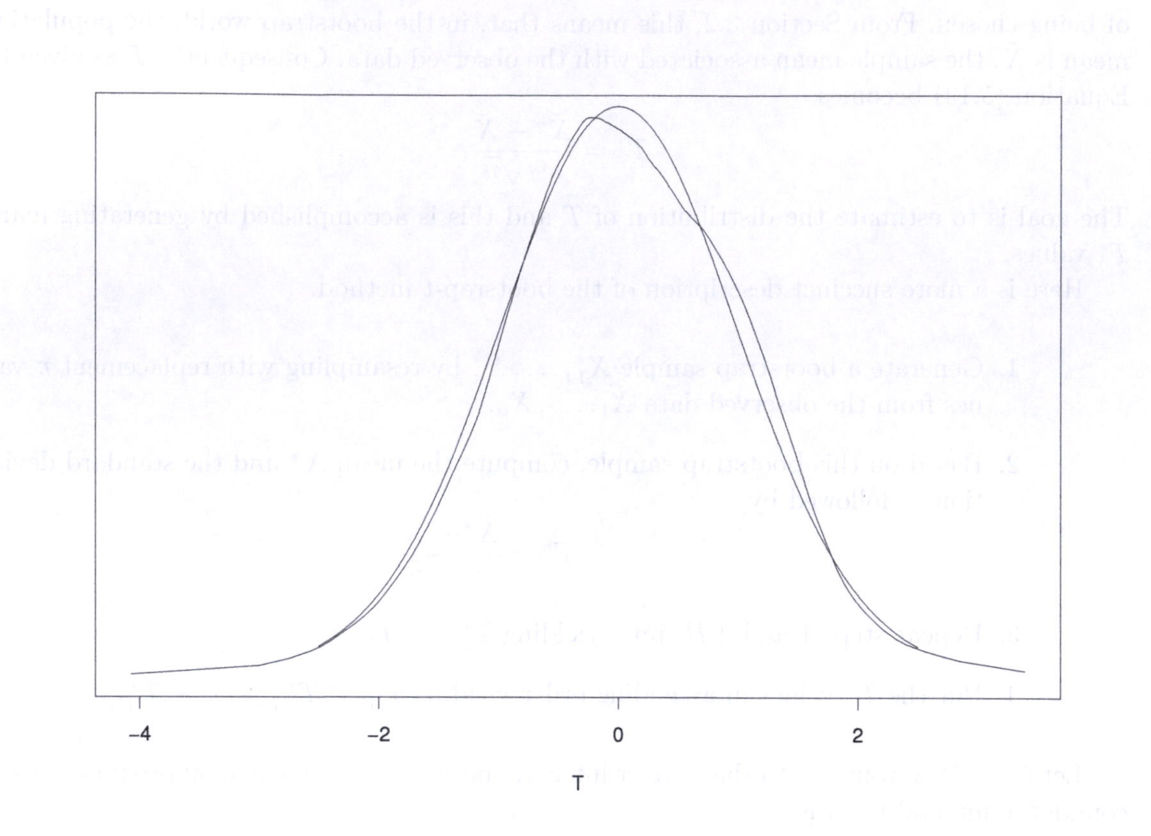

Figure 5.8 A plot of 1000 bootstrap T values based on a sample of 40 observations generated from a normal distribution. The correct estimate of the distribution of T is the symmetric smooth curve, namely a Student's T distribution with 39 degrees of freedom.

An issue is whether using a bootstrap-t method ever makes a difference, compared to using Student's T, when analyzing data from an actual study. If the answer is no, the bootstrap-t method has no practical value. But the following examples illustrate that at least in some situations, the answer is yes.

EXAMPLE

Consider data reported by Dana (1990) where the goal was to investigate issues related to self-awareness and self-evaluation. A 0.95 confidence interval based on Student's T is (161.43, 734.67). Using the bootstrap-t method via the R function `trimcibt`, the 0.95 confidence interval is (247.76, 1285.15).

EXAMPLE

In a portion of a study dealing with hangover symptoms (conducted by M. Earleywine), Student's T yields the 0.95 confidence interval (−0.62 , 6.42). The 0.95 confidence interval based on the bootstrap-t is (0.65, 25.84).

5.10.1 Symmetric Confidence Intervals

A variation of the bootstrap-t method that can be used when testing the (two-sided) hypothesis

$$H_0 : \mu = \mu_0$$

should be mentioned. Rather than use T^* as previously defined, use

$$T^* = \frac{|\bar{X}^* - \bar{X}|}{s^*/\sqrt{n}} \tag{5.15}$$

and reject $H_0 : \mu = \mu_0$ if $|T| \geq T^*_{(c)}$, where $c = (1-\alpha)B$ rounded to the nearest integer and again $T^*_{(1)} \leq \cdots \leq T^*_{(B)}$ are the B bootstrap T^* values written in ascending order. An approximate $1-\alpha$ confidence interval for μ is now given by

$$\bar{X} \pm T^*_{(c)} \frac{s}{\sqrt{n}}. \tag{5.16}$$

This is called a *symmetric* two-sided confidence interval meaning that the same quantity, namely $T^*_{(c)} \frac{s}{\sqrt{n}}$, is added and subtracted from the mean when computing a confidence interval. In contrast is the confidence interval given by Equation (5.13) which is called an *equal-tailed* confidence interval. With large sample sizes, the symmetric two-sided confidence interval enjoys some theoretical advantages over the equal-tailed confidence interval (Hall, 1988a, 1988b). The main point here is that when sample sizes are small, probability coverage and control over the probability of a Type I error can again be unsatisfactory. In some cases the actual probability coverage of these two methods differs very little, but exceptions arise.

To add perspective, Table 5.1 shows the actual Type I error probability when sampling from one of four distributions. The intention is to have a Type I error probability of 0.05, but as can be seen, this is not remotely the case in some situations. The seriousness of a Type I error will vary from one situation to the next, but some authorities would argue that when testing some hypothesis with $\alpha = 0.05$, usually the actual probability of a Type I error should not exceed 0.075 and should not drop below 0.025 (e.g., Bradley, 1978). One argument for being dissatisfied with an actual Type I error probability of 0.075 is that if a researcher believes that a Type I error probability of 0.075 is acceptable, she would have set $\alpha = 0.075$ in the first place to achieve higher power. When sampling from a skewed, relatively light-tailed distribution (the lognormal distribution indicated by LN in Table 5.2), the equal-tailed bootstrap-t (method BT in Table 5.2) performs reasonably well with a sample size of $n = 20$; the other methods are unsatisfactory. Increasing n to 100, now all four methods perform tolerably well. But when dealing with heavy-tailed distributions where outliers are common (distributions MN and SH in Table 5.1), the reverse is true. In practical terms, even with $n = 100$, there is some risk that inaccurate confidence intervals will result even with a fairly large sample size. As will be seen, if instead inferences are made with some robust measure of location, problems with controlling the probability of a Type I error can be reduced substantially.

As previously noted, Student's T test can be biased: When testing H_0: $\mu = \mu_0$, the probability of rejecting is not always minimized when $\mu = \mu_0$. In practical terms, the probability of rejecting might be higher when H_0 is true compared to certain situations where it is false. The bootstrap-t method reduces this problem but does not eliminate it.

Chapter 5 also pointed out that arbitrarily small departures from normality can destroy power when using Student's T to make inferences about the population mean. Switching to the bootstrap-t method, or any other bootstrap method, does not address this problem. But again, despite these practical problems, it turns out that the bootstrap-t method does have considerable practical value when dealing with other inferential problems, as will be seen.

Table 5.2 Actual Type I Error Probabilities for Three Methods Based on the Mean When Testing at the $\alpha = 0.05$ Level

	Dist.	Method BT	SB	T
$n = 20$	N	0.054	0.051	0.050
	LN	0.078	0.093	0.140
	MN	0.100	0.124	0.022
	SH	0.198	0.171	0.202
$n = 100$	N	0.048	0.053	0.050
	LN	0.058	0.058	0.072
	MN	0.092	0.107	0.041
	SH	0.168	0.173	0.190

N=normal; LN=lognormal; MN=mixed normal; SH=skewed, heavy-tailed; BT=equal-tailed, bootstrap-t; SB=symmetric bootstrap-t; T=Student's T

5.10.2 Exact Nonparametric Confidence Intervals for Means Are Impossible

The confidence interval for the mean, based on Student's T distribution, is parametric in the sense that it assumes sampling is from the family of normal distributions, which is characterized by two unknown parameters, namely the population mean and variance. The bootstrap-t method for computing a confidence interval is nonparametric, meaning that it is not based on a family of distributions characterized by one or more unknown parameters. Box 4.1 described a nonparametric confidence interval for the population median for which the probability coverage can be determined exactly without assuming sampling is from any particular parametric family of distributions. This confidence interval is said to be *distribution free*, meaning that the exact probability coverage can be determined assuming random sampling only. In particular, normality was not assumed. It is known, however, that the same cannot be done when working with the mean (Bahadur and Savage, 1956; cf. Donoho, 1988). The bootstrap-t, for example, provides a nonparametric confidence interval for the mean, but it is not distribution free.

5.11 THE PERCENTILE BOOTSTRAP METHOD

The percentile bootstrap method is another fundamental technique that has considerable practical value. It is stressed that it does not perform well when the goal is to make inferences about the population mean unless the sample size is very large. However, it performs very well when dealing with a variety of other problems, even with very small sample sizes, and modifications of the method have practical value as well. Although the percentile bootstrap is not recommended when working with the sample mean, it is perhaps easiest to explain in terms of the sample mean, so we start with this special case.

Again consider the goal of testing the hypothesis that the population mean has some specified value, which as usual is denoted by μ_0. Imagine that a study is repeated thousands of times and that 99% of the time, the resulting sample mean is greater than μ_0. This suggests that the hypothesis H_0: $\mu = \mu_0$ should be rejected and that the μ is greater than μ_0. Put another way, if $P(\bar{X} > \mu_0)$ is close to one, a reasonable decision is to reject the null hypothesis and decide that μ is greater than μ_0. If $P(\bar{X} > \mu_0)$ is close to zero, the reverse decision is made; otherwise, from the point of view of Tukey's three decision rule, no decision is made.

Roughly, the percentile bootstrap method uses a simulation estimate of $P(\bar{X} > \mu_0)$ based on the observed data. This is done by generating a bootstrap sample, computing the bootstrap sample $\bar{X}^*$, and repeating this process many many times, which yields an estimate of

$$p = P(\bar{X}^* > \mu_0),$$

the probability that the bootstrap sample mean is greater than the hypothesized value. In effect, p estimates $P(\bar{X} > \mu_0)$, the probability that the sample mean is greater than the hypothesized value if a study were repeated many times.

Let μ_0 be any hypothesized value. Here are the steps used by the percentile bootstrap method when dealing with the mean:

1. Generate a bootstrap sample $X_1^*, \ldots, X_n^*$ as was done in the previous section.
2. Compute the mean of this bootstrap sample, $\bar{X}^*$.
3. Repeat steps 1 and 2 B times yielding $\bar{X}_1^*, \ldots, \bar{X}_B^*$.
4. Estimate $p = P(\bar{X}^* > \mu_0)$ with $\hat{p}$, the proportion of bootstrap sample means greater than μ_0.

The percentile bootstrap p-value when testing for exact equality is

$$P = 2\text{min}(\hat{p}, 1 - \hat{p}).$$

That is, the p-value is either $2\hat{p}$ or $2(1 - \hat{p})$, whichever is smallest.

Decision Rule:

If this (estimated) p-value, P, is less than or equal to α, reject the null hypothesis. Or in the context of Tukey's three decision rule, make a decision about whether μ is greater than or less than the hypothesized value. Otherwise, no decision is made. Note that P reflects the strength of the empirical evidence that a decision can be made.

Figure 5.9 illustrates the basic idea. The left panel shows a plot of 1000 bootstrap means created in the manner just described based on a sample size of 40 generated from a normal distribution with mean $\mu = 6$. In this case, 633 of the 1000 bootstrap means are greater than 6. So in symbols, $\hat{p} = 0.633$ and the p-value is 0.734. The right panel is the same as the left, only 0.8 was added to every observation. Now $\hat{p} = 1$ and the p-value is 0.

Confidence Intervals

Now imagine the goal is to compute a 0.95 confidence interval for the population mean, μ. The strategy used by the percentile bootstrap method is to use the middle 95% of the bootstrap sample means. More generally, an approximate $1 - \alpha$ confidence interval for μ is

$$(\bar{X}_{(\ell+1)}^*, \bar{X}_{(u)}^*), \qquad (5.17)$$

where $\bar{X}_{(1)}^* \leq \cdots \leq \bar{X}_{(B)}^*$ are the B bootstrap means written in ascending order, $\ell = \alpha B/2$, rounded to the nearest integer, and $u = B - \ell$. So if $B = 20$ and $1 - \alpha = 0.8$, ℓ is 2 and u

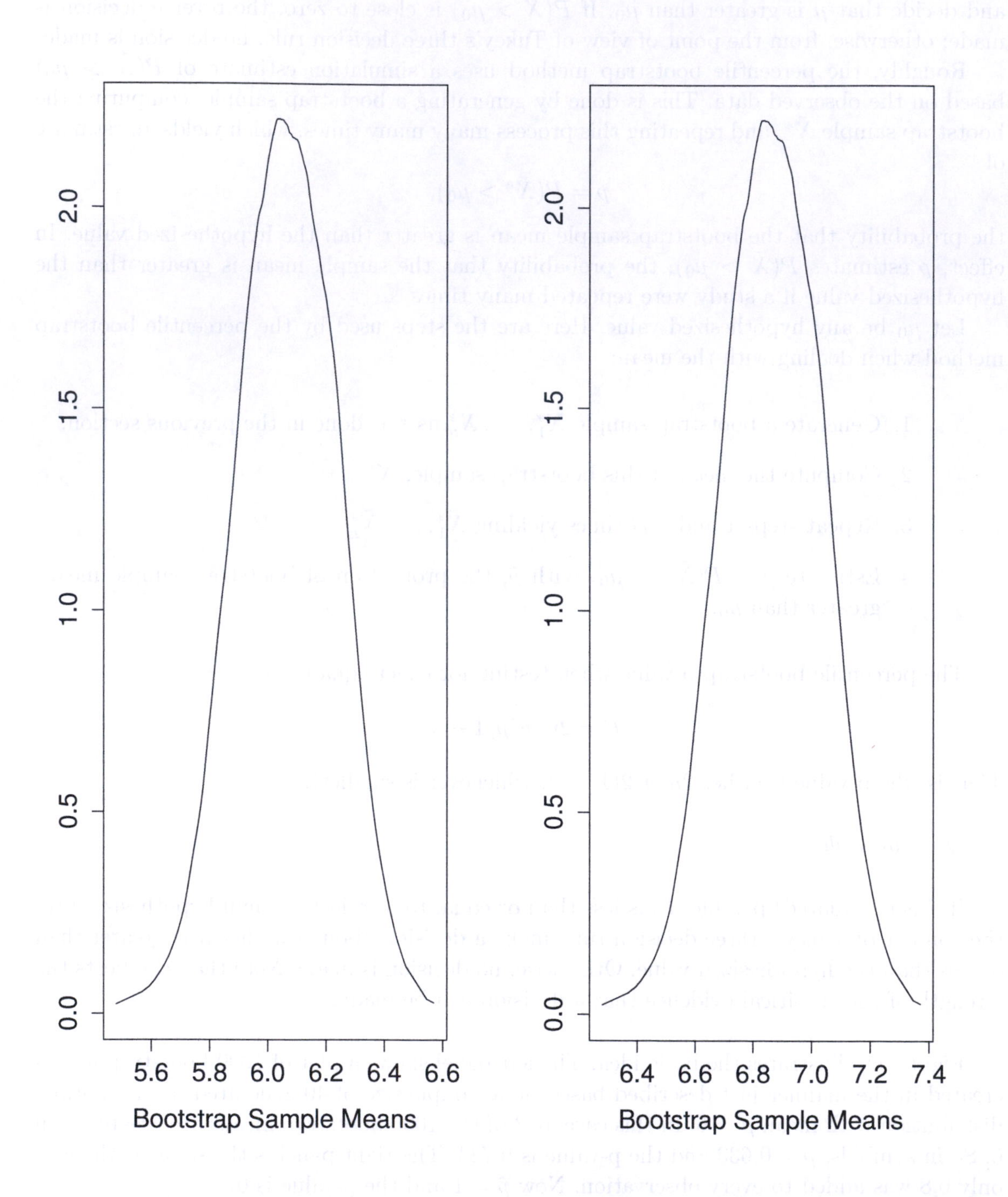

Figure 5.9 The left panel shows a plot of 1000 bootstrap means based on a sample of 40 observations generated from a normal distribution having mean $\mu = 6$. Note that the value 6 is not unusual. It is contained well within the range of bootstrap sample means, so the hypothesis H_0: $\mu = 6$ would not be rejected. The right panel is the same as the left, only observations were generated from a normal distribution with mean $\mu = 6.8$. Now the bootstrap sample means indicate that the value 6 is unusual.

is 18. For the special case $\alpha = 0.05$, $\ell = 0.025B$ (still rounding to the nearest integer), and $\bar{X}^*_{(\ell+1)}$ and $\bar{X}^*_{(u)}$ estimate the 0.025 and 0.975 quantiles of the distribution of $\bar{X}$, respectively.

Comments on Choosing B

How many bootstrap samples should be taken? That is, how large should B be? One perspective has to do with Type I errors. Would control over the Type I error probability improve by very much if B were increased? The default choices for B used by the software in this book are based on this point of view. However, there is evidence that if B is too small, this might affect power (Racine and MacKinnon, 2007; cf. Davidson and MacKinnon, 2000). So there might be practical reasons for using larger choices for B.

One final point might help before ending this section. An attempt has been made to provide some intuitive sense about the strategy used by the percentile bootstrap. But from a theoretical point of view, more needs to be done to justify this technique. Here it is merely noted that a formal justification has been derived (e.g., Hall, 1988a, 1988b; Liu and Singh, 1997).

Both theoretical and simulation studies indicate that, generally, the bootstrap-t performs better than the percentile bootstrap or Student's T when computing a confidence interval or testing some hypothesis about μ. There are exceptions, such as when sampling from a normal distribution, but to avoid poor probability coverage, the bootstrap-t method is preferable to Student's T or the percentile bootstrap. However, when working with robust measures of location, such as the median, an M-estimator or a 20% trimmed mean, we will see that typically the percentile bootstrap is preferable to the bootstrap-t. (This is particularly true when comparing medians as described in Chapter 7.)

5.12 INFERENCES ABOUT ROBUST MEASURES OF LOCATION

Generally, the percentile and bootstrap-t methods can be applied when working with any measure of location, such as a trimmed mean or an M-estimator. The details are described here along with a description of when they perform well, and when and why they are unsatisfactory.

5.12.1 Using the Percentile Method

The percentile bootstrap can be applied with any measure of location simply by replacing the sample mean with the measure of location of interest. If, for example, a 20% trimmed mean is used, generate a bootstrap sample of size n and compute the trimmed mean $\bar{X}^*_t$. If we repeat this B times yielding $\bar{X}^*_{t1}, \ldots, \bar{X}^*_{tB}$, then an approximate $1-\alpha$ confidence interval for the population trimmed mean is

$$(\bar{X}^*_{t(\ell+1)}, \bar{X}^*_{t(u)}), \tag{5.18}$$

where $\bar{X}^*_{t(1)} \leq \cdots \leq \bar{X}^*_{t(B)}$ are the B bootstrap trimmed means written in ascending order, $\ell = \alpha B/2$, rounded to the nearest integer, and $u = B - \ell$. Now $\hat{p}$ is the proportion of bootstrap sample trimmed means greater than or equal to the hypothesized value, and the p-value is computed as done in Section 5.11. Again, reject H_0: $\mu_t = \mu_0$ if the p-value is less than or equal to α.

Generally, the performance of the percentile bootstrap improves in terms of Type I errors and probability coverage as we increase the amount of trimming from 0% (the sample mean) to 20%. Moreover, with 20% trimming, it performs relatively well over a broad range of situations. But the minimum amount of trimming needed to justify the percentile bootstrap is not known. The only rule currently available is that with a minimum of 20% trimming, accurate results can be obtained even with very small sample sizes.

Note that when dealing with the median, the method in Section 4.9 can be used to compute a distribution-free confidence interval. Currently, there are no indications that the percentile bootstrap offers a practical advantage. However, in subsequent chapters where the goal is to compare two or more groups, it will be seen that when dealing with the median, there are general conditions where a percentile bootstrap is the only method known to perform well in simulations. Also, when using the one-step M-estimator or the MOM estimator and when dealing with skewed distributions, a percentile bootstrap method has been found to perform better than methods based in part on estimates of the standard error. For symmetric distributions, methods based on estimates of the standard error perform reasonably well, but all indications are that they offers little or no advantage over using a percentile bootstrap method.

5.12.2 R Functions onesampb, momci, and trimpb

The R function

```
onesampb(x, est = onestep, alpha = 0.05, nboot = 2000)
```

computes a confidence interval based on the one-step M-estimator, introduced in Chapter 2, where x is an R variable containing data, alpha is α, which defaults to 0.05, and nboot is B, the number of bootstrap samples to be used, which defaults to 2000. This function can be used with any measure of location via the argument `est`. For example, `onesampb(x, est = tmean)` would return a confidence interval based on the 20% trimmed mean. For convenience, the R function

```
momci(x,alpha=0.05,nboot=2000)
```

computes a confidence interval based on the modified one-step M-estimator and the function

```
trimpb(x,tr=0.2,alpha=0.05,nboot=2000)
```

uses a trimmed mean. The argument `tr` indicates the amount of trimming and defaults to 0.2 if not specified. Again, alpha is α and defaults to 0.05. It appears that $B = 500$ suffices, in terms of achieving accurate probability coverage with 20% trimming. But to be safe, B (`nboot`) defaults to 2000. (An argument for using $B = 2000$ can be made along the lines used by Booth and Sarker, 1998. As previously noted, if B is relatively small, this might result in relatively low power.)

EXAMPLE

For 15 law schools, the undergraduate GPA of entering students in 1973 was

3.39 3.30 2.81 3.03 3.44 3.07 3.00 3.43 3.36 3.13 3.12 2.74 2.76 2.88 2.96.

The 0.95 confidence interval returned by the R function `onesampb` is (2.95, 3.26). So among all law schools, it is estimated that the typical GPA of entering students is between 2.95 and 3.29. Using the MOM-estimator instead (the R function `momci`), the 0.95 confidence interval is (2.92, 3.35). Using the function `trimpb`, the 0.95 confidence interval for the 20% trimmed mean is (2.94, 3.25). Setting the argument tr=0 in `trimpb` results in a 0.95 confidence interval for the mean: (2.98, 3.21), illustrating that in some cases, switching from the mean to a 20% trimmed mean makes little difference.

5.12.3 The Bootstrap-t Method Based on Trimmed Means

The bootstrap-t method is readily applied when using a trimmed mean. In principle it can be applied when using an M-estimator, but this requires an estimate of its standard error. Estimates are available (e.g., Wilcox, 2005a), but studies indicate that with small to moderate sample sizes, probability coverage and control over the probability of a Type I error can be unsatisfactory. It is unclear how large the sample size must be to get accurate results. This means that working with an M-estimator, use the percentile bootstrap method.

When working with a trimmed mean, the bootstrap-t is applied in essentially the same way as was done when working with the mean. If the amount of trimming is 20% or more, it seems that using a percentile bootstrap method is best for general use. With 10% trimming, there are indications that the percentile bootstrap method continues to perform relatively well, but a definitive study has not been performed. When the amount of trimming is close to zero, it currently seems that using a bootstrap-t method is preferable. The computational steps of the most basic version of the bootstap-t method are summarized here.

Generate a bootstrap sample of size n and compute the trimmed mean and Winsorized standard deviation, which we label $\bar{X}_t^*$ and s_w^*, respectively. Let γ be the amount of trimming. With 20% trimming, $\gamma = 0.2$ and with 10% trimming, $\gamma = 0.1$. Next, compute

$$T_t^* = \frac{(1-2\gamma)(\bar{X}_t^* - \bar{X}_t)}{s_w^*/\sqrt{n}}. \tag{5.19}$$

Repeating this process B times yields B T_t^* values. Writing these B values in ascending order we get $T_{t(1)}^* \leq T_{t(2)}^* \leq \cdots \leq T_{t(B)}^*$. Letting $\ell = 0.025B$, rounded to the nearest integer, and $u = B - \ell$, an estimate of the 0.025 and 0.975 quantiles of the distribution of T_t is $T_{(\ell+1)}^*$ and $T_{(u)}^*$. The resulting 0.95 confidence interval for μ_t (the population trimmed mean) is

$$\left(\bar{X}_t - T_{t(u)}^* \frac{s_w}{(1-2\gamma)\sqrt{n}},\ \bar{X}_t - T_{t(\ell+1)}^* \frac{s_w}{(1-2\gamma)\sqrt{n}}\right). \tag{5.20}$$

Hypothesis Testing

As for testing $H_0: \mu_t = \mu_0$, compute

$$T_t = \frac{(1-2\gamma)(\bar{X}_t - \mu_0)}{s_w/\sqrt{n}}$$

and reject if

$$T_t \leq T_{t(\ell+1)}^*,$$

or if

$$T_t \geq T_{t(u)}^*.$$

The symmetric bootstrap-t method can be used as well when testing a two-sided hypothesis. Now we use

$$T_t^* = \frac{|(1-2\gamma)(\bar{X}_t^* - \bar{X}_t)|}{s_w^*/\sqrt{n}} \tag{5.21}$$

and reject H_0 if $|T_t| > T_{t(c)}^*$, where $c = (1-\alpha)B$ rounded to the nearest integer. An approximate $1-\alpha$ confidence interval for μ_t is

$$\bar{X}_t \pm T_{t(c)}^* \frac{s_w}{(1-2\gamma)\sqrt{n}}. \tag{5.22}$$

Table 5.3 Actual Type I Error Probabilities Using 20% Trimmed Means, $\alpha = 0.05$

		Method			
	Dist.	BT	SB	P	TM
$n = 20$	N	0.067	0.052	0.063	0.042
	LN	0.049	0.050	0.066	0.068
	MN	0.022	0.019	0.053	0.015
	SH	0.014	0.018	0.066	0.020

N=normal, LN=lognormal; MN=mixed normal; SH=skewed, heavy-tailed; BT=equal-tailed, bootstrap-t; SB=symmetric bootstrap-t; P=Percentile bootstrap; TM=Tukey–McLaughlin.

Table 5.2 reported the actual probability of a Type I error when using means with one of four methods. None of the methods were satisfactory for all four distributions considered, even after increasing the sample size to three hundred. Table 5.3 shows the actual probability of a Type I error when using 20% trimmed means instead. Notice that the percentile bootstrap method is the most stable; the actual probability of a Type I error ranges between 0.053 and 0.066. The other three methods do a reasonable job of avoiding Type I error probabilities above the nominal 0.05 level. But they can have actual Type I error probabilities well below the nominal level, which is an indication that their power might be less than what is obtained using the percentile method instead. With the caveat that no method is best in all situations, the percentile bootstrap with a 20% trimmed mean is a good candidate for general use.

5.12.4 R Function trimcibt

The R function

```
trimcibt(x, tr=0.2, alpha=0.05, nboot=599, side=TRUE)
```

computes a bootstrap-t confidence interval for a trimmed mean. The argument side indicates whether an equal-tailed or a symmetric confidence interval is to be computed. The default is side=TRUE resulting in a symmetric confidence interval. Using side=FALSE means that an equal-tailed confidence interval will be computed. The argument `tr` indicates the amount of trimming, which defaults to 20%. So to compute a confidence interval for the mean, set `tr=0`.

EXAMPLE

Table 3.2 reported data on the desired number of sexual partners among 105 college males. As previously indicated, these data are highly skewed with a relatively large number of outliers and this can have a deleterious effect on many methods for computing a confidence interval and testing hypotheses. If we compute the Tukey–McLaughlin 0.95 confidence interval for the 20% trimmed, described in Chapter 4, we get (1.62, 3.75). Using the R function `trimcibt` with side=F yields an equal-tailed 0.95 confidence interval of (1.28, 3.61). With side=T it is (1.51, 3.61). Using the percentile bootstrap method, the R function `trimpb` returns (1.86, 3.95). So in this particular case, the lengths of the confidence intervals do not vary that much among the methods used, but the intervals are centered around different values, which might affect any conclusions made. If `trimcibt` is used to compute a 0.95 confidence for the mean (by setting the argument tr=0), the result is (−2.46, 4704.59), which differs substantially from the confidence interval for a 20% trimmed mean.

In summary, all indications are that the percentile bootstrap is more stable (with at least 20% trimming) than the bootstrap-t method. That is, the actual Type I error probability tends to be closer to the nominal level. In addition, it has the advantage of more power, at least in some situations, compared to any other method we might choose. However, subsequent chapters will describe situations where the bootstrap-t method outperforms the percentile method as well as additional situations where the reverse is true. So both methods are important to know.

5.13 ESTIMATING POWER WHEN TESTING HYPOTHESES ABOUT A TRIMMED MEAN

An issue of fundamental importance is determining the likelihood of rejecting the null hypothesis when it is false. Put another way, if we test some hypothesis and fail to reject, this might be because the null hypothesis is true, or perhaps power is too low to detect a meaningful difference. Or in the context of Tukey's three decision rule, failing to make a decision about whether μ_t is less than or greater than μ_0 will occur if power is low. If we can estimate how much power we have based on the same data used to test some hypothesis, we are better able to discern which reason accounts for a nonsignificant result. A related issue is determining the probability of replicating the outcome of some study. As explained in Chapter 5, this issue is related to power.

Generally, it is a relatively simple matter to come up with a method for estimating power. That is, based on the data from a study, there are methods for estimating the probability that Student's T, for example, will reject if the study is replicated. But finding a method that provides a reasonably accurate estimate is another matter. Indeed, many methods have been proposed only to be discredited by subsequent studies (e.g., Wilcox and Keselman, 2002a). Currently, no method based on the mean can be recommended and even recent suggestions on how to address this problem, when using means, can be wildly inaccurate.

However, when working with a 20% trimmed mean and when testing at the $\alpha = 0.05$, all indications are that a reasonably satisfactory method of assessing power is available (Wilcox and Keselman, 2002a). There are two goals: (1) compute an (unbiased) estimate of power for some given value of $\Delta = \mu_t - \mu_0$, and (2) provide a conservative estimate of power meaning a (one-sided) confidence interval for how much power we have. Roughly, the method estimates the standard error of the trimmed mean and then, given Δ, provides an estimate of how much power we have. A possible concern, however, is that this estimate might overestimate the actual power. Based on data, we might estimate power to be 0.7, but in reality it might be 0.5 or it might be as low as 0.4. So the method also computes a (lower) 0.95 confidence interval for the actual amount of power using a percentile bootstrap technique. Briefly, for every bootstrap sample, the standard error of the trimmed mean is estimated which yields an estimate of power corresponding to whatever Δ value is of interest. Repeating this process B times yields B estimates of power which, when put into ascending order, we label $\hat{\xi}_{(1)} \leq \cdots \leq \hat{\xi}_{(B)}$. Then a conservative estimate of power is $\hat{\xi}_{(a)}$, where $a = 0.05B$ rounded to the nearest integer. So if $\hat{\xi}_{(a)} = 0.4$, say, we estimate that with probability 0.95, power is at least 0.4. If $\hat{\xi}_{(a)} = 0.6$, we estimate that with probability 0.95, power is at least 0.6.

Note that the actual power depends on Δ, which is not known. The method just outlined deals with this by estimating power for a series of Δ values. That is, the strategy is to answer the question: If $\Delta = 2$, for example, how much power would result? The actual power is not known because in particular, Δ is not known. A temptation might be to estimate Δ with $\hat{\Delta} = \bar{X}_t - \mu_0$, which yields an estimate of power and the probability of replicating a study. But this approach can be highly inaccurate and misleading and cannot be recommended. To provide a glimpse of why this approach can be highly unsatisfactory, consider the case where $\Delta = 0$ (the null hypothesis is true). Further assume that if $\alpha = 0.05$, the probability of rejecting is indeed 0.05. Now, in general, $\hat{\Delta}$ will differ from zero resulting in an estimate of power that is greater than 0.05. That is, we get an estimate of power that is biased, and often the amount of bias is severe.

5.13.1 R Functions powt1est and powt1an

The R function

```
powt1est(x, delta=0, ci=FALSE, nboot=800)
```

returns an estimate of how much power there is for some value of Δ. As usual, now x represents any R variable containing data. The argument ci defaults to F (for false) meaning that no confidence interval for power is computed. If ci=T is used, a percentile bootstrap method is used to get a conservative estimate of power. As usual, nboot indicates how many bootstrap samples are used. (That is, nboot corresponds to B.)

The R function

```
powt1an(x, ci= FALSE, plotit=TRUE, nboot=800)
```

provides a power analysis without having to specify a value for Δ. Rather, the function chooses a range of Δ values so that power will be between 0.05 and 0.9, approximately. Then it estimates how much power there is for each Δ value that it selects and plots the results when the argument plotit=TRUE. That is, the function estimates the power curve associated with the percentile bootstrap method of testing hypotheses with a 20% trimmed mean. The function also reports a lower estimate of power if ci=TRUE is used. That is, with probability 0.95, power is at least as high as this lower estimate.

EXAMPLE

Consider again the law school data used in the example at the end of Section 5.12.2 and imagine the goal is to test the hypothesis H_0: $\mu_t = 3$ with $\alpha = 0.05$. As previously indicated, the 0.95 confidence interval is (2.94, 3.29), so in particular, the null hypothesis would not be rejected. The issue here is how much power there was.

The R command

```
powt1an(x)
```

returns

```
$delta:
  0.00000  22.32513  44.65025  66.97538
  89.30051 111.62563 133.95076
 156.27589 178.60101 200.92614
 223.25127 245.57639 267.90152 290.22665
 312.55177

$power:
 0.0500 0.0607 0.0804 0.1176 0.1681
```

```
0.2353 0.3191 0.4124 0.5101 0.6117
0.7058 0.7812 0.8479 0.8984 0.9332
```

This says, for example, that when the null hypothesis is true ($\Delta = 0$), the probability of rejecting is estimated to be 0.05. When $\Delta = \mu_t - \mu_0 = 22.32$, power is estimated to be 0.0607, and when $\Delta = 44.65$, power is 0.0804.

5.14 A BOOTSTRAP ESTIMATE OF STANDARD ERRORS

Situations arise where an expression for the standard error of some estimator is not known or it takes on a rather complicated form. Examples are M-estimators and the modified one-step M-estimator (MOM) introduced in Chapter 2. So if it is desired to estimate their standard errors, it would be convenient to have a relatively simple method for accomplishing this goal. The basic bootstrap method is one way of tackling this problem.

Let $\hat{\mu}$ be any location estimator and let $\hat{\mu}^*$ be its value based on a bootstrap sample. Let $\hat{\mu}_b^*$ ($b = 1, \ldots, B$) be B bootstrap estimates of the measure of location. Then an estimate of the squared standard error of $\hat{\mu}$ is

$$S^2 = \frac{1}{B-1}\sum_{b=1}^{B}(\hat{\mu}_b^* - \bar{\mu}^*)^2$$

where $\bar{\mu}^* = \sum_{b=1}^{B} \hat{\mu}_b^*/B$.

Generally, this method for estimating the standard error of an estimator performs reasonably well. But there is at least one situation where the method performs poorly. If the goal is to estimate the standard error of the median, and if tied (duplicated) values tend to occur, a highly inaccurate estimate can result.

5.14.1 R Function bootse

The R function

```
bootse(x,nboot=1000,est=onestep)
```

computes the bootstrap estimate of the standard error of the measure of location specified by the argument `est`. By default, the one-step M-estimator is used, and the argument `nboot` corresponds to B which defaults to 1000.

5.15 EXERCISES

1. Given that $\bar{X} = 78$, $\sigma^2 = 25$, $n = 10$, and $\alpha = 0.05$, test H_0: $\mu > 80$, assuming observations are randomly sampled from a normal distribution. Also, draw the standard normal distribution indicating where Z and the critical value are located.
2. Repeat the previous problem but test H_0: $\mu = 80$.
3. For the previous problem, compute a 0.95 confidence interval and verify that this interval is consistent with your decision about whether to reject the null hypothesis.
4. For Exercise 1, determine the p-value.

5. For Exercise 2, determine the p-value.

6. Given that $\bar{X} = 120$, $\sigma = 5$, $n = 49$, and $\alpha = 0.05$, test H_0: $\mu > 130$, assuming observations are randomly sampled from a normal distribution.

7. Repeat the previous exercise but test H_0: $\mu = 130$.

8. For the previous exercise, compute a 0.95 confidence interval and compare the result with your decision about whether to reject H_0.

9. If $\bar{X} = 23$ and $\alpha = 0.025$, can you make a decision about whether to reject H_0: $\mu < 25$ without knowing σ?

10. An electronics firm mass produces a component for which there is a standard measure of quality. Based on testing vast numbers of these components, the company has found that the average quality is $\mu = 232$ with $\sigma = 4$. However, in recent years the quality has not been checked; so management asks you to check its claim with the goal of being reasonably certain that an average quality of less than 232 can be ruled out. That is, assume the quality is poor and in fact less than 232 with the goal of empirically establishing that this assumption is unlikely. You get $\bar{X} = 240$ based on a sample $n = 25$ components, and you want the probability of a Type I error to be 0.01. State the null hypothesis and perform the appropriate test assuming normality and $\sigma = 4$.

11. An antipollution device for cars is claimed to have an average effectiveness of exactly 546. Based on a test of 20 such devices, you find that $\bar{X} = 565$. Assuming normality and that $\sigma = 40$, would you rule out the claim with a Type I error probability of 0.05?

12. Comment on the relative merits of using a 0.95 confidence interval for addressing the effectiveness of the antipollution device in the previous exercise.

13. For $n = 25$, $\alpha = 0.01$, $\sigma = 5$, and H_0: $\mu \geq 60$, verify that power is 0.95 when $\mu = 56$.

14. For $n = 36$, $\alpha = 0.025$, $\sigma = 8$, and H_0: $\mu \leq 100$, verify that power is 0.61 when $\mu = 103$.

15. For $n = 49$, $\alpha = 0.05$, $\sigma = 10$, and H_0: $\mu = 50$, verify that power is approximately 0.56 when $\mu = 47$.

16. A manufacturer of medication for migraine headaches knows that its product can cause liver damage if taken too often. Imagine that by a standard measuring process, the average liver damage is $\mu = 48$. A modification of the product is being contemplated, and based on $n = 10$ trials, it is found that $\bar{X} = 46$. Assuming $\sigma = 5$, the manufacturer tests H_0: $\mu \geq 48$, the idea being that if it is rejected, there is convincing evidence that the average amount of liver damage is less than 48. Then

$$Z = \frac{46 - 48}{5/\sqrt{10}} = -1.3.$$

With $\alpha = 0.05$, the critical value is -1.645, so they do not reject because Z is not less than the critical value. What might be wrong with accepting H_0 and concluding that the modification results in an average amount of liver damage greater than or equal to 48?

17. For the previous exercise, verify that power is 0.35 if $\mu = 46$.

18. The previous exercise indicates that power is relatively low with only $n = 10$ observations. Imagine that you want power to be at least 0.8. One way of getting more power is to increase the sample size, n. Verify that for sample sizes of 20, 30, and 40, power is 0.56, 0.71, and 0.81, respectively.

19. For the previous exercise, rather than increase the sample size, what else might you do to increase power? What is a negative consequence of using this strategy?

20. Given the following values for $\bar{X}$ and s: (a) $\bar{X} = 44$, $s = 10$, (b) $\bar{X} = 43$, $s = 10$, (c) $\bar{X} = 43$, $s = 2$, test the hypothesis H_0: $\mu = 42$ with $\alpha = 0.05$ and $n = 25$.

21. For part b of the last exercise you fail to reject but you reject for the situation in part c. What does this illustrate about power?

22. Given the following values for $\bar{X}$ and s: (a) $\bar{X} = 44$, $s = 10$, (b) $\bar{X} = 43$, $s = 10$, (c) $\bar{X} = 43$, $s = 2$, test the hypothesis H_0: $\mu < 42$ with $\alpha = 0.05$ and $n = 16$.

23. Repeat the previous exercise but only test H_0: $\mu > 42$.

24. A company claims that, on average, when exposed to its toothpaste, 45% of all bacteria related to gingivitis is killed. You run 10 tests and find that the percentages of bacteria killed among these tests are 38, 44, 62, 72, 43, 40, 43, 42, 39, 41. The mean and standard deviation of these values are $\bar{X} = 46.4$ and $s = 11.27$. Assuming normality, test the hypothesis that the average percentage is 45 with $\alpha = 0.05$.

25. A portion of a study by Wechsler (1958) reports that for 100 males taking the Wechsler Adult Intelligent Scale (WAIS), the sample mean and variance on picture completion are $\bar{X} = 9.79$ and $s = 2.72$. Test the hypothesis H_0: $\mu \geq 10.5$ with $\alpha = 0.025$.

26. Given the following values for $\bar{X}_t$ and s_w: (a) $\bar{X}_t = 44$, $s_w = 9$, (b) $\bar{X}_t = 43$, $s_w = 9$, (c) $\bar{X}_t = 43$, $s_w = 3$, and assuming 20% trimming, test the hypothesis H_0: $\mu_t = 42$ with $\alpha = 0.05$ and $n = 20$.

27. Repeat the previous exercise, only test the hypothesis H_0: $\mu_t < 42$ with $\alpha = 0.05$ and $n = 16$.

28. For the data in Exercise 24, the trimmed mean is $\bar{X}_t = 42.17$ with a Winsorized standard deviation of $s_w = 1.73$. Test the hypothesis that the population trimmed mean is 45 with $\alpha = 0.05$.

29. A standard measure of aggression in 7-year-old children has been found to have a 20% trimmed mean of 4.8 based on years of experience. A psychologist wants to know whether the trimmed mean for children with divorced parents differs from 4.8. Suppose $\bar{X}_t = 5.1$ with $s_w = 7$ based on $n = 25$. Test the hypothesis that the population trimmed mean is exactly 4.8 with $\alpha = 0.01$.

30. For the following 10 bootstrap sample means, determine a p-value when testing the hypothesis H_0: $\mu = 8.5$.

7.6, 8.1, 9.6, 10.2, 10.7, 12.3, 13.4, 13.9, 14.6, 15.2.

What would be an appropriate 0.8 confidence interval for the population mean?

31. Rats are subjected to a drug that might cause liver damage. Suppose that for a random sample of rats, measures of liver damage are found to be

$$5, 12, 23, 24, 6, 58, 9, 18, 11, 66, 15, 8.$$

Verify that `trimci` returns a 0.95 confidence interval for the 20% trimmed mean, using R, equal to (7.16, 22.84). Now use `trimci` to get a confidence interval for the mean, resulting in (8.50, 33.99). Why is the confidence interval substantially longer?

32. For the data in the previous exercise, verify that the 0.95 confidence interval for the mean returned by the R function `trimcibt` is (12.40, 52.46). Note that the length of this interval is much higher than the length of the confidence interval for the mean using `trimci`.

33. Referring to the previous two exercises, which confidence interval for the mean is more likely to have probability coverage at least 0.95?

34. For the data in Exercise 30, verify that the equal-tailed .95 confidence interval for the population 20% trimmed mean using `trimcibt` is (8.17, 15.75).

35. For the data in Exercise 31, verify that the 0.95 confidence interval for the population 20% trimmed mean using `trimpb` is (9.75, 31.5).

36. Which of the two confidence intervals given in the last two exercises is likely to have probability coverage closer to 0.95?

37. For the following observations

$$2, 4, 6, 7, 8, 9, 7, 10, 12, 15, 8, 9, 13, 19, 5, 2, 100, 200, 300, 400$$

verify that the 0.95 confidence interval, based on a one-step M-estimator and returned by `onesampb`, is (7.36, 19.77).

38. For the data in the previous exercise, verify that the 0.95 confidence interval for the 20% trimmed mean returned by the R function `trimpb` is (7.25, 63). Why would you expect this confidence interval to be substantially longer than the confidence interval based on a one-step M-estimator?

39. Use `trimpb` on the data used in the previous two exercises, but this time trim 30%. Verify that the 0.95 confidence interval for the trimmed mean is (7.125, 35.500). Why do you think this confidence interval is shorter versus the confidence interval in the last exercise?

40. Repeat the last exercise, only now trim 40%. Verify that the 0.95 confidence interval is now (7.0, 14.5). Why is this confidence interval so much shorter than the confidence interval in the last exercise?

41. For the data in Exercise 37, what practical problem might occur if the standard error of the median is estimated using a bootstrap method?

42. For the data in Exercise 37, compute a 0.95 confidence interval for the median using the R function `sint` described in Chapter 4. What does this suggest about using a median versus a one-step M-estimator?

CHAPTER 6

REGRESSION AND CORRELATION

CONTENTS

A common goal is determining whether there is an association between two variables. If there is an association, there is the issue of describing it in a reasonably effective manner. Does aggression in the home have an association with the cognitive functioning of children living in the home? If yes, how might this association be described? Consider the typical exposure to solar radiation among residents living in large cities. Does knowing the amount of exposure to solar radiation provide any information about the breast cancer rate among women? In more formal terms, for any two variables, say X and Y, if we know the value of X (for example solar radiation), what does this tell us, if anything, about Y (such as breast cancer rates)? In general terms, what can be said about the conditional distribution of Y, given X. Given X, are we better able to predict the mean or median of Y? Does the variation in Y change with X? How might we characterize the strength of any association that might exist?

There is a vast literature on how to address the issues just described (e.g., Li, 1985; Montgomery and Peck, 1992; Staudte and Sheather, 1990; Hampel, Ronchetti, Rousseeuw, and Stahel, 1986; Huber, 1981; Rousseeuw and Leroy, 1987; Belsley, Kuh, and Welsch, 1980; Cook and Weisberg, 1992; Carroll and Ruppert, 1988; Hettmansperger, 1984; Hettmansperger and McKean, 1998; Wilcox, 20017; Hastie and Tibshirani, 1990) and not all issues are addressed here. One goal is to introduce two classic methods that are routinely taught and used. The first is based on what is called *least squares regression* and the second employs an estimate of what is called *Pearson's correlation.* Both least squares regression and Pearson's correlation provide useful and important tools for describing and understanding associations. But the reality is that situations arise where they are inadequate and misleading. In terms of hypothesis testing, the classic methods about to be described inherit the practical concerns associated with means that were described in Chapters 2, 4, and 5. In addition, new problems are introduced. So a goal is to develop some understanding of when these classic methods perform well as well as when and why they can be inadequate and misleading.

When classic methods are unsatisfactory, there is the issue of what might be done to correct known problems. Some alternatives to least squares regression are introduced here, but many of the improved methods are based in part on methods described in Chapters 12 and 13. Consequently, a more detailed description of robust regression methods is postponed until Chapter 14.

6.1 THE LEAST SQUARES PRINCIPLE

Imagine that we have n pairs of values

$$(X_1, Y_1), \ldots, (X_n, Y_n),$$

where, for example, X might be exposure to solar radiation and Y might be the breast cancer rate among adult women. The immediate goal is to find a straight line for predicting Y given some value for X. In more formal terms, the goal is to find a prediction rule having the form

$$\hat{Y} = b_0 + b_1 X, \tag{6.1}$$

where b_1 and b_0 are some choice for the slope and intercept, respectively. That is, given X, but not Y, we estimate that Y is equal to $\hat{Y} = b_0 + b_1 X$. The problem is determining a good choice for b_1 and b_0 based on the observed pairs of points. If, for example, the slope is positive, this suggests that as solar radiation increases, breast cancer rates tend to increase as well. And if b_1 is negative, this suggests that the opposite is true, but a point worth stressing is that this interpretation can be misleading as will be illustrated.

Note that once we choose values for the slope (b_1) and intercept (b_0), for each of the observed X values, namely $X_1, \dots, X_n$, we have a predicted Y value:

$$\hat{Y}_i = b_0 + b_1 X_i,$$

($i = 1, \dots, n$). Moreover, in general there will be some discrepancy between the observed Y_i value and its predicted value, $\hat{Y}_i$. This discrepancy can be measured with

$$r_i = Y_i - \hat{Y}_i,$$

which is called a *residual*. Residuals simply represent the error in our prediction rule based on $\hat{Y}$. The *least squares principle* for determining the slope and intercept is to determine the values for b_0 and b_1 that minimize the sum of the squared residuals. In formal terms, determine values for b_0 and b_1 that minimize

$$\sum r_i^2 = \sum (Y_i - b_0 - b_1 X_i)^2.$$

Without making any assumptions about the distribution of X or Y, the slope and intercept turn out to be

$$b_1 = \frac{\sum (X_i - \bar{X})(Y_i - \bar{Y})}{\sum (X_i - \bar{X})^2} \tag{6.2}$$

and

$$b_0 = \bar{Y} - b_1 \bar{X}, \tag{6.3}$$

respectively.

EXAMPLE

Is there some pattern to how the galaxies in the universe are moving relative to each other? Edwin Hubble collected data on two measures relevant to this issue in the hope of gaining some insight into how the universe was formed. He measured the distance of 24 galaxies from Earth plus their recession velocity. His measurements were published in 1929 and are shown in Table 6.1 where X is a galaxy's distance from Earth in megaparsecs and Y is its speed in kilometers per second. (One parsec is 3.26 light years.) For example, the first galaxy is 0.032 megaparsecs from Earth and moving away from earth at the rate of 170 kilometers per second. The third galaxy is 0.214 megaparsecs from earth and approaching Earth at the rate of 130 kilometers per second. The least squares regression line has slope $b_1 = 454.2$ and intercept $b_0 = -40.78$, which is shown in Figure 6.1. Roughly, this suggests that as the distance increases, the recession velocity increases as well.

To illustrate the notation, note that for the first galaxy in Table 6.1, which is $X_1 = 0.032$ megaparsecs from Earth, its predicted recession velocity is $\hat{Y}_1 = 454(0.032) - 40.8 = -26.3$ kilometers per second. That is, among all galaxies that are 0.032 megaparsecs from Earth, it

Table 6.1 Hubble's Data on the Distance and Recession Velocity of 24 Galaxies

Distance (X):	0.032 0.034 0.214 0.263 0.275 0.275 0.450 0.500 0.500 0.630 0.800 0.900 0.900 0.900 0.900 1.000 1.100 1.100 1.400 1.700 2.000 2.000 2.000 2.000
Velocity (Y):	170 290 −130 −70 −185 −220 200 290 270 200 300 −30 650 150 500 920 450 500 500 960 500 850 800 1090

is estimated that the average recession velocity is −26.3. But from Table 6.1 we see that for this particular galaxy, its actual recession velocity is $Y_1 = 170$, so for the first galaxy having $(X_1, Y_1) = (0.032, 170)$, there is a discrepancy between the actual and predicted velocity of $r_1 = Y_1 - \hat{Y}_1 = 170 + 26.3 = 196.3$. Table 6.2 summarizes the $\hat{Y}$ values and residuals for the data in Table 6.1.

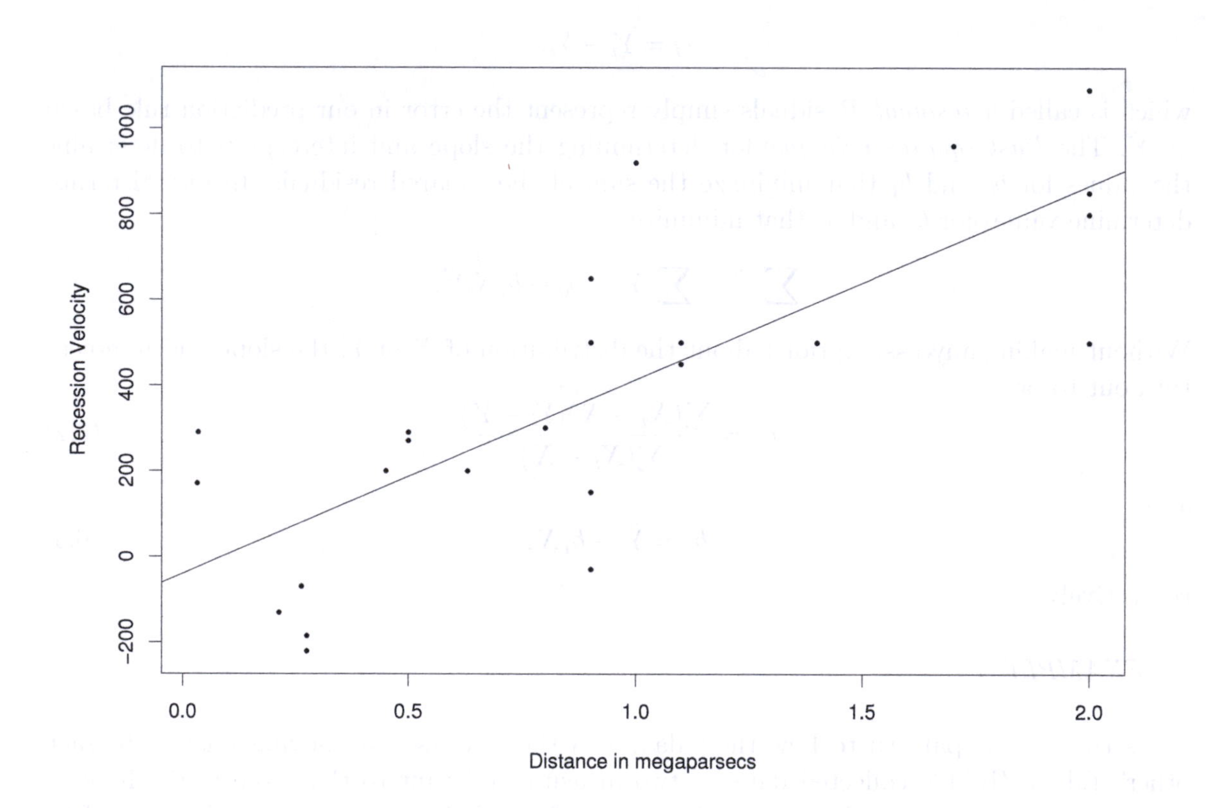

Figure 6.1 A plot of Hubble's data on the recession velocity of galaxies.

6.2 CONFIDENCE INTERVALS AND HYPOTHESIS TESTING

The least squares principle for determining the slope and intercept does not make any assumptions about the underlying distributions. In particular, normality is not required. Moreover, b_1 and b_0 provide estimates of the population slope and intercept β_1 and β_0, respectively. That is, β_1 and β_0 are the slope and intercept if all participants or objects could be measured. A common goal is testing hypotheses about β_1 and β_0, but to accomplish this goal,

Table 6.2 Fitted Values and Residuals for the Data in Table 6.1

Observation Number	Y_i	$\hat{Y}_i$	r_i
1	170	−26.3	196.3
2	290	−25.3	315.3
3	−130	56.4	−186.4
4	−70	78.7	−148.7
5	−185	84.1	−269.1
6	−220	84.1	−304.1
7	200	163.6	36.4
8	290	186.3	103.7
9	270	186.3	83.7
10	200	245.3	−45.3
11	300	322.5	−22.5
12	−30	368.0	−398.0
13	650	282	367.0
14	150	368.0	−217.0
15	500	368.0	132.0
16	920	413.4	506.6
17	450	458.8	−8.8
18	500	458.8	41.2
19	500	595.0	−95.0
20	960	731.3	228.7
21	500	867.5	−367.5
22	850	867.5	−17.5
23	800	867.5	−67.5
24	1090	867.5	222.5

Table 6.3 Measures of Marital Aggression and Recall-Test Scores

Family i	Aggression X_i	Test Score Y_i	Family i	Aggression X_i	Test Score Y_i
1	3	0	25	34	2
2	104	5	26	14	0
3	50	0	27	9	4
4	9	0	28	28	0
5	68	0	29	7	4
6	29	6	30	11	6
7	74	0	31	21	4
8	11	1	32	30	4
9	18	1	33	26	1
10	39	2	34	2	6
11	0	17	35	11	6
12	56	0	36	12	13
13	54	3	37	6	3
14	77	6	38	3	1
15	14	4	39	3	0
16	32	2	40	47	3
17	34	4	41	19	1
18	13	2	42	2	6
19	96	0	43	25	1
20	84	0	44	37	0
21	5	13	45	11	2
22	4	9	46	14	11
23	18	1	47	0	3
24	76	4			

two fundamental assumptions are typically made beyond random sampling. The first is called homoscedasticity and the second is that given X, the Y values have a normal distribution.

To describe these assumptions in a concrete manner, consider a study where the goal was to examine how children's information processing is related to a history of exposure to marital aggression. Results for two of the measures, based on data supplied by A. Medina, are shown in Table 6.3. The first, labeled X, is a measure of marital aggression that reflects physical, verbal, and emotional aggression during the last year, and Y is a child's score on a recall test. If we fit a straight line using the least squares principle, we find that $b_1 = -0.0405$ and $b_0 = 4.581$. So it is estimated, for example, that when the measure of aggression in the home is $X = 3$, that the average information processing score, among all homes with $X = 3$, is $\hat{Y} = -0.0405 + 4.581(3) = 13.7$.

Now consider all homes with an aggression score of $X = 3$, not just those shown in Table 6.3. Not all will have recall test scores of 13.7. Some will be higher and some will be lower. That is, there is variation among the recall test scores. In a similar manner, for all homes with $X = 40$ there will be variation among the recall test scores. *Homoscedasticity* refers to a situation where, regardless of what X might be, the variance of the Y values is always the same. So, for example, if among all homes with $X = 3$ the variance of the recall test scores is 14, then homoscedasticity means that the variance among all homes with $X = 50$ will again be 14. Indeed, the variance will be the same regardless of what X might be. *Heteroscedasticity* refers to a situation where the variances differ for one or more X values.

Figure 6.2 provides a graphic example of homoscedasticity. Shown is a plot of many Y values corresponding to each of three X values. For all three X values, the variation among the Y values is exactly the same. Now look at Figure 6.3 and note that the variance of the Y values differs depending on which X value we consider. That is, there is heteroscedasticity.

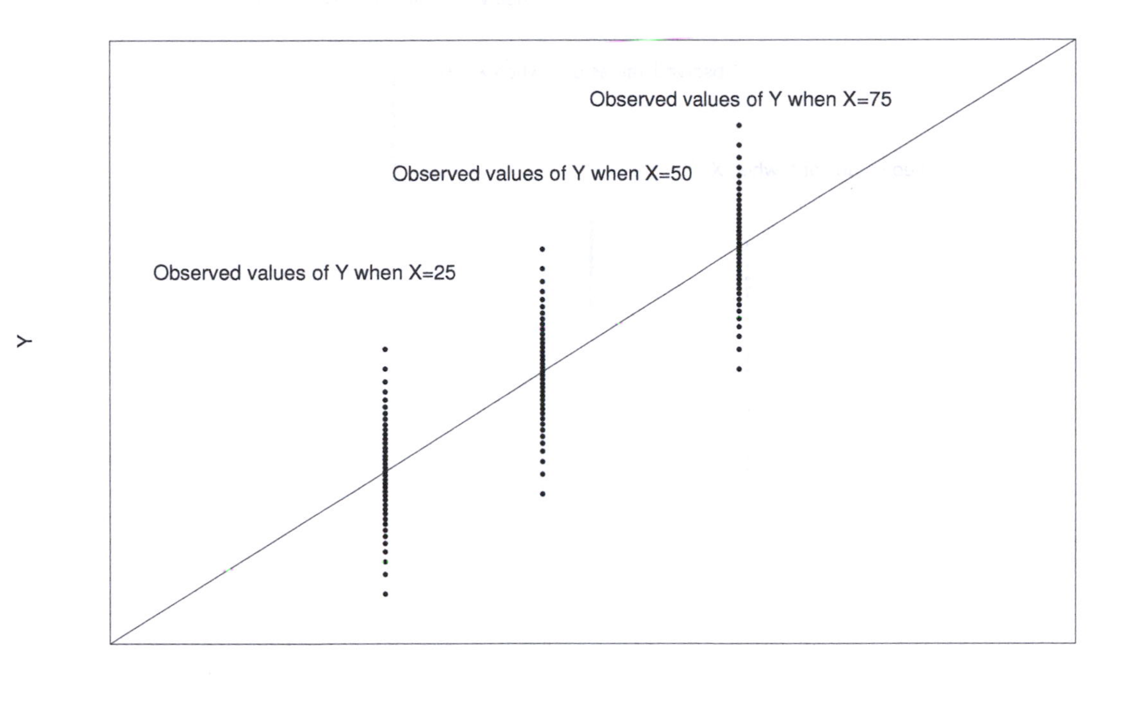

Figure 6.2 An example of homoscedasticity. The conditional variance of Y, given X, does not change with X.

The other assumption typically made when testing hypotheses is that Y, given X, has a normal distribution. In the illustration, it is assumed, for example, that among all homes with an aggression score of $X = 3$, the corresponding recall test scores have a normal distribution. The same is true, for example, among homes with an aggression score of $X = 5$.

The Classic Squares Regression Model

The assumptions just described are typically summarized in terms of the model

$$Y = \beta_0 + \beta_1 X + e, \tag{6.4}$$

where X and e are independent random variables, and e has a normal distribution with mean 0 and variance σ^2. This model says that the conditional distribution of Y, given X, has a normal distribution with mean

$$E(Y|X) = \beta_0 + \beta_1 X \tag{6.5}$$

and variance σ^2. That is, the (conditional) variance of Y, given X, does not depend in any way on the value of X.

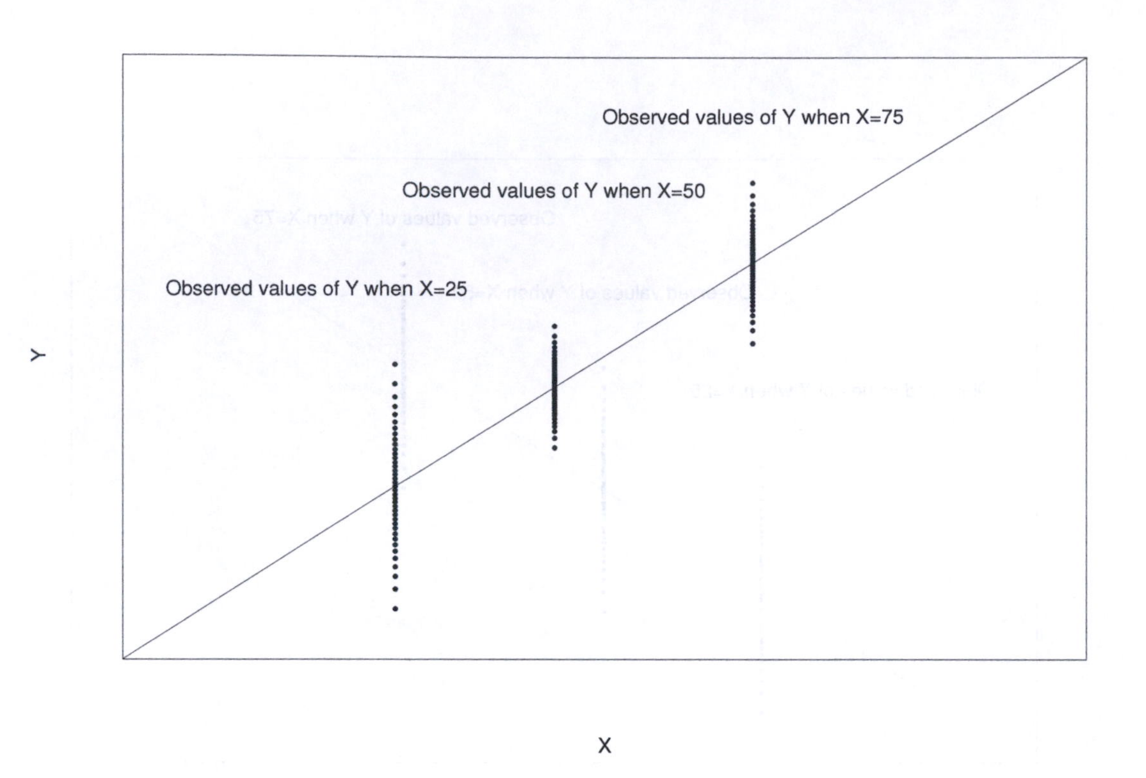

Figure 6.3 An example of heteroscedasticity. The conditional variance of Y, given X, changes with X.

6.2.1 Classic Inferential Techniques

Recall that in Chapters 4 and 5, the strategy used to compute confidence intervals and test hypotheses about the population mean was based in part on an expression for the standard error of the sample mean. As previously indicated, over many studies, the variance of the sample mean (its squared standard error) is σ^2/n, assuming random sampling. This result, coupled with assuming normality, makes it possible to test hypotheses about the population mean, μ. A similar strategy is used by the classic method for making inferences about the slope and intercept of a least squares regression line.

The Least Squares Estimate of the Slope and Intercept Has a Sampling Distribution

To begin, notice that b_1 has a sampling distribution. That is, if we were to repeat a study many times with each study based on n randomly sampled pairs of observations, we would get a collection of estimated slopes that generally differ from one another, and, of course, the same is true for the intercept. For the regression model given by Equation (6.4), it can be shown that the average value of the estimated slope, b_1, over (infinitely) many studies, is

$$E(b_1) = \beta_1$$

and the average value of the intercept is

$$E(b_0) = \beta_0.$$

That is, b_1 and b_0 are unbiased estimates of the slope and intercept. The classic approach to testing hypotheses and computing confidence intervals for the slope and intercept is based in part on expressions for the squared standard error of the least squares estimators, b_1 and b_0. To get a relatively simple expression for the standard errors, homoscedasticity is typically assumed. Momentarily assume that homoscedasticity is true and let σ^2 represent the common variance. That is, the Y values corresponding to any X value have variance σ^2 regardless of what X might be, as previously explained. In symbols, it is being assumed that

$$\mathrm{VAR}(Y|X) = \sigma^2.$$

When this assumption is true, it can be seen that the squared standard error of b_1 (the variance of b_1) is

$$\mathrm{VAR}(b_1) = \frac{\sigma^2}{\sum(X_i - \bar{X})^2} \tag{6.6}$$

and the squared standard error of b_0 is

$$\mathrm{VAR}(b_0) = \frac{\sigma^2 \sum X_i^2}{n \sum(X_i - \bar{X})^2}. \tag{6.7}$$

In practice, σ^2 is not known, but it can be estimated with

$$\hat{\sigma}^2 = \frac{1}{n-2} \sum r_i^2.$$

In words, compute the residuals as previously illustrated, in which case the estimate of the assumed common variance is the sum of the squared residuals divided by $n - 2$.

Under the assumptions that the regression model is true, and that there is random sampling, homoscedasticity, and normality, confidence intervals for the slope and intercept can be computed, and hypotheses can be tested. In particular, a $1 - \alpha$ confidence interval for the slope, β_1, is

$$b_1 \pm t\sqrt{\frac{\hat{\sigma}^2}{\sum(X_i - \bar{X})^2}}, \tag{6.8}$$

where t is the $1 - \alpha/2$ quantile of Student's t distribution with $\nu = n - 2$ degrees of freedom. (The value of t is read from Table 4 in Appendix B.) The quantity

$$\sqrt{\frac{\hat{\sigma}^2}{\sum(X_i - \bar{X})^2}}$$

is the estimated standard error of b_1. The quantity

$$\sqrt{\frac{\hat{\sigma}^2 \sum X_i^2}{n \sum(X_i - \bar{X})^2}}$$

is the estimated standard error of b_0. For the common goal of testing H_0: $\beta_1 = 0$, the hypothesis that the slope is zero, reject if the confidence interval does not contain zero. Alternatively, reject if

$$|T| \geq t,$$

where again t is the $1 - \alpha/2$ quantile of Student's t distribution with $\nu = n - 2$ degrees of freedom and

$$T = b_1\sqrt{\frac{\sum(X_i - \bar{X})^2}{\hat{\sigma}^2}}. \tag{6.9}$$

As for the intercept, β_0, a $1-\alpha$ confidence interval is given by

$$b_0 \pm t\sqrt{\frac{\hat{\sigma}^2 \sum X_i^2}{n\sum(X_i-\bar{X})^2}}. \tag{6.10}$$

It is noted that p-values can be computed when applying the hypothesis testing methods just described. As in Chapter 5, p-values refer to the probability of Type I error if the observed value of the test statistic is used as a critical value. Alternatively, the p-value is the smallest α value (Type I error probability) for which the null hypothesis would be rejected. For instance, imagine that $T=1.6$ when testing H_0: $\beta_1=0$. Using 1.6 as a critical value means that the null hypothesis is rejected if $|T|\geq 1.6$. If, for example, the degrees of freedom are $\nu=24$, it can be seen that $P(|T|\geq 1.6)=0.12$, which is the p-value. Roughly, if we reject when $|T|\geq 1.6$, the probability of a Type I error is 0.12 (under the standard assumptions previously described). Or in terms of Tukey's three decision rule, the p-value quantifies the strength of the empirical evidence that a decision can be made about whether β_1 is greater or less than zero. As will become evident, a p-value is routinely reported by the R functions described in this chapter.

EXAMPLE

Using the aggression data in Table 6.3, we test the hypothesis H_0: $\beta_1=0$ with the goal that the probability of a Type I error be 0.05, assuming normality and that the error term is homoscedastic. Because $\alpha=0.05$, $1-\alpha/2=0.975$. There are $n=47$ pairs of observations, so the degrees of freedom are $\nu=47-2=45$, and the critical value is $t=2.01$. The least squares estimate of the slope is $b_1=-0.0405$, and it can be seen that $\sum(X_i-\bar{X})^2=34659.74$ and $\hat{\sigma}^2=14.15$, so the test statistic, given by Equation (6.9), is

$$T=-0.0405\sqrt{\frac{34659.74}{14.5}}=-1.98.$$

Because $|T|=1.98<2.01$, fail to reject.

6.2.2 Multiple Regression

Multiple regression refers to a simple generalization of the regression model previously described to situations where there are two or more predictors. We observe $p+1$ variables: $Y, X_1, \ldots, X_p$ and now the typical assumption is that

$$Y=\beta_0+\beta_1 X_1+\cdots \beta_p X_p+e,$$

where e has a normal distribution and with mean zero and variance σ^2. This says that the mean of Y is assumed to be $\beta_0+\beta_1X_1+\cdots\beta_pX_p$, given $X_1,\ldots,X_p$. Homoscedasticity means that the variance of the error term, e, does not depend on the values $X_1,\ldots,X_p$. Said another way, the variance of Y does not depend on what the values $X_1,\ldots,X_p$ happen to be. In more precise terms

$$\text{VAR}(Y|X_1,\ldots,X_p)=\sigma^2.$$

That is, the (conditional) variance of Y is σ^2 regardless of what the values $X_1,\ldots,X_p$ might be.

As was the case with one predictor ($p=1$), the least squares principle is to estimate the slopes and intercept with the values $b_0, b_1, \ldots, b_p$ that minimize the sum of the squared residuals. Briefly, let

$$\hat{Y}=b_0+b_1X_1+\cdots+b_pX_p$$

be the predicted value of Y, given the value of the p predictors $X_1, \ldots, X_p$. In practice, we have n Y values, $Y_1, \ldots, Y_n$ and we denote their predicted values by $\hat{Y}_1, \ldots, \hat{Y}_n$. The least squares principle is to choose $b_0, b_1, \ldots, b_p$ so as to minimize

$$\sum (Y_i - \hat{Y}_i)^2.$$

Testing the Hypothesis That All Slopes Are Zero

A typical goal is to test

$$H_0 : \beta_1 = \cdots = \beta_p = 0, \tag{6.11}$$

the hypothesis that all p slope parameters are zero. In addition to the assumptions just described, the classic approach assumes that the error term, e, has a normal distribution, which is the same thing as saying that given values for the predictors $X_1, \ldots, X_p$, the corresponding Y values have a normal distribution.

The classic and most commonly used test of Equation (6.11) is based in part on what is called the *squared multiple correlation coefficient*:

$$R^2 = \frac{\sum (\hat{Y}_i - \bar{Y})^2}{\sum (Y_i - \bar{Y})^2}. \tag{6.12}$$

The numerator of R^2 measures the overall squared difference between the predicted Y values, $\hat{Y}$, and the mean of the Y values, $\bar{Y}$. The denominator measures the overall squared difference between the observed Y values and $\bar{Y}$. R^2 is a popular method for measuring the strength of the association between Y and the p predictors $X_1, \ldots, X_p$, but comments on the use and interpretation of R^2 are postponed until Section 6.5. The only point here is that the classic method for testing the hypothesis given by Equation (6.11) is based on the test statistic

$$F = \left(\frac{n-p-1}{p}\right)\left(\frac{R^2}{1-R^2}\right). \tag{6.13}$$

Under normality and homoscedasticity, and if the null hypothesis is true, F has what is called an F distribution with $\nu_1 = p$ and $\nu_2 = n - p - 1$ degrees of freedom. That is, the null distribution, and hence the critical value, depends on two values: the number of predictors, p, and the sample size, n. Critical values, f, for $\alpha = 0.1$, 0.05, 0.025, and 0.01 are reported in Tables 5, 6, 7, and 8, respectively, in Appendix B.

Decision Rule: Reject the hypothesis that all slopes are equal to zero if $F \geq f$.

EXAMPLE

If there are four predictors and the sample size is 20, then $p = 4$, $n = 20$, $\nu_1 = 4$, $\nu_2 = 20 - 4 - 1 = 15$, and if the Type I error probability is to be 0.05, Table 6 in Appendix B indicates that $f = 3.06$. That is, reject the hypothesis that all of the slopes are zero if $F \geq 3.06$.

6.2.3 R Functions ols and lm

R has a built-in function for testing Equation (6.11). If you have three predictors with the data stored in the R variables `x1`, `x2`, and `x3`, and the outcome predictor stored in `y`, then the built-in R command

```
summary(lm(y~x1+x2+x3))
```

will perform the F test just described, and it performs a Student's T test of H_0: $\beta_j = 0$ $(j = 0, \ldots, p)$ for each of the $p + 1$ regression parameters.

Notice the use of the R function `lm`, which stands for *linear model*. This function plays a role when dealing with a variety of commonly used models, as we shall see, particularly in Chapter 14. Here we are using an *additive model*, which is indicated by the pluses in the expression `x1+x2+x3`.

For convenience, another function (written for this book) is provided for performing the F test called

```
ols(x, y, xout = FALSE, outfun = outpro, plotit = FALSE, xlab = "X", ylab =
    "Y", zlab = "Z", RES = FALSE, ...).
```

One difference is that when dealing with a single predictor, a scatterplot of the points, which includes the least squares regression line, is created when the argument `plotit=TRUE`. Another important difference is the inclusion of the arguments `xout` and `outfun`. By default, `xout=FALSE`, meaning that no check for outliers among the `x` values is made. Setting `xout=TRUE`, outliers are removed, where the argument `outfun` indicates the method used to search for outliers. When there is one predictor only, by default a MAD-median rule is used, as described in Chapter 2. The command

```
ols(x,y,xout=TRUE,outfun=outbox)
```

would remove outliers using the boxplot rule. (With more than one predictor, the method in Section 13.1.6 is used. Consideration might also be given to using `outfun=outmgv`.)

EXAMPLE

Stromberg (1993) reports data on 29 lakes in Florida. The data were collected by the United States Environmental Protection Agency, which are reproduced in Table 6.4. There are three variables: TN, the mean annual total nitrogen concentration; NIN, the average influent nitrogen concentration; and TW, the water retention time. (The data can be downloaded from the author's web page; see Chapter 1.) In this study the two predictors of interest, X_1 and X_2, are NIN and TW, and the outcome of interest, Y, is TN. Using the built-in R function just described, it is found that $F = 0.26$, the degrees of freedom are $\nu_1 = 2$ and $\nu_2 = 26$, and if the Type I error is set at $\alpha = 0.05$, the critical value is $f = 3.37$. So the hypothesis given by Equation (6.11) is not rejected. The p-value is the probability of getting a test statistic greater than or equal to the observed value of F, 0.26, and is 0.773.

Here is how the results appear when using the R function `ols`:

```
$coef:
                 Value Std. Error    t value     Pr(>|t|)
(Intercept)  1.94674315  0.2919282  6.6685678 4.489461e-07
          x  0.02155083  0.0368536  0.5847686 5.637439e-01
          x -0.16741597  0.2678478 -0.6250414 5.373915e-01

$Ftest.p.value:
     value
 0.7730028
```

Table 6.4 Lake Data

NIN	TN	TW
5.548	2.59	0.137
4.896	3.770	2.499
1.964	1.270	0.419
3.586	1.445	1.699
3.824	3.290	0.605
3.111	.930	0.677
3.607	1.600	0.159
3.557	1.250	1.699
2.989	3.450	0.340
18.053	1.096	2.899
3.773	1.745	0.082
1.253	1.060	0.425
2.094	.890	0 0.444
2.726	2.755	0.225
1.758	1.515	0.241
5.011	4.770	0.099
2.455	2.220	0.644
.913	.590	0.266
.890	.530	0.351
2.468	1.910	0.027
4.168	4.010	0.030
4.810	1.745	3.400
34.319	1.965	1.499
1.531	2.55	0.351
1.481	.770	0.082
2.239	.720	0.518
4.204	1.730	0.471
3.463	2.860	0.036
1.727	.760	.721

Under the column headed Value we see the estimated intercept and slopes. For example, the slope associated with the first predictor (NIN) is $b_1 = 0.0216$ and the regression model is estimated to be

$$\hat{Y} = 1.94674315 + 0.02155083X_1 - 0.16741597X_2.$$

The last column gives the p-values for each individual hypothesis. For example, when testing H_0: $\beta_1 = 0$ with Student's T, the p-value is 0.5637. Note that for both predictors, no association is found.

6.3 STANDARDIZED REGRESSION

Rather than use the raw data to compute a regression line, a common strategy is to standardize the variables under study. This simply means that rather than compute the least squares estimator using the raw data, the observations are first converted to Z scores. For the aggression data in Table 6.3, for example, it can be seen that the test scores (Y) have mean $\bar{Y} = 3.4$ and standard deviation $s_y = 3.88$. The first test score is $Y_1 = 0$, and its Z score equivalent is $Z = (0 - 3.4)/3.88 = -0.88$. Of course, the remaining Y values can be converted to a Z score in a similar manner. In symbols, the Z score corresponding to Y_i is

$$Z_{yi} = \frac{Y_i - \bar{Y}}{s_y}.$$

In a similar manner, the Z score for corresponding to X_i is

$$Z_{xi} = \frac{X_i - \bar{X}}{s_x}.$$

For the aggression data, $\bar{X} = 28.5$ and $s_x = 27.45$. The first entry in Table 6.3 has $X_1 = 3$, so $Z_{x1} = -0.93$. Next, you determine the least squares estimate of the slope using the transformed X and Y values just computed. The resulting estimate of the slope will be labeled b_z, which always has a value between -1 and 1. (The quantity b_Z is equal to Pearson's correlation, which is discussed later in this chapter.) The resulting estimate of the intercept is always zero, so the regression equation takes the form

$$\hat{Z}_y = b_z Z_x.$$

For the aggression data, it can be seen that $b_z = -0.29$, so $\hat{Z}_y = -0.29(Z_x)$.

The standardized regression coefficient, b_z, can be computed in another manner. First compute the least squares estimate of the slope using the original data yielding b_1. Then

$$b_z = b_1 \frac{s_x}{s_y}.$$

EXAMPLE

For the aggression data in Table 6.3, the sample standard deviations of the X and Y values are $s_y = 3.88$ and $s_x = 27.45$, respectively. As previously indicated, the least squares estimate of the slope is $b_1 = -0.0405$. The standardized slope is just

$$b_z = -0.0405 \frac{27.45}{3.88} = -0.29.$$

A possible appeal of standardized regression is that it attempts to provide perspective on the magnitude of a predicted value for Y. Recall from Chapter 3 that for normal distributions, the value of $Z = (X - \mu)/\sigma$ has a convenient probabilistic interpretation under normality. For example, half the observations fall below a Z score of zero. A Z score of one indicates we are one standard deviation above the mean and about 84% of all observations are below this point when observations have a normal distribution. (From Table 1 in Appendix B, $Z = 1$ is the 0.84 quantile.) Similarly, $Z = 2$ refers to a point two standard deviations above the mean, and approximately 98% of all observations are below this value. Thus, for normal distributions, Z scores give you some sense of how large or small a value happens to be. Standardized regression attempts to tells us, for example, how a change of one standard deviation in X is related to changes in Y, again measured in standard deviations.

EXAMPLE

For the aggression data, assume normality and suppose we want to interpret the standardized regression estimate of the recall test when the measure of aggression is one standard deviation above or below the mean. One standard deviation above the mean of the aggression scores, X, corresponds to $Z_x = 1$, so as previously indicated, $\hat{Z}_y = (-0.29)1 = -0.29$. For a standard normal distribution, the probability of being less than -0.29 is approximately 0.39 and this provides a perspective on how the recall test is related to the measure of aggression.

Table 6.5 Z Scores for the Recall Test Scores in Table 6.3, Written in Ascending Order

−0.88	−0.88	−0.88	−0.88	−0.88	−0.88	−0.88	−0.88	−0.88	−0.88	−0.88	−0.88
−0.62	−0.62	−0.62	−0.62	−0.62	−0.62	−0.62	−0.36	−0.36	−0.36	−0.36	−0.36
−0.10	−0.10	−0.10	−0.10	0.15	0.15	0.15	0.15	0.15	0.15	0.15	0.41
0.67	0.67	0.67	0.67	0.67	0.67	1.44	1.96	2.47	2.47	3.51	

In a similar manner, one standard deviation below the mean of the aggression scores corresponds to $Z_x = -1$, so now $\hat{Z}_y = (-0.29)(-1) = 0.29$, and there is approximately a 0.61 probability that a standard normal random variable is less than 0.29.

For non-normal distributions, situations arise where Z scores can be interpreted in much the same way as when distributions are normal. But based on results in Chapters 2 and 3, this is not always the case even with a very large sample size and a very small departure from normality.

EXAMPLE

The last example illustrated how to interpret $\hat{Z}_y$ assuming normality, but now we take a closer look at the data to see whether this interpretation might be misleading. Table 6.5 shows all 47 Z_y scores for the recall test values written in ascending order. We see that 24 of the 47 values are below -0.29, so the proportion below -0.29 is $24/47 = 0.51$. This means that based on the available data, yhe estimate is that there is 0.51 probability of having a Z_y score less than -0.29. Put another way, a Z score of -0.29 corresponds, approximately, to the median. In contrast, for normal distributions, $Z_y = 0$ is the median and the probability of getting a Z_y score less than -0.29 is approximately 0.39. Thus, there is some discrepancy between the empirical estimate of the probability of getting a Z_y score less than -0.29 versus the probability you get assuming normality. The main point here is that switching to standardized regression does not necessarily provide a perspective that is readily interpretable.

A criticism of this last example is that the estimated probability of getting a Z score less than -0.29 is based on only 47 observations. Perhaps with a larger sample size, the estimated probability would be reasonably close to 0.39, the value associated with a normal distribution. However, results in Chapter 3 indicate that even with a large sample size, there can be a considerable difference, so caution is recommended when interpreting standardized regression equations.

6.4 PRACTICAL CONCERNS ABOUT LEAST SQUARES REGRESSION AND HOW THEY MIGHT BE ADDRESSED

There are a number of serious concerns regarding least squares regression in general and the classic hypothesis testing methods introduced in Section 6.2. This section summarizes some of these issues and describes how they might be addressed. Generally, there are two broad strategies that might be used. The first is to simply abandon least squares regression and replace it with an alternative regression method that deals with some of the issues illustrated here. One possibility is to use a quantile regression method, which is described later in this chapter, and another is to switch to one of the regression methods described in Chapter 14. The other strategy is to continue to use least squares regression but modify the classic methods in an attempt to deal with known problems. In some situations this latter strategy is satisfactory, but as will become evident, not all practical problems can be adequately addressed based on this approach. That is, for certain purposes to be described, there are practical reasons for abandoning least squares regression.

6.4.1 The Effect of Outliers on Least Squares Regression

Chapter 2 noted that a simple way of quantifying the sensitivity of a location estimator to outliers is with its finite sample breakdown point. The sample mean, for example, has a finite breakdown point of only $1/n$ meaning that only a single outlier is needed to render the sample mean a poor reflection of the typical value. Least squares regression line suffers from the same problem. That is, even a single outlier can have a large impact on the least squares regression line. In fact, even if no outliers are detected among the Y values (using for example the MAD-median rule or the boxplot rule in Chapter 2), and simultaneously no outliers are detected among the X values, it is still possible that a few points have an inordinate influence on the least squares regression line.

EXAMPLE

To illustrate the potential effect of a few unusual values, Figure 6.4 shows 20 points that were generated on a computer where both X and Y have normal distributions and the points are centered around the line $Y = X$. So the true slope is one ($\beta_1 = 1$). The solid straight line passing through the bulk of the points in Figure 6.4 is this least squares estimate of the regression line and has slope $b_1 = 1.01$. So in this particular case, the estimated slope is nearly equal to the true slope used to generate the data. Then two additional points were added at $X = 2.1$ and $Y = -2.4$ and are marked by the square in the lower right corner of Figure 6.4. Among all 22 X values, none are declared outliers by any of the methods in Chapter 2, and the same is true for the Y values. Yet these two additional points are clearly unusual relative to the other 20 and they have a substantial influence on the least squares estimate of the slope. (Chapter 14 describes methods that can detect the type of outliers shown in Figure 6.4.) Now the estimated slope is 0.37 and the resulting least squares regression line is represented by the dotted line in Figure 6.4. Note that the estimated intercept changes substantially as well.

EXAMPLE

Figure 6.5 shows the surface temperature (X) of 47 stars versus their light intensity. The solid line is the least squares regression line. As is evident, the regression line does a poor job of summarizing the association between the two variables under study. In this particular case, it is clear that the four points in the upper left portion of Figure 6.5 are unusual. Moreover, the R function `outbox` (described in Chapter 2) indicates that X values less than or equal to 3.84 are outliers. If we simply exclude all points with X values declared outliers and apply least squares regression to the points that remain, we get the dotted line shown in Figure 6.5, which provides a better indication of the association among the bulk of the points.

One point should be stressed. In the last example we simply restricted the range of X values to get a better fit to the bulk of the points under study. A similar strategy might be used when dealing with unusual Y values, but when restricting the range of Y values, this creates technical difficulties when testing hypotheses that require special techniques. (Some methods for dealing with this issue are described in Chapter 14.)

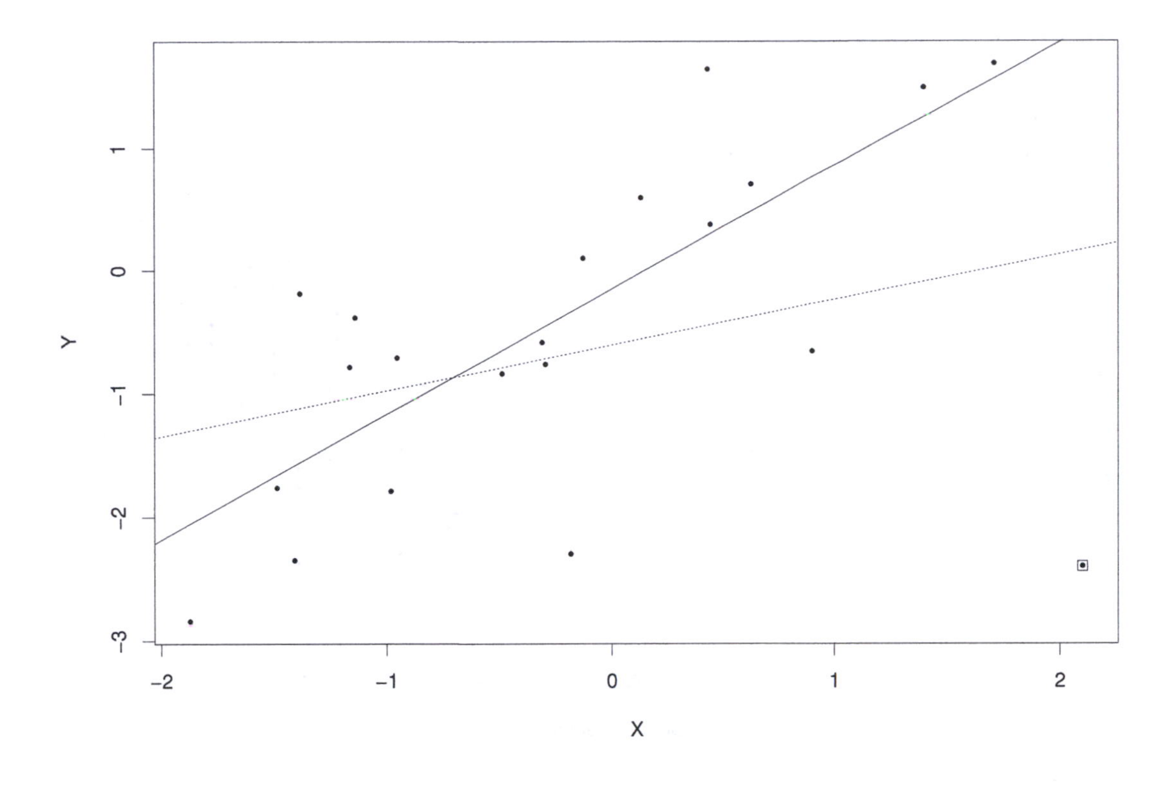

Figure 6.4 The two points marked by the square in the lower right corner have a substantial impact on the least squares regression line. Ignoring these two points, the least squares regression is given by the solid line. Including them, the least squares regression line is given by the dotted line. Moreover, none of the X or Y values are declared outliers using the methods in Chapter 2.

6.4.2 Beware of Bad Leverage Points

In regression, for any outlier among the X values, the corresponding point is called a *leverage point.* There are two kinds of leverage points: good and bad.

Bad Leverage Points

Roughly, bad leverage points are outliers that can result in a misleading summary of how the bulk of the points are associated. That is, bad leverage points are not consistent with the association among most of the data.

EXAMPLE

Consider again the example based on the lake data in Table 6.4. Recall that, based on the method in Section 6.2, no association was found when testing at the 0.05 level. Momentarily focus on the first predictor, NIN. The left panel of Figure 6.6 shows a plot of the data; the nearly horizontal line is the least squares regression line. The two largest NIN values are bad leverage points. As is fairly evident, they are not consistent with the association among the

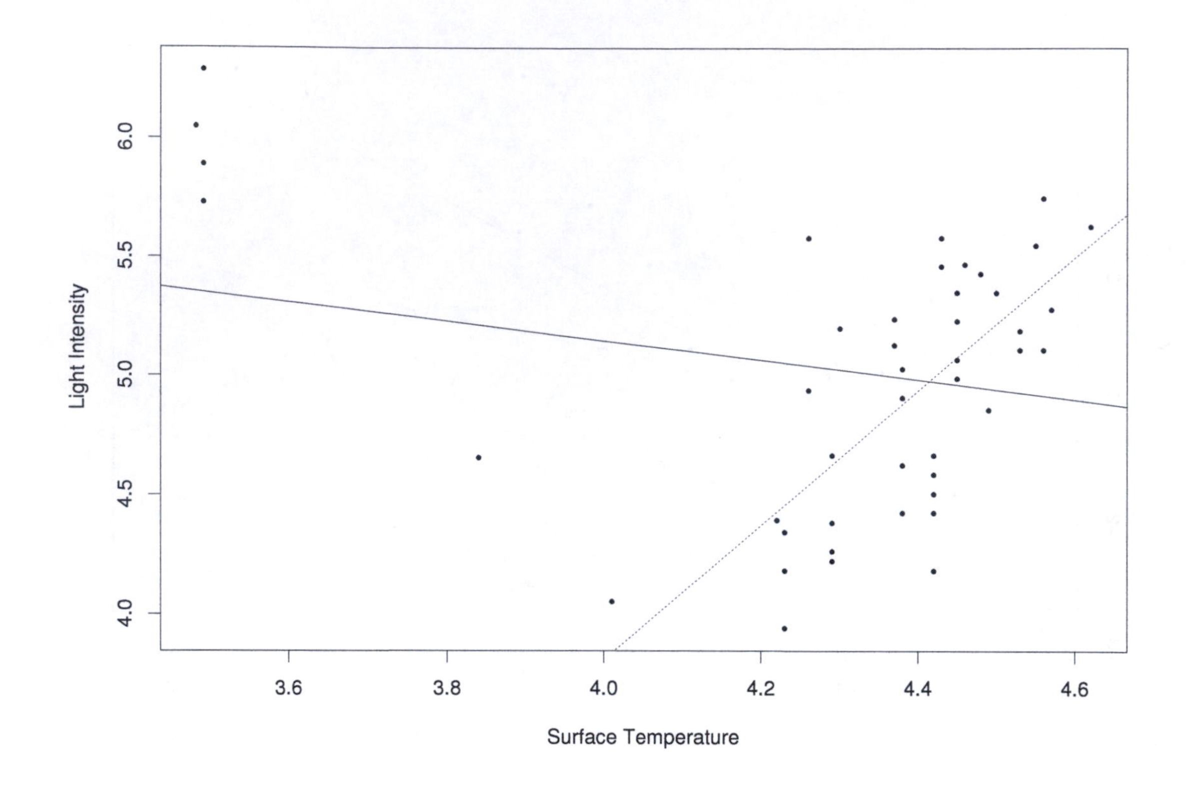

Figure 6.5 The solid line is the least squares regression line using all of the star data. Ignoring outliers among the X values, the least squares regression line is now given by the dotted line.

bulk of the points. If we eliminate them, now the regression line is given by the dotted line in the left panel of Figure 6.6 and the resulting p-value, when testing H_0: $\beta_1 = 0$, drops from 0.72 (when the outliers are included) to 0.0003. So, in particular, now the null hypothesis is rejected when testing at the 0.001 or even 0.0003 level, illustrating that bad leverage points can have a large impact when testing hypotheses.

Recall that when both predictors are included in the model, the p-value, when testing H_0: $\beta_1 = 0$, is 0.56. As just illustrated, ignoring the second predictor, the p-value is 0.72, illustrating that p-values are a function in part of which predictors are entered into the model. If we perform the analysis again with both predictors, but with outliers among the predictors removed, now the p-value when testing H_0: $\beta_1 = 0$ is 0.0009, illustrating again that a few bad leverage points can have a large impact when testing hypotheses.

EXAMPLE

L. Doi conducted a study dealing with predictors of reading ability. One portion of her study used a measure of digit naming speed (RAN1T) to predict the ability to identify words (WWISST2). Using all of the data, the slope of the least squares regression line is nearly equal to zero, which is the nearly horizontal line in the right panel of Figure 6.6. When testing H_0: $\beta_1 = 0$, the p-value is 0.76. But the scatterplot clearly reveals that the six largest

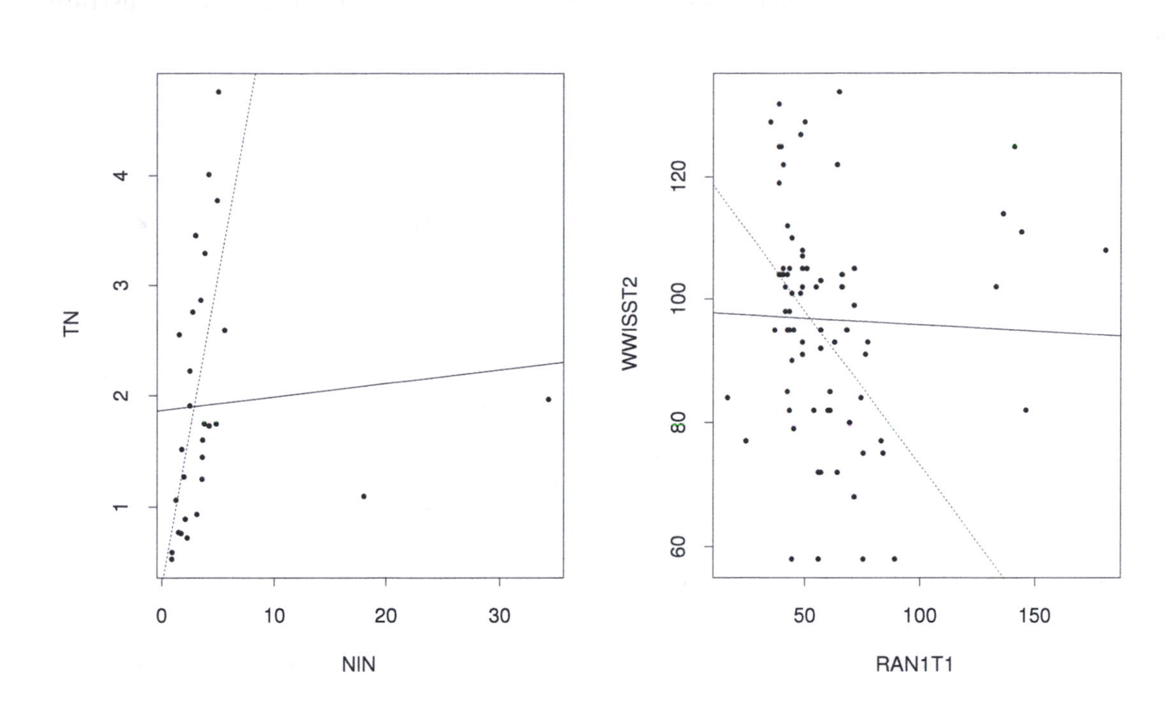

Figure 6.6 The left panel shows a scatterplot of the lake data. The bad leverage points, located in the lower right portion of the scatterplot, have a tremendous influence on the least squares estimate of the slope resulting in missing any indication of an association when using least squares regression. The right panel shows a scatterplot of the reading data. Now the bad leverage points mask a negative association among the bulk of the points.

X values are outliers. When these outliers are removed, the dotted line shows the resulting least squares regression line and now the p-value is 0.002.

Here is another way of viewing bad leverage points that might be useful. Observe that in the last two examples, bad leverage points correspond to unusually large X values. Look at the left panel of Figure 6.6 and note that all but two of the NIN values are less than 10; the other two are greater than 15. One might argue that we would like to know the association between NIN and TN when NIN has a value greater than 10. But there is almost no available information about this issue and in particular there is no compelling evidence that the association, when NIN is less than 10, is the same as the association when NIN is greater than 15. A similar concern arises when looking at the right panel of Figure 6.6.

A point worth stressing is that even if no leverage points are found using, say, the MAD-median rule, a few unusual points can still have a large impact on the least squares regression line. Figure 6.4 illustrated this point.

Good Leverage Points

Leverage points can distort the association among most of the data, but not all leverage points are bad. A crude description of a *good leverage point* is a point that is reasonably consistent with the regression line associated with most of the data. Figure 6.7 illustrates the basic idea.

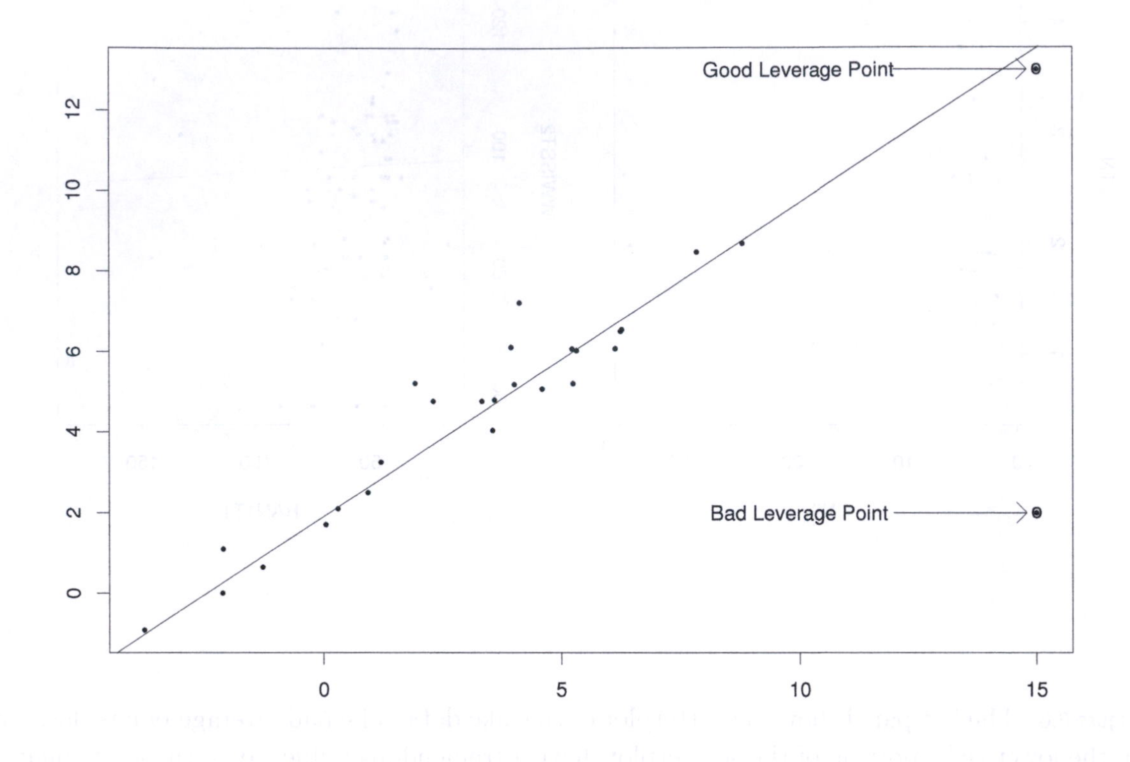

Figure 6.7 Shown are both good and bad leverage points. Good leverage points do not mask the true association among the bulk of the points and they have the practical advantage of resulting in shorter confidence intervals.

Notice that even a single leverage point can inflate $\sum(X_i - \bar{X})^2$, which is the numerator of the sample variance for the X values. (As illustrated in Chapter 2, outliers inflate the sample variance.) But $\sum(X_i - \bar{X})^2$ appears in the denominator of the expression for the standard error of b_1, so a single leverage point can cause the standard error of b_1 to be small compared to a situation where no leverage points occurred. In practical terms, we can get shorter confidence intervals, which might mean more power when there are leverage points, but caution must be exercised because leverage points can result in a poor fit to the bulk of the data as was illustrated.

6.4.3 Beware of Discarding Outliers Among the Y Values

We have seen examples where outliers among the predictors can result in a distorted sense of the association among the bulk of the points. From a technical point of view, assuming normality and homoscedasticity, it is permissible to test hypotheses using the methods in Section 6.2 after extreme X values are eliminated. But when eliminating outliers among the Y values, technical issues arise that should not be ignored. To provide at least some sense of why, recall that in Chapter 4, if we trim outliers and average the remaining data, special methods are needed to estimate the standard error. In addition, it was illustrated that if we ignore this issue, the resulting estimate of the standard error can be substantially different from an estimate that is theoretically sound. A similar issue arises here. There are theoretically sound methods for testing hypotheses using a regression estimator designed to guard against the deleterious effects of outliers among the Y values. One possibility is to switch to a quantile regression estimator, which is discussed later in this chapter, and other possibilities are described in Chapter 14.

6.4.4 Do Not Assume Homoscedasticity or that the Regression Line is Straight

In some situations, using a straight regression line provides an adequate summary of the association between two variables. But this should not be taken for granted.

EXAMPLE

Figure 6.8 shows a scatterplot of points relating the concentration of nitric oxides in engine exhaust versus its equivalence ratio, a measure of the richness of the air-ethanol mix. There is a clear indication of an association, but, clearly, fitting a straight regression line is inadequate.

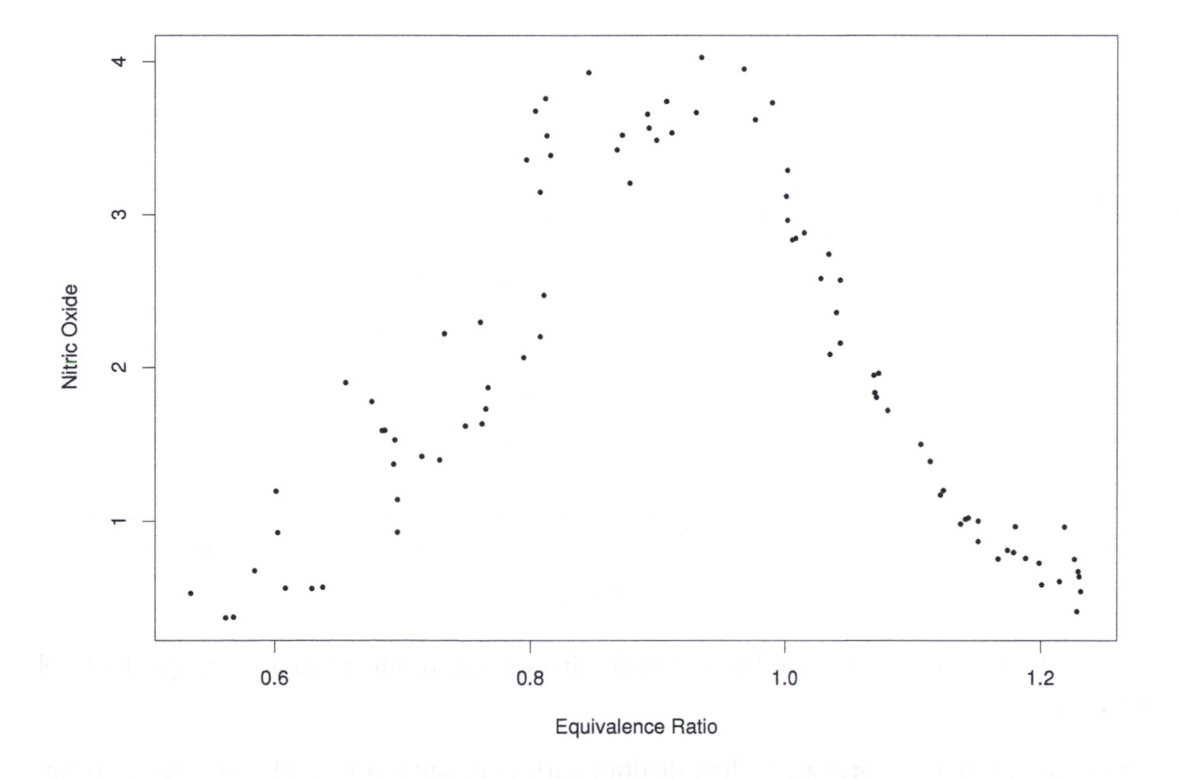

Figure 6.8 Curvature occurs in applied work as illustrated here with data from a study of how concentrations of nitric oxides in engine exhaust are related to its equivalence ratio.

EXAMPLE

This example is based on data aimed at investigating the association between a measure of a participant's quality of life and a measure of anxiety. (The data were generously supplied by D. Erceg-Hurn. The plot was created with the R function `lplot` in Section 14.4.2.) Simply fitting a straight line and testing the hypothesis that the slope is zero, the resulting p-value (using the R function `olshc4` in Section 6.4.7) is 0.0003 and the estimate of the slope is −0.7. So a superficial analysis suggests that as anxiety increases, the typical measure of the quality of life decreases. But Figure 6.9 suggests that a more accurate description is that the quality of life decreases up to a point and then levels off. (Methods in Chapter 14 lend support for this latter interpretation.)

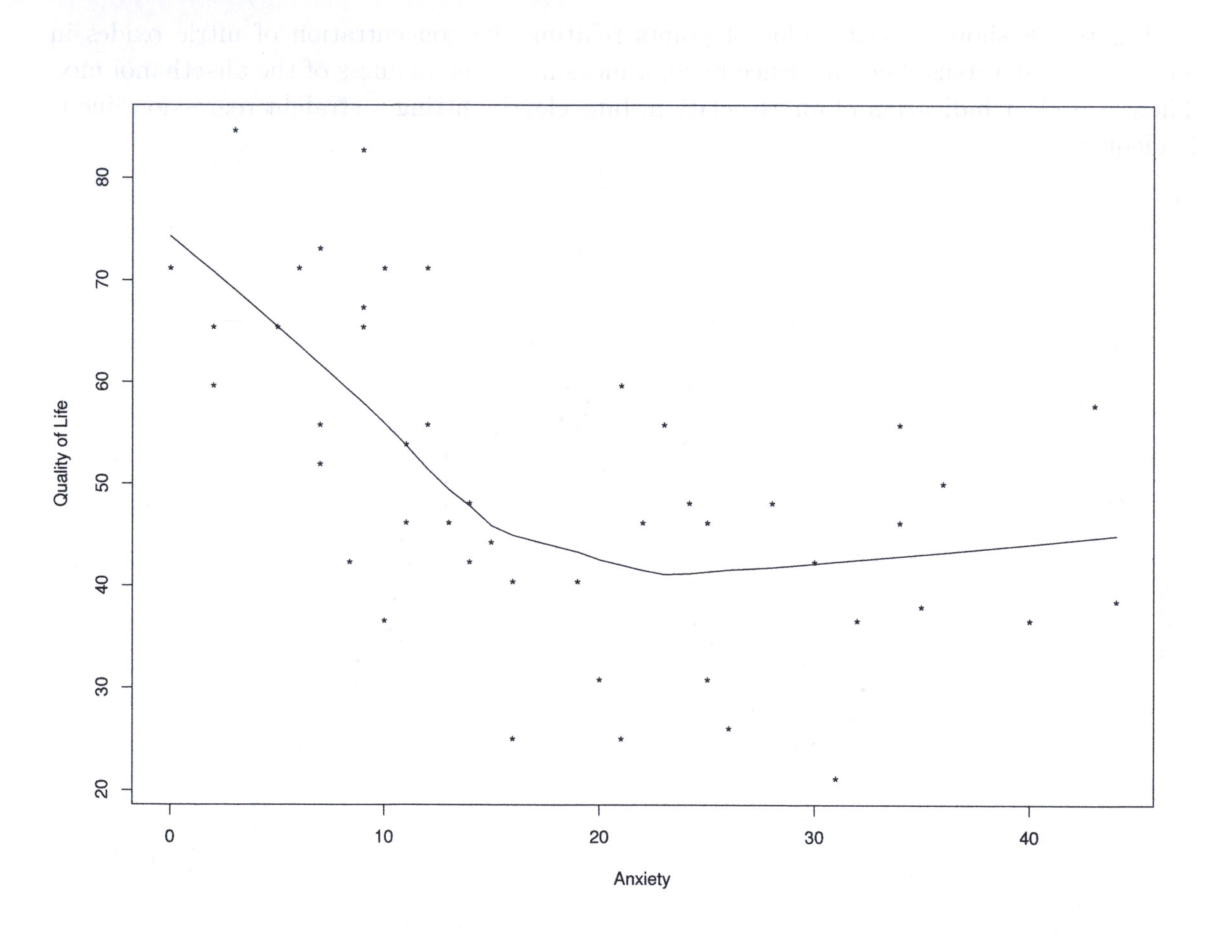

Figure 6.9 Shown is the regression line for predicting quality of life given a participant's level of anxiety.

A seemingly natural strategy when dealing with curvature is to add a quadratic term. That is, use a prediction rule having the form $\hat{Y} = \beta_0 + \beta_1 X + \beta_2 X^2$, or, more generally, include a term with X raised to some power. This might suffice in some situations, but experience with modern methods suggests that another type of nonlinearity is common: A straight line with a nonzero slope gives a reasonable prediction rule over some range of X values, but for X values outside this range, there is little or no association. That is, the slope is zero. More generally, when dealing with two or more predictors, curvature becomes a particularly difficult issue in which case methods in Chapter 14 might be considered.

6.4.5 Violating Assumptions When Testing Hypotheses

Currently, the hypothesis testing methods described in Sections 6.2.1 and 6.2.2 are routinely used. But violating the assumptions of these methods can cause serious problems.

Consider a situation where the normality assumption is valid but there is heteroscedasticity. Then the standard method for testing hypotheses about the slope, given by Equation (6.10), might provide poor control over the probability of a Type I error and poor probability coverage when computing a confidence interval, even when the sample size is large (e.g., Long and Ervin, 2000; Wilcox, 1996b). If the distributions are not normal, the situation gets worse. In some cases, the actual probability of a Type I error can exceed 0.5 when testing at the $\alpha = 0.05$ level. Perhaps an even more serious concern is that violating the homoscedasticity assumption might result in a substantial loss in power.

The homoscedasticity assumption is valid when X and Y are independent. Independence implies homoscedasticity, but $\beta_1 = 0$, for example, does not necessarily mean that there is homoscedasticity. Practical problems arise when X and Y are dependent because now there is no particular reason to assume homoscedasticity, and if there is heteroscedasticity, the wrong standard error is being used to compute confidence intervals and test hypotheses. If we could determine how $\mathrm{VAR}(Y|X)$ changes with X, a correct estimate of the standard error could be employed, but currently it seems that alternative strategies for dealing with heteroscedasticity (covered in this chapter as well as Chapter 14) are more effective.

There are methods for testing the assumption that there is homoscedasticity, one of which is described later in this chapter. But given some data, it is unknown how to tell whether any of these tests have enough power to detect situations where heteroscedasticity causes practical problems with inferential methods that assume homoscedasticity. Currently, a more effective approach appears to be to switch to some method that allows heteroscedasticity.

6.4.6 Dealing with Heteroscedasticity: The HC4 Method

Numerous methods have been proposed for testing hypotheses about the least squares regression slopes in a manner that allows heteroscedasticity. None are completely satisfactory, particularly when the sample size is small. But certain methods offer a substantial improvement over the classic methods described in Sections 6.2.1 and 6.2.2. One of the simpler techniques, which is recommended by Cribari-Neto (2004; cf. Wilcox, 1996a), is based on what is called the HC4 estimator of the standard error of b_j, say S_j^2 $(j = 0, \ldots, p)$, which is computed as described in Box 6.1, assuming familiarity with matrix algebra. (Basic matrix algebra is summarized in Appendix C.) The $1 - \alpha$ confidence interval for β_j is taken to be

$$\hat{\beta}_j \pm t S_j,$$

where t is the $1 - \alpha/2$ quantile of Student's t distribution with $\nu = n - p - 1$ degrees of freedom. As usual, reject H_0: $\beta_j = 0$ if this interval does not contain zero. This approach appears to perform relatively well when testing hypotheses about a single parameter (Ng and Wilcox, 2009). In practical terms, it can be prudent to consider the robust methods in Chapter 14.

The method can be extended to situations where the goal is to test Equation (6.11), the hypothesis that all of the slope parameters are zero (e.g., Godfrey, 2006), but no details are given here. Generally, it seems that the HC4 method controls the probability of a Type I error reasonably well under fairly severe departures from normality, even with $n = 20$. Under normality and homoscedasticity, all indications are that the classic method in Section 6.2.1 does not offer much of an advantage. That is, an argument can be made for generally using the HC4 method. However, although HC4 improves on the classic technique, there are situations where, when testing at the $\alpha = 0.05$ level, the actual Type I error probability exceeds 0.10 (Ng and Wilcox, 2009). Increasing the sample size from 20 to 100 does not correct this problem.

Box 6.1: The HC4 estimate of the standard error of the least squares estimator.

We observe

$$(Y_1, X_{11}, \ldots, X_{1p}), \ldots, (Y_n, X_{n1}, \ldots, X_{np}).$$

Let

$$\mathbf{X} = \begin{pmatrix} 1 & X_{11} \cdots & X_{1p} \\ 1 & X_{21} \cdots & X_{2p} \\ \vdots & \vdots & \vdots \\ 1 & X_{n1} \cdots & X_{np} \end{pmatrix}.$$

Let $b_0, \ldots, b_p$ be the least squares estimates of the $p + 1$ parameters. Compute

$$\mathbf{C} = (\mathbf{X}'\mathbf{X})^{-1}$$

and

$$\mathbf{H} = \text{diag}(\mathbf{XCX}')^{-1}.$$

Let $\bar{h} = \sum h_{ii}/n$, $e_{ii} = h_{ii}/\bar{h}$, and $d_{ii} = \min(4, e_{ii})$. Let $\mathbf{A}$ be the $n \times n$ diagonal matrix with the ith entry given by $r_i^2(1 - h_{ii})^{-d_{ii}}$, where r_i is the ith residual based on the least squares estimator. Let

$$\mathbf{S} = \mathbf{CX}'\mathbf{AXC}.$$

($\mathbf{S}$ is called the HC4 estimator.) The diagonal elements of the matrix $\mathbf{S}$, which we denote by $S_0^2, S_1^2, \ldots, S_p^2$, are the estimated squared standard errors of $b_0, b_1, \ldots b_p$, respectively.

6.4.7 R Functions olshc4 and hc4test

The R function

```
olshc4(x,y,alpha=0.05,xout=F,outfun=out)
```

computes $1 - \alpha$ confidence intervals using the method in Box 6.1, and p-values are returned as well. By default, 0.95 confidence intervals are returned. Setting the argument `alpha` equal to 0.1, for example, will result in 0.9 confidence intervals. The function

```
hc4test(x,y,xout=FALSE,outfun=out)
```

tests the hypothesis, given by Equation (6.11), that all of the slopes are equal to zero. Note that both functions include the option of removing leverage points via the arguments `xout` and `outfun`.

6.4.8 Interval Estimation of the Mean Response

Given some value for the independent variable, say x, of interest is computing a confidence interval for $E(Y|X = x)$. There is a classic method for accomplishing this goal assuming normality and homoscedasticity (e.g., Montgomery and Peck, 1992, p. 27). A simple modification can deal with heteroscedasticity via the HC4 estimator. (When using a robust regression estimator, see Section 14.3.2.) Let $\hat{Y} = b_0 + b_1 x$. An estimate of the standard error of $\hat{Y}$ is

$$U = \sqrt{\frac{s_y^2}{n} + S_1^2(x - \bar{X})^2},$$

where s_y^2 is the sample variance of the Y values and again S_1^2 is the HC4 estimate of the squared standard error of b_1. A $1 - \alpha$ confidence interval for $E(Y|X = x)$ is

$$\hat{Y} \pm t_{1-\alpha/2} U,$$

where $t_{1-\alpha/2}$ is the $1-\alpha/2$ quantile of Student's T distribution with $n-2$ degrees of freedom.

Now consider the goal of computing confidence intervals for $E(Y|X_i)$ $(i = 1, \ldots n)$. That is, n confidence intervals are to be computed. In the event there are only K unique values among $X_1, \ldots, X_n$, then, of course, only K confidence intervals are computed. Note that by chance, one or more of these n confidence intervals will be inaccurate. A fundamental goal is computing the confidence intervals such that the probability of one or more inaccurate confidence intervals is α. If K is small, methods in Chapter 12 can be used to accomplish this goal. Otherwise, in terms of achieving relatively high power and short confidence intervals, a much more satisfactory method is to use a simple extension of the approach in Wilcox (in press), which can be applied via the R function in the next section. (The basic strategy is outlined in Section 14.3.2.)

6.4.9 R Function olshc4band

The R function

```
olshc4band(x,y,alpha=0.05,xout=FALSE,outfun=outpro,plotit=TRUE,
  xlab="X",ylab="Y",nreps=5000,pch=".",CI=FALSE,ADJ=TRUE,SEED=TRUE)
```

computes confidence intervals for $E(Y|X = X_i)$, $i = 1, \ldots, n$. If the argument `ADJ=TRUE`, the confidence intervals are adjusted so that the probability of one or more inaccurate confidence intervals is equal to the value given by the argument `alpha`. If the argument `alpha` differs from 0.05, the adjusted critical value is estimated via a simulation. The number of replications used in the simulation is controlled via the argument `nreps`. If `ADJ=FALSE`, each confidence interval has probability coverage $1-\alpha$, in which case the probability of one or more inaccurate confidence intervals is greater than α. If `CI=TRUE`, the confidence intervals are returned. Otherwise the function simply creates a plot of the regression line plus dashed lines that indicate the confidence intervals.

EXAMPLE

Consider the Well Elderly 2 study (Clark et al., 2012) that dealt with an intervention program aimed at improving the physical and emotional well-being of older adults. The cortisol awakening response (CAR) refers to the change in the cortisol level upon awakening and measured again 30–60 minutes later. When the CAR is positive (cortisol decreases after awakening), a positive association was found between the CAR and CESD, a measure of depressive symptoms. So a simple interpretation is that when the CAR is positive, as the

CAR increases, typical depressive symptoms tend to increase as well. But to get a better understanding of the practical implications of this result, it is noted that a CESD score greater than 15 is regarded as an indication of mild depression. A score greater than 21 indicates the possibility of major depression. So an issue is whether anything can be said about whether the typical CESD score is greater than or less than 15 over the range of the observed CAR values. Removing leverage points, the individual confidence intervals indicate that for CAR between zero and 0.111, the typical CESD score is significantly less than 15 when testing at the 0.05 level. That is, the probability of one or more Type I errors is approximately 0.05. For CAR greater than 0.177, the typical CESD score is estimated to be greater than 15. But over the entire range of available CAR values, namely 0–0.358, the typical CESD scores are not significantly greater than 15.

6.5 PEARSON'S CORRELATION AND THE COEFFICIENT OF DETERMINATION

A problem of fundamental importance is quantifying the strength of the association between two variables. The most commonly used strategy is based on Pearson's correlation, which has strong ties to the least squares regression line. As usual, consider two variables X and Y. Given n pairs of observations $(X_1, Y_1), \ldots, (X_n, Y_n)$, let s_x and s_y be the sample standard deviations corresponding to the X and Y values, respectively. The *sample covariance* between X and Y is

$$s_{xy} = \frac{1}{n-1}\sum(X_i - \bar{X})(Y_i - \bar{Y}).$$

Pearson's correlation is

$$r = \frac{s_{xy}}{s_x s_y}.$$

It can be shown that $-1 \leq r \leq 1$; its value is always between -1 and 1.

Let b_1 be the least squares estimate of the slope when predicting Y, given X. A little algebra shows that Pearson's correlation is

$$r = b_1\frac{s_x}{s_y}. \tag{6.14}$$

From this last equation we have that

$$b_1 = r\frac{s_y}{s_x}. \tag{6.15}$$

So if $r > 0$, the least squares regression line has a positive slope, and if $r < 0$, the slope is negative. In standardized regression (as discussed in Section 6.2.4), the slope is equal to the correlation between X and Y. That is, the least squares regression line between

$$Z_x = \frac{X - \bar{X}}{s_x}$$

and

$$Z_y = \frac{Y - \bar{Y}}{s_y}$$

is

$$\hat{Z}_y = rZ_x.$$

The Population Pearson Correlation

There is a population analog of r, typically labeled ρ (a lower case Greek rho), which was formally introduced in Section 3.8. Roughly, it is the value of r if all individuals under study could be measured. The value of ρ always lies between -1 and 1. When X and Y are independent, it can be shown that $\rho = 0$. If $\rho \neq 0$, X and Y are dependent. However, it is possible to have dependence even when $\rho = 0$.

The Coefficient of Determination

A classic and well-known method for characterizing the strength of the association between two variables is called the coefficient of determination, which is just r^2, the square of Pearson's correlation. To provide some perspective, several descriptions are given regarding how the coefficient of determination arises.

For the first description, imagine that we ignore X in our attempts to predict Y and simply use $\hat{Y} = \bar{Y}$. Then we can measure the accuracy of our prediction rule with

$$\sum(Y_i - \bar{Y})^2,$$

the sum of the squared discrepancies between the Y values we observe and the predicted value, $\bar{Y}$. Notice that this sum is the numerator of the sample variance of the Y values. If instead we use the least squares regression line $\hat{Y} = \beta_0 + \beta_1 X$ to predict Y, then an overall measure of the accuracy of our prediction rule is

$$\sum(Y_i - \hat{Y}_i)^2.$$

The difference between these two sums measures the extent $\hat{Y}$ improves upon using $\bar{Y}$. In symbols, this difference is

$$\sum(Y_i - \bar{Y})^2 - \sum(Y_i - \hat{Y}_i)^2.$$

Finally, if we divide this difference by $\sum(Y_i - \bar{Y})^2$, we get a measure of the relative reduction in the error associated with $\hat{Y}$ versus $\bar{Y}$, which is

$$\frac{\sum(Y_i - \bar{Y})^2 - \sum(Y_i - \hat{Y}_i)^2}{\sum(Y_i - \bar{Y})^2} = r^2.$$

Here is another way of viewing the coefficient of determination. Let

$$\tilde{Y} = \frac{1}{n}\sum \hat{Y}_i$$

be the average value of the predicted Y values based on the least squares regression line. Then the variation among the predicted Y values is reflected by

$$\sum(\hat{Y}_i - \tilde{Y})^2.$$

From Chapter 2, $\sum(Y_i - \bar{Y})^2$ reflects the variation among the Y values. One way of viewing how well a least squares regression line is performing is to use the variation among the predicted Y values (the $\hat{Y}$ values) divided by the variation in the Y values. It turns out that the ratio of these two measures of variation is just r^2. That is,

$$r^2 = \frac{\sum(\hat{Y}_i - \tilde{Y})^2}{\sum(Y_i - \bar{Y})^2}. \tag{6.16}$$

This approach to measuring the strength of an association, where the variance of the predicted Y values is divided by the variance of the observed Y values, is a special case of what is called *explanatory power* (Doksum and Samarov, 1995), which is described in a broader context in Section 14.8. (One advantage of explanatory power is that it provides a way of measuring the strength of an association even when there is curvature.) Notice that if there is no association and the regression line is horizontal, all of the predicted Y values are equal to $\bar{Y}$, the average of the Y values. That is, there is no variation among the predicted Y values, in which case $r^2 = 0$. If there is a perfect association, meaning that each predicted Y value is equal to its observed value ($\hat{Y}_i = Y_i$), then $r^2 = 1$.

Another view regarding how to characterize the strength of an association is in terms of the slope of the regression line. Galton (1888) argued that this approach should be used after X and Y have been standardized. In terms of the notation introduced in the previous section, Galton argued that the least squares regression slope, based on Z_x and Z_y, be used. Recall that the least squares regression line is given by $\hat{Z}_y = rZ_x$. That is, Pearson's correlation corresponds to the slope and this provides yet another argument for using r.

Yet another way of introducing r^2 is as follows. Note that

$$Y_i - \bar{Y} = (\hat{Y}_i - \bar{Y}) + (Y_i - \hat{Y}_i).$$

It can be shown that, as a result,

$$\sum(Y_i - \bar{Y})^2 = \sum(\hat{Y}_i - \bar{Y})^2 + \sum(Y_i - \hat{Y}_i)^2.$$

The term on the left of this last equation reflects the total sum of squares and is often labeled SST. The term $\sum(\hat{Y}_i - \bar{Y})^2$ is called the model sum of squares and is labeled SSM. The term $\sum(Y_i - \hat{Y})^2$ is the error sum of squares, SSE, which is just the sum of the squared residuals. So this last equation can be written as

$$\text{SST} = \text{SSM} + \text{SSE}.$$

In other words, the overall variation in the Y values can be broken down into two pieces. The first represents the variation among the predicted Y values and the second measures the overall error when using the least squares regression line. It can be shown that

$$r^2 = \frac{\text{SSM}}{\text{SST}},$$

the proportion of the total variation "explained" by the least squares regression line.

The squared multiple correlation, R^2, introduced in Section 6.2.2 and given by Equation (6.12), is a generalization of r^2. With one predictor, $R^2 = r^2$. In general, like r^2, R^2 reflects the strength of the association *based on the least squares regression fit.* It can be shown that R is Pearson's correlation between the observed Y values and the predicted Y values, $\hat{Y}$.

6.5.1 A Closer Look at Interpreting r

Interpreting r is complicated by the fact that at least five features of the data affect its magnitude. In practical terms, caution must be exercised to not read more into the magnitude of r than is warranted.

Assuming that the regression line between X and Y is straight (there is no curvature), the first feature is the distance of the points from the line around which they are centered. That is, the magnitude of the residuals is associated with the magnitude of r. The left panel of Figure 6.10 shows a scatterplot of points with $r = 0.92$. The right panel shows another scatterplot of points which are centered around the same line as in the left panel, only they are farther from the line around which they are centered. Now $r = 0.42$.

A second feature that affects the magnitude of r is the magnitude of the slope around

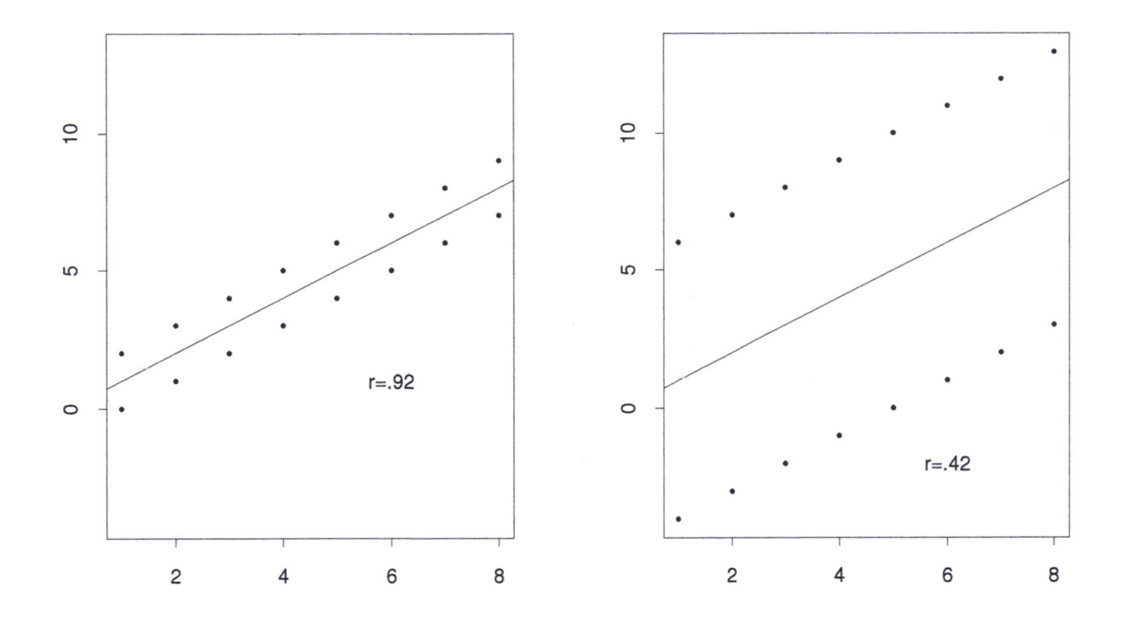

Figure 6.10 An illustration that the magnitude of the residuals affects Pearson's correlation.

which the points are centered (e.g., Barrett, 1974; Loh, 1987). Figure 6.11 shows the same points shown in the left panel of Figure 6.10, only rotated so that the slope around which they are centered has been decreased from 1 to 0.5. This causes the correlation to drop from 0.92 to 0.83. If we continue to rotate the points until they are centered around the x-axis, $r = 0$.

A third feature of data that affects r is outliers. This is not surprising because we already know that the least squares regression line has a breakdown point of only $1/n$, and we have seen how r is related to the least squares estimate of the slope, as indicated by Equation (6.13). For the star data in Figure 6.3, $r = -0.21$, which is consistent with the negative slope associated with the least squares regression line. But we have already seen that for the bulk of the points, there is a positive association. Generally, a single unusual value can cause r to be close to zero even when the remaining points are centered around a line having a nonzero slope, and one outlier can cause $|r|$ to be fairly large even when there is no association among the remaining points.

Moreover, in situations where outliers are likely to occur, even small departures from normality can greatly affect the population correlation ρ, and r can be affected as well no matter how large the sample size might be. To provide some indication of why, the left panel of Figure 6.12 shows the distribution between X and Y when both X and Y are normal and $\rho = 0.8$. In the right panel, again X and Y are normal, but now $\rho = 0.2$. So under normality, decreasing ρ from 0.8 to 0.2 has a very noticeable effect on the joint distribution of X and Y. Now look at Figure 6.13. It looks similar to the left panel of Figure 6.12 where $\rho = 0.8$, but now $\rho = 0.2$. In Figure 6.13 X is again normal, but Y has the mixed normal distribution

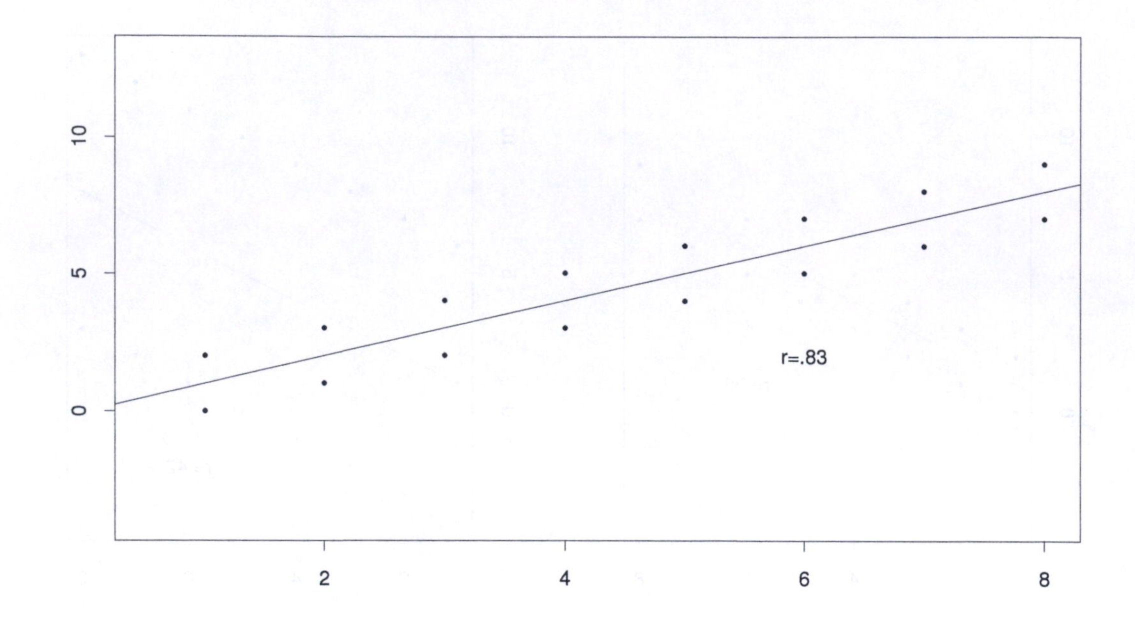

Figure 6.11 Rotating the points in the left panel of Figure 6.10 results in the plot shown here. Pearson's correlation drops from 0.92 to 0.83, which illustrates that Pearson's correlation is related to the magnitude of the slope of the line around which points are clustered.

(described Chapter 3). This demonstrates that a very small change in any distribution can have a very large impact on ρ. Also, no matter how large the sample size might be, a slight departure from normality can drastically affect r.

A fourth feature that affects the magnitude of r is any restriction in range among the X (or Y) values. To complicate matters, restricting the range of X can increase or decrease r. For example, for the data in Figure 6.1, the correlation is $r = 0.79$. Eliminating the points with $X > 0.9$, now $r = 0.38$.

The star data in Figure 6.5 illustrate that restricting the range of X (or Y) can increase r as well. If we eliminate all points having $X \leq 4.1$, r increases from -0.21 to 0.65.

A fifth feature that affects r is curvature. Consider again Figure 6.8, which shows a scatterplot of points relating the concentration of nitric oxides in engine exhaust versus its equivalence ratio, a measure of the richness of the air-ethanol mix. There is a rather obvious association, but the correlation is $r = -0.1$, a value relatively close to zero. As another example, if X is standard normal and $Y = X^2$, there is an exact association between X and Y, but for the variables X and Y, $\rho = 0$. However, based on the variables X^2 and Y, now Pearson's correlation is relatively high. (Methods for measuring the strength of an association, when there is curvature, are described in Chapter 14.)

In summary, the following features of data influence the magnitude of Pearson's correlation:

- The slope of the line around which points are clustered.

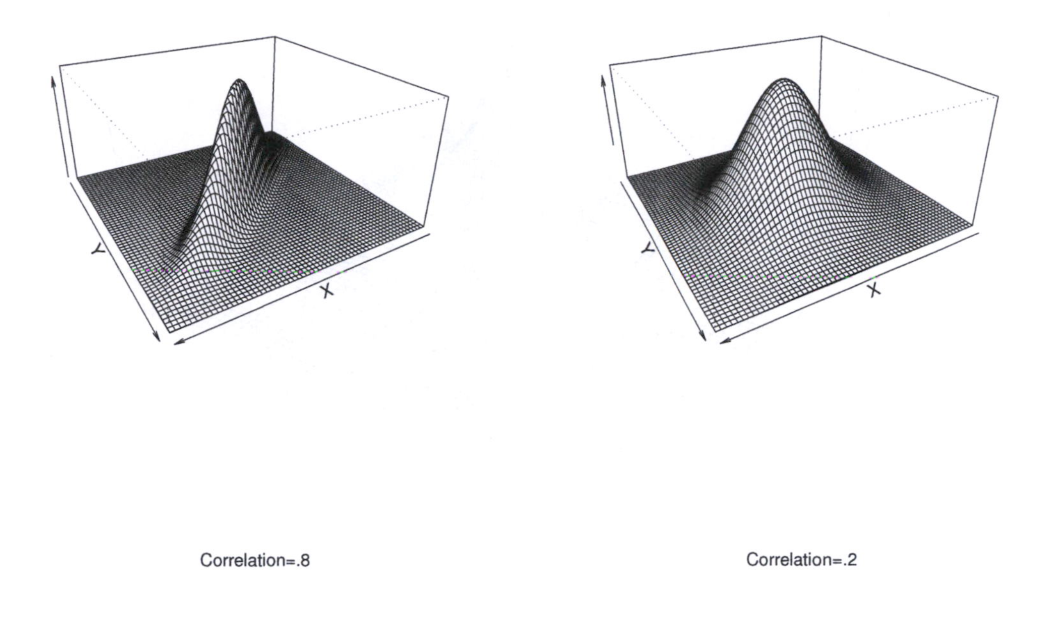

Figure 6.12 When both X and Y are normal, increasing ρ from 0.2 to 0.8 has a noticeable effect on the bivariate distribution of X and Y.

- The magnitude of the residuals.
- Outliers.
- Restricting the range of the X values, which can cause r to go up or down.
- Curvature.

A point worth stressing is that although independence implies that $\rho = 0$, $\rho = 0$ does not necessarily imply independence. In fact there are various ways in which X and Y can be dependent, yet ρ is exactly zero. For example, if X and e are independent, and $Y = |X|e$, it can be shown that the variance of Y, given X, is $X^2\sigma^2$, where again σ^2 is the variance of the error term, e. That is, X and Y are dependent because there is heteroscedasticity, yet $\rho = 0$. More generally, if there is heteroscedasticity and the least squares slope is zero, then $\rho = 0$ as well. As another example, suppose U, V, and W are independent standard normal random variables. Then it can be shown that $X = U/W^2$ and $Y = V/W^2$ are dependent (roughly because both X and Y have the same denominator), yet they have correlation $\rho = 0$.

We conclude this section by noting that the above list of factors that affect the magnitude of r is not exhaustive. Yet another feature of data that affects the magnitude of r is the reliability of the measures under study (e.g., Lord and Novick, 1968), but the details go beyond the scope of this book.

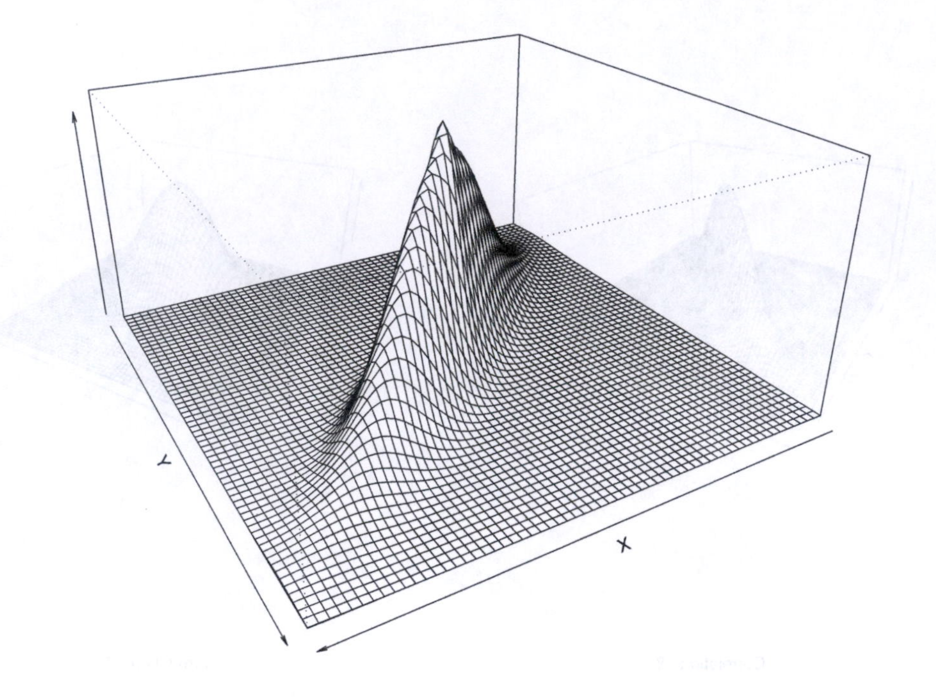

Figure 6.13 Two bivariate distributions can appear to be very similar yet have substantially different correlations. Shown is a bivariate distribution with $\rho = 0.2$, but the graph is very similar to the left panel of Figure 6.12 where $\rho = 0.8$.

6.6 TESTING H_0: $\rho = 0$

Next we describe the classic test of

$$H_0 : \rho = 0. \tag{6.17}$$

If we can reject this hypothesis, then the data suggest that X and Y are dependent.

If we assume that X and Y are independent, and if at least one of these two variables has a normal distribution, then

$$T = r\sqrt{\frac{n-2}{1-r^2}} \tag{6.18}$$

has a Student's t distribution with $\nu = n - 2$ degrees of freedom (Muirhead, 1982, p. 146; also see Hogg and Craig, 1970, pp. 339–341.) So the decision rule is to reject H_0 if $|T| \geq t$, where t is the $1 - \alpha/2$ quantile of Student's t distribution with $n - 2$ degrees of freedom, which is read from Table 4 in Appendix B.

6.6.1 R Function cor.test

The R function

```
cor.test(x,y,conf.level = 0.95)
```

tests the hypothesis of a zero correlation using Student's t distribution as just described. The function also returns a confidence interval for ρ having probability coverage indicated by the argument `conf.level`, which defaults to 0.95.

EXAMPLE

For the data in Table 6.3, $n = 47$, $r = -0.286$, so $\nu = 45$ and

$$T = -0.286\sqrt{\frac{45}{1-(-0.286)^2}} = -2.$$

With $\alpha = 0.05$, the critical value is $t = 2.01$, and because $|-2| < 2.01$, we fail to reject. That is, we are unable to conclude that the aggression scores and recall test scores are dependent with $\alpha = 0.05$.

Caution must be exercised when interpreting the implications associated with rejecting the hypothesis of a zero correlation with T. Although it is clear that T given by Equation (6.18) is designed to be sensitive to r, homoscedasticity plays a crucial role in the derivation of this test. (Recall that independence implies homoscedasticity.) When in fact there is heteroscedasticity, the derivation of T is no longer valid and can result in some unexpected properties. For instance, it is possible to have $\rho = 0$, yet the probability of rejecting H_0: $\rho = 0$ *increases* as the sample size gets large. In practical terms, a more realistic description of Student's T is that it tests the hypothesis that two random variables are independent.

For completeness, results in Bishara and Hittner (2012) indicate that when X and Y are independent, Student's T controls the Type I error fairly well when testing at the $\alpha = 0.05$ level. The primary exception was when both X and Y have a mixed normal distribution, in which case the actual Type I error probability was estimated to be between 0.066 and 0.068 for samples ranging between 10 and 160.

EXAMPLE

Figure 6.14 shows a scatterplot of 40 points generated on a computer where both X and Y have normal distributions and $\mu_x = 0$. In this particular case, Y has variance one unless $|X| > 0.5$ in which case Y has standard deviation $|X|$. So X and Y are dependent, but $\rho = 0$. For this situation, when testing at the $\alpha = 0.05$ level, the actual probability of rejecting H_0 with T is 0.098 with $n = 20$. For $n = 40$ it is 0.125, and for $n = 200$ it is 0.159. The probability of rejecting is *increasing* with n even though $\rho = 0$. When we reject, a correct conclusion is that X and Y are dependent, but it would be incorrect to conclude that $\rho \neq 0$.

Some experts might criticize this last example on the grounds that it would be highly unusual to encounter a situation where $\rho = 0$ and there is heteroscedasticity. That is, perhaps we are describing a problem that is theoretically possible but unlikely to ever be encountered. If we accept this view, then the last example illustrates that when we reject, the main reason for rejecting is unclear. That is, one might argue that surely $\rho \neq 0$, but the main reason for rejecting might be due to heteroscedasticity.

In a broader context, the last example illustrates a basic principle of fundamental importance. Classic methods that assume normality and homogeneity use correct estimates of standard errors when variables are independent, but under general conditions they use incorrect estimates when they are dependent, which can result in relatively low power and highly inaccurate confidence intervals.

A common alternative to T when making inferences about ρ is to employ what is known as *Fisher's r-to-z transformation*, but no details are provided because of results in Duncan

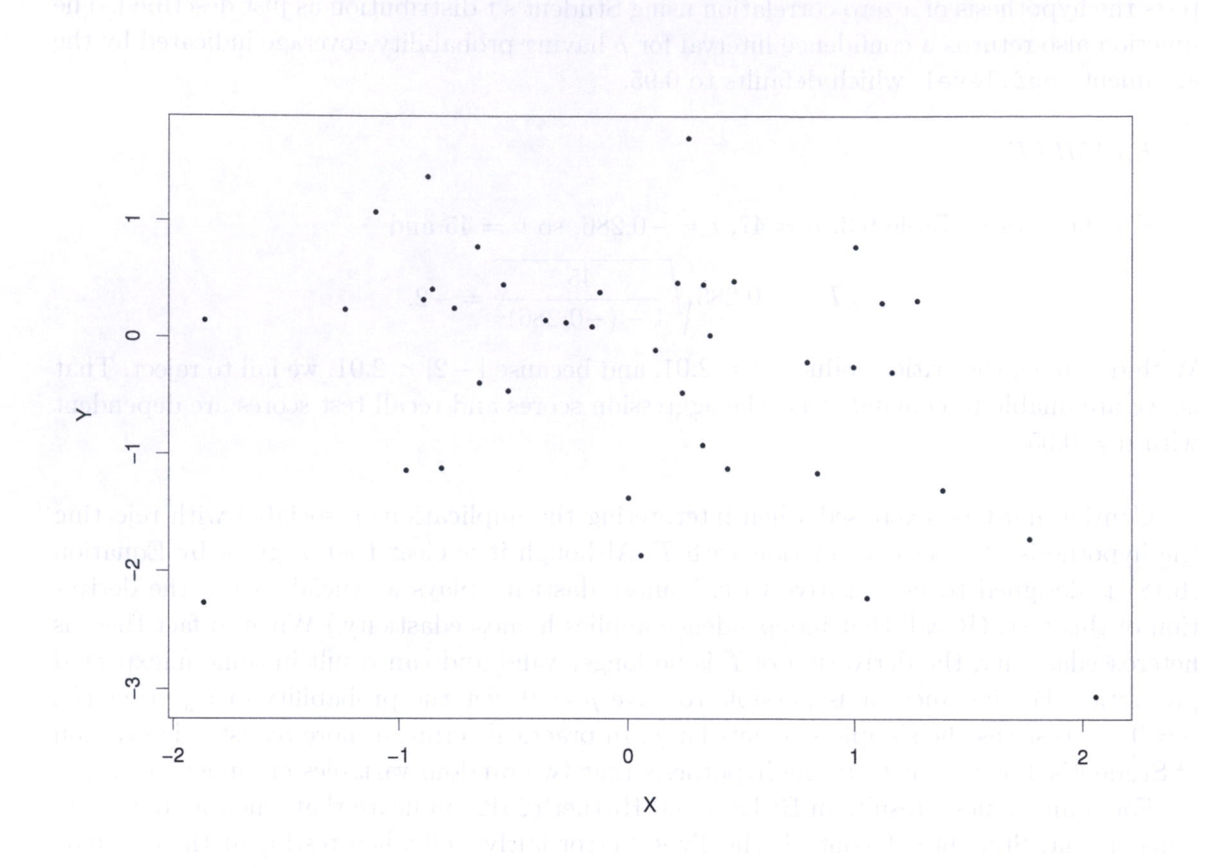

Figure 6.14 Shown are 40 points generated on a computer where Pearson's correlation is $\rho = 0$. Even though both X and Y have normal distributions, using T to test H_0: $\rho = 0$ does not control the probability of a Type I error. Indeed, control over the probability of a Type I error deteriorates as the sample size n increases because the wrong standard error is being used due to heteroscedasticity.

and Layard (1973). Briefly, Fisher's method requires normality. For non-normal distributions there are general conditions where the method does not converge to the correct answer as the sample size increases. That is, the method violates the basic principle that the accuracy of our results should increase as n gets large. Moreover, simulation studies indicate that methods based on Fisher's r-to-z transformation can result in unsatisfactory control over the Type I error probability (e.g., Hittner et al., 2003; Berry and Mielke, 2000).

6.6.2 R Function pwr.r.test

The R function

```
pwr.r.test(n = NULL, r = NULL, sig.level = 0.05, power = NULL, alternative =
                        c("two.sided", "less", "greater")),
```

which is part of the R package `pwr` and can be installed as described in Chapter 1, can be used to do a power analysis when testing the hypothesis that Pearson's correlation is equal to zero. Using the function requires specifying a value for two of the three arguments that default to NULL; the function determines the value of the argument not specified. The argument `sig.level` corresponds to α, the desired probability of a Type I error. By default, a two-sided test is assumed. The method assumes normality and is based on Fisher's r-to-z transformation. Violating the normality assumption is a serious concern. As noted in the previous section, under general conditions, methods based on Fisher's r-to-z transformation do not converge to the correct answer as the sample size gets large. Cohen (1988) suggests that, as a general guide, $\rho = 0.1$, 0.3, and 0.5 correspond to a small, medium, and large correlation, respectively.

EXAMPLE

Imagine that the goal is to test H_0: $\rho = 0$ with Student's T and that it is desired to have power equal to 0.8 when $\rho = 0.5$ and the Type I error probability is $\alpha = 0.01$. The R command `pwr.r.test(r=0.5,power=0.80,sig.level=0.01)` returns

```
          n = 41.95917
          r = 0.5
  sig.level = 0.01
      power = 0.8
alternative = two.sided
```

That is, a sample size of 42 is needed. It should be stressed, however, that the solution is based on Fisher's r-to-z transformation. As previously noted, under general conditions this transformation gives inaccurate results regardless of how large the sample size might be. In practical terms, this approach to dealing with power and the sample size can be highly inaccurate when dealing with non-normal distributions. A crude way of dealing with this problem is to determine the sample size as just illustrated, but test the hypothesis of a zero correlation using one of the robust correlations in Chapter 14.

6.6.3 Testing H_0: $\rho = 0$ When There is Heteroscedasticity

There is a simple way of testing the hypothesis that Pearson's correlation is zero in a manner that allows heteroscedasticity. First, standardize the X and Y values. That is, compute

$$x_i = \frac{X_i - \bar{X}}{s_x}$$

and

$$y_i = \frac{Y_i - \bar{Y}}{s_y},$$

where s_x and s_y are the standard deviations of X and Y, respectively. Then the slope of the least squares regression line using the x and y values estimates Pearson's correlation and the HC4 method, described in Section 6.4.5, can be used to compute a confidence interval for ρ. As usual, if this interval does not contain 0, reject H_0: $\rho = 0$. An added advantage is that it seems to perform relatively well, in terms of controlling Type I error probabilities, when dealing with nonnormal distributions. There is weak evidence that when the correlation is relatively large, method PM1 in Section 6.9 can provide a more accurate confidence interval, but this issue needs further study.

6.6.4 R Function pcorhc4

For convenience, the R function

```
pcorhc4(x,y, alpha = 0.05)
```

has been provided to compute a confidence interval for ρ using the HC4 method just described, and a p-value is returned as well.

6.6.5 When Is It Safe to Conclude that Two Variables Are Independent?

Establishing that two measures are independent is difficult. If we reject the hypothesis that Pearson's correlation is zero, we can be reasonably certain that the two measures are dependent even though this might tell us virtually nothing about the true nature of the association. But if we fail to reject, this is not remotely convincing evidence that we have independence. The basic problem is that any method that tests H_0: $\rho = 0$ might not be sensitive to the type of association that exists between the variables under study. There is a rather lengthy list of alternative methods for detecting dependence that attempt to address this problem. (See, for example, Kallenberg and Ledwina, 1999, plus the references they cite.) A few alternative techniques are described in this book.

6.7 A REGRESSION METHOD FOR ESTIMATING THE MEDIAN OF Y AND OTHER QUANTILES

About a half century prior to the derivation of least squares regression, least absolute value (LAV) regression was proposed by Roger Boscovich. That is, rather than choose the slope and intercept so as to minimize the sum of squared residuals, they are chosen to minimize the sum of the absolute values. In symbols, when there is one predictor, choose the slope b_1 and the intercept b_0 so as to minimize

$$\sum |Y_i - b_0 - b_1 X_i|. \tag{6.19}$$

An essential difference from least squares regression is that now the goal is to estimate the median of Y given X, assuming that the median is given by $b_0 + b_1 X$. In contrast, least squares regression is designed to estimate the mean of Y, given X.

A positive feature of LAV regression is that it offers protection against outliers among the Y values. This can translate into more power when testing the hypothesis that the slope is zero. (See Exercise 24 for an illustration.) Another positive feature is that when there is heteroscedasticity, its standard error can be smaller than the standard error of the least squares estimator, which also might mean more power. A negative feature is that, like least squares regression, it can be highly influenced by outliers among the X values (leverage points). That is, it guards against the deleterious effects of outliers among the Y values, but not the X values. (In fact, its breakdown point is only $1/n$.) Another concern is that when there is heteroscedasticity, its standard error (and therefore power) does not compete all that well with more modern regression estimators introduced in Chapter 14. But despite these negative characteristics, there is at least one reason for considering this approach: A generalization of the method can be used to estimate any quantile of Y given X. Situations arise where this generalization can reveal an association not detected by other techniques and it can help provide a deeper understanding of the nature of the association.

Derived by Koenker and Bassett (1978), the method assumes that the qth quantile of Y, given X, is given by $Y = \beta_0 + \beta_1 X$. A (rank inversion) method for testing hypotheses about the individual parameters was studied by Koenker (1994) and is used by the R function `rqfit` described in the next section. (For other relevant results, see Koenker and Xiao, 2002.) As for testing H_0: $\beta_1 = \cdots = \beta_p = 0$, the hypothesis that all p slopes are zero, it currently seems that a method described in Section 6.12.1 should be used. (See the R function `rqtest`.)

Here, none of the complex computational details are provided. Rather, attention is focused on easy-to-use software and some examples are included to illustrate why the method might be of interest. Readers interested in the mathematical formulation of the method, for the general case where there are p predictors, are referred to Box 6.2. (There are several numerical methods for implementing this approach.)

Box 6.2: Koenker–Bassett quantile regression method.

It is assumed that for n participants we observe

$$(Y_1, X_{11}, \ldots, X_{1p}), \ldots, (Y_n, X_{n1}, \ldots, X_{np}).$$

For some q, $0 < q < 1$, let

$$\psi_q(u) = u(q - I_{u<0}),$$

where $I_{u<0} = 1$ if $u < 0$; otherwise, $I_{u<0} = 0$. Assuming that the qth quantile of Y, given X, is given by $\beta_0 + \beta_1 X_1 + \cdots + \beta_p X_p$, the Koenker–Bassett quantile regression method estimates the unknown parameters $\beta_0, \beta_1, \ldots, \beta_p$ with the values $b_0, b_1, \ldots b_p$, respectively, that minimize

$$\sum \psi_q(r_i), \qquad (6.20)$$

where $r_i = Y_i - b_0 - b_1 X_{i1} - \cdots - b_p X_{ip}$ $(i = 1, \ldots, n)$ are the residuals.

EXAMPLE

Williams et al. (2005) conducted a study dealing generally with the Porteus Maze Test (PMT), which is used to evaluate intelligence and executive functioning and screen for intellectual deficiency. A portion of the study dealt with the association between the so-called Q score resulting from the PMT test and a measure of maladjustment for the participants in this study. The sample size is $n = 1063$. When predicting the 0.5 quantile (the median Q score), the slope is zero indicating no association. For the 0.8 and 0.9 quantiles, the slopes are estimated to be 0.0098 and 0.029, respectively. The hypothesis of a zero slope is rejected for the 0.9 quantile with $\alpha = 0.05$ suggesting that among relatively high maladjustment scores, there is an association with Q scores.

Like least absolute value (LAV) regression, a possible appeal of the quantile regression estimator is that it provides protection against the deleterious effects of outliers among the Y values. But again, outliers among the X values can be a practical concern.

EXAMPLE

Consider again the star data in Figure 6.5. As already noted, for the bulk of the data, it is evident that there is a positive association, but the least squares regression line has a negative slope. The slope of the quantile regression line, when predicting the median value of Y, is -0.69, again resulting in a misleading result. In practical terms, it can be important to check on the impact of removing outliers among the independent variable. For the situation at hand, now the slope is estimated to be 3.68.

6.7.1 R Function rqfit

R provides access to a function, called `rq`, which applies the Koenker-Bassett regression method and returns 0.95 confidence intervals for each of the individual parameters. When testing the hypothesis that a slope is zero, reject if zero is not contained in the confidence interval. As explained in Chapter 5, this means the Type I error probability is set to $\alpha = 0.05$.

Currently, the R version of this function only allows $\alpha = 0.05$. (The R function `qregci`, described in Section 6.12.1 allows $\alpha = 0.1$, 0.05, 0.025, and 0.01.) For convenience, an alternative function is supplied, called

```
rqfit(x,y,qval=0.5,xout=FALSE,outfun=out,res=FALSE),
```

which calls the function `rq`. (Both `rq` and `rqfit` assume that you have installed the R package `quantreg`. To install this package, start R and use the command `install.packages("quantreg")`.) One advantage of `rqfit` over the built-in function `rq` is that, by setting the argument `xout=TRUE`, it will remove any leverage points found by the function indicated by the argument `outfun`, which defaults to the function `out`. This means that by default, like the function `ols`, when there is one predictor only, a MAD-median rule is used. (With more than one predictor, the MVE method in Section 13.1.1 is used. Using the method in Section 13.1.6, by setting the argument `outfun=outpro`, might provide better results.) If there is more than one predictor and the goal is to test the hypothesis that all slopes are equal to zero, the hypothesis given by Equation (6.11), use the R function `rqtest` described in Section 6.12.1.

6.8 DETECTING HETEROSCEDASTICITY

This section considers the problem of testing the hypothesis that there is homoscedasticity. Many methods have been proposed, but few perform well in simulation studies. The one method found by Lyon and Tsai (1996) that performs well was derived by Koenker and Bassett (1981) and is based on a modification of a test stemming from Cook and Weisberg (1983). Another method that deserves serious consideration, and which often has more power, is described in Section 6.12.2. (For situations where there is more than one independent variable, see Zhu et al., 2016.)

The Koenker–Bassett method begins by estimating the slope and intercept using least squares regression. As usual, let r_i be the usual residuals ($i = 1, \ldots, n$). If the null hypothesis is true, then

$$\hat{\sigma}^2 = \frac{1}{n} \sum r_i^2$$

estimates the common variance. Let $A = \sum (r_i^2 - \hat{\sigma}^2)^2/n$ and $\tilde{Y} = \sum \hat{Y}_i/n$. The test statistic is

$$V = \frac{\{\sum r_i^2(\hat{Y}_i - \tilde{Y})\}^2}{A \sum (\hat{Y}_i - \tilde{Y})^2},$$

which has, approximately, a chi-squared distribution with 1 degree of freedom when the null hypothesis is true. That is, reject at the α level if $V \geq c$, where c is the $1 - \alpha$ quantile chi-squared distribution with 1 degree of freedom, which can be read from Table 3 in Appendix B. For example, with $\alpha = 0.05$, $c = 3.8415$.

As previously noted, the classic hypothesis testing methods in Section 6.2 assume that the error term is homoscedastic. Violating this assumption can affect both power and the probability of a Type I error, so a seemingly natural suggestion is to use the Koenker–Bassett method to determine whether this assumption is reasonable. Currently, however, it is unknown when this strategy has enough power to detect a situation where the methods in Sections 6.2.1 and 6.2.2 are unsatisfactory due to heteroscedasticity. Presumably, with a large enough sample size, the power of the Koenker–Bassett method will be adequate, but just how large it must be is difficult to determine. Currently, the practical value of the Koenker–Bassett method is that it might detect an association that is missed by other techniques, and it can provide empirical evidence that a particular type of dependence (heteroscedasticity) exists.

6.8.1 R Function khomreg

The R function

```
khomreg(x,y)
```

applies the Koenker–Bassett test of homoscedasticity. It returns the value of the test statistic V and a p-value.

6.9 INFERENCES ABOUT PEARSON'S CORRELATION: DEALING WITH HETEROSCEDASTICITY

As previously explained, if the classic Student's T test rejects

$$H_0 : \rho = 0,$$

the hypothesis that Pearson's correlation is zero, it is reasonable to conclude that there is dependence. But if the goal is to test the hypothesis of a zero correlation without being sensitive to heteroscedasticity, Student's T is no longer satisfactory. Section 6.4.6 also described the HC4 method for testing the hypothesis of a zero correlation in a manner that allows heteroscedasticity. The method can be used to compute a confidence interval for ρ as well, but when computing a 0.95 confidence interval and when ρ differs from zero, currently it seems that the best method for accomplishing this goal is to use the modified percentile bootstrap method described here.

The basic percentile bootstrap method described in Section 5.11 extends to the situation at hand in a simple manner. The method begins by sampling with replacement, n pairs of points from n pairs of observed values. In symbols, if we observe $(X_1, Y_1), \dots, (X_n, Y_n)$, a bootstrap sample is obtained by resampling n pairs of these points with replacement. The resulting bootstrap sample is denoted by $(X_1^*, Y_1^*), \dots, (X_n^*, Y_n^*)$.

EXAMPLE

Imagine we observe

$$(6, 2), (12, 22), (10, 18), (18, 24), (16, 29).$$

A bootstrap sample might be

$$(10, 18), (16, 29), (10, 18), (6, 2), (6, 2).$$

In the notation just introduced, $(X_1^*, Y_1^*) = (10, 18)$ and $(X_2^*, Y_2^*) = (16, 29)$.

To compute a 0.95 confidence interval for Pearson's correlation, first repeat the process of generating a bootstrap sample of size n B times yielding B bootstrap estimates of the correlation, which we label $r_1^*, \dots, r_B^*$. Then an approximate 0.95 confidence interval for Pearson's correlation, ρ, is given by the middle 95% of these bootstrap estimates. In symbols, we write these B bootstrap estimates of the slope in ascending order as $r_{(1)}^* \le r_{(2)}^* \le \dots \le r_{(B)}^*$. Letting $\ell = 0.025B$, rounded to the nearest integer, and $u = B - \ell$, an approximate 0.95 confidence interval for the slope is

$$(r_{(\ell+1)}^*, r_{(u)}^*).$$

Method PM1

The probability coverage of the confidence interval just given can differ substantially from 0.95 when n is less than 250. However, it has a property of considerable practical value: given n, the actual probability coverage is fairly stable over a relatively wide range of distributions, even when there is a fairly large degree of heteroscedasticity and the sample size is small. This suggests a method for getting a reasonably accurate confidence interval: adjust the confidence interval so that the actual probability coverage is close to 0.95 when sampling from a normal distribution and there is homoscedasticity. Then use this adjusted confidence interval for non-normal distributions or when there is heteroscedasticity. So, we adjust the percentile bootstrap method when computing a 0.95 confidence interval, based on the least squares regression estimator, in the following manner. Take $B = 599$ and for each bootstrap sample compute the least squares estimate of the slope. Next, put these 599 values in ascending order yielding $r^*_{(1)} \leq \cdots \leq r^*_{(599)}$. The 0.95 confidence interval is

$$(r^*_{(a)}, r^*_{(c)}) \tag{6.21}$$

where for $n < 40$, $a = 7$ and $c = 593$; for $40 \leq n < 80$, $a = 8$ and $c = 592$; for $80 \leq n < 180$, $a = 11$ and $c = 588$; for $180 \leq n < 250$, $a = 14$ and $c = 585$; while for $n \geq 250$, $a = 15$ and $c = 584$. Said another way, these choices for a and c stem from Gosset's strategy for dealing with small sample sizes: Assume normality and for a given sample size determine the (critical) value so that the probability of a Type I error is α, and then hope that these values continue to give good results when there is heteroscedasticity or nonnormality. This strategy performs relatively well here (Wilcox, 1996b), but it does not perform very well for other problems, such as when computing a confidence interval for the mean. Confidence intervals based on the approach just described will be called the *modified percentile* or the PM1 bootstrap method. (To get a confidence interval for the slope, based on the least squares estimator, simply replace r with the estimated slope, b_1. The computations are done by the R function `lsfitci`.)

Method PM2

For some purposes a slight variation of method PM1, just described, is needed. Rather than use $a = 15$ and $c = 584$ when $n \geq 250$, use $a = 14$ and $c = 585$ when $n \geq 180$. This will be called method *PM2.*

Hypothesis Testing. Reject H_0: $\rho = 0$ if the confidence interval does not contain zero.

When $\rho \neq 0$, the actual probability coverage remains fairly close to the 0.95 level provided ρ is not too large. But if, for example, $\rho = 0.8$, the actual probability coverage of the modified percentile bootstrap method can be unsatisfactory in some situations (Wilcox and Muska, 2001). There is no known method for correcting this problem in a satisfactory manner.

A practical issue is how this bootstrap method compares to the HC4 method for testing H_0: $\rho = 0$. Currently, in terms of controlling the probability of a Type I error, all indications are that the choice of method makes little difference. And the HC4 method has the advantage of not being limited to testing at the $\alpha = 0.05$ level. As previously noted, in terms of computing confidence intervals, it is when $\rho \neq 0$ that the bootstrap method described here is preferable. Also, in practice, the choice of method can make a practical difference as demonstrated in the example given in the next section. It should be noted, however, that an extensive comparison of the two methods has not been made for situations where $\rho \neq 0$.

For completeness, an alternative bootstrap method for computing a confidence interval for ρ was suggested by Beasley et al. (2007). The method appears to perform well when dealing with non-normal distributions and there is homoscedasticity. But when there is heteroscedasticity, it can be highly inaccurate. Moreover, the inaccuracy of the method appears to increase as the sample size gets large.

6.9.1 R Function pcorb

The R function

```
pcorb(x,y)
```

computes a 0.95 confidence interval for ρ using the modified percentile bootstrap method. Again, x and y are R variables containing vectors of observations.

EXAMPLE

For the aggression data in Table 6.3, Student's T test fails to reject the hypothesis that the correlation is zero. Using the HC4 method via the R function `pcorhc4`, the p-value is 0.092. Using the modified percentile bootstrap method instead, the R function `pcorb` returns a 0.95 confidence interval of $(-0.503, -0.008)$. So now reject $H_0 : \rho = 0$ (because the confidence interval does not contain zero) and conclude that these two variables are dependent.

6.10 BOOTSTRAP METHODS FOR LEAST SQUARES REGRESSION

This section summarizes some bootstrap methods specifically designed for making inferences about the slopes and intercepts associated with the least squares regression line. The methods described here are aimed at dealing with nonnormality and heteroscedasticity. A simple approach is to resample with replacement n pairs of values as was done in the previous section. Another approach resamples values from the residuals, but no details are given because the method deals poorly with heteroscedasticity (e.g., Wu, 1986). An alternative approach, called a wild bootstrap method, is based in part on the residuals and has been found to have practical value in simulations (e.g., Godfrey, 2006).

Momentarily focus on the case of a single predictor. Compute the least squares estimate of the slope and intercept, b_1 and b_0, respectively, and let $r_i = Y_i - b_0 - b_1 X_i$ $(i = 1, \ldots, n)$ be the residuals. A *wild bootstrap* method uses the residuals to generate bootstrap samples by multiplying the residuals by some random variable D having mean 0 and variance 1. Then this bootstrap residual is used to form a bootstrap value for Y based on X and the estimated slope and intercept. One choice for D that has received considerable attention is a discrete distribution where D has the value 1 or -1 and the probability that $D = 1$ is 0.5. That is, D is like a binomial having probability of success 0.5, only the possible outcomes are -1 and 1 rather than the usual values 0 and 1. (Another choice is to take U to have a uniform distribution over the unit interval and to use $D = \sqrt{12}(U - 0.5)$.) Imagine that n D values have been generated yielding $D_1, \ldots, D_n$. Bootstrap Y values, assuming that the hypothesis H_0: $\beta_1 = 0$ is true, are given by

$$Y_i^* = \bar{Y} + D_i r_i,$$

$i = 1, \ldots, n$. Note that there are no bootstrap X values.

As previously remarked, when estimating the squared standard error of b_1, the classic estimate used by Student's T test can be unsatisfactory when there is heteroscedasticity. As before, the HC4 estimate is used instead and is labeled V. And again, when testing the hypothesis of a zero slope, H_0: $\beta_1 = 0$, the test statistic is

$$W = \frac{|b_1|}{\sqrt{V}}.$$

The issue is how large W must be to reject. That is, the problem is determining the distribution of W when the null hypothesis is true. The strategy used by the wild bootstrap is as follows:

1. Generate bootstrap Y values using the wild bootstrap method just described, yielding n pairs of points $(Y_1^*, X_1), \ldots, (Y_n^*, X_n)$.
2. Based on the wild bootstrap values generated in step 1, compute the least squares estimate of the slope, b_1^*, the HC4 estimate of the squared standard error V^*, and the test statistic W^*.
3. Repeat steps 1 and 2 B times yielding $W_1^* \ldots, W_B^*$.

The p-value is the proportion of $W_1^*, \ldots, W_B^*$ values that are greater than or equal to W. That is, if N of the $W_1^*, \ldots, W_B^*$ values are greater than or equal to W, the p-value is N/B. To compute a $1-\alpha$ confidence interval, let c be $(1-\alpha)B$ rounded to the nearest integer. Then the confidence interval for the slope is

$$(b_1 - W_{(c)}^* \sqrt{V}, b_1 + W_{(c)}^* \sqrt{V}).$$

Positive features of the wild bootstrap method:

- In terms of Type I errors, it performs substantially better than the classic method covered in Section 6.2.1, which assumes normality and homoscedasticity, and it competes fairly well with alternative techniques that have been proposed.
- Compared to the (non-bootstrap) HC4 test (covered in Section 6.4.6) that allows heteroscedasticity, there are situations where the wild bootstrap gives better results. In particular, situations arise where HC4 is too conservative. That is, when using HC4, the actual Type I error can be substantially lower than the nominal level suggesting that it might have relatively low power. The wild bootstrap method reduces this problem.

Criticisms of the wild bootstrap method:

- Although problems with heteroscedasticity are reduced substantially, there are situations where control over the probability of a Type I error remains poor (greater than 0.1 when testing at the 0.05 level) even with a sample size of $n = 100$ (Ng and Wilcox, 2009). It is unclear the extent to which this is a practical issue. (Perhaps in applied work it is very rare to encounter a situation where the wild bootstrap gives inaccurate confidence intervals.)
- Although seemingly rare, situations can be encountered where the wild bootstrap method results in a numerical error and terminates without computing a p-value. (The covariance matrix generated by the bootstrap method can be singular.)
- The wild bootstrap method does not address the fact that the least squares estimator is not robust. Outliers remain a concern and the standard error of the least squares estimator can be high relative to alternative estimators covered in Chapter 14.

Section 6.4.6 noted that the HC4 method can be extended to situations where there are multiple predictors, $\beta_1, \ldots, \beta_p$, and the goal is to test

$$H_0 : \beta_1 = \beta_2 = \cdots = \beta_p = 0,$$

the hypothesis that all of the slopes are equal to 0. Here it is merely remarked that the wild bootstrap method just described can be used to test this hypothesis as well. The computational details are omitted but an R function described in the next section performs the calculations.

6.10.1 R Functions hc4wtest, olswbtest, and lsfitci

The R function

```
hc4wtest(x,y,nboot=500)
```

performs the wild bootstrap test of the hypothesis that the slopes of a least squares regression line are all equal to zero. The output is the (bootstrap) p-value.

The function

```
olswbtest(x,y,nboot=500)
```

also applies the wild bootstrap but rather than test the hypothesis that all of the slopes are zero, it performs a test for each slope. That is, it tests H_0: $\beta_j = 0$ ($j = 1, \ldots, p$) rather than H_0: $\beta_1 = \cdots = \beta_p = 0$.

The wild bootstrap method used by these two R functions appears to control Type I errors fairly well relative to other methods that might be used. But if these functions result in a numerical error and cannot be applied, the function

```
lsfitci(x,y,nboot=599)
```

can be used, which uses the modified bootstrap method described in the previous section where the goal was to compute a confidence interval for Pearson's correlation. This function computes 0.95 confidence intervals for each of the slopes, but when dealing with a single predictor, it does not provide a p-value and it can only be used when the Type I error is chosen to be 0.05. (It is unknown how to modify the percentile bootstrap method when α differs from 0.05.) For this reason, the emphasis here is on the wild bootstrap method. But a possible appeal of the modified bootstrap method is that execution time on a computer can be substantially lower, which might be an issue when the sample size is large. In terms of controlling the probability of Type I errors, it seems that there is little separating the two methods when using $\alpha = 0.05$. But as the next example illustrates, the choice of method can make a bit of a difference. (For a test that all slopes are zero, the function `olstest` uses a percentile bootstrap method and appears to control Type I errors reasonably well with sample sizes greater than 40.)

EXAMPLE

Using the aggression data in Table 6.3, it was illustrated in Section 6.2.1 that the hypothesis $H_0 : \beta_1 = 0$ is not rejected with $\alpha = 0.05$ using the conventional Student's T test. Using the modified percentile bootstrap method, the 0.95 confidence interval for the slope (using R) is (−0.105, −0.002); this interval does not contain zero, so you reject. The 0.95 confidence interval based on Student's T is (−0.08, 0.0002). (Using R, again `lsfitci` rejects, but it gives a slightly different confidence interval because it uses a different random number generator. Now the upper end of the confidence interval is −0.0007.)

EXAMPLE

For the selling price of homes in Table 6.8, the 0.95 confidence interval for the slope, using the modified percentile bootstrap method, is (0.166, 0.265) versus (0.180, 0.250) using Student's T. The wild bootstrap gives (0.160, 0.270). Student's T gives a shorter confidence interval, but it might be substantially less accurate because it is sensitive to violations of the assumptions of normality and homoscedasticity.

6.11 DETECTING ASSOCIATIONS EVEN WHEN THERE IS CURVATURE

As previously noted, curvature affects the magnitude of Pearson's correlation. One practical concern is that as a result, testing H_0: $\rho = 0$ might fail to reject even when there is an association. One way of dealing with this issue is to test instead the hypothesis that the regression line is both straight and horizontal, which must be the case under independence. More precisely, one might test

$$H_0 : E(Y|X) = \mu_y. \tag{6.22}$$

That is, the mean of Y, given X, does not vary with X; it is always equal to the population mean of Y. Another approach is to test

$$H_0 : M(Y|X) = \theta_y, \tag{6.23}$$

the hypothesis that the median of Y, given X, does not vary with X.

It is briefly noted that a method for testing Equation (6.22) stems from a more general method derived by Stute et al. (1998). It is based in part on a wild bootstrap technique, a version of which is described in Section 6.10. By design, the method is sensitive to any curvature, and in the event the regression line is straight, the method is designed to detect situations where the slope differs from zero. As for testing Equation (6.23), a slight modification of a method derived by He and Zhu (2003) can be used; see Wilcox (2008c) for details. For brevity, the involved computational details are not described, but R functions for applying these techniques are provided. These methods might detect dependence that is missed by other techniques. But a limitation is that when they reject, it is unclear why. That is, these tests do not provide any information about the nature of the association.

6.11.1 R Functions indt and medind

The R function

```
indt(x, y, nboot = 500, alpha = 0.05, flag = 1)
```

performs the test of independence just described, which is based on the wild bootstrap method and the mean of Y. As usual, x and y are R variables containing data, and `nboot` is B.

The function

```
medind(x,y,qval=0.5,nboot=1000,com.pval=F,alpha=0.05,xout=FALSE)
```

test the hypothesis that the median of Y does not change with X. It can also be used to test the hypothesis that the 0.75 (or 0.25) quantile of Y does not change with X by setting the argument `qval=0.75` (or 0.25). Other quantiles can be used, but now simulations are needed to determine a critical value. That is, execution time on a computer will be higher versus testing hypotheses about the median or quartiles. The number of simulations used is controlled with the argument `nboot`. Also, no p-value is provided if `com.pval=F`, in order to reduce execution time. To get a p-value, set com.pval=T. (The function also has the ability to eliminate outliers from among the X values before performing the test. This is done by setting `xout=TRUE`.) The function returns a test statistic and a critical value.

Decision Rule: Reject if the test statistic is greater than or equal to the critical value.

Table 6.6 Data on Meridian Arcs

Place	Transformed Latitude (X)	Arc Length (Y)
Quito	0.0000	56,751
Cape of Good Hope	0.2987	57,037
Rome	0.4648	56,979
Paris	0.5762	57,074
Lapland	0.8386	57,422

EXAMPLE

The data in Table 6.6 are from a study conducted about 200 years ago with the goal of determining whether the Earth bulges at the equator, as predicted by Newton, or whether it bulges at the poles. The issue was addressed by measuring latitude at various points on the Earth and trying to determine how latitude is related to a measure of arc length. The R function `indt` reports a p-value equal to 0.28. But an obvious concern is that power might be fairly low due to the very small sample size.

EXAMPLE

For the aggression data in Table 6.3, Pearson's correlation is $r = -0.286$ and we fail to reject $H_0 : \rho = 0$ at the $\alpha = 0.05$ level using Student's T. So we fail to conclude that there is an association between marital aggression and recall test scores among children. (The p-value is 0.051, so we nearly reject.) The function `indt` returns a p-value of 0.034, so, in particular, reject at the 0.05 level and conclude that marital aggression and recall test scores are dependent. That is, we have empirical evidence that aggression in the home is associated with recall test scores among children living in the home, but the function `indt` tells us nothing about what this association might be like.

In contrast, `medind` returns a test statistic equal to 0.0024 and a critical value of 0.0091, which is designed so that the Type I error probability is 0.05. Because the test statistic is less than the critical value, fail to reject. Look at the plot of the data in Figure 6.15 and note that the points in the upper left appear to be outliers. (Methods covered later in this book support this conclusion.) In particular, they reflect unusually high recall test scores. Also note that if we ignore the apparent outliers, the plot suggests that there is little or no association. Because `indt` uses a mean, a speculation is that it rejects because of the apparent outliers. The function `medind` uses a median, which will be less sensitive outliers among the recall test scores, and this might explain why the function `medind` does not reject.

6.12 QUANTILE REGRESSION

As previously indicated, quantile regression methods can be used, among other things, to fit a regression line aimed at estimating the median of Y given the values of p predictors $X_1, \ldots, X_p$. Moreover, situations arise where there is interest in estimating other quantiles of Y and a method for testing hypotheses about the individual slopes was mentioned. Here it is noted that some bootstrap methods for testing hypotheses have been examined and appear to have practical value. In simulation studies, the method mentioned in Section 6.7 seems to perform reasonably well when the Type I error probability is set at 0.05. But that method

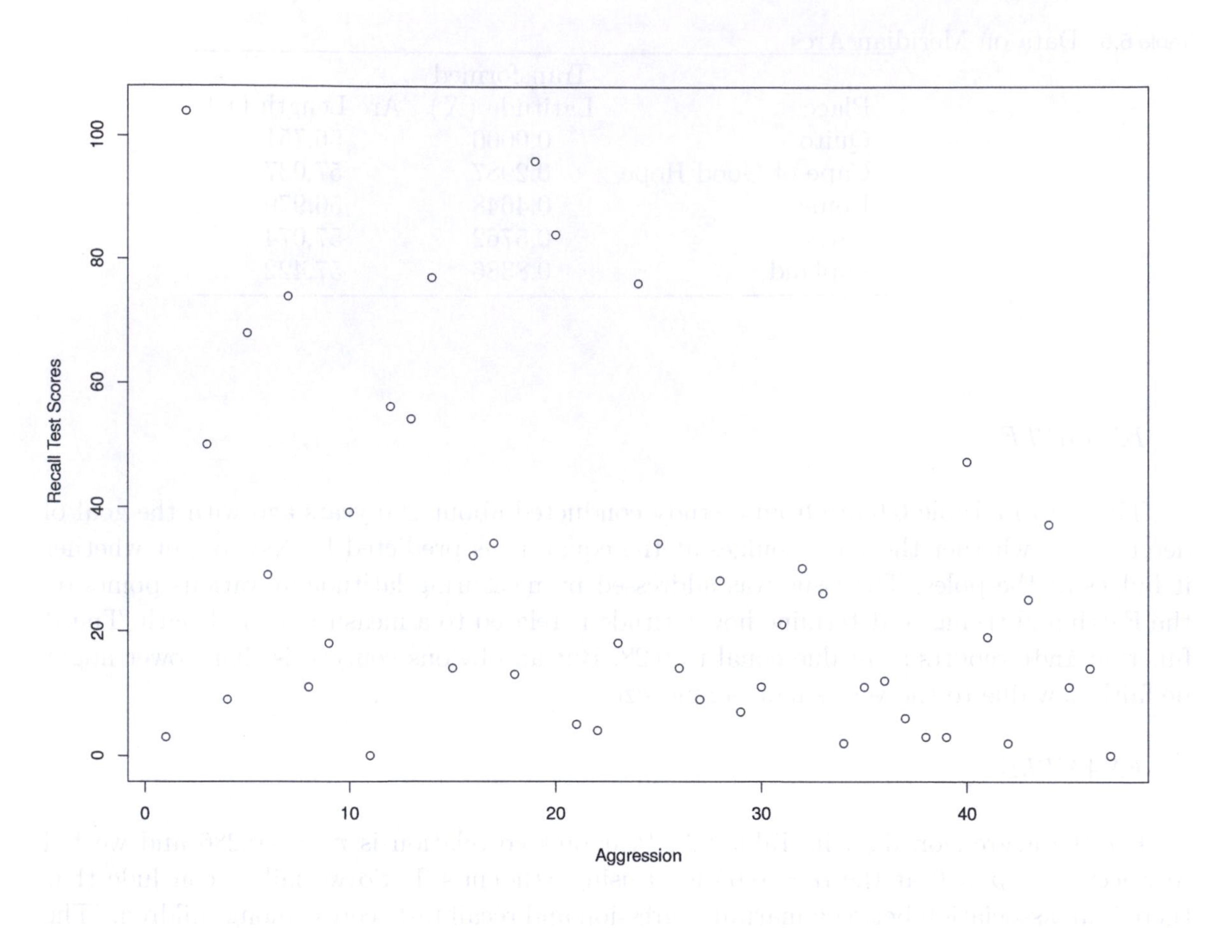

Figure 6.15 Shown is a plot of the aggression data in Table 6.3.

is limited to testing at the 0.05 level and a p-value cannot be computed. An advantage of using a bootstrap method is that it is not limited to testing at the 0.05 level. Moreover, in applied work, situations are encountered where the confidence intervals based on a particular bootstrap method give substantially different results. The precise details of the bootstrap method that is used to compute confidence intervals are not provided; interested readers can refer to Wilcox and Costa (2009) and Wilcox (2007). For details about how to test

$$H_0 : \beta_1 = \cdots = \beta_p = 0, \tag{6.24}$$

see Wilcox (2017, Section 11.1.5). R functions for applying these methods are provided in the next section.

6.12.1 R Functions qregci and rqtest

The R function

```
qregci(x, y, nboot = 100, alpha = 0.05, qval = 0.5)
```

tests the hypothesis of a zero slope for each of the p predictors when using quantile regression. This function defaults to estimating the median of Y given the values of the predictors, but other quantiles can be used via the argument `qval`. The function

```
rqtest(x, y, qval = 0.5, nboot = 200, alpha = 0.05)
```

tests the hypothesis given by Equation (6.24) that all slopes are equal to zero.

EXAMPLE

Results using the bootstrap methods just described, versus the non-bootstrap methods, can differ substantially. Consider again the lake data in Table 6.4 with the goal of predicting TN with NIN. The estimate of the slope (when predicting the median of TN with NIN) is 0.0119. Using the non-bootstrap method, the 0.95 confidence interval for the slope is (−0.046, 0.868). Notice that the estimate of the slope is not much larger than the lower end of the confidence interval, −0.046. The 0.95 confidence interval returned by the R function `qregci` is (−0.2996, 0.3234), which is roughly centered around the estimate of the slope.

6.12.2 A Test for Homoscedasticity Using a Quantile Regression Approach

Section 6.8 described a relatively simple method for testing the hypothesis of homoscedasticity within the context of regression. An alternative approach is outlined here, the practical point being that it can detect heteroscedasticity, and more generally a type of dependence that other methods can miss.

Note that one way of measuring the variation among the Y values, given X, is in terms of the interquartile range. Of course, any other two quantiles could be used and here the focus is on the 0.2 and 0.8 quantiles. For convenience, let

$$Y_{0.8} = \beta_{0,0.8} + \beta_{1,0.8}X$$

be the regression line for predicting the 0.8 quantile of Y, given X, and let

$$Y_{0.2} = \beta_{0,0.2} + \beta_{1,0.2}X$$

be the regression line for predicting the 0.2 quantile. If the variation in Y does not change with X, then the hypothesis

$$H_0 : \beta_{1,0.2} = \beta_{1,0.8} \tag{6.25}$$

must be true. That is, homoscedasticity implies that the slopes of these two regression lines must be the same. (The regression lines must be parallel). One way of testing this hypothesis is first to compute a bootstrap estimate of the standard error of $b_{1,0.8} - b_{1,0.2}$, say S_d, in which case

$$Z = \frac{b_{1,0.8} - b_{1,0.2}}{S_d}$$

and the hypothesis of identical slopes (homoscedasticity) is rejected if

$$|Z| \geq z_{1-\alpha/2},$$

the $1 - \alpha/2$ quantile of a standard normal distribution. A positive feature is that the actual Type I error probability does not appear to be very sensitive to nonnormality. But a negative feature is that when testing at the 0.05 level, the actual Type I error can be substantially smaller than 0.05 when n is small. A method for reducing this problem, when the Type I error is set at $\alpha = 0.05$, is to adjust the critical value (Wilcox and Keselman, 2006a). In particular, rather than use the 0.975 quantile of a standard normal distribution, $z_{1-\alpha/2}$, use the

$$a = 0.975 - \frac{0.104}{\sqrt{n}}$$

quantile. That is, the 0.95 confidence interval is now given by

$$(b_{0,0.8} - b_{0,0.2} - z_a S_d,\ b_{0,0.8} - b_{0,0.2} + z_a S_d).$$

(This adjustment works well when dealing with the slopes of the 0.8 and 0.2 quantile regression lines, but it is unknown how best to adjust the confidence interval when working with other quantiles.) Note that the quantile regression lines are assumed to be straight. A method that eliminates this assumption is described in Section 14.4.9.

EXAMPLE

Imagine that the sample size is $n = 25$ and the goal is to compute a 0.95 confidence interval for the difference between the slopes, $\beta_{1,0.8} - \beta_{1,0.2}$. Then the adjusted critical level is $a = 0.975 - 0.104/\sqrt{25} = 0.9542$. From Table 1 in Appendix B, $z_{0.9542} = 1.687$. So if $b_{1,0.8} = 1.5$, $b_{1,0.2} = -0.5$, and $S_d = 1.3$, the 0.95 confidence interval is

$$(2 - 1.687(1.3),\ 2 + 1.687(1.3)) = (-0.193,\ 4.193).$$

6.12.3 R Function qhomt

The R function

```
qhomt(x,y)
```

applies the test for homoscedasticity just described.

EXAMPLE

Figure 6.16 shows a scatterplot of data from a study aimed at investigating various predictors of reading ability in children. Here, a measure of sound blending (a measure of phonological awareness) is used to predict a measure of the ability to decode words. (The data were collected by L. Doi and are stored in columns 3 and 9 of the file read.dat, stored on the author's web page; see Chapter 1.) The top regression line indicates the quantile regression estimate of the 0.8 quantile of the identification score given a child's measure of phonological awareness. The bottom regression line estimates the 0.2 quantile. If there is independence, and in particular if there is homoscedasticity, these two lines should be parallel. Testing this hypothesis with the R version of the function `qhomt`, the adjusted 0.95 confidence interval is reported to be $(-2.72, -0.16)$; this interval does not contain zero, so reject and conclude there is heteroscedasticity. Using the Koenker–Bassett test of homoscedasticity instead, which was covered in Section 6.8, the p-value is 0.15. This is not to suggest that Koenker–Bassett method be excluded from consideration; it merely illustrates that the two methods can give substantially different results depending on the nature of the heteroscedasticity. An educated guess is that usually the function `qhomt` has more power than the function `khomreg`, but situations are encountered where the reverse is true.

6.13 REGRESSION: WHICH PREDICTORS ARE BEST?

A difficult problem when dealing with regression with two or more predictors is determining which predictors, or which subset of predictors, are best. Numerous methods have been proposed, most of which are known to be unsatisfactory under fairly general conditions. Methods that have been found to be unsatisfactory include stepwise regression (e.g., Montgomery and Peck, 1992, Section 7.2.3; Derksen and Keselman, 1992), a related (forward selection) method (see Kuo and Mallick, 1998; Huberty, 1989; Chatterjee and Hadi, 1988; cf. Miller, 1990), methods based on R^2 (the squared multiple correlation), and the F statistic given

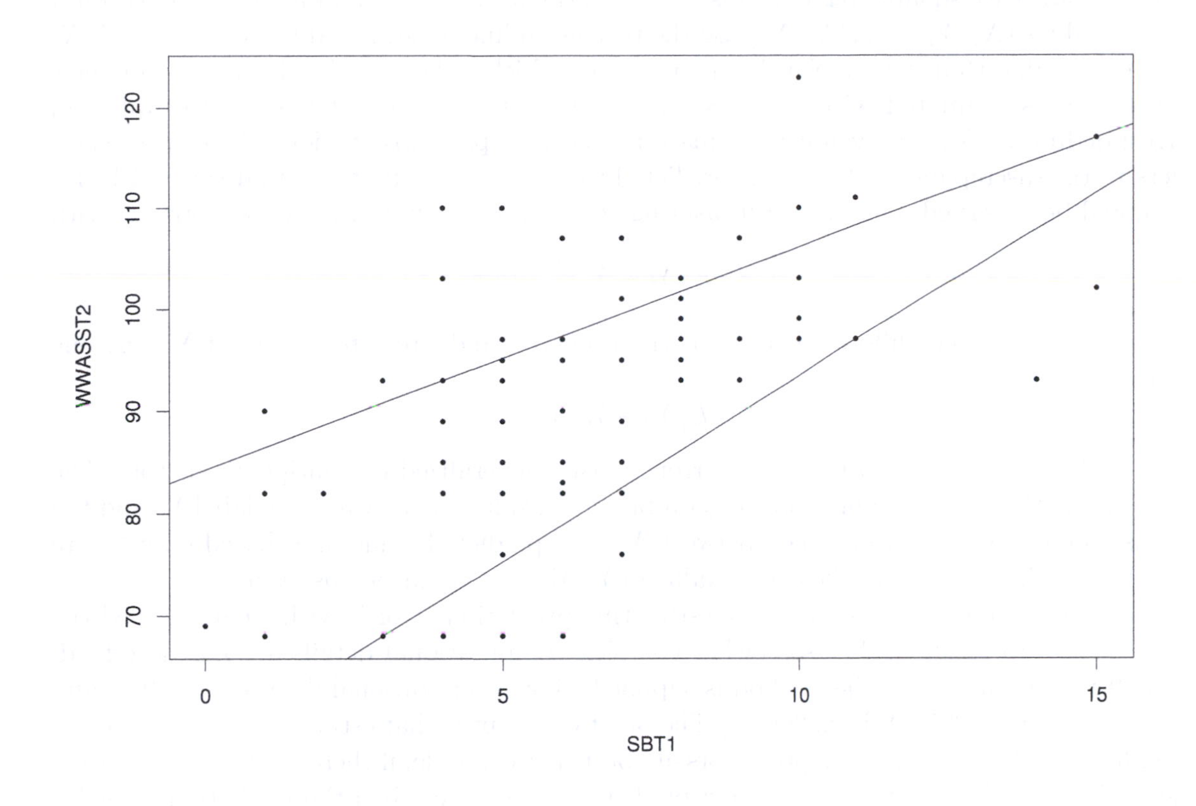

Figure 6.16 Shown is a plot of the regression lines for predicting the 0.2 and 0.8 quantiles of a measure of the ability to decode words using a measure of sound blending. Homoscedasticity means that these two regression lines should be parallel. The R function `qhomt` rejects the hypothesis that they are parallel when $\alpha = 0.05$.

by Equation (6.13), which tests the hypothesis that all slopes are zero. A homoscedastic approach based on

$$C_p = \frac{1}{\hat{\sigma}^2} \sum (Y_i - \hat{Y}_i)^2 - n + 2p,$$

called Mallow's (1973) C_p criterion, cannot be recommended either (Miller, 1990). Another approach is based on what is called ridge regression, but it suffers from problems listed by Breiman (1995). Some alternative approaches are cross-validation, bootstrap methods, namely the 0.632 estimator used here, the so-called non-negative garrote technique derived by Breiman (1995), the elastic net (Zou and Hastie, 2005), the adaptive lasso (Zou, 2006), and ordinary least squares (OLS) post-lasso (Belloni and Chernozhukov, 2013). (Breiman's method is appealing when the number of predictors is large. For an interesting variation of Breiman's method, see Tibshirani, 1996.)

Here attention is focused on two methods for estimating the relative importance of the independent variables. It is stressed that while both methods provide estimates of which independent variable are most important, they do not indicate the strength of the empirical evidence that X_1, say, is more important than X_2. For a technique aimed at addressing this issue, see method IBS in Section 14.9.3.

6.13.1 The 0.632 Bootstrap Method

Imagine that least squares regression is used to estimate the slope and intercept based on n pairs of values $(X_1, Y_1), \ldots, (X_n, Y_n)$. So the regression line is estimated to be $\hat{Y} = b_0 + b_1 X$. Further imagine that a new X value is observed, which is labeled X_0. Based on this new X value, Y is estimated with $\hat{Y}_0 = b_0 + b_1 X_0$. That is, you do not observe the value Y_0 corresponding to X_0, but you can estimate it based on past observations. *Prediction error* refers to the discrepancy between the predicted value of Y, $\hat{Y}_0$, and the actual value of Y, Y_0, if it could be observed. One way of measuring the typical amount of prediction error is with

$$E[(Y_0 - \hat{Y}_0)^2],$$

the expected squared difference between the observed and predicted value of Y. Or, one might use

$$E[|Y_0 - \hat{Y}_0)|].$$

As is evident, the notion of prediction error is easily generalized to multiple predictors. The basic idea is that via some method we get a predicted value for Y, which we label $\hat{Y}$, and the goal is to measure the discrepancy between $\hat{Y}_0$ (the predicted value of Y based on a future collection of X values) and the actual value of Y, Y_0, if it could be observed.

Currently, a relatively good way of estimating prediction error is with what is called the 0.632 bootstrap method. The somewhat complex computational details are not described, but software for applying the method is supplied. (For computational details, see Efron and Tibshirani, 1993, 1997; Wilcox, 2005a.) The practical point is that estimated prediction error provides an indication of which predictors are best. For example, if there are three predictors, and if the typical prediction error using predictor 1 is smaller than the prediction error for predictor 2, predictor 1 is estimated to be more important. Similarly, if the prediction error using predictors 1 and 2 is smaller than the prediction error using all three predictors, using predictors 1 and 2 is preferable to using all three. A criticism of R^2 is that generally it increases when adding variables, even if the variables added to the model have no relevance when predicting Y. This is roughly one of the reasons why it is ineffective as a method for identifying the best predictors. Using the 0.632 bootstrap estimate of prediction error eliminates this problem. That is, if irrelevant variables are added to the model, generally prediction error will increase. A limitation of the 0.632 bootstrap method is that it does not provide any indication of the strength of the empirical evidence that the truly best predictors have been identified.

6.13.2 R function regpre

The R function

```
regpre(x,y,regfun = lsfit, error = absfun, nboot = 100, model = NA)
```

estimates prediction error for all possible combinations of predictors when the number of predictors is less than or equal to 5. (With more than five predictors, the models to be tested must be specified with the argument `model`, which is assumed to have list mode. For example, `model`[[1]]=c(1,3,5) would estimate prediction error when using predictors 1, 3, and 5.) By default, the function uses least squares regression, but any regression estimator can be used via the argument `regfun`. For example, `regfun=rqfit` would use the quantile regression estimator. The argument error defaults to `absfun`, meaning that the average absolute prediction error is used. (Using `error=sqfun` would result in using the average squared error.

Table 6.7 Hald's Cement Data

Y	X_1	X_2	X_3	X_4
78.5	7	26	6	60
74.3	1	29	15	52
104.3	11	56	8	20
87.6	11	31	8	47
95.9	7	52	6	33
109.2	11	55	9	22
102.7	3	71	17	6
72.5	1	31	22	44
93.1	2	54	18	22
115.9	21	47	4	26
83.8	1	40	23	34
113.3	11	66	9	12
109.4	10	68	8	12

Using `regfun=tsreg` appears to be a good choice for general use. This results in using the Theil–Sen estimator described in Section 14.1.1.)

EXAMPLE

Table 6.7 shows data from a study by Hald (1952) concerning the heat evolved in calories per gram (Y) versus the amount of each of four ingredients in the mix: tricalcium aluminate (X_1), tricalcium silicate (X_2), tetracalcium alumino ferrite (X_3), and dicalcium silicate (X_4). Testing the hypothesis that all of the slopes are equal to zero using the F test in Section 6.2.2, the p-value is less than 0.001. Yet for each of the four predictors, Student's T test of H_0: $\beta_j = 0$ ($j = 1, 2, 3, 4$) has p-values 0.07, 0.5, 0.9 and 0.84, respectively. That is, we fail to reject for any specific predictor at the 0.05 level. (However, we do reject for some of the slopes when using the Theil-Sen estimator in Chapter 14.)

Here is the output from `regpre`:

```
      apparent.error    boot.est    err.632 var.used rank
 [1,]      97.360519 115.992035 135.170624        1   13
 [2,]      69.718180  77.366387  85.970872        2   10
 [3,]     149.184651 175.530506 207.953762        3   15
 [4,]      67.989763  76.842194  87.227135        4   11
 [5,]       4.454191   5.977285   7.650515       12    2
 [6,]      94.390158 134.907890 179.040382       13   14
 [7,]       5.750932   7.191115   8.736024       14    4
 [8,]      31.957133  40.487813  51.270850       23    9
 [9,]      66.836933  84.096524 104.953615       24   12
[10,]      13.518308  17.083283  21.431613       34    8
[11,]       3.700816   6.025407   8.524557      123    3
[12,]       3.690210   5.357036   7.190052      124    1
[13,]       3.910471   6.515499   9.133306      134    5
[14,]       5.678042   9.168980  12.477687      234    6
[15,]       3.681818   9.236824  14.541992     1234    7
```

```
[16,]          NA          NA 220.277075          0    16
```

What is important from an applied point of view are the last three columns. The column headed by err.632 is the 0.632 estimate of the average (squared) prediction error. The next column indicates which predictors are used. The first row shows a 1 in this column, indicating that the first predictor is used and the others are ignored. The next to last row (row 15) has 1234 in the column headed by var.used, meaning that predictors 1–4 were used. The final column is provided to help find the row with the lowest prediction error; it assigns ranks to the prediction errors, the lowest getting a rank of 1, the next lowest a rank of 2, and so on. Here, row 12 has a rank of 1 indicating that the best prediction is achieved with predictors 1, 2, and 4. The last row shows 0 under var.used, meaning that no predictors are used. (That is, ignore the predictors and simply use the regression line $\hat{Y} = \bar{Y}$.) Its rank is 16, meaning that this is the least accurate model.

6.13.3 Least Angle Regression

A positive feature of the R function `regpre` is that it automatically considers all possible subsets of predictors, provided the number of predictors is less than or equal to five. But as the number of predictors increases, it soon becomes impractical to consider all subsets of predictors when searching for the best model to predict Y. A simple strategy is to consider the prediction error associated with each predictor, ignoring the other predictors, and choose those predictors with the lowest prediction error. Yet another strategy is to use what is called least angle regression (Efron et al., 2004), which has close ties to what is called the *lasso* method. Bootstrap methods do not play a role, but least angle regression offers an alternative to prediction error when the goal is to identify the best predictors, which should be mentioned. Currently it seems that the lasso method performs relatively well in terms of identifying the most important predictors. It is unclear the extent to which it improves on estimating prediction error using the 0.632 bootstrap method just described. A possible appeal is that it can be performed quickly even with a large sample size and a large number of predictors. A criticism is that it is not robust.

6.13.4 R Function larsR

The R package `lars` applies the least angle regression method. To simplify the use of this package somewhat, the function

```
larsR(x,y,xout=FALSE,outfun=out)
```

is provided.

EXAMPLE

For the cement data in the last example, `larsR` returns

```
Sequence of LASSO moves:

Var   4  1  2  3
Step  1  2  3  4
```

The values listed after Var indicate the order of the most important variables based on least angle regression. Here, predictor 4 is identified as most important, predictor 1 is next, and predictor 3 is the least important. From the previous example, the 0.632 bootstrap method also selects predictor 4 as being the most important, but the 0.632 bootstrap method ranks predictor 2 as being more important than predictor 1, in contrast to `lars`.

6.14 COMPARING CORRELATIONS

Consider three variables X_1, X_2, and X_3 and let ρ_{jk} be the correlation between X_j and X_k ($j = 1, 2, 3$; $k = 1, 2, 3$). The goal is to test

$$H_0 : \rho_{12} = \rho_{13}. \tag{6.26}$$

Typically X_1 is some outcome variable Y while X_2 and X_3 are two predictor variables of interest. (It helps to compare the method described here with the method in Section 14.9.3.) This problem is often called comparing overlapping dependent correlations because r_{12} and r_{13} are dependent; both involve the first variable X_1. Numerous methods have been proposed for testing this hypothesis with most known to be unsatisfactory under general conditions (e.g., Hittner et al., 2003). Zou (2007) suggested a general approach to the problem. But the particular variation that was studied turns out to yield poor control over the probability of a Type I error for a wide range of situations, even when the sample size is large, because it relies on Fisher's r-to-z transformation. Here, an alternative variation is used that appears to perform relatively well (Wilcox, 2009c).

Let (l_1, u_1) and (l_2, u_2) be $1 - \alpha$ confidence intervals for ρ_{12} and ρ_{13}, respectively, based on the HC4 method described in Section 6.6.1. Let

$$L = r_{12} - r_{13} - \sqrt{(r_{12} - l_1)^2 + (u_2 - r_{13})^2 - 2\widehat{corr}(r_{12}, r_{13})(r_{12} - l_1)(u_2 - r_{13})},$$

and

$$U = r_{12} - r_{23} + \sqrt{(u_1 - r_{12})^2 + (r_{23} - l_2)^2 - 2\widehat{corr}(r_{12}, r_{13})(u_1 - r_{12})(r_{23} - l_2)},$$

where

$$\widehat{corr}(r_{12}, r_{13}) = \frac{(r_{23} - 0.5 r_{12} r_{23})(1 - r_{12}^2 - r_{13}^2 - r_{23}^2) + r_{23}^2}{(1 - r_{12}^2)(1 - r_{13})^2}$$

estimates the correlation between r_{12} and r_{13}. Then an approximate $1 - \alpha$ confidence interval for $\rho_{12} - \rho_{13}$ is

$$(L, U).$$

If this interval does not contain zero, reject the hypothesis of equal correlations.

Now consider four variables $X_1, \ldots, X_4$ and again let ρ_{jk} be the correlation between X_j and X_k ($j = 1, 2, 3, 4$; $k = 1, 2, 3, 4$). Now the goal is to compute a confidence interval for

$$\rho_{12} - \rho_{34},$$

which is sometimes called comparing non-overlapping dependent correlations. This problem arises, for example, when there are two outcome variables of interest, say Y_1 and Y_2 (labeled X_1 and X_2 here) and the issue is whether X_3 has a stronger association with Y_1 versus the association between Y_2 and X_4. Now (l_1, u_1) and (l_2, u_2) are $1 - \alpha$ confidence intervals for ρ_{12} and ρ_{34}, respectively, which are based on method PM2 described in Section 6.10. (Using HC4 results in poor control over the probability of Type I error in some situations.) The $1 - \alpha$ confidence interval for $\rho_{12} - \rho_{34}$ is (L, U), where

$$L = r_{12} - r_{34} - \sqrt{(r_{12} - l_1)^2 + (u_2 - r_{34})^2 - 2\widehat{corr}(r_{12}, r_{34})(r_{12} - l_1)(u_2 - r_{34})},$$

$$U = r_{12} - r_{34} + \sqrt{(u_1 - r_{12})^2 + (r_{34} - l_2)^2 - 2\widehat{corr}(r_{12}, r_{34})(u_1 - r_{12})(r_{34} - l_2)},$$

$$\widehat{corr}(r_{12}, r_{13}) = \frac{T1 - T2}{T3},$$

$$T1 = 0.5r_{12}r_{34}(r_{13}^2 - r_{14}^2 + r_{23}^2 + r_{24}^2) + r_{13}r_{24} + r_{14}r_{23},$$

$$T2 = r_{12}r_{13}r_{14} + r_{12}r_{23}r_{24} + r_{13}r_{23}r_{34} + r_{14}r_{24}r_{34}$$

and

$$T3 = (1 - r_{12}^2)(1 - r_{34})^2.$$

6.14.1 R Functions TWOpov and TWOpNOV

The R function

```
TWOpov(x,y)
```

returns a confidence interval for the difference between two dependent Pearson correlations when dealing with the overlapping case. The argument x is assumed to be a matrix with two columns. The correlation between y and the first column of x plays the role of r_{12}, and the correlation of y with the second column of x corresponds to r_{13}.

The function

```
TWOpNOV(x,y)
```

returns a confidence interval for the difference between two dependent correlations corresponding to the non-overlapping case. Now both x and y are assumed to be matrices with two columns. Here r_{12} is the correlation between the two variables stored in x and r_{34} is the correlation associated with the data stored in y.

EXAMPLE

Consider again the reading data introduced in the right panel of Figure 6.6, only now attention is focused on a measure of sound blending (stored in column 3 of the file read.dat) and the speed of identifying lower case letters (column 7) as predictors of comprehension (column 12). The 0.95 confidence interval for $\rho_{12} - \rho_{13}$, returned by the function TWOpov, is (0.63, 1.16), suggesting that sound blending has a stronger association with comprehension.

EXAMPLE

For the cement data, we saw that lars indicates that the fourth predictor is best for predicting the outcome variable and the third predictor is the worst. The corresponding correlations are −0.82 and −0.53, but the R function TWOpov fails to reject at the 0.05 level, casting doubt on how certain we can be that the third predictor is truly the best of the four predictors under consideration.

A related goal is to test the hypothesis

$$H_0 : \rho_{12}^2 = \rho_{13}^2. \tag{6.27}$$

That is, test the hypothesis that the strength of the association is the same. For the cement data, predictors two and four have correlations 0.816 and −0.821, respectively, with the outcome variable and we reject the hypothesis that the correlations are equal at the 0.05 level. But the strength of the associations (the squared correlations) are nearly identical. Various strategies for testing Equation (7.20) have been proposed but currently it seems that all methods can be unsatisfactory under nonnormality and heteroscedasticity.

Section 6.5.1 pointed out some concerns about Pearson's correlation: it is not robust and it can be inappropriate if a regression line is not straight. Some alternative methods for comparing measures of association, aimed at dealing with these problems, are described in Chapter 14. Included is a test of Equation (6.27) with Pearson's correlation replaced by a robust measure of association.

6.15 CONCLUDING REMARKS

The purpose of this chapter was to introduce basic concepts and to describe standard hypothesis testing methods associated with least squares regression and Pearson's correlation. Another goal was to provide some indication of what might go wrong with these standard methods and how these concerns might be addressed. It is stressed that there are a variety of alternative methods for dealing with nonnormality, heteroscedasticity, outliers, and curvature, some of which are described in Chapter 14.

6.16 EXERCISES

1. For the following pairs of points, verify that the least square regression line is $\hat{Y} = 1.8X - 8.5$.

$$X : 5, 8, 9, 7, 14$$

$$Y : 3, 1, 6, 7, 19.$$

2. Compute the residuals using the results from Exercise 1. Verify that if you square and sum the residuals, you get 47, rounding to the nearest integer.

3. Verify that for the data in Exercise 1, if you use $\hat{Y} = 2X - 9$, the sum of the squared residuals is larger than 47. Why would you expect a value greater than 47?

4. Suppose that based on $n = 25$ values, $s_x^2 = 12$, $s_y^2 = 25$, and $r = 0.6$. What is the slope of the least squares regression?

5. Verify that for the data in Table 6.3, the least squares regression line is $\hat{Y} = -0.0405X + 4.581$.

6. The following table reports breast cancer rates plus levels of solar radiation (in calories per day) for various cities in the United States. Fit a least squares regression to the data with the goal of predicting cancer rates and comment on what this line suggests.

City	Rate	Daily Calories	City	Rate	Daily Calories
New York	32.75	300	Chicago	30.75	275
Pittsburgh	28.00	280	Seattle	27.25	270
Boston	30.75	305	Cleveland	31.00	335
Columbus	29.00	340	Indianapolis	26.50	342
New Orleans	27.00	348	Nashville	23.50	354
Washington, DC	31.20	357	Salt Lake City	22.70	394
Omaha	27.00	380	San Diego	25.80	383
Atlanta	27.00	397	Los Angeles	27.80	450
Miami	23.50	453	Fort Worth	21.50	446
Tampa	21.00	456	Albuquerque	22.50	513
Las Vegas	21.50	510	Honolulu	20.60	520
El Paso	22.80	535	Phoenix	21.00	520

7. For the following data, compute the least squares regression line for predicting GPA given SAT.

SAT:	500	530	590	660	610	700	570	640
GPA:	2.3	3.1	2.6	3.0	2.4	3.3	2.6	3.5

8. For the data in the last exercise, verify that the coefficient of determination is 0.36 and interpret what this tells you.

9. For the following data, compute the least squares regression line for predicting Y from X.

X:	40	41	42	43	44	45	46
Y:	1.62	1.63	1.90	2.64	2.05	2.13	1.94

10. In Exercise 6, what would be the least squares estimate of the cancer rate given a solar radiation of 600? Indicate why this estimate might be unreasonable.

11. Maximal oxygen uptake (mou) is a measure of an individual's physical fitness. You want to know how mou is related to how fast someone can run a mile. Suppose you randomly sample six athletes and get

mou (milliliters/kilogram): 63.3 60.1 53.6 58.8 67.5 62.5
time (seconds): 241.5 249.8 246.1 232.4 237.2 238.4

Compute the correlation. Can you be reasonably certain about whether it is positive or negative with $\alpha = 0.05$?

12. Verify that for the following pairs of points, the least squares regression line has a slope of zero. Plot the points and comment on the assumption that the regression line is straight.

X: 1 2 3 4 5 6

Y: 1 4 7 7 4 1

13. Repeat Exercise 12, only for the points

X: 1 2 3 4 5 6

Y: 4 5 6 7 8 2

14. Vitamin A is required for good health. You conduct a study and find that as vitamin A intake increases, certain health problems decrease. However, for levels of vitamin A not included in your study, a sufficiently high dose can result in death. Comment on what this illustrates in the context of regression.

15. Sockett et al. (1987) report data related to patterns of residual insulin secretion in children at the time they were diagnosed with diabetes. A portion of the study was concerned with whether age can be used to predict the logarithm of C-peptide concentrations at diagnosis. The observed values are

    Age (X): 5.2 8.8 10.5 10.6 10.4 1.8 12.7 15.6 5.8 1.9 2.2 4.8 7.9 5.2 0.9 11.8 7.9 1.5 10.6 8.5 11.1 12.8 11.3 1.0 14.5 11.9 8.1 13.8 15.5 9.8 11.0 12.4 11.1 5.1 4.8 4.2 6.9 13.2 9.9 12.5 13.2 8.9 10.8

    C-peptide (Y): 4.8 4.1 5.2 5.5 5.0 3.4 3.4 4.9 5.6 3.7 3.9 4.5 4.8 4.9 3.0 4.6 4.8 5.5 4.5 5.3 4.7 6.6 5.1 3.9 5.7 5.1 5.2 3.7 4.9 4.8 4.4 5.2 5.1 4.6 3.9 5.1 5.1 6.0 4.9 4.1 4.6 4.9 5.1

    Verify that $r = 0.4$ and that you reject H_0: $\beta_1 = 0$ with $\alpha = 0.05$ and the R function `hc4test`.

16. For the data in Exercise 15, verify that a least squares regression line using only X values (age) less than 7 yields $b_1 = 0.247$ and $b_0 = 3.51$. Verify that when using only the X values greater than 7 you get $b_1 = 0.009$ and $b_0 = 4.8$. What does this suggest about using a linear rule for all of the data? (The first example in Section 14.2.2 expands on this issue.)

17. The selling price and size of a home for a suburb of Los Angeles in the year 1997 are shown in Table 6.8. At the time, even a small empty lot would cost at least \$200,000. Verify that based on the least squares regression line for these data, if we estimate the cost of an empty lot by setting the square feet of a house to $X = 0$, we get 38, 192. What does this suggest about estimating Y using an X value outside the range of observed X values?

18. For the data in Exercise 17, the sizes of the corresponding lots are:

    18,200 12,900 10,060 14,500 76,670 22,800 10,880 10,880 23,090 10,875 3498 42,689 17,790 38,330 18,460 17,000 15,710 14,180 19,840 9150 40,511 9060 15,038 5807 16,000 3173 24,000 16,600.

    Verify that the least squares regression line for estimating the selling price, based on the size of the lot, is $\hat{Y} = 11X + 436,834$.

19. Imagine two scatterplots, where in each scatterplot the points are clustered around a line having slope 0.3. If for the first scatterplot $r = 0.8$, does this mean that points are more tightly clustered around the line versus the other scatterplot where $r = 0.6$?

20. You measure stress (X) and performance (Y) on some task and get

    X : 18 20 35 16 12

    Y : 36 29 48 64 18

Table 6.8 Selling Price of Homes (divided by 1,000) Versus Size in Square Feet

Home i	Size (X_i)	Price (Y_i)	Home i	Size (X_i)	Price (Y_i)
1	2359	510	15	3883	859
2	3397	690	16	1937	435
3	1232	365	17	2565	555
4	2608	592	18	2722	525
5	4870	1125	19	4231	805
6	4225	850	20	1488	369
7	1390	363	21	4261	930
8	2028	559	22	1613	375
9	3700	860	23	2746	670
10	2949	695	24	1550	290
11	688	182	25	3000	715
12	3147	860	26	1743	365
13	4000	1050	27	2388	610
14	4180	675	28	4522	1290

Verify that you do not reject $H_0: \beta_1 = 0$ using $\alpha = 0.05$. Is this result consistent with what you get when testing $H_0: \rho = 0$? Why would it be incorrect to conclude that X and Y are independent?

21. Suppose you observe

X: 12.2, 41, 5.4, 13, 22.6, 35.9, 7.2, 5.2, 55, 2.4, 6.8, 29.6, 58.7

Y: 1.8, 7.8, 0.9, 2.6, 4.1, 6.4, 1.3, 0.9, 9.1, 0.7, 1.5, 4.7, 8.2

Verify that the 0.95 confidence interval for the slope is (0.14, 0.17). Would you reject $H_0: \beta_1 = 0$? Based on this confidence interval only, can you be reasonably certain that generally, as X increases, Y increases as well?

22. The data in Table 6.9 are from a study, conducted by L. Doi where the goal is to understand how well certain measures predict reading ability in children. Verify that the 0.95 confidence interval for the slope is (−0.16, .12) based on Equation (6.8).

23. Verify that for the data in Table 6.6, when testing the hypothesis of homoscedasticity using the R function `khomreg`, you fail to reject with $\alpha = 0.05$. Explain why this result is *not* a satisfactory reason for assuming homoscedasticity.

24. To illustrate a point, assume that for the data in Table 6.9, the goal is to predict X given Y, rather than predict Y given X. Use the R functions `ols` and `rqfit` to test the hypothesis $H_0: \beta_1 = 0$ with $\alpha = 0.05$. Verify that `ols` fails to reject, its p-value is approximately 0.97, but `rqfit` rejects. What might explain this?

25. R has a built-in dataset called leuk. The third column indicates survival times of patients diagnosed with acute myelogenous leukemia. The first column indicates the patient's white blood cell count at the time of diagnosis. Using least squares regression, assuming homoscedasticity, test the hypothesis of a zero slope using the R function `ols`. Test this hypothesis again, only now do not assume homoscedasticity; use the function `olshc4`.

Table 6.9 Reading Data

X:	34 49 49 44 66 48 49 39 54 57 39 65 43 43 44 42 71 40 41
	38 42 77 40 38 43 42 36 55 57 57 41 66 69 38 49 51 45 141
	133 76 44 40 56 50 75 44 181 45 61 15 23 42 61 146 144 89 71
	83 49 43 68 57 60 56 63 136 49 57 64 43 71 38 74 84 75 64 48
Y:	129 107 91 110 104 101 105 125 82 92 104 134 105 95 101 104 105 122 98
	104 95 93 105 132 98 112 95 102 72 103 102 102 80 125 93 105 79 125
	102 91 58 104 58 129 58 90 108 95 85 84 77 85 82 82 111 58 99
	77 102 82 95 95 82 72 93 114 108 95 72 95 68 119 84 75 75 122 127

CHAPTER 7

COMPARING TWO INDEPENDENT GROUPS

CONTENTS

Imagine that you want to compare two methods for treating headaches. One method is based on an experimental drug and the other is based on a placebo. A fundamental issue is determining whether the choice of method matters. If it does matter, how might we characterize the extent to which the experimental drug makes a practical difference? A related issue is how these two groups might be compared in terms of side effects. For example, does the experimental drug result in increased liver damage? How does the reading ability of children who watch 30 hours or more of television per week compare to children who watch 10 hours or less? How does the birth weight of newborns among mothers who smoke compare to the birth weight among mothers who do not smoke?

Roughly, there are four general approaches when comparing two groups, each of which has its own strengths and weaknesses. So a broad goal is to explain what each approach is trying to accomplish and to provide a general sense of their relative merits. The four approaches are:

1. Compare measures of location, such as the mean or median.
2. Compare measures of variation.
3. Focus on the probability that a randomly sampled observation from the first group is smaller than a randomly sampled observation from the second group.
4. Simultaneously compare all of the quantiles to get a global sense of where the distributions differ and by how much. For example, low scoring participants in group 1 might be very similar to low scoring participants in group 2, but for high scoring participants, the reverse might be true.

7.1 STUDENT'S T TEST

We begin with the classic and commonly used method for comparing two independent groups: Student's T test. The goal is to test

$$H_0 : \mu_1 = \mu_2, \tag{7.1}$$

the hypothesis that the two groups have equal means. A related goal is computing a confidence interval for $\mu_1 - \mu_2$, the difference between the means. One possible concern is that the means might not provide a reasonable measure of what is typical, as noted in previous chapters. Another concern is its standard error can be relatively high, but these issues are ignored for the moment.

The following assumptions are made:

- Random sampling.
- The observations in group 1 are independent of the observations in group 2.
- For both groups, sampling is from a normal distribution.
- The two groups have equal variances. That is, $\sigma_1^2 = \sigma_2^2$, where σ_1^2 and σ_2^2 are the variances corresponding to the groups having means μ_1 and μ_2, respectively.

The last assumption is called *homoscedasticity. Heteroscedasticity* refers to a situation where the variances differ ($\sigma_1^2 \neq \sigma_2^2$).

For convenience, let σ_p^2 represent the assumed common variance and let s_1^2 and s_2^2 be the sample variances corresponding to the two groups. Also, let n_1 and n_2 represent the corresponding sample sizes. The typical estimate of the assumed common variance is

$$s_p^2 = \frac{(n_1 - 1)s_1^2 + (n_2 - 1)s_2^2}{n_1 + n_2 - 2}, \tag{7.2}$$

where the subscript p is used to indicate that the sample variances are being pooled. Because s_1^2 and s_2^2 are assumed to estimate the same quantity, σ_p^2, a natural strategy for combining them into a single estimate is to average them, and this is exactly what is done when the sample sizes are equal. But with unequal sample sizes, a slightly different strategy is used as indicated by Equation (7.2). A weighted average is used instead where the group with the larger sample size is given more weight.

Now consider the problem of testing the null hypothesis of equal means. Simultaneously, we want a confidence interval for the difference between the population means, $\mu_1 - \mu_2$. Under the assumptions already stated, the probability of a Type I error will be exactly α if we reject H_0 when

$$|T| \geq t, \tag{7.3}$$

where

$$T = \frac{\bar{X}_1 - \bar{X}_2}{\sqrt{s_p^2\left(\frac{1}{n_1} + \frac{1}{n_2}\right)}} \tag{7.4}$$

and t is the $1 - \alpha/2$ quantile of Student's T distribution with $\nu = n_1 + n_2 - 2$ degrees of freedom. An exact $1 - \alpha$ confidence interval for the difference between the population means is

$$(\bar{X}_1 - \bar{X}_2) \pm t\sqrt{s_p^2\left(\frac{1}{n_1} + \frac{1}{n_2}\right)}. \tag{7.5}$$

(The R function `t.test`, described in Section 7.3.1, performs Student's T test.)

EXAMPLE

Salk (1973) conducted a study where the general goal was to examine the soothing effects of a mother's heartbeat on her newborn infant. Infants were placed in a nursery immediately after birth and they remained there for four days except when being fed by their mothers. The infants were divided into two groups. One group was continuously exposed to the sound of an adult's heartbeat; the other group was not. Salk measured, among other things, the weight change of the babies from birth to the fourth day. Table 7.1 reports the weight change for the babies weighing at least 3510 grams at birth. The estimate of the assumed common variance is

$$s_p^2 = \frac{(20-1)(60.1^2) + (36-1)(88.4^2)}{20+36-2} = 6335.9.$$

So

$$T = \frac{18-(-52.1)}{\sqrt{6335.9\left(\frac{1}{20}+\frac{1}{36}\right)}} = \frac{70.1}{22.2} = 3.2.$$

The degrees of freedom are $\nu = 20 + 36 - 2 = 54$. If we want the Type I error probability to be $\alpha = 0.05$, then from Table 4 in Appendix B, $t = 2.01$. Because $|T| \geq 2.01$, reject H_0 and conclude that the means differ. That is, we conclude that among all newborns we might measure, the average weight gain would be higher among babies exposed to the sound of a heartbeat compared to those that are not exposed. By design, the probability that our conclusion is in error is 0.05, assuming normality and homoscedasticity. The 0.95 confidence interval for $\mu_1 - \mu_2$, the difference between the means, is

$$[18-(-52.1)] \pm 2.01\sqrt{6335.9\left(\frac{1}{20}+\frac{1}{36}\right)} = (25.5, 114.7).$$

This interval does not contain zero, and it suggests that the difference between the means is at least 25.5, so again reject the hypothesis of equal means.

7.1.1 Choosing the Sample Sizes

Prior to collecting any data, researchers are confronted with the issue of how large of a sample size is needed. Addressing this problem in a reasonably accurate and sufficient manner—with no data—is difficult. Certainly the most common strategy is to assume normality, homoscedasticity, and that Student's T will be used when comparing groups. This approach requires making a decision about how large of a difference one wants to detect and a classic measure is the difference between the means divided by the assumed common standard deviation:

$$\delta = \frac{\mu_1 - \mu_2}{\sigma}. \tag{7.6}$$

Under normality, Cohen (1994) has suggested that small, medium, and large differences between the groups correspond to $\delta = 0.2$, 0.5, and 0.8, respectively. Once values for δ, $1-\beta$ (power), and α have been specified, it is possible to determine the sample sizes that are required using the R function `power.t.test`, described next.

7.1.2 R Function power.t.test

After issuing the R command `library(stats)`, the R function

Table 7.1 Weight Gain, in Grams, for Large Babies

Group 1 (heartbeat)			
Subject	Gain	Subject	Gain
1	190	11	10
2	80	12	10
3	80	13	0
4	75	14	0
5	50	15	−10
6	40	16	−25
7	30	17	−30
8	20	18	−45
9	20	19	−60
10	10	20	−85

$n_1 = 20$, $\bar{X}_1 = 18.0$, $s_1 = 60.1$, $s_1/\sqrt{n_1} = 13$

Group 2 (no heartbeat)							
Subject	Gain	Subject	Gain	Subject	Gain	Subject	Gain
1	140	11	−25	21	−50	31	−130
2	100	12	−25	22	−50	32	−155
3	100	13	−25	23	−60	33	−155
4	70	14	−30	24	−75	34	−180
5	25	15	−30	25	−75	35	−240
6	20	16	−30	26	−85	36	−290
7	10	17	−45	27	−85		
8	0	18	−45	28	−100		
9	−10	19	−45	29	−110		
10	−10	20	−50	30	−130		

$n_2 = 36$, $\bar{X}_2 = -52.1$, $s_2 = 88.4$, $s_2/\sqrt{n_2} = 15$

```
power.t.test(n = NULL, delta = NULL, sd = 1, sig.level = 0.05, power = NULL,
type = c("two.sample", "one.sample", "paired"), alternative = c("two.sided",
                                   "one.sided")),
```

introduced in Chapter 5, can be used to assess power and sample sizes when comparing two independent groups using Student's T test. Here, the argument `n` represents the sample size for each group. The argument `type` defaults to `"two.sample,"` and the argument `alternative` defaults to `"two.sided."` That is, unless specified otherwise, the function assumes the goal is to test H_0: $\mu_1 = \mu_2$. Here, the argument `delta` corresponds to $\delta = (\mu_1 - \mu_2)/\sigma$. As in Chapter 5, two of the arguments that default to NULL must be specified, and the function returns the value of the argument not specified.

EXAMPLE

If you want to have a 0.8 probability of rejecting when $\delta = 0.5$ and if $\alpha = 0.05$, the R command

```
power.t.test(n = NULL, delta=0.8,power=0.8)
```

indicates that for each group, the required sample size is 25.52.

7.2 RELATIVE MERITS OF STUDENT'S T

We begin by describing some positive features of Student's T. If distributions are non-normal, but otherwise *identical*, Student's T performs reasonably well in terms of controlling Type I errors in the sense that the actual Type I error probability is likely to be less than or equal to the nominal level. (Rasch et al., 2007, report relevant results when dealing with highly discrete distributions.) For example, if we test at the 0.05 level, the actual Type I error probability is probably less than or equal to 0.05. In terms of a 0.95 confidence interval, the actual probability coverage is likely to be at least 0.95. This result is somewhat expected based on features of the one-sample Student's T covered in Chapters 4 and 5. To get a rough idea of why, first note that for any two independent variables having identical distributions, their difference will have a perfectly symmetric distribution about zero, even when the distributions are skewed. For example, suppose that in Salk's study, the first group of infants have weight gains that follow the (lognormal) distribution shown in Figure 4.14, and the second group has the same distribution. If we randomly sample an observation from the first group and do the same for the second group, then the distribution of the difference will be symmetric about zero. More generally, if we sample n observations from each of two identical distributions, the difference between the sample means will have a symmetric distribution. Note that when two distributions are identical, then not only are their means equal, their variances are also equal.

Chapters 4 and 5 noted that in the one-sample case, problems with controlling Type I errors—ensuring that the actual Type I error probability does not exceed the nominal level—arise when sampling from a skewed distribution. For symmetric distributions, this problem is of little concern. So for the two-sample problem considered here, the expectation is that if there is absolutely no difference between the two groups, implying that not only do they have equal means but equal variances and the same skewness, then the actual Type I error probability will not exceed the specified α value by very much. All indications are that this argument is correct, but some simple extensions of this argument lead to erroneous conclusions. In particular, it might seem that, generally, if we sample from perfectly symmetric distributions, probability coverage and control of Type I error probabilities will be satisfactory, but we will see that this is not always the case—even under normality. In particular, perfectly symmetric distributions with unequal variances can create practical problems. Nevertheless, all indications are that generally when comparing identical distributions, so in particular the hypothesis of equal means is true and the variances are equal, Type I error probabilities will not exceed the nominal level by very much, and for this special case power is not an issue because the null hypothesis of equal means is true.

Student's T begins having practical problems when distributions differ in some manner. If sampling is from normal distributions with equal sample sizes but unequal variances, Student's T continues to perform reasonably well in terms of Type I errors (Ramsey, 1980). But when sampling from non-normal distributions, this is no longer the case. Concerns about the ability of Student's T test to control the probability of a Type I error date back to at least Pratt (1964), who established that the level of the test is not preserved if distributions differ in dispersion or shape. Even under normality there are problems when sample sizes are unequal. Basically, Type I error control, power, and probability coverage can be very poor. In fact, when sample sizes are unequal, Cressie and Whitford (1986) describe general conditions under which Student's T does not even converge to the correct answer as the sample sizes get large.

A seemingly natural suggestion for salvaging the assumption that the variances are equal is to test it. That is, test $H_0 : \sigma_1^2 = \sigma_2^2$ and if a nonsignificant result is obtained, proceed with Student's T. But this strategy has been found to be unsatisfactory, even under normality (e.g., Hayes and Cai, 2007; Markowski and Markowski, 1990; Moser, Stevens, and Watts, 1989; Wilcox, Charlin, and Thompson, 1986; Zimmerman, 2004). nonnormality makes this strategy even less satisfactory. A basic problem is that tests of the hypothesis of equal variances do not always have enough power to detect situations where the assumption of equal variances should be discarded.

Summarizing, in terms of controlling the probability of a Type I error, it appears that Student's T provides a satisfactory test of the hypothesis that two groups have identical distributions. Student's T is designed to be sensitive to the differences between the means, but in reality it is sensitive to a myriad of ways the distributions might differ such as different amounts of skewness.

EXAMPLE

To illustrate the effect of different amounts of skewness, imagine that 50 observations are sampled from the (lognormal) distribution in Figure 4.14 shifted to have a mean of zero, 25 observations are sampled from a standard normal distribution, and T is used to test the hypothesis of equal means with $\alpha = 0.05$. Then the actual probability of a Type I error is approximately 0.21. Increasing the sample sizes to 400 and 1000, respectively, the actual probability of a Type I error is approximately 0.14. Using equal sample sizes reduces this problem but does not necessarily eliminate it. If instead both sample sizes are 30, the actual Type I error probability is approximately 0.085. But increasing the standard deviation of the non-normal distribution, with the null hypothesis still true, the actual Type I error probability again exceeds 0.1.

As a method for testing the hypothesis of equal population means, one might defend Student's T in the following manner. If it rejects, we can be reasonably certain that the distributions differ. If the distributions differ, some authorities would argue that by implication, the means are not equal. That is, they would argue that in practice we never encounter situations where distributions differ in shape but have equal means. But as soon as we agree that the distributions differ, there is the possibility that the actual probability coverage of the confidence interval for the difference between the means, given by Equation (7.5), differs substantially from the $1-\alpha$ value you have specified. A crude rule is that the more the distributions differ, particularly in terms of skewness, the more inaccurate the confidence interval might be. Generally, nonnormality, outliers, and heteroscedasticity contribute to this problem. Said another way, Student's T can be used to establish that the means differ by implication, but it can be very inaccurate in terms of assessing the precision of the estimated difference.

One more problem with Student's T is that situations arise where it is biased. That is, there is a higher probability of rejecting when the means are equal compared to some situations where the means differ. If the goal is to find a method that is sensitive to differences between the means, surely this property is undesirable.

Yet another concern is that for a variety of reasons, the power of Student's T can be very low relative to many other methods that might be used. One reason is that very slight departures from normality can inflate the variances. More broadly, consider two distributions where Student's T has relatively high power. Then there are nearly identical distributions to the ones being compared where power is relatively low.

EXAMPLE

Consider the two normal distributions shown in the left panel of Figure 7.1. Both have variances one and the means differ by 1. Using Student's T with $\alpha = 0.05$, power is 0.96 with $n_1 = n_2 = 25$. But now look at the two contaminated normal distributions shown in the right panel of Figure 7.1. The difference in the means is the same as in the left panel and the plot of the distributions has an obvious similarity to the distributions in the left panel. But for the right panel, power is only 0.28. One reason power is low is that when sampling from a heavy-tailed distribution, the actual probability of a Type I error can be substantially lower than the specified α value. For example, if you use Student's T with $\alpha = 0.05$, the actual probability of a Type I error can drop below 0.01. If an adjustment could be made so that the actual probability of a Type I error is indeed 0.05, power would be better, but it would still be low relative to alternative methods that are less sensitive to outliers. The reason is that in the right panel, the variances are 10.9, compared to 1 for the distributions in the left panel. Said another way, outliers are more likely to occur when sampling from the distributions in the left panel which can inflate the standard deviations. Even when outliers are not a concern, having unequal variances or even different degrees of skewness can result in relatively poor power as well.

Finally, it is noted that when using Student's T, even a single outlier in only one group can result in a rather undesirable property. The following example illustrates the problem.

EXAMPLE

Consider the following values.

Group 1: 4 5 6 7 8 9 10 11 12 13
Group 2: 1 2 3 4 5 6 7 8 9 10

The corresponding sample means are $\bar{X}_1 = 8.5$ and $\bar{X}_2 = 5.5$ and $T = 2.22$. With $\alpha = 0.05$, the critical value is $t = 2.1$. So, assuming the underlying assumptions are reasonably accurate, reject the hypothesis of equal means and conclude that the first group has a larger population mean than the second (because the first group has the larger sample mean). Now, if we increase the largest observation in the first group from 13 to 23, the sample mean increases to $\bar{X}_1 = 9.5$. So the difference between $\bar{X}_1$ and $\bar{X}_2$ has increased from 3 to 4 and this would seem to suggest that we have stronger evidence that the population means differ and in fact

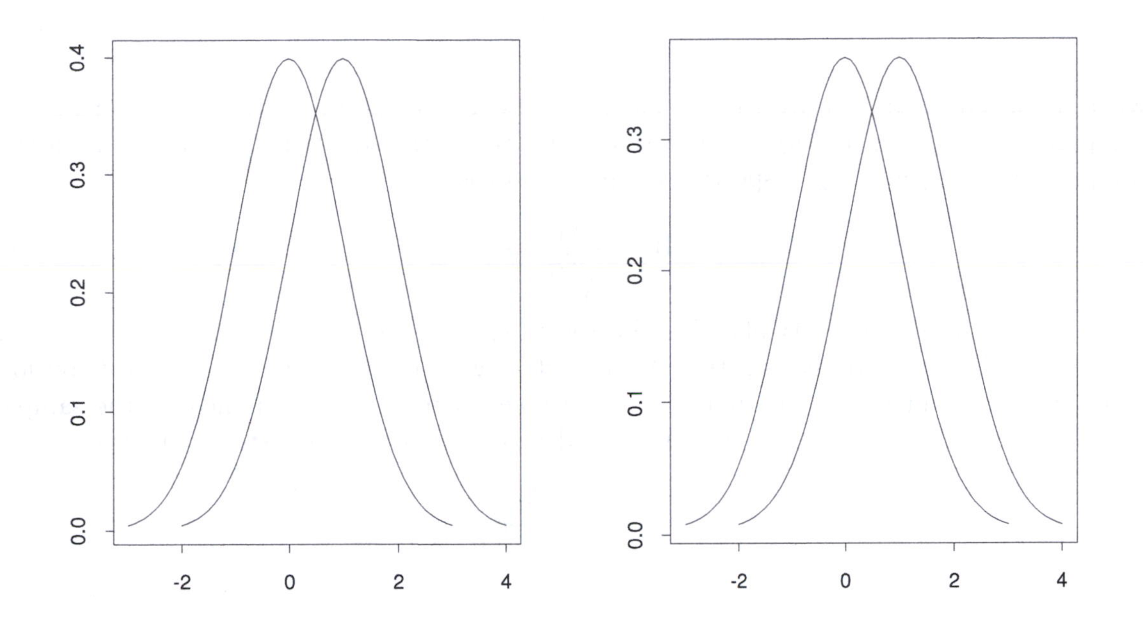

Figure 7.1 In the left panel, power is 0.96 based on Student's T, $\alpha = 0.05$. But in the left panel, power is only 0.28, illustrating the general principle that slight changes in the distributions being compared can have a large impact on the ability to detect true differences between the population means.

the first group has the larger population mean. However, increasing the largest observation in the first group also inflates the corresponding sample variance, s_1^2. In particular, s_1^2 increases from 9.17 to 29.17. The result is that T *decreases* to $T = 2.04$ and we no longer reject. That is, increasing the largest observation has more of an effect on the sample variance than the sample mean in the sense that now we are no longer able to conclude that the population means differ. Increasing the largest observation in the first group to 33, the sample mean increases to 10.5, the difference between the two sample means increases to 5, and now $T = 1.79$. So again we do not reject and in fact our test statistic is getting smaller! This illustration provides another perspective on how outliers can mask differences between population means.

7.3 WELCH'S HETEROSCEDASTIC METHOD FOR MEANS

A step toward addressing the negative features of Student's T, still assuming normality and that the goal is to compare means, is to switch to a method that allows unequal variances. Many techniques have been proposed, none of which are completely satisfactory. We describe only one such method here. One reason it was chosen is because it is a special case of a more general technique that gives more satisfactory results. That is, our goal is to build our way up to a method that performs relatively well over a broad range of situations.

Proposed by Welch (1938), we test the hypothesis of equal means (H_0: $\mu_1 = \mu_2$) as follows. For the jth group ($j = 1, 2$) let

$$q_j = \frac{s_j^2}{n_j} \tag{7.7}$$

where as usual s_1^2 and s_2^2 are the sample variances corresponding to the two groups being compared. That is, q_1 and q_2 are the usual estimates of the squared standard errors of the sample means, $\bar{X}_1$ and $\bar{X}_2$, respectively. The test statistic is

$$W = \frac{(\bar{X}_1 - \bar{X}_2)}{\sqrt{\frac{s_1^2}{n_1} + \frac{s_2^2}{n_2}}} \tag{7.8}$$

and to control the probability of a Type I error, the problem is to determine the distribution of W when the null hypothesis is true. Welch's strategy was to approximate the distribution of W using a Student's T distribution with degrees of freedom determined by the sample variances and sample sizes. In particular, the degrees of freedom are estimated with

$$\hat{\nu} = \frac{(q_1 + q_2)^2}{\frac{q_1^2}{n_1 - 1} + \frac{q_2^2}{n_2 - 1}}. \tag{7.9}$$

Decision Rule:

Reject $H_0 : \mu_1 = \mu_2$, the hypothesis of equal means, if

$$|W| \geq t, \tag{7.10}$$

where t is the $1 - \alpha/2$ quantile of Student's T distribution with $\hat{\nu}$ degrees of freedom.

One-sided tests are performed in a similar manner. To test H_0: $\mu_1 \leq \mu_2$, reject if $W \geq t$, where now t is the $1 - \alpha$ quantile of Student's T distribution with $\hat{\nu}$ degrees of freedom. As for $H_0 : \mu_1 \geq \mu_2$, reject if $W \leq t$ where t is the α quantile of Student's T.

Confidence Interval:

A $1 - \alpha$ confidence interval for $\mu_1 - \mu_2$ is

$$(\bar{X}_1 - \bar{X}_2) \pm t\sqrt{\frac{s_1^2}{n_1} + \frac{s_2^2}{n_2}}, \tag{7.11}$$

where again t is the $1 - \alpha/2$ quantile of Student's T distribution read from Table 4 in Appendix B.

The method just described is sometimes labeled the *Satterthwaite test.* The degrees of freedom used by the Welch test is a special case of general result derived by Satterthwaite (1946).

EXAMPLE

For Salk's data in Table 7.1, the value of the test statistic given by Equation (7.8) is

$$\begin{aligned} W &= \frac{18 - (-52.1)}{\sqrt{\frac{60.1^2}{20} + \frac{88.4^2}{36}}} \\ &= \frac{70.1}{19.9} \\ &= 3.52. \end{aligned}$$

To compute the estimated degrees of freedom, first compute

$$q_1 = \frac{60.1^2}{20} = 180.6,$$

$$q_2 = \frac{88.4^2}{36} = 217,$$

in which case

$$\hat{\nu} = \frac{(180.6 + 217)^2}{\frac{180.6^2}{19} + \frac{217^2}{35}} \approx 52.$$

So if the Type I error probability is to be 0.05, then for a two-sided test ($H_0 : \mu_1 = \mu_2$), $t = 2.01$, approximately, and we reject because $|W| = 3.52 > 2.01$. That is, the empirical evidence indicates that infants exposed to the sounds of a mother's heartbeat gain more weight, on average. The 0.95 confidence interval for the difference between the means can be seen to be (30, 110). In contrast, Student's T method yields a 0.95 confidence interval of (25.5, 114.7). So Welch's procedure indicates that the difference between the means of the two groups is at least 30, but Student's T leads to the conclusion that the difference between the means is at least 25.5. Generally, Welch's method provides more accurate confidence intervals than Student's T, so for the problem at hand, results based on Welch's method should be used. Also, Welch's method can provide a shorter confidence interval. There are situations where Student's T is more accurate than Welch's method, but in general the improvement is modest at best, while the improvement of Welch's method over Student's T can be substantial.

7.3.1 R function t.test

R has a built-in function called

```
t.test(x, y = NULL,var.equal=FALSE,conf.level=0.95)
```

that performs the two-sample Student's T test when values are stored in the second argument, y, and when the argument `var.equal=TRUE`. (If the second argument, y, is omitted, the one-sample Student's T test in Chapter 5 is performed.) Note that by default, `var.equal=FALSE`, meaning that the means will be compared using Welch's method described in Section 7.3, which is designed to allow unequal variances. The function also returns a confidence interval, which by default has probability coverage 0.95. To get a 0.99 confidence interval, for example, set the argument `conf.level=0.99`.

EXAMPLE

Imagine that the data in Table 7.1 are stored in the R variables salk1 and salk2. Then the command

```
t.test(salk1,salk2,var.equal=T,conf.level=0.9)
```

results in the following output:

```
        Standard Two-Sample t-Test

data:  salk1 and salk2
t = 3.1564, df = 54, p-value = 0.0026
```

```
alternative hypothesis: true difference in means is not equal to 0
90 percent confidence interval:
  32.92384 107.24282
sample estimates:
 mean of x mean of y
        18 -52.08333
```

It is common to store data with one particular variable indicating the group to which a value belongs; see Chapter 1. When this is done, the R function `t.test` can still be used using an appropriate formula. To illustrate the basic idea, imagine that the data are stored in the R variable x: the first 10 values stored in x belong to group 1 and the remaining 10 belong to group 2. The goal is to test the hypothesis of equal means, which can be done with the following R commands:

```
g=factor(c(rep(1,10),rep(2,10)))
t.test(x ~ g)
```

The R variable g is a factor variable with the first 10 values equal to 1 and the second 10 equal to 2. The R function `t.test` determines that there are two groups and essentially sorts the data in x into groups based on the values in g. Character data can used to indicate groups as well. For example, the command

```
g = factor(c(rep("A",10),rep("B",10)))
```

would indicate two groups corresponding to the labels A and B. If instead the odd numbered values in x belong to group 1 and the even numbered values belong to group 2, this could be indicated by the R command

```
g = factor(rep(c("A","B"),10)).
```

With R running, if you type g and hit return, you will see

```
 [1] A B A B A B A B A B A B A B A B A B A B
Levels: A B
```

indicating that there are two groups, called Levels, which are labeled A and B. Moreover, the first observation in x belongs to group A, the second belongs to group B, the third belongs to group A, and so on.

EXAMPLE

Chapter 1 mentioned a data set, called plasma retinol, which can be downloaded from the author's web page. It contains 14 variables. Column 2 indicates sex (1=Male, 2=Female) and column 10 indicates cholesterol consumed (mg per day). Imagine that the data are stored in the R variable plasma as described in Chapter 1. The following command would compare males to females using Welch's test:

```
t.test(plasma[,10] ~ factor(plasma[,2])).
```

The p-value is 0.0001, the sample means for males and females are 328 and 229, respectively, and so the analysis indicates that the population mean for males is higher than for females.

7.3.2 Tukey's Three-Decision Rule

Chapter 5 noted that objections have been raised about testing the hypothesis that the population mean is exactly equal to some specified value. A similar objection has been raised when testing the hypothesis that two groups have equal means (e.g., Tukey, 1991). Briefly, the criticism is that surely the means differ at some decimal place and if we could sample enough observations, the hypothesis of equal means would be rejected. One strategy for dealing with this criticism is to rely on confidence intervals.

Another strategy is to interpret results based on Tukey's three decision rule (e.g., Jones and Tukey, 2000), which was introduced in Chapter 5. Here, the three possible decisions are as follows:

1. If the hypothesis of equal means is rejected and $\bar{X}_1 < \bar{X}_2$, conclude that $\mu_1 < \mu_2$.
2. If the hypothesis of equal means is rejected and $\bar{X}_1 > \bar{X}_2$, conclude that $\mu_1 > \mu_2$.
3. If the hypothesis of equal means is not rejected, do not make a decision about which group has the larger population mean.

Said another way, the computational details when testing the hypothesis of equal means are exactly the same as before. The difference is in terms of the interpretation of the results and the stated goal. Tukey's approach is based on determining which group has the larger population mean. And he suggests making a decision about this issue if the hypothesis of equal means is rejected. Assuming the probability of a Type I error can be controlled, and that it has been set to some specified value α, the probability of erroneously concluding that the first group has the larger mean, for example, is at most $\alpha/2$.

Also, Tukey's three decision rule provides a useful perspective on p-values similar to the perspective already discussed in Chapter 5. If we decide that the group with the smallest sample mean has the smallest population mean, a p-value reflects the strength of the empirical evidence that a correct decision has been made. The closer the p-value is to zero, the stronger is the evidence that a correct decision has been made, assuming the Type I error probability can be controlled. A valid criticism is that, under general conditions, a p-value does not tell us anything about the magnitude or relative importance of any difference between the means that might exist. (This topic is discussed in more detail in Section 7.12, where an exception is indicated.) As in Chapter 5, if for example we reject the hypothesis of equal means with $\alpha = 0.05$, p-values do not tell us the likelihood of again rejecting if a study is replicated—this is a power issue.

7.3.3 Nonnormality and Welch's Method

One reason Welch's method improves upon Student's T is that under random sampling, Welch's method satisfies the basic requirement of converging to the correct answer as the sample sizes get large when randomly sampling from non-normal distributions, even when the sample sizes are unequal. That is, as the sample sizes increase, the actual probability coverage for the difference between the means will converge to $1 - \alpha$ when using Equation (7.11). For Student's T, this is true when the sample sizes are equal, but for unequal sample sizes there are general conditions where this goal is not achieved. When sampling from normal distributions, Welch's method does a better job of handling unequal variances. This can translate into more power, as well as shorter and more accurate confidence intervals. But nonnormality can be devastating in terms of both probability coverage (e.g., Algina et al., 1994) and power. Moreover, Welch's method can be biased: the probability of rejecting can be higher when the means are equal compared to situations where they are not. When the two groups being compared have identical distributions, Welch's method performs well in terms of controlling the probability of a Type I error, and because for this special case the population means are equal, power is not an issue. But when distributions differ in some manner, such as having different skewnesses, or even unequal variances, Welch's method can have poor properties under nonnormality. So although Welch's method is an improvement over Student's T, more needs to be done.

7.3.4 Three Modern Insights Regarding Methods for Comparing Means

There have been three modern insights regarding methods for comparing means, each of which has already been described. But these insights are of such fundamental importance, it is worth summarizing them here.

- Resorting to the central limit theorem in order to justify the normality assumption can be highly unsatisfactory when working with means. Under general conditions, hundreds of observations might be needed to get reasonably accurate confidence intervals and good control over the probability of a Type I error. Or in the context of Tukey's three-decision rule, hundreds of observations might be needed to be reasonably certain which group has the largest mean. When using Student's T, rather than Welch's test, concerns arise regardless of how large the sample sizes might be.
- Practical concerns about heteroscedasticity (unequal variances) have been found to be much more serious than once thought. All indications are that it is generally better to use a method that allows unequal variances.
- When comparing means, power can be very low relative to other methods that might be used. Both differences in skewness and outliers can result in relatively low power. Even if no outliers are found, differences in skewness might create practical problems. Certainly there are exceptions. But all indications are that it is prudent not to assume that these concerns can be ignored.

Despite the negative features just listed, there is one positive feature of Student's T worth stressing. If the groups being compared do not differ in any manner, meaning that they have identical distributions, so in particular the groups have equal means, equal variances, and the same amount of skewness, Student's T appears to control the probability of a Type I error reasonably well under nonnormality. That is, when Student's T rejects, it is reasonable to conclude that the groups differ in some manner, but the nature of the difference, or the main reason Student's T rejected, is unclear. Also note that from the point of view of Tukey's three-decision rule, testing and rejecting the hypothesis of identical distributions is not very interesting.

7.4 METHODS FOR COMPARING MEDIANS AND TRIMMED MEANS

As pointed out in Chapters 4 and 5, one way of dealing with outliers is to replace the mean with the median. But a concern is that the median represents an extreme amount of trimming, which might result in a relatively large standard error. In the present context, when comparing two groups, the practical implication is that power might be low relative to methods based on some alternative measure of location. As explained in Chapters 4 and 5, one strategy for dealing with this issue is to use a compromise amount of trimming. This section describes one way of comparing two groups when this strategy is followed and a method designed specifically for comparing medians is described as well.

7.4.1 Yuen's Method for Trimmed Means

Yuen (1974) derived a method for comparing the population trimmed means of two independent groups that reduces to Welch's method for means when there is no trimming. As usual, 20% trimming is a good choice for general use, but situations arise where more than 20% trimming might be beneficial (such as when the proportion of outliers exceeds 20%).

Generalizing slightly the notation in Chapter 2, and assuming 20% trimming, let $g_j = [0.2n_j]$, where again n_j is the sample size associated with the jth group ($j = 1, 2$) and $[0.2n_j]$ is the value of $0.2n_j$ rounded down to the nearest integer. (With 10% trimming, now $g_j = [0.1n_j]$.) Let $h_j = n_j - 2g_j$. That is, h_j is the number of observations left in the jth group after trimming. Let

$$d_j = \frac{(n_j - 1)s_{wj}^2}{h_j(h_j - 1)}, \tag{7.12}$$

where s_{wj}^2 is the Winsorized variance for the jth group. (The amount of Winsorizing is always the same as the amount of trimming. So for a 20% trimmed mean, Winsorize 20% as well.) Yuen's test statistic is

$$T_y = \frac{\bar{X}_{t1} - \bar{X}_{t2}}{\sqrt{d_1 + d_2}}. \tag{7.13}$$

The degrees of freedom are

$$\hat{\nu}_y = \frac{(d_1 + d_2)^2}{\frac{d_1^2}{h_1 - 1} + \frac{d_2^2}{h_2 - 1}}.$$

Confidence Interval. The $1 - \alpha$ confidence interval for $\mu_{t1} - \mu_{t2}$, the difference between the population trimmed means, is

$$(\bar{X}_{t1} - \bar{X}_{t2}) \pm t\sqrt{d_1 + d_2}, \tag{7.14}$$

where t is the $1 - \alpha/2$ quantile of Student's T distribution with $\hat{\nu}_y$ degrees of freedom.

Hypothesis Testing. The hypothesis of equal trimmed means, $H_0 : \mu_{t1} = \mu_{t2}$, is rejected if

$$|T_y| \geq t.$$

As before, t is the $1 - \alpha/2$ quantile of Student's T distribution with $\hat{\nu}_y$ degrees of freedom.

The improvement in power, achieving accurate confidence intervals, and controlling Type I error probabilities, can be substantial when using Yuen's method with 20% trimming rather than Welch's test. For instance, situations arise where Welch's test can have an actual Type I error probability more than twice the nominal level, yet Yuen's method controls the probability of a Type I error reasonably well. Wilcox (2017) describes a situation where when testing at the 0.05 level, the power of Welch's method is 0.28 but Yuen's method has power 0.78. Yuen's method is not always best in terms of Type I errors and power. Indeed, no single method is always best. But it is important to be aware that the choice of method can make a substantial difference in the conclusions reached.

7.4.2 R Functions yuen and fac2list

The R function

```
yuen(x,y,alpha=0.05,tr=0.2)
```

performs Yuen's test. The default value for the argument `alpha` is 0.05 meaning that a 0.95 confidence interval will be computed. The function also returns a p-value when testing the hypothesis of equal trimmed means. The argument `tr` indicates the amount of trimming, which defaults to 20%.

EXAMPLE

In a study of sexual attitudes, 1327 males and 2282 females were asked how many sexual partners they desired over the next 30 years.[1] We can compare the means with the R function **yuen** by setting tr=0. (This results in using Welch's test.) The 0.95 confidence interval for the difference between the means is (−1491.087, 4823.244) and the p-value is 0.30. Given the large sample sizes, a tempting conclusion might be that power is adequate and that the groups do not differ. However, the 0.95 confidence interval for the difference between the 20% trimmed means is (0.408, 2.109) with a p-value less than 0.001. In terms of Tukey's three-decision rule, there is relatively strong evidence that males have the larger population 20% trimmed mean.

The R function **yuen** does not accept a formula as is done when using the R function **t.test**. But the function

```
fac2list(x,g)
```

can be used to separate data into groups, where **x** is the R variable (often the column of a matrix or data frame) containing the data to be analyzed and **g** indicates the level of the corresponding value stored in **x**. (The built-in R function **split** can be used as well.)

EXAMPLE

Consider again the plasma retinol data used in the last example of Section 7.3.1. Again we compare males and females based on the daily consumption of cholesterol, only we use Yuen's test. Assuming the data are stored in the R variable plasma, column 2 indicates sex (male=1, female=2) and column 10 indicates the daily consumption of cholesterol. The R command

```
z=fac2list(plasma[,10],plasma[,2])
```

separates the data into groups corresponding to males and females and stores the results in z in list mode. Because male is indicated by the value 1, which is less than the value used to indicate female, z[[1]] will contain the data for the males and z[[2]] contains the data for the females. The R command

```
yuen(z[[1]],z[[2]])
```

will compare males to females using 20% trimmed means.

7.4.3 Comparing Medians

Although the median belongs to the class of trimmed means, special methods are required for comparing groups based on medians. Put another way, it might seem that Yuen's method could be used to compare medians simply by setting the amount of trimming to 0.5. (That is, when using the R function **yuen**, set tr=0.5.) When the amount of trimming is 0.2, Yuen's method performs reasonably well in terms of Type I errors and accurate confidence intervals,

[1]The data in Table 2.2 are based on the same question but are from a different study. The data used in this example, supplied by Lynn Miller, are stored in the file miller.dat and can be downloaded from the author's web page given in Chapter 1.

but as the amount of trimming gets close to 0.5, the method breaks down and should not be used. (The method used to estimate the standard error performs poorly.) If there are no tied (duplicated) values in either group, an approach that currently seems to have practical value is as follows. Let M_1 and M_2 be the sample medians corresponding to groups 1 and 2, respectively, and let S_1^2 and S_2^2 be the corresponding McKean–Schrader estimate of the squared standard errors, which was described in Section 4.9.2. Then an approximate $1 - \alpha$ confidence interval for the difference between the population medians is

$$(M_1 - M_2) \pm c\sqrt{S_1^2 + S_2^2}$$

where c is the $1 - \alpha/2$ quantile of a standard normal distribution, which can be read from Table 1 in Appendix B. Alternatively, reject the hypothesis of equal population medians if

$$\frac{|M_1 - M_2|}{\sqrt{S_1^2 + S_2^2}} \geq c.$$

But, similar to Chapter 5, if there are tied values in either group, control over the probability of a Type I error can be very poor. (A method for dealing with tied values is described in Section 7.5.3.)

7.4.4 R Function msmed

The R function

```
msmed(x,y,alpha=0.05)
```

compares medians using the McKean–Schrader estimate of the standard error. (This function contains some additional parameters that are explained in Chapter 12. If tied values are detected, this function prints a warning message.)

EXAMPLE

For the data in Table 7.1, the 0.95 confidence interval for the difference between the medians is (18.5, 91.5). This interval does not contain 0, so in particular the hypothesis of equal population medians would be rejected with $\alpha = 0.05$ indicating that infants exposed to the sound of a heartbeat have the larger population median.

7.5 PERCENTILE BOOTSTRAP METHODS FOR COMPARING MEASURES OF LOCATION

This section describes and illustrates bootstrap methods that can be used to compare measures of location. As will be seen, certain bootstrap techniques have practical advantages over non-bootstrap methods. But there are exceptions, and often the choice of which bootstrap method to use depends crucially on which measure of location is used.

The percentile bootstrap method, introduced in Section 5.11, is readily extended to the problem of comparing two groups based on any measure of location. For certain purposes, the percentile bootstrap method is the best known method for controlling the probability of a Type I error or computing accurate confidence intervals. But when the goal is to compare means, the percentile bootstrap method is not recommended.

Although the percentile bootstrap method is not recommended when comparing means, it is perhaps easiest to describe the process in terms of means with the understanding that the basic percentile bootstrap procedure remains the same when using any other measure of location. The method begins by generating a bootstrap sample from each of the two groups. This means that for the first group, n_1 observations are resampled with replacement. For the second group, n_2 observations are resampled, again with replacement. Let $\bar{X}_1^*$ and $\bar{X}_2^*$ be the bootstrap means corresponding to groups 1 and 2, respectively, and let

$$D^* = \bar{X}_1^* - \bar{X}_2^*$$

be the difference between the bootstrap means. Now suppose we repeat this process B times yielding $D_1^*, \ldots, D_B^*$. (The software written for this book uses $B = 2000$.) The middle 95% of these values, after putting them in ascending order, yields a 0.95 confidence interval for the difference between the population means. In symbols, put the values $D_1^*, \ldots, D_B^*$ in ascending order yielding $D_{(1)}^* \leq \cdots \leq D_{(B)}^*$. Then an approximate $1 - \alpha$ confidence interval for the difference between the population means, $\mu_1 - \mu_2$, is

$$(D_{(\ell+1)}^*, D_{(u)}^*), \tag{7.15}$$

where as usual $\ell = \alpha B/2$, rounded to the nearest integer, and $u = B - \ell$. So for a 0.95 confidence interval, $\ell = 0.025B$.

Hypothesis Testing. Reject the hypothesis of equal population means if the confidence interval given by Equation (7.15) does not contain zero.

Computing a p-value

A p-value can be computed and is based on the probability that a bootstrap mean from the first group is greater than a bootstrap mean from the second. In symbols, a p-value can be computed if we can determine

$$p^* = P(\bar{X}_1^* > \bar{X}_2^*). \tag{7.16}$$

The value of p^* reflects the degree of separation between the two (bootstrap) sampling distributions. If the means based on the observed data are identical, meaning that $\bar{X}_1 = \bar{X}_2$, then p^* will have a value approximately equal to 0.5. (Under normality, p^* is exactly equal to 0.5.) Moreover, the larger the difference between the sample means $\bar{X}_1$ and $\bar{X}_2$, the closer p^* will be to 0 or 1. If $\bar{X}_1$ is substantially larger than $\bar{X}_2$, p^* will be close to 1, and if $\bar{X}_1$ is substantially smaller than $\bar{X}_2$, p^* will be close to 0. (Hall, 1988a, provides relevant theoretical details and results in Hall, 1988b, are readily extended to trimmed means.) Theoretical results not covered here suggest the following decision rule when the goal is to have a Type I error probability α: reject the hypothesis of equal means if p^* is less than or equal to $\alpha/2$, or greater than or equal to $1 - \alpha/2$. Said another way, if we let

$$p_m^* = \min(p^*, 1 - p^*),$$

meaning that p_m^* is equal to p^* or $1 - p^*$, whichever is smaller, then reject if

$$p_m^* \leq \frac{\alpha}{2}. \tag{7.17}$$

We do not know p^*, but it can be estimated by generating many bootstrap samples and computing the proportion of times a bootstrap mean from the first group is greater than a bootstrap mean from the second. That is, if A represents the number of values among $D_1^*, \ldots, D_B^*$ that are greater than zero, then we estimate p^* with

$$\hat{p}^* = \frac{A}{B}. \tag{7.18}$$

Finally, reject the hypothesis of equal population means if $\hat{p}^*$ is less than or equal to $\alpha/2$ or greater than or equal to $1 - \alpha/2$. Or setting

$$\hat{p}_m^* = \min(\hat{p}^*, 1 - \hat{p}^*),$$

reject if

$$\hat{p}_m^* \leq \frac{\alpha}{2}. \tag{7.19}$$

The p-value is

$$2\hat{p}_m^*$$

(Liu and Singh, 1997).

EXAMPLE

As a simple illustration, imagine that we generate bootstrap samples from each group and compute the difference between the bootstrap sample means. Further imagine that this process is repeated 10 times (so $B = 10$ is being used) resulting in the following D^* values:

$$-2, -0.5, -0.1, -1.2, 1, -1.3, -2.3, -0.01, -1.7, -.8$$

There is one positive difference, so $\hat{p}^* = 1/10$. The smaller of the two numbers $\hat{p}^*$ and $1 - \hat{p}^*$ is 0.1. Consequently, the p-value is $2(0.1) = 0.2$.

7.5.1 Using Other Measures of Location

In principle, the percentile bootstrap method just described can be used with any measure of location. To make sure the process is clear, imagine that the goal is to compare groups using a 20% trimmed mean. Again, this means that for the first group, we generate a bootstrap sample of n_1 values by resampling with replacement n_1 values from all of the data in group 1. Based on this bootstrap sample, the 20% trimmed mean is computed yielding $\bar{X}_{t1}^*$. The same is done for the second group and the difference between these two bootstrap trimmed means is computed. In symbols, now

$$D^* = \bar{X}_{t1}^* - \bar{X}_{t2}^*.$$

This process is repeated B times and confidence intervals and p-values are computed in exactly the same manner as was done when using means.

7.5.2 Comparing Medians

The case where the goal is to compare groups using medians requires a special comment. Section 7.4.2 described a non-bootstrap method for comparing medians and it was noted that when tied values occur, it can be highly unsatisfactory in terms of Type I errors. A simple generalization of the percentile bootstrap method deals with this problem in an effective manner.

Let M_1^* and M_2^* be the bootstrap sample medians. Now, rather than using p^* as defined by Equation (7.16), use

$$p^* = P(M_1^* > M_2^*) + 0.5P(M_1^* = M_2^*).$$

In words, p^* is the probability that the bootstrap sample median from the first group is larger than the bootstrap sample median from the second, plus half the probability that they are equal. So among B bootstrap samples from each group, if A is the number of times $M_1^* > M_2^*$, and C is the number of times $M_1^* = M_2^*$, the estimate of p^* is

$$\hat{p}^* = \frac{A}{B} + 0.5\frac{C}{B}.$$

As before, the p-value is

$$2\min(\hat{p}^*, 1 - \hat{p}^*).$$

In terms of controlling the probability of a Type I error, all indications are that this method performs very well regardless of whether tied values occur (Wilcox, 2006a).

7.5.3 R Function medpb2

The R function

```
medpb2(x,y,alpha=0.05,nboot=2000,SEED=T)
```

tests the hypothesis of equal medians using the percentile bootstrap method just described. The function also returns a $1-\alpha$ confidence interval for the difference between the population medians.

7.5.4 Some Guidelines on When to Use the Percentile Bootstrap Method

As previously indicated, when using means to compare groups, the percentile bootstrap method is relatively ineffective in terms of yielding an accurate confidence interval and controlling the probability of a Type I error. However, with trimmed means, if the amount of trimming is at least 0.2, the percentile bootstrap method just described is one of the most effective methods for obtaining accurate probability coverage and achieving relatively high power. (As noted in Chapter 6, a modification of the percentile bootstrap method performs well when working with the least squares regression estimator. But this modification does not perform well when comparing the means of two independent groups.) The minimum amount of trimming needed to justify using a percentile bootstrap method, rather than some competing technique, has not been determined. For the special case where the goal is to compare medians, the percentile bootstrap method is the only known method that handles tied values in a reasonably accurate manner.

When comparing groups using the M-estimator introduced in Chapter 3, again the percentile bootstrap method appears to be best for general use. Comparing M-estimators can be done with the R function `pb2gen`, described in the next section. Many non-bootstrap methods for comparing M-estimators have been proposed, but typically they do not perform well when dealing with skewed distributions, at least with small to moderate sample sizes.

7.5.5 R Functions trimpb2, med2g, and pb2gen

The R function

```
trimpb2(x, y, tr = 0.2, alpha = 0.05, nboot = 2000)
```

compares trimmed means, using the percentile bootstrap method just described. Here `x` is any R variable containing the data for group 1 and `y` contains the data for group 2. The amount of trimming, `tr`, defaults to 20%, α defaults to 0.05, and `nboot` (B) defaults to 2000. This function returns a p-value plus a $1 - \alpha$ confidence interval for the difference between the trimmed means. For the special case there the goal is to compare medians; the function

```
med2g(x, y, alpha = 0.05, nboot = 1000)
```

can be used.

The R function

```
pb2gen(x, y, alpha = 0.05, nboot = 2000, est = onestep, ...)
```

can be used to compare groups using any measure of location in conjunction with the percentile bootstrap method. Again, `x` and `y` are any R variables containing data and `nboot` is B, the number of bootstrap samples to be used. By default, $B = 2000$. The argument `est` indicates which measure of location is to be employed. It can be any R function that computes a measure of location and defaults to the function `onestep`, which is the one-step M estimator described in Chapter 2. The argument ... can be used to reset certain default settings associated with the argument `est`. For example, if `est=mean` is used, means are compared. In contrast, the command

```
pb2gen(x, y, alpha = 0.05, nboot = 2000, est = mean,tr=0.2)
```

would compare 20% trimmed means instead. (In this case, `pb2gen` and `trimpb2` give the same results.) The command

```
pb2gen(x, y, alpha = 0.05, nboot = 2000, est = median)
```

would compare medians. (This function can handle tied values.)

EXAMPLE

A study was conducted comparing the EEG (electroencephalogram) readings of convicted murderers to the EEG readings of a control group with measures taken at various sites in the brain. For one of these sites the results were

Control Group: −0.15, −0.22, 0.07, −0.07, 0.02, 0.24, −0.60, −0.17, −0.33, 0.23, −0.69, 0.70, 1.13, 0.38

Murderers: −0.26, 0.25, 0.61, 0.38, 0.87, −0.12, 0.15, 0.93, 0.26, 0.83, 0.35, 1.33, 0.89, 0.58.

(These data were generously supplied by A. Raine.) The sample medians are −0.025 and 0.48, respectively. Storing the data in the R variables x1 and x2, the command

```
pb2gen(x1, x2, est = median)
```

returns a 0.95 confidence interval for the difference between the population medians of (−0.97, −0.085). So the hypothesis of equal population medians is rejected because this interval does not contain 0, and the data indicate that the typical measure for the control group is less than the typical measure among convicted murderers.

EXAMPLE

Table 2.2 contains data on the desired number of sexual partners over the next 30 years reported by 105 male undergraduates. The responses by 156 females are shown in Table 7.2. Does the typical response among males differ from the typical response among females? If we simply apply Student's T, we fail to reject, which is not surprising because there is an extreme outlier among the responses for males. (Note that the example given in Section 7.4.2 deals with the same issue, but with different data, and again the hypothesis of equal means was not rejected.) But if we trim only 1% of the data, Yuen's method rejects suggesting that the two distributions differ. However, with so little trimming accurate confidence intervals might be difficult to obtain. Moreover, the median response among both males and females is

Table 7.2 Desired Number of Sexual Partners for 156 Females

x:	0	1	2	3	4	5	6	7	8	10
f_x:	2	101	11	10	5	11	1	1	3	4
x:	11	12	15	20	30					
f_x:	1	1	2	1	2					

1 suggesting that in some sense the typical male and female are similar. To add perspective, another strategy is to focus on males who tend to give relatively high responses. An issue is whether they differ from females who tend to give relatively high responses. One way of addressing this issue is to compare the 0.75 quantiles of the distributions, which can be estimated with the built-in R function quantile. For example, if the responses for the males are stored in the variable `sexm`, the R command `quantile(sexm,probs=0.75)` estimates the 0.75 quantile to be 6 and for females the estimate is 3. This estimate suggests that the more extreme responses among males are typically higher than the more extreme responses among females. The command

```
pb2gen(sexm, sexf, est = quantile,probs=0.75)
```

compares the 0.75 quantiles of the two groups and returns a 0.95 confidence interval of (1, 8). So reject the hypothesis of equal 0.75 quantiles indicating that the groups differ among the higher responses. That is, in some sense the groups appear to be similar because they have identical medians, but if we take the 0.75 quantiles to be the typical response among the higher responses we might observe, the typical male appears to give higher responses than the typical female.

7.6 BOOTSTRAP-T METHODS FOR COMPARING MEASURES OF LOCATION

From Chapter 5, the basic strategy behind the bootstrap-t method, when testing hypotheses about the population mean, is to approximate the distribution of the test statistic T when the null hypothesis is true. Moreover, a fundamental difference between the bootstrap-t and the percentile bootstrap method is that the bootstrap-t is based in part on an estimate of the standard error. In contrast, standard errors play no role when using the percentile bootstrap. The immediate goal is to describe a bootstrap-t version of Welch's method for comparing means, and then generalizations to other measures of location are indicated.

7.6.1 Comparing Means

Recall that when testing the hypothesis of equal population means, Welch's test statistic is

$$W = \frac{(\bar{X}_1 - \bar{X}_2)}{\sqrt{\frac{s_1^2}{n_1} + \frac{s_2^2}{n_2}}}.$$

The probability of a Type I error can be controlled exactly if the distribution of W (over many studies) can be determined when the null hypothesis of equal means is true. Welch's strategy was to approximate the distribution of W with a Student's T distribution and the degrees of freedom estimated based on the sample variances, and sample sizes. The bootstrap-t strategy is to use bootstrap samples instead.

A bootstrap approximation of the distribution of W, when the null hypothesis is true, is obtained as follows. Generate a bootstrap sample of size n_1 from the first group and label the resulting sample mean and standard deviation $\bar{X}_1^*$ and s_1^*, respectively. Do the same for the second group and label the bootstrap sample mean and standard deviation $\bar{X}_2^*$ and s_2^*. Let

$$W^* = \frac{(\bar{X}_1^* - \bar{X}_2^*) - (\bar{X}_1 - \bar{X}_2)}{\sqrt{\frac{(s_1^*)^2}{n_1} + \frac{(s_2^*)^2}{n_2}}}. \tag{7.20}$$

Repeat this process B times yielding B W^* values: $W_1^*, \ldots, W_B^*$. Next, put these B values in ascending order, which we label $W_{(1)}^* \leq \cdots \leq W_{(B)}^*$. Let $\ell = \alpha B/2$, rounded to the nearest integer, and $u = B - \ell$. Then an approximate $1 - \alpha$ confidence interval for the difference between the means $(\mu_1 - \mu_2)$ is

$$\left((\bar{X}_1 - \bar{X}_2) - W_{(u)}^* \sqrt{\frac{s_1^2}{n_1} + \frac{s_2^2}{n_2}},\ (\bar{X}_1 - \bar{X}_2) - W_{(\ell+1)}^* \sqrt{\frac{s_1^2}{n_1} + \frac{s_2^2}{n_2}}, \right). \tag{7.21}$$

7.6.2 Bootstrap-t Method When Comparing Trimmed Means

Bootstrap-t methods for comparing trimmed means are preferable to the percentile bootstrap when the amount of trimming is close to zero. An educated guess is that the bootstrap-t is preferable if the amount of trimming is less than or equal to 10%, but it is stressed that this issue is in need of more research. The only certainty is that with no trimming, all indications are that the bootstrap-t outperforms the percentile bootstrap.

Bootstrap-t methods for comparing trimmed means are performed as follows:

1. Compute the sample trimmed means, $\bar{X}_{t1}$ and $\bar{X}_{t2}$, and Yuen's estimate of the squared standard errors, d_1 and d_2, given by Equation (7.12).

2. For each group, generate a bootstrap sample and compute the trimmed means, which we label $\bar{X}_{t1}^*$ and $\bar{X}_{t2}^*$. Also compute Yuen's estimate of the squared standard error based on these bootstrap samples, which we label d_1^* and d_2^*.

3. Compute
$$T_y^* = \frac{(\bar{X}_{t1}^* - \bar{X}_{t2}^*) - (\bar{X}_{t1} - \bar{X}_{t2})}{\sqrt{d_1^* + d_2^*}}.$$

4. Repeat steps 2 and 3 B times yielding $T_{y1}^*, \ldots, T_{yB}^*$. In terms of probability coverage, $B = 599$ appears to suffice in most situations when $\alpha = 0.05$.

5. Put the $T_{y1}^*, \ldots, T_{yB}^*$ values in ascending order yielding $T_{y(1)}^* \leq \cdots \leq T_{y(B)}^*$. The T_{yb}^* values $(b = 1, \ldots, B)$ provide an estimate of the distribution of
$$\frac{(\bar{X}_{t1} - \bar{X}_{t2}) - (\mu_{t1} - \mu_{t2})}{\sqrt{d_1 + d_2}}.$$

6. Set $\ell = \alpha B/2$ and $u = B - \ell$, where ℓ is rounded to the nearest integer.

The equal-tailed $1 - \alpha$ confidence interval for the difference between the population trimmed means, $\mu_{t1} - \mu_{t2}$, is

$$\left(\bar{X}_{t1} - \bar{X}_{t2} - T_{y(u)}^* \sqrt{d_1 + d_2},\ \bar{X}_{t1} - \bar{X}_{t2} - T_{y(\ell+1)}^* \sqrt{d_1 + d_2}\right). \tag{7.22}$$

To get a symmetric two-sided confidence interval, replace step 3 with

$$T_y^* = \frac{|(\bar{X}_{t1}^* - \bar{X}_{t2}^*) - (\bar{X}_{t1} - \bar{X}_{t2})|}{\sqrt{d_1^* + d_2^*}},$$

set $a = (1-\alpha)B$, rounding to the nearest integer, in which case a $1-\alpha$ confidence interval for $\mu_{t1} - \mu_{t2}$ is

$$(\bar{X}_{t1} - \bar{X}_{t2}) \pm T^*_{y(a)}\sqrt{d_1 + d_2}. \tag{7.23}$$

Hypothesis Testing. As usual, reject the hypothesis of equal population trimmed means ($H_0 : \mu_{t1} = \mu_{t2}$) if the $1-\alpha$ confidence interval for the difference between the trimmed means does not contain zero. Alternatively, compute Yuen's test statistic

$$T_y = \frac{\bar{X}_{t1} - \bar{X}_{t2}}{\sqrt{d_1 + d_2}}$$

and reject if

$$T_y \le T^*_{y(\ell+1)}$$

or if

$$T_y \ge T^*_{y(u)}.$$

When using the symmetric, two-sided confidence interval method, reject if

$$|T_y| \ge T^*_{y(a)}.$$

7.6.3 R Functions yuenbt and yhbt

The R function

```
yuenbt(x, y, tr = 0.2, alpha = 0.05, nboot = 599, side=FALSE)
```

uses a bootstrap-t method to compare trimmed means. The arguments are the same as those used by `trimpb2` (described in Section 7.5.4) plus an additional argument labeled `side`, which indicates whether a symmetric or equal-tailed confidence interval will be used. `Side` defaults to FALSE meaning that the equal-tailed confidence interval, given by Equation (7.22), will be computed. Setting side equal to T yields the symmetric confidence interval given by Equation (7.23).

EXAMPLE

The confidence interval based on Welch's method can differ substantially from the confidence interval based on the bootstrap-t method just described. The data for group 1 in Table 7.3 were generated from a mixed normal distribution (using R) and the data for group 2 were generated from a standard normal distribution. So both groups have population means equal to 0. Applying Welch's method, the 0.95 confidence interval for the difference between the means is (−0.988, 2.710). Using the bootstrap-t method instead, it is (−2.21, 2.24).

EXAMPLE

Table 7.4 shows data from a study dealing with the effects of consuming alcohol. (The data were supplied by M. Earleywine.) Group 1 was a control group and reflects hangover symptoms after consuming a specific amount of alcohol in a laboratory setting. Group 2 consisted of sons of alcoholic fathers. Storing the group 1 data in the R variable A1, and the group 2 data in A2, the command `trimpb2(A1,A2)` returns the following output:

Table 7.3 Data Generated from a Mixed Normal (Group 1) and a Standard Normal (Group 2)

Group 1:	3.73624506	2.10039320	−3.56878819	−0.26418493
	−0.27892175	0.87825842	−0.70582571	−1.26678127
	−0.30248530	0.02255344	14.76303893	−0.78143390
	−0.60139147	−4.46978177	1.56778991	−1.14150660
	−0.20423655	−1.87554928	−1.62752834	0.26619836
Group 2:	−1.1404168	−0.2123789	−1.7810069	−1.2613917
	−0.3241972	1.4550603	−0.5686717	−1.7919242
	−0.6138459	−0.1386593	−1.5451134	−0.8853377
	0.3590016	0.4739528	−0.2557869	

Table 7.4 The Effect of Alcohol

Group 1:	0	32	9	0	2	0	41	0	0	0
	6	18	3	3	0	11	11	2	0	11
Group 2:	0	0	0	0	0	0	0	0	1	8
	0	3	0	0	32	12	2	0	0	0

```
$p-value:
[1] 0.038

$ci:
[1] 0.1666667 8.3333333
```

This says that a 0.95 confidence interval for the difference between the population trimmed means is (0.17, 8.3). The p-value is 0.038, so in particular you would reject $H_0 : \mu_{t1} = \mu_{t2}$ at the 0.05 level. In contrast, if we use Welch's method for means, the 0.95 confidence interval is (−1.6, 10.7); this interval contains zero, so we no longer reject, the only point being that it can make a difference which method is used.

Notice that the default value for `nboot` (B) when using `yuenbt` is only 599 compared to 2000 when using trimpb2. Despite this, trimpb2 tends to have faster execution time because it is merely computing trimmed means; `yuenbt` requires estimating the standard error for each bootstrap sample, which increases the execution time considerably. When using the bootstrap-t method (the R function `yuenbt`), published papers indicate that increasing B from 599 to 999, say, does not improve probability coverage by very much, if at all, when $\alpha = 0.05$. However, a larger choice for B might result in higher power as noted in Chapter 5.

EXAMPLE

In an unpublished study by Dana (1990), the general goal was to investigate issues related to self-awareness and self-evaluation. In one portion of the study, he recorded the times individuals could keep an apparatus in contact with a specified target. The results, in hundredths of seconds, are shown in Table 7.5. Storing the data for group 1 in the R variable G1, and storing the data for group 2 in G2, the command

```
yuenbt(G1, G2)
```

Table 7.5 Self-Awareness Data

Group 1:	77 87 88 114 151 210 219 246 253
	262 296 299 306 376 428 515 666 1310 2611
Group 2:	59 106 174 207 219 237 313 365 458 497 515
	529 557 615 625 645 973 1065 3215

returns a 0.95 confidence interval of (−312.5, 16.46). This interval contains zero so we would not reject. In terms of Tukey's three-decision rule, we would make no decision about which group has the larger 20% trimmed mean. If we increase the number of bootstrap samples (B) by setting the argument `nboot` to 999, now the confidence interval is (−305.7, 10.7). We still do not reject, but increasing B alters the confidence interval slightly. In contrast, comparing medians with the method in Section 7.4.3, the 0.95 confidence interval is (−460.93, −9.07) so we reject, the only point being that even among robust estimators, the choice of method can alter the conclusions reached.

When trimming 10% or 15%, there is a modification of the bootstrap-t method that appears to improve upon the bootstrap-t methods described here (Keselman et al., 2004). Roughly, using results derived by Hall (1992), Guo and Luh (2000) derived a method that estimates the skewness of the distribution of the test statistic W and makes an appropriate adjustment based on this estimate. Keselman et al. found, however, that a bootstrap-t version of the Guo–Luh test statistic is more satisfactory in terms of controlling the probability of a Type I error. Although the computational details are not given, an R function for applying the method is supplied:

```
yhbt(x, y, tr = 0.15, alpha = 0.05, nboot = 600,PV=F).
```

By default, 15% trimming is used. The function returns a confidence interval having probability coverage specified by the argument `alpha`. If the argument `PV=T`, a p-value is reported. This method is *not* recommended when the goal is to compare means. Moreover, situations are encountered where the function terminates due to computational problems. These problems can be avoided by using a percentile bootstrap method instead. That is, use the R function `trimpb2`. Even with 10% trimming, there appears to be no practical advantage to using `yhbt` rather than a percentile bootstrap method.

7.6.4 Estimating Power and Judging the Sample Sizes

If the hypothesis of equal measures of location is not rejected, there is the issue of why. When comparing the means, one possibility is that, indeed, the population means are approximately equal. But another possibility is that the means differ and a Type II error has been committed. If, for example, the goal is to detect a difference of 0.6 between the means, and power is only 0.3 when indeed $\mu_1 - \mu_2 = 0.6$, there is a 70% chance of not rejecting. One way of trying to address this issue is estimating how much power there is. When comparing measures of location, deriving such an estimate is not difficult, but getting a reasonably accurate estimate is another matter. When comparing means, it seems that getting an estimate that remains accurate under nonnormality is very difficult or impossible, at least with small to moderate sample sizes. However, if we compare groups using the percentile bootstrap method with 20% trimmed means, a reasonably accurate estimate of power appears possible.

The basic strategy is to estimate the standard errors associated with the 20% trimmed means and then use these estimates to estimate power for a given value of the difference between the population trimmed means ($\mu_{t1} - \mu_{t2}$). The computational details are given in Wilcox and Keselman (2002a), but no details are given here. Instead, R functions are given for applying the method.

7.6.5 R Functions powest and pow2an

The R function

```
powest(x,y,delta)
```

estimates how much power there is when the difference between the population 20% trimmed means is equal to some specified value, delta. The R function

```
pow2an(x,y, ci=FALSE, plotit=TRUE, nboot=800)
```

computes a power curve using the data stored in the R variables `x` and `y`. That is, the function chooses a range of values for the difference between the population means, and for each difference it computes power. By default, the power curve is plotted. To avoid the plot and get the numerical results only, set the argument plotit=FALSE for false. Setting the argument `ci` to TRUE will result in a lower 0.95 confidence interval for the power curve to be computed using a bootstrap method based on `nboot` (B) bootstrap samples.

EXAMPLE

The R functions `powest` and `pow2an` are illustrated with data (supplied by Frank Manis) from a reading study. Theoretical arguments suggest that the groups should differ, but the hypothesis of equal 20% trimmed means (or when comparing means with any of the previously described techniques) is not rejected at the $\alpha = 0.05$ level. One possible explanation is that there is little or no difference between the groups, but another possibility is that power is low due to relatively large standard errors, meaning that detecting a substantively interesting difference is unlikely based on the sample sizes used. For a difference of 600 between the population 20% trimmed means, `powest` estimates that power is 0.8. Figure 7.2 shows the estimated power curve returned by `pow2an`. The lower dashed line is a lower 0.95 confidence interval for the actual amount of power. That is, the solid line provides an approximately unbiased estimate of power, but a possibility is that power is as low as indicated by the dashed line. Based on this analysis, it was concluded that power is low and that accepting the hypothesis of equal trimmed means is not warranted.

7.7 PERMUTATION TESTS

The so-called permutation test was introduced by R. A. Fisher in the 1930s and is sometimes recommended for comparing means and other measures of location. But in reality it is testing the hypothesis that the two groups being compared have identical distributions. Even under normality but unequal variances, the method can fail to control the probability of a Type I error when testing the hypothesis of equal means (e.g., Boik, 1987). Moreover it is known that the method is unsatisfactory for comparing the population medians (Romano, 1990). On the positive side, when testing the hypothesis of identical distributions, the probability of a Type I error is controlled exactly. An argument in favor of using the permutation test to compare means is that if distributions differ, surely the population means differ. But even if we accept this argument, the permutation test gives us little or no information about how the groups differ, let alone the magnitude of the difference between the population means, and it tells us nothing about the precision of the estimated difference between the population means based on the sample means. That is, it does not provide a confidence interval for $\mu_1 - \mu_2$. (For yet another argument in favor of the permutation test, see Ludbrook and Dudley, 1998.)

Chung and Romano (2013) summarize general theoretical concerns and limitations. Because of the known limitations of this method, for brevity, the computational details are

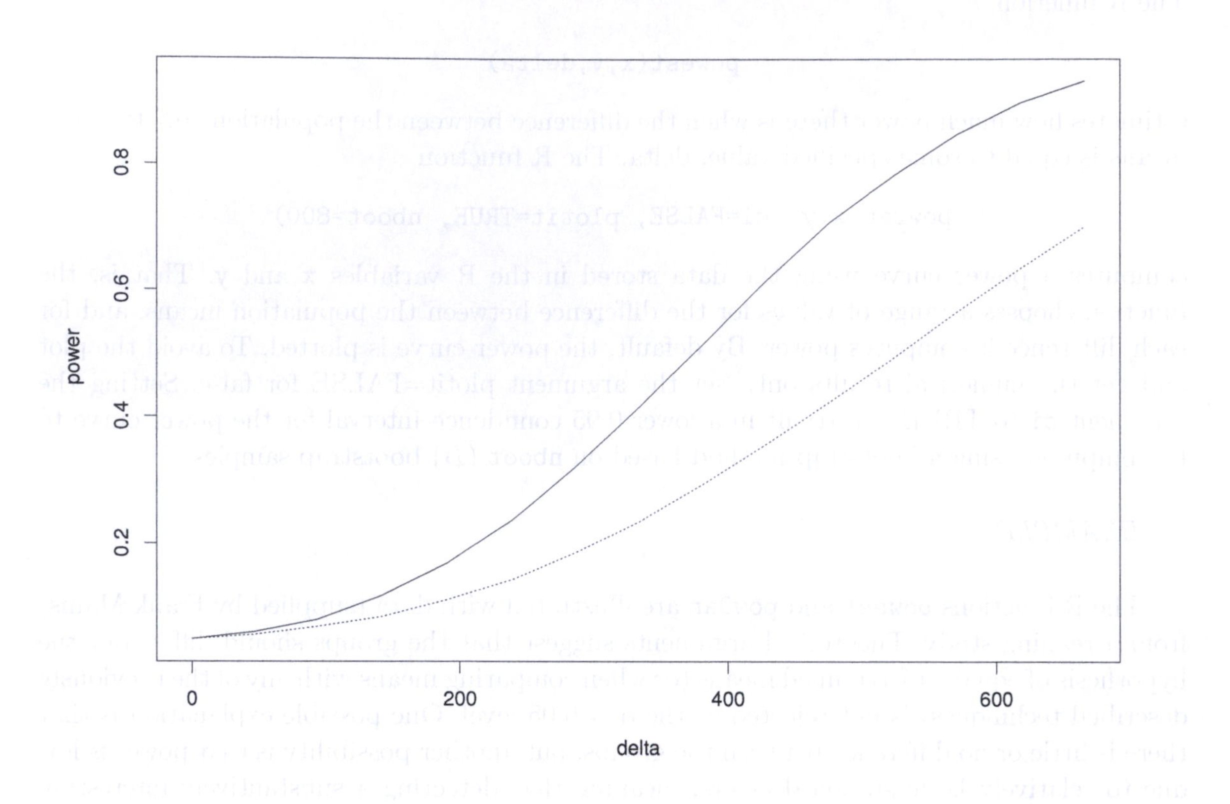

Figure 7.2 The solid line is the estimated power curve returned by pow2an. It is estimated that with 0.95 probability, power is at least as large as indicated by the lower dashed line.

omitted. Chung and Romano also indicate a variation of the permutation test that might have practical value. However, when data are generated from the distribution in Figure 4.14, and when the variances are unequal, this alternative approach (based on 1000 permutations) can perform poorly when the goal is to compare means, at least with sample sizes less than or equal to 50. The extent to which this is the case for other situations has not been determined.

It is noted that the permutation test can be applied with any measure of scale, such as the sample variances, but again the method is testing the hypothesis of equal distributions. If, for example, we use variances, examples can be constructed where we are likely to reject because the distributions differ even though the population variances are equal.

7.8 RANK-BASED AND NONPARAMETRIC METHODS

This section describes a collection of techniques for comparing two independent groups that generally fall under the rubric of rank-based or nonparametric methods. At a minimum, these methods provide a perspective that differs from any method based on comparing measures of location. Generally, the methods in this section guard against low power due to outliers. In some situations they provide more power than other methods that might be used, but the reverse is true as well. One of the methods described here, called a shift function, provides a detailed sense of how the distributions differ.

7.8.1 Wilcoxon–Mann–Whitney Test

Let p be the probability that a randomly sampled observation from the first group is less than a randomly sampled observation from the second. (Momentarily, it is assumed that tied values never occur.) If the groups do not differ, then in particular it should be the case that

$$H_0 : p = 0.5 \tag{7.24}$$

is true. The quantity p has been called a *probabilistic measure of effect size*, the *probability of concordance*, the measure of *stochastic superiority*, and the *common language measure of effect size*. This section describes a classic technique for comparing groups based on an estimate of p, called the Wilcoxon–Mann–Whitney (WMW) test. It was originally derived by Wilcoxon (1945) and later it was realized that Wilcoxon's method is the same as a procedure proposed by Mann and Whitney (1947). But as will be explained, it does not provide a satisfactory test of Equation (7.24).

First consider the problem of estimating p and for illustrative purposes suppose we observe

Group 1: 30, 60, 28, 38, 42, 54

Group 2: 19, 21, 27, 73, 71, 25, 59, 61.

Now focus on the first value in the first group, 30, and notice that it is less than four of the eight observations in the second group. So a reasonable estimate of p, the probability that an observation from the first group is less than an observation from the second, is 4/8. In a similar manner, the second observation in the first group is 60. It is less than three of the values in the second group, so a reasonable estimate of p is 3/8. These two estimates of p differ, and a natural way of combining them into a single estimate of p is to average them. More generally, if we have n_1 observations in group 1 and n_2 observations in group 2, focus on the ith observation in the first group and suppose that this value is less than V_i of the observation in group 2. So based on the ith observation in group 1, an estimate of p is V_i/n_2, and we have n_1 estimates of p: $V_1/n_2, \ldots, V_{n_1}/n_2$. To combine these n_1 estimates of p into a single estimate, average them yielding

$$\hat{p} = \frac{1}{n_1 n_2} \sum V_i. \tag{7.25}$$

As is usually done, let

$$U = n_1 n_2 \hat{p}. \tag{7.26}$$

The quantity U is called the Mann-Whitney U statistic. If $p = 0.5$, it can be shown that $E(U) = n_1 n_2/2$. More generally,

$$E\left(\frac{U}{n_1 n_2}\right) = p.$$

In other words, to estimate the probability that an observation from group 1 is less than an observation from group 2, divide U by $n_1 n_2$, the product of the sample sizes.

Next, consider the problem of estimating VAR(U), the squared standard error of U. If we assume there are no tied values and both groups have identical distributions, the classic estimate of the standard error can be derived. (Again, by no tied values is meant that each observed value occurs only once. So if we observe the value 6, for example, it never occurs again among the remaining observations.) The expression for VAR(U) is

$$\sigma_u^2 = \frac{n_1 n_2 (n_1 + n_2 + 1)}{12}.$$

This means that the null hypothesis can be tested with

$$Z = \frac{U - \frac{n_1 n_2}{2}}{\sigma_u}, \tag{7.27}$$

which has, approximately, a standard normal distribution when the assumptions are met and H_0 is true. In particular, reject if

$$|Z| \geq z_{1-\frac{\alpha}{2}},$$

where $z_{1-\alpha/2}$ is the $1-\alpha/2$ quantile of a standard normal distribution.

EXAMPLE

Continuing the illustration using the data in the last example, it can be seen that $\hat{p} = 0.479$, so $U = 23$ and

$$Z = \frac{23 - 24}{7.75} = -0.129.$$

With $\alpha = 0.05$, the critical value is 1.96; $|Z|$ is less than 1.96, so fail to reject.

Sometimes the Wilcoxon–Mann–Whitney test is described as a method for comparing medians. However, it is unsatisfactory for this purpose, as are other rank-based methods, unless certain highly restrictive assumptions are met (e.g., Fung, 1980). A crude explanation is that the Wilcoxon–Mann–Whitney test is not based on an estimate of the population medians. (See the end of Section 7.8.6 for related perspective.) One practical concern is that there are situations where power decreases as the difference between the population medians increases. Moreover, there are general conditions under which an accurate confidence interval for the difference between the medians cannot be computed based on the Wilcoxon–Mann–Whitney test (Kendall and Stuart, 1973; Hettmansperger, 1984).

In a very real sense, a more accurate description of the Wilcoxon–Mann–Whitney test is that it provides a test of the hypothesis that two distributions are identical. The situation is similar to Student's T test. When the two distributions are identical, a correct estimate of the standard error (σ_u^2) is being used. But otherwise, under general conditions, an incorrect estimate is being used, which results in practical concerns, in terms of both Type I errors and power, when using Equation (7.24) to test $H_0 : p = 0.5$.

When tied values occur with probability zero, and the goal is to test the hypothesis that the groups have identical distributions, the probability of a Type I error can be controlled exactly by computing a critical value as described, for example, in Hogg and Craig (1970, p. 373). Let

$$W = U + \frac{n_2(n_2+1)}{2} \tag{7.28}$$

and suppose the hypothesis of identical distributions H_0, given by Equation (7.24), is rejected if

$$W \leq c_L,$$

or when

$$W \geq c_U,$$

where c_L is read from Table 13 in Appendix B and

$$c_U = n_2(n_2 + n_1 + 1) - c_L.$$

Then the actual probability of a Type I error will not exceed 0.05 under random sampling. (For results on power and sample sizes, see Rosner and Glynn, 2009.)

7.8.2 R Functions wmw and wilcox.test

R has a built-in function for performing the Wilcoxon–Mann–Whitney test:

```
wilcox.test(x,y).
```

The R function

```
wmw(x,y)
```

computes a p-value based on the Wilcoxon–Mann–Whitney test using a method recommended by Hodges et al. (1990).

7.8.3 Handling Tied Values and Heteroscedasticity

A practical concern is that if groups differ, then under general circumstances the wrong standard error is being used by the Wilcoxon–Mann–Whitney test, which can result in relatively poor power and an unsatisfactory confidence interval for p. Said another way, if groups have different distributions, generally σ_u^2 is the wrong standard error for U. Another problem is how to handle tied values.

Numerous methods have been proposed for improving on the Wilcoxon–Mann–Whitney test. Currently, a method derived by Cliff (1996) appears to perform relatively well. Reiczigel et al. (2005) suggest a bootstrap method for making inferences about p and they found that it performed better than the method derived by Brunner and Munzel (2000) when sample sizes are small, say less than 30, and tied values do not occur. But when tied values do occur, their bootstrap method can perform poorly. An alternative bootstrap method could be used to deal with tied values, which is similar to the method used to compare medians, described in Section 7.5.3, but currently there is no indication that it offers a practical advantage over the method derived by Cliff. If there are no tied values, alternative heteroscedastic methods have been proposed by Mee (1990) as well as Fligner and Policello (1981). Currently it seems that for this special case, these methods offer no practical advantage. Eight other methods were compared by Newcombe (2006). Ruscio and Mullen (2012) compared twelve methods and found that if the total sample, n_1+n_2, is greater than or equal to 60, a bootstrap method not covered here tends to provide a more accurate confidence interval than Cliff's method.

7.8.4 Cliff's Method

First consider the problem of tied values and note that if we randomly sample a single observation from both groups, there are three possible outcomes: the observation from the first group is greater than the observation from the second, the observations have identical values, or the observation from the first group is less than the observation from the second. The probabilities associated with these three mutually exclusive outcomes are labeled p_1, p_2, and p_3. In symbols, if X_{ij} represents the ith observation from the jth group

$$p_1 = P(X_{i1} > X_{i2}),$$

$$p_2 = P(X_{i1} = X_{i2}),$$

and

$$p_3 = P(X_{i1} < X_{i2}).$$

Cliff (1996) focuses on testing

$$H_0 : \delta = p_1 - p_3 = 0. \tag{7.29}$$

In the event tied values occur with probability zero, in which case $p_2 = 0$, Equation (7.29) becomes H_0: $p_1 = p_3 = 0.5$. It can be shown that Equation (7.29) is tantamount to

$$H_0 : p_3 + 0.5p_2 = 0.5.$$

For convenience, let $P = p_3 + 0.5p_2$, in which case this last equation becomes

$$H_0 : P = 0.5. \tag{7.30}$$

Of course, when there are no tied values, $P = p_3 = P(X < Y)$. The parameter δ is related to P in the following manner:

$$\delta = 1 - 2P, \tag{7.31}$$

so

$$P = \frac{1-\delta}{2}. \tag{7.32}$$

Cliff's heteroscedastic confidence interval for δ is computed as follows. As usual, let X_{ij} be the ith observation from the jth group, $j = 1, 2$. For the ith observation in group 1 and the hth observation in group 2, let

$$d_{ih} = \begin{cases} -1 & \text{if } X_{i1} < X_{h2} \\ 0 & \text{if } X_{i1} = X_{h2} \\ 1 & \text{if } X_{i1} > X_{h2}. \end{cases}$$

An estimate of $\delta = P(X_{i1} > X_{i2}) - P(X_{i1} < X_{i2})$ is

$$\hat{\delta} = \frac{1}{n_1 n_2} \sum_{i=1}^{n_1} \sum_{h=1}^{n_2} d_{ih}, \tag{7.33}$$

the average of the d_{ih} values. Let

$$\bar{d}_{i.} = \frac{1}{n_2} \sum_h d_{ih},$$

$$\bar{d}_{.h} = \frac{1}{n_1} \sum_i d_{ih},$$

$$s_1^2 = \frac{1}{n_1 - 1} \sum_{i=1}^{n_1} (\bar{d}_{i.} - \hat{\delta})^2,$$

$$s_2^2 = \frac{1}{n_2 - 1} \sum_{h=1}^{n_2} (\bar{d}_{.h} - \hat{\delta})^2,$$

$$\tilde{\sigma}^2 = \frac{1}{n_1 n_2} \sum \sum (d_{ih} - \hat{\delta})^2.$$

Then

$$\hat{\sigma}^2 = \frac{(n_1 - 1)s_1^2 + (n_2 - 1)s_2^2 + \tilde{\sigma}^2}{n_1 n_2}$$

estimates the squared standard error of $\hat{\delta}$. Let z be the $1 - \alpha/2$ quantile of a standard normal distribution. Rather than use the more obvious confidence interval for δ, Cliff (1996, p. 140) recommends

$$\frac{\hat{\delta} - \hat{\delta}^3 \pm z\hat{\sigma}\sqrt{(1 - \hat{\delta}^2)^2 + z^2\hat{\sigma}^2}}{1 - \hat{\delta}^2 + z^2\hat{\sigma}^2}.$$

If the confidence interval for δ does not contain zero, reject H_0: $\delta = 0$, which means that H_0: $P = 0.5$ is rejected as well.

The confidence interval for δ is readily modified to give a confidence for P. Letting

$$C_\ell = \frac{\hat{\delta} - \hat{\delta}^3 - z\hat{\sigma}\sqrt{(1-\hat{\delta}^2)^2 + z^2\hat{\sigma}^2}}{1 - \hat{\delta}^2 + z^2\hat{\sigma}^2}$$

and

$$C_u = \frac{\hat{\delta} - \hat{\delta}^3 + z\hat{\sigma}\sqrt{(1-\hat{\delta}^2)^2 + z^2\hat{\sigma}^2}}{1 - \hat{\delta}^2 + z^2\hat{\sigma}^2},$$

a $1 - \alpha$ confidence interval for P is

$$\left(\frac{1 - C_u}{2}, \frac{1 - C_\ell}{2}\right). \tag{7.34}$$

7.8.5 R Functions cid and cidv2

The R function

```
cid(x,y,alpha=0.05,plotit=FALSE)
```

performs Cliff's method in Section 7.8.4. The function also reports a confidence interval for $P = p_3 + 0.5p_2$, which is labeled `ci.p`. The estimate of P is labeled `phat`. To get a p-value, use the function

```
cidv2(x,y,plotit=FALSE).
```

When the argument `plotit=TRUE`, these functions plot the distribution of D, where D is the difference between a randomly sampled observation from the first group, minus a randomly sampled observation from the second group. D will have a symmetric distribution around zero when the distributions are identical. The plot provides perspective on the extent to which this is the case.

7.8.6 The Brunner–Munzel Method

An alternative approach to both tied values and heteroscedasticity stems from Brunner and Munzel (2000). Their approach is based in part on what are called the midranks. To explain, first consider the values 45, 12, 32, 64, 13, and 25. There are no tied values and the smallest value is said to have *rank* 1, the next smallest has rank 2, and so on. A common notation for the rank corresponding to the ith observation is R_i. So in the example, the first observation is $X_1 = 45$ and its rank is $R_1 = 5$. Similarly, $X_2 = 12$ and its rank is $R_2 = 1$.

Now consider a situation where there are tied values: 45, 12, 13, 64, 13, and 25. Putting these values in ascending order yields 12, 13, 13, 25, 45, 64. So the value 12 gets a rank of 1, but there are two identical values having a rank of 2 and 3. The *midrank* is simply the average of the ranks among the tied values. Here, this means that the rank assigned to the two values equal to 13 would be $(2 + 3)/2 = 2.5$, the average of their corresponding ranks. So the ranks for all six values would be 1, 2.5, 2.5, 4, 5, 6.

Generalizing, consider

$$7, 7.5, 7.5, 8, 8, 8.5, 9, 11, 11, 11.$$

There are 10 values, so if there were no tied values, their ranks would be 1, 2, 3, 4, 5, 6, 7, 8, 9, and 10. But because there are two values equal to 7.5, their ranks are averaged yielding a rank of 2.5 for each. There are two values equal to 8. Their original ranks are 4 and 5, so their final ranks (their midranks) are both 4.5. There are three values equal to 11. Their original ranks are 8, 9, and 10. The average of these ranks is 9, so their midranks are all equal to 9. In summary, the ranks for the 10 observations are

$$1, 2.5, 2.5, 4.5, 4.5, 6, 7, 9, 9, 9.$$

Now consider testing H_0: $P = 0.5$, where P is defined as in Section 7.8.3. As usual, let X_{ij} be the ith observation from the jth group ($i = 1, \ldots, n_j$; $j = 1, 2$). To apply the Brunner–Munzel method, first pool all $N = n_1 + n_2$ observations and assign ranks. In the event there are tied values, ranks are averaged as just illustrated. The results for the jth group are labeled R_{ij}, $i = 1, \ldots, n_j$. That is, R_{ij} is the rank corresponding to X_{ij} among the pooled values. Let $\bar{R}_1$ be the average of the ranks corresponding to group 1 and $\bar{R}_2$ is the average for group 2. So for the jth group,

$$\bar{R}_j = \frac{1}{n_j} \sum_{i=1}^{n_j} R_{ij}.$$

Next, for the first group, rank the observations ignoring group 2 and label the results $V_{11}, \ldots V_{n_1 1}$. Do the same for group 2 (ignoring group 1) and label the ranks $V_{12}, \ldots V_{n_2 2}$.

Next, compute

$$S_j^2 = \frac{1}{n_j - 1} \sum_{i=1}^{n_j} \left(R_{ij} - V_{ij} - \bar{R}_j + \frac{n_j + 1}{2} \right)^2,$$

$$s_j^2 = \frac{S_j^2}{(N - n_j)^2}$$

$$s_e = \sqrt{N} \sqrt{\frac{s_1^2}{n_1} + \frac{s_2^2}{n_2}},$$

$$U_1 = \left(\frac{S_1^2}{N - n_1} + \frac{S_2^2}{N - n_2} \right)^2$$

and

$$U_2 = \frac{1}{n_1 - 1} \left(\frac{S_1^2}{N - n_1} \right)^2 + \frac{1}{n_2 - 1} \left(\frac{S_2^2}{N - n_2} \right)^2.$$

The degrees of freedom are

$$\hat{\nu} = \frac{U_1}{U_2}.$$

The test statistic is

$$W = \frac{\bar{R}_2 - \bar{R}_1}{\sqrt{N} s_e}.$$

Decision Rule: Reject H_0: $P = 0.5$ if $|W| \geq t$, where t is the $1 - \alpha/2$ quantile of a Student's T distribution with $\hat{\nu}$ degrees of freedom. An estimate of P is

$$\hat{P} = \frac{1}{n_1} \left(\bar{R}_2 - \frac{n_2 + 1}{2} \right) = \frac{1}{N} (\bar{R}_2 - \bar{R}_1) + \frac{1}{2}.$$

The estimate of δ is

$$\hat{\delta} = 1 - 2\hat{P}.$$

An approximate $1 - \alpha$ confidence interval for P is

$$\hat{P} \pm t s_e.$$

There is a connection between the method just described and the Wilcoxon–Mann–Whitney test that is worth mentioning:

$$U = n_2 \bar{R}_2 - \frac{n_2(n_2 + 1)}{2}.$$

That is, if you sum the ranks of the second group (which is equal to $n_2 \bar{R}_2$) and subtract $n_2(n_2 + 1)/2$, you get the Wilcoxon–Mann–Whitney U statistic given by Equation (7.26). Many books describe the Wilcoxon–Mann–Whitney method in terms of U rather than the approach used here.

Note that both the Cliff and Brunner–Munzel rank-based methods offer protection against low power due to outliers. If, for example, the largest observation among a batch of numbers is increased from 12 to one million, its rank does not change. But how should one choose between rank-based methods covered here versus methods based on robust measures of location? If our only criterion is high power, both perform well with weak evidence that, in practice, robust methods are a bit better. But the more important point is that they provide different information about how groups compare. Some authorities argue that as a measure of effect size, P or δ, as defined in this section, reflect what is most important and what we want to know. Others argue that measures of location also provide useful information; they reflect what is typical and provide a sense of the magnitude of the difference between groups that is useful and not provided by rank-based methods. The only certainty is that at present, there is no agreement about which approach should be preferred or even if it makes any sense to ask which is better.

Often Cliff's method gives similar results to the Brunner–Munzel technique. However, situations can be constructed where, with many tied values, Cliff's approach seems to be better at guaranteeing an actual Type I error probability less than the nominal α level. Based on results reported by Neuhäuser et al. (2007), with small sample sizes, Cliff's method seems preferable in terms of controlling the probability of a Type I error. (Yet another approach was derived by Chen and Luo, 2004, but results in Neuhäuser et al. do not support this method.)

Gaining Perspective

Let θ_1 and θ_2 be the population medians for groups 1 and 2, respectively. Under general conditions, both the Cliff and Bruner-Munzel methods do *not* test H_0: $\theta_1 = \theta_2$, the hypothesis that the medians are equal. However, let D be the difference between a randomly sampled observation from each group. In symbols, $D = X_1 - X_2$, where X_1 and X_2 are randomly sampled observations from groups 1 and 2, respectively. Now, the population mean of D is simply $\mu_D = \mu_1 - \mu_2$, the difference between the population means corresponding to groups 1 and 2. However, under general conditions, the median of D is not equal to the difference between the individual medians. There are exceptions, such as when both groups have symmetric distributions, but in general $\theta_D \neq \theta_1 - \theta_2$.

Here is the point: both the Cliff and Brunner–Munzel methods can be viewed as testing

$$H_0 : \theta_D = 0.$$

To see why, note that by definition, if the median of D is 0, then with probability 0.5, $D < 0$ (assuming that D is continuous). That is, $P(X_1 < X_2) = 0.5$, which is p in the notation introduced in Section 7.8.1. A confidence interval for θ_D can be computed using an extension of Cliff's method, but the details are not provided.

7.8.7 R Functions bmp and loc2dif.ci

The R function

```
bmp(x,y,alpha=0.05)
```

performs the Brunner–Munzel method. It returns a p-value when testing H_0: $P = 0.5$, an estimate of P labeled `phat`, and a confidence interval for P labeled `ci.p`. (An estimate of δ, labeled `d.hat`, is returned as well.) The R function

```
loc2dif.ci(x,y,est=median alpha=0.05)
```

computes a confidence interval for θ_D.

EXAMPLE

Table 7.4 reports data from a study of hangover symptoms among sons of alcoholics as well as a control group. Note that there are many tied values. In the second group, for example, 14 of the 20 values are zero. Welch's test for means has a p-value of 0.14, Yuen's test has a p-value of 0.076, the Brunner–Munzel method has a p-value of 0.042, and its 0.95 confidence interval for P is (0.167, 0.494). Using Cliff's method, the p-value is 0.049, the 0.95 confidence interval for δ is (0.002, 0.60), and the 0.95 confidence interval for P is (0.198, 0.490).

7.8.8 The Kolmogorov–Smirnov Test

Yet another way of testing the hypothesis that two independent groups have identical distributions is with the so-called Kolmogorov–Smirnov test. An important feature of the Kolmogorov–Smirnov test is that it can be extended to the problem of comparing simultaneously all of the quantiles using what is called the shift function described in Section 7.8.9. Exact control over the probability of a Type I error can be had assuming random sampling only. When there are no tied values, the method in Kim and Jennrich (1973) can be used to compute the exact probability of a Type I error. With tied values, the exact probability of a Type I error can be computed with a method derived by Schroër and Trenkler (1995). The R function supplied to perform the Kolmogorov–Smirnov test has an option for computing the exact Type I error probability, but the computational details are omitted. (Details can be found in Wilcox, 2017.)

To apply the Kolmogorov–Smirnov test, let $\hat{F}_1(x)$ be the proportion of observations in group 1 that are less than or equal to x, and let $\hat{F}_2(x)$ be the corresponding proportion for group 2. Let

$$U_i = |\hat{F}_1(X_{i1}) - \hat{F}_2(X_{i1})|,$$

$i = 1, \ldots, n_1$. In other words, for X_{i1}, the ith observation in group 1, compute the proportion of observations in group 1 that are less than equal to X_{i1}, do the same for group 2, take the absolute value of the difference and label the result U_i. Repeat this process for the observations in group 2 and label the results

$$V_i = |\hat{F}_1(X_{i2}) - \hat{F}_2(X_{i2})|,$$

$i = 1, \ldots, n_2$. The Kolmogorov–Smirnov test statistic is

$$KS = \max\{U_1, \ldots, U_{n_1}, V_1, \ldots, V_{n_2}\}, \quad (7.35)$$

the largest of the pooled U and V values. For large sample sizes, an approximate critical value when $\alpha = 0.05$ is

$$1.36\sqrt{\frac{n_1 + n_2}{n_1 n_2}}.$$

Reject when KS is greater than or equal to the critical value. When there are no tied values, the Kolmogorov–Smirnov test can have relatively high power, but with ties, its power can be relatively low.

7.8.9 R Function ks

The R function

```
ks(x,y,sig=TRUE,alpha=0.05),
```

written for this book, performs the Kolmogorov–Smirnov test. R has a built-in function for performing this test, but the function written for this book has the advantage of determining the exact probability of a Type I error, assuming random sampling only, even when there are tied values. (For more information, see Wilcox, 2017.) With `sig=TRUE`, the exact critical value will be used. With large sample sizes, computing the exact critical value can result in high execution time. Setting `sig=FALSE` avoids this problem by using an approximate α critical value, where α is specified by the argument `alpha`, which defaults to 0.05.

EXAMPLE

For the data in Table 7.4 the function `ks` returns

```
$test:
[1] 0.25

$critval:
[1] 0.4300698

$siglevel:
[1] 0.1165796
```

This says that the Kolmogorov–Smirnov test statistic is $KS = 0.35$; the approximate 0.05 critical value is 0.43, which is greater than KS, meaning that we would not reject at the 0.05 level. The exact p-value (labeled `siglevel`), assuming only random sampling, is 0.117.

7.8.10 Comparing All Quantiles Simultaneously: An Extension of the Kolmogorov–Smirnov Test

Roughly, when comparing medians, the goal is to compare the central values of the two distributions. But an additional issue is how low scoring individuals in the first group compare to low scoring individuals in the second. In a similar manner, how do relatively high scores within each group compare? A way of addressing this issue is to compare the 0.25 quantiles of both groups as well as the 0.75 quantiles. Or to get a more detailed sense of how the distributions differ, all of the quantiles might be compared. There is a method for comparing all quantiles in a manner that controls the probability of a Type I error exactly assuming random sampling only. The method was derived by Doksum and Sievers (1976) and is based on an extension of the Kolmogorov–Smirnov method. Complete computational details are not provided, but a function that applies the method is supplied and illustrated in the next section.

There is an alternative method for comparing quantiles that has practical value, particularly when dealing with the lower and upper quantiles. The method is based in part on a technique for controlling the probability of one or more Type I errors when performing multiple tests, which is described in Section 12.1.10. Details about this alternative method are described in Section 12.1.19. Also see the methods in Section 12.1.17.

When dealing with discrete distributions, where the possible number of responses is relatively small, there is yet another approach that can provide a more detailed indication of where and how distributions differ. For example, consumers might be asked to rate their overall satisfaction with two brands of cars. If they rate the cars on a scale from zero to 10, one could, of course, use some measure of location to compare the ratings. But it might be helpful to compare the likelihood of the individual responses. For instance, does the primary difference occur among ratings of zero? Is the likelihood of a rating 10 approximately the same for both groups? For a specific response, methods in Section 7.14 can be used. But there is a technical issue that must be addressed when dealing with two or more responses. A method aimed at addressing this issue is described and illustrated in Section 12.1.17.

7.8.11 R Function sband

The R function

```
sband(x,y, flag = FALSE, plotit = TRUE, xlab = "x (First Group)", ylab =
                                   "Delta")
```

computes confidence intervals for the difference between the quantiles using the data stored in the R variables x and y. Moreover, it plots the estimated differences as a function of the estimated quantiles associated with the first group, the first group being the data stored in the first argument, x. This difference between the quantiles, viewed as a function of the quantiles of the first group, is called a *shift function*. To avoid the plot, set the argument plotit=FALSE.

EXAMPLE

In a study by Victoroff et al. (2010), 52 fourteen-year-old refugee boys in Gaza were classified into one of two groups according to whether a family member had been wounded or killed by an Israeli. One issue was how these two groups compare based on a measure of depression. In particular, among boys with relatively high depression, does having a family member killed or wounded have more of an impact than among boys with relatively low measures of depression? Here is a portion of the output from sband:

```
           qhat lower upper
 [1,] 0.03448276    NA    18
 [2,] 0.06896552    NA    15
 [3,] 0.10344828    NA    15
 [4,] 0.13793103    NA    15
 [5,] 0.17241379    NA    16
 [6,] 0.20689655    NA    16
 [7,] 0.24137931    NA    16
 [8,] 0.27586207    NA    15
 [9,] 0.31034483    NA    16
```

```
[10,] 0.34482759    NA    16
[11,] 0.37931034    NA    16
[12,] 0.41379310   -10    19
[13,] 0.44827586    -7    20
[14,] 0.48275862    -5    26
[15,] 0.51724138    -5    26
[16,] 0.55172414    -4    26
[17,] 0.58620690    -2    34
[18,] 0.62068966    -1    NA
[19,] 0.65517241    -2    NA
[20,] 0.68965517     2    NA
[21,] 0.72413793     1    NA
[22,] 0.75862069     2    NA
[23,] 0.79310345     2    NA
[24,] 0.82758621     1    NA
[25,] 0.86206897     2    NA
[26,] 0.89655172     2    NA
[27,] 0.93103448     0    NA
[28,] 0.96551724    -3    NA
[29,] 1.00000000    -2    NA
```

The column headed by `qhat` indicates the quantile being compared. The first value listed is 0.03448276, meaning that a confidence interval for the difference between the 0.03448276 quantiles is given in the next two columns. In the column headed by `lower`, NA indicates $-\infty$. In the column headed by `upper`, NA indicates ∞. So the first row of the output says that the confidence interval for the difference between the 0.03448276 quantiles is $(-\infty, 18)$. This interval contains 0, so you would fail to conclude that the quantiles differ. The function also returns a value labeled `numsig`, which indicates how many differences among all of the estimated quantiles are significant. That is, the confidence interval does not contain 0.

Now look at row 20 of the output. This says that when comparing the 0.68965517 quantiles, the confidence interval is $(2, \infty)$. This interval does not contain 0, so reject. Looking at rows 21–26, we again reject. So no difference between the groups is found when looking at the lower quantiles, but a difference is found from the 0.69 to 0.90 quantiles. Roughly, the results indicate that among boys who have had a family member wounded or killed, the effect is more pronounced among boys with high depression scores. Moreover, the probability that all of these confidence intervals simultaneously contain the true differences is approximately 0.95. If it is desired to compute the exact probability, this can be done by setting the argument `flag=TRUE`. If `flag=TRUE` is used in the example, the output labeled pc (probability coverage) has the value 0.96762 meaning that all 29 confidence intervals contain the true differences with probability 0.96762. Said another way, the probability of making at least one Type I error among the 29 quantiles being compared is $1 - 0.96762 = 0.03238$.

7.9 GRAPHICAL METHODS FOR COMPARING GROUPS

There are several graphical techniques designed to provide perspective on how groups differ, some of which are described here. The method based on error bars, described next, should be used with caution for reasons that will be made clear. But, generally, plotting the data can be quite useful and informative.

7.9.1 Error Bars

Error bars plot the means (with the x-axis corresponding to groups) plus vertical lines above and below the sample means based on the standard errors. Figure 7.3 shows the error bars based on the data in Table 7.1. The circles indicate the values of the sample means (indicated by their position along the y-axis) and the length of the vertical lines above and below the mean is one standard error. For instance, for the first group, the estimated standard error is $s_1/\sqrt{n_1} = 13$ (as noted in Table 7.1), meaning that the two lines extending above and below the sample mean of 18 each have length 13.

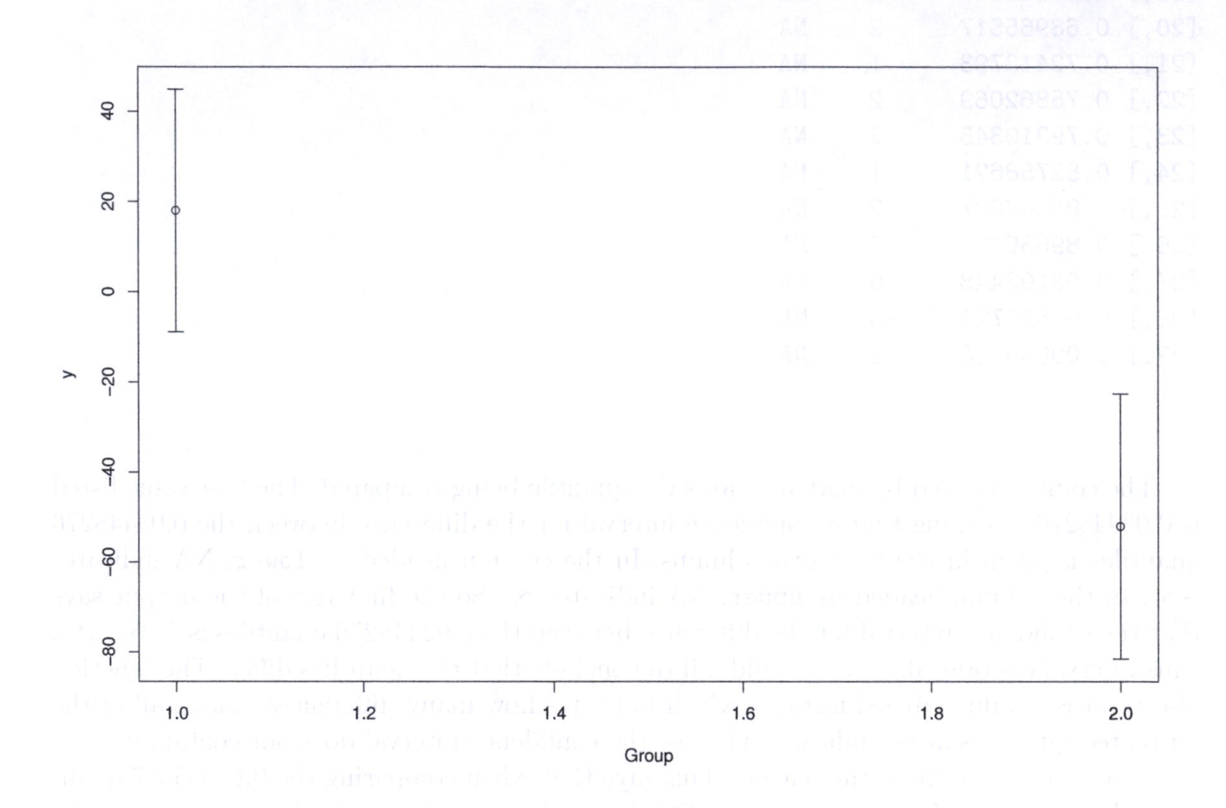

Figure 7.3 An example of error bars based on the weight-gain data in Table 7.1.

It might seem that if the intervals indicated by the error bars do not overlap, then it is reasonable to conclude that the population means, μ_1 and μ_2, differ. The bottom end of the first error bar in Figure 7.3 lies above the upper end of the other error bar indicating that the confidence interval for the first mean does not contain any of the values corresponding to the confidence interval for the second mean. Put another way, it might seem that if we reject the hypothesis of equal means when a 0.95 confidence interval for the first mean does not overlap with the 0.95 confidence interval for the second mean, the probability of a Type I error, when testing H_0: $\mu_1 = \mu_2$, is 0.05. But this is not the case (Goldstein and Healey, 1995; also see Schenker and Gentleman, 2001; cf. Afshartous and Preston, 2010; Tryon, 2001).

EXAMPLE

Thompson and Randall-Maciver (1905) recorded four measurements of male Egyptian skulls from five different time periods. For the breadth measurements of 30 skulls from the year 3300 BC and the breadth measurements of 30 skulls from the year 200 BC, Welch's test rejects with $\alpha = 0.05$. Figure 7.4 shows the error bars based on 2.67 standard errors, which corresponds to the critical value used by Welch's test when testing at the 0.01 level. The intervals do not overlap, which might seem to contradict the fact that the p-value based on Welch's test is 0.008.

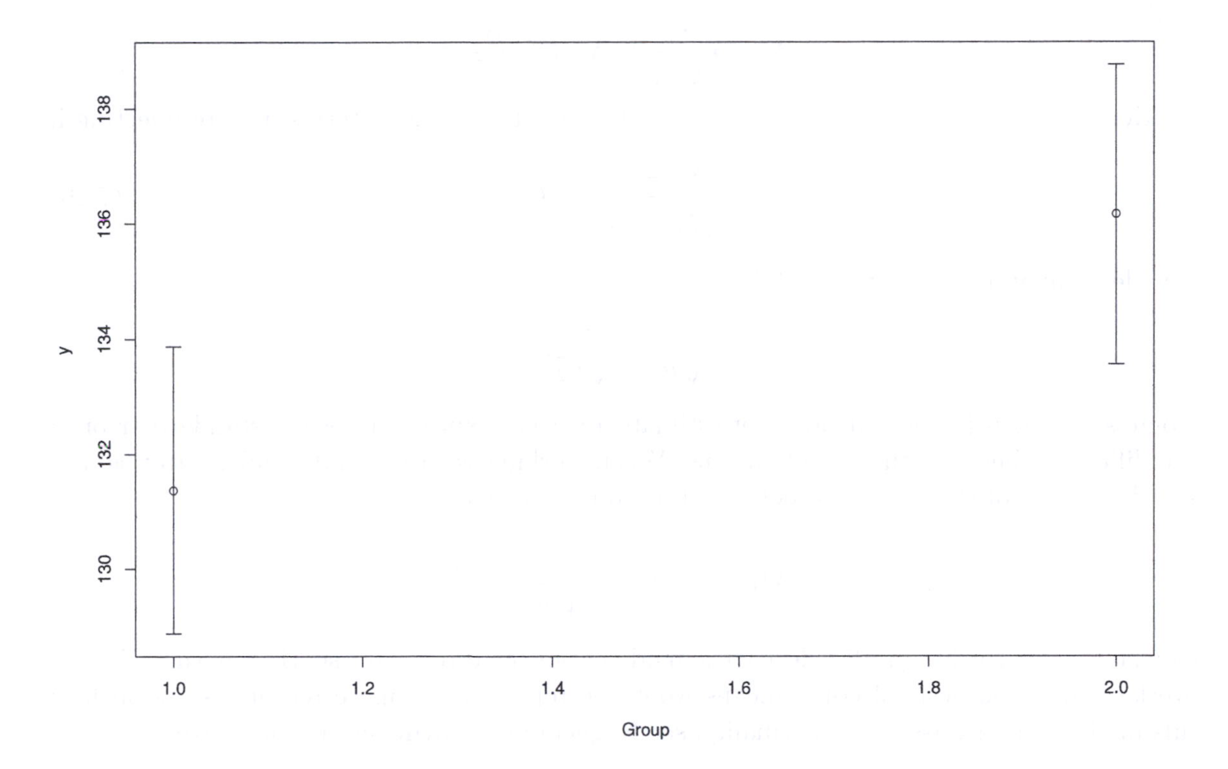

Figure 7.4 Shown are error bars for skull measurements based on Egyptian males. Group 1 corresponds to skulls from the year 3300 BC and group 2 is from the year 200 BC. Note that the intervals overlap even though Welch's test rejects the hypothesis of equal means.

Error bars are unsatisfactory when testing the hypothesis of equal means because they use the wrong standard error. To elaborate, first note that in symbols, error bars for the first mean correspond to the values

$$\bar{X}_1 - \ell \frac{s_1}{\sqrt{n_1}}$$

and

$$\bar{X}_1 + \ell \frac{s_1}{\sqrt{n_1}},$$

where ℓ is a constant chosen by the investigator. In Figure 7.4, $\ell = 2$ with the goal of plotting 0.95 confidence intervals. As for the second group, error bars correspond to the values

$$\bar{X}_2 - \ell \frac{s_2}{\sqrt{n_2}}$$

and

$$\bar{X}_2 + \ell\frac{s_2}{\sqrt{n_2}}.$$

Now consider the strategy of deciding that the population means differ if the intervals based on these error bars do not overlap. In symbols, reject the null hypothesis of equal means if

$$\bar{X}_1 + \ell\frac{s_1}{\sqrt{n_1}} < \bar{X}_2 - \ell\frac{s_2}{\sqrt{n_2}},$$

or if

$$\bar{X}_1 - \ell\frac{s_1}{\sqrt{n_1}} > \bar{X}_2 + \ell\frac{s_2}{\sqrt{n_2}}.$$

Here is a fundamental problem with this strategy. Rearranging terms, we are rejecting if

$$\frac{|\bar{X}_1 - \bar{X}_2|}{\frac{s_1}{\sqrt{n_1}} + \frac{s_2}{\sqrt{n_2}}} \geq \ell. \tag{7.36}$$

The denominator of Equation (7.36),

$$\frac{s_1}{\sqrt{n_1}} + \frac{s_2}{\sqrt{n_2}},$$

violates a basic principle—it does not estimate a correct expression for the standard error of the difference between the sample means. When working with two independent groups, the standard error of the difference between the sample means is

$$\text{VAR}(\bar{X}_1 - \bar{X}_2) = \sqrt{\frac{\sigma_1^2}{n_1} + \frac{\sigma_2^2}{n_2}}. \tag{7.37}$$

Goldstein and Healey (1995) derived a modification of error bars so that if they do not overlap, the hypothesis of equal means would be rejected. For more recent results on how this might be done, assuming normality, see Noguchi and Marmolejo-Ramos (2016).

7.9.2 R Functions ebarplot and ebarplot.med

A possible argument for error bars is that they convey useful information, despite the interpretational concern previously illustrated. The R function

```
ebarplot(x,y = NULL, nse = 1,xlab="Groups",ylab=NULL,tr=0)
```

creates error bars using the data stored in the variables `x` and `y`, assuming that you have installed the R library `plotrix`, which can be downloaded as described in Chapter 1. If the argument `y` is not supplied, the function assumes that `x` is a matrix (with columns corresponding to groups) or that `x` has list mode. If, for example, `x` is a matrix with five columns, five plots will be created for each of the five groups. If `y` is supplied, `x` must be a vector. The argument `nse` defaults to 1 indicating that one standard error is used when creating the plots. Setting `nse=2` would use two standard errors. The arguments `xlab` and `ylab` can be used to generate labels for the x-axis and y-axis, respectively. Error bars based on trimmed means can be created using the argument `tr`. For example, `tr=0.2` will use 20% trimmed means and the appropriate standard error. When dealing with medians, the R function

```
ebarplot.med(x,y=NULL,alpha=0.05,xlab="Groups",ylab=NULL)
```

can be used. It plots confidence intervals for the medians using the distribution-free method in Section 4.9.

7.9.3 Plotting the Shift Function

As already explained, a shift function indicates the difference between the quantiles of the two groups as a function of the quantiles of the first group. A plot of the shift function indicates the quantiles of the first group along the x-axis versus the estimated difference between the quantiles on the y-axis which can be done with the R function

```
sband(x,y, flag = F, plotit = T, xlab = "x (First Group)", ylab = "Delta").
```

The default label for the y-axis, indicated by the argument `ylab`, is Delta, which represents the estimated difference between the quantiles of the data stored in y and the data stored in x. The labels for both the x-axis and y-axis can be changed by putting the new label in quotes. For example, `"ylab=Difference"` will replace Delta with Difference when the plot is created.

One practical appeal of the shift function is that it provides a more detailed sense of where distributions differ and by how much. Consider, for instance, a study aimed at comparing males and females on an aptitude test. Momentarily consider the subpopulation of males and females who tend to get relatively low scores. One possibility is that these individuals differ substantially. But when we focus on the subpopulation of males and females who get relatively high scores, the reverse might be true.

EXAMPLE

Consider again the data discussed in Section 7.8.11 on depression among boys living in Gaza. Figure 7.5 shows the plot created by `sband`. The x-axis indicates measures of depression among boys with no family member wounded or killed by Israelis. The solid, nearly straight line indicates the estimated difference between all of the quantiles. In symbols, if x_q and y_q are the qth quantiles, the solid straight line is a plot of

$$\Delta(x_q) = y_q - x_q$$

as q increases from 0 to 1. We see that as we move from low measures of depression to higher measures, the effect increases in a fairly linear manner. If the groups do not differ in any manner, we should get a horizontal line at 0. That is, $\Delta(x_q)$ should be 0 for any quantile, x_q, we pick. The dashed lines above and below the plot of the estimated differences indicate a *confidence band*. This band contains *all* true differences with probability approximately equal to 0.95. The exact probability is reported when the argument `flag=TRUE` and is labeled pc. The hypothesis of equal quantiles is rejected if the lower (upper) dashed line is above (below) zero. The + indicates the location of the median for the control group (the data stored in the first argument x), and the lower and upper quartiles are marked with an o to the left and right of the +.

EXAMPLE

Figure 7.6 shows the shift function created by `sband` using the data in Table 7.1, which deals with weight gain among newborns exposed to the sound of a human heartbeat. Notice how the curve increases and then becomes a nearly horizontal line. The estimated differences are largest among infants who are relatively light at birth. This again illustrates that `sband` can reveal that the effect of some treatment can vary among subgroups of participants.

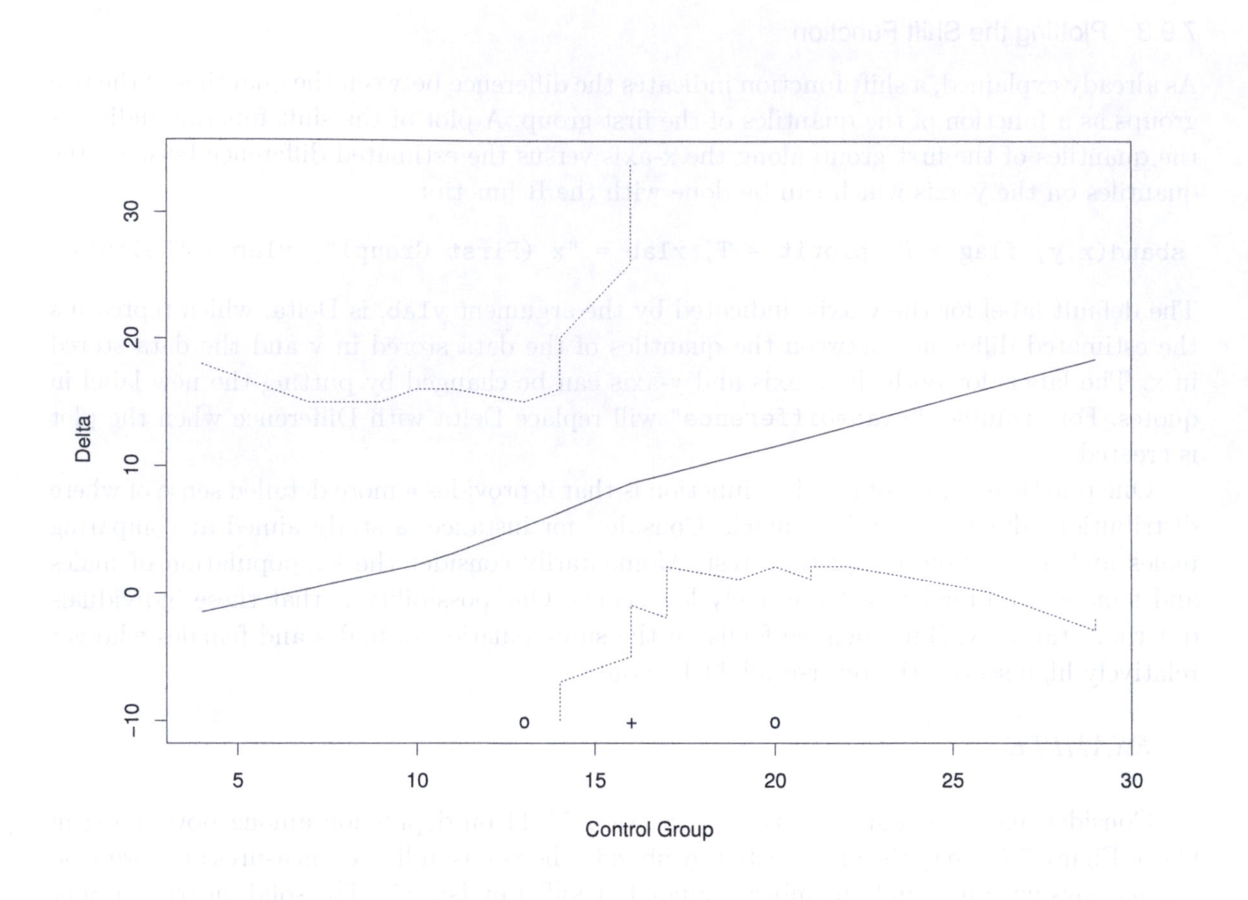

Figure 7.5 A plot of the shift function based on the Gaza data. The plot indicates that among boys with low measures of depression, there is little difference between the two groups. But as we move toward subpopulations of boys with high depression, the difference between the two groups increases.

EXAMPLE

Table 7.6 contains data from a study designed to assess the effects of ozone on weight gain in rats. (The data were taken from Doksum and Sievers, 1976.) The experimental group consisted of 22 seventy-day-old rats kept in an ozone environment for 7 days. A control group of 23 rats, of the same age, were kept in an ozone-free environment. Figure 7.7 shows the shift function. Looking at the graph as a whole suggests that the effect of ozone becomes more pronounced as we move along the x-axis up to about 19, but then the trend reverses and in fact in the upper end we see more weight gain in the ozone group.

Notice that in Figure 7.7, the left end of the lower dashed line begins at approximately $x = 22$. This is because for $x < 22$ the lower confidence band extends down to $-\infty$. That is, the precision of the estimated differences between the quantiles might be poor in this region based on the sample sizes used. Similarly, the upper dashed line terminates around $x = 27$. This is because for $X > 27$ the upper confidence band extends up to ∞.

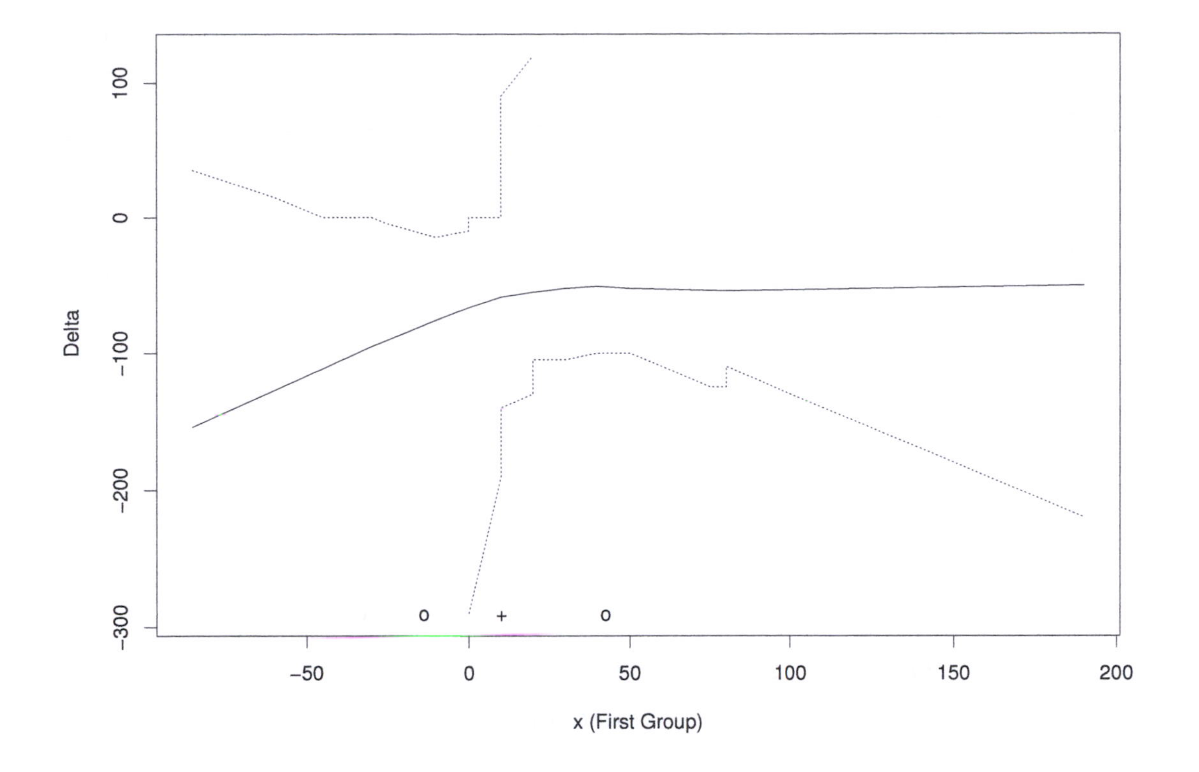

Figure 7.6 A plot of the shift function based on the data in Table 7.1. The plot indicates that weight gain differences are more pronounced among infants who have relatively low weight at birth.

7.9.4 Plotting the Distributions

Another graphical strategy is to simply plot the distributions of both groups, which can be done with the function

```
g2plot(x,y,op=4,xlab = "X", ylab = "").
```

The argument op controls the method used to estimate the distributions. By default, op=4 is used meaning that the adaptive kernel density estimator, mentioned in Chapter 2, is used. When the sample size is large, op=1, 2, or 3 might be preferable in order to avoid high execution time. The left panel of Figure 7.8 shows the plot created by this function based on the ozone data in Table 7.6. The right panel shows the corresponding boxplots, which were created with the command boxplot(x,y). It is prudent to always check the graph created

Table 7.6 Weight Gain (in Grams) of Rats in Ozone Experiment

Control:	41.0	38.4	24.4	25.9	21.9	18.3	13.1	27.3	28.5	−16.9
Ozone:	10.1	6.1	20.4	7.3	14.3	15.5	−9.9	6.8	28.2	17.9
Control:	26.0	17.4	21.8	15.4	27.4	19.2	22.4	17.7	26.0	29.4
Ozone:	−9.0	−12.9	14.0	6.6	12.1	15.7	39.9	−15.9	54.6	−14.7
Control:	21.4	26.6	22.7							
Ozone:	44.1	−9.0								

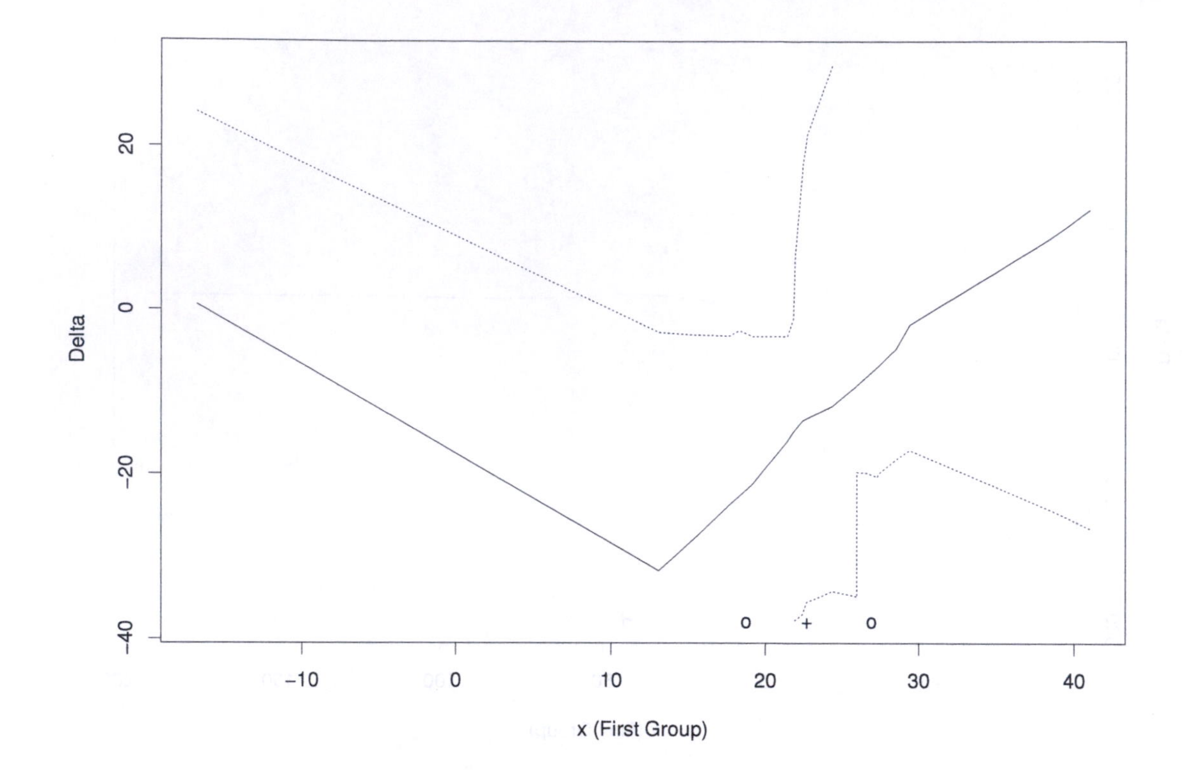

Figure 7.7 The shift function based on the data in Table 7.6. The plot indicates that for very small and very large amounts of weight gains, there is little difference between the groups. The two groups tend to differ most for weight gains just below the lower quartile of the first group.

by the function `g2plot`. For example, situations arise where methods for comparing means reject, yet the two distributions appear to be nearly identical. That is, methods based on means might reject even when the distributions differ in a trivial and uninteresting fashion.

7.9.5 R Function sumplot2g

The R function

```
sumplot2g(x,y)
```

simultaneously creates four plots: error bars, a kernel density estimates of the distributions, boxplots, and a shift function.

7.9.6 Other Approaches

The methods listed in this section are not exhaustive. Another approach is to examine a so-called *quantile-quantile* or *Q-Q plot.* (The built-in R function `qqplot` creates a Q-Q plot.) If the groups have identical quantiles, a plot of the quantiles should be close to a line having slope one and intercept zero. Another approach to measuring effect size is the so-called overlapping coefficient. You estimate the distributions associated with both groups and then compute the area under the intersection of these two curves; see Clemons and Bradley (2000)

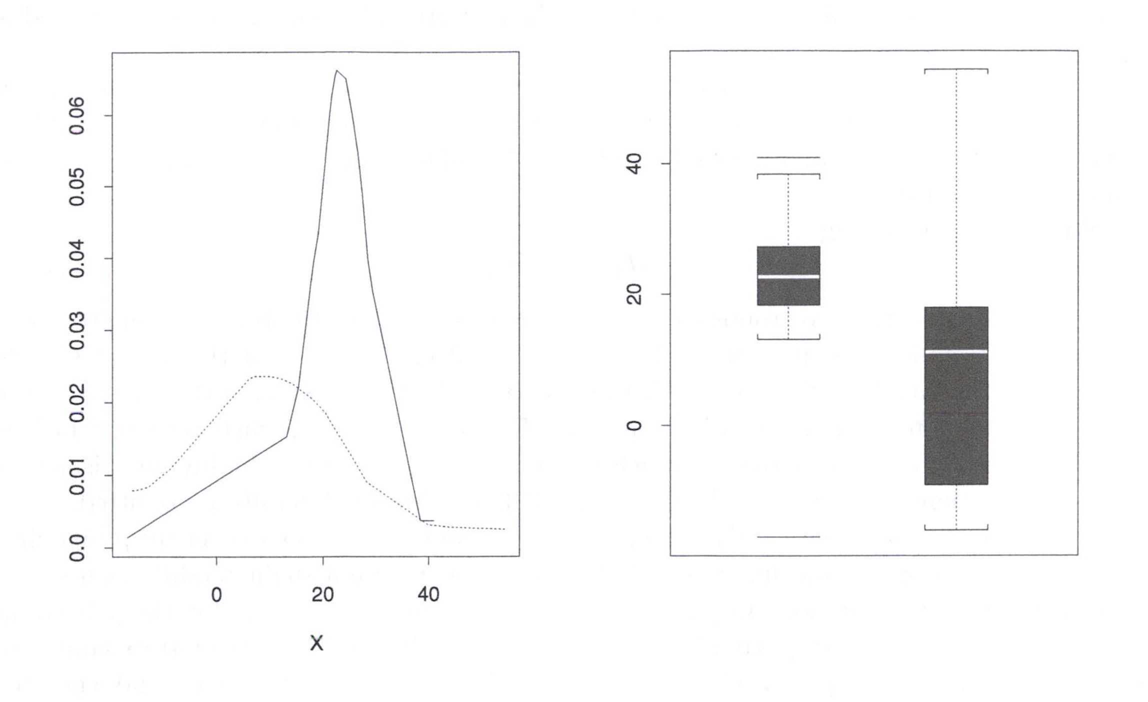

Figure 7.8 The left panel shows the graph created by the function `g2plot` using the data in Table 7.6. The right panel shows the corresponding boxplots.

for recent results on how this might be done. An area of zero corresponds to no overlap, and an area of one occurs when the distributions are identical and the groups do not differ in any manner. Yet another strategy is to plot the distribution of $D = X - Y$, the difference between a randomly sampled observation from the first and second group. If the groups do not differ, D should have a symmetric distribution about zero. (A plot of the estimated distribution of D can be created with the R function `cid`, described in Section 7.8.4, by setting the argument `plotit=TRUE`.)

7.10 COMPARING MEASURES OF VARIATION

Although the most common approach to comparing two independent groups is to use some measure of location, situations arise where there is interest in comparing variances or some other measure of scale. For example, in agriculture, one goal when comparing two crop varieties might be to assess their relative stability. One approach is to declare the variety with the smaller variance as being more stable (e.g., Piepho, 1997). As another example, consider two methods for training raters of some human characteristic. For example, raters might judge athletic ability or they might be asked to rate aggression among children in a classroom. Then one issue is whether the variance of the ratings differ depending on how the raters were trained. Also, in some situations, two groups might differ primarily in terms of the variances rather than their means or some other measure of location. To take a simple example, consider two normal distributions both having means zero with the first having variance one and the second having variance three. Then a plot of these distributions would show that they differ substantially, yet the hypotheses of equal means, equal trimmed means, equal M-estimators, and equal medians are all true. That is, to say the first group is comparable to the second is inaccurate, and it is of interest to characterize how they differ.

There is a vast literature on comparing variances and as usual not all methods are covered here. For studies comparing various methods, the reader is referred to Conover et al. (1974), Wilcox (1992), Ramsey and Ramsey (2007), and Cojbasic and Tomovic (2007) plus the references they cite.

We begin with testing

$$H_0 : \sigma_1^2 = \sigma_2^2, \tag{7.38}$$

the hypothesis that the two groups have equal variances. Many methods have been proposed. The classic technique assumes normality and is based on the ratio of the largest sample variance to the smallest. So if $s_1^2 > s_2^2$, the test statistic is $F = s_1^2/s_2^2$; otherwise, you use $F = s_2^2/s_1^2$. When the null hypothesis is true, F has a so-called F distribution, which is described in Chapter 8. But this approach has long been known to be highly unsatisfactory when distributions are non-normal (e.g., Box, 1953) so additional details are omitted.

Currently, the most successful method in terms of maintaining control over the probability of a Type I error and achieving relatively high power is to use a slight modification of the percentile bootstrap method. In particular, set $n_m = \min(n_1, n_2)$ and for the jth group ($j = 1, 2$), take a bootstrap sample of size n_m. Ordinarily we take a bootstrap sample of size n_j from the jth group, but when sampling from heavy-tailed distributions, and when the sample sizes are unequal, control over the probability of a Type I error can be extremely poor for the situation at hand. Next, for each group, compute the sample variance based on the bootstrap sample and set D^* equal to the difference between these two values. Repeat this $B = 599$ times yielding 599 bootstrap values for D, which we label $D_1^*, \ldots, D_{599}^*$. As usual, when writing these values in ascending order, we denote this by $D_{(1)}^* \leq \cdots \leq D_{(B)}^*$. Then an approximate 0.95 confidence interval for the difference between the population variances is

$$(D_{(\ell)}^*, D_{(u)}^*), \tag{7.39}$$

where for $n_m < 40$, $\ell = 7$ and $u = 593$; for $40 \leq n_m < 80$, $\ell = 8$ and $u = 592$; for $80 \leq n_m < 180$, $\ell = 11$ and $u = 588$; for $180 \leq n_m < 250$, $\ell = 14$ and $u = 585$; and for $n_m \geq 250$, $\ell = 15$ and $u = 584$. (For results on the small-sample properties of this method, see Wilcox, 2002.) Notice that these choices for ℓ and u are the same as those used in Section 7.6 when making inferences about Pearson's correlation. The hypothesis of equal variances is rejected if the confidence interval given by Equation (7.39) does not contain zero.

Using the confidence interval given by Equation (7.39) has two practical advantages over the many alternative methods one might use to compare variances. First, compared to many methods, it provides higher power. Second, among situations where distributions differ in shape, extant simulations indicate that probability coverage remains relatively accurate in contrast to many other methods one might use. If the standard percentile bootstrap method is used instead, then with sample sizes of 20 for both groups, the Type I error probability can exceed 0.1 when testing at the 0.05 level, and with unequal sample sizes it can exceed 0.15. A limitation of the method is that it is restricted to testing at the 0.05 level.

7.10.1 R Function comvar2

The R function

```
comvar2(x,y)
```

compares variances using the bootstrap method described in the previous subsection. The method can only be applied with $\alpha = 0.05$; modifications that allow other α values have not been derived. The arguments x and y are R variables containing data for group 1 and group 2, respectively. The function returns a 0.95 confidence interval for $\sigma_1^2 - \sigma_2^2$ plus an estimate of $\sigma_1^2 - \sigma_2^2$ based on the difference between the sample variances, $s_1^2 - s_2^2$, which is labeled `vardif`.

7.10.2 Brown–Forsythe Method

Chapter 3 described a measure of variation based on the average absolute distance of observations from the median. In the notation used here, if M_1 is the median of the first group, the measure of scale for the first group is

$$\hat{\tau}_1 = \frac{1}{n_1} \sum |X_{i1} - M_1|, \tag{7.40}$$

where again $X_{11}, \ldots, X_{n_1 1}$ are the observations randomly sampled from the first group. For the second group, this measure of scale is

$$\hat{\tau}_2 = \frac{1}{n_2} \sum |X_{i2} - M_2|, \tag{7.41}$$

where M_2 is the median for the second group. Notice that these measures of scale do *not* estimate the population variance (σ^2) or the population standard deviation (σ), even under normality. Some popular commercial software uses this measure of variation to compare groups, a common goal being to justify the assumption of equal variances when using Student's T. Accordingly, the method is outlined here and then comments about its relative merits are made.

For convenience, let

$$Y_{ij} = |X_{ij} - M_j|,$$

$i = 1, \ldots, n_j$; $j = 1, 2$. That is, the ith observation in the jth group (X_{ij}) is transformed to $|X_{ij} - M_j|$, its absolute distance from the median of the jth group. So the sample mean of the Y values for the jth group is

$$\bar{Y}_j = \frac{1}{n_j} \sum Y_{ij},$$

which is just the measure of scale described in the previous paragraph. Now let τ_j be the population value corresponding to $\bar{Y}_j$. That is, τ_j is the value of $\bar{Y}_j$ we would get if all individuals in the jth group could be measured. The goal is to test

$$H_0 : \tau_1 = \tau_2. \tag{7.42}$$

If we reject, we conclude that the groups differ based on this measure of dispersion.

The Brown and Forsythe (1974b) test of the hypothesis given by Equation (7.42) consists of applying Student's t to the Y_{ij} values (cf. Ramsey and Ramsey, 2007). We have already seen, however, that when distributions differ in shape, Student's T performs rather poorly, and there are general conditions under which it does not converge to the correct answer as the sample sizes get large. We can correct this latter problem by switching to Welch's test, but problems remain when distributions differ in shape. For example, suppose we sample $n_1 = 20$ observations from a normal distribution and $n_2 = 15$ observations from the distribution shown in Figure 5.6. Then when testing at the $\alpha = 0.05$ level, the actual probability of a Type I error is approximately 0.21. Like Student's T or Welch's method, the Brown–Forsythe test provides a test of the hypothesis that distributions are identical. Although it is designed to be sensitive to a reasonable measure of scale, it can be sensitive to other ways the distributions might differ. So if the goal is to compute a confidence interval for $\tau_1 - \tau_2$, the Brown–Forsythe method can be unsatisfactory if $\tau_1 \neq \tau_2$. Presumably some type of bootstrap method could improve matters, but this has not been investigated and indirect evidence suggests that practical problems will remain. Moreover, if there is explicit interest in comparing variances (σ_1^2 and σ_2^2), the Brown–Forsythe test is unsatisfactory because $\hat{\tau}_1$ and $\hat{\tau}_2$ do not estimate the population variances, σ_1^2 and σ_2^2, respectively.

7.10.3 Comparing Robust Measures of Variation

It is noted that the percentile bootstrap method can be used to compare any robust measure of variation such as the percentage bend midvariance or the biweight midvariance introduced in Chapter 2. (The measure of variation used by the Brown–Forsythe method is not robust; even a single outlier can make its value arbitrarily large.) This can be accomplished with the R function `pb2gen` used in conjunction with the functions `pbvar` and `bivar`, which compute the percentage bend midvariance and the biweight midvariance, respectively.

EXAMPLE

A sample of 25 observations was generated from the mixed normal distribution introduced in Chapter 3 and another 25 observations were generated from a standard normal. As explained in Chapter 3, the corresponding population variances differ considerably even though the corresponding probability curves are very similar. Storing the data in the R variables x and y, `comvar2(x,y)` returned a 0.95 confidence interval of (0.57, 32.2) so we correctly reject the hypothesis of equal population variances. The sample variances were $s_x^2 = 14.08$ and $s_y^2 = 1.23$. In contrast, the command `pb2gen(x,y,est=pbvar)` returns a 0.95 confidence interval of (−1.59, 11.6) for the difference between the percentage bend midvariances. The values of the percentage bend midvariances were 2.3 and 1.7. In reality, the population values of the percentage bend midvariances differ by only a slight amount and we failed to detect this due to low power.

EXAMPLE

We repeat the last example, only now we sample from the two distributions shown in Figure 7.9. One of these distributions is the contaminated normal introduced in Chapter 3 and the other is a normal distribution that has the same variance. Although there is a clear and rather striking difference between these two distributions, the population means and variances are equal. The 0.95 confidence interval returned by `comvar2` is (−13.3, 20.8) so fail to reject at the 0.05 level. Or in terms of Tukey's three decision rule, make no decision about which group has the larger variance. In contrast, comparing the percentage bend midvariances, the 0.95 confidence interval is (−28.7, −1.8). This interval does not contain zero; so reject at the 0.05 level. If the biweight midvariances are compared with the command `pb2gen(x,y,est=bivar)`, the resulting 0.95 confidence interval is (−18.8, −5.6); so again reject at the 0.05 level .

These last two examples merely illustrate that different methods can lead to different conclusions. In the first example, it is certainly true that the two distributions being compared are very similar in the sense described in Chapter 3. But the tails of the mixed normal differ from the tails of the normal, and in the context of measuring stability or reliability, there is a difference that might have practical importance. This difference happens to be detected

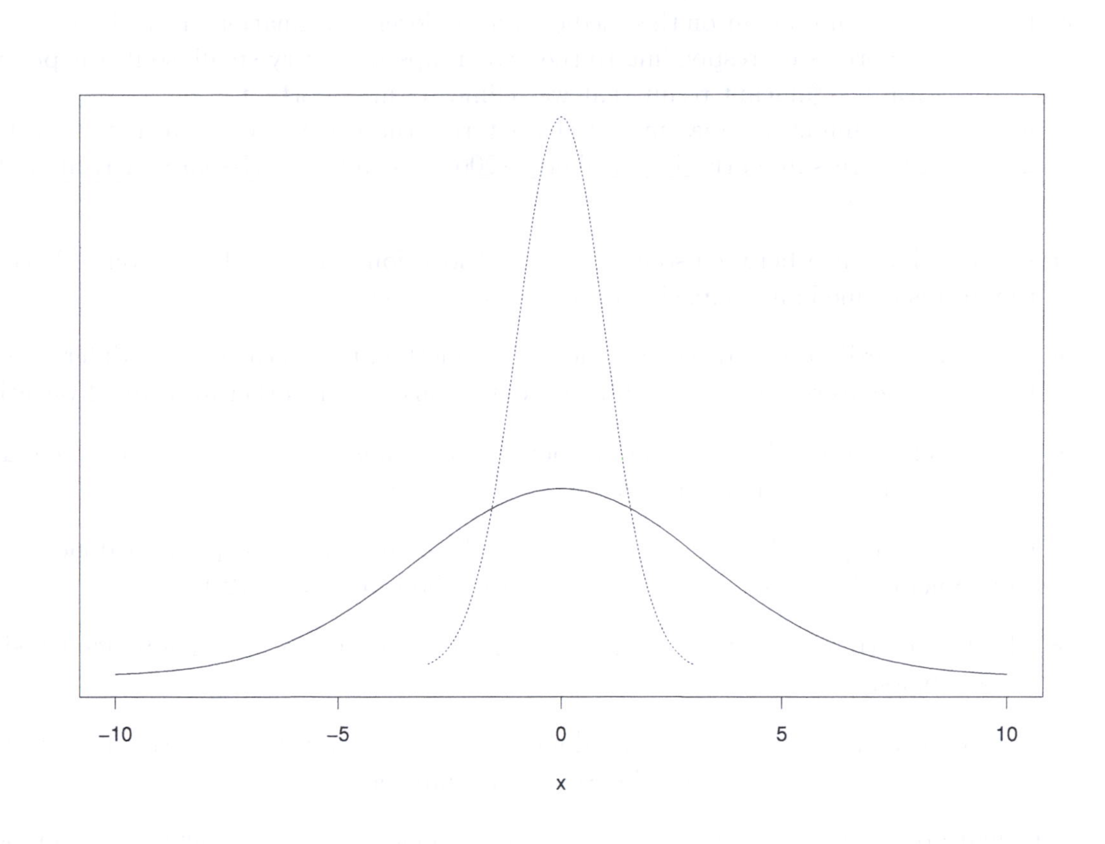

Figure 7.9 These two distributions have equal means and variances. The median and 20% trimmed means are equal to zero as well, so when comparing these groups with any of these measures of location, rejecting would mean a Type I error, yet it is clear that the distributions differ in a manner that might be considered important and substantial.

by comparing the variances but not when comparing the percentage bend or the biweight midvariances. However, in the second example, clearly the distributions differ considerably in terms of variation; comparing groups with the percentage bend or biweight midvariances detects this difference, but comparing the variances does not.

7.11 MEASURING EFFECT SIZE

It has long been recognized that, generally, merely rejecting the hypothesis of equal means (or any other measure of location) tells us virtually nothing about the extent the two groups differ (e.g., Cohen, 1994). If we reject at the 0.001 level and the first group has a larger sample mean than the second, then we conclude that the first group has the larger population mean. But generally this tells us nothing about the magnitude or importance of the difference between the groups. (An exception occurs when using Cohen's d, which is described later in this section.) An article in *Nutrition Today* (*19*, 1984, 22–29) illustrates the importance of this issue. A study was conducted on whether a particular drug lowers the risk of heart attacks. Those in favor of using the drug pointed out that the number of heart attacks in the group receiving the drug was significantly lower than the group receiving a placebo when testing at the $\alpha = 0.001$ level. However, critics of the drug argued that the difference between the number of heart attacks was trivially small. They concluded that because of the expense and side effects of using the drug, there is no compelling evidence that patients with high cholesterol levels should be put on this medication. A closer examination of the data revealed that the standard errors corresponding to the two groups were very small, so it was possible to get a statistically significant result that was clinically unimportant.

Generally, how might we measure or characterize the difference between two groups? Various strategies are summarized by Huberty (2002) as well as Grissom and Kim (2012). Some possibilities are:

- Use the difference between some measure of location, such as the difference between the means or median, coupled with a confidence interval.
- Use a standardized difference or some related method that measures the difference between the means relative to the standard deviations or some other measure of variation.
- Use a method related to explanatory power, which was mentioned in Section 6.5. This approach has connections with Pearson's correlation.
- Compare and plot the differences between all of the quantiles. (This is done by the shift function in Section 7.9.3. Also see the method in Section 12.1.19.)
- Plot the distributions. (For example, use the R function `g2plot`, already described, or use boxplots.)
- Estimate the probability that a randomly sampled observation from the first group is larger than a randomly sampled observation from the second.

The first approach has already been discussed, so no additional comments are given here. The second approach, which is commonly used, is typically implemented by assuming the two groups have a common variance, which we label σ^2. That is, $\sigma_1^2 = \sigma_2^2 = \sigma^2$ is assumed. Then the so-called standardized difference between the groups is

$$\delta = \frac{\mu_1 - \mu_2}{\sigma}. \tag{7.43}$$

Assuming normality, δ can be interpreted using results in Chapter 3. For example, if $\delta = 2$, then the difference between the means is two standard deviations, and for normal distributions we have some probabilistic sense of what this means. Cohen (1988) suggests that as a general guide, when dealing with normal distributions, $\delta = 0.2$, 0.5, and 0.8 correspond to small, medium, and large effect sizes, respectively. An estimate of δ is

$$d = \frac{\bar{X}_1 - \bar{X}_2}{s_p}, \tag{7.44}$$

which is often called *Cohen's d*, where s_p is the pooled standard deviation given by Equation (7.2). In general, determining d cannot be done based on the p-value only. However, under normality and homoscedasticity, Cohen's effect size can be determined based on the p-value and the sample sizes (Browne, 2010). A confidence interval for δ can be computed as well.

It is noted that for the probabilistic measure of effect size p, introduced in Section 7.8.1, for normal distributions with a common variance, $\delta = 0.2$, 0.5, and 0.8 corresponds to $p = 0.56$, 0.64, and 0.8, respectively (Acion et al., 2006).

One concern with δ is that if groups differ, there is no reason to assume that the variances are equal. Indeed, some authorities would argue that surely they must be unequal. We could test the hypothesis of equal variances, but how much power is needed to justify the conclusion that variances are equal if we fail to reject? A possible solution is to replace σ with the standard deviation from one of the groups. That is, we might use

$$\delta_1 = \frac{\mu_1 - \mu_2}{\sigma_1}$$

or

$$\delta_2 = \frac{\mu_1 - \mu_2}{\sigma_2}$$

which we would estimate with

$$\hat{\delta}_1 = \frac{\bar{X}_1 - \bar{X}_2}{s_1}$$

and

$$\hat{\delta}_2 = \frac{\bar{X}_1 - \bar{X}_2}{s_2},$$

respectively. A possible concern, however, is that there is the potential of getting two seemingly conflicting conclusions: δ_1, for example, might suggest a large effect size, while δ_2 indicates a small effect size. If the goal is to summarize the relative extent the means differ, and if we adopt Cohen's original suggestion that a large effect size is one that is visible to the naked eye, then the smaller of the two quantities δ_1 and δ_2 seems to be more appropriate. However, one could argue that if the variances differ substantially, there is, in some sense, a large effect, which might be reflected by the larger of two quantities δ_1 and δ_2. (The shift function, described in Section 7.8.9, provides another approach to assessing effect size in a manner that reflects both differences in measures of location and variation.)

Explanatory Measure of Effect Size

Again assuming that it is change in measures of location, relative to some measure of variation, which are of interest, there is an alternative to Cohen's d that allows heteroscedasticity (Wilcox and Tian, 2011). It is based in part on what is called explanatory power, which is a general framework for measuring the strength of an association between two variables that includes Pearson's correlation as a special case.

Note that if we are told that an observation will be sampled from the first group, a reasonable prediction of the value we would get is μ_1, the population mean (momentarily assuming that we know the value of μ_1). Similarly, if we sample from the second group, a reasonable guess at what we would observe is μ_2. The variation in the predicted values is

$$\sigma_\mu^2 = (\mu_1 - \bar{\mu})^2 + (\mu_2 - \bar{\mu})^2,$$

where $\bar{\mu} = (\mu_1 + \mu_2)/2$.

Next, momentarily imagine that an equal number of observations is sampled from each group and let σ_{pool}^2 be the (population) variance corresponding to the pooled observations. That is, we are pooling together two variables that might have different variances resulting in a variable that has variance σ_{pool}^2. Then a heteroscedastic, explanatory measure of effect size, xi, is

$$\xi = \sqrt{\frac{\sigma_\mu^2}{\sigma_{\text{pool}}^2}}.$$

To add perspective, it is noted that in the context of least squares regression, the approach leading to ξ^2 results in Pearson's squared correlation coefficient, the coefficient of determination. (Kulinskaya and Staudte, 2006, studied another approach that is somewhat related to

ξ^2.) Also, as previously noted, Cohen suggested that under normality and homoscedasticity, $\delta = 0.2$, 0.5, and 0.8 correspond to small, medium, and large effect sizes, respectively. The corresponding values of ξ are approximately 0.15, 0.35, and 0.50. Section 6.6.2 noted that small, medium, and large Pearson correlations are sometimes taken to be 0.1, 0.3, and 0.5, respectively. To the extent this seems reasonable, the same might be done when using ξ.

Estimation of ξ^2 is straightforward when there are equal sample sizes. First estimate σ^2_μ by replacing μ_1 and μ_2 with the corresponding sample means. Next, pool all $2n$ values and compute the sample variance, say s^2_{pool}, which estimates σ^2_{pool}. (Note that s^2_{pool} and s^2_p are not the same. The latter is based on a weighted average of the individual sample variances.) But when there are unequal sample sizes, this estimation method can be shown to be unsatisfactory. To deal with this, suppose the sample sizes are $n_1 < n_2$ for groups 1 and 2, respectively. If we randomly sample (without replacement) n_1 observations from the second group, we get a satisfactory estimate of ξ^2. To use all of the data in the second group, we repeat this process many times yielding a series of estimates for ξ^2, which are then averaged to get a final estimate, which we label $\hat{\xi}^2$. The estimate of ξ is just

$$\hat{\xi} = \sqrt{\hat{\xi}^2}.$$

This measure of effect size is not robust and in particular does not deal with the problem illustrated next. But a robust version is readily obtained by replacing the means and variances with some robust measure of location and scatter.

Robust Variations

A serious concern with the measure of effect size δ is nonnormality. The left panel of Figure 7.1 shows two normal distributions where the difference between the means is 1 ($\mu_1 - \mu_2 = 1$) and both standard deviations are one. So

$$\delta = 1,$$

which is often viewed as being relatively large. Now look at the right panel of Figure 7.1. As is evident, the difference between the two distributions appears to be very similar to the difference shown in the left panel, so according to Cohen we again have a large effect size. However, in the right panel, $\delta = 0.3$ because these two distributions are mixed normals with variances 10.9. This illustrates the general principle that arbitrarily small departures from normality can render the magnitude of δ meaningless. In practical terms, if we rely exclusively on δ to judge whether there is a substantial difference between two groups, situations will arise where we will grossly underestimate the degree to which groups differ, particularly when outliers occur.

Here is another concern about δ when trying to characterize how groups differ. Look at Figure 7.3. These two distributions have equal means and equal variances but they differ in an obvious way that might have practical importance. Although the difference between measures of location provide a useful measure of effect size, we need additional ways of gaining perspective on the extent to which groups differ such as the shift function, already described.

One way of dealing with nonnormality is to use a generalization of δ based on 20% trimmed means and Winsorized variances, where the Winsorized variances are rescaled so that under normality they estimate the variance (Algina et al., 2005). With 20% trimming, this means that the Winsorized variance is divided by 0.4121. That is, under normality, $s^2_w/0.4142$ estimates σ^2. If we assume the groups have equal Winsorized variances, δ becomes

$$\delta_t = 0.642\frac{\bar{X}_{t1} - \bar{X}_{t2}}{S_w},$$

where

$$S_W^2 = \frac{(n_1 - 1)s_{w1}^2 + (n_2 - 1)s_{w2}^2}{n_1 + n_2 - 2}$$

is the pooled Winsorized variance. Under normality, and when the variances are equal, $\delta = \delta_t$. If the Winsorized variances are not equal, Algina et al. suggest using both

$$\delta_{t1} = 0.642 \frac{\bar{X}_{t1} - \bar{X}_{t2}}{s_{w1}},$$

and

$$\delta_{t2} = 0.642 \frac{\bar{X}_{t1} - \bar{X}_{t2}}{s_{w2}}.$$

The measure of effect size ξ can be made more robust in a similar manner. Again replace the means with a trimmed mean, and replace the pooled sample variance with the Winsorized variance that has been rescaled to estimate the variance under normality. Henceforth, when using the measure of effect size ξ, the version based on 20% trimmed means and the 20% Winsorized variances will be assumed unless stated otherwise.

EXAMPLE

This first example illustrates that even under normality, heteroscedasticity can affect perceptions about the effect size when choosing between d and $\hat{\xi}$. With $n_1 = 80$, $n_2 = 20$, where the first group has a normal distribution with mean 0.8 and standard deviation 4, and the second group has a standard normal distribution, the median value of d (over many studies) is approximately 0.22, which is typically considered a small effect size. (The mean value of d is nearly identical to the median.) The median value of $\hat{\xi}$ is 0.40, which corresponds to $d = 0.59$ and is usually viewed as a medium effect size. So even under normality, a heteroscedastic measure of effect size can make a practical difference. If instead the first group has standard deviation 1 and the second has standard deviation 4, now the median estimates are 0.42 and 0.32, where $\hat{\xi} = 0.32$ corresponds to $d = 0.46$. That is, in contrast to the first situation, the choice between a homoscedastic and heteroscedastic measure of effect size makes little difference. If instead $n_1 = n_2 = 20$, now the median d value is 0.30 and the median $\hat{\xi}$ value is 0.34, which corresponds to $d = 0.50$. So the effect of ignoring heteroscedasticity is less of an issue with equal sample sizes, compared to the first situation considered, but it has practical consequences.

EXAMPLE

In the right panel of Figure 7.1, $\xi = 0.56$ (based on 20% trimmed means) indicating a large effect size. This is in contrast to δ, which indicates a relatively small effect size.

EXAMPLE

For the alcohol data in Table 7.4 we rejected the hypothesis of equal 20% trimmed means. If we use a standardized difference between the two groups based on the means and the standard deviation of the first group, we get $\hat{\delta}_1 = 0.4$. Using the standard deviation of the second group yields $\hat{\delta}_2 = 0.6$. So taken together, and assuming normality, these results suggest a medium effect size. The estimate of ξ is 0.44 suggesting that this robust, heteroscedastic measure of effect size is fairly large.

EXAMPLE

Consider again the first example in Section 7.4.2 dealing with the sexual attitude data. Recall that the sample sizes are $n_1 = 1327$ and $n_2 = 2282$, Welch's test returns a p-value of 0.30, but Yuen's test has a p-value less than 0.001. Cohen's effect size, d, is less than 0.0001. In contrast, $\hat{\delta}_t = 0.48$, suggesting a medium effect size and $\hat{\xi} = 0.47$, suggesting a large effect size.

A Robust, Heteroscedastic Measure of Effect Size

To get a robust version of the explanatory measure of effect size ξ, simply replace the mean with some robust measure of location and replace σ^2_{pool} with some robust measure of variation. Here, a 20% trimmed mean and a 20% Winsorized variance are used, where the Winsorized variance is rescaled to estimate the usual variance, σ^2, when sampling from a normal distribution. Again, under normality and homoscedasticity, $\delta = 0.2$, 0.5, and 0.8 roughly correspond to $\xi = 0.1$, 0.3, and 0.50, respectively.

Probabilistic Measure of Effect Size

To conclude this section, consider again p, the probability that a randomly sampled observation from the first group is larger than an observation from the second, which was discussed in Section 7.8. A criticism of effect size measures like Cohen's d is that they might not provide a relatively useful perspective regarding how the groups differ. Consider, for example, a situation where the goal is to compare two methods for treating some medical condition. Acion et al. (2006) argue that often what is needed is some sense of how likely it is that a particular treatment will be beneficial compared to some other treatment, and p provides a simple and easily understood measure that satisfies this goal. (Additional arguments for using p are given in Cliff, 1996.)

7.11.1 R Functions yuenv2 and akp.effect

The R function

```
yuenv2(x,y)
```

is exactly like the R function **yuen**, only it also reports the effect size ξ. When dealing with medians, the R function

```
med.effect(x,y)
```

can be used to compute ξ. For the computational details, see Wilcox (2017). The R function

```
akp.effect(x,y, EQVAR=TRUE, tr=0.2)
```

computes the measure of effect size δ_t, which defaults to using a 20% trimmed mean. If the argument `EQVAR=FALSE`, the function returns both δ_1 and δ_2. Setting `tr=0`, `akp.effect` returns Cohen's d when `EQVAR=TRUE`.

7.12 COMPARING CORRELATIONS AND REGRESSION SLOPES

Rather than compare measures of location or scale, situations arise where the goal is to compare correlations or regression parameters instead. That is, for every individual we have two measures and the goal is to determine whether the association for the first group differs from the association for the second. For example, in a study of schizophrenia, Dawson et al. (2000) were interested in, among other things, the association between prepulse inhibition and measures of schizophrenic symptoms. A portion of their study dealt with comparing correlations of individuals with positive symptoms to the correlation of those with negative symptoms. Also, comparing correlations or regression slopes is one strategy for determining whether a third variable modifies the association between two other variables. (See the example given at the end of this section.)

To state the goal in a more formal manner, imagine we have n_1 pairs of observations for the first group, yielding a correlation of r_1, and n_2 pairs of observations for the second group, yielding a correlation of r_2. The goal is to test

$$H_0 : \rho_1 = \rho_2,$$

the hypothesis that the two groups have equal population correlation coefficients. If we reject, this indicates that the strength of the association differs, but for reasons outlined in Chapter 6, how the associations differ is vague and unclear. Many methods for comparing correlations have been studied (e.g., Yu and Dunn, 1982; Duncan and Layard, 1973), but most methods are known to be unsatisfactory. A relatively well-known approach is based on Fisher's r-to-z transformation, but it can be highly unsatisfactory for reasons indicated in Section 6.5. Currently, one of the more effective approaches is to use a percentile bootstrap method, but adjusted to take into account the total number of observations (Wilcox and Muska, 2002; Wilcox, 2009c).

To apply the modified percentile bootstrap method, let $N = n_1 + n_2$ be the total number of pairs of observations. For the jth group, generate a bootstrap sample of n_j pairs of observations as described in Section 6.9. Let r_1^* and r_2^* represent the resulting correlation coefficients and set

$$D^* = r_1^* - r_2^*.$$

Repeat this process 599 times yielding $D_1^*, \ldots, D_{599}^*$. Then a 0.95 confidence interval for the difference between the population correlation coefficients ($\rho_1 - \rho_2$) is

$$(D_{(\ell)}^*, D_{(u)}^*),$$

where $\ell = 7$ and $u = 593$ if $N < 40$; $\ell = 8$ and $u = 592$ if $40 \le N < 80$; $\ell = 11$ and $u = 588$ if $80 \le N < 180$; $\ell = 14$ and $u = 585$ if $180 \le N < 250$; and $\ell = 15$ and $u = 584$ if $N \ge 250$. If the resulting confidence interval does not contain zero, reject the hypothesis of equal correlations. Note that this is just a simple modification of the method used to compute a confidence interval for Pearson's correlation that was described in Section 6.9.

As for testing the hypothesis of equal slopes, again numerous methods have been proposed (see, for example, Chow, 1960; Conerly and Mansfield, 1988; Oberhelman and Kadiyala, 2007; Ng and Wilcox, 2010), most of which are known to be relatively unsatisfactory. (Evidently there are no published results on how the method advocated by Oberhelman and Kadiyala, 2007, performs under nonnormality.) When using least squares regression, a method that performs relatively well and allows heteroscedasticity is the modified bootstrap method just described. That is, simply proceed as was done when working with Pearson's correlation, only replace r with the least squares estimate of the slope.

Among the methods considered by Ng and Wilcox (2010), an alternative method for comparing the slopes should be mentioned, which is based in part on a wild bootstrap method. It performs a bit better than the modified percentile bootstrap method in terms of avoiding Type I errors below the nominal level, which might translate into more power. In general the actual Type I error probability tends to be closer to the nominal than all of the other methods they considered. The approach is based on what is sometimes called the *moderated multiple regression* model, which belongs to what is known as the *general linear model*. The moderated multiple regression model is

$$Y = \beta_0 + \beta_1 X_i + \beta_2 G_i + \beta_3 X_i G_i + e, i = 1, \ldots n,$$

where X_i is the predictor variable and G_i is a *dummy variable*; $G_i = 0$ if a pair of observations (X_i, Y_i) comes from the first group, otherwise $G_i = 1$. When $G_i = 0$, the model becomes

$$Y_i = \beta_0 + \beta_1 X_i + e;$$

and when $G_i = 1$ the model is

$$Y_i = (\beta_0 + \beta_2) + (\beta_1 + \beta_3) X_i + e.$$

So if $\beta_3 = 0$, both groups have slope β_1 with different intercepts if $\beta_2 \neq 0$. That is, the hypothesis that both groups have equal slopes is

$$H_0 : \beta_3 = 0.$$

Ng and Wilcox illustrate that Student's T performs very poorly when there is heteroscedasticity. What performed best, in terms of controlling the probability of a Type I error, was the wild bootstrap method introduced in Section 6.10 used in conjunction with the HC4 estimate of the standard error.

7.12.1 R Functions twopcor, twolsreg, and tworegwb

The R function

```
twopcor(x1,y1,x2,y2)
```

computes a confidence interval for the difference between two Pearson correlations corresponding to two independent groups using the modified bootstrap method just described. The data for group 1 are stored in the R variables `x1` and `y1`, and the data for group 2 are stored in x2 and y2. The R function

```
twolsreg(x1,y1,x2,y2)
```

computes a confidence interval for the difference between the slopes based on the least squares estimator described in Chapter 6. The R function

```
tworegwb(x1,y1,x2,y2)
```

tests the hypothesis of equal slopes using the heteroscedastic wild bootstrap method in conjunction with the moderated multiple regression model.

EXAMPLE

In an unpublished study by L. Doi, there was interest in whether a measure of orthographic ability (Y) is associated with a measure of sound blending (X). Here we consider whether an auditory analysis variable (Z) *modifies* the association between X and Y. One way of approaching this issue is to partition the pairs of points (X, Y) according to whether Z is less than or equal to 14, or greater than 14, and then enter the resulting pairs of points into the R function `twopcor`. The 0.95 confidence interval for $\rho_1 - \rho_2$, the difference between the correlations, is $(-0.64, 0.14)$. This interval contains zero so we would not reject the hypothesis of equal correlations. If we compare regression slopes instead, the 0.95 confidence interval is $(-0.55, 0.18)$ and again we fail to reject. It is stressed, however, that this analysis does not establish that the association does not differ for the two groups under study. A concern is that power might be low when attention is restricted to Pearson's correlation or least squares regression. (Methods covered in subsequent chapters indicate that the measure of auditory analysis does modify the association between orthographic ability and sound blending.)

7.13 COMPARING TWO BINOMIALS

This final section considers the problem of comparing the probability of success associated with two independent binomials. For example, if the probability of surviving an operation using method 1 is p_1, and if the probability of surviving using method 2 is p_2, do p_1 and p_2 differ, and if they do differ, by how much? As another example, to what degree do men and women differ in whether they believe the President of the United States is an effective leader?

Many methods have been proposed for comparing binomials, two of which are described here. These two methods were chosen based on results in Storer and Kim (1990) and Beal (1987) where comparisons of several methods were made. It is noted, however, that competing methods have been proposed that apparently have not been compared directly to the methods covered here (e.g., Berger, 1996; Coe and Tamhane, 1993; Santner et al., 2007).

7.13.1 Storer–Kim Method

The Storer–Kim method tests H_0: $p_1 = p_2$ as follows. Imagine there are r_1 successes among n_1 trials in the first group and r_2 successes among n_2 trials in the second. So the estimates of p_1 and p_2 are $\hat{p}_1 = r_1/n_1$ and $\hat{p}_2 = r_2/n_2$. Roughly, the strategy is to determine the probability of getting an estimated difference between the probabilities greater than or equal to $|\hat{p}_1 - \hat{p}_2|$ when the null hypothesis is true. This can be done if the common probability is known. It is not known, but it can be estimated using the available data.

To elaborate, note that the possible number of successes in the first group is any integer, x, between 0 and n_1, and for the second group it is any integer, y, between 0 and n_2. For any x between 0 and n_1 and any y between 0 and n_2, set

$$a_{xy} = 1$$

if

$$\left|\frac{x}{n_1} - \frac{y}{n_2}\right| \geq \left|\frac{r_1}{n_1} - \frac{r_2}{n_2}\right|;$$

otherwise

$$a_{xy} = 0.$$

Let

$$\hat{p} = \frac{r_1 + r_2}{n_1 + n_2}.$$

The test statistic is

$$T = \sum_{x=0}^{n_1} \sum_{y=0}^{n_2} a_{xy} b(x,\ n_1,\ \hat{p}) b(y,\ n_2,\ \hat{p}),$$

where

$$b(x, n_1, \hat{p}) = \binom{n_1}{x} \hat{p}^x (1-\hat{p})^{n_1-x},$$

and $b(y, n_2, \hat{p})$ is defined in an analogous fashion. The null hypothesis is rejected if

$$T \leq \alpha.$$

That is, T is the p-value.

7.13.2 Beal's Method

Following the notation used to describe the Storer–Kim method, let $\hat{p}_1 = r_1/n_1$, $\hat{p}_2 = r_2/n_2$ and let $c = z^2_{1-\alpha/2}$ where $z_{1-\alpha/2}$ is the $1-\alpha$ quantile of a standard normal distribution. (So c is the $1-\alpha$ quantile of a chi-squared distribution with one degree of freedom.) Compute

$$a = \hat{p}_1 + \hat{p}_2$$

$$b = \hat{p}_1 - \hat{p}_2$$

$$u = \frac{1}{4}\left(\frac{1}{n_1} + \frac{1}{n_2}\right)$$

$$v = \frac{1}{4}\left(\frac{1}{n_1} - \frac{1}{n_2}\right)$$

$$V = u\{(2-a)a - b^2\} + 2v(1-a)b$$

$$A = \sqrt{c\{V + cu^2(2-a)a + cv^2(1-a)^2\}}$$

$$B = \frac{b + cv(1-a)}{1+cu}.$$

Beal's $1-\alpha$ confidence interval for $p_1 - p_2$ is

$$B \pm \frac{A}{1+cu}.$$

The choice between these two methods is not completely clear. An appeal of Beal's method is that it provides a confidence interval, but the Storer–Kim method does not. Situations arise in subsequent chapters where the Storer–Kim method has less power than Beal's method when comparing multiple groups of individuals, but when comparing two groups only, there are situations where the Storer–Kim method rejects and Beal's method does not. There are indications that the method derived by Kulinskaya, Morgenthaler and Staudte (2010), referred to as method KMS henceforth, has a practical advantage over Beal's method, but this issue needs further study. The computational details are not provided but an R function for applying it is supplied.

7.13.3 R Functions twobinom, twobici, bi2KMSv2, and power.prop.test

The R function

```
twobinom(r1 = sum(x), n1 = length(x), r2 = sum(y), n2 = length(y), x = NA, y
                                      = NA)
```

has been supplied to test H_0: $p_1 = p_2$ using the Storer–Kim method. The function can be used either by specifying the number of successes in each group (arguments `r1` and `r2`) and the sample sizes (arguments `n1` and `n2`), or the data can be in the form of two vectors containing 1s and 0s, in which case you use the arguments `x` and `y`. Beal's method can be applied with the R function

```
twobici(r1 = sum(x), n1 = length(x), r2 = sum(y), n2 = length(y), x = NA, y
                    = NA, alpha = 0.05)
```

The R function

```
bi2KMS(r1=sum(elimna(x)), n1=length(elimna(x)), r2=sum(elimna(y)),
       n2=length(elimna(y)), x=NA, y=NA, alpha=0.05)
```

applies method KMS.

EXAMPLE

If for the first group we have 7 successes among 12 observations and for the second group we have 22 successes among 25 observations, then the command `twobinom(7,12,22,25)` returns a p-value equal to 0.044. This is less than 0.05, so reject with $\alpha = 0.05$. The 0.95 confidence interval for $p_1 - p_2$ is $(-0.61, 0.048)$. This interval contains zero, so in contrast to the Storer–Kim method, fail to reject the hypothesis H_0: $p_1 = p_2$, the only point being that different conclusions might be reached depending on which method is used. The p-value returned by method KMS is 0.07.

EXAMPLE

In Table 7.2 we see that 101 of the 156 females responded that they want one sexual partner during the next 30 years. As for the 105 males in this study, 49 gave the response 1. Does the probability of a 1 among males differ from the probability among females? The R function `twobinom` returns a p-value of 0.0037 indicating that the probabilities differ even with $\alpha = 0.0037$. The command `twobici(49,105,101,156)` returns a 0.95 confidence interval of $(-0.33, -0.04)$, so again we reject, but there is some possibility that the difference between the two probabilities is fairly small.

The built-in R function

```
power.prop.test(n = NULL, p1 = NULL, p2 = NULL, sig.level = 0.05, power =
    NULL, alternative = c("two.sided", "one.sided"), strict = FALSE)
```

can be used to judge power and sample sizes, where the argument `n` is the sample size for each group, `p1` is the probability of success for group 1, and `p2` is the probability of success for group 2.

EXAMPLE

The R command

```
power.prop.test(n = 50, p1 = 0.50, p2 = 0.75)
```

indicates that if the first group has a probability of success 0.5, and the other has probability of success 0.75, and if the sample size for both groups is 50, then power is 0.75 when testing at the 0.05 level. This is a rough approximation of power in the sense that it is not based on Beal's method or the Storer–Kim method.

7.13.4 Comparing Two Discrete Distributions

For a binomial distribution, only two outcomes are possible. This section briefly comments on the situation where there are $K > 2$ possible values associated with the sample space.

Let p_{jk} denote the probability that for the jth group, the kth possible response occurs. The goal is to test

$$H_0 : p_{1k} = p_{2k}$$

for every $k = 1, \ldots, K$. Consider, for example, a Likert scale where participants rate a car's performance based on a five-point scale: 1, 2, 3, 4, 5. Now imagine that independent groups rate two different brands of cars. A possible goal is to test the global hypothesis that each possible response has the same probability of occurring. One way of accomplishing this goal is to use the Kolmogorov–Smirnov test in Section 7.8.8. It is briefly noted that an alternative approach is based on a so-called chi-squared test, but the computational details are not described. However, an R function for applying the method is provided. A more detailed understanding of how the groups compare can be achieved with the method in Section 12.1.17.

7.13.5 R Function disc2com

The R function

```
disc2com(x,y)
```

performs the chi-squared test mentioned in the previous section. (The function `binband` in Section 12.1.18 can provide a more detailed understanding of how the groups compare.)

7.14 MAKING DECISIONS ABOUT WHICH METHOD TO USE

Numerous methods have been described for comparing two independent groups. How does one choose which method to use? There is no agreed upon strategy for addressing this issue, but a few comments might help.

First, and perhaps most obvious, consider what you want to know. If, for example, there is explicit interest in knowing something about the probability that an observation from the first group is smaller than an observation from the second, use Cliff's method or the Brunner–Munzel technique. Despite the many problems with methods for comparing means, it might be that there is explicit interest in the means, as opposed to other measures of location, for reasons discussed in Chapter 2. If this is case, the R function `yuenbt` seems to be a relatively good choice for general use (setting the argument `tr=0`) with the understanding that all methods for means can result in relatively low poor, inaccurate confidence intervals, and unsatisfactory control over the probability of a Type I error. But if any method based on means rejects, it seems reasonable to conclude that the distributions differ in some manner. A possible argument for using other methods for comparing means is that they might detect differences in the distributions that are missed by other techniques (such as differences in skewness). If, for example, boxplots indicate that the groups have no outliers and a similar amount of skewness, a reasonable speculation is that using other measures of location will make little or no difference. But the only way to be sure is to actually compare groups with another measure of location.

If the goal is to maximize power, a study by Wu (2002), where data from various dissertations were reanalyzed, indicates that comparing groups with a 20% trimmed mean is likely to be best, but the only certainty is that exceptions occur. Generally, the method that has the highest power depends on how the groups differ, which is unknown. With sufficiently heavy tails, methods based on the median might have more power. To complicate matters, situations are encountered where a rank-based method has the most power. Keep in mind that for skewed distributions, comparing means is not the same as comparing trimmed means or medians.

In terms of controlling the probability of a Type I error and getting accurate confidence intervals, all indications are that when comparing identical distributions, it makes little difference which method is used. If, however, the groups differ in some manner, the choice of method might make a substantial difference in terms of both power and accurate probability coverage. If the goal is to make inferences about a measure of location, without being sensitive to heteroscedasticity and skewness, methods based on a 20% trimmed mean seem best for general use. The bootstrap method for medians also performs well based on this criterion. Generally it is better to use a method that allows heteroscedasticity rather than a method that assumes homoscedasticity. Bootstrap methods appear to have an advantage over non-bootstrap methods when sample sizes are small. Just how large the sample sizes must be so that non-bootstrap methods generally work as well as bootstrap methods is unknown. Also, keep in mind that testing assumptions, in order to justify a method based on means, is satisfactory only if the method for testing assumptions has sufficiently high power. Determining whether this is the case is difficult at best.

An issue of some importance is whether it is sufficient to use a single method for summarizing how groups differ. Different methods provide different perspectives, as was illustrated. So one strategy might be to focus on a single method for deciding whether groups differ and then use other methods, such as rank-based techniques or a shift function, to get a deeper understanding of how and where the groups differ.

A criticism of performing multiple tests is that as the number of tests performed increases, the more likely it is that at least one test will reject even when the groups do not differ. If we perform three tests at the 0.05 level, the probability of at least one Type I error is greater than 0.05. (Chapter 12 discusses this issue in more detail.) However, if several tests reject at the 0.05 level, say, and if each test controls the probability of a Type I error, then it is reasonable to conclude that the groups differ and that the probability of at least one Type I error, among all the tests performed, does not exceed 0.05. (This conclusion is justified based on sequentially rejective methods in Chapter 12.) Also, if a method fails to reject, it seems unwise to not consider whether other methods find a difference. At a minimum, using alternative methods might provide an indication of where to look for differences when conducting future investigations. Moreover, different methods provide alternative perspectives on how groups differ.

A final suggestion is to take advantage of the various plots that were described. They can be invaluable for understanding how groups differ.

7.15 EXERCISES

1. Suppose that the sample means and variances are $\bar{X}_1 = 15$, $\bar{X}_2 = 12$, $s_1^2 = 8$, $s_2^2 = 24$ with sample sizes $n_1 = 20$ and $n_2 = 10$. Verify that $s_p^2 = 13.14$, $T = 2.14$ and that Student's T test rejects the hypothesis of equal means with $\alpha = 0.05$.

2. For two independent groups of subjects, you get $\bar{X}_1 = 45$, $\bar{X}_2 = 36$, $s_1^2 = 4$, $s_2^2 = 16$ with sample sizes $n_1 = 20$ and $n_2 = 30$. Assume the population variances of the two groups are equal and verify that the estimate of this common variance is 11.25.

3. Still assuming equal variances, test the hypothesis of equal means using the data in the last exercise assuming random sampling from normal distributions. Use $\alpha = 0.05$.

4. Repeat the last exercise, only use Welch's test for comparing means.

5. Comparing the test statistics for the last two exercises, what do they suggest regarding the power of Welch's test versus Student's T test for the data being examined?

6. For two independent groups of subjects, you get $\bar{X}_1 = 86$, $\bar{X}_2 = 80$, $s_1^2 = s_2^2 = 25$, with sample sizes $n_1 = n_2 = 20$. Assume the population variances of the two groups are equal and verify that Student's T rejects with $\alpha = 0.01$.

7. Repeat the last exercise using Welch's method.

8. Comparing the results of the last two exercises, what do they suggest about using Student's T versus Welch's method when the sample variances are approximately equal?

9. If for two independent groups, you get $\bar{X}_{t1} = 42$, $\bar{X}_{t2} = 36$, $s_{w1}^2 = 25$, $s_{w2}^2 = 36$, $n_1 = 24$, and $n_2 = 16$, test the hypothesis of equal trimmed means with $\alpha = 0.05$.

10. Referring to the last exercise, compute a 0.99 confidence interval for the difference between the trimmed means.

11. For $\bar{X}_1 = 10$, $\bar{X}_2 = 5$, $s_1^2 = 21$, $s_2^2 = 29$, $n_1 = n_2 = 16$, compute a 0.95 confidence interval for the difference between the means using Welch's method and state whether you would reject the hypothesis of equal means.

12. Repeat the last exercise, only use Student's T instead.

13. Two methods for training accountants are to be compared. Students are randomly assigned to one of the two methods. At the end of the course, each student is asked to prepare a tax return for the same individual. The returns reported by the students are

Method	Returns							
1	132	204	603	50	125	90	185	134
2	92	−42	121	63	182	101	294	36

Using Welch's test, would you conclude that the methods differ in terms of the average return? Use $\alpha = 0.05$.

14. Repeat the last exercise, only compare 20% trimmed means instead.

15. You compare lawyers versus professors in terms of job satisfaction and fail to reject the hypothesis of equal means or equal trimmed means. Does this mean it is safe to conclude that the typical lawyer has about the same amount of job satisfaction as the typical professor?

16. Responses to stress are governed by the hypothalamus. Imagine you have two groups of participants. The first shows signs of heart disease and the other does not. You want to determine whether the groups differ in terms of the weight of the hypothalamus. For the first group of participants with no heart disease, the weights are

11.1, 12.2, 15.5, 17.6, 13.0, 7.5, 9.1, 6.6, 9.5, 18.0, 12.6.

For the other group with heart disease, the weights are

18.2, 14.1, 13.8, 12.1, 34.1, 12.0, 14.1, 14.5, 12.6, 12.5, 19.8, 13.4, 16.8, 14.1, 12.9.

Determine whether the groups differ based on Welch's test. Use $\alpha = 0.05$.

17. Repeat the previous exercise, only use Yuen's test with 20% trimmed means.
18. Use Cohen's d to measure effect size using the data in the previous two exercises.
19. Published studies indicate that generalized brain dysfunction may predispose someone to violent behavior. Of interest is determining which brain areas might be dysfunctional in violent offenders. In a portion of such a study conducted by Raine, Buchsbaum, and LaCasse (1997), glucose metabolism rates of 41 murderers were compared to the rates for 41 control participants. Results for the left hemisphere, lateral prefrontal region of the brain yielded a sample mean of 1.12 for the controls and 1.09 for the murderers. The corresponding standard deviations were 0.05 and 0.06. Verify that when using Student's T, $T = 2.45$, and that you reject with $\alpha = 0.05$.
20. In the previous exercise, you rejected the hypothesis of equal means. What does this imply about the accuracy of the confidence interval for the difference between the population means based on Student's T?
21. For the data in Table 7.6, if we assume that the groups have a common variance, verify that the estimate of this common variance is $s_p^2 = 236$.
22. The sample means for the data in Table 7.6 are 22.4 and 11. If we test the hypothesis of equal means using Student's T, verify that $T = 2.5$ and that you would reject with $\alpha = 0.05$.
23. Verify that the 0.95 confidence interval for the difference between the means, based on the data in Table 7.6 and Student's T, is (2.2, 20.5). What are the practical concerns with this confidence interval?
24. Student's T rejects the hypothesis of equal means based on the data in Table 7.6. Interpret what this means.
25. For the data in Table 7.6, the are 0.95 confidence interval for the difference between the means based on Welch's method is (1.96, 20.83). Check this result with the R function `yuen`.
26. In the previous exercise you do not reject the hypothesis of equal variances. Why is this *not* convincing evidence that the assumption of equal variances, when using Student's T, is justified?
27. The 20% Winsorized standard deviation (s_w) for the first group in Table 7.6 is 1.365 and for the second group it is 4.118. Verify that the 0.95 confidence interval for the difference between the 20% trimmed means, using Yuen's method, is (5.3, 22.85).
28. Create a boxplot of the data in Table 7.6 and comment on why the probability coverage, based on Student's T or Welch's method, might differ from the nominal α level.
29. For the self-awareness data in Table 7.5, verify that the R function `yuenbt`, with the argument `tr` set to 0, returns $(-571.4, 302.5)$ as a .95 confidence interval for the difference between the means.

30. For the data in Table 7.5, use the R function comvar2 to verify that the 0.95 confidence interval for the difference between the variances is (−1165766.8, 759099.7).

31. Describe a general situation where comparing medians will have more power than comparing means or 20% trimmed means.

32. For the data in Table 7.4, verify that the 0.95 confidence interval for the difference between the biweight midvariances is (−154718, 50452).

33. The last example in Section 7.5.5 dealt with comparing males to females regarding the desired number of sexual partners over the next 30 years. Using Student's T, we fail to reject which is not surprising because there is an extreme outlier among the responses given by males. If we simply discard this one outlier and compare groups using Student's T or Welch's method, what criticism might be made even if we could ignore problems with nonnormality?

34. Two methods for reducing shoulder pain after laparoscopic surgery were compared by Jorgensen et al. (1995). The data were

Group 1:	1 2 1 1 1 1 1 1 1 1 1 2 4 1 1
Group 2:	3 3 4 3 1 2 3 1 1 5 4

Verify that both the Wilcoxon–Mann–Whitney test and Cliff's method reject at the 0.05 level. Although the Kolmogorov–Smirnov test rejects with $\alpha = 0.05$, why might you suspect that the Kolmogorov–Smirnov test will have relatively low power in this particular situation?

35. Imagine that two groups of cancer patients are compared, the first group having a rapidly progressing form of the disease and the other having a slowly progressing form instead. At issue is whether psychological factors are related to the progression of cancer. The outcome measure is one where highly negative scores indicate a tendency to present the appearance of serenity in the presence of stress. The results are

Group 1:	−25	−24	−22	−22	−21	−18	−18	−18
	−18	−17	−16	−14	−14	−13	−13	−13
	−13	−9	−8	−7	−5	1	3	7
	7							
Group 2:	−21	−18	−16	−16	−16	−14	−13	−13
	−12	−11	−11	−11	−9	−9	−9	−9
	−7	−6	−3	−2	3	10		

Verify that the Wilcoxon–Mann–Whitney test rejects at the 0.05 level, but none of the other methods in Section 7.8 reject. What might explain this?

36. Two independent groups are given different cold medicines and the goal is to compare reaction times. Suppose that the decreases in reaction times when taking drug A versus drug B are as follows.

A: 1.96, 2.24, 1.71, 2.41, 1.62, 1.93

B: 2.11, 2.43, 2.07, 2.71, 2.50, 2.84, 2.88

Verify that the Wilcoxon–Mann–Whitney test rejects with $\alpha = 0.05$. Estimate the probability that a randomly sampled participant receiving drug A will have a smaller reduction in reaction time than a randomly sampled participant receiving drug B. Verify that the estimate is 0.9.

37. For the data in the previous exercise, verify that the p-value based on Cliff's method is 0.01 and for the Brunner–Munzel method the p-value is less than 0.001.

38. R contains data, stored in the R variable `sleep`, which show the effect of two soporific drugs, namely the increase in hours of sleep compared to control. Create a boxplot for both groups using the command `boxplot(sleep[[1]] ~ sleep[[2]])` and speculate about whether comparing 20% trimmed means will give a result similar to Welch's test. Use the R command `z=fac2list(sleep[,1],sleep[,2])` to separate the data into two groups and use the function `yuen` to check your answer.

CHAPTER 8

COMPARING TWO DEPENDENT GROUPS

CONTENTS

This chapter deals with the problem of comparing two dependent groups. Consider, for example, a study aimed at assessing the effect of endurance training on triglyceride levels. Imagine that the triglyceride level for each of 30 participants is measured before training begins and again after four weeks of training. This is an example of what is called a *repeated measures* or *within subjects* design, simply meaning that we repeatedly measure the same individuals over time. Let $\bar{X}_1$ and $\bar{X}_2$ be the sample means before and after training, respectively, and let μ_1 and μ_2 be the corresponding population means. One way of assessing the effects of training is to test

$$H_0 : \mu_1 = \mu_2. \tag{8.1}$$

But because the same participants are measured before and after training, it is unreasonable to assume that the sample means are independent, in which case the methods in Chapter 7 (Student's T and Welch's test) are inappropriate even under normality. They use the wrong standard error.

There are several general ways two dependent groups might be compared:

1. Compare measures of location, such as the mean or median.
2. Compare measures of variation.
3. Focus on the probability that for a randomly sampled pair of observation, the first value is less than the second.
4. Use a rank-based method to test the hypothesis that the marginal distributions are identical.
5. Simultaneously compare all of the quantiles to get a global sense of where the distributions differ and by how much.

All of the methods in Chapter 7 have analogs for the case of two dependent groups, which are summarized in this chapter.

A description of some commonly used notation is needed to illustrate some differences among the methods in this chapter. Here it is assumed that we have n randomly sampled pairs of observations:

$$(X_{11}, X_{12}), (X_{21}, X_{22}), \ldots, (X_{n1}, X_{n2}), \tag{8.2}$$

where each pair of observations might be dependent. These n pairs of values might represent, for example, n married couples, where $X_{11}, \ldots, X_{n1}$ denotes the responses given by the wives and $X_{12}, \ldots, X_{n2}$ are the responses given by their spouses. Next we form the *difference scores.* That is, for each pair of observations we take their difference and denote the results by

$$\begin{aligned}
\mathrm{D}_1 &= X_{11} - X_{12} \\
\mathrm{D}_2 &= X_{21} - X_{22} \\
\mathrm{D}_3 &= X_{31} - X_{32} \\
&\vdots \\
\mathrm{D}_n &= X_{n1} - X_{n2}.
\end{aligned}$$

Let $\bar{X}_1$ and $\bar{X}_2$ represent the sample mean of the first group (wives in the example) and the second group (husbands), respectively, and let $\bar{D}$ denote the mean of the differences. It is readily verified that

$$\bar{D} = \bar{X}_1 - \bar{X}_2.$$

Moreover, if the population means corresponding to $\bar{X}_1$, $\bar{X}_2$, and $\bar{D}$ are denoted by μ_1, μ_2, and μ_D, respectively, then

$$\mu_D = \mu_1 - \mu_2.$$

However, under general conditions, these results do not generalize to any of the robust measures of location introduced in Chapter 2. For example, if M_1, M_2, and M_D represent the sample medians of the first group, the second group, and the difference scores, respectively, then typically (but not always)

$$M_D \neq M_1 - M_2.$$

The same is true when the sample median is replaced by a 20% trimmed mean or an M-estimator, a result that will be seen to be relevant when making decisions about which method to use.

EXAMPLE

As a simple illustration, consider the following values:

i	X_{i1}	X_{i2}	$D_i = X_{i1} - X_{i2}$
1	4	6	-2
2	7	11	-4
3	6	7	-1
4	4	9	-5
5	9	9	0

Then $M_1 = 6$, $M_2 = 9$, and $M_D = -2$. So we see that $M_D \neq M_1 - M_2$.

8.1 THE PAIRED T TEST

The most common method for comparing two dependent groups is based on what is called the *paired T test.* The basic idea is that, even when the pairs of observations are dependent, testing H_0: $\mu_1 = \mu_2$, the hypothesis that the means of the marginal distributions are equal is the same as testing H_0: $\mu_D = 0$, the hypothesis that the difference scores have a mean of zero. Assuming normality, this latter hypothesis can be tested by applying Student's T using the differences $D_1, \ldots, D_n$. To make sure the details are clear, the following summary of the paired T test is provided.

Goal: For two dependent groups, test $H_0 : \mu_1 = \mu_2$, the hypothesis that they have equal means.

Assumptions: Pairs of observations are randomly sampled with each group having a normal distribution, in which case D has a normal distribution as well.

Computational Details: Compute the mean and sample variance of the D_i values:

$$\bar{D} = \frac{1}{n}\sum_{i=1}^{n} D_i$$

and

$$s_D^2 = \frac{1}{n-1}\sum_{i=1}^{n}(D_i - \bar{D})^2.$$

Next, compute the test statistic

$$T_D = \frac{\bar{D}}{s_D/\sqrt{n}}.$$

The critical value is t, the $1-\alpha/2$ quantile of Student's t distribution with $\nu = n-1$ degrees of freedom.

Decision Rule: The hypothesis of equal means is rejected if $|T_D| \geq t$.

Confidence Interval: A $1-\alpha$ confidence interval for $\mu_1 - \mu_2$, the difference between the means, is

$$\bar{D} \pm t\frac{s_D}{\sqrt{n}}.$$

As just indicated, the paired T test has $n - 1$ degrees of freedom. If the groups are independent, Student's T test (in Chapter 7) has $2(n-1)$ degrees of freedom, twice as many as the paired T test. This means that if we compare independent groups (having equal sample sizes) using the paired T test, power will be lower compared to using Student's T in Chapter 7. However, if the correlation between the observations is sufficiently high, the paired T test will have more power. That is, given a choice when designing an experiment, comparing dependent groups is preferable to comparing independent groups. To provide an indication of why, it is noted that when pairs of observations are dependent, the squared standard error of the difference between the sample means is

$$\text{VAR}(\bar{X}_1 - \bar{X}_2) = \frac{\sigma_1^2 + \sigma_2^2 - 2\rho\sigma_1\sigma_2}{n},$$

where σ_1^2 and σ_2^2 are the variances associated with groups 1 and 2, respectively. From this last equation we see that as the correlation, ρ, increases, the variance of the difference between the sample means (the squared standard error of $\bar{X}_1 - \bar{X}_2$) goes down. As noted in Chapters 5 and 7, as the standard error goes down, power goes up. In practical terms, the loss of degrees of freedom when using the paired T test will be more than compensated for if the correlation is reasonably high. Just how high it must be in order to get more power with the paired T test is not completely clear. For normal distributions, a rough guideline is that when $\rho > 0.25$, the paired T test will have more power (Vonesh, 1983). Note, however, that if the correlation is negative, the squared standard error is even higher than the situation where $\rho = 0$ meaning that comparing independent groups with Student's T in Chapter 7 is preferable to the paired T test.

8.1.1 When Does the Paired T Test Perform Well?

The good news about the paired T test is that if the observations in the first group (the X_{i1} values) have the same population distribution as the observations in the second group, so in particular they have equal means, variances, and the same amount of skewness, generally Type I error probabilities substantially higher than the nominal level can be avoided. The reason is that for this special case, the difference scores (the D_i values) have a symmetric (population) distribution in which case methods based on means perform reasonably well in terms of avoiding Type I error probabilities substantially higher than the nominal level, as indicated in Chapter 5. However, as was the case in Chapter 7, practical problems can arise when the two groups (or the two dependent variables) differ in some manner. Again, arbitrarily small departures from normality can destroy power, even when comparing groups having symmetric distributions. If groups differ in terms of skewness, the difference scores (the D_i values) have a skewed distribution, in which case the paired T test can be severely biased, meaning that power can actually decrease as the difference between the means gets large. Yet another problem is poor probability coverage when computing a confidence interval for the difference between the means. If the goal is to test the hypothesis that the two variables under study have identical distributions, the paired T test is satisfactory in terms of Type I errors. But if we reject, there is doubt as to whether this is primarily due to the difference between the means or some other way the distributions differ. With a large enough sample size, these concerns become negligible, but just how large the sample size must be is difficult to determine.

8.1.2 R Function t.test

The built-in R function

```
t.test(x)
```

can be used to perform the paired T test. If the data are stored in the R variables x and y, then the command

```
t.test(x-y)
```

accomplishes this goal. Alternatively, use the command

```
t.test(x,y,paired = TRUE).
```

EXAMPLE

R can be used to illustrate the effects of different amounts of skewness on the paired T test. Here we generate 20 observations from a standard normal distribution for the first group and another 20 observations from a lognormal distribution (shown in Figure 4.10) for the second group, shifted to have a population mean of zero. (So in this case the groups happen to be independent.) Next we apply the paired T test and note whether we reject at the 0.05 level. We repeat this 10,000 times and count the number of times we reject. If the method is performing as intended, we should reject about 500 times (5% of 10,000). Here is some R code that accomplishes this goal:

```
val=0
# Set the seed of the random number generator so
# that readers can duplicate the results.
# Or change the seed and see what happens
set.seed(2)
for(i in 1:10000){
y=rnorm(20) # Generate data from a standard normal distribution
x=rlnorm(20)-sqrt(exp(1)) # Generate data from a lognormal
# that is shifted to have mean zero.
res=t.test(x-y)
if(res[3]$p.value<=0.05)val=val+1
}
val
```

The number of rejections is indicated by the R variable `val`, which here is 865. That is, the probability of a Type I error is approximately 0.0865 even though we tested at the 0.05 level. Increasing the sample size to 30, the Type I error probability is now approximately equal to 0.0787.

Note that we can compare the means using instead the bootstrap-t method, introduced in Chapter 5. That is, use the R function `trimcibt` in conjunction with the difference scores. Theory and simulations indicate that control over the Type I error probability will tend to be better than the paired T test. (An exception is normality, where the paired T test provides exact control over the Type I error probability.)

EXAMPLE

We repeat the last example, only now we use the bootstrap-t method via the R function `trimcibt` with the argument `tr=0`. The Type I error probability is now 0.080, again with a sample size of 20.

8.2 COMPARING ROBUST MEASURES OF LOCATION

As in previous chapters, one way of dealing with the risk of low power when comparing means is to use 20% trimmed means, medians, or M-estimators instead. A possible appeal of methods for comparing 20% trimmed means is that by design, they perform about as well as the paired T test for means when the normality assumption is true. But in fairness, if a distribution is sufficiently light-tailed, comparing means might have a power advantage. If distributions are sufficiently heavy-tailed, medians or M-estimators might provide more power. As usual, if distributions are skewed, comparing means is not the same as comparing a 20% trimmed mean or other robust measures of location. Despite any concerns about the mean, there are situations where it is more appropriate than a 20% trimmed mean as discussed in Chapter 2. But to get relatively high power over a broad range of conditions, as well as improved control over the Type I error probability, using a 20% trimmed mean has appeal.

For convenience, the focus is on trimmed means, but the methods about to be described extend to medians and M-estimators as will be made clear. If two dependent groups have identical distributions, then the population trimmed mean of the difference scores is equal to the difference between the individual trimmed means, which is zero. In symbols, $\mu_{tD} = \mu_{t1} - \mu_{t2} = 0$. However, if the distributions differ, in general it will be the case that $\mu_{tD} \neq \mu_{t1} - \mu_{t2}$. In practical terms, computing a confidence interval for μ_{tD} is not necessarily the same as computing a confidence interval for $\mu_{t1} - \mu_{t2}$. The same is true when using medians or an M-estimator. So an issue is whether one should test

$$H_0 : \mu_{tD} = 0, \tag{8.3}$$

or

$$H_0 : \mu_{t1} = \mu_{t2}. \tag{8.4}$$

The latter approach is called comparing the *marginal trimmed means.* Put another way, the distributions associated with each of the dependent groups, ignoring the other group, are called *marginal distributions.* The goal is to compare the trimmed means of the marginal distributions.

To elaborate on the difference between these two hypotheses in more concrete terms, consider a study aimed at assessing feelings of life satisfaction based on randomly sampled pairs of brothers and sisters. If we test Equation (8.3), the goal in effect is to assess how the typical brother compares to his sister. Testing Equation (8.4) instead, the goal is to compare how the typical female compares to the typical male.

In terms of controlling the Type I error probability, currently it seems that there are only slight differences between the two approaches. A rough rule is that using difference scores, meaning that the goal is to test Equation (8.3), Type I errors tend be a bit higher compared to testing Equation (8.4). For example, under normality, if the pairs of observations have correlation $\rho = 0.4$, testing Equation (8.3) at the 0.05 level (using the R function `trimci`), the actual Type I error probability is approximately 0.059 compared to 0.044 when testing Equation (8.4) (using the R function `yuend` described momentarily). With $\rho = 0$, the actual Type I error probabilities are approximately 0.059 and 0.047, respectively. So in this particular case, comparing marginal trimmed means results in a Type I error probability that is closer to the nominal 0.05 level. But if the distributions are lognormal, the Type I error probabilities are approximately 0.043 and 0.033, suggesting that using difference scores might have more power. In terms of maximizing power, the optimal choice depends on how the groups differ, which, of course, is unknown. If forced to choose one approach over the other, with the goal of maximizing power, using difference scores seems to have an advantage. But perhaps situations arise where, from a substantive point of view, the choice between testing Equation (8.3) and (8.4) is relevant.

Note that testing Equation (8.3) is straightforward: compute the difference scores as done when applying the paired T test, and then apply the methods for trimmed means, or other measures of location, as described in Chapters 4 and 5. As a brief reminder, the relevant R functions are:

- `trimci` (non-bootstrap method for trimmed mean)
- `trimcibt` (bootstrap-t for trimmed mean)
- `trimpb` (percentile bootstrap for trimmed mean)
- `sintv2` (median)
- `mestci` (M-estimator)

Again assuming the data are stored in the R variables x and y, the command `sintv2(x-y)` will test the hypothesis that the difference scores have a median of zero.

There is, however, a practical appeal to testing the hypothesis of equal marginal trimmed means: When there are missing values, relatively simple but effective methods are available, which will be described in Section 8.3. There are strategies for handling missing values when using difference scores, but none seem to be satisfactory in terms of controlling Type I error probabilities. There are exceptions when comparing means, but it appears that extant simulation studies are not well designed for determining the effects of skewness and other violations of standard assumptions. Box 8.1 outlines how to test Equation (8.4) when there are no missing values.

It is noted that a bootstrap-t version of the method in Box 8.1 is available. Roughly, bootstrap samples are obtained by resampling pairs of observations as was done in Chapter 6 when dealing with Pearson's correlation. The basic idea is to determine the distribution of the test statistic T_y in Box 8.1 when the null hypothesis is true, which can be used to compute a confidence interval and a p-value. More details can be found in Wilcox (2005, section 5.9.5). A speculation is that in some situations this bootstrap-t method offers a practical advantage over the method in Box 8.1, but the extent to which this is true has not been studied.

When comparing the marginal medians, a special bootstrap method is needed that continues to perform well when tied values occur (Wilcox, 2006c). The details are not described but the R function `dmedpb` performs the calculations.

8.2.1 R Functions yuend, ydbt, and dmedpb

The R function

```
yuend(x,y,tr=0.2,alpha=0.05),
```

written for this book, compares the marginal trimmed means. That is, it tests H_0: $\mu_{t1} = \mu_{t2}$ using the method in Box 8.1. The function

```
ydbt(x,y,tr=0.2)
```

performs a bootstrap-t version. For the special case where the goal is to compare medians, use the function dmedpb(x,y, alpha=0.05, plotit = TRUE, dif = TRUE, xlab = "Group 1", ylab = "Group 2",).

(Section 12.4.4 describes some additional options when using this function.) By default, the function tests the hypothesis that the difference scores have a population median equal to zero. To compare the marginal medians, set the argument `dif=FALSE`.

Box 8.1: Comparing the marginal trimmed means.

Goal: Test H_0: $\mu_{t1} = \mu_{t2}$.
The data must be Winsorized in a manner that keeps observations paired together. In symbols, for group j ($j = 1$, 2), let $X_{(1)j} \le X_{(2)j} \le \cdots \le X_{(n)j}$ be the n values written in ascending order. Winsorizing the observations means computing

$$Y_{ij} = \begin{cases} X_{(g+1)j} & \text{if } X_{ij} \le X_{(g+1)j} \\ X_{ij} & \text{if } X_{(g+1)j} < X_{ij} < X_{(n-g)j} \\ X_{(n-g)j} & \text{if } X_{ij} \ge X_{(n-g)j}, \end{cases}$$

where as usual g is the number of observations trimmed or Winsorized from each end of the distribution corresponding to the jth group. The expression for Y_{ij} says that $Y_{ij} = X_{ij}$ if X_{ij} has a value between $X_{(g+1)j}$ and $X_{(n-g)j}$. If X_{ij} is less than or equal to $X_{(g+1)j}$, set $Y_{ij} = X_{(g+1)j}$, and if X_{ij} is greater than or equal to $X_{(n-g)j}$, set $Y_{ij} = X_{(n-g)j}$. Let

$$d_j = \frac{1}{h(h-1)} \sum_{i=1}^{n} (Y_{ij} - \bar{Y}_j)^2,$$

and

$$d_{12} = \frac{1}{h(h-1)} \sum_{i=1}^{n} (Y_{i1} - \bar{Y}_1)(Y_{i2} - \bar{Y}_2),$$

where $h = n - 2g$. The hypothesis of equal trimmed means can be tested with

$$T_y = \frac{\bar{X}_{t1} - \bar{X}_{t2}}{\sqrt{d_1 + d_2 - 2d_{12}}},$$

which is rejected if $|T_y| \ge t$, where t is the $1 - \alpha/2$ quantile of Student's t distribution with $h - 1$ degrees of freedom. A $1 - \alpha$ confidence interval for the difference between the trimmed means, $\mu_{t1} - \mu_{t2}$, is

$$(\bar{X}_{t1} - \bar{X}_{t2}) \pm t\sqrt{d_1 + d_2 - 2d_{12}}.$$

EXAMPLE

Table 8.1 reports data, taken from Rao (1948), on the weight of cork borings from 28 trees. Of specific interest was the difference in weight for the north, east, south, and west sides of the trees. For each tree, a boring was taken from all four sides, so there is some possibility that the resulting weights are dependent. That is, methods in Chapter 7 are possibly inappropriate.

Table 8.1 Cork Boring Weights for the North, East, South, and West Sides of Trees

N	E	S	W	N	E	S	W
72	66	76	77	60	53	66	63
56	57	64	58	41	29	36	38
32	32	35	36	30	35	34	26
39	39	31	27	42	43	31	25
37	40	31	25	33	29	27	36
32	30	34	28	63	45	74	63
54	46	60	52	47	51	52	53
91	79	100	75	56	68	47	50
79	65	70	61	81	80	68	58
78	55	67	60	46	38	37	38
39	35	34	37	32	30	30	32
60	50	67	54	35	37	48	39
39	36	39	31	50	34	37	40
43	37	39	50	48	54	57	43

Here we compare the east sides to the north sides as well as the south sides. Assuming the data are stored in the R variables `cork1`, `cork2`, and `cork3`, respectively, we first look at boxplots of the difference scores using the R command

```
boxplot(cork1-cork2,cork2-cork3).
```

The results are shown in Figure 8.1. Notice that the first boxplot suggests that the differences between the north and east sides have a reasonably symmetric distribution with no outliers. This might further suggest that the paired T test will give accurate probability coverage and will have relatively short confidence intervals. It is stressed, however, that the extent to which a boxplot provides a satisfactory assessment of whether accurate confidence intervals will be obtained, when working the mean, is far from clear. Using the R command `t.test(cork1-cork2)`, the 0.95 confidence interval for the population mean is (1.28, 7.43) and the p-value is 0.007. Using a 20% trimmed mean instead, with the command `trimci(cork1-cork2)`, the 0.95 confidence interval is (0.25, 7.41) and the p-value is 0.037. So both methods reject at the 0.05 level. The upper ends of the confidence intervals are similar, but the lower ends differ to a fair degree despite the apparent symmetry indicated by the boxplot.

Now consider the east and south sides. The paired T test has a p-value of 0.09. Using instead a 20% trimmed mean and testing Equation (8.3) with the R command `trimci(cork2-cork3)`, the p-value is 0.049. If instead we compare the marginal trimmed means with the R function `yuend`, the p-value is 0.12. As previously indicated, it seems that testing Equation (8.3) tends to provide the highest power among the three methods just applied, again stressing that exceptions will occur. For the situation at hand, testing Equation (8.3) is the only approach that rejects at the 0.05 level. (Using instead medians or an M-estimator, again we fail to reject at the 0.05 level.)

EXAMPLE

R has a built-in data set called Indometh. Each of the six participants was given an intravenous injection of indomethicin and the goal was to understand how plasma concentrations of indomethicin change over time. With R running, enter Indometh and hit return. The first 14 rows of data look like this:

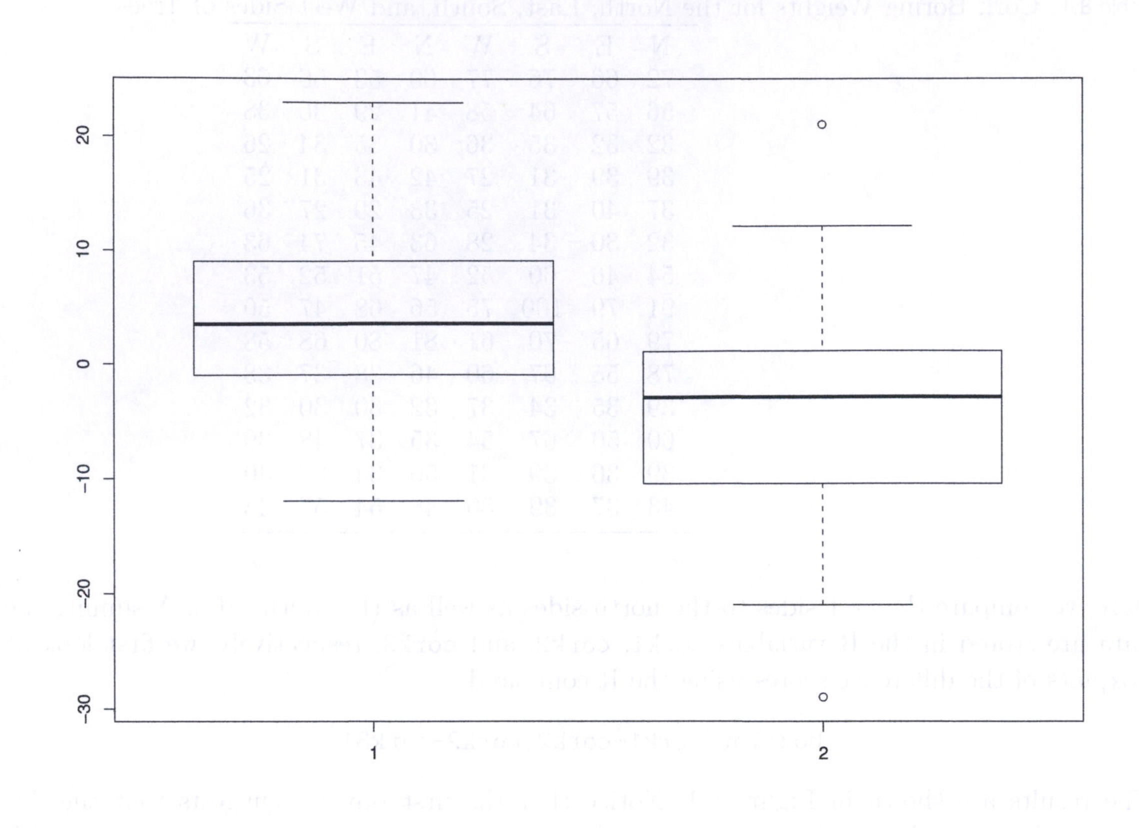

Figure 8.1 Boxplots of the difference scores based on the weight of cork borings. The left boxplot is based on the difference between north and east sides of the trees. The right boxplot is based on the east and south sides.

```
   Subject time conc
1        1 0.25 1.50
2        1 0.50 0.94
3        1 0.75 0.78
4        1 1.00 0.48
5        1 1.25 0.37
6        1 2.00 0.19
7        1 3.00 0.12
8        1 4.00 0.11
9        1 5.00 0.08
10       1 6.00 0.07
11       1 8.00 0.05
12       2 0.25 2.03
13       2 0.50 1.63
14       2 0.75 0.71
```

The first column of data, headed by Subject, contains the participant's code. The second column indicates the time at which blood samples were drawn (in hours). The third column indicates plasma concentrations of indomethicin (mcg/ml). There are various ways the data might be analyzed. One basic issue that can be addressed with the methods covered in this chapter is whether we can be reasonably certain that the typical plasma concentration at time 1 differs from the typical level at time 2. One way to accomplish this goal is to first use the R command

```
z=fac2list(Indometh[,3],Indometh[,2]).
```

This creates an R variable z having list mode, where z[[1]] indicates the plasma concentrations at time 1, which is 0.25 hours, z[[2]] contains the data on concentrations at time 2 (0.50 hours), and so on. So the concentration for the first participant, at time 1, is the first value stored in z[[1]]. The concentration for this same participant at time 2 is the first value stored in z[[2]]. One way of comparing the concentrations at times 1 and 2 is to use the R command

```
trimci(z[[1]]-z[[2]]),
```

which tests the hypothesis that the difference scores have a population trimmed mean equal to zero. The p-value is 0.028 suggesting that the typical differences between concentrations, based on a 20% trimmed mean, have changed between times 1 and 2. The command

```
yuend(z[[1]],z[[2]])
```

would compare the marginal 20% trimmed means. The resulting p-value is 0.013.

8.2.2 Comparing Marginal M-Estimators

For the special case where θ represents the population value of an M-estimator, and the goal is to test

$$H_0 : \theta_1 = \theta_2,$$

a modification of the percentile bootstrap method appears to be best for general use.

The method begins by computing a p-value in a manner similar to how p-values were computed in previous chapters when using a percentile bootstrap. For the situation at hand, generate bootstrap samples by randomly sampling, with replacement, n pairs of observations from

$$(X_{11}, X_{12}), (X_{21}, X_{22}), \ldots, (X_{n1}, X_{n2}).$$

We denote this bootstrap sample by

$$(X_{11}^*, X_{12}^*), (X_{21}^*, X_{22}^*), \ldots, (X_{n1}^*, X_{n2}^*). \tag{8.5}$$

Compute the M-estimate of location for each group, based on the bootstrap sample just generated, which we denote by $\hat{\theta}_1^*$ and $\hat{\theta}_2^*$. Repeat this B times and determine the proportion of times the bootstrap estimate from group 1 is less than the estimate for group 2. In symbols, denote the bootstrap estimates by

$$(\hat{\theta}_{11}^*, \hat{\theta}_{21}^*), \ldots, (\hat{\theta}_{1B}^*, \hat{\theta}_{2B}^*).$$

Let $I_b = 1$ if $\hat{\theta}_{1b}^* < \hat{\theta}_{2b}^*$ $(b = 1, \ldots, B)$; otherwise $I_b = 0$. Then

$$\hat{p} = \frac{1}{B} \sum I_b$$

is the proportion of times the bootstrap estimate from group 1 is less than the estimate for group 2. The usual p-value based on a percentile bootstrap method is

$$P = 2\min(\hat{p}, 1 - \hat{p}).$$

However, when comparing M-estimators using this approach the actual probability of a

Type I error can be considerably smaller than the specified level. For example, with the Type I error set at $\alpha = 0.05$, the actual level can drop below 0.01, which can result in relatively low power. Wilcox and Keselman (2002b) found that the following modification, which is also recommended when comparing groups using MOM, reduces this problem. For group 1, subtract $\hat{\theta}$ from each observation. That is, shift the first marginal distribution so that the M-estimate is zero. Do the same for the second marginal distribution (group 2). So, in particular, the data are shifted so that the null hypothesis is true. For the shifted data, take B bootstrap samples and for each sample determine whether the M-estimate from group 1 is less than the estimate from group 2. Let $\hat{q}$ be the proportion of times this is true and let

$$\hat{p}_a = \hat{p} - 0.1(\hat{q} - 0.5).$$

Then the (bias) corrected p-value is

$$P_a = 2\min(\hat{p}_a, 1 - \hat{p}_a).$$

8.2.3 R Function rmmest

The R function

```
rmmest(x, y = NA, alpha = 0.05, con = 0, est = onestep, plotit = TRUE, dif =
FALSE, grp = NA, hoch = FALSE, nboot = NA, BA = TRUE, xlab = "Group 1", ylab
                    = "Group 2", pr = TRUE, ...)
```

compares, by default, (one-step) M-estimators associated with the marginal distributions of two dependent groups. (The argument hoch refers to Hochberg's method, which is explained in Chapter 12. The argument con is explained in Chapter 12 as well.) To compare groups using MOM, set the argument `est=mom`. The argument `BA=T` means that the adjusted p-value will be used. To avoid this adjustment, use `BA=F`. The data can be stored in two R variables, `x` and `y`. Alternatively, the data can be stored in `x` only, when `x` is a matrix with columns corresponding to groups, or when `x` has list mode.

8.2.4 Measuring Effect Size

Note that when measuring effect size based on some measure of location associated with the marginal distributions, methods in Chapter 7 could be used. When using the trimmed mean associated with the difference scores, μ_{tD} can be used coupled with a confidence interval, or a simple variation of the method proposed by Algina et al. (2005) might be used:

$$d = k\frac{\bar{X}_t}{s_w}, \tag{8.6}$$

where the constant k is chosen so that, under normality, s_w/k estimates the standard deviation σ. When using a 20% trimmed mean, $k = 0.642$. When using the mean and variance, $d = \bar{X}/s$. For example, $d = 0.5$ means that the population mean is estimated to be 0.5 standard deviations from the hypothesized value, zero.

8.2.5 R Function D.akp.effect

The R function

```
D.akp.effect(x,null.value=0,tr=0.2),
```

computes the effect size d given by Eq. (8.6).

8.3 HANDLING MISSING VALUES

When dealing with randomly sampled pairs of observations, for various reasons one of the values for some pairs might be missing. For example, if the goal is to compare the cholesterol levels of married couples, it might be that the wife's cholesterol level is not available and, of course, the same might be true for some husbands. In repeated measures studies, an individual's response might be recorded at time 1, but it might not be available at time 2.

There is extensive literature on how to handle missing values (Allison, 2001; Little and Rubin, 2002; McKnight, McKnight, Sidani, and Figueredo, 2007; Molenberghs and Kenward, 2007; Daniels and Hogan, 2008). A simple strategy is to ignore any pair of values for which one of the values is missing. This is called a *complete case analysis*. But a concern is that this might result in a loss of power because some of the available data are not used. An alternative approach that has received considerable study is based on *imputation* techniques. Using the available data, missing values are filled in. When the goal is to compare the marginal means and data are imputed, a simple strategy is to then apply the paired T test, but it is known that this approach can be unsatisfactory, as noted for example by Liang et al. (2008) as well as Wang and Rao (2002). Indeed, even with no missing values, non-normality can result in relatively poor power and inaccurate confidence intervals as previously illustrated. Other methods have been proposed for handling missing values, but when testing hypotheses, it appears that none are completely satisfactory under non-normality when dealing with means.

Assuming missing responses occur at random, there is a simple method for dealing with missing values when comparing the marginal trimmed means. That is, the goal is to test the hypothesis given by Equation (8.4) using all of the available data, not just pairs of observations where both values are available. The strategy is to use a percentile bootstrap method as outlined in Chapter 5. In the present context, a bootstrap sample is obtained by randomly sampling, with replacement, n pairs of observations. In symbols, we denote this bootstrap sample by

$$(X_{11}^*, X_{12}^*), (X_{21}^*, X_{22}^*), \ldots, (X_{n1}^*, X_{n2}^*). \tag{8.7}$$

Let $\bar{X}_{1t}^*$ be the resulting trimmed mean at time 1 (or group 1) using all of the available data. That is, even if the value at time 2 is missing, the value at time 1 is used to compute the trimmed mean $\bar{X}_{1t}^*$. Similarly, $\bar{X}_{2t}^*$ is the trimmed mean at time 2. Let $D^* = \bar{X}_{1t}^* - \bar{X}_{1t}^*$. Next, repeat this process B times and denote the results by $D_1^*, \ldots, D_B^*$. Estimate $p = P(D^* > 0)$ with $\hat{p}$, the proportion of D^* values greater than 0. The p-value when testing H_0: $\mu_{t1} = \mu_{t2}$ is either $2\hat{p}$ or $2(1 - \hat{p})$, whichever is smallest.

A $1-\alpha$ confidence interval for $\mu_{t1} - \mu_{t2}$ is computed in basically the same way as outlined in Chapter 5. In the present notation, the confidence interval is given by

$$(D_{(\ell+1)}^*, D_{(u)}^*), \tag{8.8}$$

where $D_{(1)}^* \leq \cdots \leq D_{(B)}^*$ are the B bootstrap trimmed means written in ascending order, $\ell = \alpha B/2$, rounded to the nearest integer, and $u = B - \ell$.

Comments on Comparing Means

The method just described contains as a special case a method for comparing means. But a percentile bootstrap method does not perform well when using means and cannot be recommended for the problem at hand. A better approach, when using means, is to use a bootstrap-t method in conjunction with a test statistic stemming from Lin and Stivers (1974); see Wilcox (2011). The computational details are not provided, but the R function `rm2miss` described in the next section performs the calculations.

There is an R package for imputing missing values called `mice`. If `x` is a matrix, the commands

```
library(mice)
z=complete(mice(x,method="mean"))
```

will fill in missing values based on the observed values and store the results in the R variable `z`.

Imagine that Student's T is applied after imputing missing values. Under normality, if the number of missing values is small, it appears that the Type I error probability can be controlled reasonably well. But as the number of missing values increases, eventually this is no longer the case. Under non-normality, the method can be unsatisfactory, particularly when the marginal distributions differ in terms of skewness.

EXAMPLE

Imagine that both marginal distributions are standard normal, $\rho = 0$ and $n = 80$. Further imagine that the first group has 10 missing values and the same is true for the second group. When testing at the 0.05 level, the actual probability of a Type I error is approximately 0.083. If instead there are 20 missing values, the actual Type I error probability is approximately 0.144.

EXAMPLE

We repeat the first example in Section 8.2.1 where the first group has a normal distribution and the second has a lognormal distribution. Now, however, $n = 40$ pairs of observations are randomly sampled where the first value in the first group is missing, and for the second group observations 2, 3, and 4 are missing as well. When using Student's T test with $\alpha = 0.05$, with the missing values imputed via the R function `mice`, the actual Type I error probability is approximately 0.093 (based on simulations with 10,000 replications). Increasing the sample size to $n = 80$, now the actual Type I error probability is 0.073. With five missing values in the second group, again with $n = 80$, the actual Type I error probability is 0.078. Consider again the situation where the second group has three missing values, but now the second group has a skewed, heavy-tailed distribution (a g-and-h distribution with $g = h = 0.5$). The actual Type I error is 0.181, where again $n = 80$. Perhaps other R packages for handling missing values provide better results, but this remains to be determined.

8.3.1 R Functions rm2miss and rmmismcp

The R function

```
rmmismcp(x,y,est=tmean)
```

has been supplied for dealing with missing values when the goal is to test the hypothesis H_0: $\mu_{t1} = \mu_{t2}$ using the percentile bootstrap method just described. In particular, rather than using the complete case analysis strategy, it uses all of the available data to compare the marginal trimmed means, assuming any missing values occur at random. With 20% trimming or more, it appears to be one of the better methods for general use when there are missing values. By default, a 20% trimmed mean is used, but other measures of location can be used via the argument `est`. For example, `rmmismcp(x, y, est=onestep)` would compare the groups with a one-step M-estimator. The function also computes a confidence interval using Equation (8.8).

The R function

```
rm2miss(x,y,tr=0)
```

also tests H_0: $\mu_{t1} = \mu_{t2}$ assuming any missing values occur at random. It uses a bootstrap-t method in conjunction with an extension of the method derived by Lin and Stivers (1974), and by default it tests the hypothesis of equal means. It appears to be one of the better methods when the amount of trimming is small.

8.4 A DIFFERENT PERSPECTIVE WHEN USING ROBUST MEASURES OF LOCATION

Consider again a study aimed at assessing feelings of life satisfaction based on randomly sampled pairs of brothers and sisters. As previously noted, if we test Equation (8.3), the goal in effect is to assess how the typical brother compares to his sister. Comparing the marginal trimmed means instead, the goal is to compare the typical female to the typical male. There is yet another perspective that is worth mentioning: what is the typical difference between males and females? That is, rather than focus on differences between brothers and sisters, focus on the difference among all males and all females, which also differs from the difference between marginal measures of location when using a robust estimator. If this point of view is deemed interesting, estimation and hypothesis testing can be performed using an analog of Cliff's method (and the Brunner–Munzel method) as described at the end of Section 8.8.5. Roughly, begin by forming all possible differences among the participants, rather than simply taking the difference between a sister's score and her brother's score. In symbols, let

$$\mathcal{D}_{i\ell} = X_{i1} - X_{\ell 2},$$

$(i = 1, \ldots, n;\ \ell = 1, \ldots, n)$. So for the first sister, we compute the difference between her score and her brother's score as well as the difference between her score and all other males. We repeat this for all sisters resulting in n^2 differences. The goal is to determine whether the typical $\mathcal{D}$ value differs from zero. In principle, any measure of location could be applied to the $\mathcal{D}_{i\ell}$ values, but currently it seems that the best way of maximizing power is to work with the median of all n^2 differences, $M_{\mathcal{D}}$. In symbols, let $\theta_{\mathcal{D}}$ be the population median associated with $M_{\mathcal{D}}$. The goal is to test H_0: $\theta_{\mathcal{D}} = 0$, which can be done with a percentile bootstrap method. (See Wilcox, 2006b, for more details.)

8.4.1 R Functions loc2dif and l2drmci

The R function

```
loc2dif(x,y=NULL, est=median, na.rm=T, plotit=F, xlab="", ylab="",...)
```

computes $M_{\mathcal{D}}$ for the data stored in `x` (time 1, for example) and `y` (time 2). If `y` is not specified, it is assumed `x` is a matrix with two columns. The argument `na.rm=F` means that the function will eliminate any pair where one or both values are missing. If missing values occur at random and you want to use all of the available data, set `na.rm=T`. If the argument `plotit=T`, the function plots an estimate of the distribution of $\mathcal{D}$. The R function

```
l2drmci(x,y=NA,est=median,alpha=0.05,,nboot=500,SEED=TRUE,na.rm=FALSE)
```

can be used to compute bootstrap confidence interval for $\theta_{\mathcal{D}}$. The argument `na.rm` is used as was done with `loc2dif`.

EXAMPLE

Consider again the cork data in Table 8.1. For the south and east sides of the trees, the function `loc2dif` returns 1. That is, among all trees, the typical (median) difference between south and east sides is estimated to be 1. In contrast, the typical difference for a randomly sampled tree, meaning the median of the difference scores, is 3. So now we have information on how the two sides of the same tree compare, which is not the same as the difference between all the trees. Finally, the difference between the marginal medians is 1.5. This tells us something about how the typical weight for the south side of a tree compares to the typical weight of the east side. But it does not provide any direct information regarding the typical difference among all of the trees.

8.5 THE SIGN TEST

A simple method for comparing dependent groups is the so-called sign test. In essence, it is based on making inferences about the probability of success associated with a binomial distribution, which was discussed in Section 4.11. Here we elaborate a bit on its properties.

As usual, we have n randomly sampled pairs of observations:

$$(X_{11}, X_{12}), \ldots, (X_{n1}, X_{n2}).$$

That is, X_{ij} is the ith observation from the jth group. Primarily for convenience, it is temporarily assumed that tied values never occur. That is, for any pair of observations, $X_{i1} \neq X_{i2}$. Let p be the probability that for a randomly sampled pair of observations, the observation from group 1 is less than the observation from group 2. In symbols,

$$p = P(X_{i1} < X_{i2}).$$

Letting $D_i = X_{i1} - X_{i2}$, an estimate of p is simply the proportion of D_i values that are less than zero. More formally, let $V_i = 1$ if $D_i < 0$; otherwise, $V_i = 0$. Then an estimate of p is

$$\hat{p} = \frac{1}{n} \sum V_i. \tag{8.9}$$

Because $X = \sum V_i$ has a binomial probability function, results in Section 4.11 provide a confidence interval for p. If this interval does not contain 0.5, reject

$$H_0 : p = 0.5$$

and conclude that the groups differ. If $p > 0.5$, group 1 is more likely to have a lower observed value versus group 2, and if $p < 0.5$, the reverse is true.

Now consider a situation where ties can occur. Given the goal of making inferences about p, one strategy is to simply ignore or discard cases where $D_i = 0$. So if among n pairs of observations, there are N D_i values not equal to zero, then an estimate of p is

$$\hat{p} = \frac{1}{N} \sum V_i, \tag{8.10}$$

where V_i is defined as before. (For another approach for handing tied values, see Konietschke & Pauly, 2012.)

8.5.1 R Function signt

The R function

```
signt(x, y = NA, alpha = 0.05)
```

tests H_0: $p = 0.5$ with the sign test as just described. If the argument y is not specified, it is assumed that x is either a matrix with two columns corresponding to two dependent groups, or that x has list mode. The function computes the differences $X_{i1} - X_{i2}$, eliminates all differences that are equal to 0 leaving N values, determines the number of pairs for which $X_{i1} < X_{i2}$ among the N pairs that remain ($i = 1, \ldots, N$), and then calls the function `binomci`.

EXAMPLE

The output from `signt` based on the cork data is

```
$phat:
[1] 0.3076923

$ci:
[1] 0.1530612 0.5179361

$n:
[1] 28

$N:
[1] 26
```

So there are 28 pairs of points. The number of difference scores not equal to 0 is $N = 26$. The confidence interval contains 0.5, so we fail to reject H_0: $p = 0.5$.

8.6 WILCOXON SIGNED RANK TEST

The sign test provides an interesting and useful perspective on how two groups differ. However, a common criticism is that its power can be low relative to other techniques that might be used. One alternative approach is the Wilcoxon signed rank test, which tests

$$H_0 : F_1(x) = F_2(x),$$

where $F_1(x)$ is the probability that a randomly sampled observation from group 1 is less than or equal to x, and F_2 is the corresponding probability for group 2. In other words, the hypothesis is that two dependent groups have identical (marginal) distributions. To apply it, first form difference scores as was done in conjunction with the paired T test and discard any difference scores that are equal to zero. It is assumed that there are n difference scores not equal to zero. That is, for the ith pair of observations, compute

$$D_i = X_{i1} - X_{i2},$$

$i = 1, \ldots, n$ and each D_i value is either less than or greater than zero. Next, rank the $|D_i|$ values and let U_i denote the result for $|D_i|$. So, for example, if the D_i values are 6, −2, 12, 23, −8, then $U_1 = 2$ because after taking absolute values, 6 has a rank of 2. Similarly, $U_2 = 1$ because after taking absolute values, the second value, −2, has a rank of 1. Next set

$$R_i = U_i,$$

if $D_i > 0$; otherwise,

$$R_i = -U_i.$$

Positive numbers are said to have a sign of 1, and negative numbers a sign of −1, so R_i is the value of the rank corresponding to $|D_i|$ multiplied by the sign of D_i.

If the sample size (n) is less than or equal to 40 and there are no ties among the $|D_i|$ values, the test statistic is W, the sum of the positive R_i values. For example, if $R_1 = 4$, $R_2 = -3$, $R_3 = 5$, $R_4 = 2$, and $R_5 = -1$, then

$$W = 4 + 5 + 2 = 11.$$

A lower critical value, c_L, is read from Table 12 in Appendix B. So for $\alpha = 0.05$ and $n = 5$, the critical value corresponds to $\alpha/2 = 0.025$ and is 0, so reject if $W \leq 0$. The upper critical value is

$$c_U = \frac{n(n+1)}{2} - c_L.$$

In the illustration, because $c_L = 0$,

$$c_U = \frac{5(6)}{2} - 0 = 15,$$

meaning that you reject if $W \geq 15$. Because $W = 11$ is between 1 and 15, fail to reject.

If there are ties among the $|D_i|$ values or the sample size exceeds 40, the test statistic is

$$W = \frac{\sum R_i}{\sqrt{\sum R_i^2}}.$$

If there are no ties, this last equation simplifies to

$$W = \frac{\sqrt{6}\sum R_i}{\sqrt{n(n+1)(2n+1)}}.$$

For a two-sided test, reject if $|W|$ equals or exceeds $z_{1-\alpha/2}$, the $1-\alpha/2$ quantile of a standard normal distribution.

Rejecting with the signed rank test indicates that two dependent groups have different distributions. Although the signed rank test can have more power than the sign test, a criticism is that it does not provide any details about how the groups differ. For instance, in the cork boring example, rejecting indicates that the distribution of weights differs for the north versus east side of a tree, but how might we elaborate on what this difference is? One possibility is to estimate p, the probability that the weight from the north side is less than the weight from the east side. So despite lower power, one might argue that the sign test provides a useful perspective on how groups compare.

There are two other rank-based methods for comparing dependent groups that should be mentioned: Friedman's test described in Section 11.4.1 and a modern improvement on Friedman's test called method BPRM, which is described in Section 11.4.3. Both include the ability to compare more than two groups, so a description of these methods is postponed for now.

8.6.1 R Function wilcox.test

The built-in R function

```
wilcox.test(x, y, paired = FALSE, exact = TRUE)
```

performs the Wilcoxon signed rank test just described by setting the argument `paired=TRUE`. (With `paired=FALSE`, the Wilcoxon-Mann-Whitney test in Chapter 7 is performed.)

8.7 COMPARING VARIANCES

Chapter 7 noted that situations arise where the goal is to compare variances rather than some measure of location. A classic method for accomplishing this goal when dealing with dependent groups is the Morgan–Pitman test.

As usual, we have n pairs of randomly sampled observations:

$$(X_{11}, X_{12}), \ldots, (X_{n1}, X_{n2}).$$

Set

$$U_i = X_{i1} - X_{i2}$$

and

$$V_i = X_{i1} + X_{i2},$$

$(i = 1, \ldots, n)$. That is, for the ith pair of observations, U_i is the difference and V_i is the sum. Let σ_1^2 be the variance associated with the first group (the X_{i1} values) and let σ_2^2 be the variance associated with the second. The goal is to test the hypothesis that these variances are equal. It can be shown that if

$$H_0 : \sigma_1^2 = \sigma_2^2$$

is true, then Pearson's correlation between the U and V values is zero. That is, we can test the hypothesis of equal variances by testing

$$H_0 : \rho_{uv} = 0,$$

where ρ_{uv} is Pearson's correlation between U and V. The classic way of testing this hypothesis is with Student's T test as described in Section 6.6. Applying this method to the U and V values yields what is known as the *Morgan–Pitman test* for equal variances.

But we have already seen that testing the hypothesis that Pearson's correlation is equal to 0, using Student's T, can be unsatisfactory when variables are dependent, which means that the Morgan–Pitman test can be unsatisfactory when distributions differ. One basic problem is that as we move toward heavy-tailed distributions, heteroscedasticity becomes an issue when testing $H_0 : \rho_{uv} = 0$ (Wilcox, 2015a). To deal with this problem, simply use the heteroscedastic method in Section 6.6.3.

8.7.1 R Function comdvar

The R function

```
comdvar(x, y, alpha=0.05)
```

tests the hypothesis that two dependent variables have equal variances. It computes U and V as just described and then calls the R function `pcorhc4`. The function returns an estimate of the variances and a p-value. If there is interest in the confidence interval for Pearson's correlation, the command `pcorhc4(x-y, x+y)` can be used instead.

EXAMPLE

Consider again the cork boring data for the north side and east side of a tree. Again we assume the data are stored in the R variables `cork1` and `cork2`. The R command

```
comdvar(cork1, cork2, alpha=0.05)
```

tests the hypothesis of equal variances. The p-value is 0.19, so based on Tukey's three decision rule, make no decision about which variable has the larger variance when testing at the 0.05 level.

8.8 COMPARING ROBUST MEASURES OF SCALE

As for comparing dependent groups using a robust measure of scale, a basic percentile bootstrap method can be used. For example, if the goal is to test the hypothesis that two dependent groups have equal population percentage bend midvariances, generate bootstrap samples by resampling with replacement n pairs of observations. Based on this bootstrap sample, compute the percentage bend midvariance (or whichever measure of scale is of interest), and let d^* be the difference between the two estimates. Repeat this process B times yielding $d_1^*, \ldots, d_B^*$, then put these B values in ascending order and label them in the usual way, namely, $d_{(1)}^* \leq \cdots \leq d_{(B)}^*$. Then a $1-\alpha$ confidence interval for the difference between the measures of scale is $(d_{(\ell+1)}^*, d_{(u)}^*)$, where as usual $\ell = \alpha(B)/2$, rounded to the nearest integer, and $u = B - \ell$.

8.8.1 R Function rmrvar

The R function

```
rmrvar(x,y=NA,alpha=0.05, con=0, est=pbvar, hoch=TRUE, plotit=FALSE)
```

compares robust measures of variation using a percentile bootstrap method. If the argument `y` is not specified, it is assumed the data are stored in `x` in list mode or a matrix, with the columns corresponding to groups. When comparing two dependent groups, setting the argument `plotit=T` results in a scatterplot of the bootstrap estimates. By default, percentage bend midvariances are compared, but any robust measure of scatter, introduced in Chapter 2, can be used via the argument `est`. (The arguments `con` and `hoch` are explained in Chapter 12. They are relevant when comparing more than two groups.)

EXAMPLE

Chapter 7 described a study where EEG measures of convicted murderers were compared to a control group. In fact, measures for both groups were taken at four sites in the brain. For illustrative purposes, the first two sites for the control group are compared using the percentage bend midvariance described in Chapter 3. The data are

Site 1: −0.15, −0.22, 0.07, −0.07, 0.02, 0.24, −0.60, −0.17, −0.33, 0.23, −0.69, 0.70, 1.13, 0.38

Site 2: −0.05, −1.68, −0.44, −1.154, −0.16, −1.29, −2.49, −1.07, −0.84, −0.37, 0.01, −1.24, −0.33, 0.78

If the data are stored in the R variables `eeg1` and `eeg2`, the command

```
rmrvar(eeg1,eeg2)
```

returns a p-value of 0.112. So we are not able to detect a difference in the variation between these two sites when testing at the 0.05 level. Or in terms of Tukey's three-decision rule, make no decision regarding which group has the larger measure of variation. If the argument `est=pbvar` is replaced with `est=mad`, so the measure of scale MAD is used, the p-value is 0.226.

8.9 COMPARING ALL QUANTILES

Section 7.8.10 noted that it might be of interest to compare low scoring individuals in one group to low scoring participants in another. This might be done by comparing the lower quartiles. Simultaneously, one might want to compare high scoring participants using the upper quartiles. More generally, it can be informative to compare all of the quantiles, which can be done with the shift function described in Chapter 7 when comparing independent groups. Lombard (2005) derived an extension of the shift function to dependent groups (cf. Wilcox, 2006c). For an alternative technique that can have a power advantage when dealing with quantiles close to zero or one, see Section 12.4.7.

8.9.1 R Functions lband

The R function

```
lband(x,y=NA,alpha=0.05,plotit=TRUE)
```

compares all quantiles of two dependent groups using the method derived by Lombard (2005) and plots the shift function if the argument `plotit=TRUE`. If the argument `y` is not specified, it is assumed that the argument `x` is a matrix with two columns. For a method that might provide higher power, especially when dealing with tied values, see Sections 12.4.7 and 12.4.8.

EXAMPLE

Continuing the last example, `lband` finds a difference in the EEG measures for the first two sites of the control group when testing at the 0.05 level. Significant differences are found at the 0.14, 0.5, and 0.57 quantiles. That is, the 0.14 quantile for the first site differs significantly from the 0.14 quantile of the second site, and the same is true for the 0.5 and 0.57 quantiles.

8.10 PLOTS FOR DEPENDENT GROUPS

Section 7.9.3 described a plot of the shift function. The scores for the first group are plotted along the x-axis, and the y-axis indicates the difference between the estimated quantiles. More precisely, the y-axis indicates $y_q - x_q$, where x_q is the qth quantile for group 1 and y_q is the corresponding quantile for group 2. However, when creating a plot of the shift function for dependent groups, the R function `lband` should be used. Figure 8.2 shows the plot returned by `lband` based on the EEG data used in the last example. The solid line is the estimated difference between the quantiles. The dashed lines form a 0.95 confidence band for all of the differences. That is, there is a 0.95 probability that the differences among all quantiles lie between these two lines. The probability coverage can be altered via the argument `alpha` in `lband`. If the R function `sband` illustrated in Section 7.9.3 had been used, the probability coverage would no longer be 0.95, roughly because dependent variables are being compared. The R function `lband` takes the dependence into account.

Plots for the difference scores can be created using the R functions in Chapter 2. For example, the R command

```
akerd(eeg[,1]-eeg[,2])
```

would create a kernel density estimate of the distribution for the difference scores based on the EEG measures used in the previous example. If the groups do not differ, this plot should be symmetric about the value 0. When using all possible differences, as described in Section 8.4, a special R function for plotting the data has been supplied and is described in the next section.

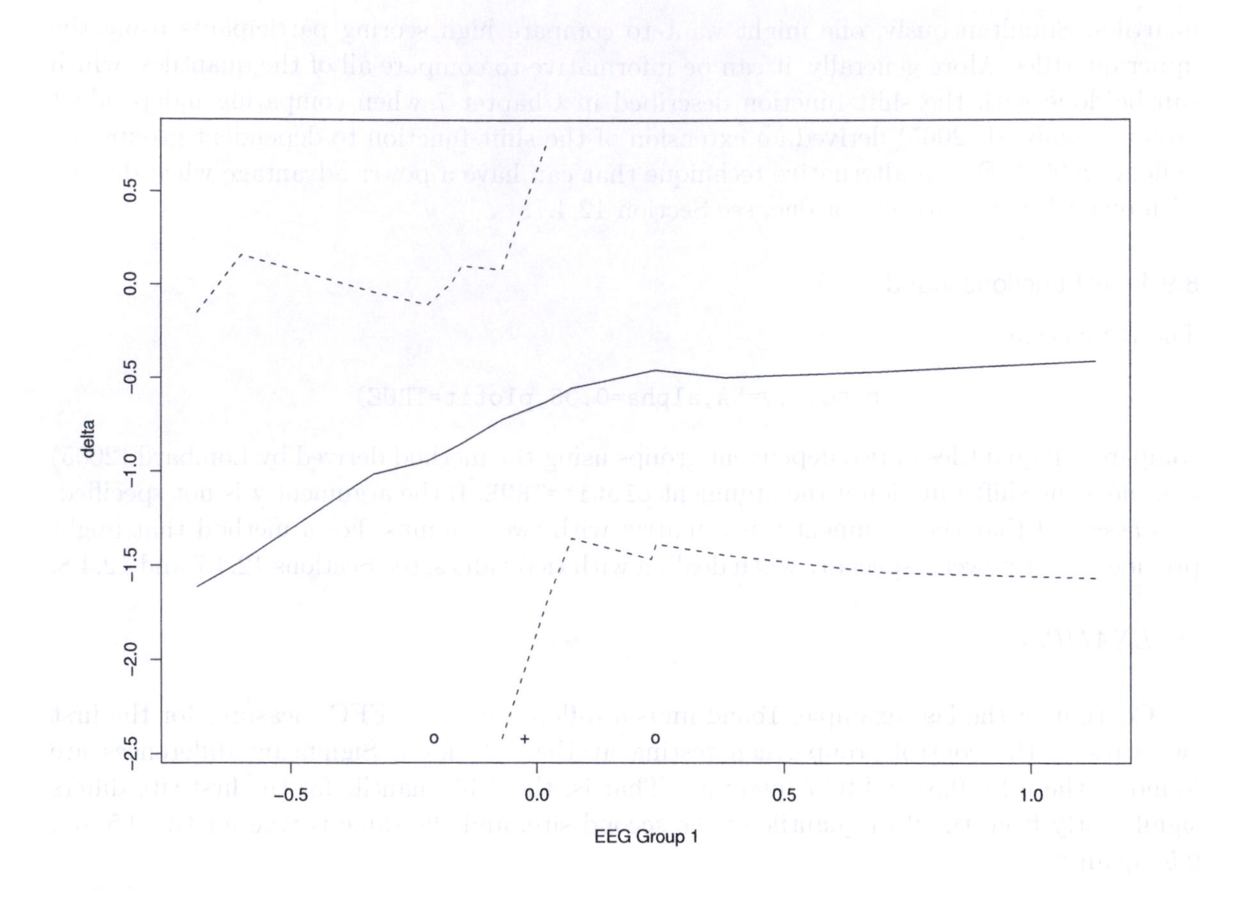

Figure 8.2 Shift function based on the EEG data from the first two sites of the control group. Note how the effect size is largest among the lower values.

8.10.1 R Function g2plotdifxy

The R function

```
g2plotdifxy(x,y)
```

creates all possible differences as described in Section 8.4 and plots the results via the adaptive kernel density estimator in Section 2.6.1. Again, if the groups do not differ in any manner, this plot should be symmetric about the value 0.

8.11 EXERCISES

1. For the data in Table 8.1, perform the paired T test for means using the weights for the east and south sides of the trees. Verify that the p-value is 0.44.

2. Repeat the previous exercise, but use 20% trimmed means instead using the difference scores in conjunction with the R function **trimci**. Verify that the p-value is 0.049.

3. If in Exercise 1 you compare the marginal 20% trimmed means with the R function **yuend**, verify that now the p-value is 0.121.

4. Generally, why is it possible to get a different p-value comparing the marginal

trimmed means versus making inferences about the trimmed mean of the difference scores?

5. Is it possible that the marginal trimmed means are equal but the trimmed mean based on the difference scores is not equal to zero? What does this indicate in terms of power?

6. Repeat Exercise 1, but now use a bootstrap-t method via the R function `trimcibt`. Verify that the p-value is 0.091.

7. Repeat the last exercise, only now use a 20% trimmed mean. Verify that the p-value is 0.049.

8. For two dependent groups you get

Group 1:	10	14	15	18	20	29	30	40
Group 2:	40	8	15	20	10	8	2	3

Compare the two groups with the sign test and the Wilcoxon signed rank test with $\alpha = 0.05$. Verify that, according to the sign test, $\hat{p} = 0.29$ and that the 0.95 confidence interval for p is (0.04, 0.71), and that the Wilcoxon signed rank test has an approximate p-value of 0.27.

9. For two dependent groups you get

Group 1:	86	71	77	68	91	72	77	91	70	71	88	87
Group 2:	88	77	76	64	96	72	65	90	65	80	81	72

Apply the Wilcoxon signed rank test with $\alpha = 0.05$. Verify that $W = 0.7565$ and that you fail to reject.

10. Section 8.2.1 analyzed the Indometh data stored in R using the R function `trimci`. Compare times 2 and 3 using means based on difference scores and verify that the p.value is 0.014.

11. Continuing the last exercise, plot the difference scores using the R function `akerd` and note that the distribution is skewed to the right, suggesting that the confidence interval for the mean might be inaccurate. Using the R function `trimcibt`, verify that the 0.95 confidence interval for the 20% of the difference scores is (0.127, 0.543). Then verify that when using the mean instead (setting the argument `tr=0`), the 0.95 confidence interval is (0.0097, 0.7969). Why is this second confidence so much longer than the first? Hint: Look at a boxplot of the difference scores.

12. The file scent_dat.txt, stored on the author's web page, contains data downloaded from a site maintained by Carnegie Mellon University. Read the file into an R variable called `scent`. The last six columns contain the time participants required to complete a pencil and paper maze when they were smelling a floral scent and when they were not. The columns headed by U.Trial.1, U.Trial.2, and U.Trial.3 are the times for no scent, which were taken taken on three different occasions. Compare scent and no scent data for occasion 1 using difference scores via the command `trimci(scent[,7]-scent[,10],tr=0)`, so the mean of the difference scores is being used. Verify that the p-value is 0.597.

13. Repeat the previous exercise, only now compare the marginal 20% trimmed means using the R function `yuend` with the argument `tr=0.2`. Verify that the p-value is 0.2209.

CHAPTER 9

ONE-WAY ANOVA

CONTENTS

This chapter extends the methods in Chapter 7 to situations where more than two independent groups are to be compared. Section 9.1 introduces a classic and commonly used method for comparing means, called the ANOVA F test, which assumes both normality and homoscedasticity. When comparing two groups only, the method reduces to Student's T test covered in Chapter 8. The method enjoys some positive features, but it suffers from the same concerns associated with Student's T test. Indeed, problems are exacerbated when comparing more than two groups and new problems are introduced. However, these problems are not completely obvious and should be described in order to motivate more modern techniques.

Numerous methods have been proposed for improving on the ANOVA F test and some of the better methods are described here. They include heteroscedastic techniques based on robust measures of location as well as rank-based or nonparametric techniques.

To help motivate the methods in this chapter, we begin with a concrete example where the goal is to compare groups using the plasma retinol data described in Chapter 1. We focus on three groups based on smoking status: never smoked, former smokers, and currently smokes. The goal is to determine whether the mean plasma retinol levels differ among these three groups. One strategy is to compare each pair of groups by testing the hypothesis of equal means using Student's T test described in Chapter 8. In symbols, test

$$H_0 : \mu_1 = \mu_2,$$

$$H_0 : \mu_1 = \mu_3,$$

$$H_0 : \mu_2 = \mu_3.$$

But even under normality and homoscedasticity, there is a technical issue that needs to be considered. Suppose there are no differences among the groups in which case none of the three null hypotheses just listed should be rejected. To keep things simple for the moment, assume all three groups have normal distributions with equal variances, in which case Student's T test provides exact control over the probability of a Type I error when testing any single hypothesis (assuming random sampling only). Further assume that all three of the hypotheses just listed are tested with $\alpha = 0.05$. So for *each* hypothesis, the probability of a Type I error is 0.05, but what is the probability of *at least one* Type I error when you perform all three tests? If you perform each of the tests with $\alpha = 0.05$, the probability of at least one Type I error will be larger than 0.05. The more tests you perform, the more likely you are to reject one or more of the hypotheses of equal means when in fact the means are equal. A common goal, then, is to test all pairs of means so that the probability of one or more Type I errors is α, where as usual α is some value you pick. Put another way, when comparing all pairs of groups, the goal is to have the probability of making no Type I errors equal to $1 - \alpha$.

Dealing with the problem just described is not immediately straightforward based on the techniques covered in previous chapters, even if we assume normality and homoscedasticity. One complication is that if we perform T tests for each pair of groups, some of these tests will be dependent. For example, when comparing group 1 to groups 2 and 3, the two test statistics (Student's T) will be dependent because both involve the sample mean corresponding to the first group. This makes it difficult to determine the probability of one or more Type I errors. (If the tests were independent, the binomial probability function could be used.)

There are two general approaches to the problem of controlling the probability of making at least one Type I error. The first and most common strategy begins by testing the hypothesis that all the groups being compared have equal means. In the illustration, the null hypothesis is written as

$$H_0 : \mu_1 = \mu_2 = \mu_3.$$

More generally, when comparing J groups, the null hypothesis is

$$H_0 : \mu_1 = \mu_2 = \cdots = \mu_J. \qquad (9.1)$$

If this hypothesis of equal means is rejected, you then make decisions about which pairs of groups differ using a method described in Chapter 12. The second general approach is to skip the methods in this chapter and use one of the appropriate techniques in Chapter 12. There are circumstances under which this latter strategy has practical advantages, as will be made evident in Chapter 12.

9.1 ANALYSIS OF VARIANCE FOR INDEPENDENT GROUPS

This section describes the most commonly used method for testing (9.1), the hypothesis of equal means. The method is called *analysis of variance* or ANOVA and was derived by Sir Ronald Fisher.

ASSUMPTIONS:

- random sampling
- independent groups
- normality
- homoscedasticity

As in Chapter 7, homoscedasticity means that all groups have a common variance. That is, if $\sigma_1^2, \ldots, \sigma_J^2$ are the population variances of the J groups, homoscedasticity means that

$$\sigma_1^2 = \sigma_2^2 = \cdots = \sigma_J^2. \tag{9.2}$$

For convenience, this common variance is labeled σ_p^2. As in previous chapters, heteroscedasticity refers to a situation where not all the variances are equal.

9.1.1 A Conceptual Overview

We begin by providing a conceptual overview of the strategy behind the ANOVA F test. This will help build a foundation for understanding its relative merits. Momentarily, primarily for notational convenience, equal sample sizes are assumed, with the common sample size denoted by n. That is, $n_1 = n_2 = \cdots n_J = n$.

Let $\bar{X}_1, \ldots, \bar{X}_J$ denote the sample means associated with the J groups being compared. The *grand mean* is the average of the J sample means:

$$\bar{X}_G = \frac{1}{J}(\bar{X}_1 + \cdots + \bar{X}_J).$$

Now, consider how we might measure the overall variation among J sample means. One approach is to use a measure similar to the sample variance, as introduced in Chapter 2. Roughly, use the typical squared difference between each sample mean and the grand mean. In symbols, use

$$\frac{1}{J-1}\sum_{j=1}^{J}(\bar{X}_j - \bar{X}_G)^2.$$

If we multiply this last equation by n, we get what is called the *mean squares between groups*, labeled here as MSBG. That is,

$$\text{MSBG} = \frac{n}{J-1}\sum_{j=1}^{J}(\bar{X}_j - \bar{X}_G)^2, \tag{9.3}$$

which can be shown to estimate the assumed common variance, σ_p^2, *if* the hypothesis of equal means is true. In fact, it can be shown that

$$E(\text{MSBG}) = \sigma_p^2.$$

That is, if the null hypothesis is true, MSBG is an unbiased estimate of the assumed common variance.

If, however, the hypothesis of equal means is false, it can be shown that

$$E(\text{MSBG}) = \sigma_p^2 + \frac{n\sum(\mu_j - \bar{\mu})^2}{J-1},$$

where

$$\bar{\mu} = \frac{1}{J}(\mu_1 + \mu_2 + \cdots + \mu_J),$$

which is called the *grand population mean.* This last equation says that the larger the variation among the population means, the larger will be the expected value of MSBG.

Note that regardless of whether the hypothesis of equal means is true or false, each of the sample variances, $s_1^2, \ldots, s_J^2$, provides an estimate of the assumed common variance, σ_p^2. A natural way of combining them into a single estimate of the assumed common variance is to average them, which is exactly what is done when the sample sizes are equal. (When sample sizes are not equal, a weighted average is used, which is described in Box 9.1 and often called the pooled variance.) This yields what is called the *mean squares within groups,* which we label MSWG. More succinctly, still assuming equal sample sizes,

$$\text{MSWG} = \frac{1}{J}\sum_{j=1}^{J} s_j^2 \tag{9.4}$$

and

$$E(\text{MSWG}) = \sigma_p^2.$$

That is, MSBG and MSWG estimate the same quantity if the null hypothesis is true. If the null hypothesis is false, on average MSBG will be larger than MSWG. It turns out to be convenient to measure the discrepancy between MSBG and MSWG with the test statistic

$$F = \frac{\text{MSBG}}{\text{MWSG}}. \tag{9.5}$$

Roughly, the larger F happens to be, the stronger the evidence that the population means differ. When the null hypothesis of equal means is true, the exact distribution of F has been derived under the assumptions of normality and equal variances. It is called an *F distribution* with degrees of freedom

$$\nu_1 = J - 1,$$

and

$$\nu_2 = N - J,$$

where $N = \sum n_j$ is the total sample size. That is, the distribution depends on two quantities: the number of groups being compared, J, and the total number of observations in all of the groups, N. The point is that you are able to determine how large F must be in order to reject H_0.

Figure 9.1 shows the distribution of F with $\nu_1 = 3$ (meaning that there are $J = 4$ groups) and $\nu_2 = 36$ degrees of freedom. The shaded region indicates the critical region when $\alpha = 0.05$; it extends from 2.87 to infinity. That is, if you want the probability of a Type I error to be 0.05, then reject the null hypothesis of equal means if $F \geq 2.87$, which is the 0.95 quantile of the F distribution. More generally, reject if $F \geq f$, where f is the $1 - \alpha$ quantile of an F distribution with $\nu_1 = J - 1$ and $\nu_2 = N - J$ degrees of freedom. Tables 5, 6, 7, and 8 in Appendix B report critical values, f, for $\alpha = 0.1$, 0.05, 0.025, and 0.01 and various degrees of freedom. For example, with $\alpha = 0.05$, $\nu_1 = 6$, $\nu_2 = 8$, Table 6 indicates that the 0.95 quantile is $f = 3.58$. That is, there is a 0.05 probability of getting a value for F that exceeds 3.58 when in fact the population means are equal. For $\alpha = 0.01$, Table 8 says that

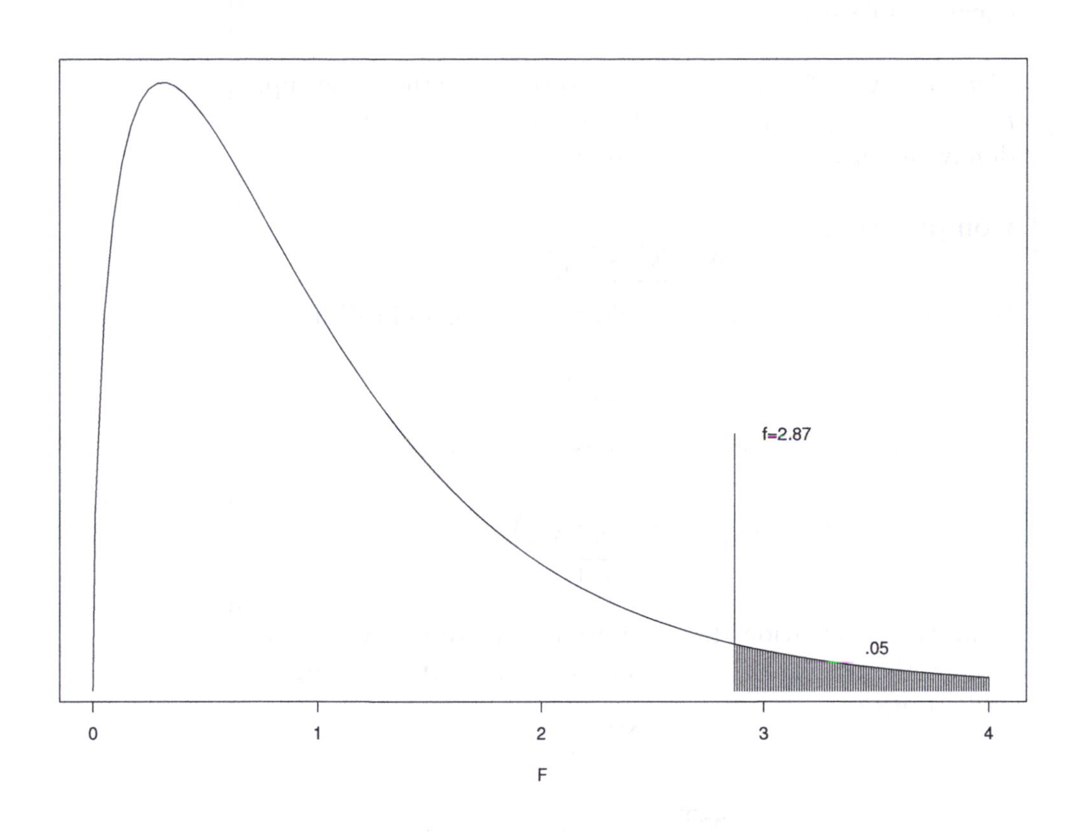

Figure 9.1 An F distribution with 3 and 36 degrees of freedom. The 0.95 quantile is $f = 2.87$ and the 0.05 critical region is indicated by the shaded portion of the plot. That is, under normality and homoscedasticity, if the hypothesis of equal means is rejected when $F \geq f$, the probability of a Type I error is exactly 0.05.

the 0.99 quantile is 6.32. This means that if you reject when $F \geq 6.32$, the probability of a Type I error will be 0.01, assuming normality and that the groups have equal variances.

The ANOVA F test can be applied when there are unequal sample sizes, still assuming normality and homoscedasticity. The computational details are relegated to Box 9.1. All of the software packages mentioned in Chapter 1 contain functions for applying this method.

EXAMPLE

Applying the ANOVA F test using computer software is fairly straightforward. But to make sure the notation is clear, the computations in Box 9.1 are illustrated with the following data.

Group 1:	7, 9, 8, 12, 8, 7, 4, 10, 9, 6
Group 2:	10, 13, 9, 11, 5, 9, 8, 10, 8, 7
Group 3:	12, 11, 15, 7, 14, 10, 12, 12, 13, 14

Box 9.1: Summary of the ANOVA F test with or without equal sample sizes.

Notation: X_{ij} refers to the ith observation from the jth group, $i = 1, \ldots, n_j$; $j = 1, \ldots, J$. (There are n_j observations randomly sampled from the jth group.)

Computations:

$$A = \sum\sum X_{ij}^2$$

(In words, square each value, add the results, and call it A.)

$$B = \sum\sum X_{ij}$$

(In words, sum all the observations and call it B.)

$$C = \sum_{j=1}^{J} \frac{1}{n_j}\left(\sum_{i=1}^{n_j} X_{ij}\right)^2$$

(Sum the observations for each group, square the result, divide by the sample size, and add the results corresponding to each group.)

$$N = \sum n_j$$

$$\text{SST} = A - \frac{B^2}{N}$$

$$\text{SSBG} = C - \frac{B^2}{N}$$

$$\text{SSWG} = \text{SST} - \text{SSBG} = A - C$$

$$\nu_1 = J - 1$$

$$\nu_2 = N - J$$

$$\text{MSBG} = \text{SSBG}/\nu_1$$

$$\text{MSWG} = \text{SSWG}/\nu_2$$

Note: SSBG stands for *sum of squares between groups*, SSWG is *sum of squares within groups*, and SST is the *total sum of squares*.

Test Statistic: $F = \frac{\text{MSBG}}{\text{MSWG}}$.

Decision Rule: Reject H_0 if $F \geq f$, the $1 - \alpha$ quantile of an F distribution with $\nu_1 = J - 1$ and $\nu_2 = N - J$ degrees of freedom.

We see that

$$A = 7^2 + 9^2 + \cdots + 14^2 = 3026,$$

$$B = 7 + 9 + \cdots + 14 = 290,$$

$$C = \frac{(7+9+\cdots+6)^2}{10} + \frac{(10+13+\cdots+7)^2}{10} + \frac{(13+11+\cdots+14)^2}{10} = 2890,$$

$$N = 10 + 10 + 10 = 30,$$

$$\text{SST} = 3026 - 290^2/30 = 222.67,$$

$$\text{SSBG} = 2890 - 290^2/30 = 86.67,$$

$$\text{SSWG} = 3026 - 2890 = 136,$$

$$\text{MSBG} = 86.67/(3-1) = 43.335,$$

$$\text{MSWG} = 136/(30-3) = 5.03,$$

so

$$F = \frac{43.335}{5.03} = 8.615.$$

The degrees of freedom are $\nu_1 = 3 - 1 = 2$ and $\nu_2 = 30 - 3 = 27$. With $\alpha = 0.01$, we see from Table 8 in Appendix B, that the critical value is $f = 5.49$. Because $8.165 \geq 5.49$, reject the hypothesis of equal means.

EXAMPLE

Is it reasonable to estimate the common variance with MSBG if the F test fails to reject? In general, the answer is no, or at least not necessarily. It might be that the means are not equal, but the power of the F test was not high enough to detect this. If the means are not equal, MSBG is estimating

$$\sigma_p^2 + \frac{n\sum(\mu_j - \bar{\mu})^2}{J-1},$$

not σ_p^2. So failing to reject is not convincing evidence that MSBG provides a reasonable estimate of the common variance.

We conclude this section by describing the conventional ANOVA model. The notation

$$\alpha_j = \mu_j - \bar{\mu}$$

represents the difference between the population mean of the jth group and the *grand population mean*,

$$\bar{\mu} = \frac{1}{J}\sum \mu_j.$$

Another common notation is

$$\begin{aligned} \epsilon_{ij} &= X_{ij} - \bar{\mu} - \alpha_j \\ &= X_{ij} - \mu_j. \end{aligned}$$

In other words, ϵ_{ij} is the difference between the ith observation in the jth group, and the corresponding mean, μ_j. That is, ϵ_{ij} is an error term: it measures the extent to which X_{ij} differs from the population mean of the jth group. Rearranging terms, this last equation becomes

$$X_{ij} = \bar{\mu} + \alpha_j + \epsilon_{ij}. \tag{9.6}$$

The ANOVA F test is obtained by assuming that ϵ_{ij} has a normal distribution with mean 0 and variance σ_p^2.

9.1.2 ANOVA via Least Squares Regression and Dummy Coding

A popular way of approaching and describing the ANOVA F test is in the context of what is called the *general linear model*, which is a framework for describing a large class of methods that contains least squares regression as a special case. A point worth stressing is that built-in R functions, based on the general linear model, assume homoscedasticity. (Even more restrictive assumptions are imposed when dealing with other classic methods described later in this book.) When dealing with ANOVA or testing hypotheses based on the least squares regression estimator, normality is assumed as well.

The main point here is that the ANOVA F test can be viewed from the perspective of least squares regression using what is called *dummy coding*. Put another way, the ANOVA F test belongs to a class of techniques based on the general linear model. As a simple example, consider the situation where the goal is to test the hypothesis that two normal distributions have equal means. Let $G = 0$ if an observation comes from group 1 and let $G = 1$ if an observation comes from group 2. A linear model that is typically used is

$$Y_i = \beta_0 + \beta_1 G_i + e_i,$$

where by assumption e has a normal distribution with mean 0 and variance σ^2. That is, homoscedasticity is assumed meaning that regardless of whether an observation is from group 1 ($G_i = 0$) or group 2 ($G_i = 1$), the variance of Y, given G, is σ^2. Note that the model is just a special case of the least squares regression model in Section 6.2. Here, the model says that if $G = 0$, the mean of Y is β_0, and if $G = 1$, the mean of Y is $\beta_0 + \beta_1$. So the two groups have equal means if $\beta_1 = 0$, and testing H_0: $\beta_1 = 0$ can be done with the method in Section 6.2, which is tantamount to using Student's T test. That is, this perspective does not deal with violations of assumptions.

Dummy coding can be extended to situations where the goal is to test the hypothesis that J independent groups have a common mean. A detailed description of this approach can be found, for example, in Cohen et al. (2003, Chapter 8). Roughly, predictors are added to the regression equation that have values 0 or 1, which makes it possible to test hypotheses about the means of the groups. A key feature of this strategy is that it designates one group as the reference group. Once this is done, the means of the remaining $(J-1)$ groups are compared to the mean of the reference group. Details are kept to a minimum, however, because the method suffers from some fundamental concerns. One is described here and another is described in Section 9.2.1.

Even when the assumptions of normality and homoscedasticity are true, there is a fundamental concern with the typical dummy coding approach to ANOVA. Imagine that we pick group 1 to be the reference group. Then the method tests the hypothesis that the mean of groups 2 through J are equal to the mean of group 1. Direct comparisons between groups 2 and 3, for example, are not made. This is reasonable when the null hypothesis of equal means is true. But if it is false, the choice of which group is the reference group can affect power.

For example, suppose that unknown to us, the means of three groups are $\mu_1 = 4$, $\mu_2 = 2$, and $\mu_3 = 6$. As is evident, the absolute difference between the first group and the other two is 2. But if we want to maximize power, a better strategy would be to compare groups 2 and 3, which is not done directly when using the typical dummy-coding scheme.

9.1.3 R Functions anova, anova1, aov, and fac2list

In its simplest form, the built-in R function

```
aov(x ~ g),
```

used in conjunction with the R function `summary` or the R function `anova`, performs the ANOVA F test. The notation `x` $\sim$ `g` indicates that the data to be analyzed, often called the values of the *dependent variable*, are stored in the R variable `x`, and the corresponding *levels* or group identifications are contained in `g`. In general, it is best to make sure that `g` is a factor variable, The reason is demonstrated in the last example in this section.

EXAMPLE

Consider the plasma retinol data in Chapter 1. Column 2 indicates the sex of the participant, stored as a 1 (male) or 2 (female), and column 10 contains information on the amount of cholesterol consumed by the participant. The first row indicates that the first participant is a female and that her daily cholesterol consumption is 170.3. If the data are stored in the R variable `plasma`, then, in the present context, the notation

```
plasma[,10] ~ plasma[,2]
```

roughly says to separate the data stored in `plasma[,10]` into groups indicated by `plasma[,2]`.

In case it helps, an alternative R function is included in the file of R functions written for this book that can be used when data are stored in a matrix, with columns corresponding to groups, or when data are stored in list mode. The function is

```
anova1(x).
```

EXAMPLE

One of the data sets stored in R is called `chickwts`. Newly hatched chicks were randomly allocated into six groups and each group was given a different feed supplement. Their weights in grams, after six weeks, are given in the first column of `chickwts` and the type of feed is indicated in the second column. The first 12 rows of this data set are:

```
   weight      feed
1     179 horsebean
2     160 horsebean
3     136 horsebean
4     227 horsebean
5     217 horsebean
6     168 horsebean
7     108 horsebean
8     124 horsebean
9     143 horsebean
10    140 horsebean
11    309   linseed
12    229   linseed
```

The R command

```
aov(chickwts[,1] ~ chickwts[,2])
```

returns

```
Terms:
                chickwts[, 2] Residuals
Sum of Squares       231129.2  195556.0
Deg. of Freedom             5        65
```

where 231129.2 is SSBG described in Box 9.1 and 195556.0 is SSWG. To apply the F test, use the command

summary(aov(chickwts[,1] ~ chickwts[,2])),

which returns

```
               Df Sum Sq Mean Sq F value    Pr(>F)
chickwts[, 2]   5 231129   46226  15.365 5.936e-10 ***
Residuals      65 195556    3009
```

So the value of the test statistic, F, is 15.365 and the p-value is 5.936e-10. That is, assuming normality and homoscedasticity, reject and conclude that the population means of the six groups are not all equal. The R command

anova(aov(chickwts[,1] ~ chickwts[,2]))

returns the same results.

To stress a feature of R, it is noted that in the last example, chickwts is not a matrix; it has list mode and it is a data frame. So the ANOVA F test could have been applied with the command

summary(aov(chickwts[[1]] ~ chickwts[[2]])).

For the situation at hand, although the data are not stored in a matrix, we are able to use `chickwts[,1]` and `chickwts[,2]` to indicate the data in columns 1 and 2, respectively. (See Chapter 1 for a discussion of data stored in a matrix.) The point being stressed is that even when the data appear to be in a matrix on the computer screen, this is not necessarily the case. The R command `is.matrix` can be used to check whether an R variable is a matrix. Here, the R command `is.matrix(chickwts)` returns false. Some R functions require that data be stored in a matrix, and applying any such function to data stored in a data frame would generate an error. The R command `chickwts=as.matrix(chickwts)` would convert the storage mode to a matrix.

The R function

```
fac2list(x,g)
```

is provided to help make various R functions easier to use when comparing two or more groups. The argument `x` is an R variable, usually some column of a matrix or data frame, containing the data to be analyzed (the dependent variable) and `g` is a column of data indicating the group to which a value, stored in `x`, belongs. When working with a data frame, this latter column of data can be a factor variable. The function sorts the data in `x` into groups indicated by the values in `g` and stores the results in list mode. The R function `split` accomplishes the same goal, but for certain purposes, it seems a bit easier to use the R function `fac2list`. The next example illustrates how this function is used and why it has practical value.

EXAMPLE

Care needs to be taken when using `aov` to make sure that the groups are generated properly. Consider the built-in data set stored in the R variable `ChickWeight`, which is a matrix containing four columns of data. The first column contains the weight of chicks, column 4 indicates which of four diets was used, and the second column gives the number of days since birth when the measurement was made, which were 0, 2, 4, 6, 8, 10, 12, 14, 16, 18, 20, 21. So for each chick, measurements were taken on 12 different days. From a statistical point of view, the following analysis is inappropriate because it involves comparing dependent groups, which requires methods covered in Chapter 11. But to illustrate a feature of R, independent groups are assumed when comparing weights corresponding to the 12 days. One way of doing the analysis is to convert the data into list mode with

```
z=fac2list(ChickWeight[,1],ChickWeight[,2]).
```

This creates 12 groups. The data for group 1 are stored in `z[[1]]` and corresponds to day 0, `z[[2]]` contains the data for day 2, and so on. If the levels of the groups are indicated by numeric values, `fac2list` puts the levels in ascending order. If the levels are indicated by a character string, the levels are put in alphabetical order. Then the command

```
anova1(z)
```

would test the hypothesis that among all 12 days, the means are equal. The results indicate that the first degrees of freedom are $\nu_1 = 11$ because there are $J = 12$ groups, and the test statistic is $F = 125.5$. But suppose we had used `aov` in the manner previously illustrated:

```
anova(aov(ChickWeight[,1] ~ ChickWeight[,2])).
```

This would return $\nu_1 = 1$ indicating there are two groups, not 12 as intended. The test statistic is now $F = 1348.7$. The following R commands would do the analysis as intended, using built-in R functions:

```
anova(aov(ChickWeight[,1] ~ as.factor(ChickWeight[,2]))).
```

So to be safe, routinely use the R command `factor` when using `aov`.

Here is another way of performing the analysis that might be helpful. First, for convenience, we store the data in another R variable:

```
z=ChickWeight.
```

Next, convert `z` to a data frame so that the columns of data can have different modes. Here the goal is to have the second column stored as a factor, while keeping the first column in its original (numeric) form, and then perform the F test. This is accomplished with the following commands:

```
z=as.data.frame(z)
z[,2]=as.factor(z[,2])
anova(aov(z[,1] ~ z[,2])).
```

A point worth stressing is that the R functions `aov`, `anova`, and `anova1` assume homoscedasticity—they have no option for handling unequal variances. This is a practical concern because violating the homoscedasticity assumption becomes an increasingly more serious problem as the number of groups increases. Sections 9.4.1 and 9.5.4 introduce R functions that are designed to handle unequal variances when comparing trimmed means that can be used to compare means as a special case.

9.1.4 Controlling Power and Choosing the Sample Sizes

When you fail to reject the hypothesis of equal means, this might be because there are no differences among the means, or any differences that exist might be difficult to detect based on the sample sizes used. Generally, failing to reject raises the fundamental concern that there might be an important difference that was missed due to low power, as stressed in previous chapters.

As in previous chapters, one strategy for trying to insure relatively high power is to choose the sample sizes, prior to collecting any data, so as to achieve some specified amount of power. This can be done if, in addition to assuming normality and homoscedasticity, the desired power and Type I error probability can be specified. Consistent with previous chapters, this means it is necessary to choose the desired probability of rejecting given some specified difference among the means.

A measure of the difference among the means that is typically used is

$$\delta^2 = \frac{\sum(\mu_j - \mu_G)^2}{\sigma_p^2}.$$

That is, we use the variation among the means relative to the assumed common variance. Of course, what constitutes a large difference will vary from one situation to the next. Cohen (1988) has suggested that as a general guide, $\delta = 0.1$, 0.25, and 0.4 correspond to small, medium, and large effect sizes, respectively.

In practical terms, the following five values depend on one another:

1. J, the number of groups
2. n, the sample size for each group
3. δ, the standardized difference among the means
4. α, the Type I error probability
5. $1 - \beta$, power

If four of these five values are specified, the fifth is determined automatically and can be computed with the R function anova.power described next.

9.1.5 R Functions power.anova.test and anova.power

Two R functions for dealing with power and sample sizes are described. The first, written for this book, is the function

```
anova.power(groups=NULL,n=NULL,delta=NULL,
        sig.level=0.05,power=NULL)
```

where the argument delta corresponds to δ^2 as used in the previous section. The function requires that the number of groups be specified via the argument groups. All but one of the remaining arguments must be specified. The function determines the value of the one unspecified argument. The argument n represents the number of observations per group. (If two or more arguments are left NULL, the function returns an error.)

The other function for dealing with power and sample sizes, which comes with R, is

```
power.anova.test(groups = NULL, n = NULL, between.var = NULL, within.var =
                    NULL, sig.level = 0.05, power = NULL)
```

The argument between.var corresponds to

$$\sigma_\mu^2 = \frac{\sum(\mu_j - \mu_G)^2}{J-1},$$

the variation among the means. The argument within.var corresponds to σ_p^2, the assumed common variance.

EXAMPLE

Imagine we have four groups and the goal is to have power equal to 0.8 when testing at the 0.05 level and $\delta = 0.4$. The command

```
anova.power(groups=4,power=0.8,delta=0.4)
```

returns $n = 28.25$. That is, a sample size of 28 for each group accomplishes this goal, assuming normality and homoscedasticity.

EXAMPLE

Imagine that for four normal distributions having means 120, 130, 140, 150 and common variance $\sigma_p^2 = 500$, the goal is to have power equal to 0.9. For illustrative purposes, suppose the values of these population means are stored in the R variable z. The command

```
power.anova.test(groups = length(z), between.var=var(z), within.var=500,
                    power=0.90)
```

returns $n = 15.19$, indicating that about 15 observations per group are needed to achieve power equal to 0.9. It can be seen that $\sigma_\mu^2 = 166.6667$, and that with $\sigma_p^2 = 500$, $\delta^2 = 1$. The command

```
anova.power(groups=4,power=0.9,delta=1)
```

also returns the value $n = 15.19$.

EXAMPLE

A researcher fails to reject the hypothesis of equal means based on a sample size of $n = 40$ for each group. This raises the issue of whether power was sufficiently high. The command

```
anova.power(groups=4,n=40,delta=0.1)
```

indicates that power is 0.35 based on $\delta^2 = 0.1$, which Cohen suggests is a relatively small effect size.

9.2 DEALING WITH UNEQUAL VARIANCES

We saw in Chapter 8 that if the population variances are unequal, but the sample sizes are equal, Student's T controls Type I errors fairly well when sampling from normal distributions except when sample sizes are very small. But with more than two groups, problems arise when using the ANOVA F statistic even when the sample sizes are equal. With unequal sample sizes, problems controlling the probability of a Type I error are exacerbated. That is, in a very real sense, as the number of groups increases, practical problems with unequal variances increase, even under normality. For example, imagine you want to compare four groups, the null hypothesis of equal means is true, and you want the probability of a Type I error to be $\alpha = 0.05$. Situations arise where the actual probability of rejecting exceeds 0.27 due to comparing normal distributions that have unequal variances. When comparing six groups, the probability of a Type I error can exceed 0.3. That is, the actual probability of a Type I error is substantially higher than the stated level of 0.05.

Another serious problem is that even if distributions are normal, but have unequal variances, the power of the F test can be low relative to more modern methods. Also, even a single outlier in one group only can result in relatively low power. Moreover, the F test can be biased, meaning that the probability of rejecting can actually decrease as the difference among the population means increases.

One suggestion for trying to salvage the F test is to first test the hypothesis of equal variances, and if you fail to reject, assume equal variances and use F. But published studies indicate that this strategy is generally unsatisfactory (e.g., Hayes and Cai, 2007; Markowski and Markowski, 1990; Moser, Stevens, and Watts, 1989; Wilcox, Charlin, and Thompson, 1986; Zimmerman, 2004). Parra-Frutos (2016) concluded that, generally, using a bootstrap version of the Brown-Forsythe (1974a) method performs better than a collection of techniques that might be used to test assumptions. (But some exceptions were found when using one of the methods that were studied.) Perhaps with sufficiently large sample sizes this approach has merit, but there are no reliable guidelines on how large the sample sizes must be. As in Chapter 8, the basic problem is that tests for equal variances do not have enough power to detect situations where violating the assumption of equal variances causes practical problems. Currently, all indications are that a better strategy is to use a method that allows heteroscedasticity and performs about as well as the ANOVA F test when in fact the homoscedasticity assumption is met. The most successful methods are based in part on robust measures of location.

A positive feature of the ANOVA F test is that if all J groups have identical distributions, it appears to control the probability of a Type I error reasonably well. Also, as in Chapter 8, some authorities would argue that it is virtually impossible for the null hypothesis of equal means to be true and simultaneously to have unequal variances. If we accept this argument, the probability of a Type I error is no longer an issue, but this does not salvage the F test because problems controlling the probability of a Type I error when variances differ reflect problems with bias. Again, power can be relatively low.

9.2.1 Welch's Test

Many methods have been proposed for testing the equality of J means without assuming equal variances (e.g., Brown and Forsythe, 1974a; Chen and Chen, 1998; Mehrotra, 1997; James, 1951; Krutchkoff, 1988; Alexander and Govern, 1994; Fisher, 1935, 1941; Cochran and Cox, 1950; Lee and Ahn, 2003; Wald, 1955; Asiribo and Gurland, 1989; Scariano and Davenport, 1986; Matuszewski and Sotres, 1986; Pagurova, 1986; Weerahandi, 1995.) Unfortunately, all of the methods just cited, plus many others, have been found to have serious practical problems (e.g., Keselman et al., 1999).

It should be noted that yet another approach to dealing with heteroscedasticity is to use the regression (or general linear model) perspective, mentioned in Section 9.1.2, in conjunction with the HC4 estimator (introduced in Chapter 6). But Ng (2009) describes situations where this approach performs poorly in terms of both control over the probability of a Type

I error and power. As usual, the risk of relatively low power is particularly high when outliers tend to be common.

Krishnamoorthy et al. (2007) review other approaches for handling unequal variances, which are known to be unsatisfactory, and they suggest using instead a parametric bootstrap method. But their method assumes normality. Under non-normality, it can perform poorly. The actual Type I error probability, when testing at the 0.05 level, can exceed 0.25 (Cribbie et al., 2010). Parra-Frutos (2014) found that bootstrap-t method coupled with the method derived by Brown and Forsythe (1974a) performs relatively well when dealing with three groups. (The simulations were limited to what Gleason, 1993, characterizes as light-tailed distributions.) The method described here performs reasonably well under normality and heteroscedasticity and it forms the basis of a technique that deals with non-normality in a relatively effective manner. The method is due to Welch (1951) and it generally outperforms the F test. The computational details are relegated to Box 9.2. (The R function t1way, described in Section 9.4.1, contains Welch's test as a special case.)

EXAMPLE

R software is readily used to perform Welch's test. But to make sure the notation is clear, Welch's test is illustrated with the data in Table 9.1, which is taken from a study dealing with schizophrenia. Referring to Table 9.1 and Box 9.2, we see that

$$w_1 = \frac{10}{0.1676608} = 59.6, w_2 = \frac{10}{0.032679} = 306.0,$$

$$w_3 = \frac{10}{0.0600529} = 166.5, w_4 = \frac{10}{0.0567414} = 176.2.$$

Therefore,

$$U = 59.6 + 306 + 166.5 + 176.2 = 708.3,$$

$$\begin{aligned} \tilde{X} &= \frac{1}{708.3}\{59.6(0.276039) + 306(0.195615) + \\ &\quad 166.5(0.149543) + 176.2(0.399699\} \\ &= 0.242, \end{aligned}$$

$$\begin{aligned} A &= \frac{1}{4-1}\{59.6(0.276039 - 0.242)^2 + 306(0.195615 - 0.242)^2 + \\ &\quad 166.5(0.149543 - 0.242)^2 + 176.2(0.399699 - 0.242)^2\} \\ &= 2.18, \end{aligned}$$

$$\begin{aligned} B &= \frac{2(4-2)}{4^2-1}\{(1 - 59.6/708.3)^2/9 + (1 - 306.0/708.3)^2/9 + \\ &\quad (1 - 166.5/708.3)^2/9 + (1 - 176.2/708.3)^2/9\} \\ &= 0.0685, \end{aligned}$$

$$F_w = \frac{2.18}{1 + 0.0685} = 2.04.$$

Box 9.2: Computations for Welch's method.

Goal: Without assuming equal variances, test $H_0 : \mu_1 = \mu_2 = \cdots = \mu_J$, the hypothesis that J independent groups have equal means.

Computations: Let

$$w_1 = \frac{n_1}{s_1^2}, w_2 = \frac{n_2}{s_2^2}, \ldots, w_J = \frac{n_J}{s_J^2}.$$

Next, compute

$$U = \sum w_j$$

$$\tilde{X} = \frac{1}{U} \sum w_j \bar{X}_j$$

$$A = \frac{1}{J-1} \sum w_j (\bar{X}_j - \tilde{X})^2$$

$$B = \frac{2(J-2)}{J^2-1} \sum \frac{(1 - \frac{w_j}{U})^2}{n_j - 1}$$

$$F_w = \frac{A}{1+B}.$$

When the null hypothesis is true, F_w has, approximately, an F distribution with

$$\nu_1 = J - 1$$

and

$$\nu_2 = \left[\frac{3}{J^2-1} \sum \frac{(1 - w_j/U)^2}{n_j - 1} \right]^{-1}$$

degrees of freedom.

Decision Rule: Reject H_0 if $F_w \geq f$, where f is the $1 - \alpha$ quantile of the F distribution with ν_1 and ν_2 degrees of freedom.

The degrees of freedom are

$$\nu_1 = 4 - 1 = 3$$

and

$$\begin{aligned} \nu_2 &= \left[\frac{3}{4^2 - 1}(0.256) \right]^{-1} \\ &= 1/.0512 \\ &= 19.5. \end{aligned}$$

If you want the probability of a Type I error to be $\alpha = 0.05$, then referring to Table 6 in Appendix B, the critical value is approximately 3.1. Because $F_w = 2.04$, which is less than 3.1, fail to reject the hypothesis of equal means.

For the data in Table 9.1, both Welch's test and the ANOVA F test fail to reject the hypothesis of equal means. It is stressed, however, that in applied work, situations arise where

Table 9.1 Measures of Skin Resistance for Four Groups

(No Schiz.)	(Schizotypal)	(Schiz. Neg.)	(Schiz. Pos.)
0.49959	0.24792	0.25089	0.37667
0.23457	0.00000	0.00000	0.43561
0.26505	0.00000	0.00000	0.72968
0.27910	0.39062	0.00000	0.26285
0.00000	0.34841	0.11459	0.22526
0.00000	0.00000	0.79480	0.34903
0.00000	0.20690	0.17655	0.24482
0.14109	0.44428	0.00000	0.41096
0.00000	0.00000	0.15860	0.08679
1.34099	0.31802	0.00000	0.87532
$\bar{X}_1 = 0.276039$	$\bar{X}_2 = 0.195615$	$\bar{X}_3 = 0.149543$	$\bar{X}_4 = 0.399699$
$s_1^2 = 0.1676608$	$s_2^2 = 0.032679$	$s_3^2 = 0.0600529$	$s_4^2 = 0.0567414$
$n_1 = n_2 = n_3 = n_4 = 10$			

Welch's test rejects and the F test does not. That is, in applied work, it can matter which test you use. The following example illustrates this point.

EXAMPLE

Consider the following data:

Group 1:	53 2 34 6 7 89 9 12
Group 2:	7 34 5 12 32 36 21 22
Group 3:	5 3 7 6 5 8 4 3

The ANOVA F test yields $F = 2.7$ with a critical value of 3.24 when $\alpha = 0.05$, so you do not reject. The p-value is 0.09. In contrast, Welch's test yields $F_w = 8$ with a critical value of 4.2, so now you reject. The p-value is 0.009, the main point being that the two methods can give decidedly different results.

9.3 JUDGING SAMPLE SIZES AND CONTROLLING POWER WHEN DATA ARE AVAILABLE

Section 5.4.3 described Stein's two-stage method, which might be used to assess power, as well as the adequacy of the sample size, when using the one-sample Student's T test. That is, given some data, you can determine how large the sample sizes must be to achieve a desired amount of power. If the sample sizes used are small compared to what is needed to achieve high power, you have empirical evidence that the null hypothesis should not be accepted. This section describes an extension of Stein's method, derived by Bishop and Dudewicz (1978), for judging the sample sizes when testing the hypothesis of equal means. Normality is assumed, but unlike the ANOVA F test, homoscedasticity is not required. In fact, under normality, the method provides exact control over both Type I error probabilities and power.

Imagine the goal is to have power $1 - \beta$ given that the difference among the means is

$$\delta = \sum(\mu_j - \bar{\mu})^2.$$

In case it helps, it is noted that if

$$\mu_1 = \cdots = \mu_{J-1},$$

but

$$\mu_J - \mu_{J-1} = a,$$

then

$$\delta = \frac{a^2(J-1)}{J}.$$

That is, if $J-1$ of the population means are equal, but the other population mean exceeds all of the other means by a, we have a simple method for determining δ which might help when trying to specify what δ should be. For example, if for three groups you want power to be high when $\mu_1 = \mu_2 = 5$ but $\mu_3 = 6$, then $a = 1$ and $\delta = 2/3$. Given δ, α, $1-\beta$, and n_j observations randomly sampled from the jth group, Box 9.3 shows how to determine N_j, the number of observations needed to achieve the desired amount of power.

EXAMPLE

Imagine the goal is to have power $1-\beta = 0.8$ when comparing three groups, where one of the groups has a population mean 2.74 larger than the other two. That is, $a = 2.74$, so $\delta = 5$. For illustrative purposes, assume $\alpha = 0.05$ and that the sample sizes are $n_1 = n_2 = n_3 = 10$, so $\nu_1 = \nu_2 = \nu_3 = 9$, and

$$\nu = \frac{3}{\frac{1}{7}+\frac{1}{7}+\frac{1}{7}} + 2 = 9.$$

The tedious calculations eventually yield a critical value of $c = 8.15$ and $d = 0.382$. If the sample variance for the first group is $s_1^2 = 5.43$, then N_1 is either equal to $n_1 + 1 = 11$ or $[s_1^2/d] + 1 = [5.43/0.382] + 1 = 15$, whichever is larger. In this particular, case $N_1 = 15$ suggesting that the original sample size of 10 is not quite satisfactory. If $N_1 = n_1 + 1$, this suggests that the available sample sizes are reasonably adequate. If the second group has a sample variance of $s_2^2 = 10$, then $N_2 = 27$ and if $s_3^2 = 20$, $N_3 = 53$. So for group 3, you have only 10 observations and about five times as many observations are required for the specified amount of power.

The method just described indicates that an additional $N_j - n_j$ observations are needed for the jth group to achieve the desired amount of power. It is noted that if these additional observations can be obtained, the hypothesis of equal means can be tested without assuming homoscedasticity. More precisely, assuming normality, the probability of a Type I error will be exactly α and power will be at least $1 - \beta$ using the Bishop-Dudewicz ANOVA method outlined in Box 9.4.

Box 9.3: Given α, δ, and n_j observations randomly sampled from the jth group, determine N_j, the number of observations for the jth group needed to achieve power $1 - \beta$.

Let z be the $1 - \beta$ quantile of the standard normal distribution. For the jth group, let $\nu_j = n_j - 1$. Compute

$$\nu = \frac{J}{\sum \frac{1}{\nu_j - 2}} + 2,$$

$$A = \frac{(J-1)\nu}{\nu - 2}, \quad B = \frac{\nu^2}{J} \times \frac{J-1}{\nu - 2},$$

$$C = \frac{3(J-1)}{\nu - 4}, \quad D = \frac{J^2 - 2J + 3}{\nu - 2},$$

$$E = B(C + D),$$

$$M = \frac{4E - 2A^2}{E - A^2 - 2A},$$

$$L = \frac{A(M-2)}{M},$$

$$c = Lf,$$

where f is the $1 - \alpha$ quantile of an F distribution with L and M degrees of freedom,

$$b = \frac{(\nu - 2)c}{\nu},$$

$$A_1 = \frac{1}{2}\{\sqrt{2}z + \sqrt{2z^2 + 4(2b - J + 2)}\},$$

$$B_1 = A_1^2 - b,$$

$$d = \frac{\nu - 2}{\nu} \times \frac{\delta}{B_1}.$$

Then

$$N_j = \max\left\{n_j + 1, \left[\frac{s_j^2}{d}\right] + 1\right\}. \tag{9.7}$$

For technical reasons, the number of observations needed for the jth group, N_j, cannot be smaller than $n_j + 1$. (The notation $[s_j^2/d]$ means you compute s_j^2/d, then round down to the nearest integer.)

Box 9.4: Bishop–Dudewicz ANOVA

Goal: Test $H_0 : \mu_1 = \cdots = \mu_J$ such that power is $1 - \beta$ and the probability of a Type I error is α.

Assumptions: Normality, random sampling.

Stage 1: Randomly sample n_j observations from the jth group, yielding X_{ij}, $i = 1, \ldots, n_j$, and compute N_j as described in Box 9.3.

Stage 2: Randomly sample $N_j - n_j$ additional observations from the jth group, which are labeled X_{ij}, $i = n_j + 1, \ldots, N_j$. For the jth group compute

$$T_j = \sum_{i=1}^{n_j} X_{ij},$$

$$U_j = \sum_{i=n_j+1}^{N_j} X_{ij}$$

$$b_j = \frac{1}{N_j}\left(1 + \sqrt{\frac{n_j(N_j d - s_j^2)}{(N_j - n_j)s_j^2}}\right),$$

$$\tilde{X}_j = \frac{T_j\{1 - (N_j - n_j)b_j\}}{n_j} + b_j U_j.$$

The test statistic is

$$\tilde{F} = \frac{1}{d}\sum(\tilde{X}_j - \tilde{X})^2,$$

where

$$\tilde{X} = \frac{1}{J}\sum \tilde{X}_j.$$

Decision Rule: Reject H_0 if $\tilde{F} \geq c$, where c is the critical value given in Box 9.3.

9.3.1 R Functions bdanova1 and bdanova2

The R function

```
bdanova1(x, alpha = 0.05, power = 0.9, delta = NA)
```

performs the calculations in Box 9.3 and returns the number of observations required to achieve the specified amount of power. The argument `power` indicates how much power you want and defaults to 0.9. The argument `delta` corresponds to δ, and the data are stored in the R variable `x` which can be an n by J matrix or it can have list mode. (In the latter case it is assumed that `x[[1]]` contains the data for group 1, `x[[2]]` contains the data for group 2, and so on.)

The R function

```
bdanova2(x1, x2, alpha = 0.05, power = 0.9, delta = NA)
```

performs the second stage analysis once the additional observations are obtained, as described in Box 9.4. Here x1 and x2 contain the first stage and second stage data, respectively.

9.4 TRIMMED MEANS

Welch's heteroscedastic method for comparing means can be extended to trimmed means. That is, the goal is to test

$$H_0 : \mu_{t1} = \mu_{t2} = \cdots = \mu_{tJ},$$

the hypothesis that J independent groups have a common population trimmed mean without assuming the groups have equal variances. (For a method based on trimmed means that assumes equal variances, see Lee and Fung, 1985.) As in Chapter 8, trimming can greatly reduce practical problems (low power, poor control over the probability of a Type I error, and bias) associated with methods for comparing means.

Compute

$$d_j = \frac{(n_j - 1)s_{wj}^2}{h_j \times (h_j - 1)},$$

$$w_j = \frac{1}{d_j},$$

$$U = \sum w_j,$$

$$\tilde{X} = \frac{1}{U} \sum w_j \bar{X}_{tj},$$

$$A = \frac{1}{J-1} \sum w_j (\bar{X}_{tj} - \tilde{X})^2,$$

$$B = \frac{2(J-2)}{J^2-1} \sum \frac{(1 - \frac{w_j}{U})^2}{h_j - 1},$$

$$F_t = \frac{A}{1+B}. \tag{9.8}$$

When the null hypothesis is true, F_t has, approximately, an F distribution with

$$\nu_1 = J - 1$$

$$\nu_2 = \left[\frac{3}{J^2-1} \sum \frac{(1 - w_j/U)^2}{h_j - 1} \right]^{-1}$$

degrees of freedom. (For $J > 2$, the expression for ν_2 reduces to $2(J-2)/(3B)$.)

Decision Rule: Reject the hypothesis of equal population trimmed means if $F_t \geq f$, the $1 - \alpha$ quantile of an F distribution with ν_1 and ν_2 degrees of freedom.

Effect Size

It is noted that the heteroscedastic measure of size, ξ, introduced in Section 8.12, is readily extended to the situation at hand. When dealing with means, now the numerator of ξ^2, the explanatory measure of effect size, is

$$\sigma_\mu^2 = \frac{1}{J-1} \sum (\mu_j - \bar{\mu})^2,$$

which measures the variation among the means. Momentarily assume that equal sample sizes are used and let σ^2 be the (population) variance of the pooled data. Then, as in Section 8.12, the explanatory measure of effect size is

$$\xi^2 = \frac{\sigma_\mu^2}{\sigma^2}.$$

An estimate of σ_μ^2 is obtained simply by replacing the population means with the corresponding sample means. The estimate of σ^2 is obtained by pooling the observations and computing the sample variance. For the case of unequal sample sizes, let m be the minimum sample size among the J groups being compared. Next, sample without replacement m observations from each group and estimate σ_μ^2. Repeat this K times and average the results yielding a final estimate of $\sigma^2(Y)$. (As in Section 8.12, if the sample sizes are unequal, simply pooling the observations and estimating σ_μ^2 with the pooled data results in a technical problem, the details of which go beyond the scope of this book.) Again, ξ will be called a heteroscedastic measure of effect size. To get a robust version of ξ, again replace the means with a 20% trimmed mean and the variance with a 20% Winsorized variance that has been rescaled to equal the population variance under normality. (Rescaling when using other amounts of trimming is automatically done by the R function `t1wayv2` described in the next section.) (An empirical likelihood method for testing the hypothesis equal methods has been derived, but Velina et al., 2016, demonstrate that the method described here performs better in terms of controlling the Type I error probability.)

9.4.1 R Functions t1way, t1wayv2, t1wayF, and g5plot

The R function

```
t1way(x,tr=0.2,grp=NA)
```

tests the hypothesis of equal trimmed means using the method just described. The argument x can have list mode or it can be a matrix. In the former case, the data for group 1 are stored in the variable `x[[1]]`, group 2 is stored in `x[[2]]`, and so on. In the latter case, `x` is an n by J matrix where column 1 contains the data for group 1, column 2 contains the data for group 2, and so forth. The argument `tr` indicates the amount of trimming and when `tr=0`, this function performs Welch's method for means described in Section 9.2.1. The argument `grp` allows you to compare a selected subset of the groups. By default all groups are used. If you set `grp=c(1,3,4)`, then the trimmed means for groups 1, 3, and 4 will be compared with the remaining data ignored. The function returns the value of the test statistic and the corresponding p-value (so specifying a value for α is not necessary). The R function

```
t1wayv2(x,tr=0.2,grp=NA)
```

is the same as `t1way`, only the measure of effect size ξ is reported as well.

EXAMPLE

For the data in Table 9.1, and using the default amount of trimming, a portion of the output from `t1way` is

```
$TEST:
[1] 5.059361
```

```
$nu1:
[1] 3

$nu2:
[1] 10.82531

$p.value:
[1] 0.01963949
```

This says that the test statistic F_t is equal to 5.06 and the p-value is 0.0194. So, in particular, you would reject the hypothesis of equal trimmed means with $\alpha = 0.05$, or even 0.02. In contrast, as previously indicated, if we compare means with Welch's method or the ANOVA F test, we fail to reject with $\alpha = 0.05$, illustrating that the choice of method can alter the conclusions reached. Setting the argument `tr` to 0, `t1way` reports the results of Welch's test to be

```
$TEST:
[1] 2.038348

$nu1:
[1] 3

$nu2:
[1] 19.47356

$siglevel:
[1] 0.1417441
```

So switching from means to 20% trimmed means, the p-value drops from 0.14 to about 0.02.

EXAMPLE

It is common to find situations where we get a smaller p-value using trimmed means rather than means. However, the reverse situation can and does occur. This point is illustrated with data taken from Le (1994) where the goal is to compare the testosterone levels of four groups of male smokers: heavy smokers (group 1), light smokers (group 2), former smokers (group 3), and non-smokers (group 4). The data are:

G1	G2	G3	G4
0.29	0.82	0.36	0.32
0.53	0.37	0.93	0.43
0.33	0.77	0.40	0.99
0.34	0.42	0.86	0.95
0.52	0.74	0.85	0.92
0.50	0.44	0.51	0.56
0.49	0.48	0.76	0.87
0.47	0.51	0.58	0.64
0.40	0.61	0.73	0.78
0.45	0.60	0.65	0.72

The p-value using Welch's method is 0.0017. Using 20% trimmed means instead, the p-value is 0.029. One reason this is not surprising is that a boxplot for each group reveals no outliers and it can be seen that the estimated standard errors for the means are smaller than the corresponding standard errors of the 20% trimmed means. However, a boxplot for the first group suggests that the data are skewed which might be having an effect on the p-value of Welch's test beyond any differences among the means.

The R function

```
t1wayF(x,fac,tr=0.2,nboot=100,SEED=T)
```

is like the R function `t1wayv2`, only `x` is a column of data and the argument `fac` is a factor variable. In essence, this function eliminates the need to use the R function `fac2list`.

EXAMPLE

Assuming the plasma retinol data (described in Chapter 1) are stored in the R variable `plasma`,

```
t1wayF(plasma[,13],plasma[,3])
```

would compare the three groups indicated by column 3 (smoking status) using 20% trimmed means and the data in column 13, which are plasma beta-carotene measures.

Boxplots of the data can be created with the R command `boxplot(x)`, where `x` is any R variable that is a matrix with column corresponding to groups, or where `x` has list mode. Kernel density plots of the distributions can be created with the R function

```
g5plot(x)
```

for up to five variables. (Also see the function `rundis` in Section 14.4.5.)

9.4.2 Comparing Groups Based on Medians

The median has a relatively large standard error under normality, and more generally when sampling from a light-tailed distribution. But with a sufficiently heavy-tailed distribution, its standard error can be relatively small. If there is a specific interest in comparing medians, the general method for trimmed means described in this section is not recommended. That is, it is not recommended that you set `tr=0.5` when using the R function `t1way` in Section 9.4.1 because this results in using a poor estimate of the standard error of the median. A better approach is to estimate the standard error of each median with the McKean–Schrader method in Chapter 4 and then use a slight modification of the method for trimmed means. In particular, let S_j^2 be the McKean-Schrader estimate of the squared standard error of M_j, the sample median corresponding to the jth group ($j = 1, \ldots, J$). Let

$$w_j = \frac{1}{S_j^2},$$

$$U = \sum w_j,$$

$$\tilde{M} = \frac{1}{U} \sum w_j M_j,$$

$$A = \frac{1}{J-1} \sum w_j (M_j - \tilde{M})^2,$$

$$B = \frac{2(J-2)}{J^2-1} \sum \frac{(1 - \frac{w_j}{U})^2}{n_j - 1},$$

$$F_m = \frac{A}{1+B}. \tag{9.9}$$

Decision Rule: Reject the hypothesis of equal population medians if $F_m \ge f$, the $1-\alpha$ quantile of an F distribution with $\nu_1 = J-1$ and $\nu_2 = \infty$ degrees of freedom.

Beware of Tied Values. The method in this section can perform poorly when there are tied values, for reasons summarized in Chapter 8. When tied values occur, currently the best strategy is to use a particular type of bootstrap method (Wilcox, 2015b) that can be applied with R function `Qanova` in Section 9.5.4. Another option is to use the R function `medpb` described in Section 12.1.12, which also performs well when there are no tied values.

9.4.3 R Function med1way

The hypothesis of equal population medians can be tested with the R function

```
med1way(x,grp=NA)
```

where the argument `grp` can be used to analyze a subset of the groups if desired. (See Section 9.4.1.) The function returns the value of the test statistic, F_m, and the p-value.

9.5 BOOTSTRAP METHODS

As stressed in previous chapters, bootstrap methods are not a panacea for the practical problems encountered when comparing measures of location. However, when comparing more than two groups they deserve serious consideration, particularly when the sample sizes are small. Indeed, certain bootstrap methods combined with certain robust measures of location appear to perform remarkably well in terms of both Type I errors and maintaining relatively high power, even with relatively small sample sizes. Said another way, if the goal is to be sensitive to how groups differ in terms of a measure of location, without being sensitive to heteroscedasticity and differences in skewness, some of the methods described here appear to be among the best possible techniques that might be used. With sufficiently large sample sizes, non-bootstrap methods based on a trimmed mean might suffice, but it remains unclear how large the sample sizes must be. An argument for using non-bootstrap methods might be that if they reject, we can be reasonably certain that the groups have different distributions suggesting that surely any measures of location we might consider differ as well. A possible concern, however, is that non-bootstrap methods are more sensitive to bias problems. As usual, relatively poor power is always a possibility when using means.

9.5.1 A Bootstrap-t Method

The bootstrap-t method in Chapter 8 can be extended to the problem of testing

$$H_0 : \mu_{t1} = \mu_{t2} = \cdots = \mu_{tJ},$$

the hypothesis of equal trimmed means. When comparing means or trimmed means with a small amount of trimming, a bootstrap-t method appears to be a relatively good method for general use. The strategy is to use the available data to estimate an appropriate critical value for the test statistic, F_t, described in Section 9.4. First, for the jth group, set

$$Y_{ij} = X_{ij} - \bar{X}_{tj}.$$

That is, for the jth group, subtract the sample trimmed mean from each of the observed values. Next, for the jth group, generate a bootstrap sample of size n_j from the Y_{ij} values which we denote by Y^*_{ij}, $i = 1, \ldots, n_j; j = 1, \ldots, J$. So, in effect, the Y^*_{ij} values represent a random sample from distributions all of which have zero trimmed means. That is, in the bootstrap world, when working with the Y_{ij} values, the null hypothesis of equal trimmed means is true. Said another way, the observations are shifted so that each has a trimmed mean of zero with the goal of empirically determining an appropriate critical value by performing simulations in the sense described in Chapter 7. The value of the test statistic F_t given by Equation (9.8) and based on the Y_{ij}^* is labeled F_t^*. In the present context, the strategy is to use a collection of F_t^* values to estimate the distribution of F_t, the test statistic based on the original observations, when the null hypothesis is true. If we can do this reasonably well, then in particular we can determine an appropriate critical value.

To estimate the critical value, repeatedly generate bootstrap samples in the manner just described, each time computing the test statistic F_t^* based on the Y_{ij}^* values. Doing this B times yields $F_{t1}^*, \ldots, F_{tB}^*$. Next, put these B values in ascending order yielding $F_{t(1)}^* \leq \cdots \leq F_{t(B)}^*$ and let u be the value of $(1-\alpha)B$ rounded to the nearest integer. Then the hypothesis of equal trimmed means is rejected if

$$F_t \geq F_{t(u)}^*. \tag{9.10}$$

9.5.2 R Functions t1waybt and BFBANOVA

The R function

```
t1waybt(x, tr = 0.2, alpha = 0.05, grp = NA, nboot = 599)
```

performs the bootstrap-t method for trimmed means that was just described. The argument x is any R variable containing data that are stored in list mode or in a matrix. In the first case x[[1]] contains the data for group 1, x[[2]] contains the data for group 2, and so on. If x is a matrix, column 1 contains the data for group 1, column 2 contains the data for group 2, and so forth. The argument grp can be used to analyze a subset of the groups. For example, grp=c(2,4,5) would compare groups 2, 4, and 5 only. As usual, alpha is α and nboot is B, the number of bootstrap samples to be used. For the special case where the goal is to compare means, results by Parra-Frutos (2014) indicate that a bootstrap version of the Brown–Forsythe (1974a) test can provide better control over the Type I error probability when comparing three groups. (No results are currently available when comparing more than three groups.) The method can be applied with the R function

```
BFBANOVA(x, nboot = 1000).
```

EXAMPLE

If the data in Table 9.1 are stored in the R variable skin, the command

```
t1waybt(skin,tr=0)
```

tests the hypothesis of equal means and returns

```
$test:
[1] 2.04

$p.value
[1] 0.2307692
```

Although not indicated, if the goal were to test the hypothesis of equal means at the 0.05 level, the critical value would have been 4.65. Note that in Section 9.2.1, the critical value, assuming normality, was 3.1.

On rare occasions, the bootstrap test statistic used by t1waybt cannot be computed. For example, if the bootstrap variance within a single group is zero, the test statistic is not defined. When this happens, the function prints a warning message as well as the actual (effective) number of bootstrap samples that were used. In the previous example, this happened once; the effective number of bootstrap samples was 598 rather than the default 599. A possible way of avoiding this problem is to use a percentile bootstrap method described in the next section. A percentile bootstrap method is not recommended when working with means, but as indicated in more detail in the next section, it does perform well when using measures of location that are reasonably robust.

9.5.3 Two Percentile Bootstrap Methods

As was the case in previous chapters, there are situations where a percentile bootstrap method is preferable to a bootstrap-t method or any non-bootstrap method that uses standard errors. If the goal is to compare M-estimators, for example, all indications are that a percentile bootstrap method is best for general use. If the amount of trimming is reasonably large, say at least 20%, again a percentile bootstrap method performs relatively well. But when comparing medians, again tied values can create serious practical problems and the methods described here are not recommended. (Use the percentile bootstrap for medians in Chapter 12 in conjunction with the R function `medpb`.)

Method SHPB

Let θ be any population measure of location, such as the 20% trimmed mean (μ_t) or the median. There are many variations of the percentile bootstrap method that can be used to test

$$H_0 : \theta_1 = \cdots = \theta_J,$$

the hypothesis that J groups have a common measure of location, but only two are described here. The first is related to a test statistic mentioned by Schrader and Hettmansperger (1980) and studied by He et al. (1990), and will be called *method SHPB.*

Let $\hat{\theta}_j$ be an estimate of θ based on data from the jth group ($j = 1, \ldots, J$). The test statistic is

$$H = \frac{1}{N} \sum n_j (\hat{\theta}_j - \bar{\theta})^2,$$

where $N = \sum n_j$, and

$$\bar{\theta} = \frac{1}{J} \sum \hat{\theta}_j.$$

To determine a critical value, shift the empirical distributions of each group so that the measure of location being used is equal to zero. That is, set $Y_{ij} = X_{ij} - \hat{\theta}_j$ as was done in Section 9.5.1. Then generate bootstrap samples from each group in the usual way from the Y_{ij} values and compute the test statistic based on the bootstrap samples yielding H^*. Repeat this B times resulting in $H_1^*, \ldots, H_B^*$, and put these B values in order yielding $H_{(1)}^* \leq \cdots \leq H_{(B)}^*$. An estimate of an appropriate critical value is $H_{(u)}^*$, where $u = (1 - \alpha)B$, rounded to the nearest integer, and H_0 is rejected if $H \geq H_{(u)}^*$. (For simulation results on how this method performs when comparing M-estimators, see Wilcox, 1993b.)

Method LSPB

The second method stems from general results derived by Liu and Singh (1997) and will be called *method LSPB.* To convey the basic strategy, momentarily consider three groups only. Let

$$\delta_{12} = \theta_1 - \theta_2$$

be the difference between the measures of location for groups 1 and 2, let

$$\delta_{13} = \theta_1 - \theta_3$$

be the difference between the measures of location for groups 1 and 3, and let

$$\delta_{23} = \theta_2 - \theta_3$$

be the difference between the measures of location for groups 2 and 3. If the null hypothesis is true, then $\delta_{12} = \delta_{13} = \delta_{23} = 0$. To simplify matters, momentarily focus on δ_{12} and δ_{13}. Further imagine that we generate bootstrap samples from each group yielding δ^*_{12} and δ^*_{13} and that we repeat this process many times for a situation where the null hypothesis is true. Then we might get the plot shown in the left panel of Figure 9.2. Note that the point (0, 0) is near the center of the cloud of points. Said another way, the point (0, 0) is deeply nested within the cloud of bootstrap values. This is what we would expect when the null hypothesis is true, which corresponds to $\delta_{12} = \delta_{13} = 0$. But if the null hypothesis is false, a plot of the bootstrap values might yield something like the right panel of Figure 9.2. Now the point (0, 0) is near the outside edge of the cloud. That is, (0, 0) is not deeply nested within the cloud of bootstrap values suggesting that the point $(\delta_{12}, \delta_{13}) = (0, 0)$ is relatively unusual and the null hypothesis should be rejected.

Generalizing, let

$$\delta_{jk} = \theta_j - \theta_k,$$

where for convenience it is assumed that $j < k$. That is, the δ_{jk} values represent all pairwise differences among the J groups. When working with means, for example, δ_{12} is the difference between the means of groups 1 and 2, and δ_{35} is the difference between groups 3 and 5. If all J groups have a common measure of location (i.e., $\theta_1 = \cdots = \theta_J$), then in particular

$$H_0 : \delta_{12} = \delta_{13} = \cdots = \delta_{J-1,J} = 0 \tag{9.11}$$

is true. It can be shown that the total number of δ's in Equation (9.11) is $L = (J^2 - J)/2$. For example, if $J = 3$, there are $L = 3$ values: δ_{12}, δ_{13}, and δ_{23}.

For each group, generate bootstrap samples from the *original* values and compute the measure of location for each group. That is, the observations are *not* centered as was done in the previous method. Said another way, bootstrap samples are *not* generated from the Y_{ij} values but rather from the X_{ij} values. Repeat this B times. The resulting estimates of location are represented by $\hat{\theta}^*_{jb}$ $(j = 1, \ldots, J; b = 1, \ldots, B)$ and the corresponding estimates of δ are denoted by $\hat{\delta}^*_{jkb}$. (That is, $\hat{\delta}^*_{jkb} = \hat{\theta}^*_{jb} - \hat{\theta}^*_{kb}$.) The general strategy is to determine how deeply $\mathbf{0} = (0, \ldots, 0)$ is nested within the bootstrap values $\hat{\delta}^*_{jkb}$ (where $\mathbf{0}$ is a vector having length L). For the special case where only two groups are being compared, this is tantamount to determining the proportion of times $\hat{\theta}^*_{1b} > \hat{\theta}^*_{2b}$, among all B bootstrap samples, which is how we proceeded in Section 8.5. But here we need special techniques for comparing more than two groups.

There remains the problem of measuring how deeply $\mathbf{0}$ is nested within the bootstrap values. Several methods have been proposed, but the involved computational details are not important here and are omitted. (Readers interested in these details are referred to Wilcox, 2017, Section 7.6.) Two methods are available via the R functions written for this book and it is the relative merits of these two methods that are important for present purposes. Briefly, the first method computes what is called *Mahalanobis distance* for each point among all of the B bootstrap points. Roughly, Mahalanobis distance is a standardized distance from the center of the data cloud that depends in part on the covariances among the $\hat{\delta}^*_{jkb}$ values. (More details about Mahalanobis distance are provided in Chapter 13.) Then the

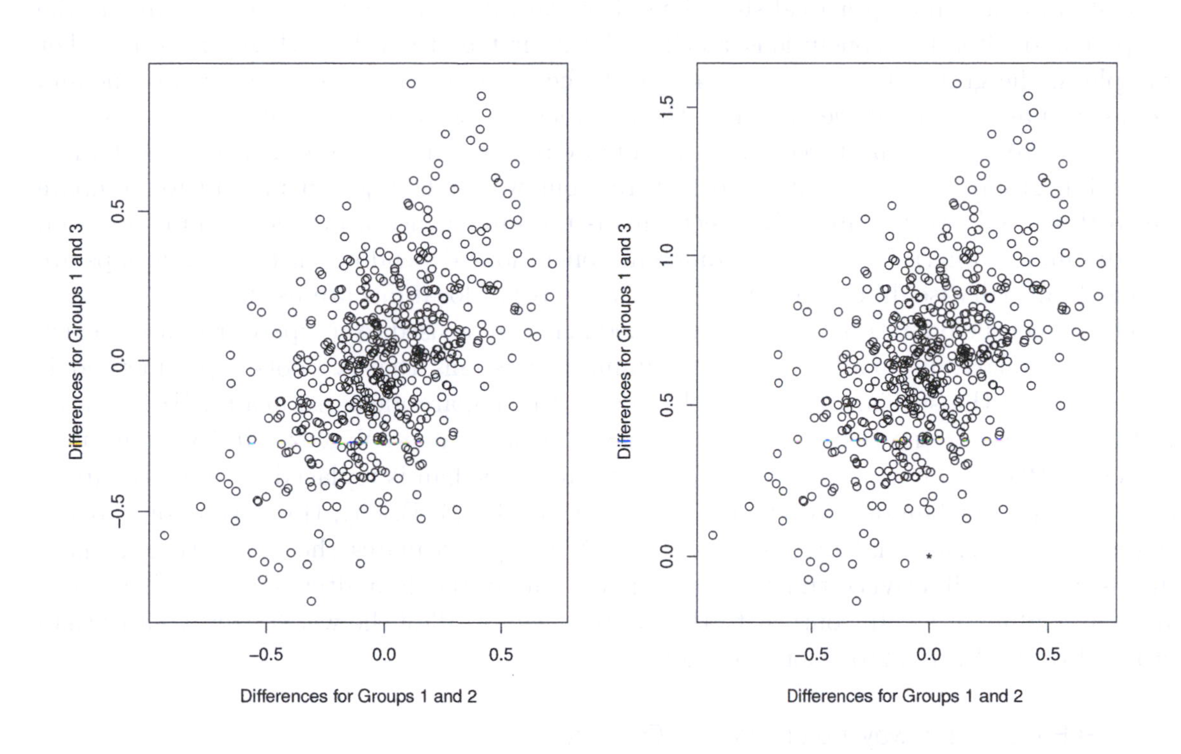

Figure 9.2 Plots of the differences between bootstrap measures of location. The left panel corresponds to a situation where the null hypothesis of equal population measures of location is true, in which case the differences should be clustered around the point (0, 0). That is, the point (0, 0) should be nested deeply within the cloud of points. The right panel corresponds to a situation where the null hypothesis is false. Note that now the point (0, 0), marked by *, is near the edge of the cloud suggesting that the point (0, 0) is relatively unusual.

distance of $\mathbf{0} = (0, \ldots, 0)$ from the center of the cloud is computed and if this distance is relatively large, reject the hypothesis that J groups have identical measures of location. When comparing M-estimators and MOM, this approach appears to be relatively effective in terms of controlling the Type I error probability. But a practical concern is that situations arise where Mahalanobis distance cannot be computed. This problem can be avoided using the second method, which is based on what is called the *projection distance* of a point. (See Wilcox, 2017, Section 6.2.5.) But a criticism of this approach is that with a large sample size, execution time can be somewhat high, particularly when the number of groups is large. If, however, R is running on a computer with a multicore processor, execution time can be reduced substantially as indicated in Section 9.5.4.

Notice that with three groups ($J = 3$), $\theta_1 = \theta_2 = \theta_3$ can be true if and only if $\theta_1 = \theta_2$ and $\theta_2 = \theta_3$. So in terms of Type I errors, *if* the null hypothesis is true, it suffices to test

$$H_0 : \theta_1 - \theta_2 = \theta_2 - \theta_3 = 0,$$

rather than testing

$$H_0 : \theta_1 - \theta_2 = \theta_2 - \theta_3 = \theta_1 - \theta_3 = 0,$$

the hypothesis that all pairwise differences are zero. An advantage of the first approach is that problems with computing Mahalanobis distance are reduced substantially and nearly eliminated among most practical situations. However, if groups differ, then rearranging the groups could alter the conclusions reached if the first of these hypotheses is tested. For example, if the groups have means 6, 4, and 2, then the difference between groups one and two, as well as two and three, is 2. But the difference between groups one and three is 4, so comparing groups one and three could mean more power. That is, we might not reject when comparing group one to two and two to three, but we might reject if instead we compare one to three and two to three. (In effect, this is the same concern mentioned in connection with dummy coding.) To help avoid different conclusions depending on how the groups are arranged, all pairwise differences among the measures of locations are used.

The percentile bootstrap methods described here can be used to compare trimmed means, but with a relatively small amount of trimming, it seems that a bootstrap-t method is preferable. For the special case where the goal is to compare medians, again SHPB can be used when there are no tied values. When there are tied values, Wilcox (2015b) found that method LSPB can be used provided the usual sample median is replaced by the estimate of the population median derived by Harrell and Davis (1982). Briefly, this estimator is based on weights $w_1, \dots w_n$ that are chosen so that $\sum w_i X_{(i)}$ estimates the population median. That is, the Harrell–Davis estimator is a weighted sum of the data after the values have been put in ascending order. In contrast to the sample median, all of the weights are greater than zero. (Also see Wilcox, 2017, for more details.)

9.5.4 R Functions b1way, pbadepth, and Qanova

Method SHPB, described in the previous section, can be applied with the R function

```
b1way(x,est=onestep,alpha=0.05,nboot=599).
```

By default it uses an M-estimator (with Huber's Ψ). The function

```
pbadepth(x, est=onestep, con=0, alpha=0.05, nboot=2000, grp=NA, op=1, MM =
      F, MC=F, SEED=T, na.rm = F, ...)
```

performs method LSPB. As usual, the argument ... can be used to reset default settings associated with the estimator indicated by the argument `est`. The argument `op` determines how the distance from the center of the bootstrap cloud is measured. By default, Mahalanobis distance is used. (The argument `con` can be used to test a hypothesis associated with a collection of linear contrasts, which are explained in Chapter 12.) To use projection distances, set `op=3`. If `op=3` and `MC=T`, the function uses a multicore processor, assuming one is available and that the R package parallel has been installed; this will reduce execution time.

For the special case where the goal is to compare the medians, the R function

```
Qanova(x,q = 0.5, op = 3, nboot = 600, MC = FALSE, SEED = TRUE)
```

can be used, which performs reasonably well even when there are tied values. The argument `q=0.5` indicates that the medians are to be compared. To compare the lower quartiles, for example, set `q=0.25`.

9.5.5 Choosing a Method

When comparing means or when using trimmed means with a small amount of trimming, the bootstrap-t method in Section 9.5.1 appears to be one of the better methods that might be used. Again it is stressed that no method for comparing means is completely satisfactory. When comparing M-estimators or MOM, there is weak evidence that method LSPB is best for general use in terms of power, but in fairness this issue is in need of more study. A possible problem with LSPB is that it uses Mahalanobis distance, which might result in a computational error. (The inverse of the covariance matrix cannot be computed.) One way of dealing with this problem is to set the argument `op=3`, which will result in using projection distance. However, particularly when comparing a large number of groups, this might result in high execution time. In this case, you have two options. If you are running R on a computer with a multicore processor, set the argument `MC=T` and again use `op=3`. (This assumes you have installed the R package parallel.) If your computer does not have a multicore processor, use method SHPB in Section 9.5.3 via the R function `b1way`.

When using trimmed means, `pbadepth` seems to be a good choice over a bootstrap-t method with a sufficient amount of trimming. But this issue is in need of further study.

9.6 RANDOM EFFECTS MODEL

The ANOVA methods covered so far deal with what is called a *fixed effect* design, roughly meaning that the goal is to compare J specific (fixed) groups. In contrast is a random effects design where groups are randomly sampled and the goal is to generalize to a larger population of groups. For example, consider a study where it is suspected that the personality of the experimenter has an effect on the results. Among all the experimenters we might use, do the results vary depending on who conducts the experiment? Here, the notion of J groups corresponds to a sample of J experimenters and for the jth experimenter we have results on n_j participants. The goal is to not only compare results among the J experimenters but to generalize to the entire population of experimenters we might use.

A study reported by Fears et al. (1996) provides another illustration where 16 estrone measures (in pg/mL) from each of five postmenopausal women were taken and found to be as shown in Table 9.2. Of interest was whether the estrone levels vary among women. That is, we envision the possibility of taking many measures from each woman, but the goal is not to simply compare the five women in the study but rather to generalize to all women who might have taken part in the study.

A study by Cronbach et al. (1972, Chapter 6) provides yet another example. The Porch index of communicative ability (PICA) is a test designed for use by speech pathologists. It is intended for initial diagnosis of patients with aphasic symptoms and for measuring the change during treatment. The oral portion of the test consists of several subtests, but to keep the illustration simple, only one subtest is considered here. This is the subtest where a patient is shown an object (such as a comb) and asked how the object is used. The response by the patient is scored by a rater on a 16-point scale. A score of 6, for example, signifies a response that is "intelligible but incorrect," and a score of 11 indicates a response that is "accurate but delayed and incomplete." A concern is that one set of objects might lead to a different rating compared to another set of objects one might use. Indeed, we can imagine a large number of potential sets of objects that might be used. To what extent do ratings differ among all of the potential sets of objects we might employ?

Let $\mu_G = E(\mu_j)$, where the expected value of μ_j is taken with respect to the process of randomly sampling a group. If all groups have a common mean, then, of course, no matter which J groups you happen to pick, it will be the case that

$$\mu_1 = \mu_2 = \cdots = \mu_J.$$

A more convenient way of describing the situation is to say that there is no variation among all the population means. A way of saying this in symbols is that $\sigma_\mu^2 = 0$, where

$$\sigma_\mu^2 = E(\mu_j - \mu_G)^2,$$

Table 9.2 Estrone Assay Measurements of a Single Blood Sample from Each of Five Postmenopausal Women

	Individuals				
Vial	P1	P2	P3	P4	P5
1	23	25	38	14	46
2	23	33	38	16	36
3	22	27	41	15	30
4	20	27	38	19	29
5	25	30	38	20	36
6	22	28	32	22	31
7	27	24	38	16	30
8	25	22	42	19	32
9	22	26	35	17	32
10	22	30	40	18	31
11	23	30	41	20	30
12	23	29	37	18	32
13	27	29	28	12	25
14	19	37	36	17	29
15	23	24	30	15	31
16	18	28	37	13	32

and again expectation is taken with respect to the process of randomly selecting μ_j. That is, among all groups of interest, σ_μ^2 is the variance of the population means. Testing the hypothesis that all groups have the same mean is equivalent to testing

$$H_0 : \sigma_\mu^2 = 0.$$

To test H_0, the following assumptions are typically made:

1. All of the (randomly) chosen groups have a normal distribution with a common variance, σ_p^2. That is, a homogeneity of variance assumption is imposed.

2. The difference, $\mu_j - \mu_G$, has a normal distribution with mean 0 and variance σ_μ^2.

3. The difference, $X_{ij} - \mu_j$, is independent of the difference $\mu_j - \mu_G$.

Let MSBG and MSWG be as defined in Section 9.1, and primarily for notational convenience, temporarily assume equal sample sizes. That is,

$$n = n_1 = \cdots = n_J.$$

Based on the assumptions just described, it can be shown that

$$E(\text{MSBG}) = n\sigma_\mu^2 + \sigma_p^2,$$

and that

$$E(\text{MSWG}) = \sigma_p^2.$$

When the null hypothesis is true, $\sigma_\mu^2 = 0$, and

$$E(\text{MSBG}) = \sigma_p^2.$$

Table 9.3 Hypothetical Data Used to Illustrate a Random Effects Model

Dosage 1	Dosage 2	Dosage 3
7	3	9
0	0	2
4	7	2
4	5	7
4	5	1
7	4	8
6	5	4
2	2	4
3	1	6
7	2	1

That is, when the null hypothesis is true, MSBG and MSWG estimate the same quantity, so the ratio

$$F = \frac{\text{MSBG}}{\text{MSWG}}$$

should have a value reasonably close to 1. If the null hypothesis is false, MSBG will tend to be larger than MSWG, so if F is sufficiently large, reject. It can be shown that F has an F distribution with $J-1$ and $N-J$ degrees of freedom when the null hypothesis is true, so reject if $F \geq f_{1-\alpha}$, where $f_{1-\alpha}$ is the $1-\alpha$ quantile of an F distribution with $\nu_1 = J-1$ and $\nu_2 = N-J$ degrees of freedom. Put more simply, the computations are exactly the same as they are for the fixed-effects ANOVA F test in Section 9.1. The only difference is how the experiment is performed. Here the levels are chosen at random, where as in Section 9.1 they are fixed.

As mentioned in Section 9.1, the fixed effect ANOVA model is often written as

$$X_{ij} = \bar{\mu} + \alpha_j + \epsilon_{ij},$$

where ϵ_{ij} has a normal distribution with mean zero and variance σ_p^2. In contrast, the random effects model is

$$X_{ij} = \mu_G + a_j + \epsilon_{ij},$$

where $a_j = \mu_j - \mu_G$. The main difference between these two models is that in the fixed effect model, α_j is an unknown *parameter*, but in the random effects model, a_j is a *random variable* that is assumed to have a normal distribution.

EXAMPLE

Suppose that for three randomly sampled dosage levels of a drug, you get the results shown in Table 9.3. To test the null hypothesis of equal means among all dosage levels you might use, compute the degrees of freedom and the F statistic as described in Box 9.1. This yields $\nu_1 = 2$, $\nu_2 = 27$, and $F = 0.53$, which is not significant at the $\alpha = 0.05$ level. That is, among all dosage levels you might have used, you fail to detect a difference among the corresponding means.

9.6.1 A Measure of Effect Size

As pointed out in Chapter 7, if you test and reject the hypothesis of equal means, there remains the issue of measuring the extent to which two groups differ. As previously explained, the p-value can be unsatisfactory. From Chapter 7, it is evident that finding an appropriate measure of effect size is a complex issue. When dealing with more than two groups, the situation is even more difficult. Measures have been proposed under the assumption of equal variances. They are far from satisfactory, but few alternative measures are available. However, measures derived under the assumption of equal variances are in common use, so it is important to discuss them here.

Suppose you randomly sample a group from among all the groups you are interested in, and then you randomly sample an individual and observe the outcome X. Let σ_X^2 be the variance of X. It can be shown that

$$\sigma_X^2 = \sigma_\mu^2 + \sigma_p^2$$

when the assumptions of the random effects model are true, where σ_p^2 is the assumed common variance among all groups we might compare. A common measure of effect size is

$$\rho_I = \frac{\sigma_\mu^2}{\sigma_\mu^2 + \sigma_p^2},$$

which is called the *intraclass correlation coefficient.* The value of ρ_I is between 0 and 1 and measures the variation among the means relative to the variation among the observations. If there is no variation among the means, in which case they have identical values, $\rho_I = 0$.

To estimate ρ_I, compute

$$n_0 = \frac{1}{J-1}\left(N - \sum \frac{n_j^2}{N}\right),$$

where $N = \sum n_j$ is the total sample size. The usual estimate of σ_μ^2 is

$$s_u^2 = \frac{\text{MSBG} - \text{MSWG}}{n_0},$$

in which case the estimate of ρ_I is

$$\begin{aligned} r_I &= \frac{s_u^2}{s_u^2 + \text{MSWG}} \\ &= \frac{\text{MSBG} - \text{MSWG}}{\text{MSBG} + (n_0 - 1)\text{MSWG}} \\ &= \frac{F-1}{F + n_0 - 1}. \end{aligned}$$

For the data in Table 9.1 it was found that $F = 1.51$, $n_0 = 10$, so

$$r_I = \frac{1.51 - 1}{1.51 + 10 - 1} = 0.05.$$

That is, about 5% of the variation among the observations is due to the variation among the means.

Donner and Wells (1986) compared several methods for computing an approximate confidence interval for ρ_I, and their results suggest using a method derived by Smith (1956). Smith's confidence interval is given by

$$r_I \pm z_{1-\alpha/2} V,$$

where $z_{1-\alpha/2}$ is the $1-\alpha/2$ quantile of the standard normal distribution, read from Table 1 in Appendix B, and

$$V = \sqrt{A(B+C+D)},$$

where

$$A = \frac{2(1-r_I)^2}{n_0^2}$$

$$B = \frac{[1+r_I(n_0-1)]^2}{N-J}$$

$$C = \frac{(1-r_I)[1+r_I(2n_0-1)]}{(J-1)}$$

$$D = \frac{r_I^2}{(J-1)^2}\left(\sum n_j^2 - \frac{2}{N}\sum n_j^3 + \frac{1}{N^2}\left(\sum n_j^2\right)^2\right).$$

For equal sample sizes an exact confidence interval is available, still assuming that sampling is from normal distributions with equal variances (Searle, 1971). Let $f_{1-\alpha/2}$ be the $1-\alpha/2$ quantile of the F distribution with $\nu_1 = J-1$ and $\nu_2 = N-J$ degrees of freedom. Similarly, $f_{\alpha/2}$ is the $\alpha/2$ quantile. Then an exact confidence interval for ρ_I is

$$\left(\frac{F/f_{1-\alpha/2}-1}{n+F/f_{1-\alpha/2}-1}, \frac{F/f_{\alpha/2}-1}{n+F/f_{\alpha/2}-1}\right),$$

where n is the common sample size. The tables in Appendix B only give the upper quantiles of an F distribution, but you need the lower quantiles when computing a confidence interval for ρ_I. To determine $f_{\alpha/2,\nu_1,\nu_2}$, you reverse the degrees of freedom, and look up $f_{1-\alpha/2,\nu_2,\nu_1}$, in which case

$$f_{\alpha/2,\nu_1,\nu_2} = \frac{1}{f_{1-\alpha/2,\nu_2,\nu_1,}}.$$

For example, if $\alpha = 0.05$, and you want to determine $f_{0.025}$ with $\nu_1 = 2$ and $\nu_2 = 21$ degrees of freedom, you first look up $f_{0.975}$ with $\nu_1 = 21$ and $\nu_2 = 2$ degrees of freedom. The answer is 39.45. Then $f_{0.025}$ with 2 and 21 degrees of freedom is the reciprocal of 39.45. That is

$$f_{0.025,2,21} = \frac{1}{39.45} = 0.025.$$

EXAMPLE

Assume that professors are rated on their level of extroversion and you want to investigate how their level of extroversion is related to student evaluations of a course. Suppose you randomly sample three professors, and their student evaluations are as shown in Table 9.4. (In reality one would, of course, want to sample more than three professors, but the goal here is to keep the illustration simple.) To illustrate how a confidence interval for ρ_I is computed, suppose $\alpha = 0.05$. Then $n = 8$, $f_{0.025} = 0.025$, $f_{0.975} = 4.42$, $F = 6.05$, and the 0.95 confidence interval for ρ_I is

$$\left(\frac{\frac{6.05}{4.42}-1}{8+\frac{6.05}{4.42}-1}, \frac{\frac{6.05}{0.025}-1}{8+\frac{6.05}{0.025}-1}\right) = (0.047,\ 0.967).$$

Hence, you can be reasonably certain that ρ_I has a value somewhere between 0.047 and 0.967. Notice that the length of the confidence interval is relatively large since ρ_I has a value

Table 9.4 Students' Ratings

Group 1	Group 2	Group 3
3	4	6
5	4	7
2	3	8
4	8	6
8	7	7
4	4	9
3	2	10
9	5	9
$\bar{X}_1 = 4.75$	$\bar{X}_2 = 4.62$	$\bar{X}_3 = 7.75$

between 0 and 1. Thus, in this case, the data might be providing a relatively inaccurate estimate of the intraclass correlation.

In some situations you might also want a confidence interval for σ_μ^2. Methods for accomplishing this goal are available, but no details are given here. For a recent discussion of this problem, see Brown and Mosteller (1991).

9.6.2 A Heteroscedastic Method

One serious concern about the random effects model just described is the assumption of equal variances. We have seen that violating this assumption can result in poor power and undesirable power properties in the fixed effect design and this problem continues for the situation at hand. This section describes a method derived by Jeyaratnam and Othman (1985) for handling unequal variances. (For an alternative approach, see Westfall, 1988.) As usual, let s_j^2 be the sample variance for the jth group, let $\bar{X}_j$ be the sample mean, and let $\bar{X} = \sum \bar{X}_j / J$ be the average of the J sample means. To test $H_0 : \sigma_\mu^2 = 0$, compute

$$q_j = \frac{s_j^2}{n_j},$$

$$\text{BSS} = \frac{1}{J-1} \sum (\bar{X}_j - \bar{X})^2,$$

$$\text{WSS} = \frac{1}{J} \sum q_j,$$

in which case the test statistics is

$$F_{jo} = \frac{\text{BSS}}{\text{WSS}}$$

with

$$\nu_1 = \frac{\left(\frac{J-1}{J} \sum q_j\right)^2}{\left(\sum \frac{q_j}{J}\right)^2 + \frac{J-2}{J} \sum q_j^2}$$

and

$$\nu_2 = \frac{(\sum q_j)^2}{\sum \frac{q_j^2}{n_j - 1}}$$

degrees of freedom. In the illustration regarding students' ratings,

$$\text{BSS} = 3.13$$

$$\text{WSS} = 0.517$$

$$F_{jo} = 6.05.$$

(The numerical details are left as an exercise.) The degrees of freedom are $\nu_1 = 1.85$ and $\nu_2 = 18.16$, and the critical value is 3.63. Because $6.05 > 3.63$, reject and conclude there is a difference among students' ratings.

When there are unequal variances, a variety of methods have been suggested for estimating σ^2_μ, several of which were compared by Rao, Kaplan, and Cochran (1981). Their recommendation is that when $\sigma^2_\mu > 0$, σ^2_μ should be estimated with

$$\hat{\sigma}^2_\mu = \frac{1}{J} \sum \ell^2_j (\bar{X}_j - \tilde{X})^2,$$

where

$$\ell_j = \frac{n_j}{n_j + 1}$$

$$\tilde{X} = \frac{\sum \ell_j \bar{X}_j}{\sum \ell_j}.$$

Evidently there are no results on how this estimate performs under non-normality.

9.6.3 A Method Based on Trimmed Means

Under normality with unequal variances, the F test can have a Type I error probability as high as 0.179 when testing at the $\alpha = 0.05$ level with equal sample sizes of 20 in each group (Wilcox, 1994a). The Jeyaratnam–Othman test statistic, F_{jo}, has a probability of a Type I error close to 0.05 in the same situation. However, when the normality assumption is violated, the probability of a Type I error using both F and F_{jo} can exceed 0.3. Another concern with both F and the Jeyaratnam–Othman method is that there are situations where power decreases even when the difference among the means increases. This last problem appears to be reduced considerably when using trimmed means. Trimmed means provide better control over the probability of a Type I error and can yield substantially higher power when there are outliers. Of course there are exceptions. Generally no method is best in all situations. But if the goal is to reduce the problems just described, an extension of the Jeyaratnam–Othman method to trimmed means has considerable practical value. (Extensions of the random effects model based on MOM have not been investigated as yet.) Readers interested in the derivation and technical details of the method are referred to Wilcox (2017, Section 7.5).

The computational details are as follows. For each of the J groups, Winsorize the observations and label the results Y_{ij}. Let h_j be the effective sample size of the jth group (the number of observations left after trimming), and compute

$$\bar{Y}_j = \frac{1}{n_j} \sum_{i=1}^{n_j} Y_{ij},$$

$$s^2_{wj} = \frac{1}{n_j - 1} \sum (Y_{ij} - \bar{Y}_j)^2,$$

$$\bar{X}_t = \frac{1}{J} \sum \bar{X}_{tj},$$

$$\text{BSST} = \frac{1}{J-1} \sum_{j=1}^{J} (\bar{X}_{tj} - \bar{X}_t)^2,$$

$$\text{WSSW} = \frac{1}{J} \sum_{j=1}^{J} \sum_{i=1}^{n_j} \frac{(Y_{ij} - \bar{Y}_j)^2}{h_j(h_j - 1)},$$

$$D = \frac{\text{BSST}}{\text{WSSW}}.$$

Let

$$q_j = \frac{(n_j - 1)s_{wj}^2}{Jh_j(h_j - 1)}.$$

The degrees of freedom are estimated to be

$$\hat{\nu}_1 = \frac{((J-1)\sum q_j)^2}{(\sum q_j)^2 + (J-2)J\sum q_j^2}$$

$$\hat{\nu}_2 = \frac{(\sum q_j)^2}{\sum q_j^2/(h_j - 1)}.$$

Decision Rule: Reject if $D \geq f$, the $1 - \alpha$ quantile of an F distribution with $\hat{\nu}_1$ and $\hat{\nu}_2$ degrees of freedom.

9.6.4 R Function rananova

The R function

```
rananova(x,tr=0.2,grp=NA)
```

performs the method described in the previous section. As usual, `x` is any R variable that has list mode or is a matrix (with columns corresponding to groups), `tr` is the amount of trimming, which defaults to 0.2, and `grp` can be used to specify some subset of the groups if desired. If `grp` is not specified, all groups stored in `x` are used. If the data are not stored in a matrix or in list mode, the function terminates and prints an error message. If the data are stored in a manner that includes a factor variable, use the R function `fac2list` to convert the data to list mode; see Section 9.1.3 for an illustration. The function `rananova` returns the value of the test statistic, D, in `rananova$teststat`, the p-value level is stored in `rananova$siglevel`, and an estimate of a Winsorized intraclass correlation is returned in the R variable `rananova$rho`. This last quantity is like the intraclass correlation ρ_I, but with the variance of the means replaced by a Winsorized variance.

EXAMPLE

Imagine that the data in Table 9.4 are stored in the R variable `data`. Then the R command `rananova(data)` returns

```
$teststat
[1] 4.911565

$df
[1]  1.850360 12.911657

$siglevel
[1] 0.02789598

$rho
[1] 0.3691589
```

So, in particular, the hypothesis of equal trimmed means is rejected at the 0.05 level. The value for `rho` indicates that about 37% of the Winsorized variance among the observations is accounted for by the Winsorized variance of the *Winsorized* means. (For technical reasons, Winsorized means are used rather than trimmed means when deriving a robust analog of the intraclass correlation coefficient.) The command `rananova(data,tr=0)` compares means instead and returns

```
$teststat
[1] 6.053237

$df
[1]  1.854842 18.157957

$siglevel
[1] 0.01087606

$rho
[1] 0.3095772
```

Now the intraclass correlation is estimated to be 0.31 meaning that the variation among the means is estimated to account for 31% of the variation among all possible observations.

9.7 RANK-BASED METHODS

As mentioned in previous chapters, a possible appeal of rank-based methods is that they are insensitive to outliers, which might translate into more power compared to a method based on means. It seems that comparing 20% trimmed means typically provides more power, but there are exceptions. Also, some researchers argue that, at least in some situations, methods based on the ordering of values (their ranks) are preferable to methods that require an interval scale (e.g., Cliff, 1996; cf. Velleman and Wilkinson, 1993; Stevens, 1951). Roughly, interval scales have units of measurement with an arbitrary zero point, such as degrees Celsius, while ordinal scales do not require a unit of measurement, only the ability to determine whether one value is larger than another.

9.7.1 The Kruskall–Wallis Test

The best-known rank-based method for comparing multiple groups is the Kruskall–Wallis test. The goal is to test

$$H_0 : F_1(x) = F_2(x) = \cdots = F_J(x), \tag{9.12}$$

the hypothesis that J independent groups have identical distributions. A crude description of the method is that it is designed to be sensitive to differences among the average ranks corresponding to the groups being compared.

The method begins by pooling all $N = \sum n_j$ observations and assigning ranks. In symbols, if X_{ij} is the ith observation in the jth group, let R_{ij} be its rank among the pooled data. When there are tied values, use midranks as described in connection with the Brunner–Munzel method in Section 8.8.6. Next, sum the ranks for each group. In symbols, compute

$$R_j = \sum_{i=1}^{n_j} R_{ij},$$

Table 9.5 Hypothetical Data Illustrating the Kruskal–Wallis Test

Group 1		Group 2		Group 3	
X_{i1}	R_{i1}	X_{i2}	R_{i2}	X_{i3}	R_{i3}
40	1	45	3	61	9
56	6	58	7	65	10
42	2	60	8	55	5
				47	4

$(j = 1, \ldots, J)$. Letting

$$S^2 = \frac{1}{N-1}\left(\sum_{j=1}^{J}\sum_{i=1}^{n_j} R_{ij}^2 - \frac{N(N+1)^2}{4}\right),$$

the test statistic is

$$T = \frac{1}{S^2}\left(-\frac{N(N+1)^2}{4} + \sum \frac{R_j^2}{n_j}\right).$$

If there are no ties, S^2 simplifies to

$$S^2 = \frac{N(N+1)}{12},$$

and T becomes

$$T = -3(N+1) + \frac{12}{N(N+1)}\sum \frac{R_j^2}{n_j}.$$

The hypothesis of identical distributions is rejected if $T \geq c$, where c is some appropriate critical value. For small sample sizes, exact critical values are available from Iman, Quade, and Alexander (1975). For large sample sizes, the critical value is approximately equal to the $1-\alpha$ quantile of a chi-square distribution with $J-1$ degrees of freedom.

EXAMPLE

Table 9.5 shows data for three groups and the corresponding ranks. For example, after pooling all $N = 10$ values, $X_{11} = 40$ has a rank of $R_{11} = 1$, the value 56 has a rank of 6, and so forth. The sums of the ranks corresponding to each group are $R_1 = 1+6+2 = 9$, $R_2 = 3+7+8 = 18$, and $R_3 = 9+10+5+4 = 28$. The number of groups is $J = 3$, so the degrees of freedom are $\nu = 2$, and from Table 3 in Appendix B, the critical value is approximately $c = 5.99$ with $\alpha = 0.05$. Because there are no ties among the N observations,

$$T = -3(10+1) + \frac{12}{10 \times 11}\left(\frac{9^2}{3} + \frac{18^2}{3} + \frac{28^2}{4}\right) = 3.109.$$

Because $3.109 < 5.99$, fail to reject. That is, you are unable to detect a difference among the distributions.

9.7.2 R Function kruskal.test

The built-in R function

```
kruskal.test(x, g, ...)
```

performs the Kruskal-Wallis test, where x contains the data and g is a factor variable. If the data are stored in x having list mode, the R command `kruskal.test(x)` is all that is required. If the data are stored in a matrix, with columns corresponding to groups, use the command

```
kruskal.test(listm(x)).
```

If there is a factor variable, a formula can be used. That is, the function can have the form

```
kruskal.test(x ~ g).
```

EXAMPLE

R has a built-in data set (a data frame) called air quality. Ozone levels are recorded for each of five months. To compare the ozone levels, using months as groups, the command

```
kruskal.test(Ozone ~ Month, data = airquality)
```

performs the Kruskal-Wallis test and returns a p-value that is less than 0.001, indicating that the groups differ.

9.7.3 Method BDM

The Kruskall–Wallis test performs relatively well when the null hypothesis of identical distributions is true, but concerns arise when the null hypothesis is false, particularly in terms of power. An improvement on the Kruskall–Wallis test, which allows tied values as well as differences in dispersion, was derived by Brunner et al. (1997) and will be called method BDM. Again, the goal is to test the hypothesis that all J groups have identical distributions. The basic idea is that if J independent groups have identical distributions and we assign ranks to the pooled data as was done in the Kruskall–Wallis test, then the average of the ranks corresponding to the J groups should be approximately equal. (This greatly oversimplifies the technical issues.) To apply it, compute the ranks of the pooled data as was done by the Kruskall–Wallis test. Again let $N = \sum n_j$ be the total number of observations and let R_{ij} be the rank of X_{ij} after the data are pooled. Let

$$\bar{R}_j = \frac{1}{n_j} \sum_{i=1}^{n_j} R_{ij}$$

be the average of the ranks corresponding to the jth group and let

$$\mathbf{Q} = \frac{1}{N}\left(\bar{R}_1 - \frac{1}{2}, \ldots, \bar{R}_J - \frac{1}{2}\right).$$

The values given by the vector $\mathbf{Q}$ are called the *relative effects* and reflect how the groups compare based on the average ranks. The remaining calculations are relegated to Box 9.5.

Box 9.5: BDM heteroscedastic rank-based method.

For the jth group, compute

$$s_j^2 = \frac{1}{N^2(n_j-1)} \sum_{i=1}^{n_j} (R_{ij} - \bar{R}_j)^2,$$

and let

$$\mathbf{V} = N\text{diag}\left\{\frac{s_1^2}{n_1}, \ldots, \frac{s_J^2}{n_J}\right\}.$$

Let $\mathbf{I}$ be a J-by-J identity matrix, let $\mathbf{J}$ be a J-by-J matrix of 1s, and set $\mathbf{M} = \mathbf{I} - \frac{1}{J}\mathbf{J}$. (The diagonal entries in $\mathbf{M}$ have a common value, a property required to satisfy certain theoretical restrictions.) The test statistic is

$$F = \frac{N}{\text{tr}(\text{M}_{11}\mathbf{V})} \mathbf{QMQ}', \tag{9.13}$$

where `tr` indicates trace and $\mathbf{Q}'$ is the transpose of the matrix $\mathbf{Q}$. (See Appendix C for how the trace and transpose of a matrix are defined.)

Decision Rule: Reject if $F \geq f$, where f is the $1-\alpha$ quantile of an F distribution with

$$\nu_1 = \frac{M_{11}[\text{tr}(\mathbf{V})]^2}{\text{tr}(\mathbf{MVMV})},$$

and

$$\nu_2 = \frac{[\text{tr}(\mathbf{V})]^2}{\text{tr}(\mathbf{V^2}\Lambda)},$$

degrees of freedom, and $\Lambda = \text{diag}\{(n_1-1)^{-1}, \ldots, (n_J-1)^{-1}\}$.

Let $F_j(x)$ be the probability that a randomly sampled observation from the jth group is less than or equal to x, and let $G(x) = \sum(F_j(x))/J$. Let P_j be the probability that a randomly sampled observation from the distribution G is less than or equal to a randomly sampled observation from the distribution associated with the jth group. It is briefly noted that in a recent technical report, Brunner et al. (2016) derived a method for testing

$$H_0 : P_1 = \cdots = P_J.$$

The method can be applied via the R package rankFD or the R function bdmP described in the next section.

9.7.4 R Functions bdm and bdmP

The R function

```
bdm(x),
```

written for this book, performs the BDM rank-based ANOVA described in Box 9.5. Here, x can have list mode or it can be a matrix with columns corresponding to groups. The function returns the value of the test statistic, the degrees of freedom, the vector of relative effects, which is labeled `q.hat`, and the p-value. The R function

```
bdmP(x)
```

performs the method derived by Brunner et al. (2016), which was described in the previous section.

EXAMPLE

For the schizophrenia data in Table 9.1, the R function `bdm` returns a p-value of 0.040 and it reports that the relative effect sizes, **Q**, are

```
q.hat:
      [,1]
[1,] 0.4725
[2,] 0.4725
[3,] 0.3550
[4,] 0.7000
```

So the conclusion is that the distributions associated with these four groups differ, and we see that the average of the ranks among the pooled data is smallest for group three and highest for group four. This is consistent with the means. That is, group three has the smallest mean and group four has the largest. The same is true when using a 20% trimmed mean or MOM.

9.8 EXERCISES

1. For the following data,

Group 1	Group 2	Group 3
3	4	6
5	4	7
2	3	8
4	8	6
8	7	7
4	4	9
3	2	10
9	5	9
$\bar{X}_1 = 4.75$	$\bar{X}_2 = 4.62$	$\bar{X}_3 = 7.75$
$s_1^2 = 6.214$	$s_2^2 = 3.982$	$s_3^2 = 2.214$

   assume that the three groups have a common population variance, σ_p^2. Estimate σ_p^2.

2. For the data in the previous exercise, test the hypothesis of equal means using the ANOVA F. Use $\alpha = 0.05$. Verify that MSBG $= 25.04$, MSWG$=4.1369$, $F = 6.05$, and that you reject at the 0.05 level. Check your results using R. The data are stored in file CH9ex1.txt on the author's web page, which can be accessed as described in Chapter 1.

3. For the data in Exercise 1, verify that Welch's test statistic is $F_w = 7.7$ with degrees of freedom $\nu_1 = 2$ and $\nu_2 = 13.4$. Using the R function `t1way`, verify that you would reject the hypothesis of equal means with $\alpha = 0.01$.

4. For the data in Exercise 1, verify that when comparing 20% trimmed means with the R function `t1way`, the p-value is 0.02.

5. For the data

Group 1	Group 2	Group 3	Group 4
15	9	17	13
17	12	20	12
22	15	23	17

store the data in an R variable having list mode and and use the R function `anova1` to verify that when using the ANOVA F test, the estimate of the assumed common variance is MSWG = 9.5. Next, compute MSWG using the R function `lapply` and the R function `list2matrix`, which converts data in list mode to a matrix.

6. Why would you not recommend the strategy of testing for equal variances, and if not significant, using the ANOVA F test rather than Welch's method?

7. For the data in Table 9.1, assume normality and that the groups have equal variances. As already illustrated, the hypothesis of equal means is not rejected. If the hypothesis of equal means is true, an estimate of the assumed common variance is MSBG = 0.12, as already explained. Describe a reason why you would prefer to estimate the common variance with MSWG rather than MSBG.

8. Using R, verify that for the following data, MSBG = 14.4 and MSWG = 12.59.

G1	G2	G3
9	16	7
10	8	6
15	13	9
	6	

9. Consider $J = 5$ groups with population means 3, 4, 5, 6, and 7, and a common variance $\sigma_p^2 = 2$. If $n = 10$ observations are sampled from each group, determine the value estimated by MSBG and comment on how this differs from the value estimated by MSWG.

10. For the following data, use R to verify that you do not reject with the ANOVA F testing with $\alpha = 0.05$, but you do reject with Welch's test.

Group 1:	10 11 12 9 8 7
Group 2:	10 66 15 32 22 51
Group 3:	1 12 42 31 55 19

What might explain the discrepancy between the two methods?

11. For the data in Table 9.1, the ANOVA F test and Welch's test were not significant with $\alpha = 0.05$. Imagine that you want power to be 0.9 if the mean of one of the groups differs from the others by 0.2. (In the notation of Section 9.3, $a = 0.2$.) Verify that according to the R function `bdanova1`, the required sample sizes for each group are 110, 22, 40, and 38. The data are stored in the file skin_dat.txt on the author's web page. Note that the file has labels in the first line and values are separated by &.

12. For the data in Table 9.1, use method LSPB to compare the groups using the modified one-step M-estimator, MOM. Compare the p-value to what you get when using a 20% trimmed mean instead. What does this illustrate?

13. Repeat the last exercise, but now use projection distances to measure the depth of the null vector. That is, set the argument `op=3`.

14. Consider the following data.

G	X	G	X	G	X
1	12	1	8	1	22
3	42	3	8	3	12
3	9	3	21	2	19
2	24	2	53	2	17
2	10	2	9	2	28
2	21	1	19	1	21
1	56	1	18	1	16
1	29	1	20	3	32
3	10	3	12	3	39
3	28	3	35	2	10
2	12				

Store this data in an R variable having matrix mode with two columns corresponding to the two variables shown. (The data are stored on the author's web page in the file ch9_ex14_dat.txt.) There are three groups with the column headed by G indicating to which group the value in the next column belongs. For example, the first row indicates that the value 12 belongs to group one and the fourth row indicates that the value 17 belongs to group two. Use the R function `fac2list` to separate the data into three groups. Now the data are stored in list mode. Then compare the three groups with Welch's test using the R function `t1way`, then compare the 20% trimmed means again using the R function `t1way`.

15. For the data in Table 9.6, verify that the p-value, based on Welch's test, is 0.98. Use the R Function `t1way`. (The data are stored on the author's web page in the file ch9_table9_6_dat.txt.)

16. For the data in Table 9.6, use `t1way` to compare 20% trimmed means and verify that the p-value is less than 0.01.

17. For the data in Table 9.6, use `pbadepth` to compare the groups using default values for the arguments. You will get the message:

```
Error in solve.default(cov, ...) :
  system is computationally singular:
   reciprocal condition number = 2.15333e-18
```

The difficulty is that the depth of the null vector is being measured by Mahalanobis distance, which requires computing the inverse of a matrix, and instances are encountered where this cannot be done when using all pairwise differences.

18. Repeat the last exercise, only now use projection distances by setting the argument `op=3`. Verify that the p-value is 0.004. Why does `op=3` avoid the error noted in the last exercise?

Table 9.6 Hypothetical Data

Group 1	Group 2	Group 3	Group 4	Group 5
10.1	10.7	11.6	12.0	13.6
9.9	9.5	10.4	13.1	11.9
9.0	11.2	11.9	13.2	13.6
10.7	9.9	11.7	11.0	12.3
10.0	10.2	11.8	13.3	12.3
9.3	9.1	11.6	10.5	11.3
10.6	8.0	11.6	14.4	12.4
11.5	9.9	13.7	10.5	11.8
11.4	10.7	13.3	12.2	10.4
10.9	9.7	11.8	11.0	13.1
9.5	10.6	12.3	11.9	14.1
11.0	10.8	15.5	11.9	10.5
11.1	11.0	11.4	12.4	11.2
8.9	9.6	13.1	10.9	11.7
12.6	8.8	10.6	14.0	10.3
10.7	10.2	13.1	13.2	12.0
10.3	9.2	12.5	10.3	11.4
10.8	9.8	13.9	11.6	12.1
9.2	9.8	12.2	11.7	13.9
8.3	10.9	11.9	12.1	12.7
93.0	110.6	119.6	112.8	112.8
96.6	98.8	113.6	108.0	129.2
94.8	107.0	107.5	113.9	124.8

CHAPTER 10

TWO-WAY AND THREE-WAY DESIGNS

CONTENTS

This chapter extends the methods in Chapter 9 to situations where the goal is to compare independent groups based on what are called two-way and three-way designs. Section 10.1 introduces some basic concepts needed to understand two-way designs. Section 10.2 describes the most commonly used approach to testing hypotheses, which assumes both normality and homoscedasticity. Then more modern methods are described including methods based on robust measures of location.

10.1 BASICS OF A TWO-WAY ANOVA DESIGN

We begin with what is called a two-way ANOVA design. When all groups are independent, this is sometimes called a between-by-between design. To convey the basics in a reasonably concrete manner, consider a study where the goal is to understand the effect of diet on weight gains in rats. Specifically, four diets are considered that differ in: (1) amount of protein (high and low) and (2) the source of the protein (beef versus cereal). The results for these four groups are reported in Table 10.1 and are taken from Snedecor and Cochran (1967). Different rats were used in the four groups, so the groups are independent. The first column gives the

Table 10.1 Weight Gains (in grams) of Rats on One of Four Diets.

Beef Low	Beef High	Cereal Low	Cereal High
90	73	107	98
76	102	95	75
90	118	97	56
64	104	80	111
86	81	98	95
51	107	74	88
72	100	74	82
90	87	67	77
95	117	89	86
78	111	58	92
$\bar{X}_1 = 79.2$	$\bar{X}_2 = 100$	$\bar{X}_3 = 83.9$	$\bar{X}_4 = 85.9$

Table 10.2 Depiction of the Population Means for Four Diets

		Source	
		Beef	Cereal
Amount	High	μ_1	μ_2
	Low	μ_3	μ_4

weight gains of rats fed a low protein diet with beef the source of protein. The next column gives the weight gains for rats on a high protein diet again with beef the source of protein, and the next two columns report results when cereal is substituted for beef.

It is convenient to depict the population means as shown in Table 10.2, which indicates, for example, that μ_1 is the population mean associated with rats receiving a high protein diet from beef. That is, μ_1 is the average weight gain if the entire population of rats were fed this diet. Similarly, μ_4 is the population mean for rats receiving a low protein diet from cereal.

A *two-way design* means that there are two *independent variables* or *factors*. In Table 10.2 the first factor is *source of protein*, which has two *levels*: beef and cereal. The second factor is *amount of protein*, which also has two levels: low and high. A more precise description is that we have a 2-by-2 design, meaning there are two factors both of which have two levels. If we compare three methods for increasing endurance and simultaneously take into account three different diets, we have a two-way design with both factors having three levels. More succinctly, this is called a 3-by-3 design. The first factor is method and the second is diet.

EXAMPLE

For the plasma retinol data described in Chapter 1, one of the variables is sex and another is smoking status: never smoked, former smoker, current smoker. Imagine we want to compare these groups in terms of their average plasma beta-carotene. Then we have a 2-by-3 ANOVA design. The first factor is sex with two levels and the second factor is smoking status with three levels. If we ignore the second factor and focus on the first factor only, we have what is called a *one-way* design with two levels.

Returning to Table 10.2, these four groups could be compared by testing

$$H_0 : \mu_1 = \mu_2 = \mu_3 = \mu_4,$$

the hypothesis that all of the means are equal. However, there are other comparisons that are often of interest. For example, a goal might be to compare the rats receiving a high versus low protein diet *ignoring* the source of the protein. To illustrate how this might be done, imagine that the values of the population means are as follows:

		Source	
		Beef	Cereal
Amount	High	$\mu_1 = 45$	$\mu_2 = 60$
	Low	$\mu_3 = 80$	$\mu_4 = 90$

For rats on a high protein diet, the mean is 45 when consuming beef compared to 60 when consuming cereal. If the goal is to characterize the typical weight gain for a high protein diet ignoring source, a natural strategy is to average the two population means yielding $(45 + 60)/2 = 52.5$. That is, the typical rat on a protein diet gains 52.5 grams. For the more general situation depicted by Table 10.2, the typical weight gain on a high protein diet would be $(\mu_1 + \mu_2)/2$. Similarly, the typical weight gain for a rat on a low protein diet would be $(\mu_3 + \mu_4)/2$, which for the situation at hand is $(80 + 90)/2 = 85$ grams. Of course, the same can be done when characterizing source of protein, ignoring amount. The typical weight gain for a rat eating beef, ignoring amount of protein, is $(45 + 80)/2 = 62.5$, and for cereal it is $(60 + 90)/2 = 75$.

One way of proceeding is to test

$$H_0 : \frac{\mu_1 + \mu_2}{2} = \frac{\mu_3 + \mu_4}{2},$$

the hypothesis that the average of the population means in the first row of Table 10.2 is equal to the average for the second row. If this hypothesis is rejected, then there is said to be a *main effect* for the amount of protein. More generally, a main effect for the first factor (amount) is said to exist if

$$\frac{\mu_1 + \mu_2}{2} \neq \frac{\mu_3 + \mu_4}{2}.$$

Similarly, a goal might be to compare source of protein ignoring amount. One way of doing this is to test

$$H_0 : \frac{\mu_1 + \mu_3}{2} = \frac{\mu_2 + \mu_4}{2},$$

the hypothesis that the average of the means in the column for beef in Table 10.2 is equal to the average for the column headed by cereal. If this hypothesis is rejected, then there is said to be a *main effect* for the source of protein. More generally, a main effect for the second factor is said to exist if

$$\frac{\mu_1 + \mu_3}{2} \neq \frac{\mu_2 + \mu_4}{2}.$$

Next, for convenience, we describe main effects using a slightly different notation. Now we let μ_{jk} be the population mean associated with the jth level of the first factor, typically called Factor A, and the kth level of the second factor, called Factor B. Furthermore, we

Table 10.3 Depiction of the Population Means for a 4-by-2 Design

		Weight	
		Obese	Not Obese
Method	1	μ_{11}	μ_{12}
	2	μ_{21}	μ_{22}
	3	μ_{31}	μ_{32}
	4	μ_{41}	μ_{42}

let J indicate the number of levels associated with Factor A and K is the number of levels associated with factor B. More succinctly, we have a J-by-K design. In the plasma retinol study, $J = 2$ (the levels are males and females) and $K = 3$ (never smoked, no longer smokes, smokes). Table 10.3 illustrates the notation for a 4-by-2 design where the goal is to compare two factors related to high blood pressure. The first factor has to do with four methods for treating high blood pressure and the second factor deals with whether a participant is considered to be obese.

The population grand mean associated with the JK groups is

$$\bar{\mu} = \frac{1}{JK}\sum_{j=1}^{J}\sum_{k=1}^{K}\mu_{jk},$$

the average of the population means. For the jth level of Factor A, let

$$\bar{\mu}_{j.} = \frac{1}{K}\sum_{k=1}^{K}\mu_{jk}$$

be the average of the K means among the levels of Factor B. Similarly, for the kth level of Factor B

$$\bar{\mu}_{.k} = \frac{1}{J}\sum_{j=1}^{J}\mu_{jk}$$

is the average of the J means among the levels of Factor A. The main effects associated with Factor A are

$$\alpha_1 = \bar{\mu}_{1.} - \bar{\mu}, \ldots, \alpha_J = \bar{\mu}_{J.} - \bar{\mu}.$$

So the main effect associated with the jth level of Factor A is the difference between the grand mean and the average of the means associated with the jth level, namely $\alpha_j = \bar{\mu}_{j.} - \bar{\mu}$. There are no main effects for Factor A if

$$\alpha_1 = \cdots = \alpha_J = 0.$$

The hypothesis of no main effects for Factor A,

$$H_0 : \alpha_1 = \cdots = \alpha_J = 0,$$

is the same as

$$H_0 : \bar{\mu}_{1.} = \bar{\mu}_{2.} = \cdots = \bar{\mu}_{J.}. \tag{10.1}$$

As for Factor B, main effects are defined by

$$\beta_1 = \bar{\mu}_{.1} - \bar{\mu}, \ldots, \beta_K = \bar{\mu}_{.K} - \bar{\mu}.$$

The hypothesis of no main effects for Factor B can be expressed as

$$H_0 : \beta_1 = \cdots = \beta_K = 0,$$

which is the same as

$$H_0 : \bar{\mu}_{.1} = \bar{\mu}_{.2} = \cdots = \bar{\mu}_{.K}. \tag{10.2}$$

10.1.1 Interactions

In addition to main effects, often researchers are interested in what is called an interaction. Consider again a 2-by-2 design where the goal is to compare high and low protein diets in conjunction with two protein sources. Suppose the *population* means associated with the four groups are:

		Source	
		Beef	Cereal
Amount	High	$\mu_1 = 45$	$\mu_2 = 60$
	Low	$\mu_3 = 80$	$\mu_4 = 90$

Look at the first row (high amount of protein) and notice that the weight gain for a beef diet is 45 grams compared to a weight gain of 60 grams for cereal. As is evident, there is an increase of 15 grams. In contrast, with a low protein diet, switching from beef to cereal results in an increase of 10 grams on average. That is, in general, switching from beef to cereal results in an increase for the average amount of weight gained, but the increase differs depending on whether we look at high or low protein. This is an example of an *interaction.*

In a 2-by-2 design, an *interaction* is said to exist if

$$\mu_{11} - \mu_{12} \neq \mu_{21} - \mu_{22}.$$

In words, for each level of Factor A, compute the difference between the means associated with Factor B. An interaction exists if these two differences are not equal. *No interaction* means that

$$\mu_{11} - \mu_{12} = \mu_{21} - \mu_{22}.$$

Various types of interactions arise and can be important when considering how groups differ. Notice that in the last illustration, there is an increase in the average weight gain when switching from beef to cereal for both high and low protein diets. For high protein there is an increase in the population mean from 45 to 60, and for low protein there is an increase from 80 to 90. In both cases, though, the largest gain in weight is associated with cereal. This is an example of what is called an *ordinal interaction.* In a 2-by-2 design as depicted in Table 10.2, an *ordinal interaction* is said to exist if

$$\mu_1 > \mu_2 \text{ and } \mu_3 > \mu_4,$$

or if

$$\mu_1 < \mu_2 \text{ and } \mu_3 < \mu_4.$$

In words, an interaction is said to be ordinal if the relative rankings remain the same. In the illustration, cereal always results in the largest weight gain regardless of whether a low or high protein diet is used.

If there is a change in the relative rankings of the means, a *disordinal interaction* is said to exist. As an illustration, imagine that the population means are as follows:

		Source	
		Beef	Cereal
Amount	High	80	110
	Low	50	30

Observe that for the first row (a high protein diet), the average weight gain increases from 80 to 110 as we move from beef to cereal. In contrast, for the low protein diet, the average weight gain decreases from 50 to 30. Moreover, when comparing beef to cereal, the relative rankings change depending on whether a high or low protein diet is used. For a high protein diet, cereal results in the largest gain, but for a low protein diet, beef results in a larger gain compared to cereal. This is an example of a disordinal interaction. In general, for the population means in Table 10.2, a disordinal interaction is said to exist if

$$\mu_1 > \mu_2 \text{ and } \mu_3 < \mu_4,$$

or if

$$\mu_1 < \mu_2 \text{ and } \mu_3 > \mu_4.$$

Research articles often present graphical displays reflecting ordinal and disordinal interactions. Figure 10.1 is an example based on the sample means just used to illustrate a disordinal interaction. Along the x-axis we see the levels of the first factor (high and low protein). The solid line extending from 110 down to 30 reflects the change in means when the source of protein is cereal. The dashed line reflects the change in means associated with beef. As is evident, the lines cross, which reflects a disordinal interaction.

Now imagine that the population means are as follows:

		Source	
		Beef	Cereal
Amount	High	50	60
	Low	80	90

Regardless of whether rats are given a high or low protein diet, cereal always results in the largest average weight gain. There is no interaction because for the first row, the means increase by 10 (from 50 to 60), and the increase is again 10 for the second row. Figure 10.2 graphically illustrates an ordinal interaction. The lines are not parallel, but they do not cross. When there is no interaction, the lines are parallel.

Notice that in the discussion of ordinal versus disordinal interactions, attention was focused on comparing means within rows. Not surprisingly, ordinal and disordinal interactions can also be defined in terms of columns. Consider again the example where the means are as follows:

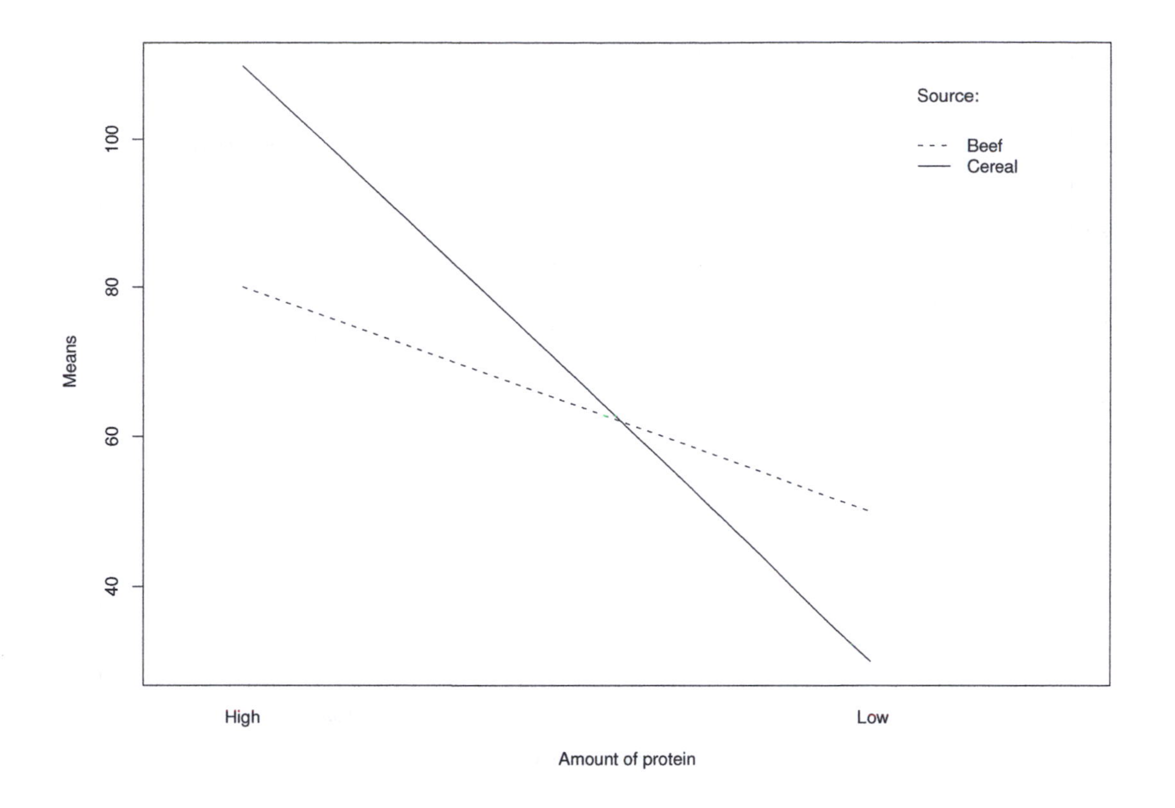

Figure 10.1 A plot often used to summarize an interaction. The dashed line shows the change in weight gain, when beef is the source of protein, as we move from high to low amounts of protein. The solid line is the change when the source is cereal.

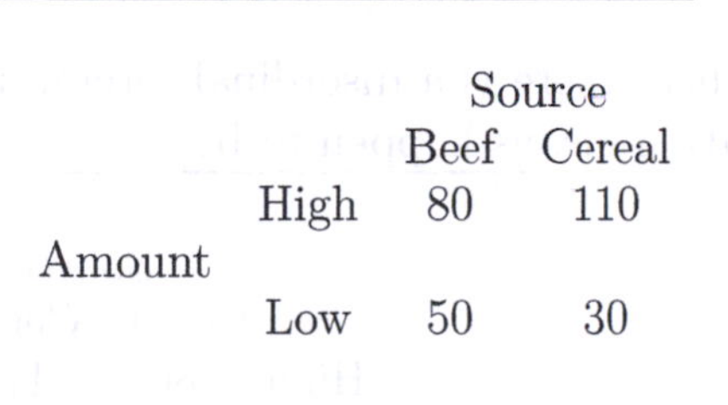

		Source	
		Beef	Cereal
Amount	High	80	110
	Low	50	30

As previously explained, there is a disordinal interaction for rows. But look at the population means in the first column and notice that 80 is greater than 50, and for the second column 110 is greater than 30. There is an interaction because $80 - 50 \neq 110 - 30$. Moreover, the interaction is ordinal because for beef, average weight gain is largest on a high protein diet, and the same is true for a cereal diet. So we can have both an ordinal and disordinal interaction depending on whether we look at differences between rows or differences between columns.

When graphing the means to illustrate a disordinal or ordinal interaction for rows, the levels of Factor A (high and low amounts of protein) were indicated by the x-axis as illustrated by Figures 10.1 and 10.2. When describing ordinal or disordinal interactions by columns, now the x-axis indicates the levels of the second factor, which in this example is source of protein (beef versus cereal). Figure 10.3 illustrates what the graph looks like for the means in Figure

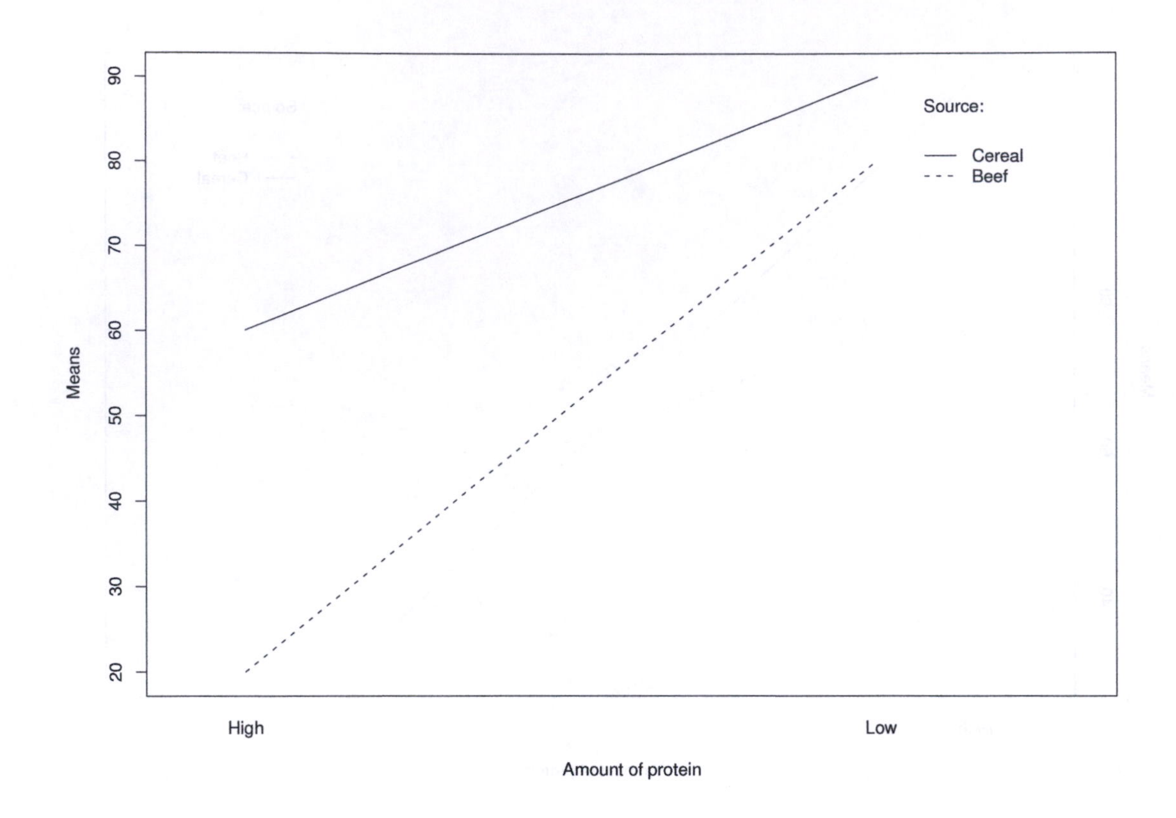

Figure 10.2 An example of an ordinal interaction. The lines are not parallel but they do not cross. Parallel lines would indicate no interaction.

10.1. Note that now the graph indicates an ordinal interaction, in contrast to the disordinal interaction in Figure 10.1.

Situations also arise where there is a disordinal interaction for both rows and columns. For example, if the population means happen to be

		Source	
		Beef	Cereal
Amount	High	80	110
	Low	100	95

then there is a disordinal interaction for rows because for the first row, cereal results in the largest gain, but for the second row (low amount of protein) the largest gain is for beef. Simultaneously, there is a disordinal interaction for columns because for the first column (beef), a low protein diet results in the largest average gain, but for the second column (cereal) a high protein diet has the largest mean.

Randomized Block Design

The term *randomized block design* refers to situations where the experimenter divides participants into subgroups called blocks with the goal that the variability within blocks

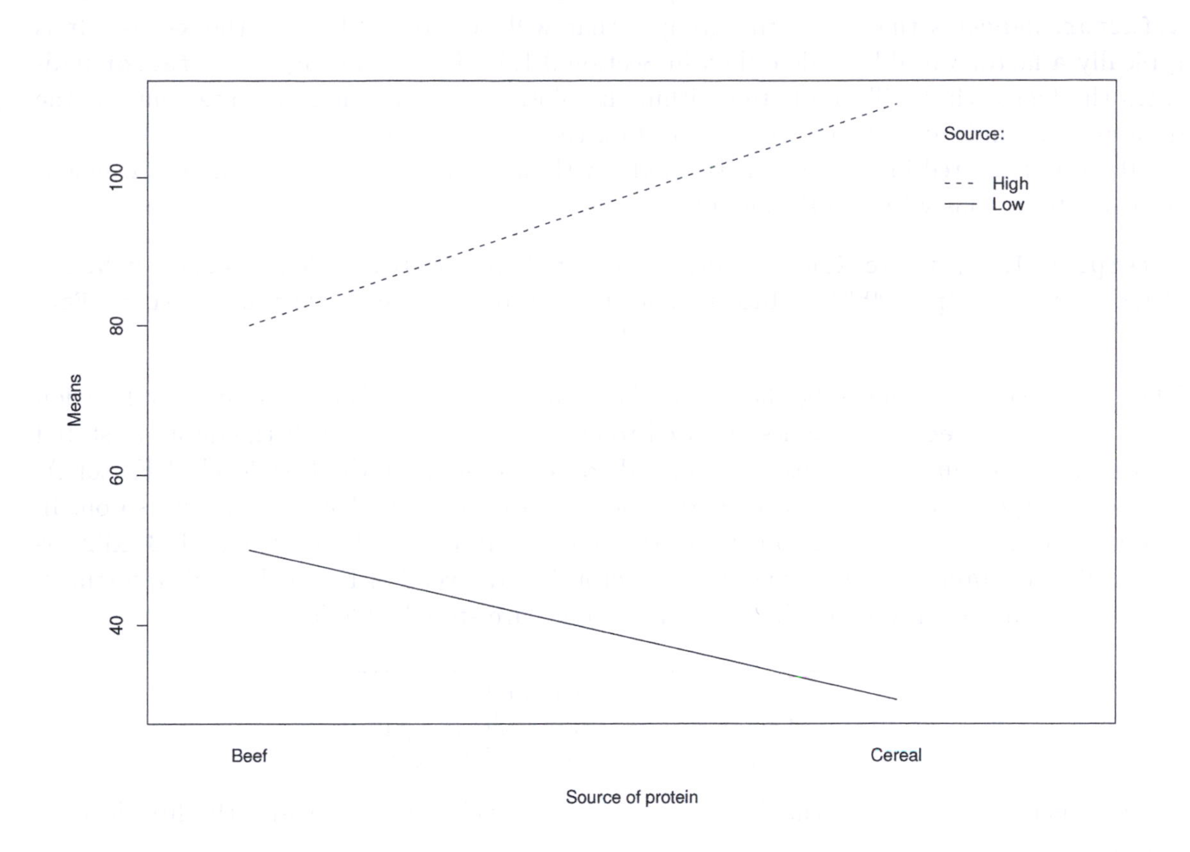

Figure 10.3 An example of an ordinal interaction when columns are used. Now the x-axis indicates the levels of the second factor.

is less than the variability between blocks. That is, the goal is to reduce the unexplained variation or experimental error by randomly assigning participants within each block to treatment conditions.

EXAMPLE

Imagine a study aimed at comparing the effectiveness of a vaccine compared to a placebo. If participants are randomly sampled and assigned to one of the treatments, we have what is called a *completely randomized design.* An alternative approach is to use a randomized block design where participants are assigned to blocks based on gender. The idea is that because men and women are physiologically different, they might react differently to medication. This randomized block design removes gender as a potential source of variability and as a potential confounding variable. The results can be analyzed with the methods in this chapter.

10.1.2 R Functions interaction.plot and interplot

R has a built-in function for plotting interactions, which has the form

```
interaction.plot(x.factor, trace.factor, response, fun = mean,trace.label =
deparse(substitute(trace.factor)), xlab = deparse(substitute(x.factor)),ylab
= ylabel).
```

(This function has some additional arguments not described here.) The first argument, `x.factor`, indicates the levels (the groups) that will be indicated along the x-axis. (It is typically a factor variable as described in Section 9.1.3.) The argument `trace.factor` indicates the factor that will be plotted within the plot. The argument `response` contains the data and `fun` indicates the measure of location that will be used.

If data are stored in a matrix or list mode, with no factor variables, it is more convenient to plot interactions with the R function

```
interplot(J, K, x, locfun = mean, locvec = NULL, g1lev = NULL, g2lev = NULL,
type = c("l", "p", "b"), xlab = "Fac 1", ylab = "means", trace.label = "Fac
2").
```

The measure of location can be chosen via the argument `locfun`. But if measures of location are stored in `locvec`, these values are used to create the plot instead. If the data are stored in list mode, the first K groups are assumed to be the data for the first level of Factor A, the next K groups are assumed to be data for the second level of Factor A, and so on. In R notation, x[[1]] is assumed to contain the data for level 1 of Factors A and B, x[[2]] is assumed to contain the data for level 1 of Factor A and level 2 of Factor B, and so forth. If, for example, a 2-by-4 design is being used, the data are stored as follows:

	Factor B			
Factor	x[[1]]	x[[2]]	x[[3]]	x[[4]]
A	x[[5]]	x[[6]]	x[[7]]	x[[8]]

For instance, x[[5]] contains the data for the second level of Factor A and the first level of Factor B.

If the data are stored in a matrix, the first K columns are assumed to be the data for the first level of Factor A, the next K columns are assumed to be data for the second level of Factor A, and so on.

EXAMPLE

Imagine that the cereal data in Table 10.1 are stored in the R variable `mm` with column 1 containing the amount of weight gained, column 2 indicating the source of protein (beef and cereal), and column 3 indicating the amount of protein (high and low). Then the R command

```
interaction.plot(mm[,2],mm[,3],mm[,1],trace.label="Amount",
    xlab="Source",ylab="Trimmed Means",fun=tmean)
```

would produce a plot using 20% trimmed means with beef and protein on the x-axis. If the data are stored in a matrix called cereal, the command

```
interplot(2,2,cereal,g1lev=c("B","C"),g2lev=c("H","L"),
xlab="Source",trace.label="Amount",ylab="Trimmed Means",locfun=tmean)
```

would produce a plot similar to Figure 10.3 but with means replaced by 20% trimmed means.

10.1.3 Interactions When There Are More Than Two Levels

So far, attention has been focused on a 2-by-2 design. What does no interaction mean when there are more than two levels for one or both factors? Basically, *no interaction* means that for any two levels of Factor A, and any two levels of Factor B, there is no interaction for the corresponding cells. As an illustration, consider the population means in Table 10.4. Now,

Table 10.4 Hypothetical Population Means Illustrating Interactions

		Factor B		
		Level 1	Level 2	Level 3
	Level 1	**10**	**20**	30
Factor A	Level 2	20	30	40
	Level 3	**30**	**40**	50

pick any two rows, say the first and third. Then pick any two columns, say the first and second. The population means for the first row and the first and second columns are 10 and 20. For the third row, the means for these two columns are 30 and 40, and these four means are printed in boldface in Table 10.4. Notice that for these four means, there is no interaction. The reason is that for the first row, the means increase from 10 to 20, and for the third row the means increase from 30 to 40. That is, for both rows, there is an increase of 10 when switching from column 1 to column 2. In a similar manner, again looking at rows 1 and 2, we see that there is an increase of 20 as we move from column 1 to column 2. That is, there is no interaction for these four means either. An interaction is said to exist among the JK means if there is an interaction for any two rows and any two columns.

EXAMPLE

Suppose the population means for a 3-by-4 design are as follows:

		Factor B			
		Level 1	Level 2	Level 3	Level 4
	Level 1	40	40	40	40
Factor A	Level 2	40	40	40	40
	Level 3	40	40	40	40

Is there an interaction? The answer is no because regardless of which two rows are picked, there is an increase of 0 as we move from any one column to another.

EXAMPLE

For the population means used in this last example, is there a main effect for Factor A or Factor B? First consider Factor A. As is evident, for any row we pick, the average of the four means is 40. That is, the average of the means is the same for all three rows, and this means there is no main effect for Factor A. In a similar manner, there is no main effect for Factor B because the average of the means in any column is again 40.

EXAMPLE

Suppose the population means for a 3-by-4 design are are as follows:

Table 10.5 Sample Means for Illustrating Main Effects and Interactions

		Source	
		Beef	Cereal
Amount	Low	$\bar{X}_{11} = 79.2$	$\bar{X}_{12} = 83.9$
	High	$\bar{X}_{21} = 100$	$\bar{X}_{22} = 85.9$

		Factor B			
		Level 1	Level 2	Level 3	Level 4
Factor A	Level 1	40	40	50	60
	Level 2	20	20	50	80
	Level 3	20	30	10	40

Is there an interaction? Looking at level 1 of Factor A, we see that the means increase by 0 as we move from Level 1 of Factor B to Level 2. The increase for Level 2 of Factor A is again 0, so there is no interaction for these four means. However, looking at Level 1 of Factor A, we see that the means increase by 10 as we move from Level 1 to Level 3 of Factor B. In contrast, there is an increase of 30 for Level 2 of Factor A, which means that there is an interaction.

10.2 TESTING HYPOTHESES ABOUT MAIN EFFECTS AND INTERACTIONS

Next, attention is focused on testing the hypotheses of no main effects and no interactions. The most commonly used method is based on the assumption that all groups have normal distributions with a common variance. As usual, the assumed common variance is labeled σ_p^2.

The computations begin by computing the sample mean corresponding to the jth level of Factor A and the kth level of Factor B. The resulting sample mean is denoted by $\bar{X}_{jk}$. That is, $\bar{X}_{jk}$ is the sample mean corresponding to the jth level of Factor A and the kth level of Factor B. For example, for the 2-by-2 study comparing high versus low protein and beef versus cereal, $\bar{X}_{11}$ is the sample mean for Level 1 of Factor A and Level 1 of Factor B. Referring to Table 10.1, $\bar{X}_{11}$ is the sample mean of the rats on a low protein diet (Level 1 of Factor A) that consume beef (level 1 of Factor B). From the first column in Table 10.1, the average weight gain for the 10 rats on this diet can be seen to be $\bar{X}_{11} = 79.2$. Level 1 of Factor A and level 2 of Factor B correspond to low protein from a cereal diet. The data are given in the third column of Table 10.1 and the sample mean is $\bar{X}_{12} = 83.9$. For Level 2 of Factor A and Level 1 of Factor B, $\bar{X}_{21} = 100$, the sample mean of the values in column 2 of Table 10.1. Finally, $\bar{X}_{22}$ is the sample mean for level 2 of Factor A and level 2 of Factor B and is 85.9. These sample means are summarized in Table 10.5.

Box 10.1: How to compute MSA, MSB, and MSINTER.

Compute:

$$\bar{X}_G = \frac{1}{JK}\sum_{j=1}^{J}\sum_{k=1}^{K}\bar{X}_{jk}$$

($\bar{X}_G$ is the average of all JK sample means.)

$$\bar{X}_{j.} = \frac{1}{K}\sum_{k=1}^{K}\bar{X}_{jk}$$

($\bar{X}_{j.}$ is the average of the sample means for the jth level of Factor A.)

$$\bar{X}_{.k} = \frac{1}{J}\sum_{j=1}^{J}\bar{X}_{jk}$$

($\bar{X}_{.k}$ is the average of the sample means for the kth level of Factor B.)

$$A_j = \bar{X}_{j.} - \bar{X}_G,\ B_k = \bar{X}_{.k} - \bar{X}_G$$

$$C_{jk} = \bar{X}_{jk} - \bar{X}_{j.} - \bar{X}_{.k} + \bar{X}_G$$

$$\text{SSA} = nK\sum A_j^2,\ \text{SSB} = nJ\sum B_k^2$$

$$\text{SSINTER} = n\sum\sum C_{jk}^2$$

$$\text{MSWG} = \frac{1}{JK}\sum\sum s_{jk}^2,\ \text{SSWG} = (n-1)JK(MSWG).$$

Then

$$\text{MSA} = \frac{\text{SSA}}{J-1},\ \text{MSB} = \frac{\text{SSB}}{K-1}$$

$$\text{MSINTER} = \frac{\text{SSINTER}}{(J-1)(K-1)}.$$

(Some books write MSINTER and SSINTER as MSAB and SSAB, respectively, where AB denotes the interaction of Factors A and B.)

Under random sampling, normality, and equal variances, a test of the hypothesis of no main effects for Factor A can be performed that provides exact control over the probability of a Type I error. Box 10.1 summarizes the bulk of the calculations when all groups have a common sample size, n. Unequal sample sizes can be used, but for present purposes the details are not important and therefore omitted. Here, J is the number of levels for Factor A and K is the number of levels for Factor B. In Box 10.1, s_{jk}^2 is the sample variance corresponding to the data used to compute $\bar{X}_{jk}$. As in the one-way design, MSWG estimates the assumed common variance, σ_p^2.

Decision Rules

The relevant hypotheses are tested as follows:

Table 10.6 Typical ANOVA Summary Table for a Two-Way Design

Source	SS	DF	MS	F
A	SSA	$J-1$	MSA = $\frac{\text{SSA}}{J-1}$	$F=\frac{\text{MSA}}{\text{MSWG}}$
B	SSB	$K-1$	MSB = $\frac{\text{SSB}}{K-1}$	$F=\frac{\text{MSB}}{\text{MSWG}}$
INTER	SSINTER	$(J-1)(K-1)$	MSINTER = $\frac{\text{SSINTER}}{(J-1)(K-1)}$	$F=\frac{\text{MSINTER}}{\text{MSWG}}$
WITHIN	SSWG	$N-JK$	MSWG = $\frac{\text{SSWG}}{N-JK}$	

- *Factor A.* The hypothesis of no main effects for Factor A is tested with

$$F=\frac{\text{MSA}}{\text{MSWG}}.$$

  The null hypothesis is rejected if $F \geq f$, the $1-\alpha$ quantile of an F distribution with $\nu_1 = J-1$ and $\nu_2 = N-JK$ degrees of freedom.

- *Factor B.* The hypothesis of no main effects for Factor B is tested with

$$F=\frac{\text{MSB}}{\text{MSWG}};$$

  reject if $F \geq f$, the $1-\alpha$ quantile of an F distribution with $\nu_1 = K-1$ and $\nu_2 = N-JK$ degrees of freedom.

- *Interactions.* The hypothesis of no interactions is tested with

$$F=\frac{\text{MSINTER}}{\text{MSWG}};$$

  reject if $F \geq f$, the $1-\alpha$ quantile of an F distribution with $\nu_1 = (J-1)(K-1)$ and $\nu_2 = N-JK$ degrees of freedom.

In terms of Type I errors, the normality assumption can be violated if all groups have identically shaped distributions and the sample sizes are not too small. This means, in particular, that the equal variance assumption is true. Put another way, if the ANOVA F test rejects, this indicates that the distributions differ, but it remains unclear how they differ and by how much. One possibility is that the F test rejects because the population means differ. Another possibility is that the F test rejects primarily because the variances differ or the amount of skewness differs. As for power, again practical problems arise under very slight departures from normality for reasons discussed in previous chapters. Unequal variances can adversely affect power as well. Generally, the more groups that are compared, the more likely the ANOVA F tests, described in this section, will be unsatisfactory when indeed groups differ.

Table 10.6 outlines a typical ANOVA summary table for a two-way design. The notation *SS* in the first row stands for sum of squares, *DF* is degrees of freedom, and *MS* is mean squares.

EXAMPLE

Consider the following ANOVA summary table:

Source	SS	DF	MS	F
A	200	1	200	1.94
B	300	2	150	1.46
INTER	500	2	250	2.42
WITHIN	620	6	103	

Referring to Table 10.6, we see that $J-1$ corresponds to the value 1 in the example, indicating that $J-1=1$, so the first factor has $J=2$ levels. Similarly, the second factor has $K=3$ levels. Table 10.6 indicates that $N-JK=6$, but $JK=6$, so $N-JK=N-6=6$, and therefore $N=12$. That is, the total number of observations among the six groups is 12. The estimate of the common variance is MSWG $=103$. If we use $\alpha=0.05$, then from Table 6 in Appendix B, the critical values for the three hypotheses are 5.99, 5.14, and 5.14. The F values are less than their corresponding critical values, so the hypotheses of no main effects for Factor A and Factor B, as well as the hypothesis of no interaction, are not rejected.

10.2.1 R function anova

The built-in R function `anova`, described in Section 9.1.3, can be used when the goal is to test the hypothesis of no main effects or no interaction in a two-way design.

EXAMPLE

Consider again the plasma retinol data and imagine that the goal is to compare groups based on two factors: sex (stored in column 2 of the R variable `plasma`) and smoking status in terms of amount of cholesterol consumed (stored in columns 3 and 10, respectively). When using the R function `aov`, connecting factors by Ş*Ť means that R will test the hypothesis of no main effects as well as no interactions. For convenience, assume the data are stored in the R variable `y`. Then the following command can be used:

```
anova(aov(y[,10]~as.factor(y[,2])*as.factor(y[,3])))
```

As just noted, the * in this last R command indicates that an interaction term in the model will be used. If we replace the * with a +, in which case the R command is now

```
anova(aov(y[,10]~as.factor(y[,2])+as.factor(y[,3]))),
```

this indicates that an *additive model* will be used. That is, main effects will be compared assuming that there is no interaction. It is sometimes recommended that if the hypothesis of no interaction is not rejected, main effects should be tested assuming that the model is additive. However, results reported by Fabian (1991) do not support this strategy. The problem is that power might not be sufficiently high to justify accepting the hypothesis of no interaction.

The output (based on the R command that includes an interaction) looks like this:

```
Analysis of Variance Table

Response: y[, 10]   Df   Sum Sq Mean Sq F value    Pr(>F)
```

```
as.factor(y[, 2])     1  355619  355619 21.9853 4.128e-06 ***
as.factor(y[, 3])     2   45637   22819  1.4107    0.2455
as.factor(y[, 2]):as.factor(y[, 3])
                      2   71014   35507  2.1951    0.1131
Residuals           309 4998170   16175
```

The line containing

```
as.factor(y[, 3])
```

reports the results for the second factor, which has a p-value of 0.2455. So, in particular, no differences are found among the three levels of smoking status when testing at the 0.05 level. The line with

```
as.factor(y[, 2]):as.factor(y[, 3])
```

reports the results when testing the hypothesis of no interaction, which has a p-value of 0.1131.

10.2.2 Inferences About Disordinal Interactions

It should be mentioned that if the hypothesis of no interactions is rejected, simply looking at the means is not enough to determine whether the interaction is ordinal or disordinal. Consider again the study on weight gain among rats and suppose that the unknown population means are as follows:

		Source	
		Beef	Cereal
Amount	Low	$\mu_1 = 60$	$\mu_2 = 60$
	High	$\mu_3 = 50$	$\mu_4 = 70$

There is an interaction because $60 - 60 \neq 50 - 70$, but the interaction is not disordinal. Further assume that the hypothesis of no interactions is correctly rejected based on the following sample means.

		Source	
		Beef	Cereal
Amount	Low	$\bar{X}_1 = 55$	$\bar{X}_2 = 45$
	High	$\bar{X}_3 = 49$	$\bar{X}_4 = 65$

Notice that the sample means suggest that the interaction is disordinal because 55 is greater than 45, but 49 is less than 65. It might be that by chance, rats on a low protein diet with beef as the source got a smaller sample mean versus rats on a low protein diet with cereal as the source. That is, it might be that the population mean corresponding to rats on a low protein diet with beef as the source is larger than the population mean for rats on a low protein diet with cereal as the source. The point is that to establish that a disordinal interaction exists for the rows, it is necessary to also reject the hypotheses $H_0 : \mu_1 = \mu_2$ and $H_0 : \mu_3 = \mu_4$. If, for example, $\mu_1 = \mu_2$, there is no disordinal interaction. Moreover, simply rejecting the hypothesis of no interaction does not indicate whether $H_0 : \mu_1 = \mu_2$ or $H_0 : \mu_3 = \mu_4$ should be rejected. Under normality, these hypotheses can be tested with

Student's T test, or Welch's test, described in Chapter 8. If both of these hypotheses are rejected, and if

$$\bar{X}_1 > \bar{X}_2 \text{ and } \bar{X}_3 < \bar{X}_4,$$

or if

$$\bar{X}_1 < \bar{X}_2 \text{ and } \bar{X}_3 > \bar{X}_4,$$

there is empirical evidence that there is a disordinal interaction.

A similar strategy is used when checking for a disordinal interaction for columns. That is, to establish that a disordinal interaction exists, it is necessary to reject both H_0: $\mu_1 = \mu_3$ and H_0: $\mu_2 = \mu_4$. If both hypotheses are rejected and the sample means satisfy

$$\bar{X}_1 > \bar{X}_3 \text{ and } \bar{X}_2 < \bar{X}_4,$$

or if

$$\bar{X}_1 < \bar{X}_3 \text{ and } \bar{X}_2 > \bar{X}_4,$$

then conclude that there is a disordinal interaction for the columns.

An illustration of how to detect a disordinal interaction is postponed until Chapter 12 because there is yet another technical issue that must be addressed: if all tests are performed at the 0.05 level, for example, the probability of making at least one Type I error can exceed 0.05.

10.2.3 The Two-Way ANOVA Model

The two-way ANOVA model can be described as follows. The population grand mean associated with the JK groups is

$$\bar{\mu} = \frac{1}{JK}\sum_{j=1}^{J}\sum_{k=1}^{K}\mu_{jk},$$

the average of the population means. Let

$$\bar{\mu}_{j.} = \frac{1}{K}\sum_{k=1}^{K}\mu_{jk}$$

be the average of the K means among the levels of Factor B associated with the jth level of Factor A. Similarly,

$$\bar{\mu}_{.k} = \frac{1}{J}\sum_{j=1}^{J}\mu_{jk}$$

is the average of the J means among the levels of Factor A associated with the kth level of Factor B. The main effects associated with Factor A are

$$\alpha_1 = \bar{\mu}_{1.} - \bar{\mu}, \ldots, \alpha_J = \bar{\mu}_{J.} - \bar{\mu}.$$

So the main effect associated with the jth level is the difference between the grand mean and the average of the means associated with the jth level, namely $\alpha_j = \bar{\mu}_{j.} - \bar{\mu}$. There are no main effects for Factor A if

$$\alpha_1 = \cdots = \alpha_J = 0.$$

As for Factor B, main effects are defined by

$$\beta_1 = \bar{\mu}_{.1} - \bar{\mu}, \ldots, \beta_K = \bar{\mu}_{.K} - \bar{\mu}.$$

The hypothesis of no main effects for Factor B can be expressed as

$$H_0 : \beta_1 = \cdots = \beta_K = 0.$$

As for interactions, let

$$\begin{aligned}\gamma_{jk} &= \mu_{jk} - \alpha_j - \beta_k - \bar{\mu} \\ &= \mu_{jk} - \bar{\mu}_{j.} - \bar{\mu}_{.k} + \bar{\mu}.\end{aligned}$$

Then no interactions means that

$$\gamma_{11} = \gamma_{12} = \cdots = \gamma_{JK} = 0.$$

Although the γ_{jk} terms are not very intuitive, they provide a convenient framework for deriving an appropriate test of the hypothesis that there are no interactions among any two levels of Factor A and Factor B.

The discrepancy between the ith observation in the jth level of Factor A and kth level of Factor B, in terms of the notation just described, is

$$\epsilon_{ijk} = X_{ijk} - \bar{\mu} - \alpha_j - \beta_k - \gamma_{jk}.$$

Rearranging terms yields

$$X_{ijk} = \bar{\mu} + \alpha_j + \beta_k + \gamma_{jk} + \epsilon_{ijk}.$$

Assuming that the error term, ϵ_{ijk}, has a normal distribution with a common variance among the groups results in the standard two-way ANOVA model that forms the basis of the hypothesis testing methods covered in this section. If there is no interaction, the model is said to be additive and reduces to

$$X_{ijk} = \bar{\mu} + \alpha_j + \beta_k + \epsilon_{ijk}.$$

Based on the model just described, it can be shown that

$$E(\text{MSWG}) = \sigma_p^2,$$

$$E(\text{MSA}) = \sigma_p^2 + \frac{nK}{J-1}\sum \alpha_j^2,$$

$$E(\text{MSB}) = \sigma_p^2 + \frac{nJ}{K-1}\sum \beta_k^2,$$

and

$$E(\text{MSINTER}) = \sigma_p^2 + \frac{n\sum\sum \gamma_{jk}^2}{(J-1)(K-1)}.$$

10.3 HETEROSCEDASTIC METHODS FOR TRIMMED MEANS, INCLUDING MEANS

A positive feature of the ANOVA F tests, described in the previous section, is that if groups have identical distributions, the probability of a Type I error seems to be controlled reasonably well under non-normality. But as was the case in Chapters 7 and 9, violating the assumption of equal variances can result in poor power properties (power can go down as the means become unequal), unsatisfactory control over the probability of a Type I error, and relatively low power compared to other methods that might be used.

Numerous methods have been proposed for dealing with heteroscedasticity when comparing measures of location. Here the focus is on comparing trimmed means, which, of course, contains comparing means as a special case. The method used here is based on an extension of a method derived by Johansen (1980), which represents a heteroscedastic approach to what is known as the general linear model.

Let $\mu_t = (\mu_{t1}, \ldots, \mu_{tJK})'$. The general form of the null hypothesis is

$$H_0 : \mathbf{C}\mu_t = 0. \tag{10.3}$$

The test statistic is

$$Q = \bar{\mathbf{X}}'\mathbf{C}'(\mathbf{CVC}')^{-1}\mathbf{C}\bar{\mathbf{X}}' \tag{10.4}$$

where the matrix $\mathbf{V}$ is computed as described in Box 10.2, assuming familiarity with basic matrix algebra. The null hypothesis is rejected if $Q \geq c_{\text{ad}}$, where c_{ad} is described in Box 10.2 as well.

To describe how $\mathbf{C}$ is constructed, let $\mathbf{C}_m$ be an $m - 1$ by m matrix having the form

$$\begin{pmatrix} 1 & -1 & 0 & 0 & \ldots & 0 \\ 0 & 1 & -1 & 0 & \ldots & 0 \\ & & & \ldots & & \\ 0 & 0 & \ldots & 0 & 1 & -1 \end{pmatrix}$$

and let $\mathbf{j}'_J$ be a $1 \times J$ matrix of ones. Then $\mathbf{C}$ is computed as described in Box 10.2.

10.3.1 R Function t2way

The R function

```
t2way(J,K,x,tr=0.2,grp=c(1:p),p=J*K,MAT=FALSE,
lev.col=c(1:2),var.col=3,pr=TRUE,IV1=NULL,IV2=NULL)
```

tests the hypotheses of no main effects and no interaction. Here, the arguments `J` and `K` denote the number of levels associated with Factors A and B. Like `t1way`, the data are assumed to be stored in `x`, which can be any R variable that is a matrix or has list mode. If factor variables are available, they can be indicated by the arguments `IV1` and `IV2`, in which case set the argument `x=data`, where `data` is a vector containing the dependent variable. That is, use of the R function `fac2list`, introduced in Section 9.1.3, is not required. The final example in this section illustrates how this is done.

If `IV1` and `IV2` are not specified, the data are assumed to be stored in the order indicated in Section 10.1.2. If they are not stored in the assumed order, the argument `grp` can be used to correct this problem. As an illustration, suppose the data are stored as follows:

	Factor B			
Factor	x[[2]]	x[[3]]	x[[5]]	x[[8]]
A	x[[4]]	x[[1]]	x[[6]]	x[[7]]

Box 10.2: Two-way ANOVA based on trimmed means.

For the hypothesis of no main effects for Factor A: $\mathbf{C} = \mathbf{C}_J \otimes \mathbf{j}'_K$, where $\otimes$ is the (right) Kronecker product defined in Appendix C.
For the hypothesis of no main effects for Factor B: $\mathbf{C} = \mathbf{j}'_J \otimes \mathbf{C}_K$.
For the hypothesis of no interaction: $\mathbf{C} = \mathbf{C}_J \otimes \mathbf{C}_K$. Let $p = JK$ and let $\mathbf{V}$ be a p by p diagonal matrix with

$$v_{jj} = \frac{(n_j - 1)s^2_{wj}}{h_j(h_j - 1)},$$

$j = 1, \ldots, p$.
If $\mathbf{C}$ has k rows, an approximate critical value is the $1 - \alpha$ quantile of a chi-squared distribution with k degrees of freedom. For small sample sizes, let

$$\mathbf{R} = \mathbf{VC}'(\mathbf{CVC}')^{-1}\mathbf{C},$$

and

$$A = \sum_{j=1}^{p} \frac{r^2_{jj}}{h_j - 1},$$

where r_{jj} is the jth diagonal element of $\mathbf{R}$. An adjusted critical value is

$$c_{\text{ad}} = c + \frac{c}{2k}\left[A\left(1 + \frac{3c}{k+2}\right)\right].$$

That is, the data for Level 1 of Factors A and B are stored in the R variable x[[2]], the data for Level 1 of A and Level 2 of B are stored in x[[3]], and so forth. To use `t2way`, first enter the R command

```
grp=c(2,3,5,8,4,1,6,7).
```

Then the command `t2way(2,4,x,grp=grp)` tells the function how the data are ordered. In the example, the first value stored in `grp` is 2, indicating that x[[2]] contains the data for level 1 of both Factors A and B, the next value is 3, indicating that x[[3]] contains the data for Level 1 of A and Level 2 of B, and the fifth value is 4, meaning that x[[4]] contains the data for Level 2 of Factor A and Level 1 of B. As usual, `tr` indicates the amount of trimming, which defaults to 0.2, and `alpha` is α, which defaults to 0.05. The function returns the test statistic for Factor A, V_a, in the R variable `t2way$test.A`, and the significance level is returned in `t2way$sig.A`. Similarly, the test statistics for Factor B, V_b, and interaction, V_{ab}, are stored in `t2way$test.B` and `t2way$test.AB`, with the corresponding significance levels stored in `t2way$sig.B` and `t2way$sig.AB`.

As a more general example, the command

```
t2way(2,3,z,tr=0.1,grp=c(1,3,4,2,5,6))
```

would perform the tests for no main effects and no interactions for a 2-by-3 design for the data stored in the R variable `z`, assuming the data for Level 1 of Factors A and B are stored in `z[[1]]`, the data for Level 1 of A and level 2 of B are stored in `z[[3]]`, and so on. The analysis would be based on 10% trimmed means.

Note that `t2way` contains an argument `p`. Generally, this argument can be ignored; it is used by `t2way` to check whether the total number of groups being passed to the function is equal to JK. If JK is not equal to the number of groups in `x`, the function prints a warning message. If, however, the goal is to perform an analysis using some subset of the groups stored in `x`, this can be done simply by ignoring the warning message. For example, suppose `x` contains data for 10 groups and it is desired to use groups 3, 5, 1, and 9 in a 2-by-2 design. That is, groups 3 and 5 correspond to Level 1 of the first factor and Levels 1 and 2 of the second. The command

t2way(2,2,x,grp=c(3,5,1,9))

accomplishes this goal.

EXAMPLE

A total of $N = 50$ male Sprague-Dawley rats were assigned to one of six conditions corresponding to a 2-by-3 ANOVA. (The data in this example were supplied by U. Hayes.) The two levels of the first factor have to do with whether an animal was placed on a fluid restriction schedule one week prior to the initiation of the experiment. The other factor has to do with the injection of one of three drugs. One of the outcome measures was sucrose consumption shortly after acquisition of a LiCl-induced conditioned taste avoidance. The output from `t2way` appears as follows:

```
$Qa
[1] 11.0931

$A.p.value
[1] 0.001969578

$Qb
[1] 3.764621

$B.p.value
[1] 0.03687472

$Qab
[1] 2.082398

$AB.p.value
[1] 0.738576
```

So based on 20% trimmed means, there is a main effect for both Factors A and B when testing at the 0.05 level, but no interaction is detected.

EXAMPLE

Using the plasma data in Chapter 1, this next example illustrates how to use `t2way` if columns of the data are used to indicate the levels of the factors. First, use the R command

```
z=fac2list(plasma[,10],plasma[,2:3])
```

to convert the data into list mode. (The R function `fac2list` was introduced in Section 9.1.3.) In effect, the data in column 10 are divided into groups according to the data stored in columns 2 and 3. Note that there is no need to indicate that the second argument is a factor variable. (The function assumes that this should be the case and converts it to a factor variable if necessary.) Then

```
t2way(2,3,z)
```

performs the analysis and returns

```
$Qa
[1] 18.90464

$A.p.value
[1] 0.001

$Qb
[1] 4.702398

$B.p.value
[1] 0.151

$Qab
[1] 2.961345

$AB.p.value
[1] 0.286
```

Alternatively, the command

t2way(x=plasma[,10],IV1=plasma[,2],IV2=plasma[,3])

could have been used.

10.4 BOOTSTRAP METHODS

As in previous chapters, a percentile bootstrap method is useful when comparing groups based on M-estimators or MOM. When comparing trimmed means with a relatively low amount of trimming, a bootstrap-t method is a good choice, particularly when the sample sizes are small.

To apply the bootstrap-t method with a trimmed mean, proceed in a manner similar to Chapter 8. That is, for the jth level of Factor A and the kth level of Factor B, subtract the trimmed mean ($\bar{X}_{tjk}$) from each of the n_{jk} observations; this is done for all JK groups. In symbols, for the jth level of Factor A and the kth level of Factor B, bootstrap samples are generated from $C_{1jk}, \dots, C_{n_{jk}jk}$, where $C_{ijk} = X_{ijk} - \bar{X}_{tjk}$. Said yet another way, center the data for each of the JK groups by subtracting out the corresponding trimmed mean in which case the empirical distributions of all JK groups have a trimmed mean of zero. That is, the distributions are shifted so that the null hypothesis is true with the goal of empirically determining an appropriate critical value. Next, generate bootstrap samples from each of these JK groups and compute the test statistics as described in Boxes 10.2 and 10.3. For the main effect associated with Factor A we label these B bootstrap test statistics as $V^*_{a1}, \dots, V^*_{aB}$; these B values provide an approximation of the distribution of V_a when the null hypothesis is true. Put these values in ascending order and label the results $V^*_{a(1)} \le \cdots \le V^*_{a(B)}$. If $V_a \ge V^*_{a(c)}$, where $c = (1-\alpha)B$, rounded to the nearest integer, reject. The hypotheses of no main effect for Factor B and no interaction are tested in an analogous manner.

Other robust measures of location can be compared with the percentile bootstrap method. Again there are many variations that might be used (which include important techniques covered in Chapter 12). Here, only one of these methods is described.

Let θ be any measure of location and let

$$\begin{aligned}\Upsilon_1 &= \frac{1}{K}(\theta_{11} + \theta_{12} + \cdots + \theta_{1K}),\\ \Upsilon_2 &= \frac{1}{K}(\theta_{21} + \theta_{22} + \cdots + \theta_{2K}),\\ &\vdots\\ \Upsilon_J &= \frac{1}{K}(\theta_{J1} + \theta_{J2} + \cdots + \theta_{JK}).\end{aligned}$$

(Υ is an uppercase Greek Upsilon.) So Υ_j is the average of the K measures of location associated with the jth level of Factor A. The hypothesis of no main effects for Factor A is

$$H_0 : \Upsilon_1 = \Upsilon_2 = \cdots = \Upsilon_J.$$

As was the case in Chapter 9, there are various ways this hypothesis might be tested. One possibility is to test

$$H_0 : \Delta_1 = \cdots = \Delta_{J-1} = 0, \tag{10.5}$$

where

$$\Delta_j = \Upsilon_j - \Upsilon_{j+1},$$

$j = 1, \ldots, J-1$. Briefly, generate bootstrap samples in the usual manner yielding $\hat{\Delta}^*_j$, a bootstrap estimate of Δ_j. Then determine how deeply $\mathbf{0} = (0, \ldots, 0)$ is nested within the bootstrap samples. If $\mathbf{0}$ is relatively far from the center of the bootstrap samples, reject.

For reasons similar to those described in Section 10.6.3, the method just described is satisfactory when dealing with the probability of a Type I error, but when the groups differ, this approach might be unsatisfactory in terms of power depending on the pattern of differences among the Υ_j values. One way of dealing with this issue is to compare all pairs of the Υ_j instead. That is, for every $j < j'$, let

$$\Delta_{jj'} = \Upsilon_j - \Upsilon_{j'},$$

and then test

$$H_0 : \Delta_{12} = \Delta_{13} = \cdots = \Delta_{J-1,J} = 0. \tag{10.6}$$

Of course, a similar method can be used when dealing with Factor B.

The percentile bootstrap method just described for main effects can be extended to the problem of testing the hypothesis of no interactions. Box 10.3 outlines how to proceed.

A criticism of the approach in Box 10.3 is that when groups differ, not all possible tetrad differences are being tested, which might affect power. One way of dealing with this problem is for every $j < j'$ and $k < k'$, set

$$\Psi_{jj'kk'} = \theta_{jk} - \theta_{jk'} + \theta_{j'k} - \theta_{j'k'}$$

and then test

$$H_0 : \Psi_{1212} = \cdots = \Psi_{J-1,J,K-1,K} = 0. \tag{10.7}$$

(Alternatively, use the R function `con2way`, described in Section 12.2.5, in conjunction with R functions in Sections 12.1.9 and 12.1.14.)

10.4.1 R Functions pbad2way and t2waybt

The R function

```
pbad2way(J,K,x,est=onestep,conall=T,alpha = 0.05,nboot=2000,grp =
                    NA,pro.dis=F,...)
```

performs the percentile bootstrap method just described, where J and K indicate the number of levels associated with Factors A and B. The argument `conall` defaults to TRUE indicating that all possible differences are used when testing hypotheses. So testing the hypothesis of no main effects for Factor A, Equation (10.6) will be tested rather than Equation (10.5). With `conall=F`, the hypothesis given by Equation (10.5) is tested. Using `conall=TRUE` can result in a numerical error as illustrated momentarily. (It uses Mahalanobis distance, which requires the inverse of a covariance matrix. But the covariance matrix can be singular.)

If this numerical error occurs, there are two options. The first is to use `conall=F`, which is acceptable if all of the hypotheses are rejected. But if one or more are not rejected, the suggestion is to use `pro.dis=T`, which avoids the numerical error by replacing Mahalanobis distance with what is called projection distance. (Projection distance does not require the inverse of a covariance matrix.) A possible appeal of using `conall=FALSE` is that it can result in faster execution time compared to using `pro.dis=TRUE`.

The R function

```
t2waybt(J, K, x, tr = 0.2, grp = c(1:p), p = J * K, nboot = 599, SEED = T)
```

compares groups using a bootstrap-t method.

EXAMPLE

For the plasma retinol data in Chapter 1, we compare plasma retinol levels (stored in column 14 of the R variable `plasma`) based on one-step M-estimators and two factors: sex and smoking status. The two R commands

```
z=fac2list(plasma[,14],plasma[,2:3])
pbad2way(2,3,z)
```

attempt to accomplish this goal using Mahalanobis distance, but this returns

```
Error in solve.default(cov, ...):
system is computationally singular:
reciprocal condition number = 0.
```

To avoid this error, use projection distance via the R command

```
pbad2way(2,3,z,pro.dis=T),
```

which reports the following p-values:

Box 10.3: How to test for no interactions using the percentile bootstrap method.

For convenience, label the JK measures of location as follows:

	Factor B			
	θ_1	θ_2	$\cdots$	θ_K
Factor	θ_{K+1}	θ_{K+2}	$\cdots$	θ_{2K}
A	$\vdots$	$\vdots$	$\cdots$	$\vdots$
	$\theta_{(J-1)K+1}$	$\theta_{(J-1)K+2}$	$\cdots$	θ_{JK}

Let $\mathbf{C}_J$ be a $J-1$ by J matrix having the form

$$\begin{pmatrix} 1 & -1 & 0 & 0 & \dots & 0 \\ 0 & 1 & -1 & 0 & \dots & 0 \\ & & & \vdots & & \\ 0 & 0 & \dots & 0 & 1 & -1 \end{pmatrix}.$$

That is, $c_{ii} = 1$ and $c_{i,i+1} = -1$; $i = 1, \dots, J-1$ and $\mathbf{C}_K$ is defined in a similar fashion. A test of no interactions corresponds to testing

$$H_0 : \Psi_1 = \cdots = \Psi_{(J-1)(K-1)} = 0,$$

where

$$\Psi_L = \sum c_{L\ell}\theta_\ell,$$

$L = 1, \dots, (J-1)(K-1)$, $\ell = 1, \cdots, J(K-1)$, and $c_{L\ell}$ is the entry in the Lth row and ℓth column of $\mathbf{C}_J \otimes \mathbf{C}_K$. That is, generate bootstrap samples yielding $\hat{\Psi}^*_L$ values, do this B times, and then determine how deeply $\mathbf{0} = (0, \dots, 0)$ is nested within these bootstrap samples.

```
$sig.levelA
[1] 0.204

$sig.levelB
[1] 0.102

$sig.levelAB
[1] 0.338
```

So in particular, no differences are found when testing at the 0.05 level.

However, comparing 20% trimmed means via a bootstrap-t method, using the R command

```
t2waybt(2,3,z)
```

returns

```
$A.p.value
[1] 0.1218698
$B.p.value
[1] 0.03505843
$AB.p.value
```

```
[1] 0.2504174
```

So, when testing at the 0.05 level, now we conclude that there are differences among the three groups of smokers, demonstrating again that the choice of method can make a practical difference. If t2way is used instead (the non-bootstrap method for comparing trimmed means), the p-value for Factor B is now 0.041.

10.5 TESTING HYPOTHESES BASED ON MEDIANS

As was the case in Chapter 9, special methods are required when the goal is to compare medians. There are non-bootstrap methods for comparing medians in a two-way design but currently it seems better to use one of two bootstrap methods. The first can be applied with the R function pbad2way with the argument est=median. The other uses a simple generalization of method SHPD in Section 10.5.3. The relative merits of these methods are unclear. Indirect evidence suggests using the method in Section 10.4, but this issue is in need of more study.

Again, tied values are an issue. The best advice is to use bootstrap methods described in Chapter 12. Indeed, a case can be made that the methods in Chapter 12 are best for general use regardless of whether tied values occur, but the details must be postponed for now. (See in particular Sections 12.1.11 and 12.1.12.)

To outline how method SHPD can be generalized to medians, let

$$\bar{M}_G = \frac{1}{JK}\sum_{j=1}^{J}\sum_{k=1}^{K} M_{jk}$$

be the average of all JK sample medians. Let

$$\bar{M}_{j.} = \frac{1}{K}\sum_{k=1}^{K} M_{jk},$$

which is the average of the sample medians for the jth level of Factor A,

$$\bar{M}_{.k} = \frac{1}{J}\sum_{j=1}^{J} M_{jk}$$

(the average of the sample means for the kth level of Factor B),

$$A_j = \bar{M}_{j.} - \bar{M}_G,$$

$$B_k = \bar{M}_{.k} - \bar{M}_G,$$

$$C_{jk} = \bar{X}_{jk} - \bar{X}_{j.} - \bar{X}_{.k} + \bar{X}_G$$

$$\text{SSA} = \sum A_j^2,$$

$$\text{SSB} = \sum B_k^2,$$

and

$$\text{SSINTER} = \sum\sum C_{jk}^2.$$

For Factor A, the goal is to determine whether SSA is sufficiently large to reject the hypothesis of no main effects based on medians. For Factor B and interactions, now the goal is to determine whether SSB and SSINTER are sufficiently large, respectively. This can be done by proceeding along the lines of Section 10.5.3.

10.5.1 R Function m2way

The computations for comparing medians using an extension of method SHPD are performed by the R function

```
m2way(J,K,x,alpha=0.05).
```

EXAMPLE

In the last example, we compared plasma retinol levels based on two factors: sex and smoking status. We rejected at the 0.05 level when analyzing smoking status based on 20% trimmed means. Comparing medians with m2way instead, the p-value is 0.088. Assuming that the data are stored in the R variable `z`, and using the R command

```
pbad2way(2,3,z,est=median,pro.dis=T)
```

the p-value for Factor B is 0.0835. So in this case, the choice between the two bootstrap methods for medians makes little difference, but in contrast to using 20% trimmed means, neither rejects at the 0.05 level. Also, there are a few tied values. Perhaps the number of tied values is small enough so as to have little impact on the methods for medians used here, but it is unclear whether this is the case.

10.6 A RANK-BASED METHOD FOR A TWO-WAY DESIGN

A rank-based method for analyzing a two-way ANOVA design was derived by Akritas et al.(1997). Roughly, the method is designed to be sensitive to differences among the average ranks of the pooled data. To describe how the notion of main effects and interactions are defined, let $F_{jk}(x)$ represent the (cumulative) probability that an observation corresponding to the jth level of Factor A and the kth level of Factor B is less than or equal to x. Main effects for Factor A are expressed in terms of

$$\bar{F}_{j.}(x) = \frac{1}{K}\sum_{k=1}^{K} F_{jk}(x),$$

the average of the distributions among the K levels of Factor B corresponding to the jth level of Factor A. Main effects for Factor B are expressed in terms of

$$\bar{F}_{.k}(x) = \frac{1}{J}\sum_{j=1}^{J} F_{jk}(x).$$

In particular, the hypothesis of no main effects for Factor A is

$$H_0 : \bar{F}_{1.}(x) = \bar{F}_{2.}(x) = \cdots = \bar{F}_{J.}(x)$$

for any x. The hypothesis of no main effects for Factor B is

$$H_0 : \bar{F}_{.1}(x) = \bar{F}_{.2}(x) = \cdots = \bar{F}_{.K}(x).$$

EXAMPLE

For the plasma retinol study, consider the two factors sex (1 = Male, 2 = Female) and smoking status (1 = Never, 2 = Former, 3 = Current Smoker). Then $F_{12}(800)$ is the probability that a randomly sampled male, who is a former smoker, has a plasma retinol level less than or equal to 800. The hypothesis of no main effects for Factor A is that the average cumulative probabilities associated with the three groups of smokers are the same for males and females.

To describe the notion of no interaction, first focus on a 2-by-2 design. Then no interaction means that for any x,

$$F_{11}(x) - F_{12}(x) = F_{21}(x) - F_{22}(x).$$

For the more general case of a J-by-K design, no interaction means that there is no interaction for any two levels of Factor A and any two levels of Factor B.

Briefly, the methods for testing the hypothesis of no main effect and no interactions are based on the ranks associated with the pooled data. So if among all JK groups there is a total of N observations, the smallest observation gets a rank of 1 and the largest gets a rank of N. The computational details for testing the hypotheses of no main effects and no interaction are not provided. Readers interested in these details are referred to Akritas et al. (1997) or Wilcox (2017, Section 7.9). Here, we simply supply an R function for applying the method, which is described next. (See Brunner et al., 2016, for an alternative approach that might be of interest.)

10.6.1 R Function bdm2way

The R function

```
bdm2way(J,K,x,alpha=0.05)
```

performs the rank-based method just described. The function returns p-values for each of the hypotheses tested as well as an estimate of what is called the relative effects. Following Akritas et al. (1997), relative effects refer to the average ranks associated with each of the JK groups minus 1/2. That is, for levels j and k of Factors A and B, respectively, the relative effect is $\bar{R}_{jk} - 0.5$, where $\bar{R}_{jk}$ is the average of the ranks. (The relative effects are related to a weighted sum of the cumulative probabilities, where the weights depend on the sample sizes.)

10.6.2 The Patel–Hoel Approach to Interactions

Patel and Hoel (1973) proposed an alternative approach to interactions in a 2-by-2 design. For Level 1 of Factor A, let $P_{11,12}$ be the probability that a randomly sampled observation from Level 1 of Factor B is less than a randomly sampled observation from a level 2 of Factor B. Similarly, for Level 2 of Factor A, let $P_{21,22}$ be the probability that a randomly sampled observation from l1 of Factor B is less than a randomly sampled observation from a Level 2 of Factor B. The Patel–Hoel definition of no interaction is that

$$p_{11,12} = p_{21,22}.$$

In the event ties can occur, let X_{ijk} be a randomly sampled observation from the jth level of Factor A and the kth level of Factor B. Now let

$$p_{11,12} = P(X_{i11} < X_{i12}) + \frac{1}{2}P(X_{i11} = X_{i12}),$$

$$p_{21,22} = P(X_{i21} < X_{i22}) + \frac{1}{2}P(X_{i21} = X_{i22}).$$

The hypothesis of no interaction is

$$H_0 : p_{11,12} = p_{21,22},$$

which can be tested using an extension of the method in Section 7.8.4. The computational details are relegated to Box 10.4. The R function `rimul`, described in Section 12.2.11, performs this test.

10.7 THREE-WAY ANOVA

Three-way ANOVA refers to a situation where there are three factors. In Section 10.3.1 we analyzed the plasma retinol data based on two factors: sex and smoking status. This data set contains another factor: vitamin use. The levels of this factor are 1 (uses vitamins fairly often), 2 (not often), and 3 (does not use vitamins). An analysis based on all three factors deals with what is called a 2-by-3-by-3 design. More generally, if three factors called A, B, and C have J, K, and L levels, respectively, we have a J-by-K-by-L design.

Main effects and two-way interactions are defined in a manner similar to a two-way design. To describe the details in a formal and precise manner, we need to generalize the notation used to describe a two-way design.

Now we let μ_{jkl} represent the mean associated with the jth level of Factor A, the kth level of Factor B, and the lth level of Factor C. Then

$$\bar{\mu}_{j..} = \frac{1}{KL}\sum_{k=1}^{K}\sum_{l=1}^{L}\mu_{jkl}$$

is the average of all population means corresponding to the jth level of Factor A. Similarly,

$$\bar{\mu}_{.k.} = \frac{1}{JL}\sum_{j=1}^{J}\sum_{l=1}^{L}\mu_{jkl}$$

and

$$\bar{\mu}_{..l} = \frac{1}{JK}\sum_{j=1}^{J}\sum_{k=1}^{K}\mu_{jkl}$$

are the averages of the population means associated with the kth and lth levels of Factors B and C, respectively.

The hypothesis of no main effects for Factor A is

$$H_0 : \bar{\mu}_{1..} = \bar{\mu}_{2..} = \cdots = \bar{\mu}_{J..}.$$

The hypotheses of no main effects for Factors B and C are

$$H_0 : \bar{\mu}_{.1.} = \bar{\mu}_{.2.} = \cdots = \bar{\mu}_{.K.}$$

and

$$H_0 : \bar{\mu}_{..1} = \bar{\mu}_{..2} = \cdots = \bar{\mu}_{..L},$$

respectively.

To describe two-way interactions, we need to extend our notation by letting

$$\bar{\mu}_{jk.} = \frac{1}{L}\sum_{l=1}^{L}\mu_{jkl},$$

$$\bar{\mu}_{j.l} = \frac{1}{K}\sum_{l=1}^{K} \mu_{jkl},$$

and

$$\bar{\mu}_{.kl} = \frac{1}{J}\sum_{l=1}^{J} \mu_{jkl}.$$

The notion of no interactions associated with Factors A and B means that for any two levels of Factor A, say j and j', and any two levels of Factor B, k and k',

$$\bar{\mu}_{jk.} - \bar{\mu}_{j'k.} = \bar{\mu}_{jk'.} - \bar{\mu}_{j'k'.}$$

No interactions associated with Factors A and C, as well as Factors B and C, are defined in an analogous fashion.

Box 10.4: Testing the hypothesis of no interaction via the Patel–Hoel approach.

For Level 1 of Factor A, let δ_1 represent δ, as described in Section 7.8.4, when comparing Levels 1 and 2 of Factor B. Let $\hat{\delta}_1$ be the estimate of δ. Its squared standard error is computed as indicated in Section 7.8.4 and will be labeled $\hat{\sigma}_1^2$. Similarly, let δ_2 be δ when focusing on Level 2 of Factor A, its estimate is denoted by $\hat{\delta}_2$, and the estimate of the squared standard error of $\hat{\sigma}_2$ is denoted by $\hat{\sigma}_2^2$. The hypothesis of no interaction corresponds to

$$H_0 : \Delta = \frac{\delta_2 - \delta_1}{2} = 0.$$

An estimate of

$$\Delta = p_{11,12} - p_{21,22}$$

is

$$\hat{\Delta} = \frac{\hat{\delta}_2 - \hat{\delta}_1}{2};$$

the estimated squared standard error of $\hat{\Delta}$ is

$$S^2 = \frac{1}{4}(\hat{\sigma}_1^2 + \hat{\sigma}_2^2).$$

A $1 - \alpha$ confidence interval for Δ is

$$\hat{\Delta} \pm z_{1-\alpha/2}S,$$

where $z_{1-\alpha/2}$ is the $1 - \alpha/2$ quantile of a standard normal distribution. The hypothesis of no interaction is rejected if this confidence interval does not contain zero.

There is also the notion of a three-way interaction. Roughly, pick any two levels of one of the factors, say Factor C. Then no three-way interaction means that the two-way interactions associated with any two levels of Factor A and any two levels of Factor B are the same.

EXAMPLE

A three-way interaction is illustrated for the simplest case: a 2-by-2-by-2 design. Focus on the first level of Factor A. For the other two factors, an interaction corresponds to situations where

$$(\mu_{111} - \mu_{112}) - (\mu_{121} - \mu_{122}) = \mu_{111} - \mu_{112} - \mu_{121} + \mu_{122} \neq 0$$

as already explained. Similarly, for the second level of Factor A, an interaction refers to a situation where

$$\mu_{211} - \mu_{212} - \mu_{221} + \mu_{222} \neq 0.$$

No three-way interaction means that the left sides of these last two equations are equal. That is, the hypothesis of no three-way interaction is

$$H_0 : \mu_{111} - \mu_{112} - \mu_{121} + \mu_{122} = \mu_{211} - \mu_{212} - \mu_{221} + \mu_{222}.$$

The ANOVA F tests for two-way designs can be extended to a three-way design, still assuming normality and homoscedasticity, but the computational details are not provided. Consistent with one-way and two-way designs, this approach seems to be satisfactory provided the groups being compared have identical distributions. But unequal variances, differences in skewness, and outliers can affect Type I errors and power. There is a heteroscedastic method for trimmed means that includes the goal of comparing means as a special case that is aimed at reducing these problems. The computational details can be found, for example, in Wilcox (2017, Section 7.3). The method is based on a generalization of a technique derived by Johansen (1980) and is similar to the method in Box 10.2. Here we merely supply appropriate software.

10.7.1 R Functions anova and t3way

The R functions `aov` and `anova` can be used to compare means in a three-way design assuming normality and homoscedasticity. Proceed as illustrated in Section 10.2.1; only now a third factor is included. If, for example, the data are stored in the R variable `y`, with the factors stored in column 2, 3, and 5, and if the outcome of interest (the dependent variable) is stored in column 10, the command

```
anova(aov(y[,10]~as.factor(y[,2])*as.factor(y[,3])*as.factor(y[,5])))
```

will perform the analysis.

To deal with heteroscedasticity, and for the more general goal of comparing trimmed means, use the R function

```
t3way(J, K, L, x, tr = 0.2, grp = c(1:p), alpha = 0.05, p = J * K * L, MAT =
FALSE, lev.col = c(1:3), var.col = 4, pr = TRUE, IV1 = NULL, IV2 = NULL, IV3
= NULL).
```

The data are assumed to be arranged such that the first L groups correspond to Level 1 of Factors A and B ($J = 1$ and $K = 1$) and the L levels of Factor C. The next L groups correspond to the first level of Factor A, the second level of Factor B, and the L levels of Factor C. If, for example, a 3-by-2-by-4 design is being used and the data are stored in list mode, it is assumed that for $J = 1$ (the first level of the first factor), the data are stored in the R variables x[[1]],...,x[[8]] as follows:

	Factor C			
Factor	x[[1]]	x[[2]]	x[[3]]	x[[4]]
B	x[[5]]	x[[6]]	x[[7]]	x[[8]]

For the second level of the first factor, $J = 2$, it is assumed that the data are stored as

	Factor C			
Factor	x[[9]]	x[[10]]	x[[11]]	x[[12]]
B	x[[13]]	x[[14]]	x[[15]]	x[[16]]

If the data are not stored as assumed by `t3way`, `grp` can be used to indicate the proper ordering. As an illustration, consider a 2-by-2-by-4 design and suppose that for $J = 1$, the data are stored as follows:

	Factor C			
Factor	x[[15]]	x[[8]]	x[[3]]	x[[4]]
B	x[[6]]	x[[5]]	x[[7]]	x[[8]]

and for $J = 2$

	Factor C			
Factor	x[[10]]	x[[9]]	x[[11]]	x[[12]]
B	x[[1]]	x[[2]]	x[[13]]	x[[16]]

Then the R command

```
grp=c(15,8,3,4,6,5,7,8,10,9,11,12,1,2,13,16)
```

and the command `t3way(2,2,3,x,grp=grp)` will test all of the relevant hypotheses at the 0.05 level using 20% trimmed means.

If the data are stored with certain columns containing information about the levels of the factors, again the R function `fac2list` can be used to convert the data into list mode so that the R function `t3way` can be used. Proceed in the manner illustrated in Section 10.3.1. Alternatively, use the arguments `IV1`, `IV2`, and `IV3` as illustrated in conjunction with the function `t2way`.

10.8 EXERCISES

1. For a 2-by-4 design with population means

		Factor B			
		Level 1	Level 2	level 3	Level 4
Factor	Level 1	μ_1	μ_2	μ_3	μ_4
A	Level 2	μ_5	μ_6	μ_7	μ_8

state the hypotheses of no main effects and no interactions.

2. Consider a 2-by-2 design with population means

		Factor B	
		Level 1	Level 2
Factor A	Level 1	$\mu_1 = 110$	$\mu_2 = 70$
	Level 2	$\mu_3 = 80$	$\mu_4 = 40$

State whether there is a main effect for Factor A, for Factor B, and whether there is an interaction.

3. Consider a 2-by-2 design with population means

		Factor B	
		Level 1	Level 2
Factor A	Level 1	$\mu_1 = 10$	$\mu_2 = 20$
	Level 2	$\mu_3 = 40$	$\mu_4 = 10$

Determine whether there is an interaction. If there is an interaction, determine whether there is an ordinal interaction for rows.

4. Make up an example where the population means in a 3-by-3 design have no interaction effect but main effects for both factors exist.

5. Imagine a study where two methods are compared for treating depression. A measure of effectiveness has been developed, the higher the measure the more effective the treatment. Further assume there is reason to believe that males might respond differently to the methods compared to females. If the sample means are

		Factor B	
		Males	Females
Factor A	Method 1	$\bar{X}_1 = 50$	$\bar{X}_2 = 70$
	Method 2	$\bar{X}_3 = 80$	$\bar{X}_4 = 60$

and if the hypothesis of no main effects for Factor A is rejected, but the hypothesis of no interactions is not rejected, what does this suggest about which method should be used? In terms of Tukey's three decision rule, what can be said about any interaction?

6. In the previous exercise, suppose the hypothesis of no interactions is rejected and in fact there is a disordinal interaction. What does this suggest about using method 1 rather than method 2?

7. Referring to Exercise 5, imagine the hypothesis of no interactions is rejected. Is it reasonable to conclude that the interaction is disordinal?

8. This exercise is based on a study by Atkinson and Polivy where the general goal was to study people's reactions to unprovoked verbal abuse. In the study, 40 participants were asked to sit alone in a cubicle and answer a brief questionnaire. After waiting far longer than it took to fill out the form, a research assistant returned to collect the responses. Half the subjects received an apology for the delay and the other half were told, among other things, that they could not even fill out the form properly. Each of these 20 subjects were divided into two groups: half got to retaliate against the research assistant by giving her a bad grade, and the other half did not get a chance to retaliate. All subjects were given a standardized test of hostility. Imagine that the sample means are

		Abuse	
		Insult	Apology
Retaliation	Yes	$\bar{X}_1 = 65$	$\bar{X}_2 = 54$
	No	$\bar{X}_3 = 61$	$\bar{X}_4 = 57$

Further assume that the hypotheses of no main effects and no interaction are rejected. Interpret this result.

9. Verify that when comparing the groups in Table 10.1 based on medians and with the R function `pbad2way` in Section 10.4.1, the p-values for Factors A and B are 0.28 and 0.059, respectively, and that for the test of no interaction, the significance level is 0.13.

10. The file CRCch10_Ex10.txt, stored on the author's web page, contains data, the first few lines of which look like this:

A	B	Outcome
1	1	32
1	1	21

The first two columns indicate the levels of two factors and the dependent variable is in column 3. Use the function `t2way` to test the hypotheses of no main effects or interactions based on means. Verify that the p-values for Factors A and B are 0.372 and 0.952, respectively.

11. For the data in the previous exercise, verify that if `t2way` is used to compare 20% trimmed means, the p-value when testing the hypothesis of no interaction is reported to be 0.408.

12. For the data in Exercise 10, verify that if `t2way` is used to compare 20% trimmed means, the p-values for Factors A and B are 0.145 and 0.93. Why does the p-value for Factor A drop from 0.372 when comparing means to 0.145 with 20% trimmed means instead?

13. Repeat the illustration in Section 10.5.1, but now compare the groups using the modified one-step M-estimator via the R function `pbad2way`. Verify that for Factor B, now the p-value is 0.059. How does this compare to the results using a 20% trimmed mean? Note: the data can be read into R as illustrated in Section 1.3.3. Gender is stored in column 2 and smoking status is in column 3. The outcome variable is in column 14.

CHAPTER 11

COMPARING MORE THAN TWO DEPENDENT GROUPS

CONTENTS

This chapter extends the methods in Chapter 8 to situations where the goal is to compare more than two dependent groups. Included are two-way and three-way designs, and what are called between-by-within designs. When attention is restricted to comparing means, it is stressed that a plethora of methods have been proposed that are not covered in this

Table 11.1 Hypothetical Blood Pressure for 36 Rabbits

Litter	Drug A	Drug B	Drug C
1	240	230	200
2	250	220	190
3	260	230	230
4	200	220	240
5	255	245	200
6	245	210	210
7	220	215	205
8	230	220	200
9	240	230	210
10	220	250	260
11	210	220	200
12	235	225	205

chapter. For book-length descriptions of these techniques, which include procedures especially designed for handling longitudinal data, see Crowder and Hand (1990), Jones (1993), and Diggle et al. (1994). This chapter covers the more basic methods for means that are typically used in applied research, and then methods that address the practical problems associated with these techniques are described. We begin with a one-way *repeated measures* design, also called a *within-subjects* design, where the goal is to compare J dependent groups. Several methods are covered with their relative merits summarized in Section 11.5. For results on a permutation-type method that appears to perform relatively well, see Friedrich et al. (2017).

11.1 COMPARING MEANS IN A ONE-WAY DESIGN

To be concrete, imagine that you are investigating three drugs for controlling hypertension, and as part of your investigation, the drugs are administered to rabbits with induced hypertension. Further imagine that littermates are used in the experiment. That is, a litter is randomly sampled, and three rabbits are randomly sampled from the litter; the first rabbit gets drug A, the second rabbit gets drug B, and the third rabbit gets drug C. Further assume that $n = 12$ randomly sampled litters are used and their blood pressure is measured after four weeks. The results are as shown in Table 11.1. A common goal is to test

$$H_0 : \mu_1 = \cdots = \mu_J, \tag{11.1}$$

the hypothesis that all J groups have a common mean. The ANOVA F test in Chapter 9 is inappropriate here because the sample means associated with the three drugs are possibly dependent due to using rabbits from the same litter.

The best-known method for testing Equation (11.1) is based on the following assumptions:

- Random sampling. In the rabbit example, litters are assumed to be randomly sampled. In more generic terms, we randomly sample n vectors of observations:

  $(X_{11}, \ldots, X_{1J}), \ldots, (X_{n1}, \ldots, X_{nJ})$.

- Normality
- Homoscedasticity
- Sphericity

Sphericity is met if, in addition to homoscedasticity, the correlations associated with any two groups have a common value. To say this in a slightly more formal way, let ρ_{jk} be Pearson's correlation corresponding to groups j and k. For Table 11.1, ρ_{23} is the correlation between blood pressure values associated with drugs B and C. It is being assumed that all ρ_{jk} $(j \neq k)$ have the same value. The sphericity assumption is met under slightly weaker conditions than those described here (e.g., Kirk, 1995), but the details are not important for present purposes for reasons that will become evident.

Let X_{ij} be the ith observation from the jth group. In the rabbit example, $X_{62} = 210$ corresponds to the rabbit from the sixth litter that received drug B. Let $\bar{X}_{i.}$ be the average of the values in the ith row, which is the ith litter in the example. For Table 11.1,

$$\bar{X}_{1.} = \frac{240 + 230 + 200}{3} = 223.3$$

and

$$\bar{X}_{2.} = \frac{250 + 220 + 190}{3} = 220.$$

Similarly, $\bar{X}_{.j}$ is the average for the jth column. In the example, $\bar{X}_{.2}$ is the average blood pressure among rabbits receiving drug B. The distribution associated with the jth group, ignoring all other groups, is called the jth *marginal distribution* and has a population mean labeled μ_j. Let

$$SSB = n\sum_{j=1}^{J}(\bar{X}_{.j} - \bar{X}_{..})^2$$

$$SSE = \sum_{j=1}^{J}\sum_{i=1}^{n}(X_{ij} - \bar{X}_{i.} - \bar{X}_{.j} + \bar{X}_{..})^2,$$

$$\text{MSB} = \frac{\text{SSB}}{J-1},$$

and

$$\text{MSE} = \frac{\text{SSE}}{(J-1)(n-1)}$$

where

$$\bar{X}_{..} = \frac{1}{Jn}\sum_{j=1}^{J}\sum_{i=1}^{n}X_{ij}$$

is the grand mean (the average of all values). The quantity SSE is called the *sum of squared residuals*. In the illustration

$$\bar{X}_{..} = 224.2$$

$$\text{SSB} = 2,787.5$$

$$\text{SSE} = 6,395.83.$$

Finally, let

$$F = \frac{\text{MSB}}{\text{MSE}}. \tag{11.2}$$

Under the assumptions previously described, if the null hypothesis of equal means is true, F has an F distribution with degrees of freedom $\nu_1 = J - 1$ and $\nu_2 = (J-1)(n-1)$. If the goal is to have a Type I error equal to α, reject if

$$F \geq f,$$

where f is the $1-\alpha$ quantile of an F distribution with $\nu_1 = J-1$ and $\nu_2 = (J-1)(n-1)$ degrees of freedom.

EXAMPLE

For the rabbit data, MSB $= 2787.5/2 = 1393.75$, MSE $= 6395.83/22 = 290.72$, so $F = 4.8$. The degrees of freedom are $\nu_1 = 2$ and $\nu_2 = 22$, and the 0.05 critical value is $f = 3.44$, so reject.

If the normality assumption is true, but the sphericity assumption is violated, the F test just described is unsatisfactory in terms of controlling the probability of a Type I error. The actual Type I error probability can exceed the nominal level. If the population variances and covariances were known, the degrees of freedom can be adjusted to deal with this problem (Box, 1954), still assuming normality. They are not known, but they can be estimated in various ways. One of the more successful estimation methods was derived by Huynh and Feldt (1976) and is used by some software packages. However, this approach can be highly unsatisfactory in terms of both Type I errors and power when sampling from non-normal distributions.

Methods for testing the sphericity assumption have been proposed, but it is unclear when such tests have enough power to detect situations where the sphericity assumption should be discarded (e.g., Boik, 1981; Keselman et al., 1980), so no details are given here. Establishing that such tests have sufficient power under non-normality is an extremely difficult problem with no adequate solution currently available. (Readers interested in more details about sphericity are referred to Kirk, 1995; as well as Rogan et al., 1979). Consistent with previous chapters, one way of reducing practical problems is to compare trimmed means instead of using a method that allows non-sphericity. Methods for accomplishing this goal are described in subsequent sections in this chapter, the first of which is covered in Section 11.2.

11.1.1 R Function aov

The built-in R function

```
aov(x ~ g),
```

introduced in Chapter 9, can be used to perform the ANOVA F test just described, again using a variation of the formula convention, which assumes certain columns of data indicate the group (or level) to which a participant belongs.

EXAMPLE

Imagine that you have data stored in the R data frame `mydata` that looks like this:

```
  dv subject myfactor
1  1      s1       f1
2  3      s1       f2
3  4      s1       f3
4  2      s2       f1
5  2      s2       f2
6  3      s2       f3
7  2      s3       f1
8  5      s3       f2
9  6      s3       f3
```

```
10  3       s4       f1
11  4       s4       f2
12  4       s4       f3
13  3       s5       f1
14  5       s5       f2
15  6       s5       f3
```

The outcome of interest is stored in the column headed by dv (dependent variable). Moreover, there are three dependent groups, which are indicated by f1, f2, and f3 in column 4. The R commands

```
blob= aov(dv ~ myfactor + Error(subject/myfactor), data=mydata)
summary(blob)
```

return the following:

```
Error: subject
          Df Sum Sq Mean Sq F value Pr(>F)
Residuals  4   12.4     3.1

Error: subject:myfactor
          Df  Sum Sq Mean Sq F value   Pr(>F)
myfactor   2 14.9333  7.4667  13.576 0.002683 **
Residuals  8  4.4000  0.5500
```

So the p-value is 0.002683.

The R function aov has no option for dealing with heteroscedasticity or non-sphericity. R has other built-in methods for comparing dependent groups, but currently it appears that they have no options for dealing with non-sphericity and robust measures of location. But R functions have been written for this book that deal with both problems, as will be seen.

11.2 COMPARING TRIMMED MEANS WHEN DEALING WITH A ONE-WAY DESIGN

Recall from Chapter 8 that when comparing two dependent groups, inferences based on difference scores are not necessarily the same as inferences based on the marginal distributions. This continues to be the case when comparing more than two groups, and again neither approach dominates the other. In terms of power, the choice of method depends on how the groups differ, which, of course, is unknown. Some of the examples given in this chapter illustrate that we can fail to reject when comparing the marginal trimmed means, but we reject when using difference scores, the point being that the choice of method can make a practical difference. This is not to suggest, however, that difference scores always give the highest power. The only certainty is that the reverse can happen, so both approaches are covered in this chapter. We begin by describing how to test

$$H_0 : \mu_{t1} = \cdots = \mu_{tJ}, \tag{11.3}$$

the hypothesis that the marginal distributions associated with J dependent groups have equal population trimmed means. In contrast to the previous section, sphericity is not assumed. So for the special case where the goal is to compare means, the method in this section appears to be a better choice for general use over the method in Section 11.1. The test statistic, labeled F_t, is computed as follows:

Winsorize the observations by computing

$$Y_{ij} = \begin{cases} X_{(g+1)j}, & \text{if } X_{ij} \le X_{(g+1)j} \\ X_{ij}, & \text{if } X_{(g+1)j} < X_{ij} < X_{(n-g)j} \\ X_{(n-g)j}, & \text{if } X_{ij} \ge X_{(n-g)j}. \end{cases}$$

Let $h = n - 2g$ be the effective sample size, where $g = [\gamma n]$, $[\gamma n]$ is γn rounded down to the nearest integer, and γ is the amount of trimming. Compute

$$\bar{X}_t = \frac{1}{J} \sum \bar{X}_{tj}$$

$$Q_c = (n - 2g) \sum_{j=1}^{J} (\bar{X}_{tj} - \bar{X}_t)^2$$

$$Q_e = \sum_{j=1}^{J} \sum_{i=1}^{n} (Y_{ij} - \bar{Y}_{.j} - \bar{Y}_{i.} + \bar{Y}_{..})^2,$$

where

$$\bar{Y}_{.j} = \frac{1}{n} \sum_{i=1}^{n} Y_{ij}$$

$$\bar{Y}_{i.} = \frac{1}{J} \sum_{j=1}^{J} Y_{ij}$$

$$\bar{Y}_{..} = \frac{1}{nJ} \sum_{j=1}^{J} \sum_{i=1}^{n} Y_{ij}.$$

The test statistic is

$$F_t = \frac{R_c}{R_e},$$

where

$$R_c = \frac{Q_c}{J-1}$$

$$R_e = \frac{Q_e}{(h-1)(J-1)}.$$

Decision Rule: Reject if $F_t \ge f$, the $1 - \alpha$ quantile of an F distribution with degrees of freedom computed as described in Box 11.1. (For simulation results on how this method performs with 20% trimmed means, see Wilcox, Keselman, Muska, and Cribbie, 2000.)

11.2.1 R Functions rmanova and rmdat2mat

The R function `rmanova`, written for this book, compares the trimmed means of J dependent groups using the calculations just described. The function has the general form

```
rmanova(x,tr=0.2,grp=c(1:length(x))).
```

The data are stored in the first argument `x`, which can either be an n-by-J matrix, the jth column containing the data for jth group, or an R variable having list mode. In the latter case, x[[1]] contains the data for group 1, x[[2]] contains the data for group 2, and so on. As usual, `tr` indicates the amount of trimming which defaults to 0.2, and `grp` can be used to compare a subset of the groups. By default, the trimmed means of all J groups are compared. If, for example, there are five groups, but the goal is to test H_0: $\mu_{t2} = \mu_{t4} = \mu_{t5}$, the command `rmanova(x,grp=c(2,4,5))` accomplishes this goal using 20% trimming.

Box 11.1: How to compute degrees of freedom when comparing trimmed means.

Let

$$v_{jk} = \frac{1}{n-1}\sum_{i=1}^{n}(Y_{ij} - \bar{Y}_{.j})(Y_{ik} - \bar{Y}_{.k})$$

for $j = 1, \ldots, J$ and $k = 1, \ldots, J$, where Y_{ij} is the Winsorized observation corresponding to X_{ij}. When $j = k$, $v_{jk} = s^2_{wj}$, the Winsorized sample variance for the jth group, and when $j \neq k$, v_{jk} is a Winsorized analog of the sample covariance.

Let

$$\bar{v}_{..} = \frac{1}{J^2}\sum_{j=1}^{J}\sum_{k=1}^{J} v_{jk}$$

$$\bar{v}_d = \frac{1}{J}\sum_{j=1}^{J} v_{jj}$$

$$\bar{v}_{j.} = \frac{1}{J}\sum_{k=1}^{J} v_{jk}$$

$$A = \frac{J^2(\bar{v}_d - \bar{v}_{..})^2}{J-1}$$

$$B = \sum_{j=1}^{J}\sum_{k=1}^{J} v^2_{jk} - 2J\sum_{j=1}^{J}\bar{v}^2_{j.} + J^2\bar{v}^2_{..}$$

$$\hat{\epsilon} = \frac{A}{B}$$

$$\tilde{\epsilon} = \frac{n(J-1)\hat{\epsilon} - 2}{(J-1)\{n-1-(J-1)\hat{\epsilon}\}}.$$

The degrees of freedom are

$$\nu_1 = (J-1)\tilde{\epsilon}$$

$$\nu_2 = (J-1)(h-1)\tilde{\epsilon},$$

where h is the effective sample size (the number of observations left in each group after trimming).

Notice that `rmanova` does not follow the R convention of using a formula. If the data are stored in a data frame, say `m`, with one of the columns indicating the levels of the groups being compared, the data can be stored in a matrix that will be accepted by `rmanova` using the R function

```
rmdat2mat(m, id.col = NULL, dv.col = NULL).
```

The argument `id.col` indicates the column containing information about the levels of the groups and `dv.col` indicates the column containing the data to be analyzed (the dependent variable).

EXAMPLE

Consider the last example where the data are stored in the R variable `mydata`. The commands

```
xx=rmdat2mat(mydata,3,1)
rmanova(xx,tr=0)
```

will perform the test based on means. The output looks like this:

```
 [1] "The number of groups to be compared is"
[1] 3
$test
[1] 13.57576
$df
[1] 1.317019 5.268076
$siglevel
[1] 0.01085563
$tmeans
[1] 2.2 3.8 4.6
$ehat
[1] 0.5752773
$etil
[1] 0.6585095
```

The value of the statistic is again $F = 13.57576$, but now the p-value is 0.01085563 rather than 0.00268 as indicated by the R function `aov`. (The last two values reported by `rmanova` are used to adjust the degrees of freedom as outlined in Box 11.1. Under sphericity, they would be equal to 1.)

EXAMPLE

Table 8.1 reports data from a study dealing with the weight of cork borings from trees. Of interest is whether the typical weights for the north, east, south, and west sides of the trees differ. Because samples taken from the same tree might be dependent, again the methods in Chapter 9 might be inappropriate. Comparing the trimmed means with the R function `rmanova`, the p-value is 0.096. Comparing means instead, the p-value is 0.007. (In Section 11.3.4 we will see that comparing trimmed means based on difference scores yields a p-value of 0.014.)

11.2.2 A Bootstrap-t Method for Trimmed Means

When comparing trimmed means, a bootstrap-t method can be applied in basically the same way as described in previous chapters, only again we generate bootstrap samples in a manner consistent with dependent groups. As usual, when working with a bootstrap-t method, we begin by centering the data. That is, we set

$$C_{ij} = X_{ij} - \bar{X}_{tj}$$

with the goal of estimating an appropriate critical value, based on the test statistic F_t, when the null hypothesis is true. The remaining steps are as follows:

1. Generate a bootstrap sample by randomly sampling, with replacement, n rows of data from the matrix

$$\begin{pmatrix} C_{11}, \ldots, C_{1J} \\ \vdots \\ C_{n1}, \ldots, C_{nJ} \end{pmatrix}$$

   yielding

$$\begin{pmatrix} C^*_{11}, \ldots, C^*_{1J} \\ \vdots \\ C^*_{n1}, \ldots, C^*_{nJ} \end{pmatrix}.$$

2. Compute the test statistic F_t based on the C^*_{ij} values generated in step 1, and label the result F^*.
3. Repeat steps 1 and 2 B times and label the results $F^*_1, \ldots, F^*_B$.
4. Put these B values in ascending order and label the results $F^*_{(1)} \leq \cdots \leq F^*_{(B)}$.

The critical value is estimated to be $F^*_{(c)}$, where $c = (1-\alpha)B$ rounded to the nearest integer. That is, reject the hypothesis of equal trimmed means if

$$F_t \geq F^*_{(c)},$$

where F_t is the test statistic in Section 11.2, which is based on the X_{ij} values.

11.2.3 R Function rmanovab

The R function

```
rmanovab(x, tr = 0.2, alpha = 0.05, grp = 0, nboot = 599)
```

performs the bootstrap-t method just described. The arguments have their usual meaning; see, for example, Section 10.5.2.

11.3 PERCENTILE BOOTSTRAP METHODS FOR A ONE-WAY DESIGN

As noted in previous chapters, when testing hypotheses based on an M-estimator, a percentile bootstrap method appears to be the best method available, and this remains the case when comparing the marginal M-estimators of multiple dependent groups. When comparing groups based on a robust estimator applied to difference scores, again a percentile bootstrap method appears to be the best choice for general use.

11.3.1 Method Based on Marginal Measures of Location

Let θ_j be any measure of location associated with the jth group and let $\hat{\theta}_j$ be an estimate of θ_j based on the available data (the X_{ij} values). The goal is to test

$$H_0 : \theta_1 = \cdots = \theta_J, \tag{11.4}$$

the hypothesis that all J dependent groups have identical measures of location. Called *Method RMPB3*, the test statistic is

$$Q = \sum(\hat{\theta}_j - \bar{\theta})^2,$$

where $\bar{\theta} = \sum \hat{\theta}_j / J$ is the average of the $\hat{\theta}_j$ values. So Q measures the variation among the estimates. If the null hypothesis is true, Q should be relatively small. The problem is determining how large Q needs to be to reject. In more formal terms, the distribution of Q needs to be determined when the null hypothesis, given by Equation (11.4), is true.

The strategy for determining a critical value is similar to how a critical value was determined when using a bootstrap-t method. Begin by shifting the empirical distributions so that the null hypothesis is true. In symbols, set $C_{ij} = X_{ij} - \hat{\theta}_j$. So for each group, the C_{ij} values have a measure of location equal to 0. Next, a bootstrap sample is obtained by resampling, with replacement, rows of data from the C_{ij} values. Again we label the results

$$\begin{pmatrix} C^*_{11}, \ldots, C^*_{1J} \\ \vdots \\ C^*_{n1}, \ldots, C^*_{nJ} \end{pmatrix}.$$

For the jth column of the bootstrap data just generated, compute the measure location that is of interest and label it $\hat{\theta}^*_j$. Compute

$$Q^* = \sum(\hat{\theta}^*_j - \bar{\theta}^*)^2,$$

where $\bar{\theta}^* = \sum \hat{\theta}^*_j / J$, and repeat this process B times yielding $Q^*_1, \ldots, Q^*_B$. Put these B values in ascending order yielding $Q^*_{(1)} \le \cdots \le Q^*_{(B)}$. Then reject the hypothesis of equal measures of location if $Q > Q^*_{(c)}$, where again $c = (1-\alpha)B$. Let P be the number of bootstrap estimates that are larger than Q. A p-value is given by P/B.

11.3.2 R Function bd1way

The R function

```
bd1way(x, est = onestep, nboot = 599, alpha = 0.05)
```

performs the percentile bootstrap method just described. By default it uses the one-step M-estimator of location, but any other estimator can be used via the argument `est`. The argument `x` is any R variable that is a matrix or has list mode, `nboot` is B, the number of bootstrap samples to be used, and `grp` can be used to analyze a subset of the groups, with the other groups ignored.

EXAMPLE

We reanalyze the data in Table 8.1 using the R function just described. Assuming the data are stored in the R matrix `cork`, the command `bd1way(cork)` returns:

```
$test
[1] 17.07862
$estimates
        N        E        S        W
48.41873 44.57139 49.25909 44.94406
$p.value
[1] 0.2020033
```

So comparing one-step M-estimators at the 0.05 level, we fail to reject the hypothesis that the typical weight of a cork boring is the same for all four sides of a tree.

11.3.3 Inferences Based on Difference Scores

Another approach is to test some appropriate hypothesis based on difference scores. One possibility is to form difference scores between the first group and the remaining $J-1$ groups and then test the hypothesis that the corresponding (population) measures of location are all equal to zero. But this approach is unsatisfactory in the sense that if we rearrange the order of the groups, power can be affected substantially.

We can avoid this problem by instead forming difference scores among all pairs of groups. There are a total of

$$L = \frac{J^2 - J}{2}$$

such differences. In symbols, we compute

$$D_{i1} = X_{i1} - X_{i2},$$
$$D_{i2} = X_{i1} - X_{i3},$$
$$\vdots$$
$$D_{iL} = X_{i,J-1} - X_{iJ}.$$

EXAMPLE

For four groups ($J = 4$), there are $L = 6$ sets of differences scores given by

$$D_{i1} = X_{i1} - X_{i2},$$
$$D_{i2} = X_{i1} - X_{i3},$$
$$D_{i3} = X_{i1} - X_{i4},$$
$$D_{i4} = X_{i2} - X_{i3},$$
$$D_{i5} = X_{i2} - X_{i4},$$
$$D_{i6} = X_{i3} - X_{i4}$$

($i = 1, \ldots n$). The goal is to test

$$H_0 : \theta_1 = \cdots = \theta_L = 0, \tag{11.5}$$

where θ_ℓ ($\ell = 1, \ldots L$) is the population measure of location associated with the ℓth set of difference scores, $D_{i\ell}$ ($i = 1, \ldots, n$). To test H_0 given by Equation (11.5), resample vectors of D values, but unlike the bootstrap-t, observations are not centered. That is, a bootstrap sample now consists of resampling with replacement n rows from the matrix

$$\begin{pmatrix} D_{11}, \ldots, D_{1L} \\ \vdots \\ D_{n1}, \ldots, D_{nL} \end{pmatrix},$$

yielding

$$\begin{pmatrix} D^*_{11}, \ldots, D^*_{1L} \\ \vdots \\ D^*_{n1}, \ldots, D^*_{nL} \end{pmatrix}.$$

For each of the L columns of the D^* matrix, compute whatever measure of location is of interest, and for the ℓth column label the result $\hat{\theta}^*_\ell$ ($\ell = 1, \ldots, L$). Next, repeat this B times yielding $\hat{\theta}^*_{\ell b}$, $b = 1, \ldots, B$ and then determine how deeply the vector $\mathbf{0} = (0, \ldots, 0)$, having length L, is nested within the bootstrap values $\hat{\theta}^*_{\ell b}$. For two groups, this is tantamount to determining how many bootstrap values are greater than zero. If most are greater than (or less than) zero, we reject. The details are relegated to Box 11.2.

11.3.4 R Function rmdzero

The R function

```
rmdzero(x,est=onestep, grp = NA, nboot = NA, ...)
```

performs the test on difference scores outlined in Box 11.2.

EXAMPLE

For the cork data in Table 8.1, setting the argument `est=tmean`, `rmdzero` returns a p-value of 0.014, so in particular reject with $\alpha = 0.05$. Recall that when these data were analyzed using `bd1way` (see the example at the end of Section 11.3.2), the p-value was 0.20. This result is in sharp contrast to the p-value obtained here, which illustrates that the choice of method can make a substantial difference in the conclusions reached.

11.4 RANK-BASED METHODS FOR A ONE-WAY DESIGN

This section describes two rank-based methods for comparing dependent groups. The first is a classic technique, called Friedman's test, which is routinely taught and used. The other is a more modern approach that has certain practical advantages over Friedman's test.

11.4.1 Friedman's Test

Friedman's test is designed to test

$$H_0 : F_1(x) = \cdots = F_J(x),$$

the hypothesis that all J dependent groups have identical distributions. The method begins by assigning ranks within rows. For example, imagine that for each individual, measures are taken at three different times yielding

Time 1	Time 2	Time 3
9	7	12
1	10	4
8	2	1
	⋮	

Box 11.2: Repeated measures ANOVA based on difference scores and the depth of 0.

Goal: Test the hypothesis, given by Equation (11.5), that all difference scores have a typical value of 0.

1. Let $\hat{\theta}_\ell$ be the estimate of θ_ℓ. Compute bootstrap estimates as described in this section and label them $\hat{\theta}^*_{\ell b}$, $\ell = 1, \dots, L$; $b = 1, \dots, B$.
2. Compute the L-by-L matrix

$$S_{\ell\ell'} = \frac{1}{B-1}\sum_{b=1}^{B}(\hat{\theta}^*_{\ell b} - \hat{\theta}_\ell)(\hat{\theta}^*_{\ell' b} - \hat{\theta}_{\ell'}).$$

Readers familiar with multivariate statistical methods might notice that $S_{\ell\ell'}$ uses $\hat{\theta}_\ell$ (the estimate of θ_ℓ based on the original difference values) rather than the seemingly more natural $\bar{\theta}^*_\ell$, where

$$\bar{\theta}^*_\ell = \frac{1}{B}\sum_{b=1}^{B}\hat{\theta}^*_{\ell b}.$$

If $\bar{\theta}^*_\ell$ is used, unsatisfactory control over the probability of a Type I error can result.

3. Let $\hat{\theta} = (\hat{\theta}_1, \dots, \hat{\theta}_L)$, $\hat{\theta}^*_b = (\hat{\theta}^*_{1b}, \dots, \hat{\theta}^*_{Lb})$ and compute

$$d_b = (\hat{\theta}^*_b - \hat{\theta})\mathbf{S}^{-1}(\hat{\theta}^*_b - \hat{\theta})',$$

where $\mathbf{S}$ is the matrix corresponding to $S_{\ell\ell'}$.

4. Put the d_b values in ascending order: $d_{(1)} \le \dots \le d_{(B)}$.
5. Let

$$\mathbf{0} = (0, \dots, 0)$$

having length L.

6. Compute

$$D = (\mathbf{0} - \hat{\theta})\mathbf{S}^{-1}(\mathbf{0} - \hat{\theta})'.$$

D measures how far away the null hypothesis is from the observed measures of location (based on the original data). In effect, D measures how deeply $\mathbf{0}$ is nested within the cloud of bootstrap values.

7. Reject if $D \ge d_{(u)}$, where $u = (1 - \alpha)B$, rounded to the nearest integer.

The ranks corresponding to the first row (the values 9, 7, and 12) are 2, 1, and 3. For the second row the ranks are 1, 3, and 2 and continuing in this manner the data become

Time 1	Time 2	Time 3
2	1	3
1	3	2
3	2	1
	⋮	

Let R_{ij} be the resulting rank corresponding to X_{ij} ($i = 1, \dots, n$; $j = 1, \dots, J$). Compute

$$A = \sum_{j=1}^{J} \sum_{i=1}^{n} R_{ij}^2$$

$$R_j = \sum_{i=1}^{n} R_{ij}$$

$$B = \frac{1}{n} \sum_{j=1}^{J} R_j^2$$

$$C = \frac{1}{4} nJ(J+1)^2.$$

If there are no ties, the equation for A simplifies to

$$A = \frac{nJ(J+1)(2J+1)}{6}.$$

The test statistic is

$$F = \frac{(n-1)(B-C)}{A-B}. \tag{11.6}$$

Reject if $F \geq f_{1-\alpha}$, or if $A = B$, where $f_{1-\alpha}$ is the $1-\alpha$ of an F distribution with $\nu_1 = J-1$ and $\nu_2 = (n-1)(J-1)$ degrees of freedom.

11.4.2 R Function friedman.test

The built-in R function

```
friedman.test(x)
```

performs Friedman's test, where the argument `x` is a matrix with columns corresponding to groups. This function also accepts a formula, and a variable indicating groups can be used. However, no illustrations are provided because it currently seems best to use an improved rank-based method described in the next section. Also, R does not use the F test statistic described here, but rather a chi-squared statistic. Results in Iman and Davenport (1980) indicate that the F statistic is better for general use.

11.4.3 Method BPRM

Numerous rank-based methods have been proposed with the goal of improving on Friedman's test in terms of both Type I errors and power. A fundamental criticism of the Friedman test is that it is based on a highly restrictive assumption regarding the covariance matrix, called *compound symmetry*, which means that all J groups have a common variance (homoscedasticity) and all pairs of groups have the same covariance. An improvement on Friedman's test that currently stands out is described by Brunner et al. (2002, Section 7.2.2) and will be called *method BPRM*. This method also improves on a technique derived by Agresti and

Pendergast (1986), which in turn improves on a method derived by Quade (1979). (R also has a built-in function for applying Quade's test.)

Roughly, method BPRM is designed to be sensitive to differences among the average ranks associated with the J groups, which can be discerned from the test statistic given by Equation (11.7). Unlike Friedman's test, ranks are assigned based on the pooled data. So if there are no ties, the smallest value among all Jn values gets a rank of 1, and the largest gets a rank of Jn. (Midranks are used if there are tied values.) Let R_{ij} be the rank of X_{ij} and as usual let $N = Jn$ denote the total number of observations. Assuming familiarity with the basic matrix operations summarized in Appendix C, let $\mathbf{R}_i = (R_{i1}, \ldots, R_{iJ})'$ be the vector of ranks for the ith participant, where $(R_{i1}, \ldots, R_{iJ})'$ is the transpose of $(R_{i1}, \ldots, R_{iJ})$. Let

$$\bar{\mathbf{R}} = \frac{1}{n}\sum_{i=1}^{n}\mathbf{R}_i$$

be the vector of ranked means, let

$$\bar{R}_{.j} = \frac{1}{n}\sum_{i=1}^{n}R_{ij}$$

denote the mean of the ranks for group j, and let

$$\mathbf{V} = \frac{1}{N^2(n-1)}\sum_{i=1}^{n}(\mathbf{R}_i - \bar{\mathbf{R}})(\mathbf{R}_i - \bar{\mathbf{R}})'.$$

The test statistic is

$$F = \frac{n}{N^2\text{tr}(\mathbf{PV})}\sum_{j=1}^{J}\left(\bar{R}_{.j} - \frac{N+1}{2}\right)^2, \tag{11.7}$$

where

$$\mathbf{P} = \mathbf{I} - \frac{1}{J}\mathbf{J},$$

$\mathbf{J}$ is a $J \times J$ matrix of all ones, and $\mathbf{I}$ is the identity matrix.

Decision Rule

Reject the hypothesis of identical distributions if

$$F \geq f,$$

where f is the $1 - \alpha$ quantile of an F distribution with degrees of freedom

$$\nu_1 = \frac{[\text{tr}(\mathbf{PV})]^2}{\text{tr}(\mathbf{PVPV})}$$

and $\nu_2 = \infty$.

11.4.4 R Function bprm

The R function

```
bprm(x)
```

performs method BPRM just described, where the argument `x` is assumed to be a matrix with J columns corresponding to groups, or it can have list mode. This function returns a p-value.

11.5 COMMENTS ON WHICH METHOD TO USE

Several reasonable methods for comparing dependent groups have been described, so there is the issue of which one to use. As usual, no method is perfect in all situations. The expectation is that in many situations where groups differ, all methods based on means perform poorly, in terms of power, relative to approaches based on some robust measure of location such as an M-estimator or a 20% trimmed mean. Currently, with a sample size as small as 21, the bootstrap-t method, used in conjunction with 20% trimmed means, appears to provide excellent control over the probability of a Type I error. Its power compares reasonably well to most other methods that might be used, but as noted in previous chapters, different methods are sensitive to different features of the data. So it is possible, for example, that the rank-based method BPRM will have more power than a method based on a 20% trimmed mean or an M-estimator.

The percentile bootstrap methods described here also do an excellent job of avoiding Type I errors greater than the nominal level, provided that the estimator being used has a reasonably high breakdown point, but there are indications that method RMPB3 can be too conservative when sample sizes are small. That is, the actual probability of a Type I error can be substantially less than α suggesting that some other method might provide better power. Nevertheless, if there is specific interest in comparing marginal distributions with M-estimators, it is suggested that method RMPB3 be used and that the sample size should be greater than 20.

Currently, among the techniques covered in this chapter, it seems that the two best methods for controlling Type I error probabilities and simultaneously providing reasonably high power are the bootstrap-t method based on 20% trimmed means and the percentile bootstrap method in Box 11.2 used in conjunction with M-estimator. (Other excellent options are covered in Chapter 12.) With near certainty, situations arise where some other technique is more optimal, but typically the improvement is small. However, comparing groups with M-estimators is not the same as comparing means, trimmed means, or medians and certainly there will be situations where some other estimator has higher power than any method based on an M-estimator or a 20% trimmed mean. If the goal is to maximize power, several methods are contenders for routine use, but, as usual, standard methods based on means are generally the least satisfactory. With sufficiently large sample sizes, trimmed means can be compared without resorting to the bootstrap-t method, but it remains unclear just how large the sample size must be.

As for the issue of whether to use difference scores rather than robust measures of location based on the marginal distributions, each approach provides a different perspective on how groups differ and they can give different results regarding whether groups are significantly different. There is some evidence that difference scores typically provide more power and better control over the probability of a Type I error, but a more detailed study is needed to resolve this issue.

11.6 BETWEEN-BY-WITHIN DESIGNS

Chapter 10 covered a two-way ANOVA design involving JK independent groups with J levels associated with the first factor and K levels with the second. A between-by-within or a split-plot design refers to a two-way design where the levels of the first factor are independent, but the measures associated with the K levels of the second factor are dependent instead.

EXAMPLE

As a simple illustration, again consider the situation where endurance is measured before and after training, but now there are two training methods. Moreover, n_1 athletes undergo training method 1 and a different, independent sample of n_2 athletes undergoes training method 2. So between methods, observations are independent, but for the two occasions, measures are possibly dependent because the same athlete is measured twice.

Main effects and interactions are defined in a manner similar to how we proceeded in Chapter 10. To make sure this is clear, consider a 2-by-2 design where the levels of Factor A are independent and the levels of Factor B are dependent. That is, we have a between-by-within design. The population means are depicted as follows:

		Factor B	
		1	2
Factor A	1	μ_{11}	μ_{12}
	2	μ_{21}	μ_{22}

The hypothesis of no main effect for Factor A is

$$H_0 : \frac{\mu_{11} + \mu_{12}}{2} = \frac{\mu_{21} + \mu_{22}}{2}$$

and the hypothesis of no interaction is

$$H_0 : \mu_{11} - \mu_{12} = \mu_{21} - \mu_{22}.$$

EXAMPLE

Imagine an intervention study aimed at dealing with depression among adults over the age of 65. One approach is to measure depression prior to treatment and 6 months later. However, it might be argued that, over 6 months, measures of depression might change even with no treatment. A way of addressing this issue is with a between-by-within design where the levels of the first factor correspond to a control group and an experimental group. Of particular interest is whether there is an interaction. That is, do changes in depression among participants in the control group differ from changes among participants in the experimental group?

11.6.1 Method for Trimmed Means

The computational details on how to compare trimmed means (including means as a special case) are tedious at best. Briefly, the hypotheses of no main effects and no interactions can be written in the form

$$H_0 : \mathbf{C}\mu_t = \mathbf{0},$$

where $\mathbf{C}$ reflects the null hypothesis of interest. The matrix $\mathbf{C}$ is constructed as described in Box 10.2. (Here, μ_t is a column vector of population trimmed means having length JK.)

For every level of Factor A, there are K dependent random variables, and each pair of these dependent random variables has a Winsorized covariance that must be estimated. For the jth level of Factor A, let g_j be the number of observations trimmed from both tails. (If the amount of trimming is γ, $g_j = [\gamma n_j]$, where $[\gamma n_j]$ is γn_j rounded down to the nearest integer.) The Winsorized covariance between the mth and ℓth levels of Factor B is estimated with

$$s_{jm\ell} = \frac{1}{n_j - 1} \sum_{i=1}^{n_j} (Y_{ijm} - \bar{Y}_{.jm})(Y_{ij\ell} - \bar{Y}_{.j\ell}),$$

where $\bar{Y}_{jm} = \sum_{i=1}^{n_j} Y_{ijm}/n_j$, and

$$Y_{ijk} = \begin{cases} X_{(g_j+1),jk} & \text{if } X_{ijk} \le X_{(g_j+1),jk} \\ X_{ijk} & \text{if } X_{(g_j+1),jk} < X_{ij} < X_{(n-g_j),jk} \\ X_{(n-g_j),jk} & \text{if } X_{ijk} \ge X_{(n_j-g_j),jk}. \end{cases}$$

The remainder of the computations are relegated to Box 11.3, where h_j is the number of observations left after trimming when dealing with the jth level of Factor A. (So $h_j = n_j - 2g_j$.) In Box 11.3, X_{ijk} represents the ith observation for level j of Factor A and level k of Factor B ($i = 1, \ldots, n_j$; $j = 1, \ldots, J$ and $k = 1, \ldots, K$). The test statistic is

$$Q = \bar{\mathbf{X}}'\mathbf{C}'(\mathbf{CVC}')^{-1}\mathbf{C}\bar{\mathbf{X}}, \tag{11.8}$$

where $\bar{\mathbf{X}}' = (\bar{X}_{t11}, \ldots, \bar{X}_{tJK})$ and the matrix $\mathbf{V}$ is computed as described in Box 11.3. Let ℓ be the number of rows associated with the matrix $\mathbf{C}$. (Technically, ℓ must be the rank of $\mathbf{C}$, which for the situation at hand corresponds to the number of rows.) Let

$$c = \ell + 2A - \frac{6A}{\ell + 2},$$

where

$$A = \frac{1}{2} \sum_{j}^{J} [\text{tr}(\{\mathbf{VC}'(\mathbf{CVC}')^{-1}\mathbf{CQ}_j\}^2) + \{\text{tr}(\mathbf{VC}'(\mathbf{CVC}')^{-1}\mathbf{CQ}_j)\}^2]/(h_j - 1)$$

and $\mathbf{Q}_j$ is computed as described in Box 11.3. When the null hypothesis is true, Q/c has, approximately, an F distribution with $\nu_1 = \ell$ and $\nu_2 = \ell(\ell + 2)/(3A)$ degrees of freedom. For Factor A, $\ell = J - 1$, for Factor B, $\ell = K - 1$, and for interactions, $\ell = (J - 1)(K - 1)$.

Decision Rule

Reject if $Q/c \ge f_{1-\alpha}$, the $1 - \alpha$ quantile of an F distribution with ν_1 and ν_2 degrees of freedom. (Also see Keselman et al., 1999.) Simulation results reported by Livacic-Rojas, Vallejo, and Fernández (2010) indicate that when comparing means, if the covariance matrices among the independent groups differ, but otherwise the marginal distributions are identical, the method in this section competes well with other techniques in terms of controlling the probability of a Type I error. But they did not consider situations where the marginal distributions differ in terms of skewness; situations can be created where, when comparing means, the method is unsatisfactory. For comparisons of many other methods when using means, see Keselman, et al. (2002), and Overall and Tonidandel (2010).

Box 11.3: Some of the computations for a between-by-within design when using trimmed means.

For fixed j, let $\mathbf{S}_j = (s_{jm\ell})$, which is the matrix of Winsorized variances and covariances for level j of Factor A. Let

$$\mathbf{V}_j = \frac{(n_j - 1)\mathbf{S}_j}{h_j(h_j - 1)}, \; j = 1, \ldots, J,$$

and let $\mathbf{V} = \mathrm{diag}(\mathbf{V}_1, \ldots, \mathbf{V}_J)$ be a block diagonal matrix. Let $\mathbf{I}_{K\times K}$ be a K-by-K identity matrix, let $\mathbf{Q}_j$ be a JK-by-JK block diagonal matrix (consisting of J blocks, each block being a K-by-K matrix), where the tth block ($t = 1, \ldots, J$) along the diagonal of $\mathbf{Q}_j$ is $\mathbf{I}_{K\times K}$ if $t = j$, and all other elements are 0. For example, if $J = 3$ and $K = 4$, then $\mathbf{Q}_1$ is a 12-by-12 block diagonal matrix where the first block is a 4-by-4 identity matrix, and all other elements are zero. As for $\mathbf{Q}_2$, the second block is an identity matrix, and all other elements are 0.

EXAMPLE

Consider a 2-by-3 design where for the first level of Factor A observations are generated from a trivariate normal distribution where all correlations are equal to zero. For the second level of Factor A, the marginal distributions are lognormal, as shown in Figure 4.10. (Here these marginal distributions are shifted so that they have means equal to zero.) Further suppose that the covariance matrix for the second level is three times larger than the covariance matrix for the first level. If the sample sizes are $n_1 = n_2 = 30$ and the hypothesis of no main effects for Factor A is tested at the 0.05 level, the actual level is approximately 0.088. If the sample sizes are $n_1 = 40$ and $n_2 = 70$ and for the second level of Factor A the marginal distributions are now skewed with heavy tails (a g-and-h distribution with $g = h = 0.5$), the probability of a Type I error, again testing at the 0.05 level, is approximately 0.188. Comparing 20% trimmed means instead, the actual type I error probability is approximately 0.035.

11.6.2 R Function bwtrim and bw2list

The R function

```
bwtrim(J, K, x, tr = 0.2, grp = c(1:p))
```

tests hypotheses about trimmed means as described in the previous section. Here, J is the number of independent groups, K is the number of dependent groups, `x` is any R variable that is a matrix or has list mode, and, as usual, the argument `tr` indicates the amount of trimming, which defaults to 0.2 if unspecified. If the data are stored in list mode, it is assumed that x[[1]] contains the data for Level 1 of both factors, x[[2]] contains the data for Level 1 of the first factor and Level 2 of the second, and so on. If the data are stored in a matrix (or a data frame), it is assumed that the first K columns correspond to Level 1 of Factor A, the next K columns correspond to Level 2 of Factor A, and so on. If the data are not stored in the proper order, the argument `grp` can be used to indicate how they are stored. For example, if a 2-by-2 design is being used, the R command

```
bwtrim(2,2,x,grp=c(3,1,2,4))
```

indicates that the data for the first level of both factors are stored in x[[3]], the data for level 1 of Factor A and level 2 of Factor B are in x[[1]], and so forth.

EXAMPLE

Consider again the data stored in the R variable ChickWeight, which is a matrix. There are four independent groups of chicks that received a different diet. The diet they received is indicated in column 4 (by the values 1, 2, 3, and 4). Each chick was measured at 12 different times, with times indicated in column 2. So we have a 4-by-12 between-by-within design. Column 1 contains the weight of the chicks. To analyze the data with the R function `bwtrim`, the data need to be stored as described in this section, which can be done with the R command

```
z=fac2list(ChickWeight[,1],ChickWeight[,c(4,2)]).
```

Then the 10% trimmed means are compared with the command

```
bwtrim(4,12,z,tr=0.1).
```

The p-value when comparing the four diets is 0.061. The p-value when comparing weight over time is extremely small, but, of course, the result that chicks gain weight over time is not very interesting. The goal here was merely to illustrate a feature of R.

Important: In this last example, look at the last argument when using the R function `fac2list`, `ChickWeight[,c(4,2)]`, and notice the use of `c(4,2)`. This indicates that the levels of the between factor are stored in column 4 and the levels of the within factor are stored in column 2. If we had used `c(2,4)`, this would indicate that the levels in column 2 correspond to the between factor, which is incorrect.

Situations are encountered where data are stored in a matrix or a data frame with one column indicating the levels of the between factor and one or more columns containing data corresponding to different times. Imagine, for example, three medications are being investigated regarding their effectiveness to lower cholesterol and that column 3 indicates which medication a participant received. Moreover, columns 5 and 8 contain the participants' cholesterol level at times 1 and 2, respectively. Then the R function

```
bw2list(x, grp.col, lev.col)
```

will convert the data to list mode so that the R function `bwtrim` can be used. The argument `grp.col` indicates the column indicating the levels of the independent groups. The argument `lev.col` indicates the K columns where the within group data are stored. For the situation just described, if the data are stored in the R variable `chol`, the R command

```
z=bw2list(chol,3,c(5,8))
```

will store the data in `z`, after which the R command

```
bwtrim(3,2,z)
```

will perform the analysis.

11.6.3 A Bootstrap-t Method

To apply a bootstrap-t method, when working with trimmed means, you first center the data in the usual way. In the present context, this means you compute

$$C_{ijk} = X_{ijk} - \bar{X}_{tjk},$$

$i = 1, \ldots, n_j$; $j = 1, \ldots, J$; and $k = 1, \ldots, K$. That is, for the group corresponding to the jth level of Factor A and the kth level of Factor B, subtract the corresponding trimmed mean from each of the observations. Next, for the jth level of Factor A, generate a bootstrap sample by resampling with replacement n_j vectors of observations from the n_j vectors of observations from the data in level j of Factor A. That is, for each level of Factor A, you have an n_j by K matrix of data, and you generate a bootstrap sample from this matrix of data as described in Section 11.4. Label the resulting bootstrap samples C^*_{ijk}. Compute the test statistic Q, given by Equation (11.8), based on the C^*_{ijk} values, and label the result Q^*. Repeat this B times yielding $Q^*_1, \ldots, Q^*_B$ and then put these B in ascending order yielding $Q^*_{(1)} \leq \cdots \leq Q^*_{(B)}$. Next, compute Q using the original data (the X_{ijk} values) and reject if $Q \geq Q^*_{(c)}$, where $c = (1 - \alpha)B$ rounded to the nearest integer.

A crude rule that seems to apply to a wide variety of situations is: the more distributions differ, the more beneficial it is to use some type of bootstrap method, at least when the sample sizes are small. However, Keselman et al. (2000) compared the bootstrap-t method just described to the non-bootstrap method for a split-plot design covered in Section 11.5.1. For the situations they examined, this rule did not apply; it was found that the bootstrap-t offered little or no advantage. Their study included situations where the correlations (or covariances) among the dependent groups differ across the independent groups being compared. However, the more complicated the design, the more difficult it becomes to consider all the factors that might influence operating characteristics of a particular method. One limitation of their study was that the differences among the covariances were taken to be relatively small. Another issue that has not been addressed is how the bootstrap-t performs when distributions differ in skewness. Having differences in skewness is known to be important when dealing with the simple problem of comparing two groups only. There is no reason to assume that this problem diminishes as the number of groups increases, and indeed there are reasons to suspect that it becomes a more serious problem. So, currently, it seems that if groups do not differ in any manner, or the distributions differ slightly, it makes little difference whether you use a bootstrap-t rather than a non-bootstrap method for comparing trimmed means. However, if distributions differ in shape, there is indirect evidence that the bootstrap-t might offer an advantage when using a split-plot design, but the extent to which this is true is not well understood.

11.6.4 R Function tsplitbt

The R function

```
tsplitbt(J,K,x,tr=0.2,alpha=0.05,grp=c(1:JK),nboot=599)
```

performs a bootstrap-t method for a split-plot (between-by-within) design as just described. The data are assumed to be arranged as indicated in conjunction with the R function `bwtrim` (as described in Section 11.5.2), and the arguments `J`, `K`, `tr`, and `alpha` have the same meaning as before. The argument `grp` can be used to rearrange the data if they are not stored as expected by the function. (See Section 11.2.1 for an illustration on how to use `grp`.)

11.6.5 Inferences Based on M-estimators and Other Robust Measures of Location

Comparing groups based on M-estimators in a between-by-within design is possible using extensions of the bootstrap methods considered in this chapter as well as Chapters 10 and 11. Roughly, bootstrap samples must be generated in a manner that reflects the dependence among the levels of Factor B, and then some appropriate hypothesis is tested using some slight variation of one of the methods already described. In fact, the problem is *not* deriving a method, but rather deciding which method should be used among the many that are available. This section briefly outlines some approaches that are motivated by published papers. Readers interested in a better understanding of the computational details are referred to Wilcox (2017, Section 8.6.5). It is noted that alternative methods might be more optimal and that for the specific situation at hand more research is needed to better understand the relative merits of different techniques. (Methods in Chapter 12 provide alternative strategies that deserve serious consideration.)

FACTOR B

Here we let θ_{jk} be any measure of location associated with level j of Factor A and level k of Factor B. For simplicity, first consider the dependent groups associated with Factor B. One approach is to compare these K groups by ignoring Factor A, forming the difference scores for each pair of dependent groups, and then applying the method in Section 11.3.3. To elaborate slightly, imagine you observe X_{ijk} ($i = 1, \ldots, n_j$; $j = 1, \ldots, J$; $k = 1, \ldots, K$). That is, X_{ijk} is the ith observation in level j of Factor A and level k of Factor B. Note that if we ignore the levels of Factor A, we can write the data as Y_{ik}, $i = 1, \ldots, N$; $k = 1, \ldots, K$, where $N = \sum n_j$. That is, we have K dependent groups that can be compared with the R function `rmdzero` in Section 11.3.4.

FACTOR A

As for Factor A, first focus on the first level of Factor B and note that the null hypothesis of no differences among the levels of Factor A can be viewed as

$$H_0 : \theta_{j1} - \theta_{j'1} = 0$$

for any $j < j'$. More generally, for any level of Factor B, say the kth, the hypothesis of no main effects is

$$H_0 : \theta_{jk} - \theta_{j'k} = 0 \tag{11.9}$$

($k = 1, \ldots, K$), and the goal is to test the hypothesis that all $C = K(J^2 - J)/2$ differences are equal to 0. Very briefly, for each level of Factor A, generate bootstrap samples as described in Section 11.3.3 and compute the measure of location for each level of Factor B. Repeat this process B times and test Equation (11.9) by determining how deeply the vector $(0, \ldots, 0)$, having length C, is nested within the B bootstrap estimates.

For Factor A, an alternative approach is to average the measures of location across the K levels of Factor B. In symbols, let

$$\bar{\theta}_{j.} = \frac{1}{K} \sum_{k=1}^{K} \theta_{jk},$$

in which case the goal is to test

$$H_0 : \bar{\theta}_{1.} = \cdots = \bar{\theta}_{J.}. \tag{11.10}$$

To test Equation (11.10) when using a bootstrap method, generate B bootstrap samples from the K dependent groups as described in Section 11.3.3. Let $\bar{\theta}^*_{j.}$ be the bootstrap estimate of $\bar{\theta}_{j.}$. For levels j and j' of Factor A, $j < j'$, set $\delta^*_{jj'} = \bar{\theta}^*_{j.} - \bar{\theta}^*_{j'.}$. The number of such differences is $(J^2 - J)/2$. Repeating this B times gives B vectors of differences, each vector having length $(J^2 - J)/2$. The strategy is to determine how deeply $\mathbf{0}$, having length $(J^2 - J)/2$, is nested within this bootstrap cloud of points.

INTERACTIONS

As for interactions, again there are several approaches one might adopt. Here, an approach based on difference scores among the dependent groups is used. To explain, first consider a 2-by-2 design, and for the first level of Factor A let $D_{i1} = X_{i11} - X_{i12}$, $i = 1, \ldots, n_1$. Similarly, for level two of Factor A let $D_{i2} = X_{i21} - X_{i22}$, $i = 1, \ldots, n_2$, and let θ_{d1} and θ_{d2} be the population measure of location corresponding to the D_{i1} and D_{i2} values, respectively. Then the hypothesis of no interaction is taken to be

$$H_0 : \theta_{d1} - \theta_{d2} = 0. \tag{11.11}$$

Again the basic strategy for testing hypotheses is generating bootstrap estimates and determining how deeply 0 is embedded in the B values that result. For the more general case of a J-by-K design, there are a total of

$$C = \frac{J^2 - J}{2} \times \frac{K^2 - K}{2}$$

equalities, one for each pairwise difference among the levels of Factor B and any two levels of Factor A.

11.6.6 R Functions sppba, sppbb, and sppbi

The R function

```
sppba(J,K,x,est=onestep, grp = c(1:JK), avg=FALSE, nboot=500,...)
```

tests the hypothesis of no main effects as given by Equation (11.9). Setting the argument `avg=TRUE` (for true) indicates that the averages of the measures of location (the $\bar{\theta}_{j.}$ values) will be used. That is, test the hypothesis of no main effects for Factor A by testing Equation (11.10). The remaining arguments have their usual meaning. The R function

```
sppbb(J,K,x,est=onestep,grp = c(1:JK),nboot=500,...)
```

tests the hypothesis of no main effects for Factor B and

```
sppbi(J,K,x,est=onestep,grp = c(1:JK),nboot=500,...)
```

tests the hypothesis of no interactions.

EXAMPLE

Section 7.5.5 mentioned a study where the goal was to compare the EEG measures for murderers to a control group. For each participant, EEG measures were taken at four sites in the brain. Imagine that the data are stored in the R variable `eeg`, a matrix with eight columns, with the control group data stored in the first four columns. Then the hypothesis of no main effects (the control group does not differ from the murderers) can be tested with the command

```
sppba(2,4,x).
```

The hypotheses of no main effects among sites as well as no interactions are tested in a similar manner using `sppbb` and `sppbi`.

11.6.7 A Rank-Based Test

A rank-based method for analyzing a between-by-within design is described by Brunner, Domhof and Langer (2002, Chapter 8). An important feature of the method is that it is designed to perform well, in terms of controlling the probability of a Type I error, when tied values occur. There are other rank-based approaches (e.g., Beasley, 2000; Beasley and Zumbo, 2003), but it seems that the practical merits of these competing methods, compared to the method described here, have not been explored.

As in Section 11.6, the notation $F_{jk}(x)$ refers to the probability that an observation corresponding to the jth level of Factor A and the kth level of Factor B is less than or equal to x. Main effects for Factor A are expressed in terms of

$$\bar{F}_{j.}(x) = \frac{1}{K}\sum_{k=1}^{K} F_{jk}(x),$$

the average of the distributions among the K levels of Factor B corresponding to the jth level of Factor A. The hypothesis of no main effects for Factor A is

$$H_0 : \bar{F}_{1.}(x) = \bar{F}_{2.}(x) = \cdots = \bar{F}_{J.}(x)$$

for any x.

EXAMPLE

The notation is illustrated with data taken from Lumley (1996) dealing with shoulder pain after surgery; the data are from a study by Jorgensen et al. (1995). Table 11.2 shows a portion of the results where two treatment methods are used and measures of pain are taken at three different times. So $F_{11}(3)$ refers to the probability of observing the value 3 or less for a randomly sampled participant from the first group (active treatment) at time 1. For the first group, $\bar{F}_{1.}(3)$ is the average of the probabilities over the three times of observing the value 3 or less. The hypothesis of no main effects for Factor A is that for any x we might pick, $\bar{F}_{1.}(x) = \bar{F}_{2.}(x)$.

Let

$$\bar{F}_{.k}(x) = \frac{1}{J}\sum_{j=1}^{J} F_{jk}(x)$$

be the average of the distributions for the kth level of Factor B. So in the shoulder example, for each of the three times shoulder pain is recorded, $\bar{F}_{.k}(x)$ is the average probability over the two groups. The hypothesis of no main effects for Factor B is

$$H_0 : \bar{F}_{.1}(x) = \bar{F}_{.2}(x) = \cdots = \bar{F}_{.K}(x).$$

As for interactions, first consider a 2-by-2 design. Then no interaction is taken to mean that for any x,

$$F_{11}(x) - F_{12}(x) = F_{21}(x) - F_{22}(x).$$

Table 11.2 Shoulder Pain Data

Active Treatment			No Active Treatment		
Time 1	Time 2	Time 3	Time 1	Time 2	Time 3
1	1	1	5	2	3
3	2	1	1	5	3
3	2	2	4	4	4
1	1	1	4	4	4
1	1	1	2	3	4
1	2	1	3	4	3
3	2	1	3	3	4
2	2	1	1	1	1
1	1	1	1	1	1
3	1	1	1	5	5
1	1	1	1	3	2
2	1	1	2	2	3
1	2	2	2	2	1
3	1	1	1	1	1
2	1	1	1	1	1
1	1	1	5	5	5
1	1	1	3	3	3
2	1	1	5	4	4
4	4	2	1	3	3
4	4	4			
1	1	1			
1	1	1			

Note: 1 = low, 5 = high

More generally, the hypothesis of no interactions among all JK groups is

$$H_0 : F_{jk}(x) - \bar{F}_{j.}(x) - \bar{F}_{.k}(x) + \bar{F}_{..}(x) = 0,$$

for any x, all j ($j = 1, \ldots, J$) and all k ($k = 1, \ldots, K$), where

$$\bar{F}_{..}(x) = \frac{1}{JK} \sum_{j=1}^{J} \sum_{k=1}^{K} F_{jk}(x).$$

Regarding how to test the hypotheses just described, complete computational details are not provided; readers interested in these details are referred to Brunner et al. (2002) or Wilcox (2017). However, an outline of the computations helps interpret the results reported by the R function in Section 11.6.8.

Let X_{ijk} represent the ith observation for level j of Factor A and level k of Factor B. As usual, the groups associated with Factor A are assumed to be independent (A is the between factor) and groups associated with Factor B are possibly dependent. The jth level of Factor A has n_j vectors of observations, each vector containing K values. If, for example, participants are measured at K times, group 1 of Factor A has n_1 participants with K measures taken at K different times. For the jth level of Factor A there are a total of n_jK observations, and among all the groups, the total number of observations is denoted by N. So the total number of vectors among the J groups is $n = \sum n_j$, and the total number of observations is $N = K \sum n_j = Kn$.

The method begins by pooling all N observations and assigning ranks. Midranks are used if there are tied values. Let R_{ijk} represent the rank associated with X_{ijk}. Let

$$\bar{R}_{.jk} = \frac{1}{n_j} \sum_{i=1}^{n_j} R_{ijk},$$

$$\bar{R}_{.j.} = \frac{1}{K} \sum_{k=1}^{K} \bar{R}_{.jk},$$

$$\bar{R}_{ij.} = \frac{1}{K} \sum_{k=1}^{K} R_{ijk},$$

$$\bar{R}_{..k} = \frac{1}{J} \sum_{j=1}^{J} \bar{R}_{.jk},$$

and

$$\bar{R}_{...} = \sum \bar{R}_{.j.}/J.$$

So, for example, $\bar{R}_{.jk}$ is the average of the ranks in level (group) j of Factor A and level k of Factor B, $\bar{R}_{.j.}$ is the average of the ranks associated with level j of Factor A, and $\bar{R}_{..k}$ is the average of the ranks associated with level k of Factor B.

FACTOR A:

The test statistic is

$$F_A = \frac{J}{(J-1)S} \sum_{j=1}^{J} (\bar{R}_{.j.} - \bar{R}_{...})^2,$$

where

$$S = \sum_{j=1}^{J} \frac{\hat{\sigma}_j^2}{n_j}$$

and

$$\hat{\sigma}_j^2 = \frac{1}{n_j - 1} \sum_{i=1}^{n_j} (\bar{R}_{ij.} - \bar{R}_{.j.})^2.$$

So, the test statistic reflects the variation among the average ranks associated with the J groups. Let

$$U = \sum_{j=1}^{J} \left(\frac{\hat{\sigma}_j^2}{n_j} \right)^2$$

and

$$D = \sum_{j=1}^{J} \frac{1}{n_j - 1} \left(\frac{\hat{\sigma}_j^2}{n_j} \right)^2.$$

The degrees of freedom are

$$\nu_1 = \frac{(J-1)^2}{1 + J(J-2)U/S^2},$$

and

$$\nu_2 = \frac{S^2}{D}.$$

Decision Rule: Reject if $F_A \geq f$, where f is the $1 - \alpha$ quantile of an F distribution with ν_1 and ν_2 degrees of freedom.

FACTOR B:

Let

$$\mathbf{R}_{ij} = (R_{ij1}, \ldots, R_{ijK})',$$

$$\bar{\mathbf{R}}_{.j} = \frac{1}{n_j} \sum_{i=1}^{n_j} \mathbf{R}_{ij}, \ \bar{\mathbf{R}}_{..} = \frac{1}{J} \sum_{j=1}^{J} \bar{\mathbf{R}}_{.j},$$

$n = \sum n_j$ (so $N = nK$),

$$\mathbf{V}_j = \frac{n}{N^2 n_j (n_j - 1)} \sum_{i=1}^{n_j} (\mathbf{R}_{ij} - \bar{\mathbf{R}}_{.j})(\mathbf{R}_{ij} - \bar{\mathbf{R}}_{.j})'.$$

So $\mathbf{V}_j$ is a K-by-K matrix of covariances based on the ranks. Let

$$\mathbf{S} = \frac{1}{J^2} \sum_{j=1}^{J} \mathbf{V}_j.$$

The test statistic is

$$F_B = \frac{n}{N^2 \text{tr}(\mathbf{P}_K \mathbf{S})} \sum_{k=1}^{K} (\bar{R}_{..k} - \bar{R}_{...})^2,$$

where $\mathbf{M}_{AB} = \mathbf{P}_J \otimes \mathbf{P}_K$ and $\mathbf{P}_J = \mathbf{I}_J - \frac{1}{J}\mathbf{H}_J$, where $\mathbf{I}_J$ is a J-by-J identity matrix and $\mathbf{H}_J$ is a J-by-J matrix of 1's. The degrees of freedom are

$$\nu_1 = \frac{(\text{tr}(\mathbf{P}_K \mathbf{S}))^2}{\text{tr}(\mathbf{P}_K \mathbf{S} \mathbf{P}_K \mathbf{S})}, \ \nu_2 = \infty.$$

Decision Rule: H_0 is rejected if $F_B \ge f$, where f is the $1-\alpha$ quantile of an F distribution with ν_1 and ν_2 degrees of freedom.

INTERACTIONS:

Let $\mathbf{V}$ be the block diagonal matrix based on the matrices $\mathbf{V}_j$, $j = 1, \dots, J$. The test statistic is

$$F_{AB} = \frac{n}{N^2 \text{tr}(\mathbf{M}_{AB}\mathbf{V})} \sum_{j=1}^{J} \sum_{k=1}^{K} (\bar{R}_{.jk} - \bar{R}_{.j.} - \bar{R}_{..k} + \bar{R}_{...})^2.$$

The degrees of freedom are

$$\nu_1 = \frac{(\text{tr}(\mathbf{M}_{AB}\mathbf{V}))^2}{\text{tr}(\mathbf{M}_{AB}\mathbf{V}\mathbf{M}_{AB}\mathbf{V})}, \ \nu_2 = \infty.$$

Decision Rule: Reject if $F_A \ge f$ (or if $F_{AB} \ge f$), where f is the $1-\alpha$ quantile of an F distribution with ν_1 and ν_2 degrees of freedom.

11.6.8 R Function bwrank

The R function

```
bwrank(J,K,x)
```

performs a between-by-within ANOVA based on ranks using the method just described. In addition to testing hypotheses as just indicated, the function returns the average ranks ($\bar{R}_{.jk}$) associated with all JK groups as well as the relative effects, $(\bar{R}_{.jk} - 0.5)/N$. The data are assumed to be stored as described in Section 11.6.2.

EXAMPLE

For the shoulder pain data in Table 11.2, the output from `bwrank` is

```
$test.A:
[1] 12.87017
$sig.A:
[1] 0.001043705
$test.B:
[1] 0.4604075
$sig.B:
[1] 0.5759393
$test.AB:
[1] 8.621151
$sig.AB:
[1] 0.0007548441
$avg.ranks:
         [,1]     [,2]     [,3]
[1,] 58.29545 48.40909 39.45455
[2,] 66.70455 82.36364 83.04545
```

```
$rel.effects:
          [,1]      [,2]      [,3]
[1,] 0.4698817 0.3895048 0.3167036
[2,] 0.5382483 0.6655580 0.6711013
```

So treatment methods are significantly different and there is a significant interaction, but no significant difference is found over time. Note that the average ranks and relative effects suggest that a disordinal interaction might exist. In particular, for group 1 (the active treatment group), time 1 has higher average ranks compared to time 2, and the reverse is true for the second group. However, the Wilcoxon signed rank test fails to reject at the 0.05 level when comparing time 1 to time 2 for both groups. When comparing time 1 to time 3 for the first group, again using the Wilcoxon signed rank test, reject at the 0.05 level, but a nonsignificant result is obtained for group 2, again comparing time 1 and time 3. So a disordinal interaction appears to be a possibility, but the empirical evidence is not compelling.

11.7 WITHIN-BY-WITHIN DESIGN

A within-by-within design refers to a two-way design where levels of both factors involve dependent groups. For instance, if the cholesterol levels of randomly sample married couples are measured at two different times, we have a within-by-within design. The hypotheses of no main effects and no interactions can be performed with trimmed means using a simple modification of the method in Box 11.3. The main difference is that all groups are possibly dependent, which alters how the covariance matrix **V** in Box 11.3 is computed. The test statistic is again given by Equation (11.8) in Box 11.3.

11.7.1 R Function wwtrim

The R function

```
wwtrim(J,K,x,grp=c(1:p),p=J*K,tr=0.2,bop=F)
```

compares trimmed means for a within-by-within design.

11.8 THREE-WAY DESIGNS

It is noted that three-way designs, as described in Section 11.7, can be extended to situations where one or more factors involve dependent groups and the goal is to compare trimmed means. The methods used here are based on an extension of the method in Johansen (1980), but the complete computational details are omitted. (Essentially, proceed as described in Wilcox, 2017, Section 7.3, only adjust the covariance matrix, which in this book corresponds to **V** in Box 11.3, to include the Winsorized covariance for any two groups that might be dependent.) The following R functions can be used to perform the analysis.

11.8.1 R Functions bbwtrim, bwwtrim, and wwwtrim

For a three-way, J-by-K-by-L design where the third factor involves dependent groups, and the other two factors deal with independent groups, called a between-by-between-by-within design, use the R function

```
bbwtrim(J,K,L,x,grp=c(1:p),tr=0.2).
```

For a between-by-within-by-within design, use

```
bwwtrim(J,K,L,x,grp=c(1:p),tr=0.2).
```

For a within-by-within-by-within design use,

```
wwwtrim(J,K,L,x,grp=c(1:p),tr=0.2).
```

11.8.2 Data Management: R Functions bw2list and bbw2list

Section 11.6.2 introduced the R function

```
bw2list(x, grp.col, lev.col)
```

for dealing with data that are stored in a matrix or a data frame with one column indicating the levels of the between factor and one or more columns containing data corresponding to different within group levels. This function can be used to deal with similar situations when using the R function `bwwtrim`.

EXAMPLE

Imagine that column 3 of the data frame `m` indicates which of two medications were used to treat some malady. Further imagine that columns 6, 7, and 8 contain measures taken at times 1, 2, and 3 for husbands, and columns 10, 11, and 12 contain the results for wives. Then the command

```
z=bw2list(m, 3, c(6,7,8,10,11,12))
```

will store the data in `z` in list mode with time the third factor. The command

```
bwwtrim(2,2,3,z)
```

will compare the groups based on a 20% trimmed mean.

The R function

```
z=bbw2list(x, grp.col, lev.col)
```

is like the function `bw2list`, only designed to handle a between-by-between-by-within design. Now the argument `grp.col` should have two values, which indicate the columns of `x` containing data on the levels of the two between factors, and `lev.col` indicates the columns containing the dependent outcomes.

EXAMPLE

The last example is repeated, only now it is assumed that independent samples of males and females are used. Moreover, column 4 is assumed to indicate whether a participant is male or female and measures, taken at three different times, are contained in columns 6, 7, and 8. Then the command

```
z=bw2list(m, c(3,4), c(6,7,8))
```

stores the data in z, in list mode, and the command

```
bbwtrim(2,2,3,z)
```

will compare the groups based on a 20% trimmed mean.

11.9 EXERCISES

1. The last example in Section 11.6.6 described a study comparing EEG measures for murderers to a control group. The entire data set, containing measures taken at four sites in the brain, can be downloaded from the author's web page as described in Chapter 1 and is stored in the file eegall.dat. The first two lines of this file described the data and need to be skipped when reading the data into R. Compare the two groups using the R function `bwtrim`. Verify that when comparing murderers to the control group, the p-value is 0.39.

2. Using the data in Exercise 1, test Equation (11.9) using a one-step M-estimator via the R function `sppba`. Verify that the p-value is 0.06.

3. In the previous exercise, you could have used averages rather than difference scores when comparing the murderers to the controls. That is, the hypothesis given by Equation (11.10) could have been tested. Using the R function `sppba` with the argument `avg` set equal to TRUE, verify that the p-value is 0.38.

4. Compare the results from the previous three exercises and comment on finding the optimal method for detecting a true difference among groups.

5. For the data used in Exercise 1, use the R function `sppbi` to test the hypothesis of no interactions based on a 20% trimmed mean. Verify that the p-value is 0.04. How does this compare to using a one-step M-esitmator?

CHAPTER 12

MULTIPLE COMPARISONS

CONTENTS

Chapters 9, 10, and 11 described methods for testing the hypothesis that more than two groups have a common measure of location. But usually a more detailed understanding of which groups differ is desired. In Chapter 9, for example, methods were described for testing

$$H_0 : \mu_1 = \cdots = \mu_J, \tag{12.1}$$

the hypothesis that J (independent) groups have equal population means. But if this hypothesis is rejected, and the number of groups is greater than two, typically it is desired to know which groups differ, how they differ and by how much. Assuming normality and homoscedasticity, a simple strategy is to ignore the methods in Chapter 9 and compare all pairs of groups using Student's T test. But as noted in Chapter 9, this raises a technical issue. Imagine that unknown to us, the hypothesis of equal means is true. Further imagine that for each pair of groups, Student's T is applied with the Type I error probability set at $\alpha = 0.05$. Then the probability of *at least one* Type I error is greater than 0.05. Indeed, the more tests that are performed, the more likely it is that at least one Type I error will occur when the hypothesis of equal means is true. When comparing all pairs of four groups, Bernhardson (1975) describes a situation where the probability of at least one Type I error can be as high as 0.29. When comparing all pairs of 10 groups this probability can be as high as 0.59.

The *familywise error rate* (FWE, sometimes called the experimentwise error rate) is the probability of making *at least one* Type I error when performing multiple tests. Determining this probability is complicated by the fact that the individual tests are not all independent. For example, if we compare the means of groups 1 and 2 with Student's T, and then compare the means of groups 1 and 3, the corresponding test statistics will be dependent because both use the sample mean from the first group. If the tests were independent, then we could use the binomial probability function to determine the probability of at least one Type I error. But because there is dependence, special methods are required.

A major goal in this chapter is to describe methods aimed at controlling FWE. But there is a broader issue that should be underscored. What role should the methods in Chapters 9 through 11 play when using the methods covered in this chapter? Currently, a common point of view is that the methods in this chapter should be applied only if some corresponding omnibus hypothesis, described in Chapters 9 through 11, rejects. One justification for this approach is that situations occur where an omnibus test correctly rejects the hypothesis of equal means, yet methods in this chapter fail to detect any differences. (Section 12.2.1 elaborates on this issue.) It is stressed, however, that there are only two methods in this chapter where the use of the methods in Chapters 9 through 11 are required in terms of controlling FWE: Fisher's method described in Section 12.1.1 and the step-down method in Section 12.1.5. And it will be seen that Fisher's method controls FWE for the special case of $J = 3$ groups only. All of the other methods in this chapter are designed to control the familywise error rate without first applying and rejecting some corresponding omnibus hypothesis covered in Chapters 9 through 11. Indeed, if the methods described in this chapter are applied only after rejecting some omnibus hypothesis, their properties have been changed: the actual probability of at least one Type I error will be less than intended, even under normality, indicating that power will be lowered. Also, when comparing means, low power due to non-normality can be exacerbated if the methods in this chapter are applied only when some method in Chapters 9–11 rejects, as will be illustrated.

12.1 ONE-WAY ANOVA AND RELATED SITUATIONS, INDEPENDENT GROUPS

This section deals with the one-way ANOVA design described in Chapter 9. Included are some methods for getting more details about how the distributions of two independent groups differ beyond looking at some measure of location.

The immediate goal is to compare the means of all pairs of groups rather than simply test the hypothesis that all J groups have a common mean. Simultaneously, the goal is to have the probability of at least one Type I error approximately equal to some specified value, α. Included in this chapter are some classic, commonly used methods for comparing means: Fisher's least significant difference method, the Tukey–Kramer method and Scheffé's method, which assume both normality and equal variances. Problems with these classic methods have long been established (e.g., Wilcox, 1996c), but they are commonly used, so they are important to know.

12.1.1 Fisher's Least Significant Difference Method

Assuming normality and homoscedasticity, Fisher's least significant difference (LSD) method begins by performing the ANOVA F test in Section 9.1. If it rejects at the α level, the means for each pair of groups are compared using an extension of Student's T, again at the α level.

To elaborate, imagine that the ANOVA F test rejects the hypothesis of equal means based on some chosen value for α. Let MSWG (described in Box 9.1) be the estimate of the assumed common variance, which is based on the data from all of the groups. To test

$$H_0 : \mu_j = \mu_k, \tag{12.2}$$

the hypothesis that the mean of the jth group is equal to the mean of the kth group, compute

$$T = \frac{\bar{X}_j - \bar{X}_k}{\sqrt{MSWG\left(\frac{1}{n_j} + \frac{1}{n_k}\right)}}. \tag{12.3}$$

When the assumptions of normality and homoscedasticity are met, T has a Student's T

Table 12.1 Hypothetical Data for Three Groups

G1	G2	G3
3	4	6
5	4	7
2	3	8
4	8	6
8	7	7
4	4	9
3	2	10
9	5	9

distribution with $\nu = N - J$ degrees of freedom, where J is the number of groups being compared and $N = \sum n_j$ is the total number of observations in all J groups.

Decision Rule: Reject the hypothesis of equal means if

$$|T| \geq t_{1-\frac{\alpha}{2}},$$

where $t_{1-\alpha/2}$ is the $1 - \frac{\alpha}{2}$ quantile of Student's T distribution with $N - J$ degrees of freedom.

EXAMPLE

For the data in Table 12.1, it can be seen that MSWG = 4.14, the sample means are $\bar{X}_1 = 4.75$, $\bar{X}_2 = 4.62$, and $\bar{X}_3 = 7.75$, the F test is significant with the Type I error probability set at $\alpha = 0.05$, and so, according to Fisher's LSD procedure, proceed by comparing each pair of groups with Student's T test. For the first and second groups,

$$T = \frac{|4.75 - 4.62|}{\sqrt{4.14(\frac{1}{8} + \frac{1}{8})}} = 0.128.$$

The degrees of freedom are $\nu = 21$, and with $\alpha = 0.05$, Table 4 in Appendix B says that the critical value is 2.08. Therefore, you fail to reject. That is, the F test indicates that there is a difference among the three groups, but Student's T suggests that the difference does not correspond to groups 1 and 2. For groups 1 and 3,

$$T = \frac{|4.75 - 7.75|}{\sqrt{4.14(\frac{1}{8} + \frac{1}{8})}} = 2.94,$$

and because 2.94 is greater than the critical value, 2.08, reject. That is, conclude that groups 1 and 3 differ. In a similar manner, conclude that groups 2 and 3 differ as well because $T = 3.08$.

When the assumptions of normality and homoscedasticity are true, Fisher's method controls FWE when $J = 3$. That is, the probability of at least one Type I error will be less than or equal to α. However, when there are more than three groups ($J > 3$), this is not necessarily the case (Hayter, 1986). To illustrate why, suppose four groups are to be compared; the first three have equal means, but the mean of the fourth group is so much larger than the other three that power is close to one. That is, with near certainty, we reject with the ANOVA F test and proceed to compare all pairs of means with Student's T at the α level. This means that we will test

$$H_0 : \mu_1 = \mu_2,$$

$$H_0 : \mu_1 = \mu_3,$$

$$H_0 : \mu_2 = \mu_3,$$

each at the α level, and the probability of at least one Type I error among these three tests will be greater than α.

12.1.2 The Tukey–Kramer Method

Tukey was the first to propose a method that controls FWE. Called Tukey's *honestly significance difference* (HSD), he assumed normality and homoscedasticity and obtained an exact solution when all J groups have equal sample sizes and the goal is to compare the means for each pair of groups. That is, the method guarantees that the probability of at least one Type I error is exactly equal to α. Kramer (1956) proposed a generalization that provides an approximate solution when the sample sizes are unequal and Hayter (1984) showed that when there is homoscedasticity and sampling is from normal distributions, Kramer's method guarantees that FWE will be less than or equal to α.

When comparing the jth group to the kth group, the Tukey–Kramer $1 - \alpha$ confidence interval for $\mu_j - \mu_k$ is

$$(\bar{X}_j - \bar{X}_k) \pm q\sqrt{\frac{\text{MSWG}}{2}\left(\frac{1}{n_j} + \frac{1}{n_k}\right)}, \tag{12.4}$$

where n_j is the sample size of the jth group and MSWG is the mean square within groups, which estimates the assumed common variance. (See Box 9.1.) The constant q is read from Table 9 in Appendix B and depends on the values of α, J (the number of groups being compared), and the degrees of freedom,

$$\nu = N - J,$$

where again N is the total number of observations in all J groups. Under normality, equal variances, and equal sample sizes, the *simultaneous probability coverage* is exactly $1 - \alpha$. That is, with probability $1 - \alpha$, it will be simultaneously true that the confidence interval for $\mu_1 - \mu_2$ will indeed contain $\mu_1 - \mu_2$, the confidence interval for $\mu_1 - \mu_3$ will indeed contain $\mu_1 - \mu_3$, and so on.

Decision Rule: Reject $H_0 : \mu_j = \mu_k$ if

$$\frac{|\bar{X}_j - \bar{X}_k|}{\sqrt{\frac{\text{MSWG}}{2}\left(\frac{1}{n_j} + \frac{1}{n_k}\right)}} \geq q.$$

Equivalently, reject if the confidence interval given by Equation (12.4) does not contain 0. In terms of Tukey's three-decision rule, if we fail to reject, make no decision about whether μ_j is greater than μ_k.

EXAMPLE

Table 12.2 shows some hypothetical data on the ratings of three brands of cookies. Each brand is rated by a different sample of individuals. There are a total of $N = 23$ observations, so the degrees of freedom are $\nu = 23 - 3 = 20$, the sample means are $\bar{X}_1 = 4.1$, $\bar{X}_2 = 6.125$, and $\bar{X}_3 = 7.2$, the estimate of the common variance is MSWG $= 2.13$, and with $\alpha = 0.05$,

Table 12.2 Ratings of Three Types of Cookies

Method 1	Method 2	Method 3
5	6	8
4	6	7
3	7	6
3	8	8
4	4	7
5	5	
3	8	
4	5	
8		
2		

Table 9 in Appendix B indicates that $q = 3.58$. The confidence interval for $\mu_1 - \mu_3$ is

$$(4.1 - 7.2) \pm 3.58\sqrt{\frac{2.13}{2}\left(\frac{1}{10} + \frac{1}{5}\right)} = (-5.12, -1.1).$$

This interval does not contain 0, so reject the hypothesis that the mean ratings of brands 1 and 3 are the same. Comparing brand 1 to brand 2, and brand 2 to brand 3 can be done in a similar manner, but the details are left as an exercise.

12.1.3 R Function TukeyHSD

The R function

```
TukeyHSD(x,conf.level = 0.95) ,
```

in conjunction with the R function `aov`, can be used to apply Tukey's method. (The R function `TukeyHSD` contains some additional arguments not discussed here.)

EXAMPLE

The R function `TukeyHSD` is illustrated with the data frame stored in R called warp breaks, which reports the number of warp breaks per loom, where a loom corresponds to a fixed length of yarn. One of the factors is tension, which is stored in the third column and has three levels labeled H, L, and M. The R commands

```
fm1=aov(breaks ~ tension, data = warpbreaks)
TukeyHSD(fm1)
```

compare the means corresponding to the three levels of the factor tension and returns

```
$tension
           diff       lwr        upr     p adj
M-L -10.000000 -19.55982 -0.4401756 0.0384598
H-L -14.722222 -24.28205 -5.1623978 0.0014315
H-M  -4.722222 -14.28205  4.8376022 0.4630831.
```

For example, the sample mean for tension M, minus the mean for tension L, is −10. The column headed "p adj" is the adjusted p-value based on the number of tests to be performed. For instance, when comparing groups M and L, p adj is 0.0384598. That is, if Tukey's test is performed with the probability of a least one Type I error set at 0.0384598, reject the hypothesis of equal means when comparing groups M and L.

12.1.4 Tukey–Kramer and the ANOVA F Test

As alluded to in the introduction to this chapter, it is common practice to use the Tukey–Kramer method only if the ANOVA F test rejects the hypothesis of equal means. As previously explained, when using the Tukey–Kramer method with equal sample sizes, the probability of at least one Type I error is exactly α under normality and homoscedasticity. But if the Tukey–Kramer method is used contingent on rejecting with the ANOVA F, this is no longer true—FWE will be less than α indicating that power will be reduced (Bernhardson, 1975).

12.1.5 Step-Down Methods

All-pairs power refers to the probability of detecting all true differences among all pairwise differences among the means. For example, suppose the goal is to compare four groups where $\mu_1 = \mu_2 = \mu_3 = 10$, but $\mu_4 = 15$. In this case, all-pairs power refers to the probability of rejecting $H_0 : \mu_1 = \mu_4$, and $H_0 : \mu_2 = \mu_4$, and $H_0 : \mu_3 = \mu_4$. Still assuming normality and homoscedasticity, it is possible to achieve higher all-pairs power than the Tukey–Kramer method using what is called a step-down technique while still achieving the goal of having the probability of at least one Type I error less than or equal to α. One price for this increased power is that confidence intervals for the differences among the pairs of means can no longer be computed. A summary of the method is as follows:

1. Test H_0: $\mu_1 = \cdots = \mu_J$ at the $\alpha_J = \alpha$ level of significance. Assuming normality and homoscedasticity, this is done with the ANOVA F test in Chapter 9. If H_0 is not rejected, stop and fail to find any differences among the means. (In the context of Tukey's three-decision rule, make no decisions about which groups have the larger mean.) Otherwise, continue to the next step.

2. For each subset of $J-1$ means, test the hypothesis that these means are equal at the $\alpha_{J-1} = \alpha$ level of significance. If all such tests are nonsignificant, stop. Otherwise continue to the next step.

3. For each subset of $J-2$ means, test the hypothesis that they are equal at the $\alpha_{J-2} = 1-(1-\alpha)^{(J-2)/J}$ level of significance. If all of these tests are nonsignificant, stop; otherwise, continue to the next step.

4. In general, test the hypothesis of equal means, for all subsets of p means, at the $\alpha_p = 1-(1-\alpha)^{p/J}$ level of significance, when $p \leq J-2$. If all of these tests are nonsignificant, stop and fail to detect any differences among the means; otherwise, continue to the next step.

5. The final step consists of testing all pairwise comparisons of the means at the $\alpha_2 = 1-(1-\alpha)^{2/J}$ level of significance. In this final step, when comparing the jth group to the kth group, either fail to reject, fail to reject by implication from one of the previous steps, or reject.

EXAMPLE

Consider $J = 5$ methods designed to increase the value of a client's stock portfolio which we label methods A, B, C, D, and E. Further assume that when comparing these five methods, you are willing to sacrifice confidence intervals to enhance your all-pairs power. Assume that you want the FWE to be $\alpha = 0.05$. The first step is to test

$$H_0 : \mu_1 = \mu_2 = \mu_3 = \mu_4 = \mu_5$$

at the $\alpha_5 = \alpha = 0.05$ level of significance, where the subscript 5 on α_5 indicates that in the first step, all $J = 5$ means are being compared. If you fail to reject H_0, stop and decide that there are no pairwise differences among the five methods. If you reject, proceed to the next step which consists of testing the equality of the means for all subsets of four groups. In the illustration, suppose the F test for equal means is applied as described in Chapter 9 yielding $F = 10.5$. Assuming the critical value is 2.6, you would reject and proceed to the next step. That is, you test

$$H_0 : \mu_1 = \mu_2 = \mu_3 = \mu_4$$

$$H_0 : \mu_1 = \mu_2 = \mu_3 = \mu_5$$

$$H_0 : \mu_1 = \mu_2 = \mu_4 = \mu_5$$

$$H_0 : \mu_1 = \mu_3 = \mu_4 = \mu_5$$

$$H_0 : \mu_2 = \mu_3 = \mu_4 = \mu_5.$$

In this step you test each of these hypotheses at the $\alpha_4 = \alpha = 0.05$ level of significance, where the subscript 4 indicates that each test is comparing the means of four groups. Note that both the first and second steps use the same significance level, α. If in the second step, all five tests are nonsignificant, you stop and fail to detect any pairwise differences among the five methods; otherwise, you proceed to the next step. In the illustration, suppose the values of your test statistic, F, are 9.7, 10.2, 10.8, 11.6, and 9.8 with a critical value of 2.8. So you reject in every case, but even if you reject in only one case, you proceed to the next step.

The third step consists of testing all subsets of exactly three groups, but this time you test at the

$$\alpha_3 = 1 - (1 - \alpha)^{3/5}$$

level of significance, where the subscript 3 is used to indicate that subsets of three groups are being compared. In the illustration, this means you test

$$H_0 : \mu_1 = \mu_2 = \mu_3$$

$$H_0 : \mu_1 = \mu_2 = \mu_4$$

$$H_0 : \mu_1 = \mu_3 = \mu_4$$

$$H_0 : \mu_1 = \mu_2 = \mu_5$$

$$H_0 : \mu_1 = \mu_3 = \mu_5$$

$$H_0 : \mu_1 = \mu_4 = \mu_5$$

$$H_0 : \mu_2 = \mu_3 = \mu_4$$

$$H_0 : \mu_2 = \mu_3 = \mu_5$$

$$H_0 : \mu_2 = \mu_4 = \mu_5$$

$$H_0 : \mu_3 = \mu_4 = \mu_5$$

using $\alpha_3 = 1 - (1 - 0.05)^{3/5} = 0.030307$. If none are rejected, stop and fail to detect any pairwise difference among all pairs of methods; otherwise, continue to the next step.

Table 12.3 Illustration of the Step-Down Procedure

Groups	F	α	ν_1	Critical value	Decision
ABCDE	11.5	$\alpha_5 = 0.05$	4	2.61	Sig.
ABCD	9.7	$\alpha_4 = 0.05$	3	2.84	Sig.
ABCE	10.2				Sig.
ABDE	10.8				Sig.
ACDE	11.6				Sig.
BCDE	9.8				Sig.
ABC	2.5	$\alpha_3 = 0.0303$	2	3.69	Not Sig.
ABD	7.4				Sig.
ACD	8.1				Sig.
ABE	8.3				Sig.
ACE	12.3				Sig.
ADE	18.2				Sig.
BCD	2.5				Not Sig.
BCE	9.2				Sig.
BDE	8.1				Sig.
CDE	12.4				Sig.
AB	5.1	$\alpha_2 = 0.0203$	1	5.85	Not Sig. by implication
AC	6.0				Not Sig. by implication
AD	19.2				Sig.
AE	21.3				Sig.
BC	1.4				Not Sig. by implication
BD	6.0				Not Sig. by implication
BE	15.8				Sig.
CD	4.9				Not Sig. by implication
CE	13.2				Sig.
DE	3.1				Not Sig.

The final step is to compare the jth group to the kth group by testing

$$H_0 : \mu_j = \mu_k$$

at the

$$\alpha_2 = 1 - (1 - \alpha)^{2/5}$$

level. In the illustration, $\alpha_2 = 0.020308$. In this final stage, make one of three decisions: fail to reject H_0, fail to reject H_0 due to the results from a previous step, or reject.

To clarify the second decision, suppose that H_0: $\mu_1 = \mu_3 = \mu_4$ is not rejected. Then by implication, the hypotheses H_0: $\mu_1 = \mu_3$, H_0: $\mu_1 = \mu_4$, and H_0: $\mu_3 = \mu_4$ would not be rejected *regardless* of what happens in the final step. That is, H_0: $\mu_1 = \mu_3$, H_0: $\mu_1 = \mu_4$, and H_0: $\mu_3 = \mu_4$ would be declared not significant by implication, even if they were rejected in the final step. This might seem counterintuitive, but it is necessary in order to control the familywise Type I error probability. Table 12.3 summarizes the results.

As stressed in Chapter 9, the ANOVA F test performs rather poorly when the normality or homoscedasticity assumption is violated. One particular problem is low power under arbitrarily small departures from normality. When comparing means with a step-down procedure, there is a sense in which this problem is exacerbated. To illustrate why, imagine you are comparing four groups. All of the groups have unequal means. The first three groups have normal distributions, but the fourth has the mixed normal described in Section 3.7. Then the ANOVA F test, applied to all four groups, can have low power, which in turn can mean that the step-down method described here has low power as well. In fact, even a single outlier in one group can mask substantial differences among the other groups being compared. (Dunnett and Tamhane, 1992, describe results on a step-up method that assumes normality and homoscedasticity, but it suffers from the same problem just described.) A simple way of addressing this concern is to replace the ANOVA F test with the global test for trimmed means described in Section 9.4.

Method Fpmcp

Consider a situation where K tests are performed, where each test compares the trimmed means of two independent groups based on Yuen's method. When these K tests are independent, a simpler step-down method can be used that provides relatively high power (Wilcox and Clark, 2015). Denote the resulting p-values by $p_1, \ldots, p_K$. Results in Fisher (1932) can be used to test the hypothesis that all K tests are not significant. The test statistic is

$$F = -2\sum \log(p_i),$$

which has a chi-squared distribution with $2K$ degrees of freedom when all K hypotheses are true. (Here, log indicates the natural logarithm.) That is, reject at the α level if F exceeds the $1 - \alpha$ quantile of a chi-squared distribution with $2K$ degrees of freedom. Let $p_{(1)} \leq \cdots \leq p_{(K)}$ denote the ordered p-values and suppose the goal is to have the probability of one or more Type I errors equal to α. First, perform Fisher's global test based on all K p-values. If not significant at the α level, stop. If significant, reject the hypothesis associated with the smallest p-value and then apply Fisher's method based on $p_{(2)} \leq \cdots \leq p_{(K)}$ at the $\alpha/2$ level. If not significant, stop. If significant, reject the hypothesis associated $p_{(2)}$ and then apply Fisher's method based on $p_{(3)} \leq \cdots \leq p_{(K)}$ at the $\alpha/3$ level. Continue in this fashion until a non-significant result is obtained or all K hypotheses are rejected. (Wilcox and Clark also report results based on alternatives to Fisher's method.) It is stressed that this method requires independent tests. If the tests are dependent, such as when performing all pairwise comparisons, control over the probability of one or more Type I errors can be unsatisfactory.

12.1.6 Dunnett's T3

For multiple comparison procedures based on means, Dunnett (1980a, 1980b) documented practical problems with methods that assume homoscedasticity and then compared several heteroscedastic methods, two of which stood out when sampling from normal distributions. Although non-normality can ruin these methods, they are important because they provide a basis for deriving substantially improved techniques.

Dunnett's T3 procedure is just Welch's method described in Section 7.3, but with the critical value adjusted so that FWE is approximately equal to α when sampling from normal distributions. Let s_j^2 be the sample variance for the jth group, again let n_j be the sample size, and set

$$q_j = \frac{s_j^2}{n_j}, \; j = 1, \ldots, J.$$

When comparing group j to group k, the degrees of freedom are

$$\hat{\nu}_{jk} = \frac{(q_j + q_k)^2}{\frac{q_j^2}{n_j - 1} + \frac{q_k^2}{n_k - 1}}.$$

The test statistic is

$$W = \frac{\bar{X}_j - \bar{X}_k}{\sqrt{q_j + q_k}}.$$

Decision Rule: Reject $H_0 : \mu_j = \mu_k$ if $|W| \geq c$, where the critical value, c, is read from Table 10 in Appendix B. (This table provides the 0.05 and 0.01 quantiles of what is called the *Studentized maximum modulus distribution.*) When using Table 10, the critical value depends on the total number of comparisons to be performed. When performing all pairwise comparisons, the total number of comparisons is

$$C = \frac{J^2 - J}{2}.$$

So with three groups ($J = 3$), the total number of tests is

$$C = \frac{3^2 - 3}{2} = 3.$$

If you have $J = 4$ groups *and* you plan to perform all pairwise comparisons, $C = (4^2 - 4)/2 = 6$.

EXAMPLE

Suppose the goal is to compare five groups to a control group and that only these five comparisons are to be done. That is, the goal is to test $H_0 : \mu_j = \mu_6$, $j = 1, \ldots, 5$, so the number of comparisons is $C = 5$. If $\alpha = 0.05$ and the degrees of freedom are 30, the critical value is $c = 2.73$. If you have five groups and plan to do all pairwise comparisons, $C = 10$, and with $\alpha = 0.01$ and $\nu = 20$, the critical value is 3.83.

A confidence interval for $\mu_j - \mu_k$, the difference between the means of groups j and k, is given by

$$(\bar{X}_j - \bar{X}_k) \pm c\sqrt{\frac{s_j^2}{n_j} + \frac{s_k^2}{n_k}}.$$

By design, the simultaneous probability coverage will be approximately $1 - \alpha$, under normality, when computing C confidence intervals and c is read from Table 10. (The R function `lincon`, described in Section 12.1.9, can be used to apply Dunnett's T3 method by setting the argument `tr`=0.)

EXAMPLE

Table 9.1 reports skin resistance for four groups of individuals. If the goal is to compare all pairs of groups with $\alpha = 0.05$, then $C = 6$, and the confidence interval for the difference between the means of the first two groups is $(-0.35, 0.52)$; this interval contains 0, so you fail to detect a difference. The degrees of freedom are 12.3, the critical value is $c = 3.07$, the test statistic is $W = 0.56$, and again fail to reject because $|W| < 3.07$. It is left as an exercise to show that for the remaining pairs of means, we again fail to reject.

Stressing a point made in the introduction to this chapter, should Dunnett's T3 method be applied only if Welch's test in Chapter 9 rejects? In terms of controlling the probability of at least one Type I error, all indications are that the answer is no. The T3 method is designed to control FWE. There is nothing in the derivation of the method that requires first testing the hypothesis that all J groups have equal means.

12.1.7 Games–Howell Method

An alternative to Dunnett's T3 is the Games and Howell (1976) method. When comparing the jth group to the kth group, you compute the degrees of freedom, $\hat{\nu}_{jk}$, exactly as in Dunnett's T3 procedure, and then you read the critical value, q, from Table 9 in Appendix B. (Table 9 reports some quantiles of what is called the Studentized range distribution.) The $1-\alpha$ confidence interval for $\mu_j - \mu_k$ is

$$(\bar{X}_j - \bar{X}_k) \pm q\sqrt{\frac{1}{2}\left(\frac{s_j^2}{n_j} + \frac{s_k^2}{n_k}\right)}.$$

If this interval does not contain 0, reject H_0: $\mu_j = \mu_k$, which is the same as rejecting if

$$\frac{|\bar{X}_j - \bar{X}_k|}{\sqrt{\frac{1}{2}\left(\frac{s_j^2}{n_j} + \frac{s_k^2}{n_k}\right)}} \geq q.$$

Under normality, the Games–Howell method appears to provide more accurate probability coverage, compared to Dunnett's T3 method, when all groups have a sample size of at least 50. A close competitor under normality is Dunnett's (1980b) C method, but no details are given here.

EXAMPLE

Imagine there are three groups with $\bar{X}_1 = 10.4$, $\bar{X}_2 = 10.75$,

$$\frac{s_1^2}{n_1} = 0.11556,$$

$$\frac{s_2^2}{n_2} = 0.156.$$

Then $\hat{\nu} = 19$ and with $\alpha = 0.05$, $q = 3.59$, so the confidence interval for $\mu_1 - \mu_2$ is

$$(10.4 - 10.75) \pm 3.59\sqrt{\frac{1}{2}(0.11556 + 0.156)} = (-0.167, 0.97).$$

This interval contains 0, so fail to reject the hypothesis of equal means.

12.1.8 Comparing Trimmed Means

Dunnett's T3 method is readily extended to trimmed means. Briefly, for each pair of groups, apply Yuen's method in Section 7.4.1 and control FWE by reading a critical value from Table 10 in Appendix B rather than Table 4. As was the case when using the T3 method, the critical value depends on C, the number of tests to be performed, as well as the degrees of freedom associated with the two groups being compared, which is estimated as indicated in Section 7.4.1.

12.1.9 R Functions lincon, stepmcp and twoKlin

The R function

```
lincon(x,con=0,tr=0.2,alpha=0.05)
```

compares all pairs of groups using the extension of the T3 method to trimmed means. The argument `x` is assumed to be a matrix with J columns or to have list mode. The argument `tr` controls the amount of trimming and defaults to 20%. Unlike Table 10 in Appendix B, the function can be used with any valid value for α and it can handle situations where the total number of tests is greater than 28. (The argument `con` is explained in Section 12.2.5.) The R function

```
stepmcp(x,con=0,tr=0.2,alpha=0.05)
```

can used to apply the step-down method in Section 12.1.5 when dealing with five groups or less. But rather than use the ANOVA F test, this function uses the global test for trimmed means described in Section 9.4.

The R function

```
twoKlin(x=NULL,x1=NULL,x2=NULL,tr=0.2,alpha=0.05,pr=TRUE)
```

performs method Fpmcp, which is based on trimmed means and is also described in Section 12.1.5. If the data are stored in `x`, `x` is assumed to have $2K$ columns if it is a matrix, or length $2K$ if it has list mode, in which case the first test of the K independent tests is based on the data in columns 1 and $K+1$, the second test is based on columns 2 and $K+2$, and so on. If data are stored in `x1` and `x2`, the function performs a test based on the data in column 1 of `x1` and `x2`, followed by a test using the data in column 2 of `x1` and `x2`, and so forth.

12.1.10 Alternative Methods for Controlling FWE

There are several alternative methods for controlling FWE that have practical value in certain settings (particularly when comparing dependent groups). This section describes these methods for future reference.

Bonferroni Method

The simplest method is based on the Bonferroni inequality. If C hypotheses are to be tested, test each hypothesis at the α/C level. Provided the probability of a Type I error can be controlled for each of the individual tests, FWE will be at most α.

Rom's Method

Several improvements on the Bonferroni method have been published that are designed to ensure that the probability of at least one Type I error does not exceed some specified value, α. One that stands out is a *sequentially rejective* method derived by Rom (1990), which has been found to have good power relative to several competing techniques (e.g., Olejnik et al., 1997). To apply it, compute a p-value for each of the C tests to be performed and label them $P_1, \ldots, P_C$. Next, put the p-values in descending order, which are now labeled $P_{[1]} \geq P_{[2]} \geq \cdots \geq P_{[C]}$. Proceed as follows:

1. Set k = 1.

2. If $P_{[k]} \leq d_k$, where d_k is read from Table 12.4, stop and reject all C hypotheses; otherwise, go to step 3.

3. Increment k by 1. If $P_{[k]} \leq d_k$, stop and reject all hypotheses having a significance level less than or equal to d_k.

Table 12.4 Critical Values, d_k, for Rom's Method

k	$\alpha = 0.05$	$\alpha = 0.01$
1	0.05000	0.01000
2	0.02500	0.00500
3	0.01690	0.00334
4	0.01270	0.00251
5	0.01020	0.00201
6	0.00851	0.00167
7	0.00730	0.00143
8	0.00639	0.00126
9	0.00568	0.00112
10	0.00511	0.00101

4. If $P_{[k]} > d_k$, repeat step 3.

5. Continue until you reject or all C hypotheses have been tested.

An advantage of Rom's method is that its power is greater than or equal to the Bonferroni method. In fact, Rom's method always rejects as many or more hypotheses. A negative feature is that confidence intervals, having simultaneous probability coverage $1 - \alpha$, are not readily computed. A limitation of Rom's method is that it was derived assuming that independent test statistics are used, but various studies suggest that it continues to control the Type I error reasonably well when the test statistics are dependent.

Hochberg's Method

A method similar to Rom's method was derived by Hochberg (1988) where, rather than use the d_k values in Table 12.4, use $d_k = \alpha/k$. For $k = 1$ and 2, d_k is the same as in Rom's method. An advantage of Hochberg's method is that it can be used when the number of tests to be performed exceeds 10 or when α is not equal to 0.05 or 0.01. Here, when using Rom's method, the d_k values in Table 12.4 are used if $k \leq 10$. If the number of tests exceeds 10, the d_k values are determined via Hochberg's method. If α is not equal to 0.05 or 0.01, Rom's method cannot be applied, and so Hochberg's method is used instead. Like Rom's method, the probability of at least one Type I error will not exceed α provided that the probability of a Type I error can be controlled for each of the individual tests. Rom's method offers a slight advantage over Hochberg's method in terms of power. But Hochberg's method relaxes the assumption that the test statistics are independent.

Hommel's Method

Yet another method for controlling the probability of one or more Type I errors was derived by Hommel (1988). Let $p_1, \ldots, p_C$ be the p-values associated with C tests and let

$$j = \max\{i \in \{1, \cdots, C\} : p_{(n-i+k)} > k\alpha/i \text{ for } k = 1, \ldots, i\},$$

where $p_{(1)} \leq \cdots \leq p_{(C)}$ are the p-values written in ascending order. If the maximum does not exist, reject all C hypotheses. Otherwise, reject all hypotheses such that $p_i \leq \alpha/j$. Adjusted p-values based on Hommel's method can be computed via the R function `p.adjust` in Section

12.1.12, which also computes an adjusted p-value based on Hochberg's method. It seems that typically there is little or no difference between the adjusted p-values based on the Hochberg and Hommel methods, but situations are encountered where Hommel's method rejects and Hochberg's method does not, and Hommel's method is based on slightly weaker assumptions.

Benjamini and Hochberg Method

Benjamini and Hochberg (1995) proposed a technique similar to Hochberg's method, only in step 1 of Hochberg's method, $P_{[k]} \leq \frac{\alpha}{k}$ is replaced by

$$P_{[k]} \leq \frac{(C-k+1)\alpha}{C}.$$

Results reported by Williams et al. (1999) support the use of the Benjamini–Hochberg method over Hochberg.

EXAMPLE

Suppose six hypotheses are to be tested with the Benjamini–Hochberg method based on the following results:

Number	Test	p-value	
1	$H_0: \mu_1 = \mu_2$	$P_1 = 0.010$	$P_{[5]}$
2	$H_0: \mu_1 = \mu_3$	$P_2 = 0.015$	$P_{[3]}$
3	$H_0: \mu_1 = \mu_4$	$P_3 = 0.005$	$P_{[6]}$
4	$H_0: \mu_2 = \mu_3$	$P_4 = 0.620$	$P_{[1]}$
5	$H_0: \mu_2 = \mu_4$	$P_5 = 0.130$	$P_{[2]}$
6	$H_0: \mu_3 = \mu_4$	$P_6 = 0.014$	$P_{[4]}$

Because $P_{[1]} > 0.05$, fail to reject the fourth hypothesis. Had it been the case that $P_{[1]} \leq 0.05$, you would stop and reject all six hypotheses. Because you did not reject, set $k = 2$, and because $C = 6$, we see that $(C-k+1)\alpha/C = 5(0.05)/6 = 0.0417$. Because $P_{[2]} = 0.130 > 0.0417$, fail to reject the fifth hypothesis and proceed to the next step. Incrementing k to 3, $(C-k+1)\alpha/C = 4(0.05)/6 = 0.0333$, and because $P_{[3]} = 0.015 \leq 0.0333$, reject this hypothesis and the remaining hypotheses having p-values less than or equal to 0.0333. That is, reject hypotheses 1, 2, 3, and 6 and fail to reject hypotheses 4 and 5. If the Bonferroni inequality had been used instead, we see that $0.05/6 = 0.00833$, so only hypothesis 3 would be rejected.

A criticism of the Benjamini–Hochberg method is that situations can be constructed where some hypotheses are true, some are false, and the probability of at least one Type I error will exceed α among the hypotheses that are true (Hommel, 1988; cf. Keselman et al., 1999). In contrast, Hochberg's method does not suffer from this problem. In practical terms, the Benjamini–Hochberg method relaxes somewhat the control over the probability of at least one Type I error with the benefit of possibly increasing power. When performing many tests, the increase in power can be of practical importance.

Although the Benjamini–Hochberg method does not necessarily control the probability of at least one Type I error, it does have the following property. When testing C hypotheses, let Q be the proportion of hypotheses that are true and rejected. That is, Q is the proportion of Type I errors among the null hypotheses that are correct. If all hypotheses are false, then of course $Q = 0$, but otherwise Q can vary from one experiment to the next. That is, if we repeat a study many times, the proportion of erroneous rejections will vary. The *false discovery rate* is the expected value of Q. That is, if a study is repeated infinitely many times, the false discovery rate is the average proportion of Type I errors among the hypotheses that are true. Benjamini and Hochberg (1995) show that their method ensures that the false discovery rate is less than or equal to α. For results on how the Benjamini–Hochberg method performs in terms of controlling the false discovery rate, even when the test statistics are dependent, see Benjamini and Yekutieli (2001). The Benjamini–Hochberg method can be improved if the number of true hypotheses is known. Of course it is not known how many null hypotheses are in fact correct, but Benjamini and Hochberg (2000) suggest how this number might be estimated which can result in higher power. (For related results, see Finner and Roters, 2002; Finner and Gontscharuk, 2009; Sarkar, 2002.)

12.1.11 Percentile Bootstrap Methods for Comparing Trimmed Means, Medians, and M-estimators

As explained in Chapter 7, the only known method for comparing medians that remains reasonably accurate when tied values occur is based on a percentile bootstrap method. Also, when there are no tied values, there seems to be no compelling reason to use a non-bootstrap method. When the goal is to compare all pairs of independent groups via medians, it currently appears that the best approach, in terms of controlling the probability of at least one Type I error, is to use Rom's method in conjunction with the percentile bootstrap method in Chapter 7. That is, use the percentile bootstrap method to determine a p-value for each pair of groups and then proceed as indicated in the previous section. If the number of tests to be performed exceeds 10, use Hochberg's method instead to control FWE. This method also appears to perform well when using trimmed means provided the amount of trimming is not too small. Hommel's method is another option.

When comparing M-estimators or MOM, the same method used to compare medians appears to be best for general use when the sample sizes are large. That is, for each group, an M-estimator is computed rather than a median; otherwise, the computations are the same. But with small sample sizes, roughly meaning that all sample sizes are less than or equal to 80, the method is too conservative in terms of Type I errors: the actual level can drop well below the nominal level. An adjustment that appears to be reasonably effective is applied as follows.

For each group, generate B bootstrap samples and for each sample compute the M-estimator of location (or MOM). For each pair of groups, determine the proportion of bootstrap estimates from the first group that are larger than the estimates from the second. That is, proceed as described in Section 7.5 and compute $\hat{p}^*$ as given by Equation (7.16). For the situation here where the goal is to compare all pairs of groups, the total number of $\hat{p}^*$ values is $C = (J^2 - J)/2$. Put the $\hat{p}^*$ values in *descending* order yielding $\hat{p}^*_{[1]} \geq \hat{p}^*_{[2]} \geq \cdots \geq \hat{p}^*_{[C]}$. So, for example, $\hat{p}^*_{[1]}$ is the largest of the C values just computed and $\hat{p}^*_{[C]}$ is the smallest. Decisions about the individual hypotheses are made as follows. If $\hat{p}^*_{[1]} \leq \alpha_1$, where α_1 is read from Table 12.5, reject all C hypotheses. If $\hat{p}^*_{[1]} > \alpha_1$, but $\hat{p}^*_{[2]} \leq \alpha_2$, fail to reject the hypothesis associated with $\hat{p}^*_{[1]}$, but the remaining hypotheses are rejected. If $\hat{p}^*_{[1]} > \alpha_1$ and $\hat{p}^*_{[2]} > \alpha_2$, but $\hat{p}^*_{[3]} \leq \alpha_3$, fail to reject the hypotheses associated with $\hat{p}^*_{[1]}$ and $\hat{p}^*_{[2]}$, but reject the remaining hypotheses. In general, if $\hat{p}^*_{[c]} \leq \alpha_c$, reject the corresponding hypothesis and all other hypotheses having smaller $\hat{p}^*_m$ values. For other values of α (assuming $c > 1$) or for $c > 10$, use

$$\alpha_c = \frac{\alpha}{c}.$$

This method is a slight modification of Hochberg's method and is called *method SR*. If any group has a sample size greater than 80, Hochberg's method is used instead.

Table 12.5 Values of α_c for $\alpha = 0.05$ and 0.01

c	$\alpha = 0.05$	$\alpha = 0.01$
1	0.02500	0.00500
2	0.02500	0.00500
3	0.01690	0.00334
4	0.01270	0.00251
5	0.01020	0.00201
6	0.00851	0.00167
7	0.00730	0.00143
8	0.00639	0.00126
9	0.00568	0.00112
10	0.00511	0.00101

It is briefly noted that there is a percentile bootstrap method designed specifically for comparing 20% trimmed means (Wilcox, 2001d). The extent it improves on using the basic percentile bootstrap method described here is unknown.

12.1.12 R Functions medpb, linconpb, pbmcp, and p.adjust

The R function

```
medpb(x,alpha=0.05, nboot=NA, grp=NA, est=median, con=0, bhop=F)
```

performs all pairwise comparisons of J independent groups using the method for medians described in the previous section. The argument `nboot` determines how many bootstrap samples will be used. By default, `nboot=NA`, meaning that the function will choose how many bootstrap samples will be used depending on how many groups are being compared.

The R function

```
linconpb(x,alpha=0.05,nboot=NA,grp=NA,est=tmean,con=0,bhop=FALSE,...)
```

is exactly the same as `mdepb`, only it defaults to using a 20% trimmed mean. (The R function `tmcppb` is identical to `linconpb`.)

The R function

```
pbmcp(x,alpha=0.05,nboot=NA,grp=NA,est=onestep,con=0,bhop=F)
```

can be used to compare measures of location indicated by the argument `est`, which defaults to a one-step M-estimator. If the largest sample size is less than 80 and `bhop=F`, method SR is used to control FWE. But if the largest sample size exceeds 80, Hochberg's method is used, still assuming that `bhop=F`. If `bhop=T`, the Benjamini–Hochberg method is used. With small sample sizes, this function appears to be best when comparing groups based on an M-estimator or MOM. If the goal is to compare groups using MOM, set the argument `est=mom`.

The R function

```
p.adjust(p, method = p.adjust.methods, n = length(p))
```

adjusts a collection of p-values based on the Bonferroni method and related techniques that include the Hochberg and Hommel methods. Setting the argument `method = "hochberg"`, for example, would adjust the p-values based on Hochberg's method. Other methods include: `"hommel"`, `"bonferroni"`, and `"BH"`, where BH indicates the Benjamini–Hochberg method. The default method is `"holm"`, a method not covered here.

12.1.13 A Bootstrap-t Method

When comparing all pairs of means, or trimmed means with a small amount of trimming, a bootstrap-t version of the method in Section 12.1.6 and 12.1.8 appears to be a relatively good choice for general use. Roughly, for each pair of groups, compute the absolute value of Yuen's test statistic. Of course, some of these test statistics will be larger than others. Denote the largest value by $T_{\max}$. For the case where all groups have the same (population) trimmed mean, if we could determine the distribution of $T_{\max}$, meaning that we could compute $P(T_{\max} \leq c)$ for any value of c we might choose, then FWE (the probability of at least one Type I error) could be controlled. The strategy is to use bootstrap samples to estimate this distribution. Here is a more detailed outline of the method.

Goal: Test H_0: $\mu_{tj} = \mu_{tk}$, for all $j < k$ so that the probability of at least one Type I error is approximately α. That is, compare all pairs of groups with the goal that FWE is α.

1. Compute the sample trimmed means for each group, namely $\bar{X}_{t1}, \ldots, \bar{X}_{tJ}$, and then Yuen's estimate of the squared standard errors, $d_1, \ldots, d_J$; see Equation (8.12).

2. For each group, generate a bootstrap sample and compute the trimmed means, which we label $\bar{X}^*_{t1}, \ldots, \bar{X}^*_{tJ}$. Also compute Yuen's estimate of the squared standard error yielding $d^*_1, \ldots, d^*_J$.

3. For groups j and k ($j < k$), compute

$$T^*_{yjk} = \frac{|(\bar{X}^*_{tj} - \bar{X}^*_{tk}) - (\bar{X}_{tj} - \bar{X}_{tk})|}{\sqrt{d^*_1 + d^*_2}}.$$

4. Let $T^*_{\max}$ be the largest of the values just computed.

5. Repeat steps 2 through 4 B times yielding $T^*_{\max 1}, \ldots, T^*_{\max B}$.

6. Put the $T^*_{\max 1}, \ldots, T^*_{\max B}$ values in ascending order yielding $T^*_{\max(1)} \leq \cdots \leq T^*_{\max(B)}$.

7. Let $u = (1 - \alpha)B$, rounded to the nearest integer. Then the confidence interval for $\mu_{tj} - \mu_{tk}$ is

$$(\bar{X}_{tj} - \bar{X}_{tk}) \pm T^*_{\max(u)}\sqrt{d_j + d_k}.$$

8. The hypothesis H_0: $\mu_{tj} = \mu_{tk}$ is rejected if

$$\frac{|\bar{X}_{tj} - \bar{X}_{tk}|}{\sqrt{d_j + d_k}} \geq T^*_{\max(u)}.$$

12.1.14 R Function linconbt

The R function

```
linconbt(x,con=0,tr=0.2,alpha = 0.05)
```

compares all pairs of groups using the bootstrap-t method just described. (The argument con is explained in Section 12.2.5.)

12.1.15 Rank-Based Methods

Consider two independent groups and let P be the probability that a randomly sampled observation from the first group is less than a randomly sampled observation from the second. Sections 7.8.4 and 7.8.6 described rank-based methods for testing the hypothesis

$$H_0 : P = 0.5. \tag{12.5}$$

both of which can handle tied values as well as heteroscedasticity. It is noted that both methods are readily extended to the problem of comparing all pairs of groups in a manner that controls FWE. Simply proceed as before only read the critical value from Table 10 in Appendix B with infinite degrees of freedom. Another approach is to use Hochberg's method to control FWE. It appears that generally Hochberg's method is preferable, although situations can be constructed where Table 10 in Appendix B is preferable in terms of power. For instance, when the goal is to have FWE equal to 0.05, and the number of tests is $C \leq 28$, Hochberg's method is preferable. But if, for example, the goal is to have FWE equal to 0.01, with C sufficiently large, using Table 10 to control FWE is preferable.

12.1.16 R Functions cidmul, cidmulv2, and bmpmul

The R function

```
cidmul(x,alpha=0.05)
```

performs all pairwise comparisons given the goal of testing Equation (12.5) for each pair of groups. For each pair of groups, the function applies Cliff's method in Section 7.4 and controls the probability of making at least one Type I error via the argument `alpha`. The R function

```
cidmulv2(x,alpha=0.05,g=NULL,dp=NULL,CI.FWE=FALSE)
```

uses Cliff's method in conjunction with Hochberg's method for controlling FWE and seems to be preferable to the R function `cidmul` in most situations. If the argument `g=NULL`, then `x` is assumed to be a matrix or have list mode. If `g` is specified, it is assumed that column `g` of `x` is a factor variable and that the dependent variable of interest is in column `dp` of `x`. (The argument `x` can be a matrix or data frame.) If the argument `CI.FWE=FALSE`, then confidence intervals are computed based on Cliff's method. That is, for each pair of groups, the goal is to compute a $1-\alpha$ confidence interval, which means that the simultaneous probability coverage is not designed to be $1-\alpha$. With `CI.FWE=TRUE`, the probability coverage for each confidence interval is based on the Hochberg critical p-value. For example, if when comparing groups 2 and 4, the Hochberg critical p-value is 0.0125, then the function returns a 0.0125 confidence interval for P, rather than a 0.05 confidence interval (assuming the argument `alpha=0.05`).

The function

```
bmpmul(x,alpha=0.05)
```

is the same as `cidmul`, only now each pair of groups is compared using the Brunner–Munzel method in Section 7.7.6.

12.1.17 Comparing the Individual Probabilities of Two Discrete Distributions

Section 7.8.10 noted that when comparing discrete distributions, alternatives to the shift function might have practical value. One such method is described here. The goal is to test

$$H_0 : P(X = x) = P(Y = x) \tag{12.6}$$

for each possible value x in a manner that controls the familywise error rate. Imagine, for example, the possible values are 0, 1, 2, 3, 4 and 5. Then the goal is to test six hypotheses such that the probability of one or more Type I errors is approximately equal to α. Note that for a specific value for x, testing Equation (12.6) corresponds to testing the hypothesis that two independent binomial distributions have the same probability of success. That is, either $X = x$ or $X \neq x$. Letting $p_1 = P(X = x)$ and $p_2 = P(Y = x)$, the goal is to test H_0: $p_1 = p_2$, which can be done with methods in Section 7.13. Here, the Storer–Kim method is used unless stated otherwise. When testing multiple hypotheses, the probability of one or more Type I errors is controlled using Hochberg's method in Section 12.1.10. The main point here is that testing Equation (12.6) for each possible x value might yield details of practical importance that are missed when focusing on a measure of location only. A limitation of the Storer–Kim method is that it does not provide confidence intervals. A way of dealing with this issue is to replace the Storer–Kim method with the method derived by Kulinskaya et al. (2010).

12.1.18 R Functions binband, splotg2, cumrelf, and cumrelfT

The R function

```
binband(x,y,KMS=FALSE,alpha=0.05)
```

performs the method described in the previous section. By default, the individual probabilities are compared using the Storer–Kim method. Setting the argument KMS=TRUE, the method derived by Kulinskaya et al. (2010) is used, which provides a confidence interval for $p_1 - p_2$ for each possible x value. Again, Hochberg's method controls FWE.

The R function

```
splotg2(x,y,op=TRUE,xlab="X",ylab="Rel. Freq.")
```

plots the relative frequencies for all distinct values found in each of two groups. With op=TRUE, a line connecting the points corresponding to the relative frequencies is formed. (Starting with Rallfun-v32, binband contain the argument plotit=TRUE, meaning that by default, a plot is created via the R function splotg2.)

The R function

```
cumrelf(x, y=NA, xlab="X",ylab="CUM REL FREQ")
```

plots the cumulative relative frequency of two or more groups. The argument can be a matrix, with columns corresponding to groups, or x can have list mode. If data are stored in the argument y, it is assumed that there are two groups only with the data for the second group stored in y. Suppose it is desired to test the hypothesis

$$P(X \leq x) = P(Y \leq x).$$

That is, the goal is to determine whether the value x corresponds to the same quantile in two independent groups, which can done using methods for comparing the two independent binomials described in Section 7.13. This is in contrast to testing the hypothesis that the 0.25, for example, is the same quantile for both groups. The R function

```
cumrelfT(x, y, pts, xlab = "X", ylab = "CUM REL FREQ", plotit = TRUE, op =
                                  1)
```

tests this hypothesis for every value stored in the argument `pts`. For instance, `pts=c(0,1)` would test both $H_0 : P(X \leq 0) = P(Y \leq 0)$ and $H_0 : P(X \leq 1) = P(Y \leq 1)$. The argument `op=1` means that the Storer–Kim method will be used. Setting `op=2`, method KMS is used instead.

EXAMPLE

Erceg-Hurn and Steed (2011) investigated the degree to which smokers experience negative emotional responses (such as anger and irritation) upon being exposed to anti-smoking warnings on cigarette packets. Smokers were randomly allocated to view warnings that contained only text, such as "Smoking Kills, " or warnings that contained text and graphics, such as pictures of rotting teeth and gangrene. Negative emotional reactions to the warnings were measured on a scale that produced a score between 0 and 16 for each smoker, where larger scores indicate greater levels of negative emotions. (The data are stored on the author's web page in the file smoke.csv.) The means and medians differ significantly. But to get a deeper understanding of how the groups differ, look at Figure 12.1, which shows plots of the relative frequencies based on the R function splotg2. Note that the plot suggests that the main difference between the groups has to do with the response zero: the proportion of participants in the text-only group responding zero is 0.512 compared to 0.192 for the graphics group. Testing the hypothesis that the corresponding probabilities are equal, based on the R function binband, the p-value is less than 0.0001. The probabilities associated with the other possible responses do not differ significantly at the 0.05 level except for the response 16; the p-value is 0.0031. For the graphics group, the probability of responding 16 is 0.096 compared to 0.008 for the text-only group. So a closer look at the data, beyond comparing means and medians, suggests that the main difference between the groups has to do with the likelihood of giving the most extreme responses possible, particularly the response zero.

12.1.19 Comparing the Quantliles of Two Independent Groups

Section 7.8.10 described a method for comparing all of the quantiles of two independent distributions. Assuming random sampling only, the method controls the probability of one or more Type I errors. However, power might be relatively poor when there is interest in the lower and upper quantiles, particularly when dealing with tied (duplicated) values. A method for dealing with these concerns is to use a percentile bootstrap method in conjunction with the quantile estimator derived by Harrell and Davis (1982), with the probability of one or more Type I errors controlled using one of the methods in Section 12.1.10 (Wilcox et al., 2014). When dealing with tied values, a crucial feature of the Harrell–Davis estimator is that it is a weighted sum of all of the observed values, with all of the weights getting some positive value. That is, the Harrell–Davis estimator chooses positive weights $w_1, \dots w_n$ such that $\sum w_i X_{(i)}$ estimates the quantile of interest. When estimating the population median, for example, this in contrast to the usual sample median that uses only the middle one or two values.

To elaborate, bootstrap samples are generated as described in Section 7.5. Let $\hat{d}^*_q = \hat{\theta}^*_{1q} - \hat{\theta}^*_{2q}$ be the estimated difference between qth quantiles based on this bootstrap sample. Repeat this process B times yielding $\hat{d}^*_{q1}, \dots, \hat{d}^*_{qB}$. Put these B values in ascending order yielding $\hat{d}^*_{q(1)} \leq \cdots, \hat{d}^*_{q(B)}$. Then a $1 - \alpha$ confidence interval for $q_1 - q_2$ and a p-value can be computed as described in Section 7.5. Here, the probability of one or more Type I errors is controlled using Hochberg's method, described in Section 12.1.10, unless stated otherwise.

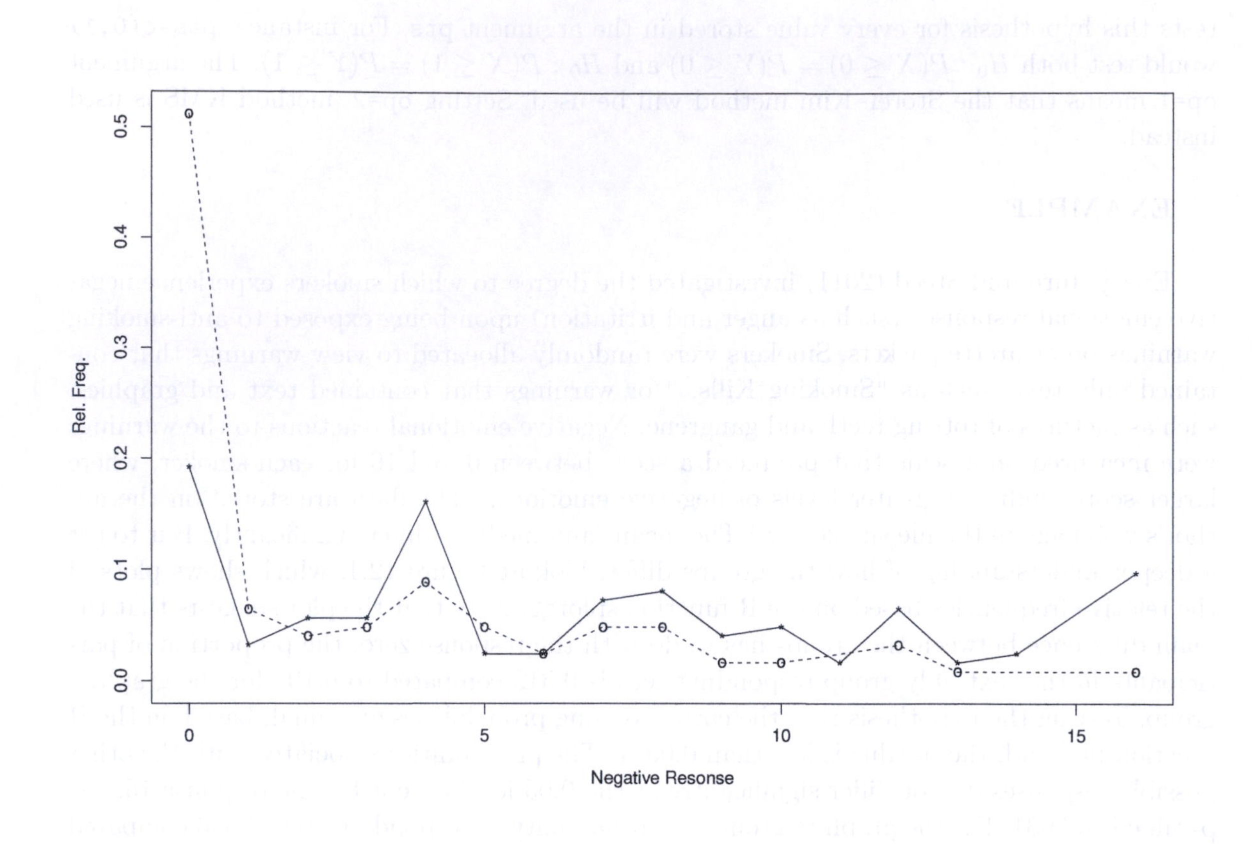

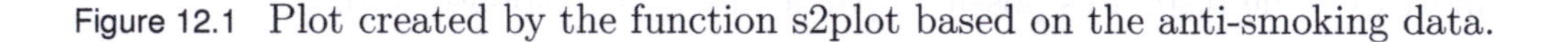
Figure 12.1 Plot created by the function s2plot based on the anti-smoking data.

12.1.20 R Functions qcomhd and qcomhdMC

The R function

```
qcomhd(x, y, q = c(0.1, 0.25, 0.5, 0.75, 0.9), nboot = 2000, plotit = TRUE,
        SEED = TRUE, xlab = "Group 1", ylab = "Est.1-Est.2", tr=0.2)
```

compares quantiles using the method described in the previous section. By default, the 0.1, 0.25, 0.5, 0.75, 0.9 quantiles are compared, but this can be altered using the argument q. The R function

```
qcomhdMC(x, y, q = c(0.1, 0.25, 0.5, 0.75, 0.9), nboot = 2000, plotit =
   TRUE, SEED = TRUE, xlab = "Group 1", ylab = "Est.1-Est.2", tr=0.2)
```

is exactly the same as qcomhd, only it takes advantage of a multicore processor if one is available.

EXAMPLE

This example stems from a study aimed at improving the physical and emotional well-being of older adults (Clark, et al., 2011). A portion of the study measured depressive symptoms for a control group and a group that received intervention. Comparing 20% trimmed means or medians found no significant differences when testing at the 0.05 level. However, the bulk of the participants were not overly depressed, so it does not seem surprising that no significant difference was found. However, there is the issue of comparing the participants who had relatively high measures of depression. A way of doing this is to compare the 0.8, 0.85, and 0.9 quantiles. The corresponding p-values, using the method in the previous section, are 0.061, 0.006, and 0.012, respectively, suggesting that among participants who have relatively high measures of depression, intervention can be beneficial. In contrast, comparing all of the quantiles using the method in Section 7.8.10, no significant differences are found.

12.1.21 Multiple Comparisons for Binomial and Categorical Data

For J independent binomial distributions, let p_j be the probability of success associated with the jth group, and consider the goal of testing

$$H_0 : p_j = p_k$$

for every $j < k$ such that FWE is equal to α. A simple method is to use the Storer–Kim method and control FWE with Hochberg's method.

Section 7.13.4 mentioned a classic method for testing the hypothesis that two independent discrete distributions are identical. That is, the sample space contains $K > 2$ possible values and the goal is to test the hypothesis that the probabilities associated with each possible value are identical for both groups. One way of comparing J independent groups is to compare each pair of groups using the method in Section 7.13.4 and control FWE using Hochberg's method.

12.1.22 R Functions skmcp and discmcp

The R function

```
skmcp(x,alpha=0.05)
```

performs all pairwise comparisons of J independent binomial distributions using the Storer–Kim method as described in the previous section. The argument `x` can be a matrix with J columns containing 0's and 1's, or it can have list mode. When the sample space contains $K > 2$ possible values, the R function

```
discmcp(x,alpha=0.05)
```

can be used.

12.2 TWO-WAY, BETWEEN-BY-BETWEEN DESIGN

When dealing with a two-way ANOVA design, it is convenient to describe relevant multiple comparison procedures in the context of what are called linear contrasts. For a J-by-K design, let $L = JK$ represent the total number of groups. In this section it is assumed all L groups are independent. As noted in Chapter 10, this is often referred to as a between-by-between design.

By definition, a *linear contrast* is any linear combination of the means having the form

$$\Psi = \sum_{\ell=1}^{L} c_\ell \mu_\ell, \tag{12.7}$$

where $c_1, \ldots, c_L$, called *contrast coefficients*, are constants that sum to 0. (Ψ is an upper case Greek psi.) In symbols, Ψ is a linear contrast if

$$\sum c_\ell = 0.$$

EXAMPLE

Section 10.1 described a 2-by-2 design dealing with weight gain in rats. The two factors were source of protein (beef versus cereal) and amount of protein (low versus high). Here the population means are represented as follows:

		Source	
		Beef	Cereal
Amount	Low	μ_1	μ_2
	High	μ_3	μ_4

Consider the hypothesis of no main effect for the first factor, amount of protein. As explained in Chapter 10, the null hypothesis is

$$H_0 : \frac{\mu_1 + \mu_2}{2} = \frac{\mu_3 + \mu_4}{2}.$$

Rearranging terms in this last equation, the null hypothesis can be written as a linear contrast, namely,

$$H_0 : \mu_1 + \mu_2 - \mu_3 - \mu_4 = 0.$$

That is $\Psi = \mu_1 + \mu_2 - \mu_3 - \mu_4$, the contrast coefficients are $c_1 = c_2 = 1$, $c_3 = c_4 = -1$, and the null hypothesis is H_0: $\Psi = 0$.

EXAMPLE

Now consider a 3-by-2 design:

		Factor B	
		1	2
	1	μ_1	μ_2
Factor A	2	μ_3	μ_4
	3	μ_5	μ_6

An issue that is often of interest is not just whether there is a main effect for Factor A, but which levels of Factor A differ and by how much. That is, the goal is to compare Level 1 of Factor A to Level 2, Level 1 to Level 3, and Level 2 to Level 3. In symbols, the three hypotheses of interest are

$$H_0 : \frac{\mu_1 + \mu_2}{2} = \frac{\mu_3 + \mu_4}{2},$$

$$H_0 : \frac{\mu_1 + \mu_2}{2} = \frac{\mu_5 + \mu_6}{2},$$

and

$$H_0 : \frac{\mu_3 + \mu_4}{2} = \frac{\mu_5 + \mu_6}{2}.$$

In terms of linear contrasts, the goal is to test

$$H_0 : \Psi_1 = 0,$$

$$H_0 : \Psi_2 = 0,$$
$$H_0 : \Psi_3 = 0,$$

where

$$\Psi_1 = \mu_1 + \mu_2 - \mu_3 - \mu_4,$$
$$\Psi_2 = \mu_1 + \mu_2 - \mu_5 - \mu_6,$$

and

$$\Psi_3 = \mu_3 + \mu_4 - \mu_5 - \mu_6.$$

For Ψ_1, the linear contrast coefficients are $c_1 = c_2 = 1$, $c_3 = c_4 = -1$, and $c_5 = c_6 = 0$. For Ψ_2, the linear contrast coefficients are $c_1 = c_2 = 1$, $c_3 = c_4 = 0$, and $c_5 = c_6 = -1$.

As for main effects associated with Factor B, because there are only two levels, there is a single hypothesis to be tested, namely

$$H_0 : \mu_1 + \mu_3 + \mu_5 - \mu_2 - \mu_4 - \mu_6 = 0.$$

So the linear contrast is

$$\Psi_1 = \mu_1 + \mu_3 + \mu_5 - \mu_2 - \mu_4 - \mu_6$$

and the contrast coefficients are now $c_1 = c_3 = c_5 = 1$ and $c_2 = c_4 = c_6 = -1$.

The hypotheses of no interactions can be written as linear contrasts as well. For example, the hypothesis of no interaction for the first two rows corresponds to

$$H_0 : \mu_1 - \mu_2 = \mu_3 - \mu_4,$$

which is the same as testing

$$H_0 : \Psi_4 = 0,$$

where

$$\Psi_4 = \mu_1 - \mu_2 - \mu_3 + \mu_4.$$

So now the linear contrast coefficients are $c_1 = c_4 = 1$, $c_2 = c_3 = -1$, and $c_5 = c_6 = 0$. Similarly, for rows 1 and 3, the hypothesis of no interaction is

$$H_0 : \Psi_5 = 0,$$

where

$$\Psi_5 = \mu_1 - \mu_2 - \mu_5 + \mu_6.$$

In general, there are a collection of C linear contrasts that one might want to test, and the goal is to devise a method for performing these C tests in a manner that controls FWE. (An R function designed to generate contrast coefficients relevant to main effects and interactions is described in Section 12.2.5.)

12.2.1 Scheffé's Homoscedastic Method

Assuming normality and homoscedasticity, Scheffé's classic method can be used to test C hypotheses about C linear contrasts such that FWE is less than or equal to α—regardless of how large C might be. Let Ψ be any specific linear contrast and let

$$\hat{\Psi} = \sum_{\ell=1}^{L} c_\ell \bar{X}_\ell. \tag{12.8}$$

That is, estimate the population mean μ_ℓ of the ℓth group with $\bar{X}_\ell$, the sample mean of the ℓth group, and then plug this estimate into Equation (12.7) to get an estimate of Ψ. The $1-\alpha$ confidence interval for Ψ is

$$(\hat{\Psi} - S, \hat{\Psi} + S), \tag{12.9}$$

where

$$S = \sqrt{(L-1)f_{1-\alpha}\text{MSWG}\sum \frac{c_\ell^2}{n_\ell}},$$

MSWG is the mean square within groups (described in Chapter 9) that estimates the assumed common variance, and $f_{1-\alpha}$ is the $1-\alpha$ quantile of an F distribution with $\nu_1 = L-1$ and $\nu_2 = N - L$ degrees of freedom, where $N = \sum n_j$ is the total number of observations in all L groups. (For a one-way design with J levels, $L = J$, and for a J-by-K design, $L = JK$.) In particular, H_0: $\Psi = 0$ is rejected if the confidence interval given by Equation (12.9) does not contain 0. Alternatively, reject H_0: $\Psi = 0$ if

$$\frac{|\hat{\Psi}|}{\sqrt{(L-1)(\text{MSWG})\sum \frac{c_\ell^2}{n_\ell}}} \geq \sqrt{f_{1-\alpha}}.$$

EXAMPLE

For the special case where all pairwise comparisons of J independent groups are to be performed, Scheffé's confidence interval for the difference between the means of the jth and kth groups, $\mu_j - \mu_k$, is

$$(\bar{X}_j - \bar{X}_k) \pm S,$$

where

$$S = \sqrt{(J-1)f_{1-\alpha}(\text{MSWG})\left(\frac{1}{n_j} + \frac{1}{n_k}\right)},$$

and $f_{1-\alpha}$ is the ANOVA F critical value based on $\nu_1 = J - 1$ and $\nu_2 = N - J$ degrees of freedom.

There is an interesting connection between Scheffé's multiple comparison procedure and the ANOVA F test in Chapter 9. Imagine that rather than perform all pairwise comparisons, the goal is to test the hypothesis that all linear contrasts among J independent groups are equal to 0. Assuming normality and homoscedasticity, the ANOVA F test in Section 9.1 accomplishes this goal. So a possible argument for using the ANOVA F test, as opposed to ignoring it and simply applying a multiple comparison procedure aimed at all pairwise comparisons of means, is that it tests a broader range of hypotheses that includes all pairwise comparisons as a special case. If the F test rejects, this means that one or more linear contrasts differ significantly from 0.

Scheffé's method is one of the best-known multiple comparison procedures, but it suffers from the same problems associated with other homoscedastic methods already covered. Even under normality, if there is heteroscedasticity, problems arise in terms of Type I errors (Kaiser and Bowden, 1983). Also, under normality and homoscedasticity, the Tukey–Kramer method will yield shorter confidence intervals than Scheffé's method when the goal is to perform all pairwise comparisons. But for a sufficiently large collection of linear contrasts, the reverse is true (Scheffé, 1959, p. 76). It can be applied via the R function `Scheffe`, but more modern methods are recommended.

12.2.2 Heteroscedastic Methods

Two heteroscedastic methods for linear contrasts are covered in this section.

Welch–Šidák Method

The first contains Dunnett's T3 method as a special case and is called the Welch–Šidák method. (The computations are performed by the R function `lincon` described in Section 12.1.9.) Again let L represent the total number of groups, let Ψ be any linear contrast, and let C represent the total number of contrasts to be tested. An expression for the squared standard error of $\hat{\Psi}$ is

$$\sigma^2_{\hat{\Psi}} = \sum \frac{c^2_\ell \sigma^2_\ell}{n_\ell},$$

where σ^2_ℓ and n_ℓ are the variance and sample size of the ℓth group. An estimate of this quantity is obtained simply by replacing σ^2_ℓ with s^2_ℓ, the sample variance associated with the ℓth group. That is, estimate $\sigma^2_{\hat{\Psi}}$ with

$$\hat{\sigma}^2_{\hat{\Psi}} = \sum \frac{c^2_\ell s^2_\ell}{n_\ell}.$$

Let

$$q_\ell = \frac{c^2_\ell s^2_\ell}{n_\ell}.$$

The degrees of freedom are estimated to be

$$\hat{\nu} = \frac{(\sum q_\ell)^2}{\sum \frac{q^2_\ell}{n_\ell - 1}}.$$

The test statistic is

$$T = \frac{\hat{\Psi}}{\hat{\sigma}_{\hat{\Psi}}}.$$

The critical value, c, is a function of $\hat{\nu}$ and C (the total number of hypotheses you plan to perform) and is read from Table 10 in Appendix B. Reject if $|T| \geq c$, and a confidence interval for Ψ is

$$\hat{\Psi} \pm c\hat{\sigma}_{\hat{\Psi}}. \tag{12.10}$$

Kaiser–Bowden Method

A heteroscedastic analog of Scheffé's method was derived by Kaiser and Bowden (1983). In principle, if enough linear contrasts are tested, it will have shorter confidence intervals than those based on the Welch–Šidák method. But at what point this is the case is unclear and it appears that the number of hypotheses to be tested must be quite large before the Kaiser–Bowden method offers a practical advantage. For brevity, the computational details of the Kaiser–Bowden are omitted, but an R function for applying it is provided.

EXAMPLE

Consider a 2-by-2 design and suppose the sample means are $\bar{X}_1 = 10$, $\bar{X}_2 = 14$, $\bar{X}_3 = 18$, $\bar{X}_4 = 12$; the sample variances are $s^2_1 = 20$, $s^2_2 = 8$, $s^2_3 = 12$, $s^2_4 = 4$; and the sample sizes are $n_1 = n_2 = n_3 = n_4 = 4$. Further assume the goal is to test three hypotheses with the

Welch–Šidák method: no main effects for Factor A, no main effects for Factor B, and no interaction. In terms of linear contrasts, the goal is to test

$$H_0 : \Psi_1 = 0,$$

$$H_0 : \Psi_2 = 0,$$

$$H_0 : \Psi_3 = 0,$$

where

$$\Psi_1 = \mu_1 + \mu_2 - \mu_3 - \mu_4,$$

$$\Psi_2 = \mu_1 + \mu_3 - \mu_2 - \mu_4,$$

$$\Psi_3 = \mu_1 - \mu_2 - \mu_3 + \mu_4.$$

Moreover, the probability of at least one Type I error is to be 0.05 among these three tests. For the first hypothesis, the estimate of Ψ_1 is

$$\hat{\Psi}_1 = 10 + 14 - 18 - 12 = -6.$$

The estimate of the squared standard error is

$$\begin{aligned}\hat{\sigma}^2_{\hat{\Psi}_1} &= \frac{1^2(20)}{4} + \frac{1^2(8)}{4} + \frac{(-1)^2(12)}{4} + \frac{(-1)^2(4)}{4} \\ &= 11,\end{aligned}$$

the degrees of freedom for the Welch–Šidák method can be seen to be 9.3, and with $C = 3$, the critical value is approximately $c = 2.87$. The test statistic is

$$T = \frac{-6}{\sqrt{11}} = -1.8,$$

and because $|T| < 2.87$, fail to reject. The other two hypotheses can be performed in a similar manner, but the details are left as an exercise. If instead the Kaiser–Bowden method is used, it can be seen that the length of the confidence intervals are considerably longer than those based on the Welch–Šidák method. Again, when performing a large number of tests, the Kaiser–Bowden method might offer a practical advantage, but clear guidelines are not available.

EXAMPLE

Chapter 10 noted that when investigating the possibility of a disordinal interaction, it can be necessary to compare pairs of groups associated with the interaction of interest. For instance, in the previous example we have that $\bar{X}_1 < \bar{X}_2$ but that $\bar{X}_3 > \bar{X}_4$. Establishing that there is a disordinal interaction requires rejecting both $H_0 : \mu_1 = \mu_2$ and $H_0 : \mu_3 = \mu_4$. One way of controlling FWE when testing these two hypotheses is with the Welch–Šidák method.

12.2.3 Extension of Welch–Šidák and Kaiser–Bowden Methods to Trimmed Means

The Welch–Šidák and Kaiser–Bowden methods are readily extended to trimmed means. The focus here is on a generalization of the Welch–Šidák method, which can be applied via the R function `lincon` in Section 12.1.9. Details about a generalization of the Kaiser–Bowden are described in Wilcox (2017).

First consider the goal of computing a confidence interval for the difference between the trimmed means for each pair of groups. That is, compute a confidence interval $\mu_{tj} - \mu_{tk}$, for all $j < k$, such that the simultaneous probability coverage is approximately $1 - \alpha$. The confidence interval for $\mu_{tj} - \mu_{tk}$ is

$$(\bar{X}_{tj} - \bar{X}_{tk}) \pm c\sqrt{d_j + d_k},$$

where

$$d_j = \frac{(n_j - 1)s_{wj}^2}{h_j(h_j - 1)},$$

h_j is the number of observations left in group j after trimming, and c is read from Table 10 in Appendix B with

$$\hat{\nu} = \frac{(d_j + d_k)^2}{\frac{d_j^2}{h_j - 1} + \frac{d_k^2}{h_k - 1}}$$

degrees of freedom. As for linear contrasts, let

$$\Psi = \sum_{\ell=1}^{L} c_\ell \mu_{t\ell}.$$

It is assumed that there are a total of C such linear contrasts and for each the goal is to test $H_0 : \Psi = 0$ with FWE equal to α. Compute

$$\hat{\Psi} = \sum_{\ell=1}^{L} c_\ell \bar{X}_{t\ell},$$

$$d_\ell = \frac{c_\ell^2 (n_\ell - 1) s_{w\ell}^2}{h_\ell(h_\ell - 1)},$$

where $s_{w\ell}^2$ and h_ℓ are the Winsorized variance and effective sample size (the number of observations left after trimming) of the ℓth group, respectively. An estimate of the squared standard error of $\hat{\Psi}$ is

$$S_e = \sum d_\ell.$$

Letting

$$G = \sum \frac{d_\ell^2}{h_\ell - 1},$$

the estimated degrees of freedom are

$$\hat{\nu} = \frac{S_e^2}{G}.$$

Then a confidence interval for Ψ, based on a trimmed analog of the Welch–Šidák method, is

$$\hat{\Psi} \pm c\sqrt{S_e},$$

where c is read from Table 10 in Appendix B (and is again a function of C and $\hat{\nu}$).

12.2.4 R Function kbcon

The R function

```
Scheffe(x,con=0,alpha=0.05)
```

performs Scheffé's method and the R function

```
kbcon(x,con=0,tr=0.2,alpha=0.05)
```

applies the Kaiser–Bowden method. By default these functions perform all pairwise comparisons. As usual, the argument `x` is any R variable that contains the data, which can have list mode or it can be a matrix with J columns.

12.2.5 R Functions con2way and conCON

The R functions `lincon` and `linconbt` introduced in Sections 12.1.9 and 12.1.14, respectively, as well as `kbcon`, contain an argument `con` that can be used to specify the linear contrast coefficients relevant to the hypotheses of interest. To help analyze a J-by-K design, the R function

```
con2way(J,K)
```

is supplied, which generates the linear contrast coefficients needed to compare all main effects and interactions. The contrast coefficients for main effects associated with Factor A and Factor B are returned in `conA` and `conB`, respectively, which are matrices. The columns of the matrices correspond to the various hypotheses to be tested. The rows contain the contrast coefficients corresponding to the groups. Contrast coefficients for all interactions are returned in `conAB`. When using `con2way` in conjunction with the R function `lincon`, it is assumed that the groups are arranged as in Section 10.1.2. The following example provides an illustration.

EXAMPLE

Consider again a 3-by-2 design where the means are arranged as follows:

		Factor B	
		1	2
	1	μ_1	μ_2
Factor A	2	μ_3	μ_4
	3	μ_5	μ_6

The R command

```
con2way(3,2)
```

returns

```
$conA
     [,1] [,2] [,3]
[1,]    1    1    0
[2,]    1    1    0
[3,]   -1    0    1
[4,]   -1    0    1
[5,]    0   -1   -1
[6,]    0   -1   -1
```

```
$conB
     [,1]
[1,]    1
[2,]   -1
[3,]    1
[4,]   -1
[5,]    1
[6,]   -1

$conAB
     [,1] [,2] [,3]
[1,]    1    1    0
[2,]   -1   -1    0
[3,]   -1    0    1
[4,]    1    0   -1
[5,]    0   -1   -1
[6,]    0    1    1
```

Each matrix has 6 rows because there are a total of 6 groups. The first column of the matrix `conA` contains 1, 1, −1, −1, 0, 0, which are the contrast coefficients for comparing the first two levels of Factor A, as explained in the second example at the beginning of Section 12.2. The second column contains the contrast coefficients for comparing Levels 1 and 3. The final column contains the coefficients for comparing Levels 2 and 3. In a similar manner, `conB` contains the contrast coefficients for the two levels of Factor B and `conAB` contains the contrast coefficients for all interactions.

EXAMPLE

Chapter 10 analyzed the plasma retinol data, stored in R, based on two factors: sex and smoking status. Here the goal is to perform all multiple comparisons based on 20% trimmed means. As previously indicated, first use the R command

```
z=fac2list(plasma[,10],plasma[,2:3])
```

to convert the data into list mode and store it in the R variable `z`. We have a 2-by-3 design, and so the command

```
w=con2way(2,3)
```

creates the contrast coefficients and stores them in `w`. Then the R command

```
lincon(z,con=w$conA)
```

will compare the two levels of Factor A (male and female). The command

```
lincon(z,con=w$conB)
```

will perform all pairwise comparisons based on the three levels of Factor B (smoking status) and

```
lincon(z,con=w$conAB)
```

will perform all interactions associated with any two rows and columns.

While the types of comparisons just illustrated are often of interest, others arise. But now it is necessary to specify the relevant contrast coefficients.

EXAMPLE

A classic problem is comparing treatment groups to a control group. That is, rather than perform all pairwise comparisons, the goal is to compare group 1 to group J, group 2 to group J, and so on. In symbols, the goal is to test

$$H_0\colon \mu_{t1} = \mu_{tJ},$$

$$H_0\colon \mu_{t2} = \mu_{tJ},$$

$$\vdots$$

$$H_0\colon \mu_{t,J-1} = \mu_{tJ}$$

with FWE equal to α. For illustrative purposes, assume there are four groups and the goal is to compare the first three groups to the control group (group 4) with the R function `lincon`. Store the matrix

$$\begin{pmatrix} 1 & 0 & 0 \\ 0 & 1 & 0 \\ 0 & 0 & 1 \\ -1 & -1 & -1 \end{pmatrix}$$

in some R variable, say `mmat`. This can be done with the R function

```
conCON(J,conG=1),
```

where the argument `conG` indicates which of the J groups is the control group. For the situation at hand, the command

```
nmat=conCON(4,conG=4)
```

would store the linear contrast coefficients in `nmat`. Then the command

```
lincon(x,con=mmat)
```

will perform the relevant comparisons. For the schizophrenia data in Table 10.1, the results are

```
$test:
     con.num      test     crit         se       df
           1 -2.221499 2.919435 0.08703807 8.550559
           2 -1.582364 2.997452 0.10097658 7.593710
           3 -4.062500 2.830007 0.06688001 9.999866

$psihat:
     con.num     psihat   ci.lower   ci.upper
           1 -0.1933550 -0.4474570  0.0607470
           2 -0.1597817 -0.4624542  0.1428908
           3 -0.2717000 -0.4609709 -0.0824291
```

So again the conclusion is that groups 3 and 4 differ and that group 3 has a smaller population trimmed mean.

12.2.6 Linear Contrasts Based on Medians

The method for testing linear contrasts based on trimmed means should not be used for the special case where the goal is to compare medians. But a modification of the method for trimmed means appears to perform relatively well when the goal is to compare medians and there are no tied values. Again let L indicate the number of groups to be compared and let M_ℓ represent the sample median of the ℓth group. Let $\theta_1, \ldots, \theta_L$ be the population medians, and now let

$$\Psi = \sum_{\ell=1}^{L} c_\ell \theta_\ell$$

be some linear contrast of interest. Compute

$$\hat{\Psi} = \sum_{\ell=1}^{L} c_\ell M_\ell,$$

and let

$$S_e^2 = \sum c_\ell^2 S_\ell^2,$$

where S_ℓ^2 is the McKean–Schrader estimate of the squared standard error of M_ℓ. Then an approximate $1-\alpha$ confidence interval for Ψ is

$$\hat{\Psi} \pm c S_e,$$

where c is read from Table 10 in Appendix B with degrees of freedom $\nu = \infty$ and where C is again the number of hypotheses to be tested. As usual, this method is designed so that FWE will be approximately α.

12.2.7 R Functions msmed and mcp2med

The R function

```
msmed(x, y = NA, con = 0, alpha = 0.05)
```

can be used just like the function `lincon` to test hypotheses about linear contrasts based on the median. (When doing so, ignore the second argument y.) So if data are stored in a matrix called `mat` having six columns corresponding to six groups,

```
msmed(mat)
```

will perform all pairwise comparisons using medians. (If only two groups are being compared, the data for the second group can be stored in the second argument, y.)

For convenience, the R function

```
mcp2med(J, K, x, con = 0, alpha = 0.05, grp = NA, op = F)
```

is supplied, which by default performs all (pairwise) multiple comparisons among main effects, and it tests the hypothesis of no interaction for all relevant 2-by-2 cells. The function calls `con2way` to create the contrast coefficients and then it uses `msmed` to test the corresponding hypotheses.

If, however, there are tied values, use the R function `medpb` in Section 12.1.12. It contains an argument `con` that can be used to specify the linear contrasts of interest. In fact, using `medpb` is a reasonable choice even when there are no tied values.

12.2.8 Bootstrap Methods

The bootstrap-t method is readily extended to situations where the goal is to test hypotheses about C linear contrasts. Again let $\Psi_1, \ldots, \Psi_C$ indicate the linear contrasts of interest, which includes all pairwise comparisons of the groups as a special case. A bootstrap-t method for testing

$$H_0 : \Psi = 0$$

for each of C linear contrasts, such that FWE is α, is applied as follows:

1. For each of the L groups, generate a bootstrap sample, $X^*_{i\ell}$, $i = 1, \ldots, n_\ell$; $\ell = 1, \ldots, L$. For each of the L bootstrap samples, compute the trimmed mean, $\bar{X}^*_{t\ell}$,

$$\hat{\Psi}^* = \sum c_\ell \bar{X}^*_{t\ell},$$

$$d^*_\ell = \frac{c^2_\ell (n_\ell - 1)(s^*_{w\ell})^2}{h_\ell (h_\ell - 1)},$$

   where $(s^*_{w\ell})^2$ is the Winsorized variance based on the bootstrap sample taken from the ℓth group, and h_ℓ is the effective sample size of the ℓth group (the number of observations left after trimming).

2. Compute

$$T^* = \frac{|\hat{\Psi}^* - \hat{\Psi}|}{\sqrt{A^*}},$$

   where $\hat{\Psi} = \sum c_\ell \bar{X}_{t\ell}$ and $A^* = \sum d^*_\ell$. The results for each of the C linear contrasts are labeled $T^*_1, \ldots, T^*_C$.

3. Let

$$T^*_q = \max\{T^*_1, \ldots, T^*_C\}.$$

   In words, T^*_q is the maximum of the C values $T^*_1, \ldots, T^*_C$.

4. Repeat steps 1 through 3 B times yielding T^*_{qb}, $b = 1, \ldots, B$.

Let $T^*_{q(1)} \leq \cdots \leq T^*_{q(B)}$ be the T^*_{qb} values written in ascending order, and let $u = (1 - \alpha)B$, rounded to the nearest integer. Then the confidence interval for Ψ is

$$\hat{\Psi} \pm T^*_{q(u)} \sqrt{A},$$

where $A = \sum d_\ell$,

$$d_\ell = \frac{c^2_\ell (n_\ell - 1) s^2_{w\ell}}{h_\ell (h_\ell - 1)},$$

and the simultaneous probability coverage is approximately $1 - \alpha$. This method can be applied with the R function `linconbt` described in Section 12.1.14; the linear contrast coefficients are specified via the argument `con`.

As for the percentile bootstrap method, simply proceed along the lines in Section 12.1.11. Main effects and interactions can be tested using appropriately chosen linear contrasts, as previously indicated. Some R functions designed specifically for a two-way design are described in the next section.

12.2.9 R Functions mcp2a, bbmcppb, bbmcp

For convenience, when dealing with a two-way ANOVA design, the R function

```
mcp2a(J,K,x,est=onestep,con=0,alpha=0.05,nboot=NA,grp=NA,...)
```

is supplied that performs all pairwise multiple comparisons among the rows and columns, and it tests all linear contrasts relevant to interactions. Method SR is used to control the probability of at least one Type I error if the largest sample size is less than 80. Otherwise, Rom's method is used. By default, `mcp2a` uses the one-step M-estimator of location.

The R function

```
bbmcppb(J, K, x, est=tmean, JK = J * K, alpha = 0.05, grp = c(1:JK), nboot =
                    500, bhop = FALSE, SEED = TRUE)
```

performs multiple comparisons for a two-way ANOVA design based on trimmed means in conjunction with a percentile bootstrap method. It uses the linear contrasts for main effects and interactions as previously described. Rom's method is used to control the probability of one or more Type I errors. For $C > 10$ hypotheses, or when the goal is to test at some level other than 0.05 and 0.01, Hochberg's method is used. Setting the argument `bhop=T`, the Benjamini–Hochberg method is used instead. The R function

```
bbmcp(J, K, x, tr = 0.2, alpha = 0.05, grp = NA, op = FALSE)
```

is like the R function `bbmcppb`, only it is based on the method in Section 12.2.3.

12.2.10 The Patel–Hoel Rank-Based Interaction Method

Section 11.6.2 described the Patel–Hoel approach to interactions when dealing with a 2-by-2 design. For a J-by-K design, imagine that all possible interactions are to be tested using the Patel–Hoel method with the goal that the probability of at least one Type I error is α. A strategy for achieving this goal is to read a critical value from Table 10 in Appendix B with $\nu = \infty$ degrees of freedom, noting that the total number of hypotheses to be tested is

$$C = \frac{J^2 - J}{2} \times \frac{K^2 - K}{2}.$$

12.2.11 R Function rimul

The R function

```
rimul(J,K,x,alpha=0.05,p=J*K,grp=c(1:p))
```

tests all hypotheses relevant to interactions using the Patel–Hoel approach with the goal that FWE be equal to α.

12.3 JUDGING SAMPLE SIZES

As noted in Chapters 7 and 8, failing to reject some hypothesis might be because there is little or no difference between the groups, or there might be an important difference that was missed due to low power. So issues of general importance are determining whether the sample size was sufficiently large to ensure adequate power, how large the sample size should have been if power is judged to be inadequate, and the related problem of controlling the length of confidence intervals. This section describes two methods for accomplishing the last of these goals when working with linear contrasts.

12.3.1 Tamhane's Procedure

The first of the two methods is due to Tamhane (1977). A general form of his results can be used to deal with linear contrasts, but here attention is restricted to all pairwise comparisons among J independent groups. The goal is to compute confidence intervals having simultaneous probability coverage equal to $1-\alpha$, such that the length of the confidence intervals have some specified value, $2m$, given s_j^2, the sample variance from the jth group based on n_j observations ($j=1,\ldots,J$). The constant m is called the *margin of error* and reflects the desired level of accuracy when estimating a linear contrast. The computational details are as follows:

Compute

$$A=\sum\frac{1}{n_j-1}$$

$$\nu=\left[\frac{J}{A}\right],$$

where the notation $[J/A]$ means you compute J/A and round down to the nearest integer. Next, determine h from Table 11 in Appendix B with ν degrees of freedom. (Table 11 gives the quantiles of the range of independent Student T variates.) Note that Table 11 assumes $\nu\le 59$. For larger degrees of freedom, use Table 9 in Appendix B instead. Let

$$d=\left(\frac{m}{h}\right)^2.$$

Letting s_j^2 be the sample variance for the jth group, the total number of observations required from the jth group is

$$N_j=\max\{n_j+1,\left[\frac{s_j^2}{d}\right]+1\}.$$

Once the additional observations are available, compute the generalized sample mean, $\tilde{X}_j$, for the jth group as described and illustrated in Box 9.4. The confidence interval for $\mu_j-\mu_k$ is

$$(\tilde{X}_j-\tilde{X}_k-m,\tilde{X}_j-\tilde{X}_k+m).$$

EXAMPLE

Assume that for three groups, $n_1=11$, $n_2=21$, $n_3=41$, and the goal is to compute confidence intervals having length 4 and having simultaneous probability coverage $1-\alpha=0.95$. So,

$$A=\frac{1}{10}+\frac{1}{20}+\frac{1}{40}=0.175,$$

$$\nu=\left[\frac{3}{0.175}\right]=[17.14]=17,$$

and $h=3.6$. Confidence intervals with length 4 means that $m=2$, so

$$d=\left(\frac{2}{3.6}\right)^2=0.3086.$$

If the sample variance for the first group is $s_1^2=2$,

$$\begin{aligned}N_1&=\max\{11+1,\left[\frac{2}{0.3086}\right]+1\}\\&=\max(12,[6.4]+1)\\&=12.\end{aligned}$$

Because $n_1 = 11$ observations are already available, $12 - 11 = 1$, one more observation is required. If you get $\tilde{X}_1 = 14$ and $\tilde{X}_2 = 17$, the confidence interval for the difference between the means is

$$(14 - 17 - 2, 14 - 17 + 2) = (-5, \ -1).$$

12.3.2 R Function tamhane

Tamhane's two-stage multiple comparison procedure can be applied with the R function

```
tamhane(x, x2 = NA, cil = NA, crit = NA).
```

The first stage data are stored in x (in a matrix or in list mode) and the second stage data in x2. If x2 contains no data, the function prints the degrees of freedom, which can be used to determine h as described in the previous section. Once h has been determined, store it in the argument crit. If no critical value is specified, the function terminates with an error message; otherwise it prints the total number of observations needed to achieve confidence intervals having length cil (given by the third argument). (So in the notation of Section 12.3.1, the value of cil divided by 2 corresponds to m.) If data are stored in the argument x2, confidence intervals are computed as described in the previous section and returned by the function tamhane in the R variable ci.mat; otherwise, the function returns ci.mat with the value NA.

EXAMPLE

For the data in Table 10.1, if the goal is to perform all pairwise comparisons such that the confidence intervals have length 1 (so $m = 0.5$) and FWE is to be 0.05, then the degrees of freedom are 9 and from Table 11 in Appendix B, $h = 4.3$. If the data are stored in the R variable skin, the command

```
tamhane(skin,cil=1,crit=4.3)
```

returns

```
$n.vec:
[1] 13 11 11 11
```

indicating that the required sample sizes are 13, 11, 11, and 11, respectively.

12.3.3 Hochberg's Procedure

This section describes another two-stage procedure that controls the length of the confidence interval, which was derived by Hochberg (1975). In contrast to Tamhane's procedure, Hochberg's method allows the possibility of no additional observations being required in the second stage, and it has been generalized to situations where the goal is to compare trimmed means. Hochberg's method ensures that the lengths of the confidence intervals are at most $2m$, in contrast to Tamhane's methods for which the confidence intervals have lengths exactly equal to $2m$. (As before, m is some constant chosen by the investigator.) A summary of the computations for the general case where C linear contrasts are to be tested is described as follows. Here, n_j represents the sample size for the jth group in the first stage, and s_{jw}^2 is the sample Winsorized variance for the jth group based on these n_j observations. (Hochberg's original derivation assumed equal sample sizes in the first stage, but the adjustment for unequal sample sizes used here appears to perform well.)

As in Tamhane's procedure, read the critical value h from Table 11 in Appendix B with the degrees of freedom

$$\nu = J\left(\sum \frac{1}{h_j - 1}\right)^{-1},$$

where h_j is the number of observations in the jth group left after trimming. If $\nu > 59$, use Table 9 in Appendix B instead. Compute

$$d = \left(\frac{m}{h}\right)^2.$$

The total number of observations needed from the jth group is

$$N_j = \max(n_j, \left[\frac{s_{jw}^2}{(1-2\gamma)^2 d}\right] + 1). \tag{12.11}$$

So if $N_j - n_j$ is large, and if we fail to reject any hypothesis involving μ_{tj}, it seems unreasonable to accept the null hypothesis. Or in the context of Tukey's three-decision rule, make no decision about which group has the largest population trimmed mean.

If the additional $N_j - n_j$ observations can be obtained, confidence intervals are computed as follows. Let $\bar{X}_{jt}$ be the trimmed mean associated with the jth group based on all N_j values. For all pairwise comparisons, the confidence interval for $\mu_{jt} - \mu_{kt}$, the difference between the population trimmed means corresponding to groups j and k, is

$$(\bar{X}_{jt} - \bar{X}_{kt}) \pm hb,$$

where

$$b = \max\left(\frac{s_{jw}}{(1-2\gamma)\sqrt{N_j}}, \frac{s_{kw}}{(1-2\gamma)\sqrt{N_k}}\right).$$

As for the linear contrast $\Psi = \sum c_j \mu_j$, sum the positive c_j values and label the result a. N_j is computed as before, but now

$$d = \left(\frac{m}{ha}\right)^2.$$

Let

$$b_j = \frac{c_j s_{jw}}{\sqrt{N_j}}.$$

Let A be the sum of the positive b_j values, C be the sum of the negative b_j values, and compute

$$D = \max(A, -C),$$

$$\hat{\Psi} = \sum c_j \bar{X}_{jt},$$

in which case the confidence interval for Ψ is

$$\hat{\Psi} \pm hD.$$

EXAMPLE

To illustrate Hochberg's method, imagine that with four groups, each having a sample size of $n = 25$, the goal is to compute a confidence interval for $H_0 : \Psi = 0$ with $\alpha = 0.05$, where

$$\Psi = \mu_1 + \mu_2 - \mu_3 - \mu_4,$$

and the length of the confidence interval is to be at most $2m = 8$. So $\nu = 24$ and from Table 11 in Appendix B, $h = 3.85$. We see that the sum of the positive contrast coefficients is $a = 2$, so

$$d = \left(\frac{4}{3.85(2)}\right)^2 = 0.2699.$$

If $s_1^2 = 4$, $s_2^2 = 12$, $s_3^2 = 16$, and $s_4^2 = 20$, then $N_1 = \max(25,\ [4/0.2699] + 1) = 25$. Hence, no additional observations are required. The sample sizes for the other three groups are $N_2 = 45$, $N_3 = 60$, and $N_4 = 75$.

Once the additional observations are available, compute

$$b_1 = \frac{1 \times 2}{\sqrt{25}} = 0.4.$$

Similarly, $b_2 = 0.5164$, $b_3 = -0.5164$, and $b_4 = -0.5164$. The sum of the positive b_j values is $A = b_1 + b_2 = 0.4 + 0.5164 = 0.9164$. The sum of the negative b_j values is $C = -1.0328$, so $-C = 1.0328$,

$$D = \max(0.9164, 1.0328) = 1.0328$$

$$hD = 3.85 \times 1.0328 = 3.976.$$

If after the additional observations are sampled, the sample means are $\bar{X}_1 = 10$, $\bar{X}_2 = 12$, $\bar{X}_3 = 8$, and $\bar{X}_4 = 18$, then

$$\hat{\Psi} = 10 + 12 - 8 - 18 = -4,$$

and the confidence interval is

$$-4 \pm 3.976 = (-7.976,\ -0.024),$$

so reject H_0.

12.3.4 R Function hochberg

The R function

```
hochberg(x, x2 = NA, cil = NA crit = NA, con = 0, tr = 0.2, alpha = 0.05,
                              iter = 10000, SEED = T )
```

performs Hochberg's two-stage procedure. The first four arguments are used in the same manner as in Section 12.3.2. The argument `con` can be used to specify linear contrasts of interest. If `con=0` is used, all pairwise comparisons are performed. Otherwise, `con` is a J-by-C matrix where each column of `con` contains the contrast coefficients for the C tests to be performed. The default amount of trimming is 0.2. If no value for the argument `crit` is specified, an appropriate critical value is determined based on a simulation where the number of replications used is determined by the argument `iter`.

12.4 METHODS FOR DEPENDENT GROUPS

Multiple comparison methods for independent groups typically take advantage of the independence among the groups in some manner. Examples are the Tukey–Kramer method, Scheffé's method, as well as Dunnett's T3 and the Games–Howell methods, which require that the groups being compared are independent. When comparing dependent groups instead, generally some modification of the methods for independent groups must be used. The main point here is that the methods in Section 12.1.10 can be used when comparing dependent groups. Briefly, consider any method in Chapter 8 aimed at comparing two dependent groups. Further assume that the method controls the probability of a Type I error and imagine that we compute p-values for each hypothesis that is tested. Then the probability of at least one Type I error is controlled (it will not exceed the nominal level) when testing multiple hypotheses if we use Rom's method, or Hochberg's method or the Bonferroni method, which are described in Section 12.1.10.

EXAMPLE

Consider J dependent groups and imagine that all pairwise comparisons are to be performed based on difference scores. Then the number of hypotheses to be tested is $C = (J^2 - J)/2$. So if for each pair of groups, the hypothesis that the trimmed mean of the difference scores is equal to zero is tested as described in Section 9.2, and if the goal is to have FWE less than or equal to 0.05, the Bonferroni method performs each test at the $0.05/C$ level. Alternatively, compute a p-value for each test that is performed. Then Rom's method can be applied and will guarantee that the probability of at least one Type I error is at most α. The same is true when using Hochberg's methods. These latter two methods will have power greater than or equal to the power of the Bonferroni method.

12.4.1 Linear Contrasts Based on Trimmed Means

A collection of linear contrasts can be tested when working with trimmed means corresponding to dependent groups. First consider a single linear contrast based on the marginal trimmed means of L groups:

$$\Psi = \sum_{\ell=1}^{L} c_\ell \mu_{t\ell}.$$

Then H_0: $\Psi = 0$ can be tested as outlined below. Alternatively, a generalization of difference scores can be used instead. That is, set

$$D_i = \sum_{\ell=1}^{L} c_\ell X_{i\ell}$$

and test the hypothesis that the population trimmed mean of the D_i values is 0, which can be done with the method in Section 5.6 and the R function `trimci`. When testing C such hypotheses, FWE can be controlled with Rom's method.

How to test a linear contrast based on the marginal trimmed means of dependent groups.

Goal: Test H_0: $\Psi = 0$, for each of C linear contrasts such that FWE is α. (There are L groups.)

Let Y_{ij} ($i = 1, \ldots, n$; $j = 1, \ldots, L$) be the Winsorized values, which are computed as described in Box 8.1. Let

$$A = \sum_{j=1}^{L} \sum_{\ell=1}^{L} c_j c_\ell d_{j\ell},$$

where

$$d_{j\ell} = \frac{1}{h(h-1)} \sum_{i=1}^{n} (Y_{ij} - \bar{Y}_j)(Y_{i\ell} - \bar{Y}_\ell),$$

and h is the number of observations left in each group after trimming. Let

$$\hat{\Psi} = \sum_{\ell=1}^{L} c_\ell \bar{X}_{t\ell}.$$

Test statistic:

$$T = \frac{\hat{\Psi}}{\sqrt{A}}.$$

Decision Rule: Reject if $|T| \geq t$, where t is the $1-\alpha/2$ quantile of a Student's T distribution with $\nu = h - 1$ degrees of freedom. When testing more than one linear contrast, FWE can be controlled with Rom's method.

12.4.2 R Function rmmcp

The R function

```
rmmcp(x,con = 0, tr = 0.2, alpha = 0.05, dif=TRUE)
```

performs multiple comparisons among dependent groups using trimmed means and Rom's method for controlling FWE. By default, difference scores are used and all pairwise comparisons are performed. Setting `dif=FALSE` results in comparing marginal trimmed means. Linear contrasts can be tested via the argument `con`. (When α differs from both 0.05 and 0.01, FWE is controlled with Hochberg's method as described in Section 12.1.10.)

12.4.3 Comparing M-estimators

When the goal is to compare dependent groups via M-estimators, a simple generalization of the percentile bootstrap method in Section 8.2.2 can be used. In the present situation, generate bootstrap samples by randomly sampling, with replacement, n rows of data from the matrix

$$\begin{pmatrix} X_{11}, \ldots, X_{1J} \\ \vdots \\ X_{n1}, \ldots, X_{nJ} \end{pmatrix}$$

yielding

$$\begin{pmatrix} X^*_{11}, \ldots, X^*_{1J} \\ \vdots \\ X^*_{n1}, \ldots, X^*_{nJ} \end{pmatrix}.$$

An M-estimator of location is estimating some unknown measure of location, θ . First consider the goal of comparing all pairs of groups by testing

$$H_0 : \theta_j = \theta_k \qquad (12.12)$$

for all $j < k$. As usual, the goal is to have FWE equal to α. For the two groups being compared, compute a p-value. That is, compute the proportion of bootstrap values from the first group that are greater than the bootstrap values from the second group, and label the result $\hat{p}^*$. The p-value is

$$2\min(\hat{p}^*, 1 - \hat{p}^*).$$

One strategy for controlling FWE is to use Rom's method. But when dealing with M-estimators, if the sample sizes are small, this approach is too conservative in terms of Type

I errors. That is, the actual probability of a Type I error tends to be substantially less than the nominal level with small to moderate sample sizes. A method for correcting this problem is to use the bias correction method in Section 8.2.2 and use Rom's method to control the probability of at least one Type I error.

Difference scores can be analyzed as well using a simple extension of the percentile bootstrap method in Chapter 8. Here, there is the issue of controlling FWE, which can be done with the methods in Section 12.1.10.

12.4.4 R Functions rmmcppb, dmedpb, dtrimpb, and boxdif

The R function

```
rmmcppb(x,y=NA,alpha=0.05,con=0,est=onestep,plotit=TRUE,dif=TRUE,
grp=NA,nboot=NA,BA=FALSE,hoch=FALSE,xlab="Group 1",ylab="Group 2",pr=TRUE,
SEED=TRUE,...)
```

performs all pairwise comparisons using the percentile bootstrap method just described. The argument `y` defaults to NA meaning that the argument `x` is assumed to be a matrix with J columns or to have list mode. If data are stored in the argument `y`, then it is assumed that two dependent groups are to be compared with the data in the first group stored in `x`. To test Equation (12.12), set the argument `dif=FALSE`; otherwise, difference scores are used. If `dif=FALSE` and `hoch=FALSE`, method SR in Section 12.1.11 is used to control FWE; otherwise Hochberg's method in Section 12.1.10 is used. The R function

```
dmedpb(x,y=NA, alpha=0.05, con=0, est=median, plotit=TRUE, dif=TRUE, grp=NA,
hoch=TRUE, nboot=NA,xlab="Group 1", ylab="Group 2", ylab.ebar = NULL,
pr=TRUE, SEED=TRUE, PCI = FALSE,...)
```

is the same as `rmmcppb`, only it defaults to comparing medians, and it is designed to handle tied values. Hochberg's method is used to control the probability of one or more Type I errors. If the argument `dif=TRUE` and `est=median`, confidence intervals based on the medians of the difference scores are created if `PCI=TRUE`. The y-axis can be labeled with the argument `ylab.ebar`.

The R function

```
dtrimpb(x,y=NA,alpha=0.05,con=0,est=tmean,plotit=TRUE,dif=TRUE,grp=NA,
hoch=TRUE, nboot=NA, xlab="Group 1", ylab="Group 2", pr=TRUE, SEED=TRUE,...)
```

is exactly the same as `dmedpb`, only it defaults to using 20% trimmed means.

Another goal might be to create boxplots for all pairwise difference scores among J dependent groups. The R function

```
boxdif(x,names)
```

accomplishes this goal. The argument `x` is assumed to be a matrix or a data frame. The boxplots can be labeled via the optional argument `names`.

12.4.5 Bootstrap-t Method

When dealing with dependent groups, a bootstrap-t method appears to be one of the best methods when comparing linear contrasts based on means, or when using trimmed means with a relatively small amount of trimming. The strategy is similar to other bootstrap-t methods previously described in this chapter. Readers interested in more precise details are referred to Wilcox (2017, Section 8.2.3).

12.4.6 R Function bptd

The R function

```
bptd(x, tr = 0.2, alpha = 0.05, con = 0, nboot = 599)
```

tests hypotheses about linear contrasts associated with dependent groups using a bootstrap-t method. The argument `con` can be used to specify the set of linear contrasts that are of interest. Its default value is 0 meaning that all pairwise comparisons are performed.

12.4.7 Comparing the Quantiles of the Marginal Distributions

Section 8.9 described a method for comparing all of the quantiles of two dependent variables based on the marginal distributions. The method is distribution free, meaning that assuming random sampling only, the probability of one or more Type I errors can be determined exactly. But a negative feature is that power can be relatively low when dealing with the lower and upper quantiles, particularly when there are tied values. A method for dealing with these issues was suggested by Wilcox and Erceg-Hurn (2012). It uses a bootstrap method in conjunction with the Harrell–Davis estimator mentioned in Section 12.1.19. Briefly, proceed as described in Section 9.5, only now the Harrell–Davis estimator is used and the probability of one or more Type I errors is controlled with Hochberg's method described in Section 12.1.10.

It is noted that there are some additional methods for comparing the quantiles of dependent variables that might be of interest. Included is a method based on the quantiles of the difference scores. Details are described in Wilcox (2017).

12.4.8 R Function Dqcomhd

The R function

```
Dqcomhd(x,y,q = c(1:9)/10, nboot = 1000, plotit = TRUE, SEED = TRUE, xlab =
   "Group 1", ylab = "Est.1-Est.2", na.rm = TRUE, alpha=0.05,ADJ.CI=TRUE).
```

compares the quantile of the marginal distributions of two dependent variables as described in the previous section. By default, all of the deciles are compared, which are estimated via the Harrell–Davis estimator. The argument q can be used to specify the quantiles to be compared. The argument `na.rm=TRUE` means that vectors with missing values will be removed; otherwise all of the available data are used. The argument `ADJ.CI=TRUE` means that the confidence intervals are based on the critical p-values used by Hochberg's method.

12.5 BETWEEN-BY-WITHIN DESIGNS

There are various ways of performing multiple comparisons when dealing with a between-by-within (or split-plot) design using a combination of methods already described. A few approaches are summarized here, assuming that levels of Factor B correspond to dependent groups, the levels of Factor A are independent, and the goal is to compare trimmed means.

Method BWMCP

One approach is to define main effects and interactions as described in Section 12.2 and then use a slight modification of the method in Section 12.4.1 to compute a test statistic. The modification consists of taking into account whether groups are independent when estimating the standard error of an appropriate linear contrast. (If groups j and ℓ are independent, set

$d_{j\ell} = 0$ when computing A in Section 12.4.1.) Note, however, that when some groups are independent, the degrees of freedom used in Section 12.4.1 are no longer appropriate. One way of dealing with this technical issue is to use a bootstrap-t method, which is the approach used by method BWMCP.

Method BWAMCP: Comparing Levels of Factor A for Each Level of Factor B

To provide more detail about how groups differ, another strategy is to focus on a particular level of Factor B and perform all pairwise comparisons among the levels of Factor A. Of course, this can be done for each level of Factor B.

EXAMPLE

Consider again a 3-by-2 design where the means are arranged as follows:

		Factor B	
		1	2
	1	μ_1	μ_2
Factor A	2	μ_3	μ_4
	3	μ_5	μ_6

For Level 1 of Factor B, method BWAMCP would test H_0: $\mu_1 = \mu_3$, H_0: $\mu_1 = \mu_5$, and H_0: $\mu_3 = \mu_5$. For Level 2 of Factor B, the goal is to test H_0: $\mu_2 = \mu_4$, H_0: $\mu_2 = \mu_6$, and H_0: $\mu_4 = \mu_6$. These hypotheses can be tested by creating the appropriate linear contrasts and using the R function `lincon`, which can be done with the R function `bwamcp` described in the next section.

Method BWBMCP: Dealing with Factor B

When dealing with Factor B, there are four variations of method BWMCP that might be used, which here are lumped under what will be called method BWBMCP. The first two variations ignore the levels of Factor A and test hypotheses based on the trimmed means corresponding to the difference scores or in terms of the marginal trimmed means. That is, the data are pooled over the levels of Factor A. The other two variations do not pool the data over the levels of Factor A, but rather perform an analysis based on difference scores or the marginal trimmed means for each level of Factor A. In more formal terms, consider the jth level of Factor A. Then there are $(K^2 - K)/2$ pairs of groups that can be compared. If for each of the J levels of Factor A, all pairwise comparisons are performed, the total number of comparisons is $J(K^2 - K)/2$.

EXAMPLE

Consider a 2-by-2 design where the first level of Factor A has 10 observations and the second has 15. For convenience we imagine that the levels of Factor B correspond to times 1 and 2. So we have a total of 25 pairs of observations with the first 10 corresponding to Level 1 of Factor A. When analyzing Factor B, the pooling strategy just described means the goal is to compare either the difference scores corresponding to all 25 pairs of observations, or to compare the marginal trimmed means, again based on all 25 observations. That is, time 1 is compared to time 2 without paying attention to the two levels of Factor A. Not pooling means that for Level 1 of Factor A either compare difference scores corresponding to times 1 and 2, or compare the marginal trimmed means. The same could be done for Level 2 of Factor A.

Method BWIMCP: Interactions

As for interactions, for convenience we focus on a 2-by-2 design with the understanding that the same analysis can be done for any two levels of Factor A and any two levels of Factor B. To begin, focus on the first level of Factor A. There are n_1 pairs of observations corresponding to the two levels of Factor B, say times 1 and 2. Form the difference scores, which for level j of Factor A are denoted by D_{ij} $(i = 1, \ldots, n_j)$, and let μ_{tj} be the population trimmed means associated with these difference scores. Then one way of stating the hypothesis of no interaction is

$$H_0 : \mu_{t1} = \mu_{t2}.$$

In other words, the hypothesis of no interaction corresponds to the trimmed means of the difference scores associated with Level 1 of Factor A being equal to the trimmed means of the differences scores associated with Level 2 of Factor A. Of course, the hypothesis of no interaction can be stated in a similar manner for any two levels of Factors A and B. The goal is to test all such hypotheses in a manner that controls FWE, and again Rom's method might be used.

Methods SPMCPA, SPMCPB, and SPMCPI

Methods SPMCPA, SPMCPB, and SPMCPI are essentially the same as method BWAMCP, BWBMCP, and BWIMCP, respectively, only a percentile bootstrap method is used rather than a bootstrap-t. In practical terms, groups can be compared based on an M-estimator or MOM.

12.5.1 R Functions bwmcp, bwamcp, bwbmcp, bwimcp, spmcpa, spmcpb, spmcpi, and bwmcppb

The R function

```
bwmcp(J, K, x, tr = 0.2, alpha = 0.05,con=0, nboot=599)
```

performs method BWMCP described in the previous section. By default, it creates all relevant linear contrasts for main effects and interactions, as described in the beginning of Section 12.2. (It calls the R function `con2way` and then applies the method in Section 12.4.1, taking into account which pairs of groups are independent.)

The R function

```
bwamcp(J, K, x, tr = 0.2, alpha = 0.05)
```

performs multiple comparisons associated with Factor A using the method BWAMCP, described in the previous section. The function creates the appropriate set of linear contrasts and calls the R function `lincon`. The function returns three sets of results corresponding to Factor A, Factor B, and all interactions. The critical value reported for each of the three sets of tests is designed to control the probability of at least one Type I error.

The R function

```
bwbmcp(J, K, x, tr = 0.2, con = 0, alpha = 0.05, dif = TRUE, pool=FALSE)
```

uses method BWBMCP to compare the levels of Factor B. If `pool=TRUE` is used, it simply pools the data for you and then calls the function `rmmcp`. If `dif=FALSE` is used, the marginal trimmed means are compared instead. By default, `pool=FALSE` meaning that

$$H_0 : \mu_{tjk} = \mu_{tjk'}$$

is tested for all $k < k'$ and $j = 1, \ldots, J$. For each level of Factor A, the function simply selects data associated with the levels of Factor B and sends it to the R function `rmmcp`.

EXAMPLE

Here is an example of the output from `bwbmcp` when using default settings for the arguments `pool` and `dif` when dealing with a 2-by-2 design.

```
[1] "For level  1  of Factor A:"
     Group Group       test   p.value p.crit        se
[1,]     1     2 -0.6495768 0.5246462   0.05 0.2388609
     Group Group     psihat  ci.lower  ci.upper
[1,]     1     2 -0.1551585 -0.6591109 0.3487939
[1] "For level  2  of Factor A:"
     Group Group       test   p.value p.crit        se
[1,]     1     2 -0.3231198 0.7505451   0.05 0.2257491
     Group Group    psihat   ci.lower  ci.upper
[1,]     1     2 -0.072944 -0.5492329 0.4033449
```

For Level 1 of Factor A, ignoring Level 2 of Factor A, the p-value for comparing Levels 1 and 2 of Factor B is 0.5246. On the next line we see −0.1551585 under `psihat`. That is, the estimate of the linear contrast, which in this case is based on the 20% trimmed mean of the difference scores, is −0.1551585.

As for interactions, the R function

```
bwimcp(J, K, x, tr = 0.2, alpha = 0.05)
```

can be used.

EXAMPLE

Imagine that the goal is to compare a control group to an experimental group at time 1 and then again at times 2, 3, and 4. That is, we have a 2-by-4, between-by-within design and the goal is to compare the two levels of Factor A for each of the four levels of Factor B. Simultaneously, the goal is to have FWE equal to α. If the data are stored in the R variables `x`, the R command

```
bwamcp(2,4,x,tr=0.1)
```

will perform the analysis based on 10% trimmed means.

EXAMPLE

Another example stems from an illustration in Chapter 11 where EEGs for murderers were measured at four sites in the brain. The same was done for a control group, in which case the goal might be, among other things, to determine which sites differ between the two groups. When working with means under the assumption of normality, the problem is simple: Compare each site with Welch's method and control FWE with Rom's procedure. To compare groups based on an M-estimator of location, use the R function `spmcpa`. If the EEG data are stored in the R function `eeg`, the command

```
spmcpa(2, 4, eeg,est=onestep)
```

returns

```
$output
     con.num    psihat p.value  p.sig  ci.lower   ci.upper
[1,]       1 -0.498333   0.010 0.0254 -0.871071 -0.0567815
[2,]       2 -0.073500   0.846 0.0500 -0.634454  0.5099546
[3,]       3  0.129390   0.510 0.0338 -0.487223  0.7832449
[4,]       4 -0.055310   0.956 0.0500 -0.683182  0.5334029

$con
     [,1] [,2] [,3] [,4]
[1,]    1    0    0    0
[2,]    0    1    0    0
[3,]    0    0    1    0
[4,]    0    0    0    1
[5,]   -1    0    0    0
[6,]    0   -1    0    0
[7,]    0    0   -1    0
[8,]    0    0    0   -1

$num.sig
[1] 1
```

The contrast coefficients are reported in `$con`. For example, among the eight groups, the first column indicates that measures of location corresponding to the first and fifth group are compared. (That is, for Level 1 of Factor B, compare Levels 1 and 2 of Factor A.) The value of `num.sig` is 1, meaning that one significant difference was found based on whether the value listed under `sig.test` (the p-value) is less than or equal to corresponding values listed under `crit.sig`, where `crit.sig` is chosen to control FWE. Here, the data indicate that the typical EEG for murderers differ from the control group at the first site where measures were taken.

The R function

```
bwmcppb(J, K, x, tr = 0.2, alpha = 0.05, nboot = 500, bhop = F)
```

simultaneously performs all multiple comparisons related to all main effects and interactions using a percentile bootstrap method. Unlike `spmcpa`, `spmcpb`, and `spmcpi`, the function `bwmcppb` is designed for trimmed means only and has an option for using the Benjamini–Hochberg method via the argument `bhop`.

12.6 WITHIN-BY-WITHIN DESIGNS

A within-by-within design refers to situations where all groups are possibly dependent, in contrast to a between-by-within design where the levels of Factor A are independent. Multiple comparisons are readily performed using methods already described for handling linear contrasts among dependent groups.

EXAMPLE

Imagine a two-way ANOVA design where husbands and wives are measured at two different times. Then the number of groups is $L = 4$, all four groups are dependent, and this is an example of a two-way ANOVA with a within-by-within design. That is, the levels of Factor A are possibly dependent as are the levels of Factor B. Assume that for wives, the trimmed means are μ_{t1} and μ_{t2} at times one and two, and the trimmed means at times one and two for the husbands are μ_{t3} and μ_{t4}. If the goal is to detect an interaction by testing

$$H_0 : \mu_{t1} - \mu_{t2} = \mu_{t3} - \mu_{t4},$$

the linear contrast is

$$\Psi = \mu_{t1} - \mu_{t2} - \mu_{t3} + \mu_{t4}.$$

To use `rmmcp`, first store the contrast coefficients in some R variable. One way to do this is with the R command

```
mat=matrix(c(1, -1, -1, 1)).
```

Alternatively, use the command

```
mat=con2way(2,2)$conAB.
```

If the data are stored in the R matrix `m1`, then the command

```
rmmcp(m1,con=mat,dif=FALSE)
```

will perform the computations.

12.6.1 Three-Way Designs

It is briefly noted that the multiple comparison procedures for two-way designs, previously described, can be extended to three-way designs, including situations where one or more factors involve dependent groups. The basic strategies used with a two-way design are readily extended to a three-way design, but the details are omitted. Instead, R functions for applying the methods are supplied and illustrated. For situations where all groups are independent, or all groups are dependent, R functions already described can be used. The main difficulty is generating all of the linear contrast coefficients, which can be done with the R function `con3way` described in the next section.

12.6.2 R Functions con3way, mcp3atm, and rm3mcp

This section describes three R functions aimed at facilitating the analysis of a three-way design. The first function

```
con3way(J,K,L)
```

generates all of the linear contrast coefficients needed to test hypotheses about main effects and interactions.

EXAMPLE

We illustrate how to use `con3way` when dealing with a between-by-between-by-between design and the goal is to test the hypothesis of no three-way interaction. (Main effects and two-way interactions are handled in a similar manner.) Consider again the weight-gain illustration in Section 11.1 where there are two factors: amount of protein (high and low) and source of protein (beef and cereal). Here we imagine that there is a third factor, type of medication, having two levels, namely, placebo and experimental. It is convenient to represent the means as follows:

Factor A: Placebo

		Source	
		Beef	Cereal
Amount	High	μ_1	μ_2
	Low	μ_3	μ_4

Factor A: Experimental

		Source	
		Beef	Cereal
Amount	High	μ_5	μ_6
	Low	μ_7	μ_8

For the first level of Factor A (placebo), interaction refers to

$$(\mu_1 - \mu_2) - (\mu_3 - \mu_4).$$

For the second level of Factor A, interaction refers to

$$(\mu_5 - \mu_6) - (\mu_7 - \mu_8).$$

The hypothesis of no three-way interaction corresponds to

$$(\mu_1 - \mu_2) - (\mu_3 - \mu_4) = (\mu_5 - \mu_6) - (\mu_7 - \mu_8).$$

So in terms of a linear contrast, the hypothesis of no three-way interaction is

$$H_0 : \mu_1 - \mu_2 - \mu_3 + \mu_4 - \mu_5 + \mu_6 + \mu_7 - \mu_8 = 0. \qquad (12.13)$$

What is needed is a convenient way of generating the linear contrast coefficients, and this is accomplished with the R function `con3way`. For example, the R command

```
m=con3way(2,2,2)
```

stores in `m$conABC` the linear contrast coefficients for testing the hypothesis of no three-way interaction. When the factors have more than two levels, `con3way` generates all sets of linear contrasts relevant to all three-way interactions and again returns them in `con3way$conABC`. In a similar manner, all linear contrasts relevant to main effects for Factor A are stored in `m$conA`.

The next step is to test the hypothesis of no three-way interaction with an R function described in Section 12.1 or 12.2 that is designed for independent groups. For example, if the data are stored in the R variable `dat`, the command

```
lincon(dat,con=m$conABC)
```

would test the hypothesis of no three-way interaction based on a 20% trimmed mean.

For convenience, the R function

```
mcp3atm(J, K, L, x, tr = 0.2, con = 0, alpha = 0.05, grp = NA, pr = TRUE)
```

is provided, which performs all of the multiple comparisons based on the linear contrast coefficients generated by `con3way` when dealing with a between-by-between-by-between design.

A within-by-within-by-within design is handled in a manner similar to a between-by-between-by-between design. The only difference is that now we use a method for linear contrasts that is designed for dependent groups, which were described in Section 12.4. Assuming that all linear contrasts generated by `con3way` are of interest, the R function

```
rm3mcp(J, K, L, x, tr = 0.2, alpha = 0.05, dif = TRUE, op = FALSE, grp = NA)
```

can be used.

As for situations where there are both between and within factors, use a bootstrap method via the R functions described in Section 12.6.4.

EXAMPLE

The last example is repeated assuming that all groups are dependent. Again use the R function `con3way` to generate the linear contrast coefficients. If the contrast coefficients are stored in `m$conABC`, the R command

```
rmmcp(dat,con=m$conABC)
```

tests the hypothesis of no three-way interaction.

12.6.3 Bootstrap Methods for Three-Way Designs

Bootstrap methods are readily extended to three-way designs, including situations where one or more factors involve dependent groups. The next section summarizes some R functions that are available.

12.6.4 R Functions bbwmcp, bwwmcp, bwwmcppb, bbbmcppb, bbwmcppb, bwwmcppb, and wwwmcppb

Bootstrap-t Methods

Section 12.5 described how to perform multiple comparisons, using a bootstrap-t method, when dealing with a between-by-within design. The same strategy is readily extended to linear contrasts relevant to a three-way design. The R function

```
bbwmcp(J, K, L, x, tr = 0.2, JKL = J * K * L, con = 0, alpha = 0.05, grp =
       c(1:JKL), nboot = 599, SEED = TRUE, ...)
```

performs all multiple comparisons associated with main effects and interactions using a bootstrap-t method in conjunction with trimmed means when analyzing a between-by-between-by-within design. The function uses `con3way` to generate all of the relevant linear contrasts and then uses the function `lindep` to test the hypotheses. The critical value is designed to control the probability of at least one Type I error among all the linear contrasts associated with Factor A. The same is done for Factor B and Factor C.

The R function

```
bwwmcp(J, K, L, x, tr = 0.2, JKL = J * K * L, con = 0, alpha = 0.05, grp =
                    c(1:JKL), nboot = 599, SEED = TRUE, ...)
```

handles a between-by-within-by-within design. (For a within-by-within-by-within design, the R function `rmmcp` in Section 12.4.2 can be used in conjunction with the R function `con3way`. For a between-by-between-by-between design, use the R function `lincon` or the percentile bootstrap method that is performed by the R function `bbbmcppb` described next.)

Percentile Bootstrap Methods

Percentile bootstrap methods can be used as well and appear to perform relatively well when using a trimmed mean with at least 20% trimming; for smaller amounts of trimming, a bootstrap-t method seems preferable. For a between-by-between-by-between design, assuming the goal is to compare trimmed means, use the R function

```
bbbmcppb(J, K, L, x, tr = 0.2, JKL = J * K * L, con = 0, alpha = 0.05, grp =
                    c(1:JKL), nboot = 599, SEED = TRUE, ...).
```

For a between-by-between-by-within design, use the R function

```
bbwmcppb(J, K, L, x, tr = 0.2, JKL = J * K * L, con = 0, alpha = 0.05, grp =
                    c(1:JKL), nboot = 599, SEED = TRUE, ...).
```

As for a between-by-within-by-within and within-by-within-by-within design, now use the functions

```
bwwmcppb(J, K, L, x, tr = 0.2, JKL = J * K * L, con = 0, alpha = 0.05, grp =
                    c(1:JKL), nboot = 599, SEED = TRUE, ...)
```

and

```
wwwmcppb(J, K, L, x, tr = 0.2, JKL = J * K * L, con = 0, alpha = 0.05, grp =
                    c(1:JKL), nboot = 599, SEED = TRUE, ...),
```

respectively.

12.7 EXERCISES

1. Assuming normality and homoscedasticity, what problem occurs when comparing multiple groups with a Student's T test and the groups have a common mean?
2. For five independent groups, assume that you plan to do all pairwise comparisons of the means and you want FWE to be 0.05. Further assume that $n_1 = n_2 = n_3 = n_4 = n_5 = 20$, $\bar{X}_1 = 15$, $\bar{X}_2 = 10$, $s_1^2 = 4$, and $s_2^2 = 9$, $s_3^2 = s_4^2 = s_5^2 = 15$. Test $H_0: \mu_1 = \mu_2$ using (a) Fisher's method (assuming the ANOVA F test rejects), (b) Tukey–Kramer, (c) Dunnett's T3, (d) Games–Howell, and (e) Scheffé's method,
3. Repeat the previous exercise, but now with $n_1 = n_2 = n_3 = n_4 = n_5 = 10$, $\bar{X}_1 = 20$, $\bar{X}_2 = 12$, $s_1^2 = 5$, $s_2^2 = 6$, $s_3^2 = 4$, $s_4^2 = 10$, and $s_5^2 = 15$.
4. You perform six tests and get the p-values 0.07, 0.01, 0.40, 0.001, 0.1, and 0.15. Based on the Bonferroni inequality, which would be rejected with FWE equal to 0.05?
5. For the previous exercise, if you use Rom's method, which tests would be rejected?

6. You perform five tests and get p-values 0.049, 0.048, 0.045, 0.047, and 0.042. Based on the Bonferroni inequality, which would be rejected with FWE equal to 0.05?

7. Referring to the previous exercise, which would be rejected with Rom's procedure?

8. Imagine you compare four groups with Fisher's method and you reject the hypothesis of equal means for the first two groups. If the largest observation in the fourth group is increased, what happens to MSWG? What does this suggest about power when comparing groups one and two with Fisher's method?

9. Repeat the previous exercise but with the Tukey–Kramer and Scheffé methods instead.

10. For the data in Table 9.1, each group has 10 observations, so when using Tamhane's method to compute confidence intervals for all pairwise differences, the degrees of freedom are 9 and the value of h in Table 11 of Appendix B is 4.3. Verify that if the goal is to have confidence intervals with FWE equal to 0.05 and lengths 0.5, the required sample sizes for each group are 50, 11, 18, and 17.

11. Repeat the previous exercise using the R function `hochberg` in Section 12.3.4. Verify that the required sample sizes are 50, 10, 18, and 17 when using means. Compare this to using 20% trimmed means.

12. Use the R function `lincon` to verify that for the data in Table 10.1, the hypothesis of equal 20% trimmed means is rejected for groups 3 and 4, with FWE equal to 0.05, but otherwise no differences are found.

13. Use the R function `msmed` to compare all pairs of groups based on medians using the data in Table 10.1. Note that the function prints a warning that there are tied values in group 3, and suggests using the R function `medpb`. Use this function and compare the results to those returned by `msmed`.

14. For the four groups in Table 10.1, use the R functions `apply` and `bootse` to compute the standard errors for the Harrell–Davis estimator, which is computed with the R function `hd`. Compare the results to the estimated standard error for the 20% trimmed means returned by `trimse`.

15. Perform all pairwise comparisons of the groups in Table 10.1 using the R function `rmmcppb`. First use the default estimator, which is a one-step M-estimator and then use the MOM estimate of location. Use difference scores. Note that when using a one-step M-estimator, NA is returned for the p-values and some confidence intervals. Why does this happen? Verify that when using MOM, a difference between groups 2 and 3 is found with FWE set at 0.05.

16. Thompson and Randall-Maciver (1905) report four measurements of male Egyptian skulls from five different time periods. The first was maximal breadth of skull and the five time periods were 4000 BC, 3300 BC, 1850 BC, 200 BC, and 150 AD. The data are stored on the author's web site in the file skull_data.txt. The first four columns contain the skull measurements and the fifth column contains the time periods. For the data in column 1 (maximal breadth), perform all pairwise comparisons using means via the R function `lincon`. Hint: use the R

function `fac2list` to divide the data in groups based on the time periods. Next, compare the groups using `tmcppb` when using a 20% trimmed mean and compare the results to those based on means compared via `lincon`.

17. Among the five groups in the previous exercise, only one group was found to have an outlier based on the boxplot rules in Chapter 2. What might explain why more significant differences are found when comparing groups based on 20% trimmed means versus the mean?

CHAPTER 13

SOME MULTIVARIATE METHODS

CONTENTS

Roughly, the multivariate methods in this chapter deal with situations where two or more outcome variables are of interest. In the context of regression, for example, a goal might be to estimate simultaneously the typical cholesterol and the typical triglyceride level given that an individual is 20 pounds over his or her ideal weight. More generally, *multivariate regression* refers to a situation where there are $p \geq 2$ outcome (dependent) variables of interest. In the notation of Chapter 6, Y is now a matrix having n rows and p columns rather than a single vector having length n. Of course, a separate regression analysis could be done for each outcome variable. But there are circumstances where dealing with all outcome variables simultaneously might make a practical difference, as we shall see.

As another example, Raine et al. (1997) were interested in comparing EEG measures of convicted murderers to a control group. But rather than measure EEG at a single site in the brain, four sites were used. One strategy is to perform a separate analysis for each site. However, new insights might result if we take into account the overall structure of the data when comparing the groups.

To elaborate slightly, momentarily assume normality and homoscedasticity and that Student's T is used to compare the EEG measures using the data from each site. So four Student's T tests are performed. It is possible that all four tests fail to reject, yet when we compare the groups in a manner that takes the overall pattern of results into account, a true difference between the groups is detected. (An illustration is given in Section 13.2.) Not all multivariate methods take into account the overall structure of the data, as will become evident. But many do, particularly various robust methods developed in recent years.

It is stressed that there are many multivariate statistical methods beyond those covered here. The only goal here is to touch on some of the more basic methods and issues, particularly methods relevant to the topics covered in previous chapters. For books dedicated to multivariate methods, see, for example, Anderson (2003), Grimm and Yarnold (1995), Hand and Taylor (1987), Krzanowski (1988), Manly (2004), Rencher (2002), Tabachnick and Fidell (2006).

As previously noted, basic matrix algebra is summarized in Appendix C. Here, some computational details are provided in case they help, but the focus is on conceptual issues aimed at helping the reader choose an appropriate method for achieving some desired goal. As usual, R functions are available for performing the methods to be described. R provides an excellent set of functions for performing arithmetic operations relevant to matrices, including the basic operations summarized in Appendix C, but for present purposes these details do not play an important role and are not discussed.

13.1 LOCATION, SCATTER, AND DETECTING OUTLIERS

When dealing with multivariate data, the basic issue of detecting outliers might seem trivial: apply an outlier detection rule to each of the variables under study. There is, however, a concern, which is roughly illustrated by an example given by Comrey (1985): It is not unusual to be young, and it is not unusual for someone to have hardening of the arteries. But it is unusual to be both young and have hardening of the arteries.

What is needed is a method for detecting outliers that takes into account the overall structure of the data. One way of stating this goal in a slightly more formal manner is that rotating the points should not alter our conclusions about whether a pair of points (or more generally a vector of p points) is unusual. To elaborate, consider the plot of points shown in Figure 13.1. Note that the point indicated by the arrow certainly seems unusual in the sense that it is clearly separated from the other points. (And indeed it is unusual based on how the data were generated.) But look at the right panel, which shows a boxplot for each of the two variables and note that no outliers are found. However, suppose we rotate the points 45 degrees, which results in the scatterplot shown in the left panel of Figure 13.2. The arrow points to the apparent outlier noted in Figure 13.1, which is now flagged as an outlier by the first boxplot in the right panel of Figure 13.2. Moreover, the second boxplot flags two additional points as outliers.

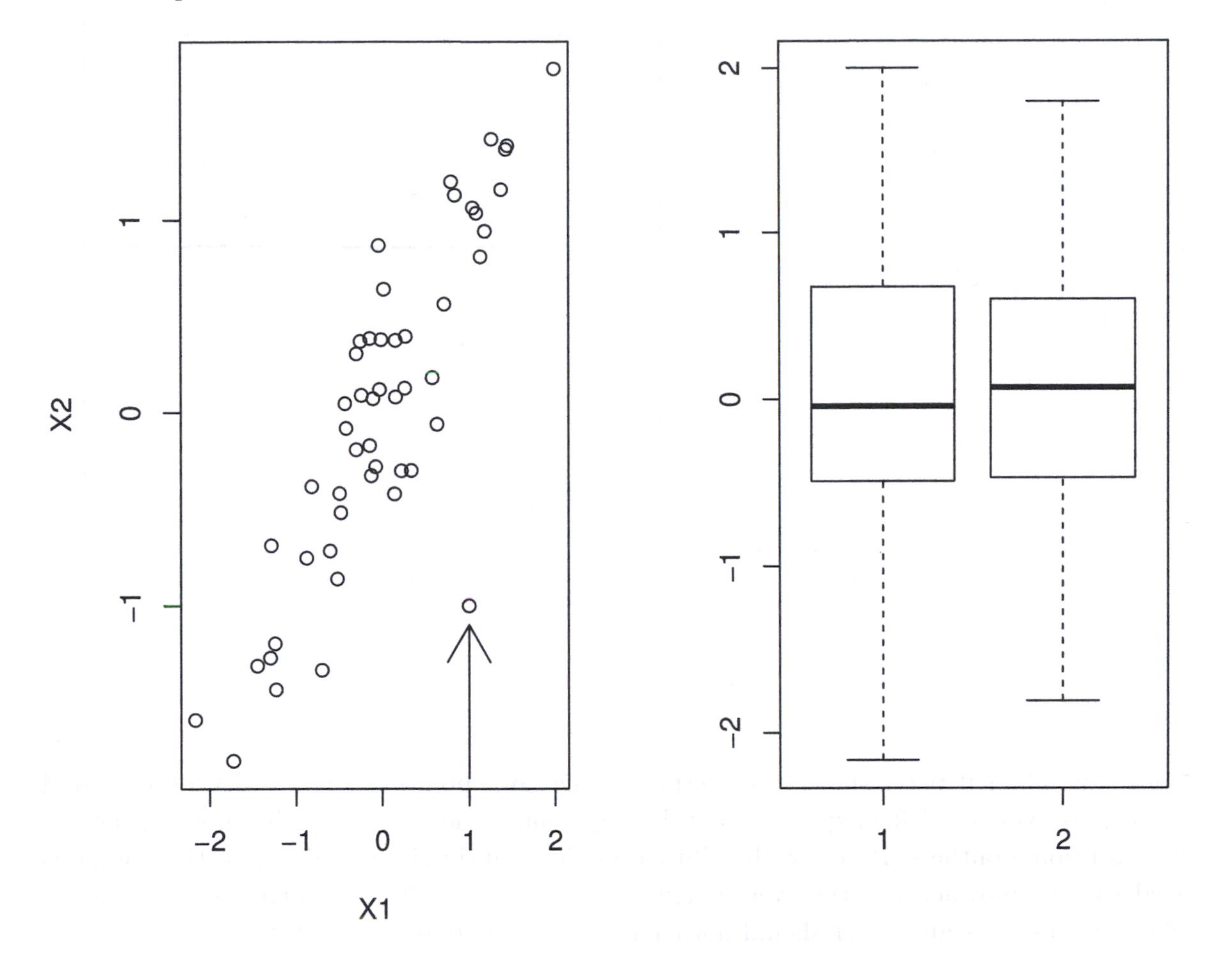

Figure 13.1 The left panel shows a scatterplot of points with one point, indicated by the arrow, appearing to be a clear outlier. But based on boxplots for the individual variables, no outliers are found.

One strategy for detecting outliers among multivariate data is to rotate the points in a variety of ways, each time using a boxplot to check for outliers among the marginal distributions. But this strategy is rather impractical, particularly when dealing with three or more variables. A classic method for dealing with this issue is based on what is called Mahalanobis distance. Imagine that for each participant, we have p measures (e.g., height, weight, cholesterol level, blood pressure, and so on). For the ith participant, the p responses are denoted by

$$\mathbf{X}_i = (X_{i1}, \ldots X_{ip}).$$

That is, $(X_{i1}, \ldots, X_{ip})$ is a row vector having length p. Let s_{jk} be the sample covariance between the jth and kth variables. It might help to recall that s_{jj}, the covariance of a variable with itself, is just the sample variance s_j^2. Denote the matrix of covariances by

$$\mathbf{S} = (s_{jk}).$$

So the diagonal elements of $\mathbf{S}$ are the variances of the p variables under study. The Maha-

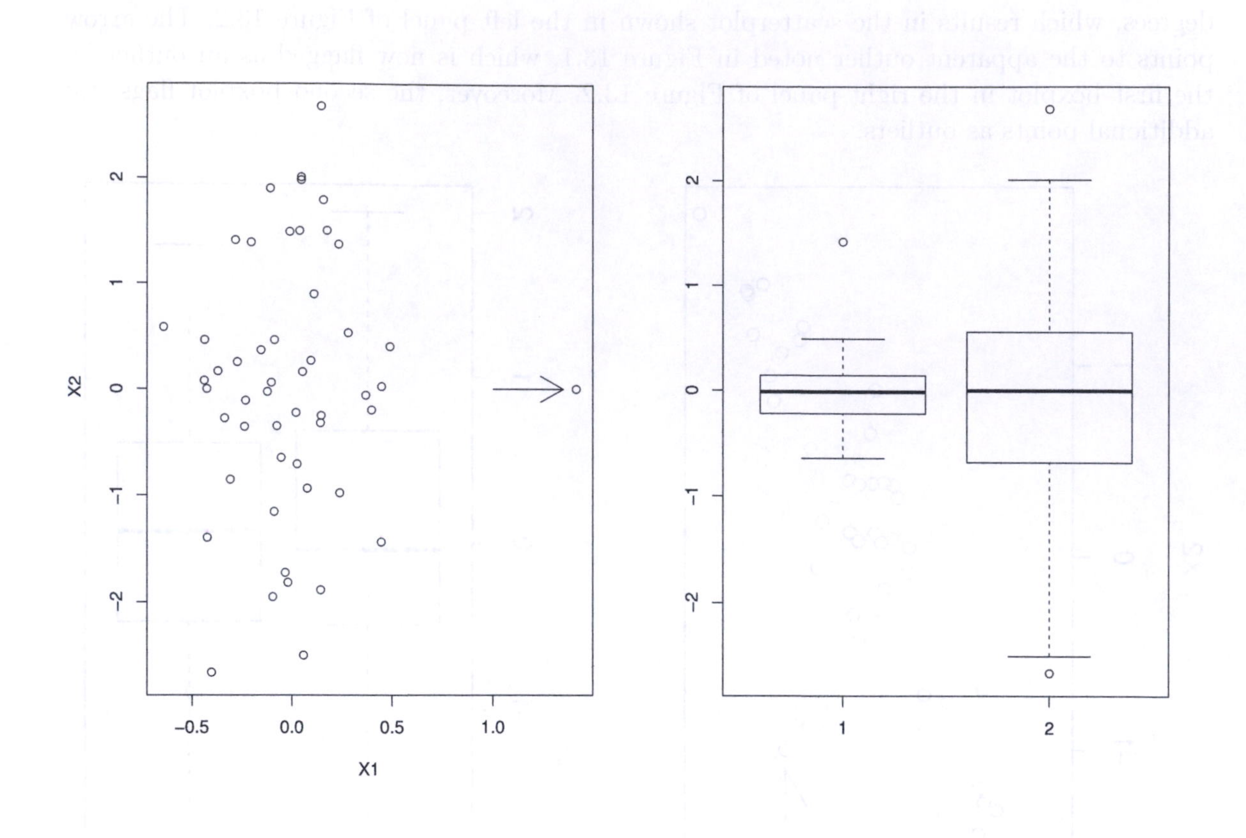

Figure 13.2 The left panel shows a scatterplot of the same points in Figure 13.1, only rotated 45 degrees. Now the left boxplot detects the apparent outlier, and the other boxplot detects two additional outliers. Roughly, this illustrates that multivariate outlier detection methods need to take into account the overall structure of the data. In particular, decisions about whether a point is an outlier should not change when the points are rotated.

lanobis distance of $\mathbf{X}_i$ from the sample mean,

$$\bar{\mathbf{X}} = \frac{1}{n}\sum \mathbf{X}_i,$$

is

$$D_i = \sqrt{(\mathbf{X}_i - \bar{\mathbf{X}})\mathbf{S}^{-1}(\mathbf{X}_i - \bar{\mathbf{X}})'} \tag{13.1}$$

where $(\mathbf{X}_i - \bar{\mathbf{X}})'$ is the transpose of $(\mathbf{X}_i - \bar{\mathbf{X}})$. (See Appendix C for a description of the transpose of a matrix.) Roughly, D_i reflects the standardized distance of a point from the sample mean that takes into account the structure of the data.

To add perspective, look at the inner ellipse shown in Figure 13.3. The points on this ellipse all have the same Mahalanobis distance from the mean. This particular ellipse was constructed so that it contains about half of the points. Points on the outer ellipse also have the same Mahalanobis distance. The outer ellipse was constructed with the property that under normality, it will contain about 97.5% of the points in a scatterplot. This suggests declaring points outside the outer ellipse to be outliers and this represents a multivariate generalization of the outlier detection method introduced in Section 2.4.1, which was based on the mean and variance. The points in Figure 13.3 are the same as the points in Figure

13.1, so now we would declare the point outside the outer ellipse an outlier even though boxplots based on the marginal distributions find no outliers, as indicated in Figure 13.1.

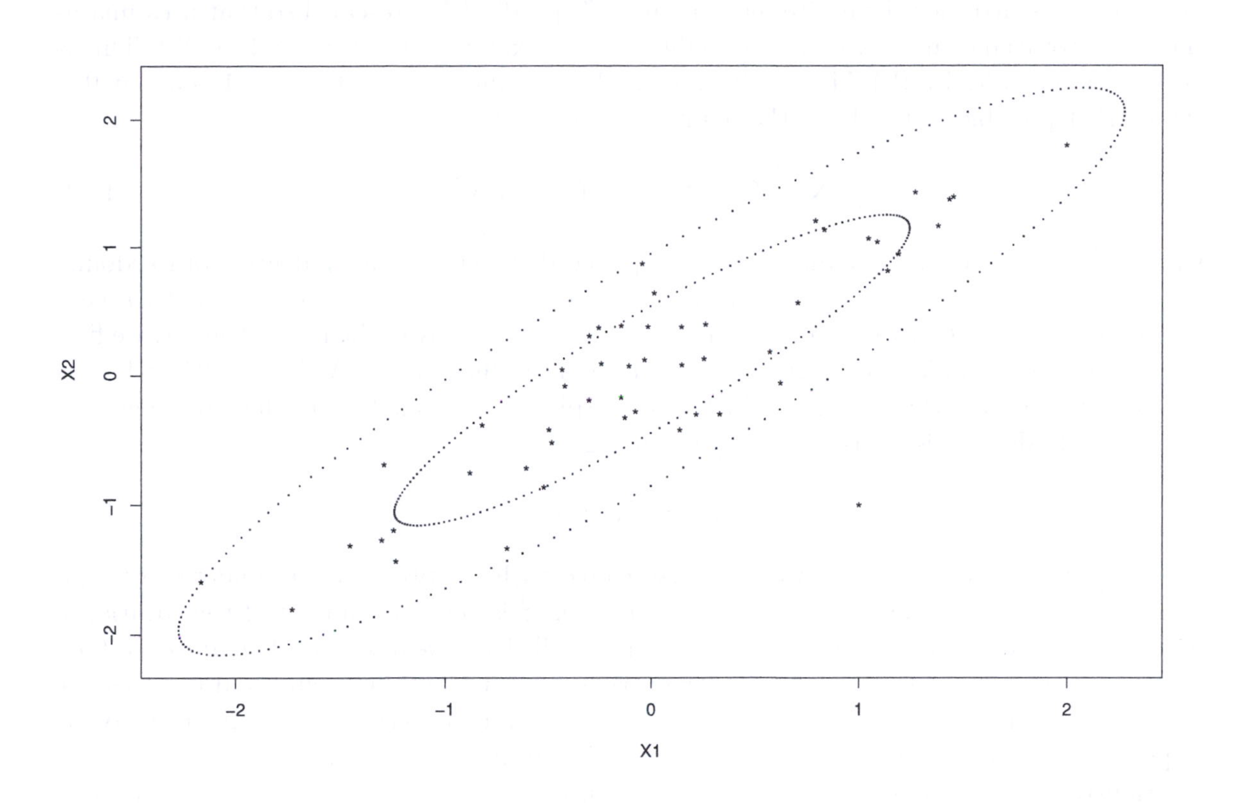

Figure 13.3 All of the points on the inner ellipse have the same Mahalanobis distance from the center. The same is true for the outer ellipse.

There are, however, two concerns with using Mahalanobis distance as an outlier detection method. The first is that it suffers from masking. That is, like the outlier detection method in Section 2.4.1, the very presence of outliers can result in finding no outliers at all. The reason is that the mean and particularly the covariance matrix are sensitive to outliers. (Their breakdown point is only $1/n$.) The second concern is that Mahalanobis distance implies that points that are equidistant from the mean form an ellipse as indicated in Figure 13.3. In terms of detecting outliers, it is being assumed that points outside some properly chosen ellipse are declared outliers. When dealing with elliptically shaped distributions, this would seem to suffice. But otherwise it might be preferable to use a method that is more flexible regarding how it determines whether a point is an outlier.

13.1.1 Detecting Outliers Via Robust Measures of Location and Scatter

One general strategy for detecting multivariate outliers in a manner that avoids masking is to use a modification of Mahalanobis distance where the mean and covariance matrix are replaced by some robust analog. Many possibilities have been proposed. One that has received considerable attention is based on the minimum volume ellipsoid (MVE) estimator (Rousseeuw and Leroy, 1987). To convey the basic idea, consider the bivariate case. As previously noted, the inner ellipse in Figure 13.3 contains about half the data. Many ellipses can be constructed that contain half of the data. The MVE estimator searches among all such ellipses for the one that has the smallest area. (When there are more than two variables, it searches for the ellipsoid having the smallest volume.) Then the mean and covariance of the points in this ellipse, which we label **C** and **M**, respectively, are taken to be the measures of location and scatter for the entire set of points. (Typically **M** is rescaled so that it estimates the usual covariance matrix under normality; see for example Marazzi, 1993, p. 254. This is done automatically by R.) The resulting breakdown point for both **C** and **M** is about 0.5, the highest possible value. Then the point $\mathbf{X}_i$ is declared an outlier if

$$\sqrt{(\mathbf{X}_i - \mathbf{C})\mathbf{M}^{-1}(\mathbf{X}_i - \bar{\mathbf{C}})'} > \sqrt{\chi^2_{0.975,p}}, \tag{13.2}$$

where $\chi^2_{0.975,p}$ is the 0.975 quantile of a chi-squared distribution with p degrees of freedom.

An alternative to the MVE estimator is the so-called *minimum covariance determinant* (MCD) estimator (Rousseeuw and van Driessen, 1999). To convey the basic strategy, we first must describe the notion of a *generalized variance* (introduced by S. Wilks in 1934), which is intended to measure the extent to which a scatterplot of points is tightly clustered together. For bivariate data, this measure of dispersion is given by

$$s_g^2 = s_1^2 s_2^2(1 - r^2), \tag{13.3}$$

where s_1^2 and s_2^2 are the sample variances associated with the two variables under study, and r is Pearson's correlation. (In the multivariate case, s_g^2 is the determinant of the covariance matrix.) The (positive) square root of s_g^2 will be called the *generalized standard deviation.* Recall from Chapter 3 that the smaller the variance, the more tightly clustered together are the values. Moreover, a single outlier can inflate the sample variance tremendously. What is particularly important here is that the sample variance can be small only if all values are tightly clustered together with no outliers. In Chapter 6 we saw that the correlation is sensitive to how far points happen to be from the (least squares) regression line around which they are centered. When the sample variances are small and Pearson's correlation is large, the generalized variance will be small.

The left panel of Figure 13.4 shows a scatterplot of 100 points for which $s_g = 1.303$. The right panel shows a scatterplot of another 100 points, only now they are more tightly clustered around the line having slope 1 and passing through the origin, which results in a smaller generalized variance, namely $s_g^2 = 0.375$.

Now consider any subset of the data containing half of the points. The strategy behind the MCD estimator is to search among all such subsets and identify the one with the smallest generalized variance. The MCD measure of location and covariance is just the mean and covariance of these points. (For results supporting the use of the MCD estimator over the MVE estimator, see Woodruff and Rocke, 1994.) Like the MVE estimator, computing the MCD measure of location and covariance is a nontrivial, computer-intensive problem, but R has a function that performs the calculations, which is described in the next section. (For a description of the algorithm used, see Rousseeuw and van Driessen, 1999.) Moreover, R automatically rescales the MCD covariance matrix so that it estimates the usual covariance matrix under normality. Once these measures of location and scale are available, the relative distance of a point from the center of a data cloud can be measured again using a simple modification of Mahalanobis distance.

13.1.2 R Functions cov.mve and cov.mcd

The R function

```
covmve(m)
```

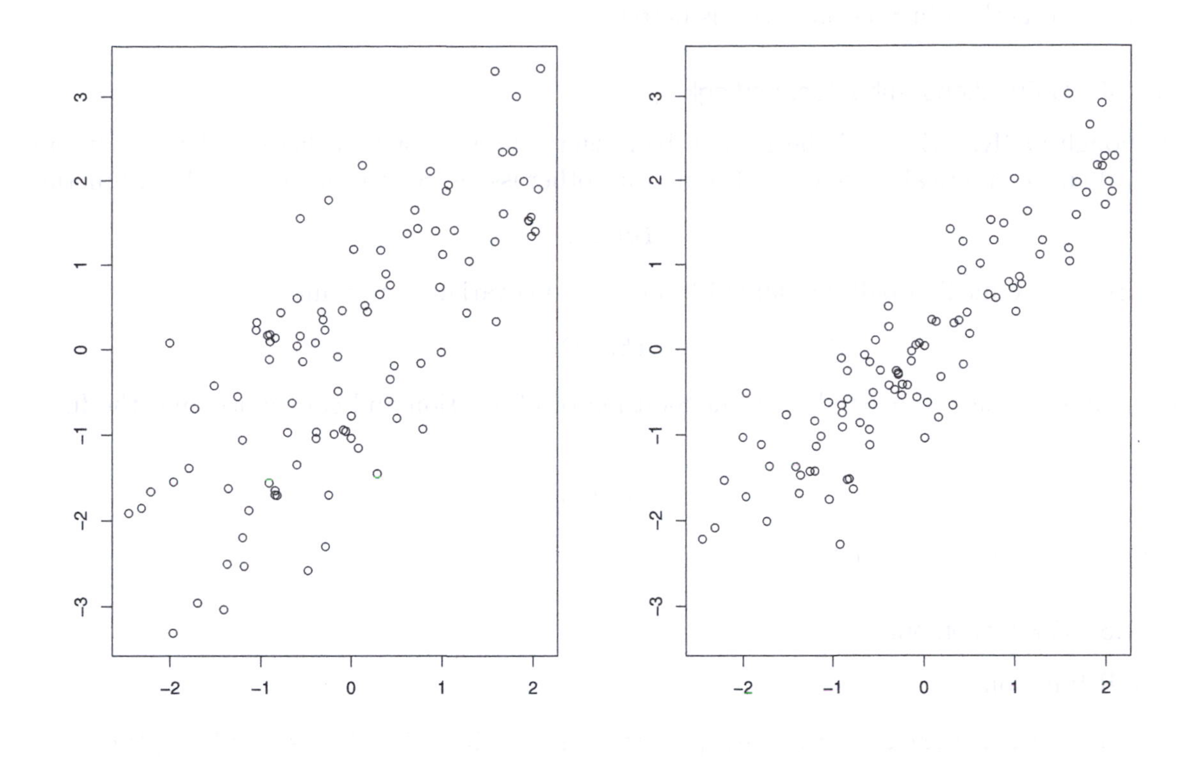

Figure 13.4 The generalized variance reflects the extent to which points are tightly clustered together. In the left panel, the generalized variance is $s_g = 1.303$. In the right panel, the generalized variance is $s_g^2 = 0.375$.

computes the minimum volume ellipsoid estimate of location and scatter for the data stored in `m`, which is assumed to be a matrix with n rows and p columns. (It calls the R function `cov.mcd` after attaching the R package MASS.) The function

```
covmcd(m)
```

computes the MCD estimate of location and scatter. (It calls the R function `cov.mcd` after attaching the R package MASS.)

It should be noted that, by default, `cov.mve` does not return the MVE measure of location and scatter. Rather, it removes points flagged as outliers based on Equation (13.2) and returns the mean and covariance matrix based on the remaining data. The R function `cov.mcd` does the same thing.

13.1.3 More Measures of Location and Covariance

For completeness, there are other robust measures of location and covariance that could be used when checking for outliers. Examples are the *translated biweight S-estimator* (Rocke, 1996; cf. Rocke and Woodruff, 1996.), the *median ball algorithm* (Olive, 2004) and the *orthogonal Gnanadesikan–Kettenring* (OGK) estimator. From Wilcox (2008a), using the OGK estimator to detect outliers is not recommended. (A modification of this estimator performs tolerably well but other methods seem better for general use.) An approach based on the median ball algorithm appears to be unsatisfactory as well. Using TBS to detect outliers performs relatively well if $n \geq 20$ and the number of variables is $p \leq 5$. For $p > 5$, the projection method in Section 13.1.5 is better.

13.1.4 R Functions rmba, tbs, and ogk

Although OGK, TBS, and the median ball algorithm are not used here to detect outliers, they do have practical value when dealing with other issues. So it is noted that the R function

```
rmba(m)
```

computes the median ball measure of location and covariance, the function

```
tbs(m)
```

computes Rocke's translated biweight S-estimator of location and covariance, and the function

```
ogk(m)
```

computes the OGK estimator.

13.1.5 R Function out

The R function

```
out(m,x, cov.fun = cov.mve, plotit = T, SEED = T, xlab = ''X'', ylab =
                    ''Y'', crit = NULL)
```

detects outliers using Equation (13.2). The measure of location **C** and the covariance matrix **M** used by Equation (13.2) are indicated by the argument `cov.fun`, which defaults to the MVE estimator. Setting `cov.fun=out.mcd` results in using the MCD estimator. (Other options for this argument are `ogk`, `tbs`, and `rmba`, which result in using the OGK, TBS, and median ball algorithm, respectively, but except for `tbs`, it seems that these options are relatively unsatisfactory.)

13.1.6 A Projection-Type Outlier Detection Method

This section outlines a method for detecting outliers that does not require the use of a covariance matrix, which is a feature that has practical value in a variety of settings. Moreover, with the understanding that no single method is always best, the method in this section performs relatively well compared to competing techniques (Wilcox, 2008a) based on two criteria. The first is that under normality, the proportion of points declared outliers should be relatively small. The second is that if a point is truly unusual relative to the distribution that generated the data, it should have a high probability of being declared an outlier.

The method is based on what are called (orthogonal) projections, but the computational details are not given. (They can be found in Wilcox, 2017, Section 6.4.9.) The goal here is merely to explain the basic idea. To do this, momentarily focus on the bivariate case ($p = 2$) and look at Figure 13.5, which shows a scatterplot of points with a line through the center. Call this line L. An *orthogonal projection* of a point, say **A**, onto line L means that we draw a line through point **A** that is perpendicular to line L. The point where this line intersects L is the projection of point **A** onto line L. The arrow in Figure 13.5 points to the projection of the point at the other end of the arrow.

Note that once all of the data are projected onto a line, we have, in effect, reduced the p-variate data to the univariate case. This means we can use a boxplot rule or the MAD-median rule to check for outliers. The basic idea is that if a point is an outlier based on some

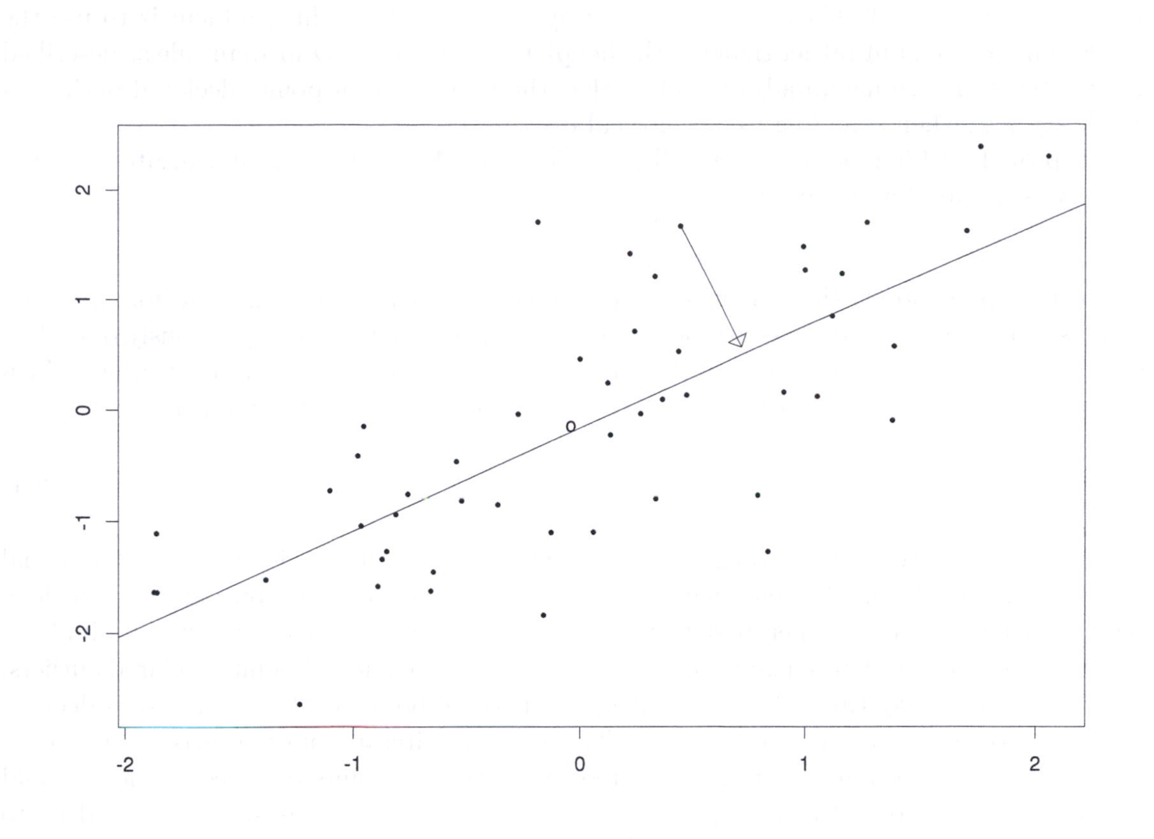

Figure 13.5 This figure illustrates the projection of a point onto a line. The head of the arrow points to the projection of the point located at the other end of the arrow.

projection of the points onto a line, then it is an outlier. But there are infinitely many lines onto which the points might be projected. To make this approach practical, a finite number of projections must be used. Currently, a strategy that performs well begins by finding the center of the data cloud. The marginal medians could be used and several other possibilities are available, including the MVE, MCD, OGK, TBS, and the median ball algorithm method previously mentioned. Next, for each point, draw a line through it and the center of the data cloud. There are n such lines. For each line, project all of the data onto this line and check for outliers using the MAD-median rule or the boxplot rule described in Sections 2.4.2 and 2.4.4, respectively. A point is declared an outlier if it is an outlier based on any of these n projections.

Another outlier detection method that performs about as well as the projection method is described in Wilcox (2017, Section 6.4.7) and is called the *minimum generalized variance* (MGV) method. It is based in part on a covariance matrix, but the inverse of the covariance matrix is not used, which has practical advantages. In some situations it seems to perform a bit better than the projection method, but in other situations it seems to be a bit worse. Given some data, it is difficult to determine whether the MGV method should be preferred over the projection method, so further details are omitted.

Filzmoser et al. (2008) noted that under normality, if the number of variables is large, the proportion of points declared outliers by the better known outlier detection methods can be relatively high. This concern applies to all the methods covered here with the projection method seemingly best at avoiding this problem. But with more than nine variables ($p > 9$), it breaks down as well. Currently, the best way of dealing with this problem is to use the projection method, but rather than use the boxplot rule or the MAD-median rule as described in Chapter 2, determine an adjustment so that the proportion of points declared outliers is small, say 5%, when sampling from a normal distribution.

To provide a bit more detail, recall from Section 2.4.2 that in the univariate case, the value X is declared an outlier if

$$\frac{|X - M|}{\text{MADN}} > 2.24.$$

In the present context, the n points are projected onto a line and this rule for detecting outliers is applied. This process is repeated for each of the n projections previously described and a point is declared an outlier if it is flagged an outlier based on any projection. When the number of variables (p) is large, the strategy is to declare a point an outlier if

$$\frac{|X - M|}{\text{MADN}} > c, \tag{13.4}$$

where c is chosen so that the proportion of points declared outliers is approximately equal to some (specified) small value when sampling from a normal distribution. This is done via simulations. That is, generate data from a p-variate normal distribution where all of the variables have correlations equal to zero, determine the proportion of points declared outliers, and repeat this many times. If the average proportion of points declared outliers is deemed too large, say greater than 0.05, increase c. This process is iterated until a satisfactory choice for c is obtained, meaning that the expected proportion of points declared outliers should be approximately equal to some specified value. A refinement of this strategy would be to generate data from a multivariate normal distribution that has the same covariance matrix as the data under study. Currently, this does not seem necessary or even desirable, but this issue is in need of further study.

13.1.7 R Functions outpro, outproMC, outproad, outproadMC, and out3d

The R function

```
outpro(m,gval = NA, center = NA, plotit = T, op = T, MM = F, cop = 3, xlab
        ="VAR 1", ylab = "VAR 2")
```

checks for outliers using the projection method described in the previous section. By default, the method uses the boxplot rule on each projection. To use the MAD-median rule, set the argument `MM=T`. The argument `cop` determines how the center of the data cloud is computed. By default, the marginal medians are used. The options `cop=`2, 3, 4, 5, 6, 7 correspond to MCD, marginal medians, MVE, TBS, RMBA, and a measure of location called the *spatial* (L1) *median*, respectively. These choices can make a difference in the results, but from a practical point of view it is unclear which method is best for routine use. An alternative choice for the center of the data can be specified with the argument center. (The argument `op` has to do with how the data are plotted; see Wilcox, 2017, for more information.) The argument `gval` corresponds to c in Equation (13.4). The function chooses a value for `gval` if one is not specified.

The R function

```
outproad(m,gval = NA, center = NA, plotit = T, op = T, MM = F, cop = 3, xlab
        = "VAR 1", ylab = "VAR 2",rate = 0.05)
```

is the same as `outpro`, only the decision rule is adjusted so that the expected proportion of points declared outliers, under normality, is approximately equal to the value indicated by the argument `rate`, which defaults to 0.05. Use this function if the number of variables is greater than 9.

The R functions

```
outproMC(m, gval,center = NA, plotit = T, op = T, MM = F, cop = 3, xlab =
                    "VAR 1", ylab = "VAR 2")
```

and

```
outproadMC(m, center = NA, plotit = T, op = T, MM = F, cop = 3, xlab = "VAR
                    1", ylab = "VAR 2",rate = 0.05)
```

are the same as outpro and outproad, respectively, but they take advantage of a multicore processor if one is available. (They require the R package parallel.)

When working with three dimensional data, the R function

```
out3d(x,outfun=outpro,xlab="Var 1",ylab="Var 2",zlab="Var 3",
                    reg.plane=F,regfun=tsreg)
```

will plot the data and indicate outliers with an *. If the argument `reg.plane =T`, this function also plots the regression plane based on the regression estimator indicated by the argument `regfun`, assuming that `regfun` returns the intercept and slope estimates in `regfun$coef`. When plotting the regression surface, the function assumes the outcome variable is stored in the third column of the argument `x`. The two predictors are stored in the first two columns. To plot the least squares regression surface, use `regfun=lsift`. By default, the function uses the Theil–Sen regression estimator described in Section 15.1.1. (The function `out3d` requires the R package `scatterplot3d`, which can be installed as described in Chapter 1.)

13.1.8 Skipped Estimators of Location

Skipped estimators of location refer to the simple strategy of checking for outliers, removing any that are found, and averaging the values that remain. Examples are the measures of location returned by the R functions `cov.mve` and `cov.mcd`. If the projection method for detecting outliers is used, we get what is called the *OP estimator*. Various alternative estimators, which are not skipped estimators, have been proposed. An example is a generalization of trimmed means, which include as a special case Tukey's median. In the univariate case, they reduce to the trimmed mean and median described in Chapter 2. (Details about these and other estimators can be found in Wilcox, 2017.) A recent comparison of various estimators, in terms of their overall accuracy (measured by the generalized variance of an estimator), suggests that the OP estimator performs relatively well (Ng and Wilcox, 2010). This is not to suggest that all competing estimators be ruled out. The only point is that the OP estimator appears to have practical value and to warrant serious consideration when analyzing data.

13.1.9 R Function smean

The R function

```
smean(x, MM=FALSE, MC=FALSE)
```

computes the skipped estimator of location based on the projection method for detecting outliers. With `MM=FALSE`, a boxplot rule is applied to each projection when checking for outliers; otherwise, a MAD-median rule is used. With `MC=TRUE`, multiple cores are used to reduce execution time, assuming the R package multicore has been installed.

13.2 ONE-SAMPLE HYPOTHESIS TESTING

Chapter 4 described methods for testing the hypothesis that a measure of location is equal to some specified value. For the multivariate case, there is a classic method based on means that should be mentioned. Called Hotelling's T^2 test, the goal is to test

$$H_0 : \mu = \mu_0$$

where here μ represents a vector of p population means and μ_0 is a vector of specified constants. The test statistic is

$$T^2 = n(\bar{\mathbf{X}} - \mu_0)\mathbf{S}^{-1}(\bar{\mathbf{X}} - \mu_0)',$$

where $\mathbf{S}$ is the variance-covariance matrix corresponding to the p measures under study. Let f be the $1-\alpha$ quantile of an F distribution with degrees of freedom $\nu_1 = p$ and $\nu_2 = n-p$. Under (multivariate) normality, the probability of a Type I error will be α if the null hypothesis is rejected when

$$T^2 \geq \frac{p(n-1)}{n-p}f.$$

But it is unknown how to get good control over the probability of a Type I error for a reasonably wide range of non-normal distributions. There is the usual concern that power might be relatively low due to outliers. Consequently, the focus here is on a method where the mean is replaced by the OP estimator of location.

To help convey the basic idea, we illustrate the method for the bivariate case, with the understanding that the same method can be used when dealing with $p > 2$ variables. Let θ represent the population analog of the OP estimator of location and imagine the goal is to test

$$H_0 : \theta = \theta_0,$$

where θ_0 is some specified vector. To be a bit more concrete, suppose the goal is to test H_0: θ=(0, 0) based on the data in Figure 13.6. A method that has been found to perform well in simulations is based on a percentile bootstrap method. As was done in Chapter 6 when dealing with regression, imagine that the data are stored in a matrix having n rows and p columns. In symbols, the bootstrap sample is generated by randomly sampling with replacement n rows from

$$\begin{array}{ccc} (X_{11}, & \ldots, & X_{1p}) \\ & \vdots & \\ (X_{n1}, & \ldots, & X_{np}), \end{array}$$

yielding

$$\begin{array}{ccc} (X^*_{11}, & \ldots, & X^*_{1p}) \\ & \vdots & \\ (X^*_{n1}, & \ldots, & X^*_{np}). \end{array}$$

Next, compute the OP estimate of location based on this bootstrap sample yielding $\hat{\theta}^*$. Repeat this B times yielding $\hat{\theta}^*_1, \ldots, \hat{\theta}^*_B$. Figure 13.7 shows the resulting estimates when B = 500 (using the R function smeancrv2 described in the next section). The issue is whether the point $\theta = (0, 0)$, which is marked by the + in Figure 13.7 and corresponds to the null hypothesis, is deeply nested within the cloud of bootstrap sample estimates. If the answer is yes, we fail to reject. But if the point $\theta = (0, 0)$ is sufficiently far from the center of the cloud of points, we reject the null hypothesis. (When using a percentile bootstrap method, the center of the bootstrap data cloud is taken to be $\hat{\theta}$, the OP estimate of location based on the original data, $\mathbf{X}_1, \ldots, \mathbf{X}_n$.)

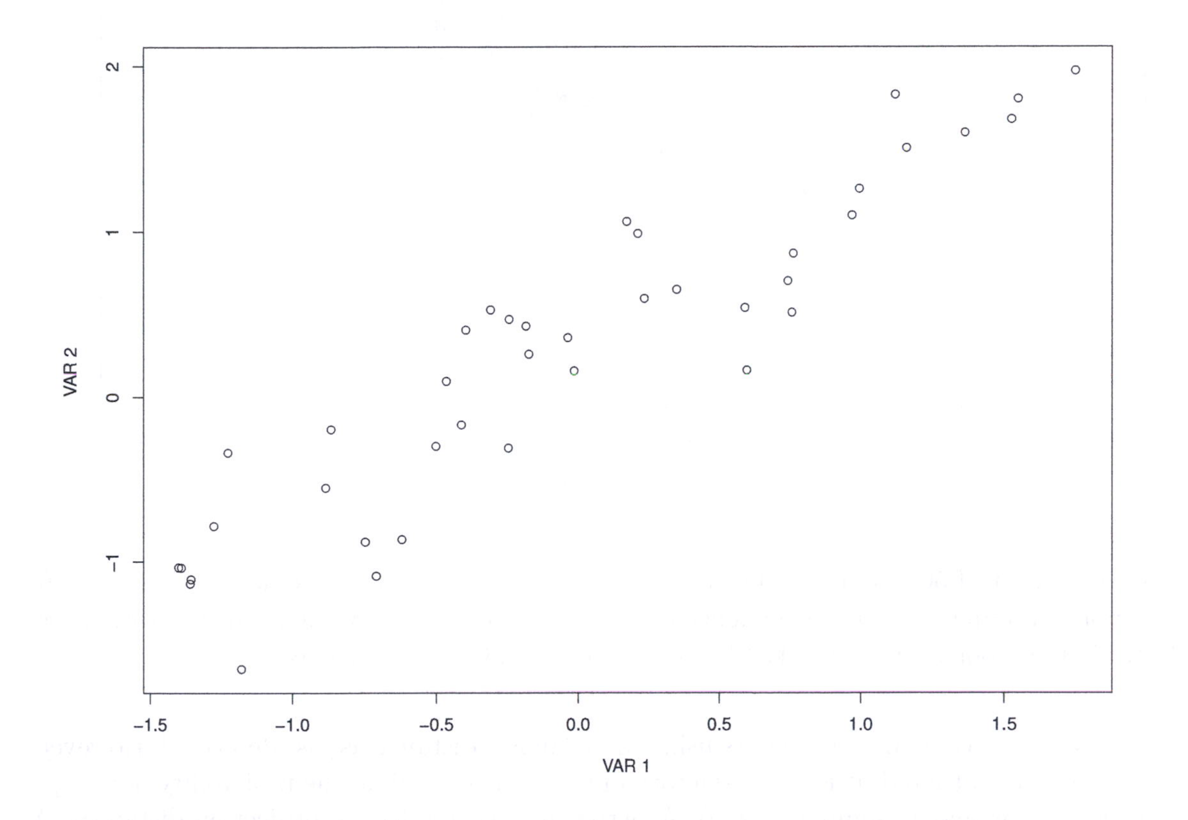

Figure 13.6 A scatterplot of data, generated from a bivariate normal distribution having mean $(0, 0.25)$, $n = 40$.

One way of computing a p-value is to first compute the Mahalanobis distance of every point from the center of the bootstrap cloud. Imagine that this is done resulting in the distances $d_1^*, \ldots, d_B^*$. Next, compute the Mahalanobis distance of the hypothesized value, $(0, 0)$, from the center of the bootstrap cloud, which we label d_0^*. Then from Liu and Singh (1997), a p-value is given by the proportion of $d_1^*, \ldots, d_B^*$ that are greater than or equal to d_0^*. Moreover, the closest 95% of the points to the OP estimate of location (meaning the points having the smallest distances) form a 0.95 confidence region for the population measures of location, θ. This confidence region, based on the data in Figure 13.6, is indicated by the polygon in Figure 13.7. The + in Figure 13.7 indicates the location of the hypothesized values. It lies outside this polygon indicating that the null hypothesis should be rejected.

It is noted that the data in Figure 13.6 were generated from a normal distribution having means equal to $(0, 0.25)$, so rejecting the hypothesis H_0: $\theta = (0, 0)$ was the correct decision. If, however, we test the hypothesis that the second variable has a population OP measure of location equal to 0, ignoring the first variable, we fail to reject.

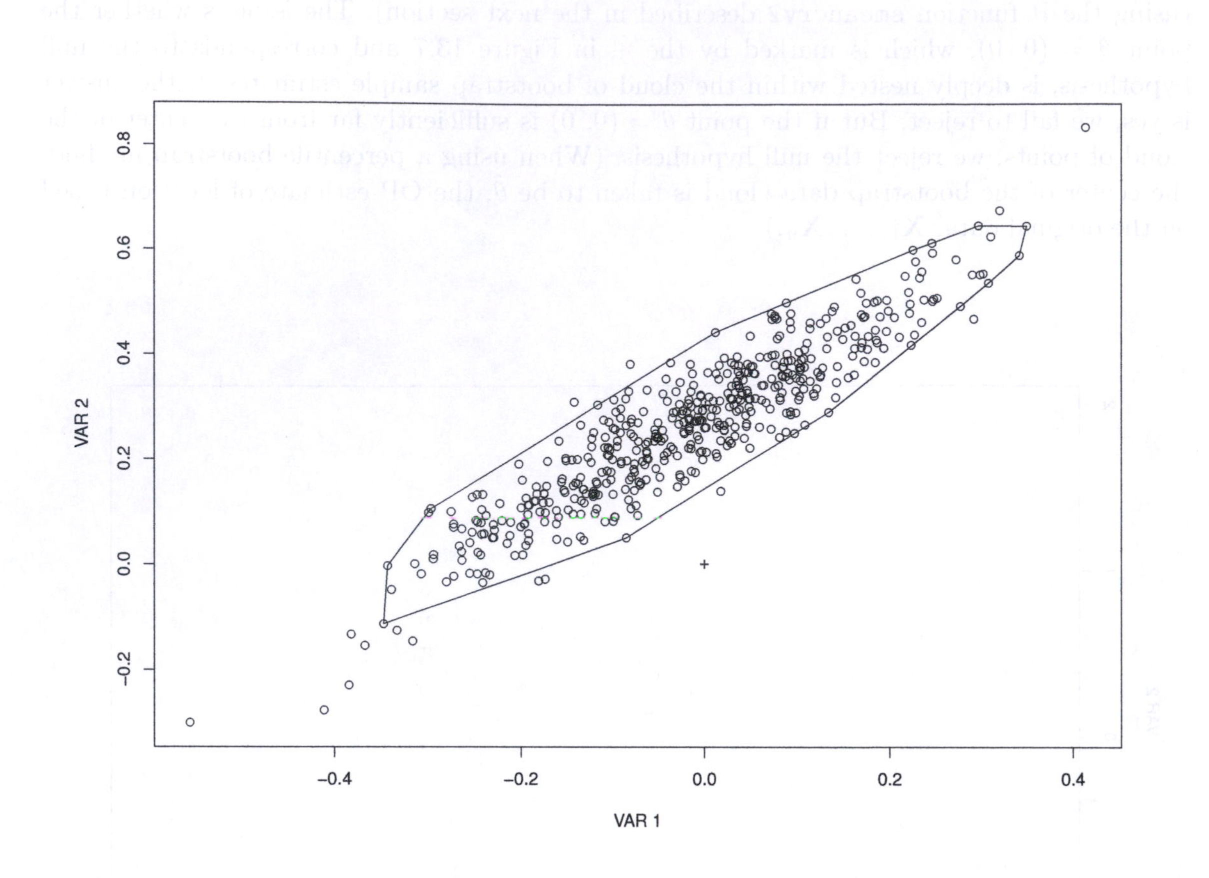

Figure 13.7 Plot of 500 bootstrap OP estimates of location based on the data in Figure 13.6. The polygon indicates a 0.95 confidence region for the population value of the OP estimator. Note that the point (0, 0), marked by the +, lies outside the confidence region.

Basic statistical training suggests using Mahalanobis distance as just described. However, for the problem at hand, it is unsatisfactory in terms of controlling the probability of a Type I error (Wilcox and Keselman, 2006b). A better approach is to use projection distances. A rough sketch of how to compute projection distances is relegated to Box 13.1. More precise computational details can be found in Wilcox (2017, Section 6.2.5). The polygon in Figure 13.7 is based on the points closest to the center of the bootstrap cloud, based on the projection distances among all B points.

Box 13.1: An outline of how to compute projection distances.

Consider a data cloud having n points. Momentarily focus on the first point, $\mathbf{X}_1$, and consider the problem of measuring how far it is from the center of the cloud. Its projection distance can be roughly described as follows. Recall that when using the projection method for detecting outliers, a portion of the process consisted of projecting all points onto n lines. That is, for each data point $\mathbf{X}_i$, draw a line through it and the center of the data, and then project all points onto this line. Once the first point, $\mathbf{X}_1$, is projected onto a line corresponding to $\mathbf{X}_i$, as illustrated, for example, in Figure 13.5, it will have some distance from the center of the cloud. That is, $\mathbf{X}_1$ will have n distances corresponding to the n lines formed by $\mathbf{X}_i$ ($i = 1, \ldots, n$) and the center of the data cloud. Let d_1 be the maximum distance of among all n projections. Then d_1 is said to be the projection distance of $\mathbf{X}_1$ from the center. This process can be repeated for the remaining $n - 1$ points, yielding their corresponding projection distances $d_2, \ldots, d_n$.

13.2.1 Comparing Dependent Groups

Previously covered methods provide a multivariate extension of the paired T test introduced in Chapter 8. They also provide an alternative approach to analyzing more than two dependent groups. This section illustrates how these methods might be used.

EXAMPLE

Imagine an intervention study where the goal is to assess a method for treating depression and feelings of life satisfaction. Further imagine that prior to receiving treatment, measures of depression and life satisfaction are taken for each participant, and again after 6 months of treatment. One way of analyzing the data is to compare measures of depression using the paired T test, or some robust analog, and do the same using a measure of life satisfaction. That is, difference scores are computed for both variables. We test the hypothesis that the difference scores associated with depression have a typical value of 0, and we repeat this for the measure of life satisfaction. An alternative strategy is to test the hypothesis that the OP measure of location, associated with the bivariate distribution of difference scores, is 0. Depending on the nature of the association between depression and life satisfaction, this might result in more power.

EXAMPLE

Imagine an endurance training study where endurance is measured prior to training and again every 4 weeks for a total of 12 weeks. So participants are measured four times. The methods in Section 11.3.3 can be used to compare changes in endurance. That is, we have a repeated measures design with four levels. One approach is to test the hypothesis given by Equation (11.4), as described in Section 11.3.3. That is, form the difference scores for time 1 and time 2, time 1 and time 3, and so on. Because there are four dependent levels, this results in six sets of difference scores. An issue is whether the typical difference is 0 for all possible time pairs. In Section 11.3.3, a measure of location is computed for each set of difference scores without taking into account the overall structure of the data. Another approach is to compare all six sets of differences via the OP estimator. That is, the typical changes in endurance are based on a measure of location that takes into account the overall structure of the data when dealing with outliers.

13.2.2 R Functions smeancrv2, hotel1, and rmdzeroOP

The R function

```
smeancrv2(m, nullv = rep(0, ncol(m)), nboot = 500, plotit = TRUE, MC =
            FALSE, xlab = "VAR 1", ylab = "VAR 2")
```

tests the hypothesis that the population OP measure of location has some specified value. The data are assumed to be stored in a matrix having n rows and p columns, which is denoted here by the argument `m`. The null value, indicated by the argument `nullv`, defaults to (0, 0, ..., 0) having length p. When the number of variables is $p = 2$, and if the argument `plotit=TRUE`, the function plots the bootstrap estimates of location and indicates the central 95% as illustrated by the polygon in Figure 13.7. This polygon represents a 0.95 confidence region. That is, there is a 0.95 probability that the population OP measure of location is within this polygon. Setting `MC=TRUE` can reduce execution time assuming your computer has a multicore processor and that the R package parallel has been installed.

The R function

```
rmdzeroOP(m, nullv = rep(0, ncol(m)), nboot = 500, MC = FALSE)
```

tests the hypothesis that all pairwise differences among J dependent groups have an OP measure of location equal to $\mathbf{0}$, a vector of 0's having length $(J^2 - J)/2$. This function simply forms all relevant difference scores and calls the function `smeancrv2`; it can be used to deal with the situation depicted in the last example. In essence, the function provides a robust, multivariate extension of the methods in Chapter 12.

EXAMPLE

Exercise 12 in Chapter 8 described a data set dealing with scents. The last six columns of the data matrix contain the time participants required to complete a pencil and paper maze when they were smelling a floral scent and when they were not. The columns headed by U.Trial.1, U.Trial.2, and U.Trial.3 are the times for no scent, which correspond to measures taken on three different occasions. Here it is assumed the data are stored in the R variable `scent` having 12 columns of data, with columns 7 through 9 giving the times corresponding to no scent and corresponding to occasions 1 through 3, respectively. Similarly, columns 10 through 12 correspond to the three occasions when participants were smelling a floral scent. For each of the three occasions, we compute the difference between the scented and unscented times using the R command

```
dif=scent[,7:9]-scent[,10:12].
```

So, for example, column 1 of the R variable `dif` contains the differences corresponding to the two conditions no scent and scent, on occasion 1. The R command

```
smeancrv2(dif)
```

tests the hypothesis that the typical value of these three differences are all equal to zero and returns a p-value of 0.272.

EXAMPLE

Continuing the last example, imagine that for the unscented condition, the goal is to test the hypothesis that the difference scores associated with the three occasions typically have a value equal to 0. That is, difference scores are being formed between occasions 1 and 2, 1 and 3, and 2 and 3, and the goal is to determine whether the population OP measure of location associated with these three difference scores is equal to (0, 0, 0). The command

```
rmdzeroOP(scent[,7:9])
```

accomplishes this goal and returns a p-value of 0.03.

The R function

```
hotel1(x,null.value=0,tr=0)
```

performs Hotelling's T^2 test that the vector of trimmed means is equal to the value specified by the argument `null.value`. By default ,`tr=0`, meaning that means are used. Setting the argument `tr=0.2`, for example, 20% trimmed means will be used instead. The default null value is 0. The argument `x` is assumed to be a data frame or a matrix.

13.3 TWO-SAMPLE CASE

There is a classic multivariate method for comparing two or more independent groups based on means. An outline of the method is given in Section 13.4 along with illustrations regarding how to apply it using built-in R functions. The focus here is on comparing two independent groups using the robust OP measure of location in conjunction with a percentile bootstrap method.

The method in Section 13.2 is readily extended to the two-sample case. That is, for two independent groups, θ_j represents the value of θ (the population OP measure of location) associated with the jth group ($j = 1, 2$). The goal is to test

$$H_0 : \theta_1 - \theta_2 = 0. \tag{13.5}$$

Now we generate bootstrap samples from each group, compute the corresponding OP estimates of location, label the results $\hat{\theta}_1^*$ and $\hat{\theta}_2^*$, and set

$$d^* = \hat{\theta}_1^* - \hat{\theta}_2^*.$$

Repeat this process B times yielding $d_1^*, \ldots, d_B^*$. Then H_0 is tested by determining how deeply the vector $(0, \ldots, 0)$ is nested within the cloud of d_b^* values, $b = 1, \ldots, B$, again using projection depth. Here, the center of the bootstrap cloud is $(\hat{\theta}_1 \ , \hat{\theta}_2)$, the estimate of location based on the original data. So roughly, when testing Equation (13.5), the goal is to determine how close the null vector is to the center of the bootstrap cloud. If its depth is low, meaning that its standardized distance from the center of the cloud is high, reject the hypothesis given by Equation (13.5).

More precisely, let D_b be the OP distance associated with the bth bootstrap sample and let D_0 be the distance associated with null vector $(0, \ldots, 0)$. Set $I_b = 1$ if $D_b > D_0$, otherwise $I_b = 0$. Then the p-value is

$$\hat{p} = \frac{1}{B}\sum_{b=1}^{B} I_b.$$

However, an adjusted critical p-value, say p_c, is required for n small. That is, reject if the p-value is less than or equal to p_c. A limitation of the method is that p_c is limited to testing at the $\alpha = 0.05$ level. Another limitation is that situations are encountered where the method cannot be applied due to division by zero when computing the OP measure of location based on a bootstrap sample.

13.3.1 R Functions smean2, mat2grp, matsplit, and mat2list

The R function

```
smean2(m1, m2, nullv = rep(0, ncol(m1)), nboot = 500, plotit = T)
```

performs the method just described. Here, the data are assumed to be stored in the matrices m1 and m2, each having p columns. The argument `nullv` indicates the null vector and defaults to a vector of zeros. If the number of variables is $p = 2$ and the argument `plotit=T`, the function plots a confidence region for the OP (population) measure of location.

EXAMPLE

Thompson and Randall-Maciver (1905) report four measurements for male Egyptian skulls from five different time periods: 4000 BC, 3300 BC, 1850 BC, 200 BC, and 150 AD. There are 30 skulls from each time period and four measurements: maximal breadth, basibregmatic height, basialveolar length, and nasal height. Here, the first and last time periods are compared, based on the OP measures of location. Figure 13.8 shows the plot created by `smean2` when using the first two variables only. The + in the lower right corner indicates the null vector (0, 0), which clearly is outside the 0.95 confidence region. The p-value is 0.002.

13.3.2 R functions matsplit, mat2grp, and mat2list

This section describes some R functions for manipulating data files. For data stored in a matrix, say `m`, they divide the data into groups based on the values stored in some column of `m`.

The R function

```
matsplit(m,coln)
```

splits the matrix `m` into two matrices based on the values in the column of `m` indicated by the argument `coln`. This column is assumed to have two values only. Results are returned in `$m1` and `$m2`.

The R function

```
mat2grp(m,coln)
```

also splits the data in a matrix into groups based on the values in column `coln` of the matrix `m`. Unlike `matsplit`, `mat2grp` can handle more than two values (i.e., more than two groups), and it stores the results in list mode.

EXAMPLE

For the plasma data in Chapter 1, column 2 contains two values: 1 for male and 2 for female. The R command

```
y=mat2grp(plasma,2)
```

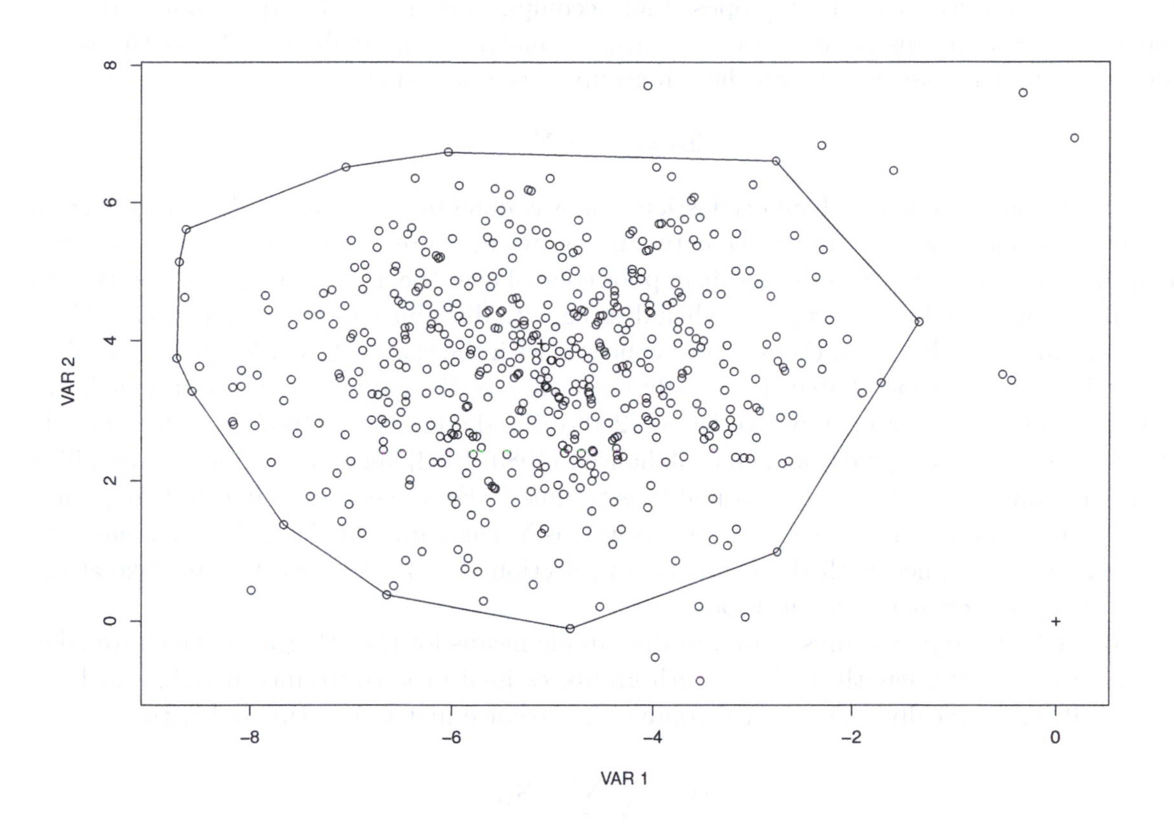

Figure 13.8 The polygon indicates a 0.95 confidence region based on the skull data. That is, there is a 0.95 probability that the population measures of location are located inside this region. The null vector is indicated by a +.

will store the plasma data for males in `y[[1]]` as a matrix, and the data for females will be stored in `y[[2]]`, again as a matrix. Imagine, for instance, we want to compare these two groups based on the data in columns 11 through 13. This would be accomplished by the command

```
smean2(y[[1]][,11:13],y[[2]][,11:13]).
```

The R function

```
mat2list(m,grp.dat)
```

splits data into groups based on the values stored in the columns of `m` indicated by the corresponding values stored in argument `coln`. This function is potentially more convenient than `mat2grp(m,coln)` because it can work on a subset of the variables stored in `m`. In contrast, `mat2grp(m,coln)` uses all of the variables stored in `m`.

13.4 MANOVA

Imagine that for each participant we have p measures and the goal is to compare J independent groups . For the jth group ($j = 1, \ldots, J$), let μ_j be the vector of means having length p. There is a classic *multivariate analysis of variance* method, generally known as MANOVA, where the goal is to test

$$H_0 : \mu_1 = \cdots = \mu_J, \tag{13.6}$$

the hypothesis that the vectors of (population) means are identical among all J groups. Several test statistics have been proposed for accomplishing this goal that assume normality and that the J groups have identical covariance matrices. In symbols, if $\mathbf{\Sigma}_j$ is the $p \times p$ covariance matrix associated with the jth group, it is assumed that

$$\mathbf{\Sigma}_1 = \cdots = \mathbf{\Sigma}_J.$$

So in addition to assuming homoscedasticity, as was done by the ANOVA F test in Section 9.1, it is assumed that the covariance between any two measures is the same for all J groups. Violating these assumptions can result in poor control over the Type I error probability and, as usual, power can be relatively low when dealing with distributions where outliers are likely to be encountered. In effect, the practical concerns with the classic ANOVA F test, described in Section 9.2, are exacerbated. Here, only an outline of the computations is provided. For complete details, see for example Anderson (2003), Hand and Taylor (1987), and Krzanowski (1988). Gamage et al. (2004) as well as Johansen (1980, 1982) suggest methods for handling unequal covariance matrices, but normality is assumed. (For a method limited to two groups, again assuming normality, see Kim, 1992, as well as Yanagihara and Yuan, 2005.) Johansen's method can be applied with the R function in Section 13.4.3. (Also see Konietschke et al., 2015, for results on bootstrap methods.)

For each of the p measures, compute the sample means for the jth group, which we label $\bar{\mathbf{X}}_j$, a vector having length p. For the jth group, estimate the covariance matrix and label the result $\mathbf{S}_j$. Typically, the assumed common covariance matrix is estimated with

$$\mathbf{W} = \frac{1}{N} \sum n_j \mathbf{S}_j,$$

where $N = \sum n_j$. It roughly plays the role of the within group sum of squares introduced in Chapter 9. An analog of the between group sum of squares is

$$\mathbf{B} = \sum n_j (\bar{\mathbf{X}}_j - \bar{\mathbf{X}})(\bar{\mathbf{X}}_j - \bar{\mathbf{X}})',$$

where $\bar{\mathbf{X}}$ is the grand mean. A popular test statistic, among the many that have been proposed, is known as the Pillai–Bartlett trace statistic. (It is the sum of the eigenvalues of a matrix based on a ratio involving $\mathbf{B}$ and $\mathbf{W}$, the details of which are not given here.) Denoting this statistic by F_m, it rejects if $F_m \geq f_{1-\alpha}$, the $1 - \alpha$ quantile of an F distribution with $\nu_1 = J - 1$ and $\nu_2 = N - p - 1$ degrees of freedom. R uses an F distribution to get an approximate critical value, but it uses adjusted degrees of freedom.

Concerns about Type I error control and power, described in Chapter 9 regarding the ANOVA F test, extend to the method just described. An extension of the test statistic F_m to trimmed means was studied by Wilcox, Keselman, Muska, and Cribbie (2000), but it was judged to be unsatisfactory in simulations. A better approach was studied by Wilcox (1995) and is described in Section 13.4.2.

13.4.1 R Function manova

R has built-in functions for performing the MANOVA method just outlined that are formulated in terms of what is called the *general linear model.* Here, the model assumes the groups have identical covariance matrices. To apply the classic MANOVA method, first use the R function `manova` to specify the model. For example, imagine that the data are stored in the R variable `xx`, a matrix with four columns. The first three columns contain the outcome variables of interest and the last column indicates the levels (groups) that are to be compared. Then the R command

```
fit=manova(xx[,1:3] ~ as.factor(xx[,4]))
```

specifies the model. To perform the F test, use the command

```
summary(fit).
```

A portion of the output will look like this:

```
                    Df Pillai  approx F     Pr(>F)
as.factor(xx[, 4])  2 0.2221   3.9970     0.000855 ***
Residuals           97
```

The first line indicates Pillai, which means that an F distribution with adjusted degrees of freedom was used when testing the null hypothesis. Here, the critical value is taken from an F distribution with 6 and 192 degrees of freedom rather than 2 and 97 degrees of freedom as is done in the previous section. (The output labels the estimated degrees of freedom 6 and 192 as num Df and den Df, respectively.) Wilks' test statistic is the most popular in the literature, but the default PillaiŰ–Bartlett statistic is recommended by Hand and Taylor (1987).

There is a simple way of performing a separate (univariate) ANOVA F test for each of the three outcome variables: use the R command

```
summary.aov(fit)).
```

13.4.2 Robust MANOVA Based on Trimmed Means

Johansen (1980) derived an alternative to the classic MANOVA method that allows the covariance matrices among the J groups to differ. Johansen's method can be generalized to trimmed means and has been found to compare well to other methods that have been derived (Wilcox, 1995).

For the jth group, let $\bar{\mathbf{X}}_j = (\bar{X}_{tj1}, \ldots, \bar{X}_{tjp})$ denote the vector of trimmed means and let $\mathbf{V}_j$ be the Winsorized covariance matrix. Let

$$\tilde{R}_j = \frac{n_j - 1}{(n_j - 2g_j)(n_j - 2g_j - 1)} \mathbf{V}_j,$$

where $g_j = \gamma n_j$, rounded down to the nearest integer, and γ is the amount of trimming. The remaining calculations are relegated to Box 13.2.

Box 13.2: Robust MANOVA using trimmed means.

Let

$$\mathbf{W}_j = \tilde{R}_j^{-1},$$

$$\mathbf{W} = \sum \mathbf{W}_j,$$

and

$$A = \frac{1}{2}\sum_{j=1}^{J}[\{tr(\mathbf{I} - \mathbf{W}^{-1}\mathbf{W}_j)\}^2 + tr\{(\mathbf{I} - \mathbf{W}^{-1}\mathbf{W}_j)^2\}]/f_j,$$

where $f_j = n_j - 2g_j - 1$. The estimate of the population trimmed means, assuming H_0 is true, is

$$\hat{\mu}_t = \mathbf{W}^{-1}\sum \mathbf{W}_j\bar{\mathbf{X}}_j.$$

The test statistic is

$$F = \sum_{j=1}^{J}\sum_{k=1}^{p}\sum_{m=1}^{p} w_{mkj}(\bar{X}_{mj} - \hat{\mu}_m)(\bar{X}_{kj} - \hat{\mu}_k), \qquad (13.7)$$

where w_{mkj} is the mkth element of $\mathbf{W}_j$, $\bar{X}_{mj}$ is the mth element of $\bar{\mathbf{X}}_j$, and $\hat{\mu}_m$ is the mth element of $\hat{\mu}_t$.
Decision Rule: Reject if

$$F \geq c + \frac{c}{2p(J-1)}\left\{A + \frac{3cA}{p(J-1)+2}\right\},$$

where c is the $1 - \alpha$ quantile of a chi-squared distribution with $p(J - 1)$ degrees of freedom.

Notice that for a between-by-within design, which was described and illustrated in Chapter 11, MANOVA provides another way of comparing the J independent groups. As a simple illustration of why MANOVA might make a difference, consider a 2-by-3, between-by-within design and imagine that for the first level of Factor A, the population means are (2, 0, 0); and for the second level of Factor A the means are (0, 0, 2). So based on the approach in Chapter 11, the levels of Factor A do not differ; each has an average population mean of 2/3. But from a MANOVA perspective, the levels of Factor A differ. (Of course, for appropriate multiple comparisons, again there are differences between the two levels of Factor A that might be detected using methods in Chapter 12.) A limitation of the robust MANOVA method is that situations are encountered where the inverse of the matrices in Box 13.2 cannot be computed. (The covariance matrices are singular.)

13.4.3 R Functions MULtr.anova and MULAOVp

The R function

```
MULtr.anova(x, J = NULL, p = NULL, tr = 0.2,alpha=.05)
```

performs the robust MANOVA method based on trimmed means. The argument J defaults to NULL, meaning that x is assumed to have list mode with length J, where x[[j]] contains a matrix with n_j rows and p columns, $j = 1, \ldots, J$. If the arguments J and p are specified, the data can be stored as described in Section 11.6.2. The R function

```
MULAOVp(x, J = NULL, p = NULL, tr = 0.2)
```

also performs the robust MANOVA method based on trimmed means, only it returns a p-value.

EXAMPLE

Imagine that two independent groups are to be compared based on measures taken at three different times. One way of comparing the groups is with a robust MANOVA method. If the data for the first group are stored in the R variable m1, a matrix having three columns, and if the data for the second group are stored in m2, also having three columns, the analysis can be performed as follows:

```
x=list()
x[[1]]=m1
x[[2]]=m2
MULtr.anova(x).
```

The function returns the test statistic and a critical value.

EXAMPLE

Section 11.6.2 illustrated how to analyze the data stored in the R variable ChickWeight. There were four independent groups of chicks that received a different diet and each chick was measured at 12 different times. To analyze the data with the R function MULAOVp, the data can again be stored as described Section 12.6.2. That is, use the R command

```
z=fac2list(ChickWeight[,1],ChickWeight[,c(4,2)]).
```

The command

```
MULAOVp(z,4,12)
```

would attempt to compare the groups, but in this case R reports that the inverse of the matrices used by this robust MANOVA method cannot be computed. Switching to means by setting the argument tr=0 produces the same error.

13.5 A MULTIVARIATE EXTENSION OF THE WILCOXON–MANN–WHITNEY TEST

Given the goal of comparing two independent groups, there are several extensions of rank-based methods to multivariate data, with each answering a different question. So, to some extent, the choice of method depends on which perspective seems most meaningful for the data being analyzed. And, of course, multiple perspectives might be desirable.

Consider the line connecting the center of the data of the first group to the center of the second group. Figure 13.9 illustrates the basic idea, which shows a scatterplot of data generated from two bivariate normal distributions. The points indicated by o denote an observation from the first group, and the points indicated by + are from the second group. The line passes through the sample means of the two groups. An approach to generalizing the

Wilcoxon–Mann–Whitney test is to first project all of the points onto this line. Roughly, once this is done, test the hypothesis that the projected points corresponding to the first group are less than the points corresponding to the second group. There are some computational and technical details that require attention, which can be found in Wilcox (2017, Section 6.9). Here, we merely describe and illustrate an R function that performs the calculations; see Section 13.5.2.

13.5.1 Explanatory Measure of Effect Size: A Projection-Type Generalization

When dealing with multiple outcome variables, there is the issue of assessing effect size. When dealing with two independent groups, one way of generalizing the methods in Section 7.11 is to use the projection of the data just described. One particular possibility is to plot the projected points for both groups, the point being that even with more than three variables for each group, we are able to get some visual sense of the extent to which the groups differ. A numeric measure of the separation of the groups, based on their measures of location, is obtained by computing the explanatory measure of effect size, which was introduced in Section 7.11, using the projected points. Another measure of effect size is the probability that based on the projections, an observation from the first group is less than an observation from the second.

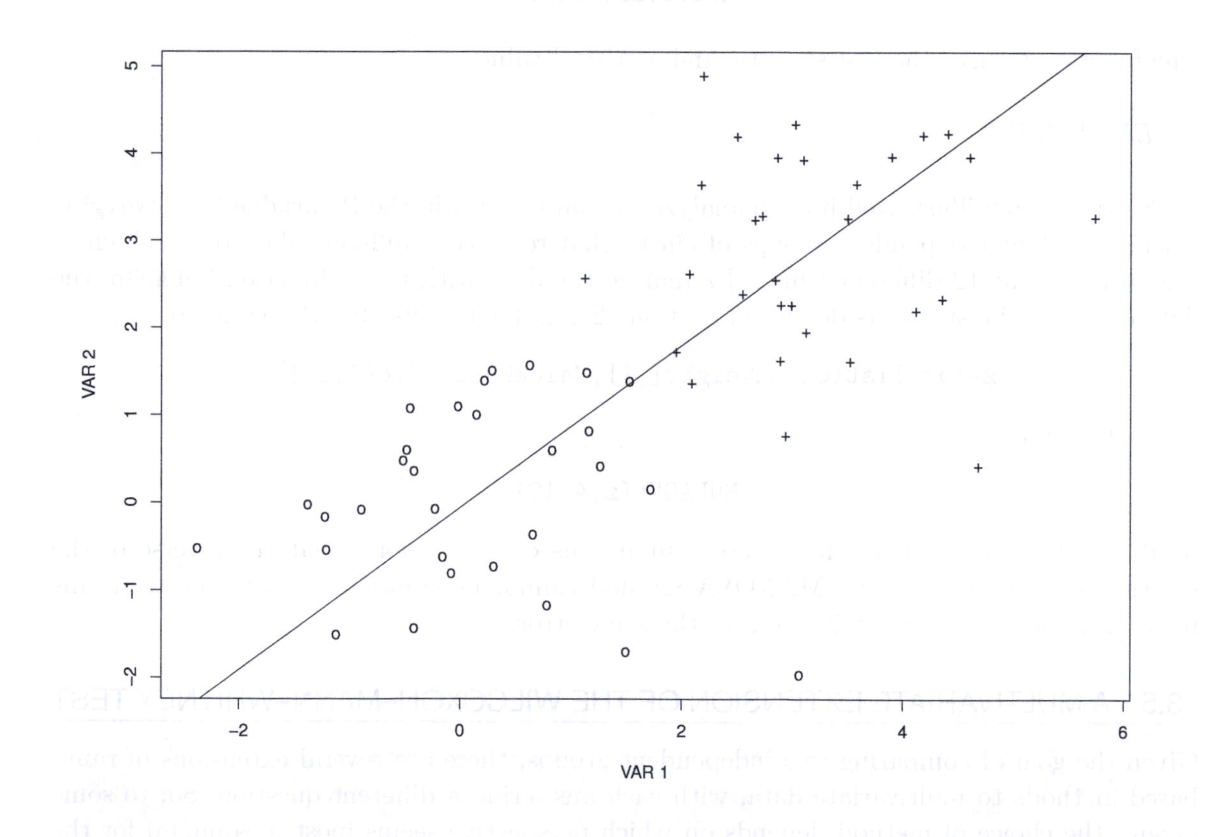

Figure 13.9 The line passes through the sample means of the two groups, with points from the first group indicated by o. A way of visualizing the difference between the groups is to project all points onto this line.

13.5.2 R Function mulwmwv2

The R function

```
mulwmwv2(m1, m2, plotit=TRUE, cop=3, alpha=0.05, nboot=1000, pop=4)
```

performs the generalization of the Wilcoxon–Mann–Whitney test outlined in the previous section. The argument cop determines which measure of location will be used for each group. By default the marginal medians are used. Setting cop=1 results in the Donoho–Gasko median, and 2 will result in the MCD measure of location. The argument pop determines the type of plot that will be used when the argument plotit=TRUE. By default, an adaptive kernel density estimator is used. (The function computes the projected points for each group and plots the results by calling the R function g2plot.) Setting pop=2 results in boxplots. The Type I error probability can be set via the argument alpha; the function does not return a p-value.

EXAMPLE

Consider again the skull data used in the previous example, only now the groups are compared using all four skull measurements. The R function mulwmwv2 returns

```
$phat
[1] 0.1372222

$lower.crit
[1] 0.2494444

$upper.crit
[1] 0.7033333

$Effect.Size
[1] 0.8319884
```

The plot of the projected points is shown in Figure 13.10. The solid line is the plot of the projected points associated with the first group, and the dashed line is the plot for the second group. The function estimates that the probability of a projected point from the first group being less than a projected point from the second group is 0.137, which corresponds to the output under phat. If this value is less than the value indicated by lower.crit, or greater than the value under upper.crit, reject the hypothesis that the probability is 0.5. So here this hypothesis is rejected and the explanatory measure of effect size is 0.832, suggesting a relatively large separation between the corresponding measures of location.

13.6 RANK-BASED MULTIVARIATE METHODS

There are several rank-based methods for dealing with multivariate data where the goal is to compare J independent groups (e.g., Choi and Marden, 1997; Liu and Singh, 1993; Hettmansperger et al. 1997; Munzel and Brunner, 2000b). The theoretical details of the method developed by Hettmansperger et al. assume tied values never occur and that distributions are symmetric, so their method is not considered here. Here the focus is on the Munzel–Brunner method and the Choi–Marden method.

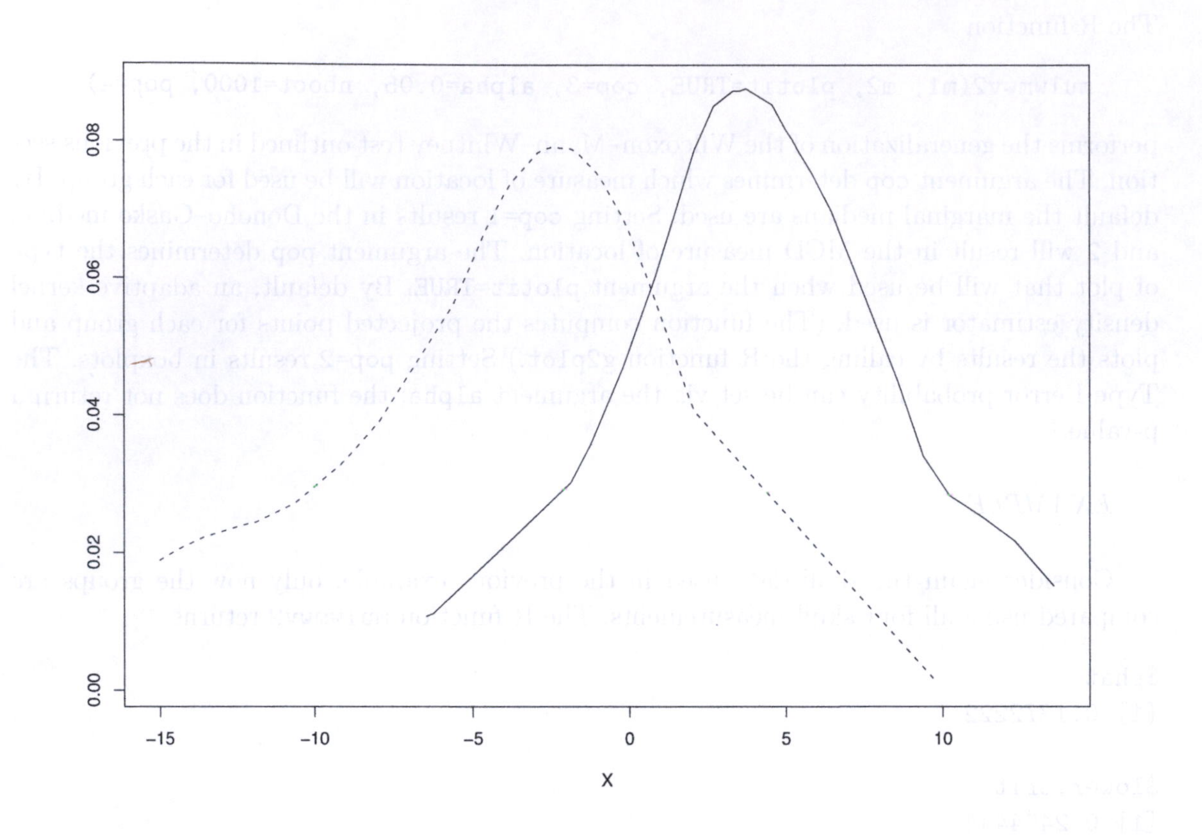

Figure 13.10 Shown is the plot returned by the R function mulwmwv2 based on the skull data. The solid line is the plot of the projected points associated with the first group. The dashed line is the plot for the second group. Notice that the plot suggests that there is a relatively low probability that a projected point from the first group is less than a projected point from the second.

13.6.1 The Munzel–Brunner Method

This section describes a one-way multivariate method derived by Munzel and Brunner (2000b). For the jth group, there are n_j randomly sampled vectors of observations, with each vector containing K measures. Let $F_{jk}(x) = P(X_{jk} \le x)$ be the distribution associated with the jth group and kth measure ($k = 1, \ldots, K$). For example, $F_{32}(6)$ is the probability that for the third group, the second variable will be less than or equal to 6 for a randomly sampled individual. For the kth measure, the goal is to test the hypothesis that all J groups have identical distributions. The more general goal is to test the hypothesis that simultaneously, all groups have identical distributions for each of the K measures under consideration. That is, the goal is to test

$$H_0 : F_{1k}(x) = \cdots = F_{Jk}(x) \tag{13.8}$$

for all possible values of x and all $k = 1, \ldots, K$.

The method begins by pooling the first of the K measures of all J groups and assigning ranks. Ties are handled in the manner described in Section 7.8.6. (Midranks are used.) Repeat this process for the remaining $K - 1$ measures and label the results R_{ijk}. That is, R_{ijk} is the rank of the ith observation in the jth group and for the kth measure. Let

$$\bar{R}_{jk} = \frac{1}{n_j}\sum_{i=1}^{n_j} R_{ijk}$$

be the average of the ranks for the jth group corresponding to the kth measure. Set

$$\hat{Q}_{jk} = \frac{\bar{R}_{jk} - .5}{n},$$

where $n = \sum n_j$ is the total number of randomly sampled vectors among the J groups. The remaining calculations are summarized in Box 13.3. The $\hat{Q}$ values are called the *relative effects* and reflect the ordering of the average ranks. If, for example, $\hat{Q}_{11} < \hat{Q}_{21}$, the typical rank for variable one in group one is less than the typical rank for variable one in group two. More generally, if $\hat{Q}_{jk} < \hat{Q}_{j'k}$, then based on the kth measure, the typical rank (or observed value) for group j is less than the typical rank for group j'.

Box 13.3: The Munzel–Brunner one-way multivariate method

Let $\mathbf{I}_J$ be a J-by-J identity matrix, let $\mathbf{H}_J$ be a J-by-J matrix of ones, and compute

$$\mathbf{P}_J = \mathbf{I}_J - \frac{1}{J}\mathbf{H}_J,\ \mathbf{M}_A = \mathbf{P}_J \otimes \frac{1}{K}\mathbf{H}_K,$$

$$\hat{\mathbf{Q}} = (\hat{Q}_{11}, \hat{Q}_{12}, \ldots, \hat{Q}_{1K}, \hat{Q}_{21}, \ldots, \hat{Q}_{JK})',$$

$$\mathbf{R}_{ij} = (R_{ij1}, \ldots, R_{ijK})',\ \bar{\mathbf{R}}_j = (\bar{R}_{j1}, \ldots, \bar{R}_{jK})',$$

$$\mathbf{V}_j = \frac{1}{nn_j(n_j-1)} = \sum_{i=1}^{n_j}(\mathbf{R}_{ij} - \bar{\mathbf{R}}_j)(\mathbf{R}_{ij} - \bar{\mathbf{R}}_j)',$$

where $n = \sum n_j$ and let

$$\mathbf{V} = \mathrm{diag}\{\mathbf{V}_1, \ldots, \mathbf{V}_J\}.$$

The test statistic is

$$F = \frac{n}{\mathrm{tr}(\mathbf{M}_A\mathbf{V})}\hat{\mathbf{Q}}' M_A \hat{\mathbf{Q}}.$$

Decision Rule: Reject if $F \geq f$, where f is the $1-\alpha$ quantile of an F distribution with

$$\nu_1 = \frac{(\mathrm{tr}(\mathbf{M}_A\mathbf{V}))^2}{\mathrm{tr}(\mathbf{M}_A\mathbf{V}\mathbf{M}_A\mathbf{V})},$$

and $\nu_2 = \infty$ degrees of freedom.

13.6.2 R Function mulrank

The R function

```
mulrank(J, K, x)
```

Table 13.1 CGI and PGI scores after four weeks of treatment

Exercise		Clomipramine		Placebo	
CGI	PGI	CGI	PGI	CGI	PGI
4	3	1	2	5	4
1	1	1	1	5	5
2	2	2	0	5	6
2	3	2	1	5	4
2	3	2	3	2	6
1	2	2	3	4	6
3	3	3	4	1	1
2	3	1	4	4	5
5	5	1	1	2	1
2	2	2	0	4	4
5	5	2	3	5	5
2	4	1	0	4	4
2	1	1	1	5	4
2	4	1	1	5	4
6	5	2	1	3	4

performs the one-way multivariate method in Box 13.3. The data are stored in `x` which can be a matrix or have list mode. If `x` is a matrix, the first K columns correspond to the K measures for group 1, the second K correspond to group 2, and so forth. If stored in list mode, `x[[1]], ..., x[[K]]` contain the data for group 1, `x[[K+1]], ..., x[[2K]]` contain the data for group 2, and so on.

EXAMPLE

Table 13.1 summarizes data (reported by Munzel and Brunner, 2000) from a psychiatric clinical trial where three methods are compared for treating individuals with panic disorder. The three methods are exercise, clomipramine, and a placebo. The two measures of effectiveness are a clinical global impression (CGI) and the patient's global impression (PGI). The test statistic returned by `mulrank` is $F = 12.7$ with $\nu_1 = 2.83$ and a p-value less than 0.001. The relative effects are:

```
$q.hat:
          [,1]      [,2]
[1,] 0.5074074 0.5096296
[2,] 0.2859259 0.2837037
[3,] 0.7066667 0.7066667
```

So among the three groups, the second group, clomipramine, has the lowest relative effects. That is, the typical ranks were lowest for this group, and the placebo group had the highest ranks on average.

13.6.3 The Choi–Marden Multivariate Rank Test

This section describes a multivariate analog of the Kruskal–Wallis test derived by Choi and Marden (1997). There are, in fact, many variations of the approach they considered, but here attention is restricted to the version they focused on. As with the method in Section 13.5.1, we have K measures for each individual and there are J independent groups. For the jth group and any vector of constants $\mathbf{x} = (x_1, \ldots, x_K)$, let

$$F_j(\mathbf{x}) = P(X_{j1} \le x_1, \ldots, X_{jK} \le x_K).$$

So, for example, $F_1(\mathbf{x})$ is the probability that for the first group, the first of the K measures is less than or equal to x_1, the second of the K measures is less than or equal to x_2, and so forth. The null hypothesis is that for any $\mathbf{x}$,

$$H_0 : F_1(\mathbf{x}) = \cdots = F_J(\mathbf{x}), \tag{13.9}$$

which is sometimes called the *multivariate hypothesis* to distinguish it from Equation (13.8), which is called the *marginal hypothesis.* The multivariate hypothesis is a stronger hypothesis in the sense that if it is true, then by implication the marginal hypothesis is true as well. For example, if the marginal distributions for both groups are standard normal distributions, the marginal hypothesis is true, but if the groups have different correlations, the multivariate hypothesis is false.

The Choi–Marden method represents an extension of a technique derived by Möttönen and Oja (1995) and is based on a generalization of the notion of a rank to multivariate data. (Also see Chaudhuri, 1996, Section 4.) First, consider a random sample of n observations with K measures for each individual or thing and denote the ith vector of observations by

$$\mathbf{X}_i = (X_{i1}, \ldots, X_{iK}).$$

Let

$$A_{ii'} = \sqrt{\sum_{k=1}^{K} (X_{ik} - X_{i',k})^2}.$$

Here, the "rank" of the ith vector is itself a vector (having length K) given by

$$\mathbf{R}_i = \frac{1}{n} \sum_{i'=1}^{n} \frac{\mathbf{X}_i - \mathbf{X}_{i'}}{A_{ii'}},$$

where

$$\mathbf{X}_i - \mathbf{X}_{i'} = (X_{i1} - X_{i'1}, \ldots, X_{iK} - X_{i'K}).$$

The remaining calculations are summarized in Box 13.4. All indications are that this method provides good control over the probability of a Type I error when ties never occur. There are no known problems when there are tied values, but this issue is in need of more research.

Box 13.4: The Choi–Marden method.

Pool the data from all J groups and compute rank vectors as just described in the text. The resulting rank vectors are denoted by $\mathbf{R}_1, \ldots, \mathbf{R}_n$, where $n = \sum n_j$ is the total number of vectors among the J groups. For each of the J groups, average the rank vectors and denote the average of these vectors for the jth group by $\bar{\mathbf{R}}_j$.

Next, assign ranks to the vectors in the jth group, ignoring all other groups. We let $\mathbf{V}_{ij}$ (a column vector of length K) represent the rank vector corresponding to the ith vector of the jth group ($i = 1, \ldots, n_j$; $j = 1, \ldots, J$) to make a clear distinction with the ranks based on the pooled data. Compute

$$\mathbf{S} = \frac{1}{n-J} \sum_{j=1}^{J} \sum_{i=1}^{n_j} \mathbf{V}_{ij} \mathbf{V}'_{ij},$$

where $\mathbf{V}'_{ij}$ is the transpose of $\mathbf{V}_{ij}$ (so $\mathbf{S}$ is a K-by-K matrix). The test statistic is

$$H = \sum_{j=1}^{J} n_j \bar{\mathbf{R}}'_j \mathbf{S}^{-1} \bar{\mathbf{R}}_j. \tag{13.10}$$

(For $K = 1$, H does not quite reduce to the Kruskall–Wallis test statistic. In fact, H avoids a certain technical problem that is not addressed by the Kruskall–Wallis method.)

Decision Rule: Reject if $H \geq c$, where c is the $1-\alpha$ quantile of a chi-squared distribution with degrees of freedom $K(J-1)$.

13.6.4 R Function cmanova

The R function

```
cmanova(J,K,x)
```

performs the Choi–Marden method just described. The data are assumed to be stored in x as described in Section 13.6.2.

13.7 MULTIVARIATE REGRESSION

Consider a regression problem where there are p predictors $\mathbf{X} = (X_1, \ldots, X_p)$ and q responses $\mathbf{Y} = (Y_1, \ldots, Y_q)$. For example, in the reading study mentioned in Section 6.4.2, there were four outcome variables of interest (measures of reading ability), so here $q = 4$, and the goal was to investigate how well seven variables ($p = 7$) are able to predict the four measures of reading ability. The usual multivariate regression model is

$$\mathbf{Y} = \mathbf{B}'\mathbf{X} + \mathbf{b}_0 + \epsilon, \tag{13.11}$$

where $\mathbf{B}$ is a $(p \times q)$ slope matrix, $\mathbf{b}_0$ is a q-dimensional intercept vector, and the errors $\epsilon = (\epsilon_1, \ldots, \epsilon_q)$ are independent and identically distributed with mean $\mathbf{0}$ and covariance matrix $\mathbf{\Sigma}_\epsilon$, a square matrix of size q.

Here is the typical way of describing an extension of the least squares regression estimator in Section 6.1 (where there is $q = 1$ outcome variable) to the situation at hand, where $q > 1$. To begin, let

$$\mu = \begin{pmatrix} \mu_x \\ \mu_y \end{pmatrix} \quad \text{and} \quad \mathbf{\Sigma} = \begin{pmatrix} \mathbf{\Sigma}_{xx} & \mathbf{\Sigma}_{xy} \\ \mathbf{\Sigma}_{yx} & \mathbf{\Sigma}_{yy} \end{pmatrix}.$$

So μ is a vector of means, with the first p means corresponding to the means of the p predictors, and the remaining means are the q means associated with the dependent variable $\mathbf{Y}$. As for $\mathbf{\Sigma}$, it is composed of four covariance matrices. The first is the covariance matrix associated with the p predictors, which is denoted by $\mathbf{\Sigma}_{xx}$. The notation $\mathbf{\Sigma}_{xy}$ refers to the covariances between the predictors and the outcome variable. The estimates of $\mathbf{B}$ (the slope parameters) and $\mathbf{b}_0$ are

$$\hat{\mathbf{B}} = \hat{\mathbf{\Sigma}}_{xx}^{-1}\hat{\mathbf{\Sigma}}_{xy} \tag{13.12}$$

and

$$\hat{\mathbf{b}}_0 = \hat{\mu}_y - \hat{\mathbf{B}}'\hat{\mu}_x, \tag{13.13}$$

respectively. The estimate of the covariance matrix associated with the error term, ϵ, is

$$\hat{\mathbf{\Sigma}}_\epsilon = \hat{\mathbf{\Sigma}}_{yy} - \hat{\mathbf{B}}'\hat{\mathbf{\Sigma}}_{xx}\hat{\mathbf{B}}. \tag{13.14}$$

This standard way of describing multivariate regression might seem to suggest that it takes into account the dependence among the outcome variables. However, this is not the case in the following sense. The estimation procedure just described is tantamount to simply applying the least squares estimator to each of the dependent variables (e.g., Jhun and Choi, 2009). Imagine, for example, there are three outcome variables ($q = 3$), and say four predictors ($p = 4$). If we focus on the first outcome variable, ignoring the other two, and apply least squares regression, the resulting estimates of the slopes and intercept would be identical to the values returned by the multivariate regression estimator just described.

13.7.1 Multivariate Regression Using R

Multivariate regression, using R in conjunction with the estimator just described, can be done using commands very similar to those illustrated in Chapter 6.

EXAMPLE

As a simple illustration, imagine that there are three predictors stored in the R matrix x and two outcome variables stored in y. Then the R command

```
model=lm(y ~ x)
```

specifies the model given by Equation (13.10). To test hypotheses about the slope parameters being equal to zero, assuming normality and homoscedasticity, use the command

```
summary.aov(model).
```

The output will look like this:

```
 Response 1 :
            Df Sum Sq Mean Sq F value Pr(>F)
x            3  1.418   0.473   0.362 0.7808
Residuals   36 47.008   1.306

 Response 2 :
            Df  Sum Sq Mean Sq F value  Pr(>F)
x            3  3.8862  1.2954  2.4642 0.07805 .
Residuals   36 18.9248  0.5257
```

So for the first dependent variable, labeled Response 1, the hypothesis that all three slopes are zero has a p-value of 0.7808. For the second outcome variable, the p-value is 0.07805.

13.7.2 Robust Multivariate Regression

The estimates of the slopes and intercepts given by Equations (13.12) and (13.13) are not robust, for reasons outlined in Chapter 6. Numerous methods have been proposed for dealing with this problem. Two recent examples are a robust multivariate regression estimator derived by Rousseeuw, et al. (2004) and another proposed by Bai et al. (1990), both of which take into account the dependence among the outcome variables, in contrast to most other estimators that have been proposed (cf. Ben et al., 2006). This is of practical importance because when the outcome variables are positively correlated, the overall estimate of the parameters can be more accurate compared to methods that do not take into account the association among the outcome variables (e.g., Jhun and Choi, 2009). Another advantage of the Rousseeuw et al. estimator is that it performs well, in terms of achieving a relatively small standard error, when there is heteroscedasticity (Wilcox, 2009a). Whether the Bai estimator has a similar advantage has not been investigated.

Another approach to multivariate regression is to simply apply some robust estimator to each of the outcome variables under study. In terms of achieving a relatively small standard error, even when there is heteroscedasticity, the Theil–Sen estimator (which is introduced in Section 14.1.1) appears to be a good choice (Wilcox, 2009a). But like the method in the previous section, this approach does not take into account the dependence among the outcome variables. This issue can be addressed somewhat by first removing any outliers found by the projection method in Section 13.1.6 and then applying the Theil–Sen estimator to the data that remain. That is, combine both the predictors $\mathbf{X}$ and outcome variables $\mathbf{Y}$ into a single matrix and then check for outliers. If, for example, row 9 of the combined matrix is declared an outlier, remove row 9 from both $\mathbf{X}$ and $\mathbf{Y}$. The same is done for all other rows declared outliers, and then the Theil–Sen estimator is applied using the remaining data. In terms of testing hypotheses about the slopes and intercepts, a possible appeal of the Theil–Sen estimator is that published papers indicate that accurate confidence intervals and good control over the probability of a Type I error can be achieved even under heteroscedasticity. For the situation at hand, one can simply test hypotheses focusing on the first outcome (dependent) variable, ignoring the others, using the R functions in Sections 14.3.1. The same can be done with the remaining outcome variables. Currently, however, given the goal of testing hypotheses, there are no published papers on how best to proceed when the Theil–Sen estimator is applied after removing outliers via the projection outlier method as just described. There are no results regarding testing hypotheses based on the estimator derived by Rousseeuw et al. (2004).

13.7.3 R Function mlrreg and mopreg

The R function

```
mlrreg(x,y).
```

performs the multivariate regression estimator derived by Rousseeuw et al. (2004). Currently the best method for testing H_0: $\mathbf{B} = 0$ is based in part on a bootstrap estimate of the standard errors. The method can be applied via the R function

```
mlrreg.Stest(x,y,nboot=100,SEED=TRUE).
```

The function

```
mopreg(x,y,regfun=tsreg,cop=3,KEEP=TRUE, MC=FALSE)
```

estimates the slopes and intercepts for each outcome variable using the method specified by the argument `regfun`, which defaults to the Theil–Sen estimator. Setting the argument `KEEP=FALSE`, any outliers detected by the projection method will be eliminated. As usual, `MC=TRUE` will reduce execution time, assuming a multicore processor is available.

13.8 PRINCIPAL COMPONENTS

Imagine a situation involving a large number of variables. For various reasons, it might be desired to reduce the number of variables while simultaneously retaining a large proportion of the information that is available. For example, suppose numerous measures are taken for each participant and that five of the variables are used to measure a participant's cognitive ability. Further imagine that the five cognitive variables are highly correlated. Then in some sense these five cognitive variables provide redundant information, in which case it might be desired to reduce the five variables to one or two variables only. *Principal components analysis* (PCA) is one way of approaching this problem.

The basic idea is perhaps easiest to explain if we momentarily focus on the goal of reducing p variables down to one variable only. The principal component strategy consists of using a linear combination of the p variables. For the ith participant, the goal is to replace the p variables $X_{i1}, \ldots, X_{ip}$ with

$$V_i = w_1 X_{i1} + \cdots + w_p X_{ip},$$

where the constants $w_1, \ldots, w_p$ are to be determined, and this is done for each participant ($i = 1, \ldots, n$). In effect, the $n \times p$ matrix of data, $\mathbf{X}$, represents n vectors, each having length p, which has been reduced to the n values $V_1, \ldots, V_n$. The strategy used by principal components is to choose $w_1, \ldots, w_p$ so that the variance among the values $V_1, \ldots, V_n$ is relatively large. To give some graphical sense of why, look at Figure 13.5 and imagine that all the points are projected onto the line shown. In the notation used here, these projections are obtained using a particular choice for the weights $w_1, \ldots, w_p$. Once all points are projected onto this line, the variance of the resulting points can be computed. But suppose instead the points had been projected onto a line that is perpendicular to the one shown in Figure 13.5. Notice now that the variance of the resulting points will be smaller, roughly suggesting that more information has been lost in some sense. The principal component strategy is to retain as much information as possible by choosing the weights so as to maximize the variance.

To make progress, a restriction must be imposed on the weights $w_1, \ldots, w_p$. To explain why, consider the case where all of the variables have a positive correlation. Then the larger the weights, the larger will be the variance. That is, the solution is to take $w_1 = \cdots = w_p = \infty$, which has no practical value. To deal with this issue, the convention is to require that

$$\sum_{j=1}^{p} w_j^2 = 1.$$

Now consider the problem of reducing the p variables down to two variables rather than just one. So for the ith participant, the goal is to compute two linear combinations of the p variables based on two sets of weights:

$$V_{i1} = w_{11}X_{i1} + \cdots + w_{1p}pX_{ip},$$

and

$$V_{i2} = w_{21}X_{i1} + \cdots + w_{2p}X_{ip}$$

($i = 1, \ldots, n$), which are generally called the *principal component scores.* For both weighted sums, again the goal is to maximize their variance. That is, the goal is to maximize the variance of the values $V_{11}, V_{21}, \ldots, V_{1n}$, as well as the values $V_{12}, V_{22}, \ldots, V_{2n}$. But now an additional condition is imposed: the weights are chosen so that Pearson's correlation between $V_{11}, V_{21}, \ldots, V_{1n}$ and $V_{12}, V_{22}, \ldots, V_{2n}$ is 0. This process can be extended so that the original data are transformed into an $n \times p$ matrix $\mathbf{V}$ with the property that the p variables have zero (Pearson) correlations and the variances of each of the p variables have been maximized. Moreover, if we let $s_1^2, \ldots, s_p^2$, represent the sample variances associated with columns $1, \ldots, p$ of $\mathbf{V}$, respectively, the method is constructed so that

$$s_1^2 \geq s_2^2 \geq \cdots \geq s_p^2. \tag{13.15}$$

More theoretically oriented books provide a formal mathematical description of principal component analysis (e.g., Mardia et al., 1992, Section 8.2.1). Box 13.5 provides a brief outline, one point being that the *eigenvalues* (defined in Appendix C) of the covariance (or correlation) matrix play a role; they correspond to the variances of the principal components. This helps explain why modern robust methods sometimes report eigenvalues associated with some robust analog of the covariance matrix.

Box 13.5: Outline of a principal component analysis.

Briefly, consider any $p \times p$ matrix $\mathbf{\Sigma}$. Here, $\mathbf{\Sigma}$ is either a (population) correlation matrix or a covariance matrix based on the p variables under study. Let $\lambda_p \geq \cdots \geq \lambda_1 \geq 0$ be the eigenvalues (described in Appendix C) associated with $\mathbf{\Sigma}$ and let $\mathbf{\Lambda}$ be the diagonal matrix corresponding to $\lambda_p, \ldots, \lambda_1$, where $\mathbf{\Lambda}'\mathbf{\Lambda} = \mathbf{I}$. (That is, $\mathbf{\Lambda}$ is orthogonal.) Then it is possible to determine a matrix $\mathbf{\Gamma}$ such that $\mathbf{\Gamma}'\mathbf{\Sigma}\mathbf{\Gamma} = \mathbf{\Lambda}$ using what is called a *spectral decomposition.* The principal component transformation of $\mathbf{X}$ is

$$\mathbf{P} = \mathbf{\Gamma}'(\mathbf{X} - \mu). \tag{13.16}$$

It can be shown that the (population) variances corresponding to the p columns of $\mathbf{P}$ are equal to the eigenvalues $\lambda_p, \ldots, \lambda_1$. This helps explain why some recently developed robust generalizations of principal component analysis report eigenvalues associated with some robust covariance matrix. The idea is that the transformed variables with the larger eigenvalues account for more of the variability among the original data, $\mathbf{X}$. The sample variances indicated by Equation (13.15) are the sample eigenvalues based on an estimate of the covariance matrix $\mathbf{\Sigma}$.

Using the correlations, rather than the covariances, becomes an issue when dealing with the common goal of deciding how many of the p principal components are needed to capture the bulk of the variability. A common strategy is to use the first m principal components, with m chosen so as to account for a relatively large proportion of the total variance. But as illustrated in the next section, differences in the scale of the variables can have a large impact on the choice of m when using the covariance matrix. A method for dealing with this issue is to use a correlation matrix instead.

13.8.1 R Functions prcomp and regpca

The built-in R function

```
prcomp(x,cor=FALSE)
```

in conjunction with the R function `summary`, can be used to perform the classic principal component analysis just outlined. The argument `cor` indicates whether the analysis will be done on the covariance matrix or the correlation matrix. By default, `cor=FALSE`, meaning that the covariance matrix will be used. The R function `predict` returns the principal component scores. (The function `prcomp` has many additional arguments that are not explained here. Use the R command `?prcomp` for more information.)

The R function `prcomp` returns an error if any values are missing. If it is desired to simply remove any rows of data with missing values, the R function

```
regpca(x,cor=T,SCORES = FALSE, scree=TRUE, xlab="Principal
        Component",ylab="Proportion of Variance"),
```

will do this automatically and then call the R function `prcomp`. The function `regpca` also creates a scree plot, which is a line segment that shows the fraction of the total variance as a function of the number of components. Scree plots are sometimes used to help judge where the most important components cease and the least important components begin. The point of separation is called the "elbow." Setting the argument `SCORES=TRUE`, the function returns the principal component scores. Unlike `prcomp`, the function `regpca` defaults to using the correlation matrix rather than the covariance matrix. To use the covariance matrix, set the argument `cor=FALSE`.

EXAMPLE

As an illustration, 200 observations were generated from a multivariate normal distribution with $p = 4$ variables and all correlations equal to 0. Storing the data in the R variable `x`, the command `summary(prcomp(x))` returned

```
Importance of components:
                         PC1   PC2   PC3   PC4
Standard deviation     1.113 0.963 0.959 0.914
Proportion of Variance 0.316 0.236 0.235 0.213
Cumulative Proportion  0.316 0.552 0.787 1.000
```

The first line, headed by "Standard deviation," indicates the standard deviation of the principal component scores. The next line reports the proportion of variance. That is, for each component, divide the variance of the principal component scores by the total variance (the sum of the variances). The next line reports the cumulative proportions based on the proportion of variance values listed in the previous line.

Notice that the standard deviations for each of the principal components (labeled PC1, PC2, PC3, and PC4) are approximately equal, suggesting that all four variables are needed to capture the variation in the data, a result that was expected because all pairs of variables have zero correlation. The cumulative proportion associated with the first three principal components is 0.787. A scree plot is shown in Figure 13.11. The bottom (solid) line shows the variance associated with the principal components. The upper (dashed) line is the cumulative proportion. Note that the lower line is nearly horizontal with no steep declines, suggesting that all four components should be used to capture the variability in the data. Another rule that is sometimes used is to retain those components for which the proportion of variance is greater than 0.1. When the proportion is less than 0.1, it has been suggested that the corresponding principal component rarely has much interpretive value. (Assuming multivariate normality, methods for testing the equality of the eigenvalues are available. See, for example, Muirhead, 1982, Section 9.6.)

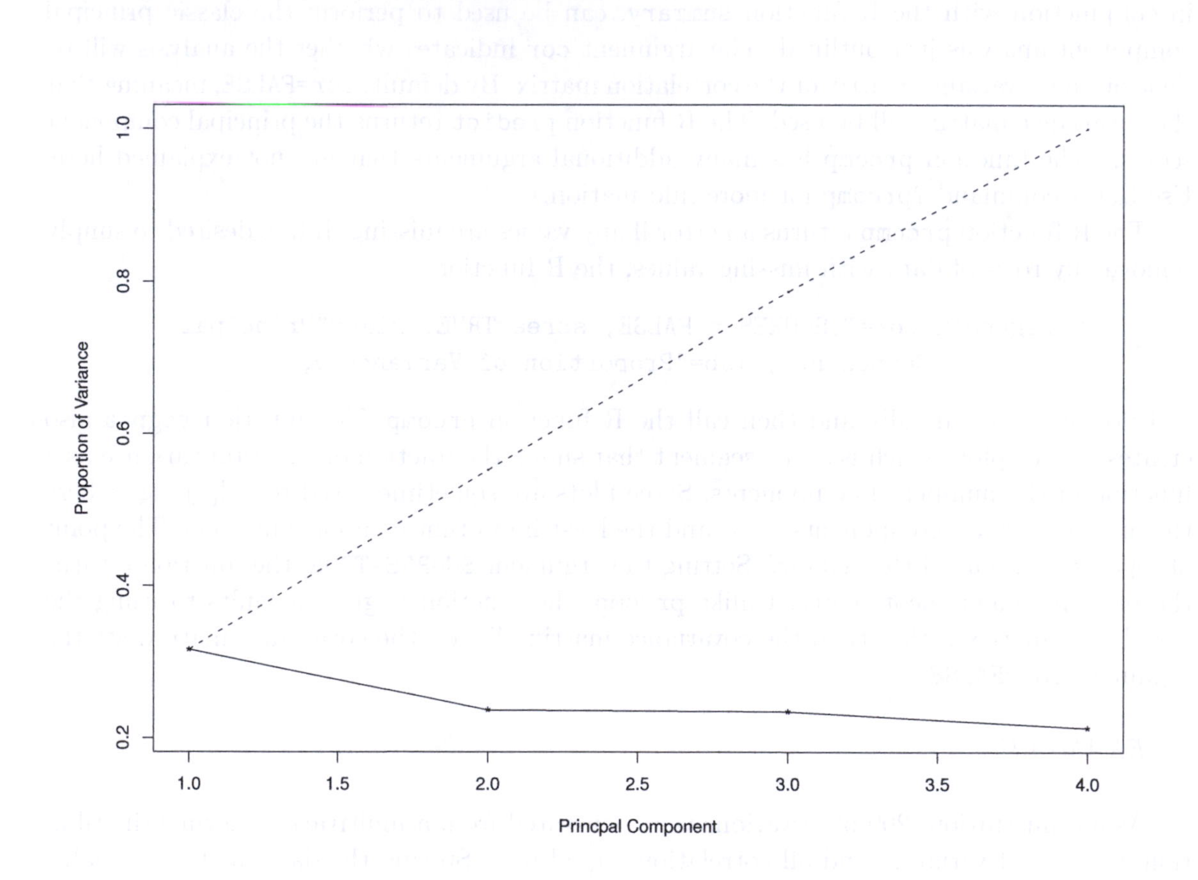

Figure 13.11 The scree plot corresponding to the first illustration in this section where all variables have normal distributions with correlations all equal to zero. The bottom (solid) line indicates the variance associated with the principal components. The upper (dashed) line is the cumulative proportion.

EXAMPLE

We repeat the last example, only now the correlations among all four variables is 0.9. Now the output is

```
Importance of components:
                         PC1    PC2    PC3    PC4
Standard deviation     1.869 0.3444 0.3044 0.2915
```

```
Proportion of Variance 0.922 0.0313 0.0244 0.0224
Cumulative Proportion  0.922 0.9531 0.9776 1.0000
```

Note that the first principal component has a much larger standard deviation than the other three principal components. The proportion of variance accounted for by PC1 is 0.922, suggesting that it is sufficient to use the first principal component only to capture the variability in the data. Figure 13.12 shows the scree plot. Note the steep decline going from the first to the second principal component, after which the curve is nearly horizontal. This suggests that a single principal component is sufficient for capturing the variability in the data.

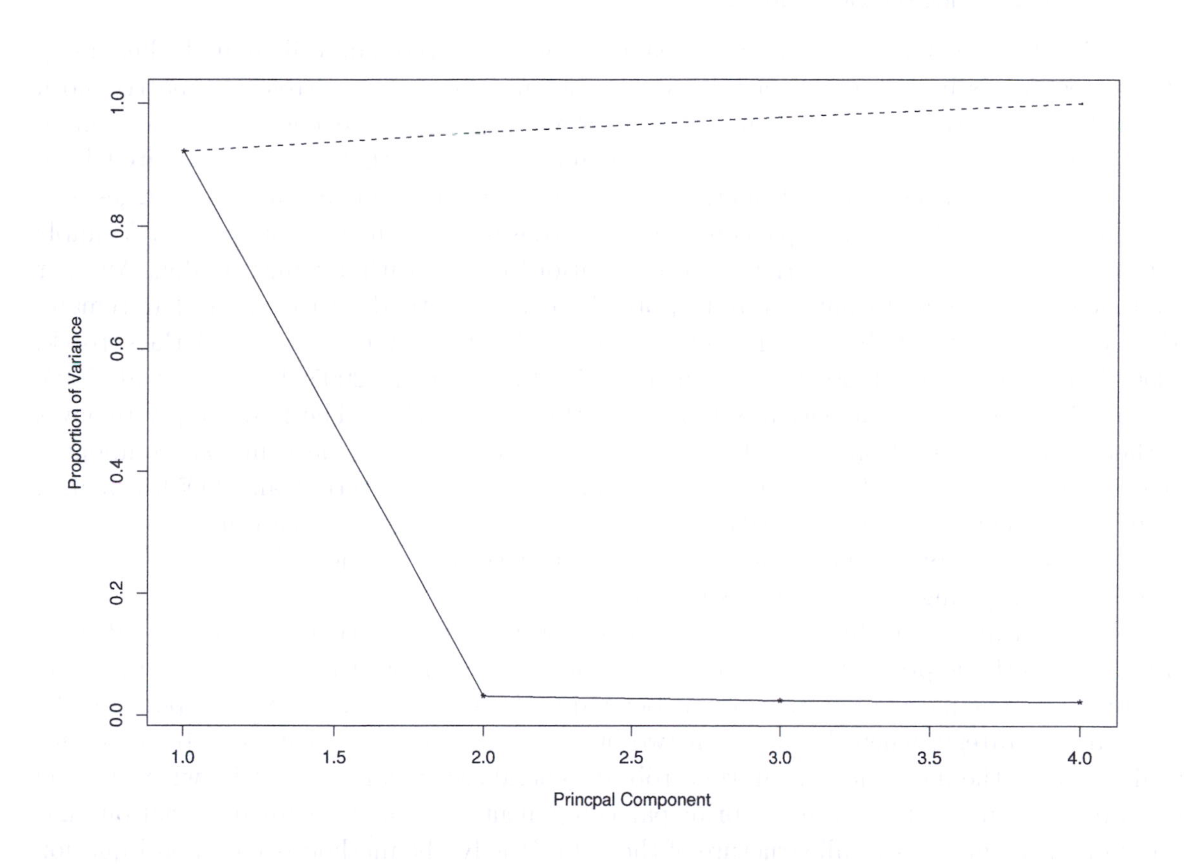

Figure 13.12 The scree plot corresponding to the second illustration in this section where all variables have normal distributions with correlations all equal to 0.9. Note that in contrast to Figure 13.11, for the solid line, there is a distinct bend at component 2, which is taken to suggest that only the first component is important.

An important point is that changing the scale of one of the variables can alter the conclusions reached regarding how many components are needed to capture the bulk of the variability.

EXAMPLE

Consider again the first example in this section where all of the correlations are zero. As was noted, the output suggests that all four variables are needed to capture the bulk of the variability among the data. Here, the same data are used, only the first variable is multiplied by 12, as might be done when switching from feet to inches. Now the output looks like this:

```
Importance of components:
                          Comp.1   Comp.2   Comp.3   Comp.4
Standard deviation      12.35025 1.025970 0.959234 0.91773695
Proportion of Variance   0.98119 0.006776 0.005923 0.00542179
Cumulative Proportion    0.98119 0.988655 0.994578 1.00000000
```

which suggests that a single principal component is needed rather than four. For this reason, when variables are measured on different scales, it is often suggested that principal component analysis be done on the correlation matrix of the variables rather the covariance matrix.

13.8.2 Robust Principal Components

As previously indicated, the classic principal components method, described at the beginning of this section, is based on the variance-covariance matrix, or the correlation matrix, both of which are sensitive to outliers. In practical terms, situations are encountered where, upon closer scrutiny, the resulting components explain a structure that has been created by a mere one or two outliers (e.g., Huber, 1981, p. 199). This has led to numerous suggestions regarding how the classic principal component analysis might be made more robust. A simple approach is to replace the covariance (or correlation) matrix with a robust analog. Another approach is to eliminate outliers and apply the classic method to the data that remain. More complex methods have been proposed (e.g., Maronna, 2005; Croux and Haesbroeck, 2000; Locantore et al., 1999; Li and Chen, 1985; Hubert et al., 2005; Engelen et al. 2005; She et al., 2016). Here, attention is limited to three approaches. The first simply removes outliers via the projection method in Section 13.1.5 followed by the principal component analysis in Section 13.8. The second is a method proposed by Hubert et al. (2005). Roughly, their method consists of a combination of down weighting outliers and certain projections of the data. The computational details are rather involved and not described here. But an R function for applying the method is supplied.

The third approach differs in a fundamental way from the other methods listed here. All of the methods previously cited have a property in common: they are based in part on maximizing some measure of variation associated with each of the principal components. An alternative strategy, when dealing with two or more principal components, is to search for projections of the data that maximize a robust generalized variance. That is, when judging the variance among two or more principal components, use a measure of variation that takes into account the overall structure of the data. Briefly, the method is based on Equation (13.16) with the mean replaced by the MCD measure of location. Simultaneously, the method chooses the matrix $\mathbf{\Gamma}$ in Equation (13.16) so as to maximize the generalized variance of $\mathbf{P}$ associated with some robust correlation (or covariance) matrix. (This is done using a general algorithm derived by Nelder and Mead, 1965, for maximizing a function. See Olsson and Nelson, 1975, for a discussion of the relative merits of the Nelder–Mead algorithm.)

13.8.3 R Functions outpca, robpca, robpcaS, Ppca, and Ppca.summary

The R function

```
outpca(x,cor=TRUE, SCORES=FALSE, ADJ=FALSE, scree=TRUE, xlab="Principal
           Component", ylab="Proportion of Variance")
```

eliminates outliers via the projection method and applies the classic principal component analysis to the remaining data. In contrast to the convention used by R, the correlation matrix, rather than the covariance matrix, is used by default. To use the covariance matrix, set the argument `cor=FALSE`. Setting `SCORES=TRUE`, the principal component scores are returned.

If the argument ADJ=TTUE, the R function outproad is used to check for outliers rather than the R function outpro, which is recommended if the number of variables is greater than 9. By default, the argument scree=TRUE, meaning that a scree plot will be created.

The function

```
robpcaS(x, SCORES=FALSE)
```

provides a summary of the results based on the method derived by Hubert et al. (2005). A more detailed analysis is performed by the function

```
robpca(x, scree=TRUE, xlab = "Princpal Component", ylab = "Proportion of
                        Variance"),
```

which returns the eigenvalues and other results discussed by Hubert et al. (2005), but these details are not discussed here.

The R function

```
Ppca(x, p=ncol(x)-1, SCORES=FALSE, SCALE=TRUE, gcov=rmba, MC=F)
```

applies the method aimed at maximizing a robust generalized variance. (The algorithm for applying this method represents a substantial improvement on the algorithm used in Wilcox, 2008b; see Wilcox, 2010a.) This particular function requires the number of principal components to be specified via the argument p, which defaults to $p-1$. The argument SCALE=TRUE means that the marginal distributions will be standardized based on the measure of location and scale corresponding to the argument gvoc, which defaults to the median ball algorithm.

When using the method aimed at maximizing a robust generalized variance, there is the issue of how different choices for the number of components compare. The R function

```
Ppca.summary(x, MC=FALSE, SCALE=TRUE)
```

is aimed at addressing this issue. It calls Ppca using all possible choices for the number of components, computes the resulting generalized standard deviations, and reports their relative size. If you have access to a multicore processor, setting the argument MC=TRUE will reduce execution time.

The output from the function Ppca.summary differs in crucial ways from the other functions described here. To illustrate it, again consider the same data used in the first example of Section 13.7.1. So we have multivariate normal data with all correlations equal to 0.0. The output from Ppca.summary is

```
                   [,1]      [,2]      [,3]      [,4]
Num. of Comp. 1.0000000 2.000000 3.0000000 4.0000000
Gen.Stand.Dev 1.1735029 1.210405 1.0293564 1.0110513
Relative Size 0.9695129 1.000000 0.8504234 0.8353002
```

The second line indicates the (robust) generalized standard deviation given the number of components indicated by the first line. So when using two components, the generalized standard deviation is 1.210405. Note that the generalized standard deviations are not in descending order. Using two components results in the largest generalized standard deviation. But observe that all four generalized standard deviations are approximately equal, which is what we would expect for the situation at hand. The third line of the output is obtained by dividing each value in the second line by the maximum generalized standard deviation. Here, reducing the number of components from 4 to 2 does not increase the generalized standard deviation by very much, suggesting that 4 or maybe 3 components should be used. Also observe that there is no proportion of variance used here, in contrast to classic PCA.

In classic PCA, an issue is how many components must be included to capture a reasonably large proportion of the variance. When using the robust generalized variance, one first looks at the relative size of the generalized standard deviations using all of the components. If the relative size is small, reduce the number of components. In the example, the relative size using all four components is 0.835 suggesting that perhaps all four components should be used.

Now we consider the data set in the second example of Section 13.7.1 where all of the correlations are 0.9. The output from `Ppca.summary` is

```
                    [,1]      [,2]      [,3]       [,4]
Num. of Comp. 1.000000 2.0000000 3.0000000 4.00000000
Gen.Stand.Dev 2.017774 0.6632588 0.2167982 0.05615346
Relative Size 1.000000 0.3287082 0.1074442 0.02782942
```

As indicated, a single component results in a relatively large generalized standard deviation suggesting that a single component suffices. The relative sizes corresponding to 3 and 4 components are fairly small suggesting that using 3 or 4 components be ruled out. Even with 2 components the relative size is fairly small.

EXAMPLE

Consider again the reading study mentioned in Section 6.4.2. Here the focus is on five variables: two measures of phonological awareness, a measure of speeded naming for digits, a measure of speeded naming for letters, and a measure of the accuracy of identifying lower case letters. Using the classic principal component analysis based on the correlation matrix, the R function `regpca` returns

```
Importance of components:
                          Comp.1 Comp.2 Comp.3 Comp.4 Comp.5
Standard deviation        1.4342 1.0360 0.9791 0.7651 0.57036
Proportion of Variance 0.4114 0.2146 0.1917 0.1170 0.06506
Cumulative Proportion  0.4114 0.6260 0.8178 0.9349 1.00000
```

The R function `robpcaS` returns

```
                          [,1]  [,2]  [,3]  [,4]   [,5]
Number of Comp.          1.000 2.000 3.000 4.000 5.0000
Robust Stand Dev         2.239 1.265 1.219 0.977 0.6069
Proportion Robust var 0.531 0.169 0.157 0.101 0.0390
Cum. Proportion          0.531 0.701 0.859 0.960 1.0000
```

So both functions suggest that three or four components are needed to capture the bulk of the variability. A similar result is obtained with the R function `Ppca.summary` *if* the variables are not standardized. However, if they are standardized, `Ppca.summary` returns

```
                 [,1]      [,2]   [,3]      [,4]      [,5]
Num. of Comp. 1.0000 2.0000 3.0000 4.00000 5.00000
Gen.Stand.Dev 1.7125 1.5155 0.7229 0.47611 0.31127
Relative Size 1.0000 0.8849 0.4221 0.27801 0.18176
```

The second line shows the robust generalized standard deviations based on the number of components used. Because the relative sizes using three, four, or five components are rather small, the results suggest that two components suffice.

13.9 EXERCISES

1. Consider a bivariate data set and imagine that the data are projected onto the x-axis and then checks for outliers are made using the MAD-median rule. The same is done when projecting the data onto the y-axis and both times no outliers are found. Why might the conclusion of no outliers be unsatisfactory?

2. Repeat the example in Section 13.3.1 based on the skull data, only use the final skull measures rather than the first two.

3. Using R, generate data and store it in the R variable `x` using the commands

   ```
   set.seed(3)
   x=rmul(100,p=5,mar.fun=ghdist,h=0.2,rho=0.7)
   x[,1]=rlnorm(100)
   ```

   This creates five random variables, the last four of which have symmetric, heavy-tailed distributions with a common correlation of 0.7. The first variable has a skewed distribution with relatively light tails and is independent of the other four variables. Use `regpca` to perform the classic PCA and verify that for two and three components, the cumulative proportion of variances are 0.824 and 0.893, respectively, suggesting that two or three components are needed to capture the variability in the data.

4. Using the data generated in the previous exercise, compute the principal component scores and verify that the corresponding correlation matrix is approximately equal to the identity matrix.

5. For the data used in the previous two exercises, verify that the R functions `outpro` and `outproad` identify the same points as outliers. Why was this expected?

6. For the plasma data in Chapter 1, use the function `mat2grp` to sort the data into two groups based on whether a participant is male or female. Compare these two groups in terms of dietary beta-carotene consumed (labeled BETADIET) and dietary retinol consumed (labeled RETDIET) using the R function `smean2`. Verify that the p-value is 0.258.

CHAPTER 14

ROBUST REGRESSION AND MEASURES OF ASSOCIATION

CONTENTS

Regression is an extremely vast topic that is difficult to cover in a single book let alone a few chapters. Moreover, many major advances that have considerable practical value have occurred during the last quarter century, including new and improved methods for measuring the strength of an association. The goal in this chapter is to cover some of the modern methods that are particularly relevant in applied work with the understanding that many issues and techniques are not discussed. For more information, see for example Maronna et al. (2006) and Wilcox (2017).

14.1 ROBUST REGRESSION ESTIMATORS

Chapter 6 noted that least squares regression can yield highly inaccurate results regarding the association among the bulk of the points due to even a single outlier. The quantile regression estimator mentioned in Chapter 6, which contains least absolute value regression as a special case, guards against outliers among the Y values, but leverage points (outliers among the predictor variables) can again result in a misleading and distorted sense about the association among the bulk of the points. Many alternative regression estimators have been proposed for dealing with this problem (e.g., Wilcox, 2017, Chapter 10). Some of these alternative estimators have an advantage beyond guarding against outliers: when there is heteroscedasticity, their standard errors can be substantially smaller than the standard error of the least squares regression estimator. In practical terms, some robust regression estimators might have relatively high power.

This section summarizes several robust regression estimators that seem to be particularly useful, with the understanding that no single estimator is always optimal. In particular, it is not claimed that estimators not listed here have no practical value. Various illustrations are included to help convey some sense of when certain estimators perform well, and when and why they might be unsatisfactory. The methods described in this chapter deal with heteroscedasticity in a relatively effective manner. Atkinson et al. (2016) describe an approach to heteroscedasticity where the error term is taken to be $\lambda(X)e$, where $\lambda(X)$ is assumed to be a particular function. This is in contrast to the methods here that make no parametric assumption about the nature of the heteroscedasticity. The relative merits of the method derived by Atkinson et al., compared to the estimators described here, have not been studied.

As in Chapter 6, this section assumes that for the population of individuals or things under study, the typical value of Y, given $X_1, \dots, X_p$, is given by

$$Y = \beta_0 + \beta_1 X_1 + \cdots + \beta_p X_p.$$

The goal is to estimate $\beta_0, \beta_1 \dots \beta_p$ with $b_0, \dots, b_p$, respectively. The problem is finding a good choice for $b_0, \dots, b_p$ based on the available data. Following the notation in Chapter 6, for the ith vector of observations $(Y_i, X_{i1}, \dots, X_{ip})$, we let

$$\hat{Y}_i = b_0 + b_1 X_{i1} + \cdots + b_p X_{ip}$$

be the predicted value of Y and

$$r_i = Y_i - \hat{Y}_i$$

be the corresponding residual.

14.1.1 The Theil–Sen Estimator

Momentarily consider a single predictor and imagine that we have n pairs of values:

$$(X_1, Y_1), \dots, (X_n, Y_n).$$

Consider any two pairs of points for which $X_i > X_j$. The slope corresponding to the two points (X_i, Y_i) and (X_j, Y_j) is

$$b_{1ij} = \frac{Y_i - Y_j}{X_i - X_j}. \tag{14.1}$$

The Theil (1950) and Sen (1968) estimate of β_1 is the median of all the slopes represented by b_{1ij}, which is labeled $b_{1\text{ts}}$. The intercept is estimated in one of two ways. The first is to use

$$b_{0\text{ts}} = M_y - b_{1\text{ts}} M_x,$$

where M_y and M_x are the sample medians corresponding to the Y and X values, respectively. The second uses the median of the residuals. (The latter method appears to have a slight advantage from a robustness point of view.)

When there is one predictor, the finite sample breakdown point of the Theil–Sen estimator is approximately 0.29 (Dietz, 1989) meaning that about 29% of the data must be altered to make the resulting estimate of the slope and intercept arbitrarily large or small. A negative feature is that the finite sample breakdown point decreases as the number of predictors, p, gets large. Another positive feature is that its standard error can be tens even hundreds of times

Table 14.1 Boscovich's Data on Meridian Arcs

X:	0.0000	0.2987	0.4648	0.5762	0.8386
Y:	56,751	57,037	56,979	57,074	57,422

smaller than the ordinary least squares estimator when the error term is heteroscedastic, even under normality.

EXAMPLE

The computation of the Theil–Sen estimator is illustrated with the data in Table 14.1, which were collected about 200 years ago and analyzed by Roger Boscovich. The data deal with determining whether the Earth bulges at the center, as predicted by Newton, or whether it bulges at the poles as suggested by Cassini. Here X is a transformed measure of latitude and Y is a measure of arc length. (Newton's prediction was that $\beta_1/(3\beta_0) \approx 1/230$.) For the first two pairs of points, the estimated slope is

$$\frac{57037 - 56751}{0.2987 - 0} = 1560.1.$$

Computing the slope for the remaining nine pairs of points and taking the median yields 756.6. It is left as an exercise to verify that the intercept is estimated to be 56,685. Interestingly, $b_{1\text{ts}}/(3b_{0\text{ts}}) = 0.0044$ which is fairly close to Newton's prediction: $1/230 = 0.0043$. (Least squares gives a very similar result.)

Several methods have been proposed for extending the Theil–Sen estimator to $p > 1$ predictors. Here an approach described in Wilcox (2017, Section 10.2) is used. The computational details are omitted but an R function for applying the method is supplied.

A point worth stressing is that even when a regression estimator has a high breakdown point, this does not necessarily mean that leverage points have little or no influence on the estimated slope. It merely means that leverage points cannot make the estimated slope arbitrarily large or small. In practical terms, it is advisable to consider what happens to the estimated slope if outliers among the predictors are removed using one of the methods in Chapter 13 (or Chapter 2 in the case of a single predictor). The example in the next section underscores this point.

14.1.2 R Functions tsreg, tshdreg, and regplot

The R function

```
tsreg(x,y, xout=FALSE, outfun=outpro, iter=10, varfun=pbvar, corfun=pbcor,
   plotit=FALSE, WARN=TRUE, HD=FALSE, OPT=FALSE, xlab="X",ylab="Y",...)
```

computes the Theil–Sen estimator. Setting the argument `xout=TRUE`, leverage points are removed based on the outlier detection method specified by the argument `outfun`. The argument `outfun` defaults to the R function `outpro`, which applies the projection method in Section 13.1.6 when there is more than one independent variable. The argument `OPT=FALSE` means that the intercept is estimated using the median of the residuals; `OPT=TRUE` means that the intercept is estimated with $M_y - b_{1\text{ts}}M_x$. (With `HD=TRUE`, the usual sample median is replaced by the Harrell–Davis estimator.) When there is a single predictor, the R function

```
regplot(x,y,regfun=tsreg,xlab="X",ylab="Y")
```

creates a scatterplot of the data that includes the regression line specified by the argument `regfun`, which defaults to the Theil–Sen estimator. (The function assumes that the slope and intercept, specified by the argument `regfun`, are returned in `$coef`.) The R function

```
tshdreg(x, y, xout=FALSE, outfun=out, iter=10, varfun=pbvar,
                        corfun=pbcor,...)
```

computes another version of the Theil–Sen estimator that might provide more power when there are tied values among the dependent variable (Wilcox and Clark, 2013). When estimating the slopes, it replaces the usual sample median with the Harrell–Davis estimator, which was briefly mentioned in Section 9.5.3.

EXAMPLE

For the star data in Figure 6.5, `tsreg` returns

```
$coef:

 Intercept
 -2.623636 1.727273
```

Removing leverage points by setting the argument `xout=T`, now the estimated slope is 3.22, the point being that even with a reasonably high breakdown point, a robust regression estimator might be highly influenced by leverage points.

14.1.3 Least Median of Squares

The *least median of squares* (LMS) regression estimator appears to have been first proposed by Hampel (1975) and further developed by Rousseeuw (1984). The strategy is to determine the slopes and intercept $(b_0, \ldots, b_p)$ that minimize

$$\mathrm{MED}(r_1^2, \ldots, r_n^2),$$

the median of the squared residuals. The breakdown point is approximately 0.5, the highest possible value. The LMS regression estimator has some negative theoretical properties, which are summarized in Wilcox (2017, Section 10.5). However, it is often suggested as a diagnostic tool or as a preliminary fit to data, and on occasion it does perform well relative to other methods that might be used.

14.1.4 Least Trimmed Squares and Least Trimmed Absolute Value Estimators

Rousseeuw's (1984) *least-trimmed squares* (LTS) estimator is based on minimizing

$$\sum_{i=1}^{h} r_{(i)}^2,$$

where $r_{(1)}^2 \leq \ldots \leq r_{(h)}^2$ are the squared residuals written in ascending order. With $h = [n/2]+1$, where $[n/2]$ is $n/2$ rounded down to the nearest integer, the same breakdown point as the LMS estimator is achieved. However, $h = [n/2] + [(p+1)/2]$ is often used for technical reasons summarized in Wilcox (2017, Section 10.4).

A close variation of the LTS estimator is the *least trimmed absolute* (LTA) estimator. Now the strategy is to choose the intercept and slope so as to minimize

$$\sum_{i=1}^{h} |r|_{(i)}. \tag{14.2}$$

Like LTS, the LTA estimator can have a much smaller standard error than the least squares estimator, but its improvement over the LTS estimator seems to be marginal at best, at least based on what is currently known (Wilcox, 2001.)

14.1.5 R Functions lmsreg, ltsreg, and ltareg

The R function

```
lmsreg(x,y),
```

which is stored in the R package MASS, computes the least median of squares regression estimator. The R command `library(MASS)` provides access to the function `lmsreg`.

The R function

```
ltsreg(x,y)
```

computes the LTS regression estimate. Like `lmsreg`, access to the function `ltsreg` is obtained via the R command `library(MASS)`.

The R function

```
ltareg(x,y,tr=0.2,h=NA)
```

computes the LTA estimator. If no value for `h` is specified, the value for h is taken to be $n-$[tr], where [tr] is tr rounded down to the nearest integer. By default, `tr` is equal to 0.2.

14.1.6 M-estimators

There are several variations of what are called regression M-estimators, which represent generalizations of the M-estimator of location introduced in Chapter 2. Generally, M-estimators can be written in terms of what is called weighted least squares. Weighted least squares means that the slopes and intercept are determined by minimizing

$$\sum w_i r_i^2,$$

for some choice of the weights $w_1, \ldots, w_n$. M-estimators choose values for the weights that depend in part on the values of the residuals. If $w_1 = \cdots = w_n = 1$, we get the ordinary least squares estimator. Readers interested in computational details are referred to Wilcox (2017, Sections 10.5-10.9). One of the better M-estimators was derived by Coakley and Hettmansperger (1993), which can be applied with the R function described in the next section. It has a breakdown point of 0.5 and its standard error competes well with the least squares estimator under normality. Briefly, the Coakley–Hettmansperger uses LTS to get an initial fit and then it adjusts the estimate based in part on the resulting residuals. (The minimum volume covariance matrix in Chapter 13 plays a role as well.) Yet another regression estimator that enjoys excellent theoretical properties is the MM-estimator derived by Yohai (1987).

14.1.7 R Function chreg

The R function

```
chreg(x,y)
```

computes the Coakley–Hettmansperger regression estimator and

```
MMreg(x,y)
```

computes the MM-estimator.

14.1.8 Deepest Regression Line

The sample median, as described in Chapter 2, can be viewed as the deepest point among the n values available when the sample size is odd. If the sample size is even, the median is the average of the two deepest values. This point of view can be extended to regression lines. In particular, Rousseeuw and Hubert (1999) derived a numerical measure of how deeply a regression line is nested within a scatterplot of points. They proposed estimating the slope and intercept of a regression based on the line that is deepest among the n points that are observed. In effect, their method provides an alternative to the quantile regression estimator described in Section 6.7 where the goal is to estimate the median of Y given X. Unlike the quantile regression line in Chapter 6, the deepest regression line provides protection against bad leverage points. For a single predictor, if all predictor values X are distinct, the breakdown point is approximately 0.33.

14.1.9 R Function mdepreg

The R function

```
mdepreg(x,y)
```

computes the deepest regression line.

14.1.10 Skipped Estimators

Skipped regression estimators remove outliers using one of the methods in Chapter 13 and then fit a regression line to the remaining data. The projection outlier detection method appears to be a relatively good choice followed by the Theil–Sen estimator. This is called the OP regression estimator. A natural guess is to apply least squares regression after any outliers are removed, but in terms of achieving a relatively low standard error, the Theil–Sen estimator is a better choice (Wilcox, 2003).

14.1.11 R Functions opreg and opregMC

The R function

```
opreg(x,y)
```

computes the skipped regression estimator using the Theil–Sen estimator after removing outliers via the projection method. It performs well under normality and homoscedasticity, in terms of achieving a relatively small standard error, and its standard error might be substantially smaller than the least squares estimator when there is heteroscedasticity. The R function

```
opregMC(x,y)
```

is the same as `opreg`, only it takes advantage of a multicore processor, assuming one is available and that the R package parallel has been installed.

14.1.12 S-estimators and an E-type Estimator

S-estimators represent yet another general approach to estimating the slopes and intercept. The basic strategy is to take the slopes and intercept to be the values that minimize some robust measure of variation applied to the residuals (Rousseeuw and Yohai, 1984). For example, rather than minimize the variance of the residuals, as done by least squares, minimize the percentage bend midvariance. Many variations and extensions have been proposed (e.g.,

Wilcox, 2017, Section 10.13.1). One that seems to perform relatively well begins by estimating the parameters by minimizing the percentage bend midvariance associated with the residuals. Next, compute the residuals based on the resulting fit and then eliminate any point (X_i, Y_i) for which the corresponding residual is an outlier based on the MAD-median rule. Finally, re-estimate the slopes and intercepts with the Theil–Sen estimator. This is called the TSTS estimator and belongs to what is called the class of *E-type* estimators.

14.1.13 R Function tstsreg

The R function

```
tstsreg(x,y)
```

computes the TSTS estimator just described. The R package WRScpp contains a C++ version of this function that can reduce execution time considerably when using a bootstrap method. Install WRScpp as described in Chapter 1. Then use the command `library(WRScpp)`, after which the function `tstsreg_C(x,y)` can be used.

14.2 COMMENTS ON CHOOSING A REGRESSION ESTIMATOR

Many regression estimators have been proposed that can have substantial advantages over the ordinary least squares estimator. But no single estimator dominates and there is the practical issue that the choice of estimator can make a practical difference in the conclusions reached. One point worth stressing is that even when an estimator has a high breakdown point, only two outliers, properly placed, can have a large impact on some of the robust estimates of the slopes. Examples are the Coakley–Hettmansperger estimator, which has the highest possible breakdown point, the Theil–Sen estimator, and the MM-estimator derived by Yohai (1987), as well as other robust regression estimators not covered here. Estimators that seem to be particularly good at avoiding this problem are the LTS, LTA, TSTS, and OP regression estimators. Among these estimators, the OP and TSTS estimators can have an advantage in terms of achieving a relatively low standard error, and so they have the potential of providing relatively high power. A negative feature of the OP and TSTS estimators is that with large sample sizes, execution time might be an issue. The R function `opregMC` is supplied to help deal with this when a multicore processor is available. As for TSTS, the R function `tstsreg_C(x,y)` can reduce execution time substantially, assuming the R package WRScpp has been installed.

In terms of achieving a relatively small standard error when dealing with heteroscedasticity, certain robust regression estimators can offer a distinct advantage compared to the least squares estimator. (For more details, see Table 10.3 in Wilcox, 2017.) Ones that perform well include the Theil–Sen estimator, the OP regression estimator, the MM-esitmator, and the TSTS estimator.

EXAMPLE

Using data from actual studies, the examples in Section 6.4.2 illustrated that a few outliers can have a large impact on the least squares regression estimator. For the lake data shown in the left panel of Figure 6.6, all of the robust regression estimators described in this chapter give similar estimates of the slope, which differ substantially from the least squares estimate. But for the reading data in the right panel, this is not the case as illustrated in Figure 14.1. Based on the LTS, Coakley–Hettmansperger, LTA, Theil–Sen, and OP estimators, the estimates of the slope are 0.072, 0.023, −0.275, −0.282, and −0.6, respectively. It is noted that if the Theil–Sen estimator is applied via the function `tsreg`, and if the argument `xout=T` is used in order to remove leverage points, the estimated slope is −0.6, the same estimate returned by the OP estimator. In this particular case, LTS and Coakley–Hettmansperger are highly influenced by the largest six RAN1T1 values, which are clearly leverage points. If these six points are eliminated, the estimated slopes are now −0.484 and −0.361, respectively. The MM-estimator, not shown in Figure 14.1, estimates the slope to be −0.0095. In practical terms, even when using a robust regression estimator, it seems prudent to check the effect of removing leverage points. When using the Theil–Sen estimator via the R function `tsreg`, this can be done by setting the argument `xout=TRUE`. The OP and TSTS estimators can be used as well.

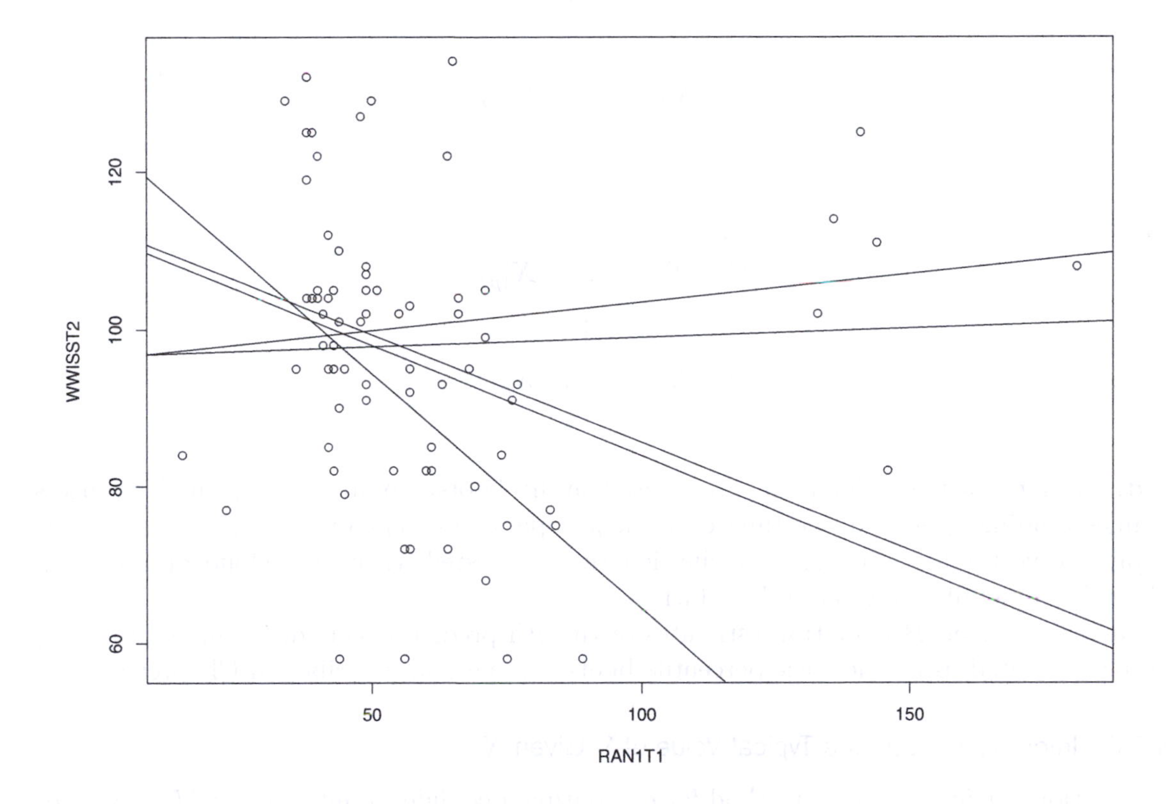

Figure 14.1 Robust estimators can differ substantially. Here, the slopes based on LTS, Coakley–Hettmansperger, LTA, Theil–Sen and OP are 0.072, .023, −0.275, −0.282, and −0.6, respectively. This illustrates that it is prudent to check what happens when leverage points are removed.

14.3 INFERENCES BASED ON ROBUST REGRESSION ESTIMATORS

This section takes up the problem of making inferences about regression parameters when using a robust regression estimator. Included is a technique for computing a confidence band for $M(Y|X)$, some conditional measures of location associated with Y, given X. That is, for each value among $X_1, \ldots, X_n$, a confidence interval for $M(Y|X_i)$ is computed $(i = 1, \ldots, n)$ with the property that the simultaneous probability coverage is approximately equal to $1-\alpha$.

In the event there are only K unique values among $X_1, \ldots, X_n$, then, of course, only K confidence intervals are computed.

14.3.1 Testing Hypotheses About the Slopes

Generally, given the goal of testing

$$H_0 : \beta_1 = \cdots = \beta_p = 0 \tag{14.3}$$

a basic percentile bootstrap method performs relatively well in terms of controlling the probability of a Type I error. Briefly, generate a bootstrap sample by resampling with replacement n rows from

$$\begin{array}{ccc} (Y_1, X_{11}, & \ldots, & X_{1p}) \\ & \vdots & \\ (Y_n, X_{n1}, & \ldots, & X_{np}), \end{array}$$

yielding

$$\begin{array}{ccc} (Y_1^*, X_{11}^*, & \ldots, & X_{1p}^*) \\ & \vdots & \\ (Y_n^*, X_{n1}^*, & \ldots, & X_{np}^*). \end{array}$$

Next, estimate the regression parameters based on this bootstrap sample. Repeat this process B times yielding B bootstrap estimates of the p slope parameters and then check to see how deeply the vector $\mathbf{0} = (0, \ldots, 0)$, having length p, is nested within the cloud of bootstrap values. More details are given in Box 14.1.

Note that given B bootstrap estimates of the jth predictor, confidence intervals for β_j can be computed using the basic percentile bootstrap method described in Chapter 6.

14.3.2 Inferences About the Typical Value of Y Given X

This section briefly outlines a method for computing a confidence interval for $M(Y|X_i)$, for each X_i ($i = 1, \ldots, n$), such that the simultaneous probability coverage is approximately $1 - \alpha$. Letting θ_0 be some specified value, there is the related goal of testing

$$M(Y|X_i) = \theta_0, \tag{14.4}$$

($i = 1, \ldots, n$) such the probability of one or more Type I errors is approximately α.

Section 6.4.8 mentioned how to deal with these goals when using the least squares regression estimator. The main point here is that the same goals can be achieved when using a robust regression estimator. The basic strategy is to estimate $M(Y|X_i)$ with $\hat{Y}_i = b_0 + b_1 X_i$, where as usual b_0 and b_1 are estimates of the intercept and slope, respectively. Next, compute a bootstrap estimate of the standard error of $\hat{Y}_i$ yielding say $\hat{\tau}_i$. The goal, then, is to choose a critical value, say c, such that the n confidence intervals

Box 14.1: Test (14.3), the hypothesis that all slope parameters are zero.

The B bootstrap estimates of the p regression parameters are denoted by $\hat{\beta}^*_{jb}$, $j = 1, \ldots, p$; $b = 1, \ldots, B$. Let

$$s_{jk} = \frac{1}{B-1}\sum_{b=1}^{B}(\hat{\beta}^*_{jb} - \hat{\beta}_j)(\hat{\beta}^*_{kb} - \hat{\beta}_k),$$

where $\hat{\beta}_j$ is the estimate of β_j based on the original data. Compute

$$d_b = (\hat{\beta}^*_b - \hat{\beta})\mathbf{S}^{-1}(\hat{\beta}^*_b - \hat{\beta})',$$

where $\mathbf{S}$ is the matrix corresponding to s_{jk}, $\hat{\beta} = (\hat{\beta}_1, \ldots, \hat{\beta}_p)$, and $\hat{\beta}^*_b = (\hat{\beta}^*_{1b}, \ldots, \hat{\beta}^*_{pb})$. The value of d_b measures how far away the bth bootstrap vector of estimated slope parameters is from the center of all B bootstrap values. Put the d_b values in ascending order yielding $d_{(1)} \leq \cdots \leq d_{(B)}$. The test statistic is

$$D = (\mathbf{0} - \hat{\beta})\mathbf{S}^{-1}(\mathbf{0} - \hat{\beta})'$$

and measures how far away the null hypothesis is from the estimated slope parameters. Reject if $D \geq d_{(u)}$, where $u = (1-\alpha)B$, rounded to the nearest integer.

$$\hat{Y}_i \pm c\hat{\tau}_i \tag{14.5}$$

($i = 1, \ldots, n$) have simultaneous probability coverage $1 - \alpha$. If there are only K unique values among $X_1, \ldots, X_n$, and if K is relatively small, methods in Chapter 12 can be used. But otherwise, in terms of achieving relatively high power and relatively short confidence intervals, the methods in Chapter 12 are generally unsatisfactory. An alternative method for determining an appropriate critical value is much more satisfactory (Wilcox, in press a).

Briefly, momentarily assume that there is homoscedasticity and that for any i,

$$\frac{\hat{Y}_i - \theta_0}{\hat{\tau}_i}$$

has a standard normal distribution when $\beta_1 = \beta_0 = 0$. Then, for a given sample size, n, a simulation is performed for determining the distribution of the minimum p-value when the n hypotheses given by Equation (14.4) are true. The α quantile of this distribution is then used to compute confidence intervals. More precisely, if p_c is the α quantile of the minimum p-value when the n hypotheses are true, taking c to be the $1 - p_c/2$ quantile of a standard normal distribution results in confidence intervals that have, approximately, simultaneous probability coverage $1 - \alpha$. Simulations indicate that the method continues to perform well under non-normality and when there is heteroscedasticity. The R function `regYci`, described in the next section, applies the method.

14.3.3 R Functions regtest, regtestMC, regci, regciMC, regYci, and regYband

The R function

```
regtest(x, y, regfun = tsreg, nboot = 600, alpha = 0.05, plotit = TRUE, grp
```

Table 14.2 Hald's Cement Data

Y	X_1	X_2	X_3	X_4
78.5	7	26	6	60
74.3	1	29	15	52
104.3	11	56	8	20
87.6	11	31	8	47
95.9	7	52	6	33
109.2	11	55	9	22
102.7	3	71	17	6
72.5	1	31	22	44
93.1	2	54	18	22
115.9	21	47	4	26
83.8	1	40	23	34
113.3	11	66	9	12
109.4	10	68	8	12

```
= c(1:ncol(x)), nullvec= c(rep(0, length(grp))), xout = FALSE, outfun =
outpro)
```

tests the hypothesis that the regression coefficients are all equal to 0. By default, the Theil–Sen estimator is used but other regression estimators can be used via the argument `regfun` provided the estimated slopes are returned in `regfun$coef`. The function can be used to test the hypothesis that the slopes are equal to some specified set of constants, other than 0, via the argument `nullvec`. For example, if there are two predictors, the R command `regtest(x,y,nullvec=c(2,4))` will test the hypothesis that $\beta_1 = 2$ and $\beta_2 = 4$. The argument `grp` can be used to indicate that a subset of the parameters is to be tested, which can include the intercept term. For example, setting `grp=c(0,3)` will test H_0: $\beta_0 = \beta_3 = 0$, assuming the argument `nullvec` is not specified. The command

```
regtest(x,y,grp=c(2,4,7))
```

will test H_0: $\beta_2 = \beta_4 = \beta_7 = 0$. The R function

```
regtestMC(x, y, regfun = tsreg, nboot = 600, alpha = 0.05, plotit = TRUE,
grp = c(1:ncol(x)), nullvec = c(rep(0, length(grp))), xout = FALSE, outfun =
out)
```

is the same as `regtest`, only it takes advantage of a multicore processor if one is available.

The R function

```
regci(x, y, regfun = tsreg, nboot = 599, alpha = 0.05, SEED = TRUE, pr =
TRUE, xout = FALSE, outfun = out, ...)
```

tests hypotheses about the individual parameters.

EXAMPLE

This example illustrates that it is possible to reject the hypothesis that the slope parameters are all equal to 0, but fail to reject for any of the individual slope parameters. Table 14.2 shows data from a study by Hald (1952) concerning the heat evolved in calories per gram (Y) versus the amount of each of four ingredients in the mix: tricalcium aluminate (X_1), tricalcium silicate (X_2), tetracalcium alumino ferrite (X_3), and dicalcium silicate (X_4). Consider the first and third predictors and suppose we test H_0: $\beta_1 = \beta_3 = 0$ with the R function **regtest** in conjunction with the least squares estimator. The p-value is 0.047, so in particular, reject with $\alpha = 0.05$. However, if we test the individual slope parameters with the R function **regci**, the 0.95 confidence intervals for β_1 and β_3 are (−0.28, 5.93) and (−2.3, 3.9), respectively, so we fail to reject for either of the predictor variables. This phenomenon, where the omnibus test is significant but the individual tests are not, is known to occur when using the conventional F test (described in Section 6.2.2) as well (e.g., Fairley, 1986). The reason is that when two estimators have a reasonably strong association, the resulting confidence region for the two parameters is a relatively narrow ellipse. Figure 14.2 shows a plot of the bootstrap estimates of the slopes, which provides some indication of why this phenomenon occurs. The square, where the horizontal and vertical lines intersect, corresponds to the hypothesized value. As is evident, it is relatively far from the bulk of the bootstrap estimates. However, in order to reject H_0: $\beta_1 = 0$ at the 0.05 level, which corresponds to parameter 1 in Figure 14.2, 97.5% of the bootstrap estimates would need to be either above or below the horizontal line. To reject H_0: $\beta_3 = 0$, 97.5% of the bootstrap estimates would need to be to the right or to the left of the vertical line. Said another way, computing separate confidence intervals is essentially computing a rectangular confidence region for the two parameters under investigation. When the two estimators are approximately independent, this tends to give similar results to those obtained with the confidence region used by the R function **regtest**, but otherwise it is possible for one method to reject when the other does not.

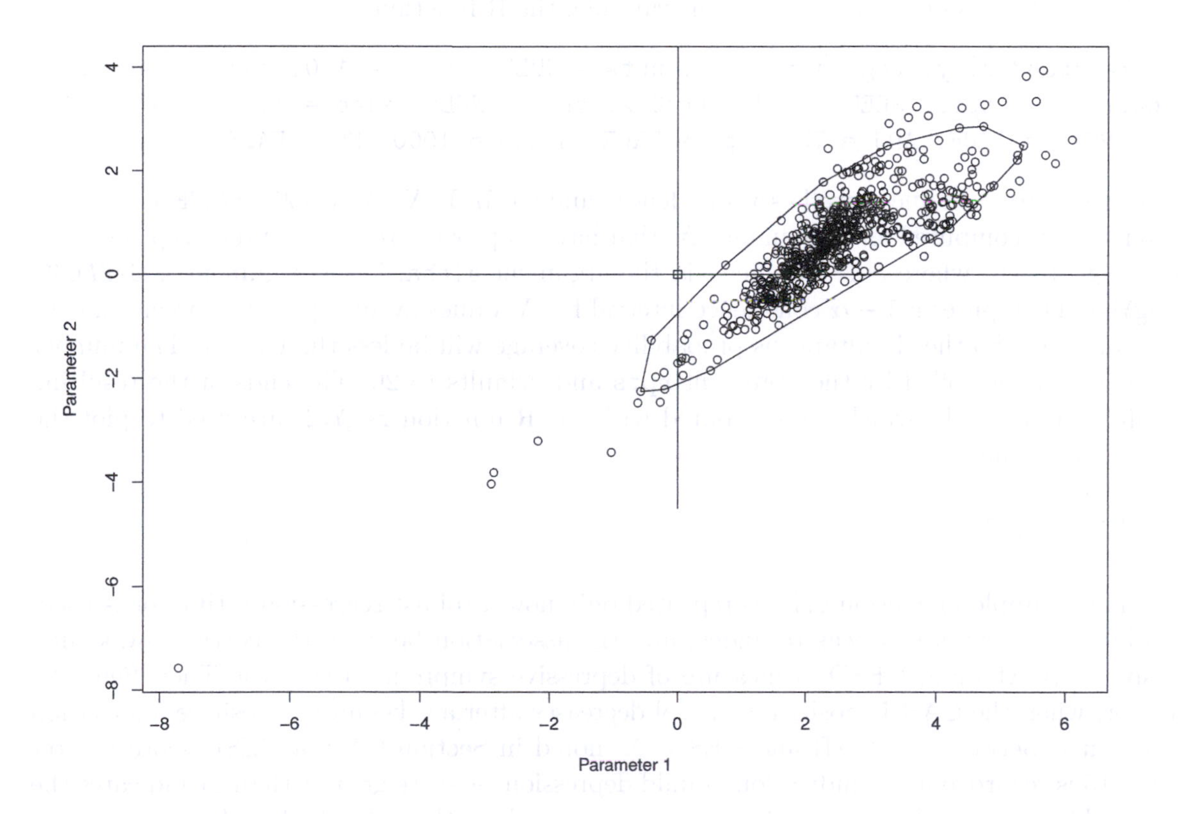

Figure 14.2 It is possible to reject the hypothesis that all slopes are equal to 0, yet fail to reject for each individual slope, as illustrated here.

The R function

```
regYci(x,y,regfun=tsreg, pts=unique(x), nboot=100, ADJ=FALSE, xout=FALSE,
outfun=out, SEED=TRUE, alpha=0.05, crit=NULL, null.value=0, plotPV=FALSE,
scale=TRUE, span=0.75, xlab="X", xlab1="X1", xlab2="X2", ylab="p-values",
    zlab="p-values", theta=50, phi=25, MC=FALSE, nreps=1000, SM=FALSE,
                              pch="*",...)
```

computes confidence intervals for $M(Y|X_i)$, $i = 1,\ldots,n$, as outlined in Section 14.3.2. If the argument ADJ=TRUE, the confidence intervals are adjusted so that the simultaneous probability coverage is approximately equal to $1-\alpha$, where α is specified via the argument alpha. And in terms of testing (14.4), the function computes a critical p-value so that the probability of one or more Type I errors does not exceed the value given by the argument alpha. That is, imagine that K hypotheses are tested and let p_k be the p-value associated with the kth hypothesis. The function computes p_c with the property that if the kth hypothesis is rejected when $p_k \le p_c$, the probability of one or more Type I errors is less than or equal to the value given by the argument alpha. When testing at the 0.05 level, this critical p-value can be computed quickly. Otherwise, execution time increases substantially. Setting the argument MC=TRUE can reduce execution time considerably. Note that by default the function uses the Theil–Sen estimator, but simulations suggest that, even using the least squares estimator, control over the probability of a Type I error is reasonably good in most situations. The function can be used with more than one independent variable. If the number of independent variables is one or two, setting the argument plotPV=TRUE results in a plot of the p-values as a function of the independent variables.

When there is only one independent variable, the R function

```
regYband(x, y, regfun = tsreg, npts = NULL, nboot = 100, xout = FALSE,
outfun = outpro, SEED = TRUE, tr=0.2, crit = NULL, xlab = "X", ylab = "Y",
  SCAT = TRUE, ADJ = TRUE, pr = TRUE, nreps = 1000, MC = FALSE, ...)
```

plots the regression line as well as a confidence band for $M(Y|X)$. When ADJ=TRUE, confidence intervals are computed for each unique X_i that have, approximately, simultaneous probability coverage $1-\alpha$, where α is specified via the argument alpha. If the argument ADJ=FALSE, regYband computes a $1-\alpha$ confidence interval for X values evenly spaced between $\min(X)$ and $\max(X)$. So the simultaneous probability coverage will be less than $1-\alpha$. The number of points is controlled by the argument npts and defaults to 20. The ends of the resulting confidence intervals, which are computed with the R function regYci, are used to plot the confidence band.

EXAMPLE

The example in Section 6.4.9 is repeated only now a robust regression estimator is used. To briefly review, a goal was to understand the association between the cortisol awakening response (CAR) and CESD, a measure of depressive symptoms. Using the Theil–Sen estimator, when the CAR is positive (cortisol decreases after awakening), a positive association was found between the CAR and CESD. As noted in Section 6.4.9, a CESD score greater than 15 is regarded as an indication of mild depression. A score greater than 21 indicates the possibility of major depression. Removing leverage points, the individual confidence intervals indicate that for CAR between zero and 0.10, the typical CESD score is significantly less than 15 when testing at the 0.05 level. That is, the probability of one or more Type I errors is approximately 0.05. For CAR greater than 0.23, the typical CESD score is estimated to be greater than 15. But over the entire range of available CAR values, namely 0−0.358, the typical CESD scores are not significantly greater than 15.

14.3.4 Comparing Measures of Location via Dummy Coding

Section 9.1.2 noted that methods for comparing means can be viewed from a regression perspective where groups are indicated via dummy coding. In particular, means can be compared using methods associated with least squares regression that were covered in Chapter 6. As previously stressed, this approach does not represent an effective method for dealing with non-normality when the goal is to compare means. In principle, dummy coding can be used when the goal is to compare robust measures location by replacing the least squares estimator with one of the robust regression estimators covered in this chapter. However, when using the Theil–Sen estimator, simulations do not support this approach: control over the Type I error probability can be poor (Ng, 2009). Whether a similar problem occurs when using some other robust regression estimator has not been investigated.

14.4 DEALING WITH CURVATURE: SMOOTHERS

Situations are encountered where using a straight regression line appears to provide a reasonable summary of the association between two variables. But exceptions occur and so methods for dealing with curvature are required. Momentarily consider a single predictor. A simple strategy is to include values of the predictor raised to some power. For example, rather than use $\hat{Y} = \beta_0 + \beta_1 X$, one might use $\hat{Y} = \beta_0 + \beta_1 X + \beta_2 X^2$ or even $\hat{Y} = \beta_0 + \beta_1 X + \beta_2 X^2 + \beta_3 X^3$. But even these more flexible models can prove to be inadequate when trying to understand the nature of the association. Many methods have been derived with the goal of providing an even more flexible approach to studying curvature that are generally known as *nonparametric regression* estimators or *smoothers*. (R has at least six built-in smoothers.) Several smoothers have considerable practical value and so the goal in this section is to outline a few that deserve serious consideration. Readers interested in more details about smoothers are referred to Hastie and Tibshirani (1990), Efromovich (1999), Eubank (1999), Fan and Gijbels (1996), Fox (2001), Green and Silverman (1993), Gyöfri et al. (2002), and Härdle (1990). Wilcox (2017) summarizes many more smoothers not covered here. (For the special case where Y is binary, see Section 14.5.4.)

A crude description of the strategy used by smoothers is as follows. Suppose we want to estimate the mean of Y, given that $X = 6$, based on n pairs of observations: $(X_1, Y_1), \ldots, (X_n, Y_n)$. Robust measures of location might be of interest as well, but for the moment attention is restricted to the mean of Y given X. The strategy is to focus on the observed X values close to 6 and use the corresponding Y values to estimate the mean of Y. Typically, smoothers give more weight to Y values for which the corresponding X values are close to 6. For pairs of points for which the X value is far from 6, the corresponding Y values are ignored. The general problem, then, is determining which values of X are close to 6 and how much weight the corresponding Y values should be given. Many methods have been proposed.

All of the smoothers described here can be used when there are multiple predictors. So it is possible to estimate Y, given values for the predictors, without assuming a particular parametric form for the regression surface. For two predictors ($p = 2$), plots of the regression surface can be created as will be illustrated. There is a concern, however, generally known as the curse of dimensionality: as p increases, neighborhoods with a fixed number of points become less local (Bellman, 1961). In practical terms, as p increases, the expectation is that increasingly large sample sizes are needed to get reasonably accurate estimates. For $p = 2$, all indications are that the smoothers described here perform reasonably well provided the sample size is not overly small. The extent this is the case for $p > 2$ is unclear.

14.4.1 Cleveland's Smoother

One of the earliest suggestions for implementing the strategy just outlined is due to Cleveland (1979). Again imagine that the goal is to estimate the mean of Y given that $X = 6$. Cleveland fits a weighted least squares regression line to the data. For X_i values close to 6, the weight (roughly the importance) attached to the point (X_i, Y_i) is relatively high. For X_i values sufficiently far from 6, the weights are set equal to zero. So we have a regression estimate of the mean of Y given that $X = 6$. Of course we can repeat this process for a range of X values, which provides an overall estimate of the true regression line. The idea is that, although a regression line might not be linear over the entire range of X values, it will be approximately linear over small intervals of X. That is, for X values close to 6, say, a linear regression line might perform reasonably well and can be used to estimate Y given that $X = 6$. But for X values far from 6, some other regression estimate of Y might be necessary. Cleveland's method is known as a *locally weighted running-line smoother* or *LOESS*.

The details of Cleveland's method are instructive. Given the goal of estimating the mean of Y corresponding to some specific value for X, measure the distance of X from each of the observed X_i values with

$$\delta_i = |X_i - X|.$$

For example, if $X = 10$, and $X_1 = 5.2$, $X_2 = 8.8$, $X_3 = 10.5$, it follows that $\delta_1 = |5.2 - 10| = 4.8$, $\delta_2 = |8.8 - 10| = 1.2$, $\delta_3 = |10.5 - 10| = 0.5$, and so forth. Next, sort the δ_i values and retain the pn pairs of points that have the smallest δ_i values, where p is a number between 0 and 1. The value of p represents the proportion of points used to predict Y and is generally referred to as the *span*. For the moment, suppose $p = 1/2$. If, for example, $n = 43$, this means that you retain 22 of the pairs of points that have X_i values closest to $X = 10$. These 22 points are the *nearest neighbors* to $X = 10$. Let δ_m be the maximum value of the δ_i values that are retained. Set

$$Q_i = \frac{|X - X_i|}{\delta_m},$$

and if $0 \leq Q_i < 1$, set

$$w_i = (1 - Q_i^3)^3,$$

otherwise set

$$w_i = 0.$$

Finally, use *weighted least squares* to predict Y using w_i as weights (cf. Fan, 1992). That is, determine the values b_1 and b_0 that minimize

$$\sum w_i(Y_i - b_0 - b_1 X_i)^2$$

and estimate the mean of Y corresponding to X to be $\hat{Y} = b_0 + b_1 X$. Because the weights (the w_i values) change with X, generally a different regression estimate of Y is used when the value of X is altered.

Let $\hat{Y}_i$ be the estimated mean of Y given that $X = X_i$ based on the method just described. Then an estimate of the regression line is obtained by the line connecting the points $(X_i, \hat{Y}_i)$ $(i = 1, \ldots, n)$ and is called a *smooth*. The span, p, controls the raggedness of the smooth. If p is close to 1, we get a straight line even when there is curvature. If p is too close to 0, an extremely ragged line is obtained instead. By choosing a value for p between 0.2 and 0.8, usually curvature can be detected and a relatively smooth regression line is obtained.

This section described the basic features of Cleveland's smoother. But it should be noted that a more complex algorithm is used by R. Included are weights (the w_i values) based in part on what is called Tukey's biweight. These weights help guard against the deleterious impact of outliers among the dependent variable. Moreover, a confidence band can be computed for the regression line in a manner that allows heteroscedasticity, and which has simultaneous probability coverage equal to $1-\alpha$ (Wilcox, in press c).

14.4.2 R Functions lowess, lplot, lplot.pred, and lplotCI

The built-in R function

```
lowess(x,y,p=2/3)
```

computes Cleveland's smoother. (A closely related R function is `loess`, which has options not available when using `lowess`.) The value for p, the span, defaults to 2/3. You can create a scatterplot of points that contains this smooth with the R commands

```
plot(x,y)
lines(lowess(x,y)).
```

If the line appears to be rather ragged, try increasing p to see what happens. If the line appears to be approximately horizontal, indicating no association, check to see what happens when p is lowered.

The R function

```
lplot(x,y, span=0.75, pyhat=FALSE, eout=F, xout=F, outfun=out, plotit=TRUE,
   expand=0.5, low.span=2/3, varfun=pbvar,family="gaussian", scale=TRUE,
         xlab="X", ylab="Y", zlab="", theta=50, phi=25),
```

written for this book, plots Cleveland's smoother automatically and provides a variety of useful options. For example, it will remove all outliers if `eout=TRUE`. To remove leverage points only, use `xout=TRUE`. If `pyhat=TRUE`, the function returns $\hat{Y}_i$, the estimate of Y given that $X = X_i$. To suppress the plot, use `plotit=FALSE`. The argument `low.span` is the value of p, the span, when using a single predictor. The arguments `xlab` and `ylab` can be used to label the x-axis and y-axis, respectively. (When there are two or more independent variables, the argument `family="gaussian"` means fitting is by least squares. If `family="symmetric"` is used, the function uses an estimation procedure that deals with outliers among the dependent variable.) The arguments `theta` and `phi` can be used to rotate three-dimensional plots. The argument `scale` is important when plotting a three-dimensional plot. If there is no association, `scale=FALSE` typically yields a good plot. But when there is an association, `scale=TRUE` is usually more satisfactory. (The argument `varfun` is explained in Section 14.5.)

Imagine that a regression line is fitted using the R function `lplot` and the data stored in the R variables `x` and `y`. Further imagine that some new values for the independent variable are stored in `z`, but the corresponding values for the dependent variable are not available. For example, university admissions officers have data on undergraduate GPAs, GRE scores, and graduate GPAs for former students. Of interest is estimating the graduate GPA of a student applying for admission based on her undergraduate GPA and GRE scores. A way of estimating the typical value of the dependent variable, based on `z`, is to use the estimate of the regression line based on `x` and `y`. The R function

```
lplot.pred(x, y, pts = x, xout = FALSE, outfun = outpro, span = 2/3, ...)
```

accomplishes this goal by setting the argument `pts=z`.

The R function

```
lplotCI(x, y, plotit = TRUE, xlab = "X", ylab = "Y", p.crit = NULL, alpha =
  0.05, span = 2/3, CIV = FALSE, xout = FALSE, outfun = outpro, pch = ".",
    SEED = TRUE, nboot = 100, pts = NULL, npts = 25, nreps = 2000,...)
```

plots the regression plus a confidence band having probability coverage `1-alpha`, where the argument `alpha` defaults to 0.05. More precisely, let L be the median of the values stored in `x` minus 1.5 times the value of MADN, where MADN is also based on the values in `x`. Let U be the median of the values stored in `x` plus 1.5 times the value of MADN. The function determines `npts` points evenly spaced between L and U, inclusive, and computes confidence intervals such that the simultaneous probability coverage is approximately `1-alpha`. The default value for the argument `npts` is 25. Setting the argument `CIV=TRUE`, the function returns the confidence intervals; otherwise it simply plots the results. The default for the argument `span` is `NULL`, meaning that the function chooses the span p based on the sample size. Execution time is low when `alpha=0.05` is used. Otherwise it can be relatively high because the function must determine how to adjust the confidence intervals so that the simultaneous probability coverage is $1 - \alpha$.

EXAMPLE

For the diabetes data in Exercise 15 of Chapter 6, the goal was to understand the association between the age of children at diagnosis and their C-peptide levels. As noted there, the hypothesis of a zero slope is rejected with the R function `hc4test`; the p-value is 0.034. Student's T test of a zero slope has a p-value of 0.008. So a temptation might be to conclude that as age increases, C-peptide levels increase as well. But look at Figure 14.3, which shows Cleveland's smooth. Note that for children up to about the age of 7, there seems to be a positive association. But after the age of 7, it seems that there is little or no association at all.

EXAMPLE

The plasma retinol data described in Chapter 1 contains data on the amount of fat and fiber consumed per day, which are used here to predict plasma retinol levels. Assuming the data for fat and fiber are stored in the R variable `x`, and plasma retinol measures are stored in `y`, the R command

```
lplot(x,y,scale=TRUE,xlab="FAT",ylab="FIBER",zlab="PLASMA RETINOL")
```

creates the plot shown in Figure 14.4. The plot suggests that for low fat consumption, as the amount of fiber consumed increases, the typical plasma retinol level increases as well. But for high fat consumption, it appears that as fiber consumption increases, there is relatively little or no increase in plasma retinol levels.

14.4.3 Smoothers Based on Robust Measures of Location

A natural guess at how to extend Cleveland's method to robust measures of location is to replace the weighted least squares estimator with one of the robust regression estimators covered in Section 14.1. However, this strategy is known to be highly unsatisfactory. The reason is that robust regression estimators can be insensitive to curvature, so the resulting smooth often misses curvature when in fact it exists. A better strategy is to use what is called the *running-interval smoother*. To estimate some measure of location for Y, given some value for X, a running-interval smoother searches for *all* points close to the value of X that is of

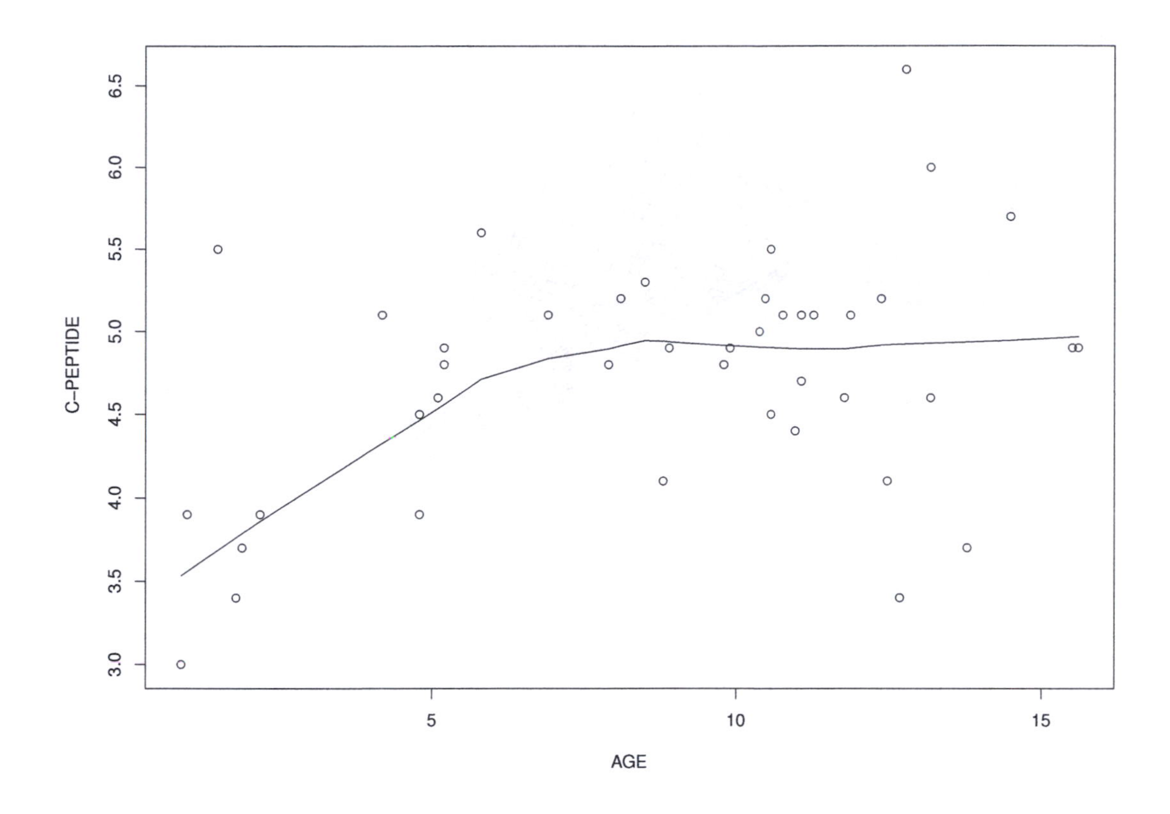

Figure 14.3 A smooth created by the R function `lplot` using the diabetes data in Exercise 15 of Chapter 6. Note that there seems to be a positive association up to about the age of 7, after which there is little or no association.

interest and then it computes a measure of location based on the corresponding Y values. Note that the number of X_i values close to X will depend on the value of X. In contrast, Cleveland's method uses the k nearest points with k fixed and chosen in advance.

To elaborate, compute MAD based on the X values and label it MAD_x. Let f be some constant that is chosen in a manner to be described and illustrated. Then a specific value X is said to be close to X_i if

$$|X_i - X| \le f\left(\frac{\text{MAD}_x}{0.6745}\right).$$

So for normal distributions, X is close to X_i if X is within f standard deviations of X_i. Now consider all of the Y_i values corresponding to the X_i values that are close to X. Then an estimate of the typical value of Y, given X, is the estimated measure of location based on the Y values just identified. For example, if six X_i values are identified as being close to $X = 22$, and the corresponding Y values are 2, 55, 3, 12, 19, and 21, then the estimated mean of Y, given that $X = 22$, would be the average of these six numbers: 18.7. The estimated 20% trimmed mean of Y, given that $X = 22$, would be the trimmed mean of these six values, which is 13.75.

A running-interval smoother is created as follows. For each X_i, determine which of the X_j values are close to X_i, compute a measure of location associated with the corresponding Y_j values, and label this result $\hat{Y}_i$. So we now have the following n pairs of numbers: $(X_1, \hat{Y}_1), \ldots, (X_n, \hat{Y}_n)$. The running-interval smooth is the line formed by connecting these

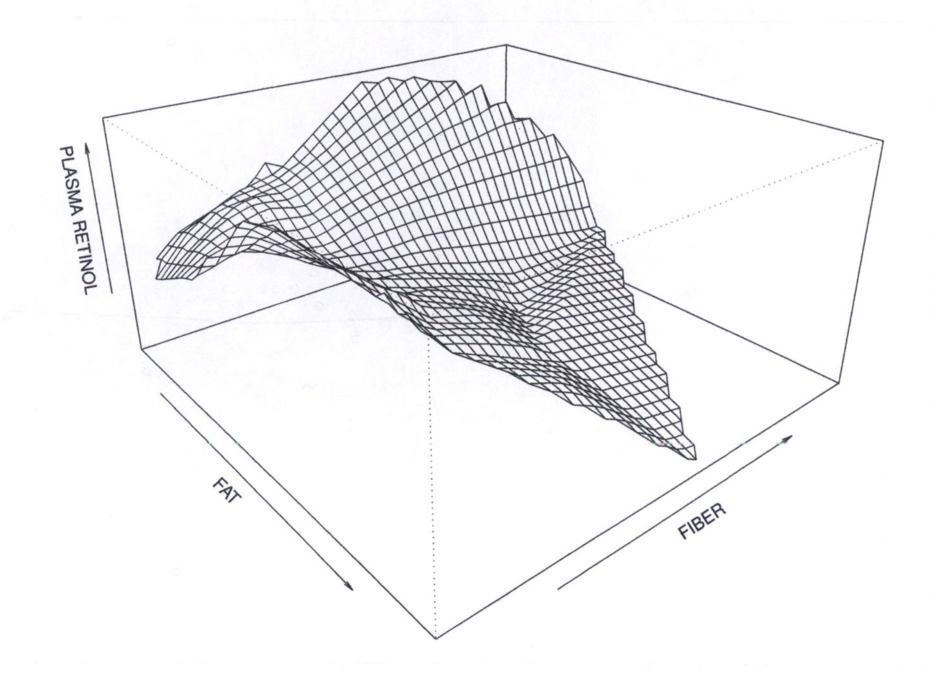

Figure 14.4 The smooth created by `lplot` when predicting plasma retinol given fat and fiber consumption.

points. The span, f, controls how ragged the line will be. As with Cleveland's method, if f is sufficiently large, the smooth will be a straight, horizontal line, even when there is curvature, and if f is too close to zero, the result is a very ragged line. The basic idea behind the running-interval smoother can be extended to multiple predictors (e.g., Wilcox, 2017), but the computational details are not described.

The description of the running-interval just given is aimed at estimating the typical value of Y given X. But in various situations it can be useful and informative to study the quantiles of Y given X. For example, given X, what is the 0.75 quantile associated with Y? Quantile regression lines can be estimated via the Harrell–Davis estimator. That is, apply the Harrell–Davis estimator using all of the Y_i values for which the corresponding X_i values satisfy $|X_i - X| \leq f\text{MAD}_x/0.6745$. From an aesthetic point of view, it helps to smooth the initial fit again using LOESS. It is noted that this approach to estimating a quantile regression line avoids certain practical concerns associated with the constrained B-spline smoother (COBS) that was derived by Koenker and Ng (2005); see Wilcox (2016a).

14.4.4 R Functions rplot, rplotCIS, rplotCI, rplotCIv2, rplotCIM, rplot.pred, qhdsm, and qhdsm.pred

The R function

```
rplot(x, y, est = tmean, scat = TRUE, fr = NA, plotit = TRUE, pyhat = FALSE,
  efr = 0.5, theta = 50, phi = 25, scale = TRUE, expand = 0.5, SEED = TRUE,
```

```
varfun = pbvar, outfun = outpro, nmin = 0, xout = FALSE, out = FALSE, eout =
   FALSE, xlab = "X", ylab = "Y", zscale = FALSE, zlab = " ", pr = TRUE,
duplicate = "error", ticktype = "simple", LP = TRUE, OLD = FALSE, pch = ".",
                              prm = TRUE, ...)
```

computes a running-interval smooth. The argument `est` determines the measure of location that is used and defaults to a 20% trimmed mean. The argument `fr` corresponds to the span, f, and defaults to 0.8. The function returns the $\hat{Y}_i$ values if the argument `pyhat=TRUE`. By default, a scatterplot of the points is created with a plot of the smooth. To avoid the scatterplot, set `scat=FALSE`. The argument `LP=TRUE` means that the original smooth is smoothed again using LOESS. The function

```
rplot.pred(x,y,pts=x,est=tmean,fr=1,nmin=1,xout=FALSE,outfun=outpro,...)
```

uses the data in `x` and `y` to estimate the typical value of Y, given that X has the values stored in the argument `pts`, using the running-interval smooth. (The R function `runYhat` accomplishes the same goal.)

The R function

```
rplotS(x,y,tr=0.2,fr=0.8, plotit=TRUE, scat=TRUE, pyhat=FALSE, SEED=TRUE,
dfmin=8, eout=FALSE, xout=FALSE, xlab="X", ylab="Y", outfun=out, LP=TRUE,
                        alpha=0.05, pch=".",...)
```

computes confidence intervals for $M(Y|X_i)$ for each X_i via the Tukey–McLaughlin method, provided the degrees of freedom are at least 8, and then it plots the results. By default, $M(Y|X_i)$ is the 20% trimmed mean of Y given that $X = X_i$. Another goal is to compute confidence intervals such that with probability 0.95 say, all of the confidence intervals contain the true value of $M(Y|X_i)$. A criticism of this function is that it does not achieve this goal.

There are three functions aimed at addressing this issue. Details about the underlying methods are summarized in Wilcox (2016b). The first is

```
rplotCI(x,y,tr=0.2,fr=0.5, p.crit=NA, plotit=TRUE, scat=TRUE, SEED=TRUE,
pyhat=FALSE, pts=NA, npts=25, xout=FALSE, xlab="X", ylab="Y", low.span=2/3,
     nmin=12, outfun=out, LP=TRUE, LPCI=FALSE, MID=TRUE, alpha=0.05,
                             pch=".",...).
```

If picks covariate values and computes confidence intervals using the Tukey–McLaughlin method, but confidence intervals are adjusted so that the simultaneous probability coverage is approximately 0.95. The method requires a sample size of at least 50. The argument `npts` controls how many covariate values are used, which defaults to 25. Execution time remains low when using `npts=10`, but otherwise it can be fairly high. (The probability coverage can be altered via the argument `alpha` at the expense of substantially higher execution time.) Roughly, the function determines how to adjust the confidence intervals when dealing with a normal distribution. For example, with $n = 50$, if a $1 - 0.004846$ confidence interval is computed for each of the 25 covariate values, the probability that all 25 confidence intervals contain $M(Y|X_i)$ $(i = 1, \ldots, 25)$ is approximately 0.95. Using a span `fr=0.5` or smaller is crucial. Using `fr=0.8`, for example, the simultaneous probability coverage is reasonably close to the nominal level when there is a very weak association, but otherwise it can drop below 0.90. Using `fr=0.5` generally suffices, but when there is a strong association, such as a correlation greater than 0.5, `fr=0.2` should be used. Also, to be safe, if $n > 500$, again `fr=0.2` is a good choice unless the strength of the association is fairly weak, in which case `fr=0.5` performs reasonably well.

The R function

```
rplotCIv2(x,y,tr=0.2, fr=0.5, p.crit=NA, plotit=TRUE, scat=TRUE, SEED=TRUE,
pyhat=FALSE, pts=NA, npts=25, xout=FALSE, xlab="X", ylab="Y", low.span=2/3,
nmin=12, outfun=out, LP=TRUE, LPCI=FALSE, MID=TRUE, alpha=0.05, pch=".",...)
```

is like the previous function, only it computes confidence intervals for every X_i, provided the number of X_j values that are close to X_i is reasonably large, say 12. (This can be altered via the argument `nmin`.) The argument `p.crit` indicates the adjustment that is made for each confidence interval. For example, `p.crit=0.001` means that 0.999 confidence intervals would be computed. The function determines `p.crit` so that the simultaneous probability coverage is approximately equal to 1-`alpha`, where the argument `alpha` defaults to 0.05.

The R function

```
rplotCIM(x, y, est = hd, fr = 0.5, p.crit = NA, plotit = TRUE, scat = TRUE,
 pyhat = FALSE, pts = NA, npts = 25, xout = FALSE, xlab = "X", ylab = "Y",
  low.span = 2/3, nmin = 16, outfun = out, LP = FALSE, LPCI = FALSE, MID =
                TRUE, alpha = 0.05, pch = ".", ...)
```

is like the function `rplotCI`, only it is designed for medians with the individual confidence intervals based on the method in Section 4.9. (The R function `sint` is used.)

The function

```
qhdsm(x, y, qval = 0.5, q = NULL, pr = FALSE, xout = FALSE, outfun = outpro,
plotit = TRUE, xlab = "X", ylab = "Y", zlab = "Z", pyhat = FALSE, fr = NULL,
  LP = FALSE, theta = 50, phi = 25, ticktype = "simple", nmin = 0, scale =
                TRUE, pr.qhd = TRUE, pch = ".", ...)
```

estimates the regression line aimed at predicting the qth quantile of Y, given X. It is essentially the running-interval smoother based on the Harrell–Davis estimator. The argument `qval` (or `q`) controls which quantile will be used and defaults to 0.5, the median. Setting the argument `q=c(0.25,0.75)`, for example, the function will plot the regression lines for the lower and upper quartiles. Setting the argument `LP=TRUE` results in a smoother looking curve, but when plotting two or more regression lines, it can result in estimates for the lower quantile being greater than estimates of the upper quantile. (The R package Rallfun-v31 uses `LP=FALSE` by default. Earlier versions used `LP=TRUE` by default.) The R function

```
qhdsm.pred(x,y,pts=x,q=0.5,fr=1,nmin=1,xout=FALSE,outfun=outpro,...)
```

predicts the qth quantile of Y for the data in `pts`, using the smooth based on the data in `x` and `y`.

14.4.5 Prediction When X Is Discrete: The R Function rundis

Consider a situation where there are multiple Y values for each X value and where X is discrete with a relatively small sample space. In this case it might be of interest to compute a measure of location for Y corresponding to each X value and plot the results. For example, for all Y values corresponding to $X = 1$, say, compute some measure of location, do the same for $X = 2$, and so on. This is in contrast to the smooths previously described where you search for all X values close to 2, for example, and compute a measure of location based on the corresponding Y values. For convenience, the R function

```
rundis(x,y,est=mom,plotit=TRUE,pyhat=FALSE,...)
```

has been supplied to perform this task.

14.4.6 Seeing Curvature with More Than Two Predictors

Most smoothers can be used with more than two predictors, but as is evident, visualizing curvature cannot be done in a simple manner. A variety of techniques have been proposed for dealing with this problem, and a comparison of several methods was made by Berk and Booth (1995). They focused on predicting means, but the strategies they considered are readily extended to robust measures of location. Generally, these methods can help, but situations can be created where any one method fails.

A simple strategy, sometimes called the *partial response plot*, is to check a smooth for each individual predictor ignoring the others. An alternative approach is to plot the residuals versus the predicted values. Experience suggests that often this strategy can be unsatisfactory. Another strategy is based on what is called a partial residual plot. The idea dates back to Ezekiel (1924, p. 443) and was named a partial residual plot by Larsen and McCleary (1972). To explain it, imagine there are p predictors and that you want to check for curvature associated with predictor j. Assuming that the other predictors have a linear association with Y, fit a regression plane to the data ignoring the jth predictor. The *partial residual plot* simply plots the resulting residuals versus X_j. A smoother applied to this plot can be used to check for curvature.

14.4.7 R Function prplot

The R function

```
prplot(x,y,pvec=ncol(x),regfun=tsreg,...)
```

creates a partial residual plot assuming that curvature is to be checked for the predictor indicated by the argument `pvec`. The argument `x` is assumed to be an n-by-p matrix. By default, it is assumed that curvature is to be checked using the data stored in the last column of the matrix `x`. The argument `regfun` indicates which regression method will be used and defaults to Theil–Sen. The function uses the smoother LOESS described in Section 14.4.1.

EXAMPLE

To provide a simple illustration of `prplot`, data were generated according to the model $Y = X_1 + X_2 + X_3^2 + e$ where all three predictors and the error term have standard normal distributions. Assuming the data are stored in the R variables `x` and `y`, the R command

```
prplot(x[,1:2],y)
```

creates the partial residual plot in the left panel of Figure 14.5. That is, fit the model $Y = \beta_0 + \beta_1 X_1 + e$, compute the residuals, and use LOESS to predict the residuals based on X_2. Here, X_3 is being ignored. The plot suggests that the association between Y and X_2 is approximated reasonably well with a straight line. The right panel of Figure 14.5 is the partial residual plot for X_3 based on the residuals associated with the model $Y = \beta_0 + \beta_1 X_1 + \beta_2 X_2 + e$, which results from the R command

```
prplot(x,y).
```

As evident, the plot suggests that there is curvature.

14.4.8 Some Alternative Methods

The methods already covered for detecting and describing curvature, when there are multiple predictors, are far from exhaustive. Although complete details about other methods are not provided, it might help to mention some of the alternative strategies that have been proposed.

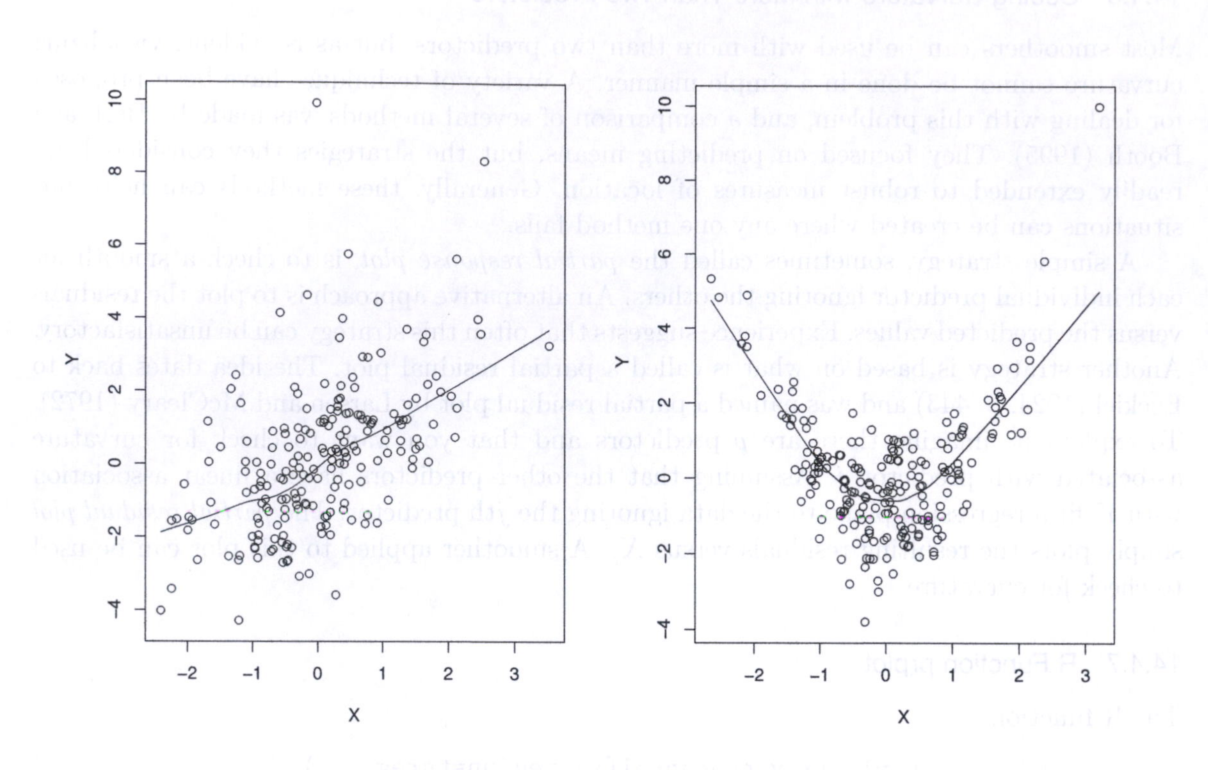

Figure 14.5 The left panel shows the plot created by prplot based on X_1 and X_2, ignoring X_3. That is, fit the model $Y = X_1 + e$ with the goal of checking for curvature associated with X_2. The plot suggests that there is little or no curvature, which is correct. The right panel shows the plot when the model $Y = X_1 + X_2 + e$ is fit to the data and the goal is to check for curvature associated with X_3. The plot suggests that there is curvature, which is true.

One approach is to assume that for some collection of functions $f_1, \ldots, f_p$,

$$Y = \beta_0 + f_1(X_1) + \cdots + f_p(X_p) + e, \tag{14.6}$$

and then try to approximate these functions using some type of smoother. This is called a *generalized additive model.* Details can be found in Hastie and Tibshirani (1990) who described an algorithm for applying it to data. (R has built-in functions for applying this technique; see the methods listed under generalized additive models in the R *Guide to Statistics.*)

The augmented partial residual plot (Mallows, 1986) is like the partial residual plot, only it includes a quadratic term for the predictor being investigated. For a generalization of this method, see Cook (1993).

Another approach is to search for nonlinear transformations of both Y and the predictors that result in an additive model. That is, search for functions $f_y, f_1, \ldots, f_p$ such that

$$f_y(Y) = f_1(X_1) + \cdots + f_p(X_p) + e$$

provides a good fit to data. An algorithm for implementing the method was derived by Breiman and Friedman (1985) and is called *alternating conditional expectations* or *ace.* (R has built-in functions for applying the technique.) For some refinements and extensions, see Tibshirani (1988).

14.4.9 Detecting Heteroscedasticity Using a Smoother

Sections 6.8 and 6.12.2 described methods for testing the hypothesis that there is homoscedasticity. Yet another way of testing this hypothesis is via the running-interval smoother (Wilcox, 2006e). Let $r_i = Y_i - \hat{Y}_i$ $(i = 1, \ldots, n)$ denote the residuals where $\hat{Y}_i$ is the estimate of the typical value of Y based on X_i and the running-interval smoother. Consider the regression line where X is the independent variable and $|r_i|$ is the dependent variable. Homoscesdaticity means that the slope of this regression line is zero, which can be tested with the method in Section 14.3.1. Note that the method in Section 6.12.2 assumes that the 0.2 and 0.8 quantile regression lines are straight. The method in this section does not require this assumption. For variations of the method just described, see Wilcox (2017).

14.4.10 R Function rhom

The R function

```
rhom(x,y, op=1, op2=FALSE, tr=0.2,plotit=TRUE, xlab="NA", ylab="NA",
    zlab="ABS(res)", est=median, sm=FALSE, SEED=TRUE, xout=FALSE,
                         outfun=outpro,...)
```

tests the hypothesis that there is homoscedasticity using the method described in the previous section. (The arguments op and op2 refer to variations of the method that are described in Wilcox, 2017.)

14.5 SOME ROBUST CORRELATIONS AND TESTS OF INDEPENDENCE

There are two general approaches to measuring the strength of an association in a robust manner. The first is to use a robust analog of Pearson's correlation that is not based on any particular regression estimator. The other approach first fits a regression line to the data, and then the strength of the association is measured based on this fit. For the first approach, there are two types of robust correlations. The first, sometimes called an M-type correlation, guards against outliers among each variable, but these correlations do not deal with outliers in a manner that takes into account the overall structure of the data. The other, sometimes called an O-type correlation, down weights or eliminates outliers using a technique that takes into account the overall structure. This section summarizes methods based on the first approach. (For a summary of some additional measures not covered here, see Shevlyakov and Smirnov, 2011, as well as Wilcox, 2017.) An example of the second approach is described in Section 14.5.5. Methods for testing the hypothesis of a zero correlation are described and provide a relatively simple robust alternative to the hypothesis testing methods based on Pearson's correlation, which were described in Chapter 6.

14.5.1 Kendall's tau

Kendall's tau is based on the following idea. Consider two pairs of observations: (X_1, Y_1) and (X_2, Y_2). For convenience, assume that $X_1 < X_2$. If $Y_1 < Y_2$, then these two pairs of numbers are said to be concordant. That is, if Y increases as X increases, or if Y decreases as X decreases, we have concordant pairs of observations. If two pairs of observations are not concordant, they are said to be discordant.

If the ith and jth pairs of points are concordant, let $K_{ij} = 1$. If they are discordant, let $K_{ij} = -1$. Kendall's tau is the average of all K_{ij} values for which $i < j$. More succinctly, Kendall's tau is estimated with

$$\hat{\tau} = \frac{2\sum_{i<j} K_{ij}}{n(n-1)}, \tag{14.7}$$

which has a value between −1 and 1. If $\hat{\tau}$ is positive, there is a tendency for Y to increase with X—possibly in a nonlinear fashion—and if $\hat{\tau}$ is negative, the reverse is true. If as X increases, Y always increases as well, $\hat{\tau} = 1$. If as X increases, Y always decreases, $\hat{\tau} = -1$.

The population analog of $\hat{\tau}$ is labeled τ and can be shown to be zero when X and Y are independent. The classic test of

$$H_0 : \tau = 0 \tag{14.8}$$

is based on

$$Z = \frac{\hat{\tau}}{\sigma_\tau},$$

where

$$\sigma_\tau^2 = \frac{2(2n+5)}{9n(n-1)}.$$

The null hypothesis is rejected if

$$|Z| \geq z_{1-\frac{\alpha}{2}},$$

where $z_{1-\alpha/2}$ is the $1 - \alpha/2$ quantile of a standard normal distribution, which can be read from Table 1 in Appendix B.

Section 6.6.1 described an R function for testing the hypothesis that Pearson's correlation is equal to 0.0. The same function can be used to test Equation (14.8).

EXAMPLE

If the data are stored in the R variables x and y, the R command

```
cor.test(x,y,"kendall")
```

computes Kendall's tau and tests the hypothesis given by Equation (14.8).

EXAMPLE

Imagine we observe the 10 values $X = 0.1, 0.2, \ldots, 1$ and that $Y = X^2$. Then there is a perfect monotonic increasing relationship between X and Y, $\hat{\tau} = 1$ and Pearson's correlation is $r = 0.975$. So in this particular case there is little separating the two coefficients. However, Kendall's tau provides some protection against missing an association due to one or more outliers. For example, if the largest Y value is increased from 1 to 10, again $\hat{\tau} = 1$ but now $r = 0.59$ with a p-value (based on Student's T) equal to 0.07. Increasing the largest Y value to 50, $r = 0.54$.

14.5.2 Spearman's rho

Spearman's rho, labeled r_s, is just Pearson's correlation based on the ranks associated with the two variables under study. That is, if we observe $(X_1, Y_1), \ldots, (X_n, Y_n)$, convert $X_1, \ldots, X_n$ to ranks as was done, for example, in Section 8.8.5. Do the same for the values $Y_1, \ldots, Y_n$, and compute Pearson's correlations based on these ranks. Under independence, the population analog of r_s, ρ_s, is zero. Also, like Kendall's tau, Spearman's rho is exactly equal to 1 if there is a monotonic increasing relationship between X and Y. That is, Y never decreases as X increases. In addition, $\rho_s = -1$ if the association is monotonic decreasing instead.

The usual approach to testing

$$H_0 : \rho_s = 0$$

is based on

$$T = r_s \sqrt{\frac{n-2}{1-r_s^2}}.$$

When there is independence, T has, approximately, a Student's t distribution with $\nu = n-2$ degrees of freedom. So reject and conclude there is an association if $|T| \geq t$, where t is the $1-\alpha/2$ quantile of a Student's t distribution with $n-2$ degrees of freedom.

Like Kendall's tau, Spearman's rho provides protection against outliers among the X values, or among the Y values, but it does not take into account the overall structure of the data. That is, a few unusual points, properly placed, can have a substantial influence on its value. De Winter et al. (2016) compared Pearson's correlation to Spearman's correlation and concluded that Pearson's correlation is suitable for light-tailed distributions, whereas Spearman's correlation is preferable when variables feature heavy-tailed distributions or when outliers are present,

14.5.3 Winsorized Correlation

The Winsorized correlation coefficient is obtained by Winsorizing the n pairs of observations as described in Box 9.1 of Section 9.2. (For technical reasons, trimming is a less satisfactory approach to defining a robust correlation.) The Winsorized correlation between X and Y is just Pearson's correlation applied to the Winsorized values. The resulting correlation coefficient will be labeled r_w.

Letting ρ_w be the population analog of r_w, it can be shown that when X and Y are independent, $\rho_w = 0$. If we assume independence—implying homoscedasticity—a simple test of

$$H_0 : \rho_w = 0$$

is based on the test statistic

$$T_w = r_w \sqrt{\frac{n-2}{1-r_w^2}}. \tag{14.9}$$

Let

$$\nu = n - 2g - 2,$$

where $g = [\gamma n]$, and γ is the amount of Winsorizing. (Remember that $[\gamma n]$ means to compute γn, and then round down to the nearest integer.) Reject if $|T_w| \geq t_{1-\alpha/2}$, the $1-\alpha/2$ quantile of Student's t distribution with ν degrees of freedom. Setting the amount of Winsorizing to 0 (i.e., using $\gamma = 0$), T_w reduces to the test statistic covered in Section 6.6.

14.5.4 R Function wincor

The R function

```
wincor(x,y,tr=0.2)
```

computes the Winsorized correlation and tests the hypothesis H_0: $\rho_w = 0$ using the Student's T test described in the previous section. The amount of Winsorization is controlled by the argument `tr` and defaults to 0.2.

14.5.5 OP or Skipped Correlation

Kendall's tau, Spearman's rho, and the Winsorized correlation all offer protection against the deleterious effects of outliers among the marginal distributions. But a criticism is that they do not deal with outliers in a manner that takes into account the overall structure of the data. Section 13.1 summarized some robust measures of covariance that do take into account the overall structure of the data when dealing with outliers. Each is readily converted to a robust analog of Pearson's correlation. Here, attention is restricted to what is called the OP correlation coefficient, which belongs to the class of skipped correlations. The strategy is to check for outliers using the projection method in Section 13.1.6, remove any outliers that are found, and compute Pearson's correlation using the data that remain. This process is often called a skipped correlation even though, strictly speaking, the term skipped correlation refers to any strategy for removing outliers, after which some correlation coefficient is computed. It seems that in terms of detecting an association between two variables, the OP correlation performs relatively well (Stuart, 2009). Letting r_p indicate the OP correlation, the hypothesis of a zero correlation can be tested with

$$T_p = r_p \sqrt{\frac{n-2}{1-r_p^2}}.$$

The null hypothesis is rejected at the $\alpha = 0.05$ level if $|T_p| \geq c$, where

$$c = \frac{6.947}{n} + 2.3197$$

(Wilcox, 2004a). However, this method assumes homoscedasticity, so a more accurate description is that it tests the hypothesis that two random variable are independent. To deal with heterocedasticity, a percentile bootstrap method can be used via the R functions described in Section 14.5.8.

14.5.6 R Function scor

The R function

```
scor(x,y, plotit = TRUE, xlab = "VAR 1", ylab = "VAR 2", MC = FALSE)
```

computes the OP correlation. By default, it creates a scatterplot with the central half of the data indicated by a polygon. To avoid the plot, set the argument `plotit=FALSE`. If you have access to a multicore processor, setting the argument `MC=TRUE` can reduce execution substantially if the sample size is large. The function tests the hypothesis that the OP correlation is equal to zero, but only at the $\alpha = 0.05$ level. More precisely, it reports a test statistic and critical value, and the hypothesis of a zero correlation is rejected if the test statistic exceeds the critical value. To get a p-value, or to deal with heterocedasticity, use the R function `corb` or the R function `scorci` described in Section 14.5.8, which use a percentile bootstrap method.

EXAMPLE

Using R, data were generated based on the model $Y = X + e$, where both X and e have standard normal distributions. The sample size was $n = 20$. The values of Pearson's correlation, Kendall's tau, Spearman's rho, the 20% Winsorized correlation, and the OP correlation were 0.598, 0.40, 0.571, 0.467, and 0.598, respectively. All five correlations are significant at the 0.05 level. But if two outliers are added at $(X, Y) = (2.1, -2.4)$, now Pearson's correlation, Kendall's tau, Spearman's rho, and the 20% Winsorized correlation are equal to 0.011, 0.17, 0.225, and 0.181, respectively, none of which reject at the 0.05 level. However, the OP correlation is again 0.598 and rejects. The reason is that the first four correlations are sensitive to the two outliers at $(2.1, -2.4)$, roughly because they do not take into account the overall structure of the data. In contrast, the OP correlation detects the two outliers at (2.1, −2.4) and removes them, the result being that they do not alter the correlation.

14.5.7 Inferences about Robust Correlations: Dealing with Heteroscedasticity

All of the hypothesis testing methods associated with the robust correlations, previously described in this chapter, are sensitive to heteroscedasticity. That is, when they reject, the main reason could be due to heteroscedasticity. If the goal is to test the hypothesis of independence, this is not a practical concern. But if the goal is to test the hypothesis of a zero correlation in a manner that is insensitive to heteroscedasticity, currently the best strategy is to use a basic percentile bootstrap method.

14.5.8 R Functions corb and scorci

The R function

```
corb(x,y, corfun = pbcor, nboot = 599,...)
```

can be used to test hypotheses and compute confidence intervals for any of the robust correlations covered in this chapter. By default, the function uses the percentage bend correlation, which is not covered here. (See Wilcox, 2017, Section 9.3.1 for details.) Setting the argument `corfun=wincor`, for example, would compute a confidence interval based on the Winsorized correlation. When using the OP correlation, the skipped correlation in Section 14.5.5 that takes into account the overall structure of the data when removing outliers, it is suggested that the command

```
corb(x,y, corfun = scor, plotit=FALSE)
```

be used. Otherwise, the function will create a scatterplot for each bootstrap sample, which can increase execution time considerably. Alternatively, use the R function

```
scorci(x,y,nboot=1000,alpha=0.05, SEED=TRUE, plotit=TRUE, STAND=TRUE,
            corfun=pcor, cop=3,...).
```

The argument `corfun=pcor` means that Pearson's correlation will be used after outliers are removed. The function `scorciMC` takes advantage of a multicore processor if one is available.

14.6 MEASURING THE STRENGTH OF AN ASSOCIATION BASED ON A ROBUST FIT

Given a fit to the data based on a robust regression estimator or smoother, how should the strength of the association be measured? The approach used here is based on simple generalizations of the notion of explanatory power, which was studied in a general context by Doksum and Samarov (1995). The generalization used here consists of simply replacing the usual variance with some robust analog and taking $\tilde{Y}$ to be the predicted value of Y based on any regression estimator or smoother. In symbols, let $\tau^2(Y)$ be any measure of variation. Then a robust analog of *explanatory power* is

$$\eta^2 = \frac{\tau^2(\tilde{Y})}{\tau^2(Y)}, \tag{14.10}$$

eta squared. The *explanatory strength of association* is the (positive) square root of explanatory power, η.

To put η^2 in perspective, if $\tilde{Y}$ is the predicted value of Y based on the least squares regression line, and τ^2 is the usual variance, then η^2 reduces to the squared multiple correlation coefficient given by Equation (6.12) in Section 6.2.2. In the case of a single predictor, still using least squares regression, η is just Pearson's correlation, ρ, assuming that the sign of η is taken to be the sign of the slope of the least squares regression line.

To make the explanatory strength of an association practical, a choice for the measure of variation, τ^2, must be made. Here, τ^2 is taken to be the percentage bend midvariance, which is computed as described in Section 2.3.10, or the 20% Winsorized variance. For completeness, other robust measures of variation represent reasonable choices (e.g., Wilcox, 2017), but the extent to which they offer a practical advantage, when measuring the strength of an association, is unknown.

In principle, explanatory power can be estimated when using any regression method or smoother. First, compute the percentage bend midvariance based on predicted Y values, say $\hat{\tau}^2(\tilde{Y})$, and then compute the percentage bend midvariance based on the observed Y values, $\hat{\tau}^2(Y)$, in which case the estimate of η^2 is

$$\hat{\eta}^2 = \frac{\hat{\tau}^2(\tilde{Y})}{\hat{\tau}^2(Y)}. \tag{14.11}$$

There is a fundamental issue: does the choice of method for obtaining the predicted Y values make a practical difference when estimating η^2? The answer has been found to be yes, at least with small to moderate sample sizes (e.g., Wilcox, 2010c). Two regression estimators that seem to perform relatively well, given the goal of estimating η^2, are the Theil–Sen estimator when the regression surface is a plane, and Cleveland's smoother (LOESS), described in Section 14.4.1, when there is curvature.

Section 14.4.2 described an R function, `lplot`, for plotting Cleveland's nonparametric regression line (LOESS). One of the arguments is `varfun`, which can now be explained. It indicates the measure of variation used when estimating explanatory power and defaults to the percentage bend midvariance.

Renaud and Victoria-Feser (2010) compared several other robust analogs of R^2, the coefficient of determination, which represent alternatives to η^2. Their results suggest using a measure of association based in part on a fit to the data obtained via the MM-estimator, which is not covered here.

14.7 COMPARING THE SLOPES OF TWO INDEPENDENT GROUPS

Consider two independent groups and imagine that the goal is to test

$$H_0 : \beta_{11} = \beta_{12}, \tag{14.12}$$

the hypothesis that the two groups have equal slopes. Chapter 7 noted that a modified percentile bootstrap method, as well as a wild bootstrap method, performs reasonably well when using least squares regression. When comparing slopes based on robust regression estimators, it suffices to use the basic percentile bootstrap method. That is, generate bootstrap samples from each group as described in Section 7.6, compute the slope for each group based on these bootstrap samples, and label them b_{11}^* and b_{12}^*. Next, repeat this process B times, let d_b^* be the difference between the bootstrap estimate of the slopes based on the bth bootstrap sample ($b = 1, \dots, B$), and let $\hat{p}^*$ be the proportion of times the bootstrap estimate for the first group is less than the bootstrap estimate from the second. That is, $\hat{p}^*$ is the proportion of d_b^* values less than zero. The p-value for the hypothesis given by Equation (14.12) is

$$2\min(\hat{p}^*, 1 - \hat{p}^*).$$

As usual, reject the null hypothesis if the p-value is less than or equal to α. Putting the d_b^* values in ascending order, the $1-\alpha$ confidence interval for the difference between the slopes, $\beta_{11}-\beta_{12}$, is $(d^*_{(\ell+1)}, d^*_{(u)})$, where as usual $\ell=\alpha B/2$, rounded to the nearest integer, and $u=B-\ell$. Ng and Wilcox (2010) found that this method, used in conjunction with the Theil–Sen estimator, compares very well to several others that have been proposed.

14.7.1 R Function reg2ci

The R function

```
reg2ci(x1, y1, x2, y2, regfun=tsreg, nboot=599, alpha=0.05, plotit=T)
```

compares the slopes of two groups using the method just described. The data for group 1 are stored in `x1` and `y1`, and for group 2 they are stored in `x2` and `y2`. As usual, `nboot` is B, the number of bootstrap samples, `regfun` indicates which regression estimator is to be used and defaults to the Theil–Sen estimator, and `plotit=T` creates a plot of the bootstrap estimates.

EXAMPLE

This example is based on the Well Elderly 2 data described in Section 14.3.3. A portion of the study dealt with the CAR (described in Section 14.3.3) and a measure of meaningful activities (MAPA). Here the focus is on two ethnic groups: Latino/Hispanic and blacks. An issue is whether these groups differ in regard to the nature of the association between MAPA and the CAR. Figure 14.6 shows the plot created by the R function `reg2ci`. The p-value when comparing the slopes is 0.048.

14.8 TESTS FOR LINEARITY

Smoothers provide an informal check on curvature. This section describes two methods that can be used to establish curvature in a more formal manner. (A third method was derived by Wang and Qu, 2007, but it is unknown how it compares to the two methods given here.)

The first strategy is to split the data into two groups based on the X values. For example, one might split the data based on the median. If there is no curvature, then these two sets of observations should have identical slopes which can be tested as described in Section 14.7.

An alternative approach is to test the hypothesis that there is a linear association between Y and some set of predictors. That is, the goal is to test the hypothesis that for some $\beta_0,\ldots,\beta_p$, $Y=\beta_0+\beta_1X_1+\cdots+\beta_pX_p+e$. Here, it is *not* assumed that the error term is homoscedastic. The theoretical justification for the method in this section is due to Stute, Manteiga, and Quindimil (1998). For simplicity, the method is described for the case of a single predictor only, but multiple predictors can be handled as well.

Let $\hat{Y}$ be some regression estimate of Y. Least squares could be used, but it has been shown that this can lead to problems in terms of controlling the probability of a Type I error. So it is suggested that some robust estimator be used instead. Theil–Sen seems to be a good choice. For fixed j $(1\le j\le n)$, set $I_i=1$ if $X_i\le X_j$; otherwise $I_i=0$, and let

$$\begin{aligned} R_j &= \frac{1}{\sqrt{n}}\sum I_i(Y_i-\hat{Y}_i) \\ &= \frac{1}{\sqrt{n}}\sum I_i r_i, \end{aligned} \tag{14.13}$$

where $r_i=Y_i-\hat{Y}_i$ are the usual residuals. The (Kolmogorov) test statistic is the maximum absolute value of all the R_j values. That is, the test statistic is

$$D=\max|R_j|, \tag{14.14}$$

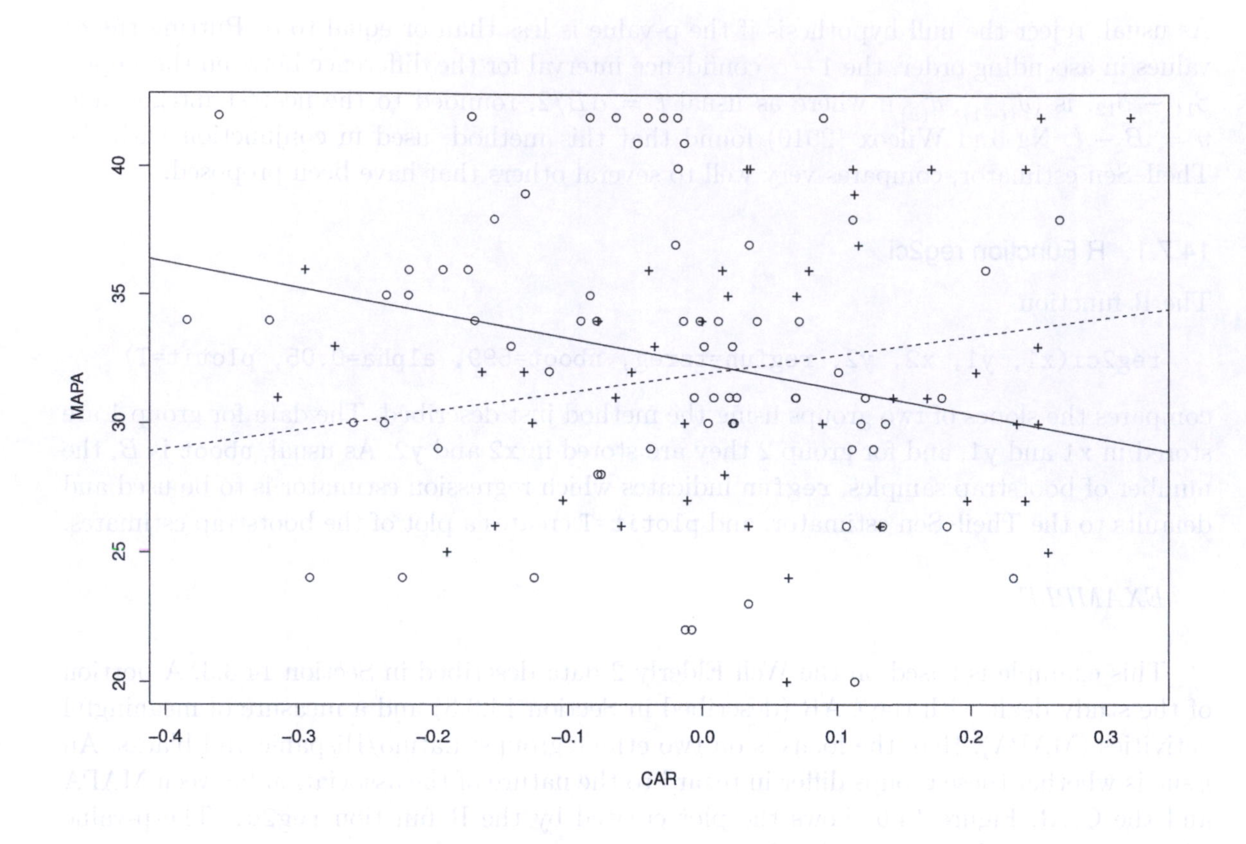

Figure 14.6 Shown are the regression lines when predicting MAPA scores based on the CAR. The solid line is for Latino/Hispanics and the dashed line is for blacks.

where max means that D is equal to the largest of the $|R_j|$ values. A Cramér–von Mises test statistic can be used instead where now

$$D = \frac{1}{n} \sum R_j^2. \tag{14.15}$$

A critical value is determined as described in Box 14.2.

Box 14.2: How to determine the critical value when testing the hypothesis of a straight line.

A critical value is determined using a wild bootstrap method. Generate n observations from a uniform distribution and label the results $U_1, \ldots, U_n$. Next, for $i = 1, \ldots, n$, set

$$V_i = \sqrt{12}(U_i - 0.5),$$

$$r_i^* = r_i V_i,$$

and

$$Y_i^* = \hat{Y}_i + r_i^*.$$

Then based on the n pairs of points $(X_1, Y_1^*), \ldots, (X_n, Y_n^*)$, compute the test statistic and label it D^*. Repeat this process B times and label the resulting test statistics $D_1^*, \ldots, D_B^*$. Finally, put these B values in ascending order yielding $D_{(1)}^* \leq \cdots \leq D_{(B)}^*$. The critical value is $D_{(u)}^*$, where $u = (1 - \alpha)B$ rounded to the nearest integer. That is, reject if

$$D \geq D_{(u)}^*.$$

Notice that if this method rejects, this indicates in particular that there is an association. Generally, there are many ways of testing the hypothesis of no association and the method used can make a practical difference as illustrated in the next example. A practical problem is that the best method, in terms of power, for testing the hypothesis of no association is difficult to determine because it depends on the true nature of the association, which is unknown.

14.8.1 R Functions lintest, lintestMC, and linchk

The R function

```
lintest(x,y,regfun=tsreg,nboot=500,alpha=0.05)
```

tests the hypothesis that a regression surface is a plane using the method outlined in the previous section. (Execution time is fairly fast with one predictor, but it might be slow when there are multiple predictors instead.) When reading the output, the Kolmogorov test statistic is labeled `dstat` and its critical value is labeled `critd`. The Cramér–von Mises test statistic is labeled `wstat`. The default regression method (indicated by the argument `regfun`) is Theil–Sen. The R function

```
lintestMC(x,y,regfun=tsreg,nboot=500,alpha=0.05)
```

is the same as `lintest`, only it takes advantage of a multicore processor, if one is available, with the goal of reducing execution time.

The function

```
linchk(x,y,sp,pv=1,regfun=tsreg,nboot=599,alpha=0.05)
```

splits the data into two groups according to whether predictor `pv` has a value less than the value stored in the argument `sp`. For example,

```
linchk(x,y,sp=10,pv=3)
```

would split the data into two groups based on whether predictor 3 has a value less than 10. Then it compares the regression parameters for these two groups with the R function `reg2ci`.

EXAMPLE

For the plasma retinol data shown in Figure 14.4, the R function `lintest` rejects the hypothesis that the regression surface is a plane; the reported p-value is 0 (meaning that it is less than 0.002). This indicates that there is curvature, meaning in particular there is an association. Various methods that assume there is no curvature fail to find an association. For example, using the classic F test in Chapter 6 in conjunction with the least squares regression estimator, the p-value is 0.257. Using the R function `hc4test`, which uses the HC4 estimator to deal with heteroscedasticity, again the p-value is 0.257. With the bootstrap method used by the R function `olstest`, which allows heteroscedasticity, the p-value is 0.644. Testing the hypothesis of zero slopes via the R function `regtest`, using the Theil–Sen estimator, the p-value is 0.417.

14.9 IDENTIFYING THE BEST PREDICTORS

A problem that has received considerable attention is identifying a subset of predictors that might be used in place of the p predictors that are available. In particular, which predictors are best at predicting some outcome variable? This is an extremely difficult problem with many proposed solutions that have been found to be relatively unsatisfactory.

Section 6.13 summarized various strategies for identifying the best predictors. Among the methods discussed, two seem to perform relatively well: least angle regression and the 0.632 bootstrap estimate of prediction error. But they do not address an important issue: how compelling is the evidence that predictor 1, say, is better than predictor 2? Inferential methods, when using the lasso or least angle regression, have been derived assuming homoscedasticity and normality (Tibshirani et al., 2016; Lee et al., 2016). Here the focus is on robust methods for determining the relative importance of the independent variables that do not assume homoscedasticity or normality.

Two basic strategies for answering this question are described in this section. The first, described in Section 14.9.1, is based on robust generalizations of the methods in Section 6.14. There are in fact two variations of this approach as will be seen.The second approach, which is described in Section 14.9.3, deals with a limitation associated with the first approach.

14.9.1 Inferences Based on Independent Variables Taken in Isolation

To simplify matters, imagine that there are two independent variables: X_1 and X_2. The methods in this section are based on the following strategy. Compare the strength of the association between Y and X_1, ignoring X_2, to the strength of the association between Y and X_2, ignoring X_1. This is in contrast to the method in Section 14.9.3, where both X_1 and X_2 are entered into the model simultaneously. There are two variations of the approach used here. The first is to use robust analogs of the coefficient of determination (the squared Pearson correlation) without using any particular regression estimator. That is, use one of the measures of association outlined in Section 14.5. Another strategy is to measure the strength of the association based on some robust regression fit to the data. There are two variations of this latter approach. The first is to fit a regression line, assuming the regression line is straight, using for example the Theil–Sen estimator. The other uses a smoother.

Method SA1

The first general approach in this section, called method *SA1*, simply replaces Pearson's correlation with a robust analog without fitting any particular regression model to the data. There are two related goals when comparing predictors based on Pearson's correlation. The first is to test

$$H_0 : \rho_{y1} = \rho_{y2},$$

where ρ_{y1} is the correlation between the outcome variable of interest Y and the first predictor, and ρ_{y2} is defined in an analogous fashion. The other is to test

$$H_0 : \rho^2_{y1} = \rho^2_{y2}. \tag{14.16}$$

The focus here is on testing a robust analog of Equation (14.16). As noted in Chapter 6, despite many attempts at finding a method for testing Equation (14.16), extant techniques are known to be unsatisfactory under general conditions. Moreover, as noted in Chapter 13, there are many robust analogs of the usual covariance matrix, with each yielding a robust alternative to Pearson's correlation. Here the median ball algorithm is used primarily because there are published results indicating that good control over the Type I error probability can be achieved. (Perhaps other robust measures of association can have a practical advantage in some situations, but this issue is in need of further study.) So, letting ζ^2 represent the correlation stemming from the median ball algorithm, the goal is to test

$$H_0 : \zeta^2_{y1} = \zeta^2_{y2}. \tag{14.17}$$

Let τ^2_y be the median ball measure of variation among the Y values and let $\psi^2_{yj} = \tau^2_y(1 - \zeta^2_{yj})$, $j = 1$, 2. The method used here tests

$$H_0 : \psi^2_{y1} = \psi^2_{y2}, \tag{14.18}$$

which is the same as testing Equation (14.17). Although ζ^2_{y1} and ζ^2_{y2} are readily estimated, there is no known method for controlling the Type I error probability if these estimates are used. But if we test Equation (14.18) instead, a method has been found that performs well in simulations (Wilcox, 2010d). The method computes a bootstrap estimate of the standard error of

$$D = \hat{\psi}^2_{y1} - \hat{\psi}^2_{y2},$$

yielding say S_D, where $\hat{\psi}^2_{y1}$ and $\hat{\psi}^2_{y2}$ are estimates of ψ^2_{y1} and ψ^2_{y2}, respectively. The test statistic is

$$U = \frac{D}{S_D},$$

and a critical value, when testing at the $\alpha = 0.05$ level, is

$$c = 2.09 - \frac{5.596}{\sqrt{n}}.$$

The null hypothesis is rejected if $|U| \geq c$.[1] Currently, adjusted critical values are not available when $\alpha \neq 0.05$ and the method does not provide a p-value.

Method SA2

Method SA1 uses a robust measure of scatter without any reference to a particular regression estimator. An alternative approach begins by applying some regression estimator, measuring the strength of association for each predictor based on this fit, and then testing the hypothesis that the strength of the association is the same for both predictors. This is the strategy used by *Method SA2*; it is based on the Theil–Sen estimator with the strength of the association measured via explanatory power.

Roughly, take independent bootstrap samples for each of the predictors under investigation. That is, take a bootstrap sample from (X_{i1}, Y_i) and compute a bootstrap estimate of η_1^2, say $\tilde{\eta}_1^2$, and then take a new, independent bootstrap sample from (X_{i2}, Y_i) yielding $\tilde{\eta}_2^2$, and let $D^* = \tilde{\eta}_1^2 - \tilde{\eta}_2^2$. Repeating this process B times yields $D_1^*, \ldots, D_B^*$. Letting $P = P(D^* < 0)$, a (generalized) p-value is

$$p = 2\min(P, 1 - P) \tag{14.19}$$

(Liu and Singh, 1997). If we use the usual estimate of p (as described, for example, in Section 5.11) and reject if $p \leq \alpha$, simulations indicate that the actual Type I error will be less than or equal to α. But a negative feature is that the actual Type I error probability can be substantially less than α when the sample size is small, indicating that power can be relatively low. There is an adjustment for dealing with this problem when testing at the $\alpha = 0.05$ level (Wilcox, 2010d), but the involved computational details are not described. (All indications are that for $n \geq 100$, an adjustment is no longer needed.)

Method SA3

Method SA2 can be extended to situations where Cleveland's (LOESS) smoother, described in Section 14.4.1, is used in place of the Theil–Sen estimator. Again, an adjustment is needed to control the Type I error probability reasonably well when the sample size is small. The extensive computations are not described, but an R function is provided that performs the calculations.

14.9.2 R Functions regpord, ts2str, and sm2strv7

The R function

```
regpord(x,y)
```

performs method SA1 for all pairs of predictors. That is, for predictors j and k, which are assumed to be stored in a matrix x having p columns, it tests the hypothesis that they have the same strength of association with the outcome variable stored in y. This is done for all $j < k$. As previously indicated, method SA1 is limited to testing at the $\alpha = 0.05$ level.

Here is an example of how a portion of the output is reported.

```
$crit.value
[1] 1.5004
```

[1]For sufficiently large n, theory suggests that when computing the critical value, the 2.09 should be replaced by 1.96, but it is unclear at what point this should be done.

```
$results
  Pred. Pred  test.stat     Decision
1     1    2  0.2781442 fail to reject
2     1    3 -0.3680995 fail to reject
3     2    3 -0.6204060 fail to reject
```

The last column indicates that for each pair of predictors, no difference in the strength of the association was found, meaning that the absolute value of the test statistic did not exceed the critical value.

The R function

```
ts2str(x,y)
```

performs method SA2. The current version is designed to compare two predictors only; the argument `x` is assumed to be a matrix with two columns. The function returns a p-value as well a value labeled `p.crit`. Reject at the $\alpha = 0.05$ level if the p-value is less than or equal to `p.crit`.

The R function

```
sm2strv7(x,y,regfun=tsreg,nboot=500,alpha=0.05)
```

performs method SA3.

EXAMPLE

In an unpublished study by L. Doi, a general goal was to identify good predictors of reading ability in children. Two of the predictors were a measure of sound blending and the speed of identifying lower case letters. One of the outcome variables was the ability to decode words. The estimated explanatory power for the first predictor, which is returned by `lplot`, is 0.384. For the other predictor, explanatory power is 0.07. Both methods SA2 and SA3 reject at the 0.05 level, indicating that sound blending has a stronger association with the ability to decode words. In contrast, method SA1 fails to reject.

14.9.3 Inferences When Independent Variables Are Taken Together

A concern about the methods in Section 14.9.1 is that the strength of the association between Y and X_1 can depend on whether X_2 is included in the model. In particular, the estimated slope associated with X_1, b_1, can be positive and differ significantly from zero when ignoring all other independent variables, yet when some additional independent variable is added to the model, now b_1 does not differ significantly from zero and can be relatively small.

EXAMPLE

The phenomenon just described can happen when, for example, X_1 and X_2 are correlated and data are generated from the model $Y = \beta_1 X_1 + \beta_2 X_2 + \epsilon$, where $\beta_1 = 1$ and $\beta_2 = 0$. The following R commands illustrate this point.

```
set.seed(1) # Set the seed of the random number generator so that readers
can duplicate the results
x=rmul(50,rho=0.6)
y=rnorm(50,mean=x[,1])
olshc4(x[,2],y)
olshc4(x,y)
```

The second command generates a matrix of data with 50 rows and two columns. Each column of the data is generated from a standard normal distribution and Pearson's correlation between these two variables is $\rho = 0.6$. The command `rnorm(50,mean=x[,1])` generates 50 values from a normal distribution having means equal to the values in `x[,1]` and variance one. For example, `x[1,1]` is equal to -0.468, so `y[1]` contains a value generated from a normal distribution with mean -0.468 and variance one. In a similar manner, `x[2,1]` is equal to -0.019, so `y[2]` contains a value generated from a normal distribution with mean -0.019. More formally, the command generates data for which the mean of Y is $\beta_1 X_1 + \beta_2 X_2$, where $\beta_1 = 1$ and $\beta_2 = 0$. The command `olshc4(x[,2],y)` tests H_0: $\beta_2 = 0$, the hypothesis that the second independent variable has a zero slope using the data in `x[,2]`, ignoring the data in `x[,1]`. The estimate of the slope is $b_2 = 0.52$ and the p-value is 0.003. Roughly, even though the values for the dependent variable were generated in a manner that ignores the second independent variable, an association is found due to the correlation between the second independent variable and the first independent variable. The final command includes both independent variables when testing for a zero slope. Now the estimate of β_2 is -0.006, very close to the true value, and the p-value is 0.976. As for the slope associated with the first independent variable, the estimate is $b_1 = 1.03$, very close to the true value, and the p-value is less than 0.001.

When assessing the relative importance of two independent variables, what is needed is a method that deals with situations where both independent variables are included in the model. First focus on the model

$$Y = \beta_0 + \beta_1 X_1 + \beta_2 X_2 + (\lambda_1(X_1) + \lambda_2(X_2))\epsilon,$$

where ϵ has some unknown variance σ^2, $E(\epsilon) = 0$, and ϵ is independent of X_1 and X_2. The functions λ_1 and λ_2 are used to model heteroscedasticity. For convenience, let

$$M(X_1, X_2) = \beta_1 X_1 + \beta_2 X_2$$

denote some conditional measure of location given X_1 and X_2. In the notation used in Section 14.6, $M(X_1, X_2) = \tilde{Y}$. From Equation (14.10), the numerator of explanatory power is based on some measure of dispersion associated with $M(X_1, X_2)$. Let $M_j(X_j) = \beta_j X_j$, where it is stressed that β_j is the slope associated with X_j when the other independent variable is included in the model. Let η_j^2 be some measure of variation associated with $M_j(X_j)$ over the range of possible X_j values. That is, η_j^2 is the numerator of explanatory power given by Equation (14.10) based on the slope associated with the jth independent variable. The goal is to test

$$H_0 : \eta_1^2 = \eta_2^2. \tag{14.20}$$

Now consider the linear model

$$Y = \beta_0 + \beta_1 X_1 + \beta_2 X_2 + \beta_3 X_3 + (\lambda_1(X_1) + \lambda_2(X_2) + \lambda_2(X_3))\epsilon.$$

Let $M_{12}(X_1, X_2) = \beta_1 X_1 + \beta_2 X_2$ and let $M_3(X_3) = \beta_3 X_3$, where β_1, β_2, and β_3 are the slopes when all three independent variables are included in the model. Let η_{12}^2 be some measure of dispersion associated with $M_{12}(X_1, X_2)$ and let η_3^2 be some measure of dispersion associated with $M_3(X_3)$. Here, the measure of dispersion is taken to be the 20% Winsorized variance. To determine whether the first two independent variables are more important than the third independent variable, one can test

$$H_0 : \eta_{12}^2 = \eta_3^2. \tag{14.21}$$

Or in terms of Tukey's three decision rule, is it reasonable to make a decision about whether

η_{12}^2 is less than or greater than η_3^2 based on the available data? As is evident, this approach is readily generalized to more than three independent variables.

There is, however, a limitation that should be stressed. While the method in this section provides a valid way of making inferences about the relative importance of X_1 and X_2 taken together, versus X_3, the method is meaningless regarding the relative importance of X_1 and X_2 taken together, versus X_2. That is, it is pointless testing $H_0 : \eta_{12}^2 = \eta_2^2$ because it is generally the case that $\eta_{12}^2 > \eta_2^2$.

Method IBS

One way of testing the relevant hypotheses is to use a basic percentile bootstrap method that generates bootstrap samples as described in Section 14.3.1. However, when using the quantile regression estimator, this approach performs poorly in terms of controlling the Type I error probability. When testing at the 0.05 level, and when the slopes differ from zero, the actual level can exceed 0.10.

Consider the goal of testing Eq. (14.20). Wilcox (in press b) considered a slight modification of the standard percentile bootstrap method that uses independent bootstrap samples. That is, generate a bootstrap sample and estimate η_1^2 yielding $\hat{\eta}_2^2$. Then generate a new bootstrap sample and estimate η_2^2 yielding $\hat{\eta}_2^2$. Repeat this B times and estimate $P(\hat{\eta}_1^2 < \hat{\eta}_2^2)$ using the method Section 7.8.1, which is denoted by $\hat{p}$. So in effect, B^2 pairs of values are used when computing $\hat{p}$. Then a p-value is $2\min(\hat{p}, 1-\hat{p})$ (cf. Racine and MacKinnon, 2007b). Extant simulations indicate that now, even when the slopes differ from zero, the actual Type I error probability is substantially smaller than the nominal level, still using the quantile regression estimator. Wilcox (in press b) found, however, that using instead the Theil–Sen estimator, reasonably good control over the Type I error probability is achieved. (Some simulation results are also reported in Wilcox, 2017). Limited simulations suggest that when using the MM-estimator, again control over the Type I error probability is good. As is evident, a similar approach can be used when testing Eq. (14.21).

14.9.4 R Function regIVcom

The R function

```
regIVcom(x,y, IV1=1, IV2=2, regfun=tsreg, nboot=200, xout=FALSE,
         outfun=outpro, SEED=TRUE, MC=FALSE, tr=0.2,...)
```

performs method IBS. The arguments `IV1` and `IV2` indicate which independent variables are to be compared. For example, setting the arguments `IV1=c(1,2)` and `IV2=3` would compare the strength of the association between independent variables 1 and 2 and the dependent variable, to the strength of the association between the third independent variable and the dependent variable. That is, the function would test Eq. (14.21). The argument `tr` indicates the amount of Winsorizing. The function returns a p-value and estimates of η^2 for the independent variables associated with `IV1` and `IV2`, respectively, which are labeled est.1 and est.2. The function also returns their ratio, estimates of explanatory power (η^2 divided by the variance of Y) labeled e.pow1 and e.pow2, the square root of e.pow1 and e.pow2 labeled strength.assoc.1 and strength.assoc.2 (analogs of Pearson's correlation), and the ratio of these values, which is labeled strength.ratio.

EXAMPLE

Data from the Well Elderly 2 study (Clark et al., 2011) are used to illustrate method IBS. Generally, the Well Elderly 2 study was designed to assess the effectiveness of an intervention program aimed at improving the physical and emotional well-being of older adults. A portion of the study was aimed at understanding the association between a measure of life satisfaction (LSIZ) and two independent variables: a measure of depressive symptoms (CESD) and a measure of interpersonal support (PEOP). Here the focus is on measures taken after intervention. Least angle regression indicates that CESD is more important than PEOP. But this does not indicate the strength of the empirical evidence that this is the case. Using method IBS, and letting η_1^2 denote the explanatory power associated with CESD, the p-value is $p = 0.012$ and η_1/η_2 was estimated to be 2 indicating that, indeed, a measure of depressive symptoms is more important than a measure of interpersonal support.

EXAMPLE

Consider again the Well Elderly 2 study used in the previous example. Another explanatory variable of interest was a measure of meaningful activities (MAPA). Now the issue is the extent CESD continues to be more important when CESD, PEOP, and MAPA are used to predict LSIZ. Again, least angle regression indicates that CESD is the most important explanatory variable followed by PEOP. But comparing the explanatory power associated with CESD to the explanatory power associated with PEOP and MAPA the p-value is 0.44 and η_{12}/η_3 was estimated to be 0.83. So now the relative importance of a measure of depressive symptoms (CESD) is less striking. Finally, least angle regression indicates that MAPA is the least important explanatory variable when both CESD and PEOP are included in the model. To add perspective, the explanatory power associated with MAPA was compared to the explanatory power associated with both CESD and PEOP. Now $p = 0.002$ and η_1/η_{23} was estimated to be 0.148. So method IBS lends support to the indication that MAPA is indeed less important than CESD and PEOP, taken together, when all three of these explanatory variables are included in the model.

14.10 INTERACTIONS AND MODERATOR ANALYSES

This section discusses the issue of detecting interactions within the context of regression. Consider some outcome variable, Y, and two predictors, X_1 and X_2. Roughly, the issue is whether knowing the value of X_2 modifies the association between Y and X_1. For example, for the reading study mentioned in the last example, there was interest in how a measure of orthographic ability (Y) is related to a measure of auditory analysis X_1. A third variable in this study was a measure of sound blending (X_2). Does knowing the value of this third variable alter the association between Y and X_1, and if it does, how? More generally, there is interest in knowing whether a particular factor affects the magnitude of some effect size. Such factors are called *moderators* (e.g., Judd et al., 2001).

Graphically, an interaction, in the context of regression, can be described roughly as follows. Let X_2 be any value of the second predictor variable, X_2. For example, X_2 could be the median of the X_2 values, but any other value could be used in what follows. Now consider the outcome variable Y and the first predictor (X_1), and imagine that we split the n pairs of points $(Y_1, X_{11}), \ldots, (Y_n, X_{n1})$ into two groups: those pairs for which the corresponding

X_2 value is less than X_2 and those for which the reverse is true. No interaction means that the regression lines corresponding to these two groups are parallel. For example, if for the first group $Y = X_1^2 + e$ and for the second $Y = X_1^2 + 6 + e$, these regression lines are parallel and there is no interaction. But if for the second group $Y = X_1^2 + 8X_1 + 3 + e$, say, then the regression lines intersect (at $X_1 = -3/8$) and we say that X_2 modifies the association between Y and X_1.

A popular method for checking and modeling interaction assumes that

$$Y = \beta_0 + \beta_1 X_1 + \beta_2 X_2 + \beta_3 X_1 X_2 + e. \tag{14.22}$$

That is, use the product of the two predictors to model interaction and conclude that an interaction exists if

$$H_0 : \beta_3 = 0 \tag{14.23}$$

is rejected. (Saunders, 1955, 1956, appears to be the first to suggest this approach to detecting interactions in regression; cf. Cronbach, 1987; Baron and Kenny, 1986.) The model given by Equation (14.22) often plays a role in what is called a *moderator analysis*, roughly meaning that the goal is to determine the extent to which knowing the value of one variable, X_2, alters the association between Y and X_1. The hypothesis given by Equation (14.23) can be tested using methods already covered. That is, for the observations $(Y_1, X_{11}, X_{12}), \ldots, (Y_n, X_{n1}, X_{n2})$, set $X_{i3} = X_{i1}X_{i2}$ and for the model $Y = \beta_0 + \beta_1 X_1 + \beta_2 X_2 + \beta_3 X_3 + e$ test $H_0 : \beta_3 = 0$. However, it currently seems that a collection of tools is needed to address the issue of interactions in an adequate manner.

To add perspective on the product term just described, suppose we fix (or condition on) X_2. That is, we treat it as a constant. Then a little algebra shows that Equation (14.22) can be written as

$$Y = (\beta_0 + \beta_2 X_2) + (\beta_1 + \beta_3 X_2) X_1 + e.$$

So the intercept term becomes $(\beta_0 + \beta_2 X_2)$ and the slope for X_1 changes as a linear function of X_2. If $\beta_3 = 0$, then

$$Y = (\beta_0 + \beta_2 X_2) + \beta_1 X_1 + e.$$

That is, knowing X_2 alters the intercept term but not the slope. Said another way, if we split the data into two groups according to whether X_2 is less than or greater than some constant c, the corresponding regression lines will be parallel, consistent with the description given previously.

It is briefly mentioned that one way of investigating the extent to which the variable X_2 is a modifier variable is to compute the strength of the association between Y and X_1 as a function of X_2. A method for doing this was derived by Doksum et al. (1994), but it seems that a very large sample size is needed when applying this technique. (In addition, software for implementing the method is not readily available.)

A fundamental concern is whether the model given by Equation (14.22) provides a reasonably satisfactory method for detecting and describing an interaction. Perhaps using the product term, X_1X_2, is too simple. If, for example, the true regression model involves $X_1X_2^3$, modeling interaction with Equation (14.22) might result in relatively low power when testing the hypothesis of no interaction because the interaction is modeled in an inaccurate fashion. A more flexible approach for establishing that there is an interaction is to test

$$H_0 : Y = \beta_0 + f_1(X_1) + f_2(X_2) + e, \tag{14.24}$$

the hypothesis that for some unknown functions f_1 and f_2, a generalized additive model fits the data, versus the alternative hypothesis

$$H_1 : Y = \beta_0 + f_1(X_1) + f_2(X_2) + f_3(X_1, X_2) + e.$$

This can be done using an extension of a method derived by Dette (1999) in conjunction with the running-interval smoother. Details are summarized in Wilcox (2017, Section 11.6.3; cf. Barry, 1993; Samarov, 1993).

14.10.1 R Functions olshc4.inter, ols.plot.inter, regci.inter, reg.plot.inter and adtest

The R function

```
olshc4.inter(x,y, alpha = 0.05, xout = FALSE, outfun = out, ...)
```

tests hypotheses about the slope parameters in Equation (14.22) using the least squares estimator. Heteroscedasticity is addressed with the HC4 estimator mentioned in Section 6.4.6. As usual, leverage points can be eliminated by setting the argument `xout=TRUE`. To use a robust regression estimator, use the R function

```
regci.inter(x,y, regfun = tsreg, nboot = 599, alpha = 0.05, SEED = TRUE, pr
    = TRUE, xout = FALSE, outfun = out, ...).
```

The argument `regfun` indicates the regression estimator that will be used, which defaults to the Theil–Sen estimator. The R function

```
ols.plot.inter(x,y, pyhat = FALSE, eout = FALSE, xout = FALSE, outfun = out,
 plotit = TRUE, expand = 0.5, scale = FALSE, xlab = "X", ylab = "Y", zlab =
      "", theta = 50, phi = 25, family = "gaussian", duplicate =
                 "error",ticktype="simple",....)
```

plots an estimate of the regression surface using the least squares estimator and assuming the model given Equation (14.22). The R function

```
reg.plot.inter(x,y, regfun=tsreg, pyhat = FALSE, eout = FALSE, xout = FALSE,
outfun = out, plotit = TRUE, expand = 0.5, scale = FALSE, xlab = "X", ylab =
   "Y", zlab = "", theta = 50, phi = 25, family = "gaussian", duplicate =
                 "error",ticktype="simple",...)
```

plots the regression surface again based on the model given by Equation (14.22), only now a robust regression estimator is used. It can be interesting and informative to compare the plots created by `ols.plot.inter` and `reg.plot.inter` to the plots created by `lplot` and `rplot`.

The R function

```
adtest(x, y, nboot=100, alpha=0.05, xout=F, outfun=out, SEED=TRUE, ...)
```

tests the hypothesis given by Equation (14.24). There are two test statistics for accomplishing this goal. Both are reported by the function along with their corresponding p-values.

A point worth stressing is that when checking for an interaction, the model given by Equation (14.22), which is commonly used, might not provide a satisfactory approximation of the regression surface—a smoother that provides a more flexible approach to curvature might be needed.

EXAMPLE

A portion of a study conducted by Shelley Tom and David Schwartz is used to illustrate that Equation (14.22) can be unsatisfactory when modeling an interaction. The dependent variable, labeled the Totagg score, is a sum of peer nomination items that were based on an inventory that included descriptors focusing on adolescents' behaviors and social standing. (The peer nomination items were obtained by giving children a roster sheet and asking them to nominate a certain amount of peers who fit particular behavioral descriptors.) The independent variables were grade point average (GPA) and a measure of academic engagement. The sample size is $n = 336$. Assuming that the model given by Equation (14.22) is true, the hypothesis of no interaction (H_0: $\beta_3 = 0$) is not rejected using the least squares estimator. The p-value returned by the R function `olswbtest` is 0.6. (The p-value returned by `olshc4` is 0.64.) But look at the left panel of Figure 14.7, which shows the plot of the regression surface assuming Equation (14.22) is true. (This plot was created with the R function `ols.plot.inter`.) Compare this to the right panel, which is an estimate of the regression surface using LOESS (created by the R function `lplot`). This suggests that using the usual interaction model is unsatisfactory for the situation at hand. The R function `adtest` returns a p-value less than 0.01 indicating that an interaction exists.

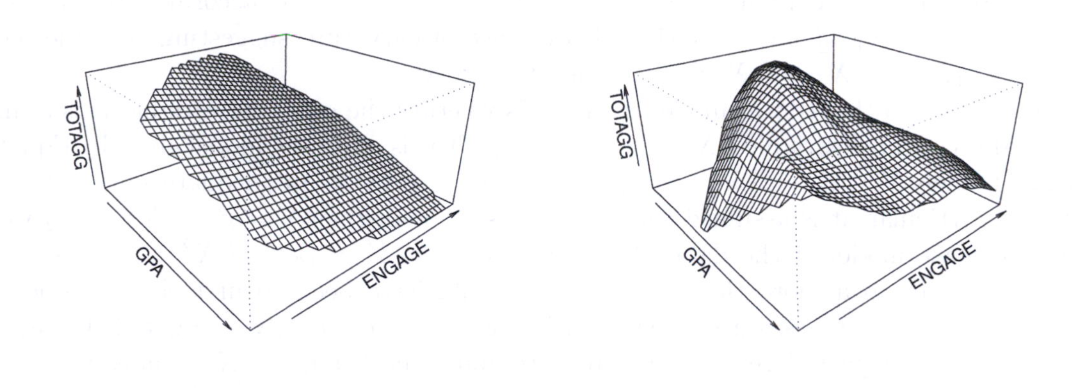

Figure 14.7 The left panel shows the plot created by `ols.plot.inter`, which assumes that an interaction can be modeled with $Y = \beta_0 + \beta_1 X_1 + \beta_2 X_2 + \beta_3 X_1 X_2 + e$ and where the least squares estimate of the parameters is used. The right panel shows an approximation of the regression surface based on the R function `lplot`.

14.10.2 Graphical Methods for Assessing Interactions

Some alternative graphical methods might be useful when studying interactions. The first simply plots a smooth of Y versus X_1 given particular values for X_2. So if there is no interaction, and this plot is created at say $X_2 = c_1$ and $X_2 = c_2$, $c_1 \neq c_2$, the regression lines should be parallel. The R function `kercon`, described in the next section, creates such a plot. More precisely, three smooths are created corresponding to three choices for X_2: the lower quartile, the median, and the upper quartile. Computational details can be found in Wilcox (2017, Section 10.6).

Another possibility is to split the data into two groups according to whether X_2 is less than some specified value and plot a nonparametric regression line for both groups. No interaction means that the resulting regression lines will be approximately parallel. The R function `runsm2g`, described in Section 14.9.3, creates such a plot by splitting the data into two groups according to whether X_2 is less than its median. Partial residual plots might also be used as illustrated in the next example.

EXAMPLE

Two hundred vectors of observations were generated according to the model $Y = X_1 + X_2 + X_1X_2 + e$ where X_1, X_2, and e are independent standard normal random variables. Based on how the data were generated, a partial residual plot based on the term X_1X_2 should produce a reasonably straight line. Applying the R function `prplot` to the data creates the smooth shown in the left panel of Figure 14.8, and we see that we get a reasonably straight line that has a slope approximately equal to the true slope, 1. The right panel shows the plot returned by `prplot`, only now the data were generated according to the model $Y = X_1 + X_2 + X_1X_2^2 + e$. Now there is evidence of curvature suggesting that the model $Y = \beta_0 + \beta_1X_1 + \beta_2X_2 + \beta_3X_1X_2 + e$ is unsatisfactory.

Notice that in this last example, the issue is whether the outcome variable Y is a linear combination of three variables: X_1, X_2, and X_1X_2. That is, does the model given by Equation (14.22), namely $Y = \beta_0 + \beta_1X_1 + \beta_2X_2 + \beta_3X_1X_2 + e$, provide an adequate summary of the data? In particular, it is assumed that if X_1X_2 is ignored, then $Y = \beta_0 + \beta_1X_1 + \beta_2X_2 + e$ is an adequate model of the data. But if, for example, $Y = \beta_0 + \beta_1X_1^2 + \beta_2X_2 + e$, this is not the case. So a more thorough analysis would have been to first check a smoother to see whether the regression line between Y and X_1 is reasonably straight. If the answer is yes, next use a partial residual plot to determine whether using X_2 is adequate. If the answer is yes, then the strategy used by the partial residual plot, consisting of fitting the model $Y = \beta_0 + \beta_1X_1 + \beta_2X_2 + e$ and then plotting the resulting residuals versus X_1X_2, is a reasonable strategy for checking the model $Y = \beta_0 + \beta_1X_1 + \beta_2X_2 + \beta_3X_1X_2 + e$.

14.10.3 R Functions kercon, runsm2g, regi

The R function

```
kercon(x, y, pyhat = FALSE, cval = NA, plotit = TRUE, eout = FALSE, xout =
   FALSE, outfun = out, iran = 0.05, xlab = "X", ylab = "Y", pch = "." )
```

creates the first of the plots mentioned in the previous section. The argument `x` is assumed to be a matrix with two columns. By default, three plots are created: a smooth of Y and X_1, given that X_2 is equal to its lower quartile, its median, and its upper quartile. Different choices for the X_2 values can be specified via the argument `cval`.

The R function

```
runsm2g(x1, y1, x2, val = median(x2), est = tmean, sm = FALSE, fr = 0.8,
                      xlab = "X", ylab = "Y", ...)
```

splits the data in `x1` and `y` into two groups based on the value in the argument `val` and the data stored in the argument `x2`. By default, a median split is used. The function returns a

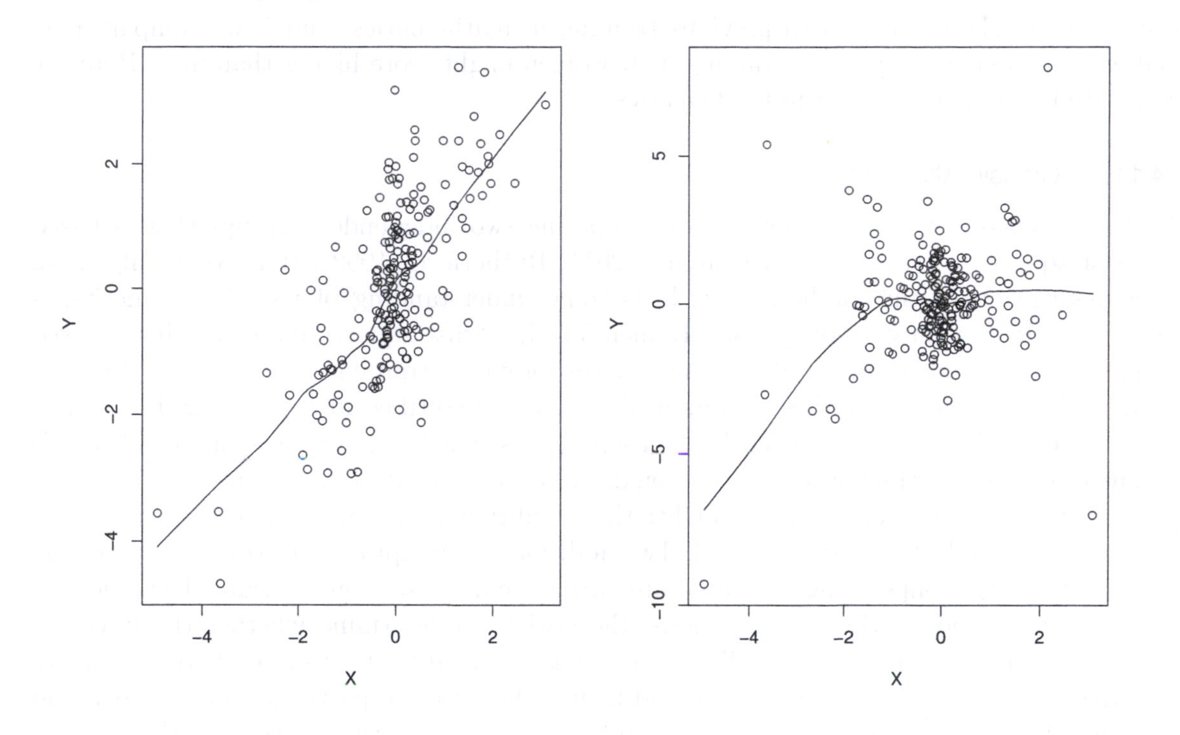

Figure 14.8 The left panel shows the plot created by prplot when indeed the interaction model $Y = X_1 + X_2 + X_1X_2 + e$ is correct. The sample size is $n = 200$ and the smooth is reasonably straight as it should be. The right panel shows the plot when the model $Y = X_1 + X_2 + X_1X_2 + e$ is incorrect; the true model is $Y = X_1 + X_2 + X_1X_2^2 + e$.

smooth for both groups. If there is no interaction, these two smooths should be approximately parallel. The smooths are based on the goal of estimating the trimmed mean of the outcome variable. But other measures of location can be used via the argument `est`.

The R function

```
regi(x, y, z, pt=median(z), est=onestep, regfun=tsreg, testit=FALSE,...)
```

creates two smooths in a manner similar to the function `runsm2g`. In fact this function simply calls the function `runsm2g`, only it uses by default a one-step M-estimator rather than a 20% trimmed mean. An advantage of this function is that it also has an option for replacing the smoother with a robust regression estimator, via the argument `regfun`, which by default is the Theil–Sen estimator. This is done by setting the argument `testit=TRUE`, in which case the function also tests the hypothesis that the two regression lines have equal slopes. Rejecting the hypothesis of equal slopes indicates an interaction.

14.11 ANCOVA

Roughly, *analysis of covariance* (ANCOVA) refers to methods aimed at comparing groups based on some measure of location in a manner that takes into account the information provided by some predictor variable called a *covariate*. As a simple illustration, imagine that men and women are compared in terms of their typical score on some mathematics aptitude test and it is found that the typical male scores higher than the typical female. However, test scores might be related to previous training in mathematics, and if we compare men and women having comparable training, now women might score higher than men. Here the covariate is previous training in mathematics.

14.11.1 Classic ANCOVA

There is a classic ANCOVA method for comparing two independent groups that is based on least squares regression (e.g., Huitema, 2011; Rutherford, 1992). It is commonly used, so a description of the method, as well as some understanding of its relative merits, is important. In the illustration previously mentioned, a least squares regression line is used to predict mathematics aptitude for men, given previous training. The same is done for women. When testing hypotheses, homoscedasticity, as described in Section 6.2, is assumed for both regression lines. Moreover, both groups are assumed to have the same (conditional) variance. In the illustration, the (conditional) variance of aptitude tests among men, given previous training, is assumed to be equal to the (conditional) variance among women. So two types of homoscedasticity are assumed. Two additional assumptions are that the regression lines for the two groups being compared are parallel and the outcome variable has a normal distribution. Based on these assumptions, the goal is to determine whether the intercept terms differ. For the mathematics illustration, it is assumed that given any two amounts of training, the difference between women and men is the same. In particular, among men and women with three semesters of training, differences in their aptitude scores is the same as among participants with 12 semesters of training. More formally, it is assumed that for the jth group,

$$Y_j = \beta_{0j} + \beta_1 X_j + e, \tag{14.25}$$

where e has a normal distribution with mean 0 and variance σ^2. That is, both groups are assumed to have the same slope β_1 but possibly different intercepts and the variance of the error term, e, is assumed to be the same for both groups as well. The goal is to test

$$H_0 : \beta_{01} = \beta_{02}, \tag{14.26}$$

the hypothesis that the intercepts are equal. (This ANCOVA model belongs to the class of *general linear models* first mentioned in Chapter 9.) Violation of any of these assumptions can result in poor power and misleading results. Violating two or more of these assumptions exacerbates practical concerns. And as usual, all indications are that testing assumptions is generally ineffective because such tests might not have enough power to detect violations of assumptions that have practical consequences. A better strategy is to use a method designed to perform well when the classic assumptions are true, but which continues to perform well when these assumptions are violated.

EXAMPLE

Using the plasma retinol data in Chapter 1, we illustrate a commonly used approach to ANCOVA using built-in R functions, which assumes normality, both types of homoscedasticity, and equal slopes, as summarized by Equation (14.25). The goal is to compare males and females, indicated by values in column 2 of the R variable `plasma`, in terms of their plasma retinol, stored in column 14, taking into account grams of fat consumed per day, which is stored in column 7. A two-step process is used. The first step is to fit a least squares regression model to both groups with the goal of testing H_0: $\beta_{11} = \beta_{12}$, the hypothesis that the two groups have equal slopes. The R commands are

```
model=lm(plasma[,14] ~ as.factor(plasma[,2])*plasma[,7]).
summary.aov(model)
```

Here, the * in the term `as.factor(plasma[,2]*plasma[,7]` indicates that the model includes an interaction term for the group variable `plasma[,2]` and the covariate `plasma[,7]`. For the situation at hand, this means that the slopes of the regression lines depend on whether we are dealing with males or females. The output returned by R is

```
                        Df   Sum Sq  Mean Sq F value  Pr(>F)
as.factor(plasma[, 2])   1   464927   64927 11.16 0.00093 ***
plasma[, 7]              1   229799   29799  5.51 0.01946 *
as.factor(plasma[, 2]):plasma[, 7]
                         1    51989   51989  1.24 0.26479
Residuals              311 12955403   41657
```

What is important here is the third line, which indicates that when testing the hypothesis of equal slopes, the p-value is 0.2647935. The hypothesis of equal slopes is not rejected, which is taken to mean that one can proceed to compare the intercepts assuming that the regression lines are parallel. Now the R command

```
model=lm(plasma[,14] ~ as.factor(plasma[,2])+plasma[,7])
summary.aov(model)
```

is used, where the + in the term `as.factor(plasma[,2])+plasma[,7]` indicates that no interaction term is used. That is, the regression lines are taken to be parallel. The output returned by R is

```
                        Df   Sum Sq   Mean Sq F value  Pr(>F)
as.factor(plasma[, 2]) 1   464927   464927  11.152 0.00094 ***
plasma[, 7]            1   229799   229799   5.512 0.01951 *
Residuals            312 13007392    41690
```

What is important now is the first line. The p-value, when testing the hypothesis of equal intercepts, is 0.0009412, indicating that plasma retinol levels among females differ from males, taking into account the grams of fat consumed per day. (The second line indicates that when testing the hypothesis of a zero slope, assuming the groups have identical slopes, the p-value is 0.0195104.)

Important Point: Be sure to notice that in these last two R commands, the group variable `plasma[,2]` appears before the covariate variable `plasma[,7]`. The R commands

```
model=lm(plasma[,14] ~ as.factor(plasma[,7])*plasma[,2])
summary.aov(model)
```

are incorrect and would alter the results.

A criticism of the two-step process just described is that power, when testing the hypothesis of equal slopes, might not be sufficiently high to justify accepting the null hypothesis (cf. Fabian, 1991). A more fundamental concern is that violating any of the classic assumptions just described can create serious practical problems in terms of Type I errors, power, and the goal of getting a reasonably accurate sense about how the groups compare. Testing assumptions is an option. But as previously noted, it is unknown how to determine whether such tests have enough power to detect situations where violating the assumptions has practical consequences. Some least squares methods that allow non-parallel regression lines are available, a classic example being the Johnson–Neyman method (Johnson and Neyman, 1936). But concerns remain regarding non-normality, outliers, curvature, and heteroscedasticity.

Henceforth, attention is focused on modern robust methods. But due to space limitations, a complete description of all modern methods is not provided. Readers interested in a more complete description of these techniques, as well as the computational details associated with some of the methods to be described, are referred to Wilcox (2017, Chapter 12). Here attention is focused on some of the methods that currently seem to have practical importance. Two basic types are covered. The first is based on a parametric regression model and the second is based on a robust smoother.

14.11.2 Robust ANCOVA Methods Based on a Parametric Regression Model

This section focuses on ANCOVA methods for comparing two independent groups that are based on the following model:

$$Y_{ij} = \beta_{0j} + \beta_{1j}X + \lambda(X_j)\epsilon_j \tag{14.27}$$

($j = 1, 2$), where ϵ_j has variance σ_j^2 and $\lambda(X_j)$ is some unknown function that models how the conditional variance of Y, given X, varies with X. In contrast to the classic ANCOVA method, it is not assumed that the regression lines are parallel; it is not assumed that the error term $\lambda(x_j)\epsilon_j$ is homoscedastic, between group homoscedasticity, $\sigma_1^2 = \sigma_2^2$, is not assumed; and it is not assumed that the error terms, ϵ_1 and ϵ_2, have normal distributions. Let x denote some value of the covariate X that is of interest. So for the jth group, it is assumed that the typical value of Y, given that $X = x$, is

$$M_j(x) = \beta_{0j} + \beta_{1j}x.$$

For example, for the first group, $M_1(6)$ might represent the population mean of Y given that $X = 6$, or it could be the population value corresponding to a 20% trimmed mean or median. The estimate of $M_j(x)$ is simply

$$\hat{M}_j(x) = b_{0j} + b_{1j}x,$$

where b_0 and b_1 are estimates of the intercept and slope, respectively. The goal is to test

$$H_0 : M_1(x) = M_2(x), \tag{14.28}$$

for one or more values of the covariate X. Currently, the best approach appears to be one that is based in part on a bootstrap estimate of the standard error of $\hat{M}_j(x)$ (Wilcox, 2013).

For fixed j, generate a bootstrap sample yielding

$$(X_{1j}^*, Y_{1j}^*), \ldots, (X_{n_jj}^*, Y_{n_jj}^*).$$

Estimate the slope and intercept based on this bootstrap sample yielding b_{1j}^* and b_{0j}^*, respectively. For x specified, let $\hat{M}_j^*(x) = b_{0j}^* + b_{1j}^*x$. Repeat this process B times yielding $\hat{M}_{jb}^*(x)$ ($b = 1, \ldots, B$). Then, from basic principles (e.g., Efron and Tibshirani, 1997), an estimate of the squared standard error of $\hat{M}_j(x)$ is

$$\hat{\tau}_j^2 = \frac{1}{B-1}\sum(\hat{M}_{jb}^*(x) - \bar{M}_j^*(x))^2, \tag{14.29}$$

where $\bar{M}_j^*(x) = \sum \hat{M}_{jb}^*(x)/B$. In terms of controlling the probability of a Type I error, $B = 100$ appears to suffice. Letting z be the $1 - \alpha/2$ quantile of a standard normal distribution, an approximate $1 - \alpha$ confidence interval for $M_1(x) - M_2(x)$ is

$$\hat{M}_1(x) - \hat{M}_2(x) \pm z\sqrt{\hat{\tau}_1^2 + \hat{\tau}_2^2}. \tag{14.30}$$

Method S1

There remains the issue of choosing the covariate points where the regression lines are to be compared. A simple approach is to choose $K = 5$ points evenly spaced between the smallest and largest covariate values. When using the Theil–Sen estimator, confidence intervals that have simultaneous probability approximately equal to $1 - \alpha$ can be computed by using a critical value reported in Table 10 in Appendix B (the Studentized maximum modulus distribution) with infinite degrees of freedom. Following Wilcox (2017, Section 12.1.1), this will be called method S1. A possible criticism of this approach is that using a relatively small number of covariate points might miss details that have practical importance. But if K is increased, low power becomes an issue.

Method S2

Consider the situation where covariate points are chosen as done by method S1, only K is relatively large. Method S2 computes confidence intervals in a manner that provides more power than what would be obtained using the Studentized maximum modulus distribution. Here, the default choice is $K = 25$, but even larger values for K can be used.

Let $\hat{p}_k$ be the p-value when testing (14.28) based on x_k ($k = 1, \ldots, K$). Let

$$p_{\min} = \min\{\hat{p}_1, \ldots, \hat{p}_K\}.$$

Let p_c be the α quantile associated with the distribution of $p_{\min}$. Then the probability of one or more Type I errors is α if (14.28) is rejected for any covariate value x_k for which $p_k \leq p_c$.

The strategy used to determine p_c is to momentarily assume normality, homoscedasticity, and that there is no association. Then an estimate of p_c is computed via a simulation. All indications are that when using the Theil–Sen estimator or the quantile regression estimator, control over the probability of one or more Type I errors is very good when data are generated from a non-normal distribution and when there is heteroscedasticity. When using the Theil–Sen estimator with $n = 30$ and when $\alpha = 0.05$, estimates of the actual probability of one or more Type I errors ranged between 0.02 and 0.05 among the situations described in Wilcox (2017, Section 12.1.2). Moreover, method S2 was found to offer a distinct power advantage over method S1 for a range of situations.

It is briefly noted that Method S2 has been extended to two covariates (Wilcox, 2017, Section 12.1.3). The rather involved details are not covered here. But an R function (`ancJNmp`) is described in the next section for applying the method.

14.11.3 R Functions ancJN, ancJNmp, anclin, reg2plot, and reg2g.p2plot

The R function

```
ancJN(x1,y1,x2,y2, pts=NULL, Dpts=FALSE,regfun=tsreg, fr1=1, fr2=1, SCAT =
    TRUE, pch1 = "+", pch2 = "o", alpha=0.05, plotit=TRUE, xout=FALSE,
        outfun=out, nboot=100, SEED=TRUE, xlab="X", ylab="Y",...)
```

performs method S1 described in the previous section. By default, the Theil–Sen estimator is used but other estimators can be used via the argument `regfun`. Least squares regression can be used by setting the argument `regfun=ols`. The covariate values, for which the hypothesis given by Eq. (14.28) is to be tested, can be specified by the argument `pts`. By default, the function picks five covariate values evenly spaced between the smallest and largest covariate values stored in the arguments x1 and x2. Confidence intervals are adjusted so that the simultaneous probability coverage is approximately equal to $1 - \alpha$, where α is specified by the argument alpha. The function reports p-values, which can be adjusted to control the probability of one or more Type I errors via the R function p.adjust. The arguments pch1 and pch2 determine the symbols used when creating a scatterplot. Setting SCAT=FALSE, no scatterplot is created; only the regression lines are plotted.

The R function

```
anclin(x1,y1,x2,y2,regfun=tsreg, pts=NULL, ALL=FALSE, npts=25, plotit=TRUE,
SCAT=TRUE, pch1="*", pch2="+", nboot=100, ADJ=TRUE, xout=FALSE, outfun=out,
SEED=TRUE, p.crit=0.015, alpha=0.05, crit=NULL, null.value=0, plotPV=FALSE,
        scale=TRUE, span=0.75, xlab="X",xlab1="X1", xlab2="X2"
    ylab="p-values",ylab2="Y", theta=50, phi=25, MC=FALSE, nreps=1000,
                                pch="*",...)
```

applies method S2. The argument ALL=FALSE means that the covariate values are chosen to be values evenly spaced between the minimum value and maximum values observed. The number of covariate values is controlled by the argument npts. If ALL=TRUE, the hypothesis given by (14.28) is tested for each of the unique values among all of the covariate values. If the desired probability of one or more Type I errors, indicated by the argument alpha, differs from 0.05, the function computes an estimate of p_c with the number of replications used in the simulation controlled by the argument nreps. If a multicore processor is available, setting MC=TRUE can reduce execution time considerably. By default, the regression lines are plotted with the labels for the x-axis and y-axis controlled by the arguments xlab and ylab2, respectively. If the argument plotPV=TRUE and plot=FALSE, the p-values are plotted and now the argument ylab controls the label for the y-axis. The arguments pch1 and pch2 control the symbol used when creating a scatterplot for group 1 and 2, respectively. If the argument SCAT=FALSE, no scatterplot is created.

The R function

```
ancJNmp(x1, y1, x2, y2, regfun = qreg, p.crit = NULL, DEEP = TRUE, plotit =
 TRUE, xlab = "X1", ylab = "X2", null.value = 0, FRAC = 0.5, cov1 = FALSE,
SMM = TRUE, tr=0.2, nreps = 1000, MC = FALSE, pts = NULL, SEED = TRUE, nboot
                  = 100, xout = FALSE, outfun = out, ...)
```

is like the R functions ancJN and anclin, only it is designed for two or more covariates and it has no option for plotting the regression line for the special case where only one covariate is being used. In contrast to the function ancJN, a quantile regression estimator is used by default.

The R function

```
reg2plot(x1, y1, x2, y2, regfun = tsreg, xlab = "X", ylab = "Y", xout =
                 FALSE, outfun = out, STAND = TRUE, ...)
```

plots two regression lines, the first is based on the data in x1 and y1, and the second is based on the data in x2 and y2. When there are two independent variables, the R function

```
reg2g.p2plot(x1,y1,x2,y2,x out=FALSE, outfun=out, xlab="Var 1", ylab="Var
      2", zlab="Var 3", regfun=tsreg,COLOR=TRUE,STAND=TRUE,
                tick.marks=TRUE,type="p",pr=TRUE,...)
```

plots the regression planes.

14.11.4 ANCOVA Based on the Running-interval Smoother

This section describes some ANCOVA methods that are designed to deal with curvature. Again, the regression lines are not assumed to be parallel. Moreover, homoscedasticity is not assumed.

For the jth group, again let $M_j(x)$ be some population measure of location associated with Y given that $X = x$. A general goal is to test

$$H_0 : M_1(x) = M_2(x), \text{ for all } x. \tag{14.31}$$

That is, the regression lines do not differ in any manner. Methods aimed at testing Equation (14.31) are called *global nonparametric ANCOVA* methods.

Another goal is determining where the regression lines differ and by how much. For the Well Elderly 2 study described in Section 14.3.2, the goal might be to determine how the ethnic groups compare based on the 20% trimmed mean taking into account a participant's cortisol awakening response. Put more formally, the goal is to compare the regression lines at design points (X values) that are of interest. This will be called *local nonparametric ANCOVA*, to distinguish it from the class of global nonparametric ANCOVA methods described in the previous paragraph.

A method for testing Equation (14.31) was derived by Srihera and Stute (2010), which is based on a smoother designed specifically for means. The method allows heteroscedasticity but it is unknown how well it performs under non-normality or how it might be generalized to robust measures of location. (Many related methods based on means are cited by Srihera and Stute as well as Wilcox, 2017. Included is a method derived by Young and Bowman, 1995, as well as Bowman and Young, 1996, which assumes both normality and homoscedasticity.) Here, a sketch of a method is provided that is based on the running-interval smoother and a 20% trimmed mean. Readers interested in more details are referred to Wilcox (2010b).

The method begins by fitting the running-interval smoother, described in Section 14.4.3, to the first group and measuring how deeply this line is nested within the scatterplot for the first group as well as the second group. The method for measuring the depth of the smoother is based on a generalization of the notion of depth used by the regression estimator in Section 14.1.8. If the regression lines are identical for both groups, their depths should be approximately the same. Let D_1 be the difference between these two depths. In a similar manner, if we first fit a smoother using the data from the second group and measure its depth relative to both groups, the difference between the two depths, say D_2, should be relatively small. The test statistic is the larger of the two values $|D_1|$ and $|D_2|$, say D_m. The hypothesis of identical distributions is rejected if $D_m \geq c$, where the critical value c is determined via a bootstrap method. The choice of smoother matters in terms of controlling the probability of a Type I error. Two smoothers that perform well in simulations are the running-interval smoother in Section 14.4.3 and the quantile smoother (COBS) mentioned in Section 14.4.5. However, the running interval smoother is recommended because COBS might indicate a substantial amount of curvature when none exists (Wilcox, 2016a).

Method Y

Next, a local nonparametric ANCOVA method is described that will be called method Y. This alternative approach is aimed at determining where the regression lines differ and by how much. In terms of both Type I errors and power, the method about to be described compares well to the classic ANCOVA method in Section 14.11.1, when the normality and homoscedasticity assumptions are met (Wilcox, 1997). To describe the method, first assume that x has been chosen with the goal of computing a confidence interval for $M_1(x) - M_2(x)$.

For the jth group, let X_{ij}, $i = 1, \ldots, n_j$ be values of the predictors that are available. The value $M_j(x)$ is estimated as described in Section 14.4.3. That is, for fixed j, estimate $M_j(x)$ using the Y_{ij} values corresponding to the X_{ij} values that are close to x. Let $N_j(x)$ be the number of observations used to compute the estimate of $M_j(x)$. That is, $N_j(x)$ is the number of points in the jth group that are close to x, which in turn is the number of Y_{ij} values used to estimate $M_j(x)$. Provided that both $N_1(x)$ and $N_2(x)$ are not too small, a reasonably accurate confidence interval for $M_1(x) - M_2(x)$ can be computed using methods already covered, which includes Yuen's method for trimmed means, described in Section 7.4.1, or a bootstrap method could be used. When comparing the regression lines at more than one design point, confidence intervals for $M_1(x) - M_2(x)$, having simultaneous probability coverage approximately equal to $1 - \alpha$, can be computed as described in Section 12.1.6. By default, the R function ancova, described in the next section, picks five design points where the regression lines will be compared based in part on whether both $N_1(x)$ and $N_2(x)$ are sufficiently large to get good control over the Type I error probability. The function has an option for letting the user pick the x values of interest. Note that rather than compare trimmed means, modern improvements on the Wilcoxon–Mann–Whitney test could be used that were described in Sections 7.8.4 and 7.8.6.

Method SPB

Note that a basic bootstrap method can be used in place of Yuen's test. That is, bootstrap samples are generated from the Y_{ij} values used to estimate $M_j(x)$. For convenience, this will be called method SPB.

Method UB

Currently there are two methods that might provide more power compared to methods Y and SPB. The first stems from a simple modification of a method designed to compare dependent groups, which was suggested by Wilcox (2014). The method uses a percentile bootstrap method, but unlike method SPB, bootstrap samples are generated from the entire set of observed values. That is, for the jth group, generate a bootstrap sample from $(X_{1j}, Y_{1j}), \ldots, (X_{n_j j}, Y_{n_j j})$. Then apply the running-interval smoother to estimate $M_j(x)$. Otherwise, the details are the same as a standard percentile bootstrap method. Currently, this method is based on five covariate values and FWE is controlled via Hochberg's method.

Method TAP

Another way of possibly improving power is to use a larger number of covariate values and control FWE along the lines indicated in Section 14.11.2 in connection with method S2. (Complete computational details are described in Wilcox, 2017, Section 12.2.4.) The current default is to use 25 covariate values.

14.11.5 R Functions ancsm, Qancsm, ancova, ancovaWMW, ancpb, ancovaUB, ancboot, ancdet, runmean2g, qhdsm2g, and l2plot

The R function

```
ancsm(x1, y1, x2, y2, crit.mat = NULL, nboot = 200, SEED = TRUE, REP.CRIT =
FALSE, LP = TRUE, est = tmean, fr = NULL, plotit = TRUE, sm = FALSE, xout =
        FALSE, outfun = out, xlab = "X", ylab = "Y", ...)
```

tests the hypothesis that two independent groups have identical regression lines. That is, it performs a global nonparametric ANCOVA method aimed at testing Equation (14.31) based on the running-interval smoother in conjunction with trimmed means. The arguments `x1` and `y1` contain the data for group 1, and the group 2 data are assumed to be stored in the next two arguments, `x2` and `y2`. As usual, leverage points are removed if the argument `xout=TRUE`.

The R function

```
Qancsm(x1, y1, x2, y2, nboot = 200, qval = 0.5, xlab = "X", ylab = "Y",
                plotit = TRUE, xout=FALSE, outfun=out,...)
```

is another global method. The only difference from `ancsm` is that it is based on a quantile regression smoother. By default, the smoother is aimed at estimating the (conditional) median of Y, given X. Other quantiles can be used via the argument `qval`, but for quantiles other than the median, it is unknown how well the method controls the probability of a Type I error.

The R function

```
ancova(x1,y1,x2,y2, fr1=1, fr2=1, tr=0.2, alpha=0.05, plotit=TRUE, pts = NA,
                         xout=FALSE, outfun=out,...)
```

performs a local nonparametric ANCOVA method based on the running-interval smoother, method Y in the previous section, which defaults to estimating the 20% trimmed mean of Y given X. The arguments `x1, y1, x2, y2, tr`, and `alpha` have their usual meanings. The arguments `fr1` and `fr2` are the spans used for groups 1 and 2, respectively. The argument `pts` can be used to specify the X values at which the two groups are to be compared. For example, `pts=12` will result in comparing the trimmed mean for group 1 (based on the values stored in `y1`) to the trimmed mean of group 2 given that $X = 12$. If there is no trimming, the null hypothesis is $H_0 : E(Y_1|X = 12) = E(Y_2|X = 12)$, where Y_1 and Y_2 are the outcome variables of interest corresponding to the two groups. Using `pts=c(22,36)` will result in testing two hypotheses. The first is $H_0 : M_1(22) = M_2(22)$ and the second is $H_0 : M_1(36) = M_2(36)$. If no values for `pts` are specified, then the function picks five X values and performs the appropriate tests. The values that it picks are reported in the output as illustrated below. Generally, this function controls FWE (the probability of one or more Type I errors) using Hochberg's method in Section 12.1.10. If `plotit=TRUE` is used, the function also creates a scatterplot and smooth for both groups with a + and a dashed line indicating the points and the smooth, respectively, for group 2. If the argument `xout=TRUE`, leverage points are removed before creating the plot.

The function

```
ancovaWMW(x1,y1,x2,y2, fr1=1, fr2=1, alpha=0.05, plotit=TRUE,
pts=NA,xout=FALSE, outfun=out,LP=TRUE, sm=FALSE, est=hd,...)
```

is like the function ancova, only it compares groups in terms of the likelihood that a randomly sampled observation from the first group is less than a random sampled observation from the second group. This is done via Cliff's method in Section 7.8.4, which is one way of improving the Wilcoxon–Mann–Whitney test.

The function

```
ancpb(x1,y1,x2,y2, est=hd, pts=NA, fr1=1, fr2=1, nboot=599, plotit=TRUE,...)
```

applies method SPB described in the previous section. That is, the function is like the function `ancova` only a percentile bootstrap method is used to test hypotheses and by default the Harrell–Davis estimate of the median is used. FWE is controlled using Hochberg's method in Section 12.1.10. In essence, the function creates groups based on the values in pts; in conjunction with the strategy behind the smooth, it creates the appropriate set of linear contrasts, and then it calls the function pbmcp described in Section 12.1.12.

The function

```
ancboot(x1,y1,x2,y2,fr1=1,fr2=1,tr=0.2,nboot=599,pts=NA,plotit = T)
```

compares trimmed means using a bootstrap-t method. Now FWE is controlled as described in Section 12.1.13. Bootstrap samples are generated in the same manner as done by method SPB.

The R function

```
ancovaUB(x1=NULL,y1=NULL,x2=NULL,y2=NULL, fr1=1, fr2=1, p.crit=NULL,
    padj=FALSE, pr=TRUE, method="hochberg", FAST=TRUE, est=tmean,
  alpha=0.05,plotit=TRUE, xlab="X", ylab="Y", pts=NULL, qpts=FALSE,
 qvals=c(0.25,0.5,0.75), sm=FALSE, xout=FALSE, eout=FALSE, outfun=out,
 LP=TRUE, nboot=500, SEED=TRUE, nreps=2000, MC=FALSE, nmin=12, q=0.5,
                  SCAT=TRUE, pch1="*", pch2="+",...)
```

applies method UB. By default, FAST=TRUE, meaning that the critical p-value, $\hat{p}_c$, will be computed quickly if in addition the argument alpha=0.05. Otherwise $\hat{p}_c$ must be estimated via a simulation, which can require a relatively high execution time. The argument nreps controls how many replications are used when computing $\hat{p}_c$. Setting the argument MC=TRUE can reduce execution time considerably if a multicore processor is available. If the argument qpts=TRUE, the covariate points are chosen to be the quantiles associated with the data in x1; the quantiles used are controlled by the argument qvals. If qpts=FALSE, the covariate values are chosen as done by method Y described in the previous section.

If the argument padj=TRUE, the function reports adjusted p-values using the method indicated by the argument method. Hochberg's method is used by default. Hommel's method can be used by setting the argument method="hommel". A possible appeal of using padj=TRUE is that execution time is very low when the goal is to test hypotheses at some level other than 0.05, but power might be reduced. If qpts=TRUE, covariate values are chosen based on the quantiles indicated by the argument qvals in conjunction with the data in the argument x1.

The R function

```
ancdet(x1,y1,x2,y2,fr1=1,fr2=1,tr=0.2,
alpha=0.05,plotit=TRUE,plot.dif=FALSE,pts=NA,sm=FALSE,
pr=TRUE,xout=FALSE,outfun=out,MC=FALSE,
npts=25,p.crit=NULL,nreps=5000,SEED=TRUE,EST=FALSE,
SCAT=TRUE,xlab="X",ylab="Y",pch1="*",pch2="+",...)
```

applies method TAP. The argument npts indicates how many covariate values will be used, which defaults to 25. With plotit=TRUE, the function plots a smooth for both regression lines. Setting plot.dif=TRUE, the function plots the estimated difference between the regression lines, $\hat{M}_1(x) - \hat{M}_2(x)$, based on the covariate values that were used. A confidence band is also plotted based on the adjusted p-value, $\hat{p}_c$. That is, the simultaneous probability coverage among the K confidence intervals is approximately $1-\alpha$, where α is specified via the argument alpha, which defaults to 0.05. If EST=TRUE, or alpha differs from 0.05, a simulation estimate of $\hat{p}_c$ is used where the number of replications is controlled by the argument nreps. Setting MC=TRUE can reduce execution if a multicore processor is available. The confidence intervals that are returned are adjusted so that the simultaneous probability coverage is approximately equal to alpha. The last column of the output indicates whether a significant result is obtained based on the corresponding confidence interval.

Some additional R functions for plotting the smooths corresponding to two groups are provided in case they help. The R function

```
runmean2g(x1, y1, x2, y2, fr=0.8, est=tmean, xlab="X", ylab="Y", ...)
```

creates a scatterplot for both groups (with a + used to indicate points that correspond to the second group) and it plots an estimate of the regression lines for both groups using the running-interval smoother in Section 14.4.3. By default, it estimates the 20% trimmed mean of Y, given X. But some other measure of location can be used via the argument `est`. The smooth for the first group is indicated by a solid line, and a dashed line is used for the other group. This function can be used to plot quantile regression lines by setting the argument `est=hd` and specifying the quantile of interest via the argument `q`. For example, `q=0.75` would plot the 0.75 quantile regression lines. For convenience, the R function

```
qhdsm2g(x1, y1, x2, y2,q=0.5, qval=NULL, LP=TRUE, fr=0.8, xlab="X",
            ylab="Y", xout=FALSE, outfun=outpro,....)
```

is provided, which accomplishes the same goal. The R function

```
l2plot(x1, y1, x2, y2,span=2/3,xlab = "X", ylab = "Y")
```

also plots smoothers for each group, only it uses LOESS to estimate the regression lines.

EXAMPLE

Consider again the Well Elderly 2 study described in Section 14.3.2. After six months of intervention, one of the goals was to compare males and females in terms of a life satisfaction measure (SF36) using the cortisol awakening response as a covariate. First, suppose the groups are compared using method S1 in conjunction with the least squares regression estimator. Assuming that the data for males are stored in x1 and y1, and the data for females are stored in x2 and y2, this was accomplished with the command

```
ancJN(x1,y1,x2,y2, xout=TRUE, outfun=outbox, xlab="CAR", ylab="SF36",
                  plotit=TRUE, regfun=ols).
```

Here is a portion of the output:

```
$output
              X    Est1     Est2       DIF      TEST       se     ci.low
[1,] -0.32160000 47.47286 40.78677 6.6860850 2.4272872 2.754550 -0.3931095
[2,] -0.14490072 45.75017 40.79015 4.9600156 2.8930128 1.714481  0.5537994
[3,] -0.01307176 44.46493 40.79267 3.6722575 2.4353167 1.507918 -0.2030914
[4,]  0.08277492 43.53049 40.79451 2.7359887 1.5076466 1.814741 -1.9278967
[5,]  0.26549698 41.74909 40.79800 0.9510864 0.3216437 2.956956 -6.6482914
        ci.hi     p.value
[1,] 13.765279 0.015212206
[2,]  9.366232 0.003815657
[3,]  7.547606 0.014878761
[4,]  7.399874 0.131644995
[5,]  8.550464 0.747722624
```

So based on the confidence intervals, and assuming a linear model, a significant result is obtained at the 0.05 level for one covariate value only, namely the second value, −0.14490072. As can be seen, the corresponding p-value is 0.0038. The adjusted p-value (via the R function `p.adjust` using Hochberg's method) is 0.019. All of the other four adjusted p-values are greater than 0.059. Using instead the Theil–Sen estimator, a significant result is obtained for the first four covariate values.

Now the groups are compared using method UB by setting the argument `padj=TRUE` when using the R function `ancovaUB`. Here are the results:

```
$output
               X    p.values p.adjusted
[1,] -0.32160000 0.009029345 0.02708804
[2,] -0.14490072 0.000000000 0.00000000
[3,] -0.01307176 0.004000000 0.01600000
[4,]  0.08277492 0.088000000 0.17600000
[5,]  0.26549698 0.952380952 0.95238095
```

So for the first three covariate values, a significant result is obtained at the 0.05 level, in contrast to the results based on the least squares estimator. From the point of view of Tukey's three decision rule, decide that for the negative CAR values, males have higher SF36 scores compared to females. For the positive CAR values, no decision is made. Figure 14.9 shows the plot created by `ancovaUB`. As can be seen, for negative CAR values, the two regression lines appear to be nearly parallel and nearly horizontal. But when the CAR is greater than zero, differences between the two regression lines diminish as the CAR increases. Method TAP, applied via the R function `ancdet`, indicates a significance difference for CAR values ranging between −0.32 to −0.077.

As a final note, robust methods for handling more than one covariate have been derived. Interested readers are referred to Wilcox (2017).

14.11.6 R Functions Dancts, Dancols, Dancova, Dancovapb, DancovaUB, and Dancdet

The methods used by the R functions `ancts, ancols, ancova, ancpb, ancovaUB`, and `ancdet` are readily extended to situations where dependent groups are being compared (Wilcox, 2017). The corresponding R functions are `Dancts, Dancols, Dancova, Dancovapb, DancovaUB`, and `Dancdet`, respectively. The arguments to these latter functions are virtually the same as those functions aimed at comparing independent groups; so for brevity, further details are omitted. The only difference is that these latter functions include an argument `DIF` indicating whether difference scores are to be used. By default, `DIF=FALSE` meaning that measures of location associated with the marginal distributions will be used.

EXAMPLE

Another goal in the Well Elderly 2 study was to assess the impact of intervention on a measure of perceived health (SF36) before and after intervention using the cortisol awakening response (CAR) as a covariate. Figure 14.10 shows a plot of the smooths. Prior to intervention, the regression line appears to be reasonably straight, but after intervention this is no longer the case. Assuming the data prior to intervention are stored in `xx1` (CAR) and `yy1` (SF36), and the data after intervention are stored in `xx2` and `yy2`, the command

```
DancovaUB(xx1,yy1,xx2,yy2)
```

performs method DUB and creates the plot shown in Figure 14.10. Like the function

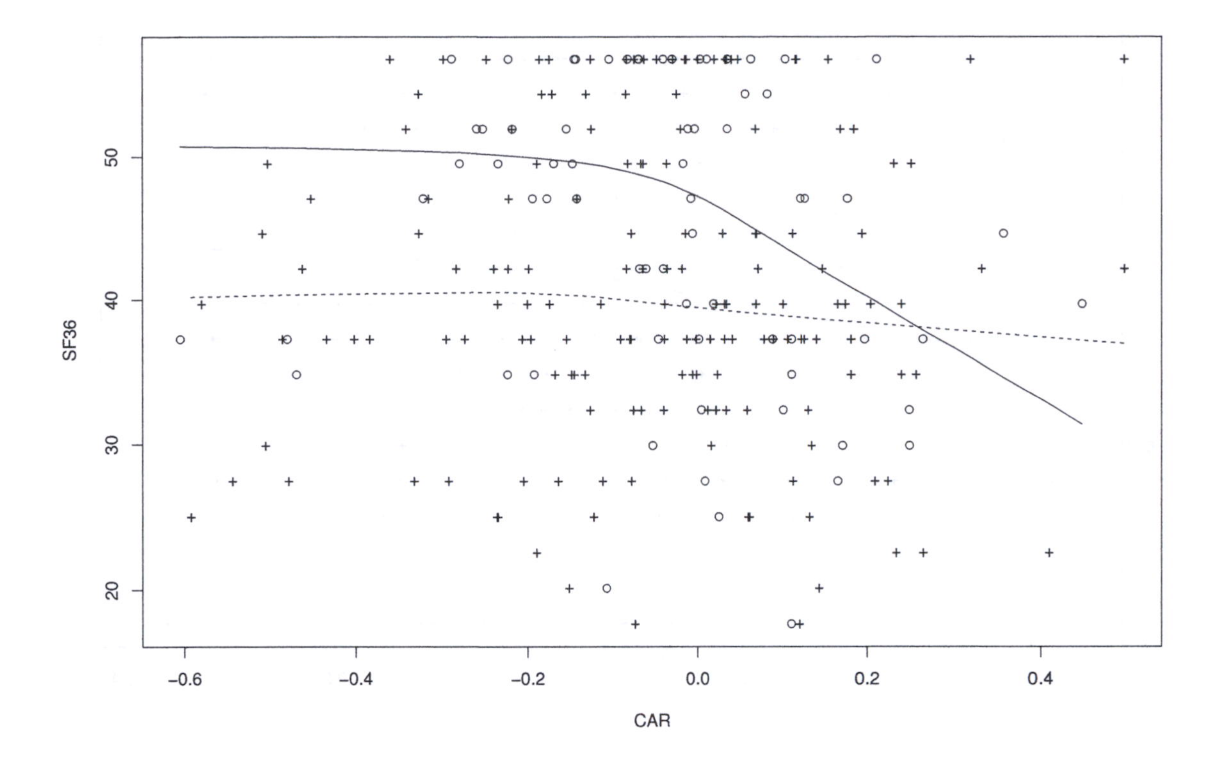

Figure 14.9 The smooths for males (solid line) and females where the goal is to estimate the median SF36 score given a value for the CAR

ancovaUB, by default the regression lines are compared at the 0.25, 0.5, and 0.75 quantiles of the covariate. (Alternative quantiles can be used via the argument qval.) For the data at hand, comparisons are made at CAR=−0.099, 0.054, and 0.317. (More detailed comparisons can be performed using alternative functions as previously explained.) When testing at the 0.05 level, the function returns a significant result for CAR=−0.099 and 0.317. So the results suggest that for the point where CAR is negative, perceived health is higher after intervention. For CAR positive, at some point the reverse is true. In contrast, assuming the regression lines are straight and using the R function Dancts, no significant differences are found.

14.12 EXERCISES

1. For the predictor X_2 in Table 14.2, if the R function runYhat is used to estimate the typical value of Y given that $X_2 = 250$, based on a 20% trimmed mean, the function returns NA. Explain why this occurs.

2. Table 6.4 reports data on 29 lakes in Florida and are stored in the file lake_dat.txt on the author's web page. Assuming that you want to predict Y (the mean annual nitrogen concentration stored in column two) given TW (water retention time stored in column three), plot a smooth using the R function lplot. Test the hypothesis that the slope is zero using regci, first with xout=FALSE and then with xout=TRUE. Comment on what the results illustrate.

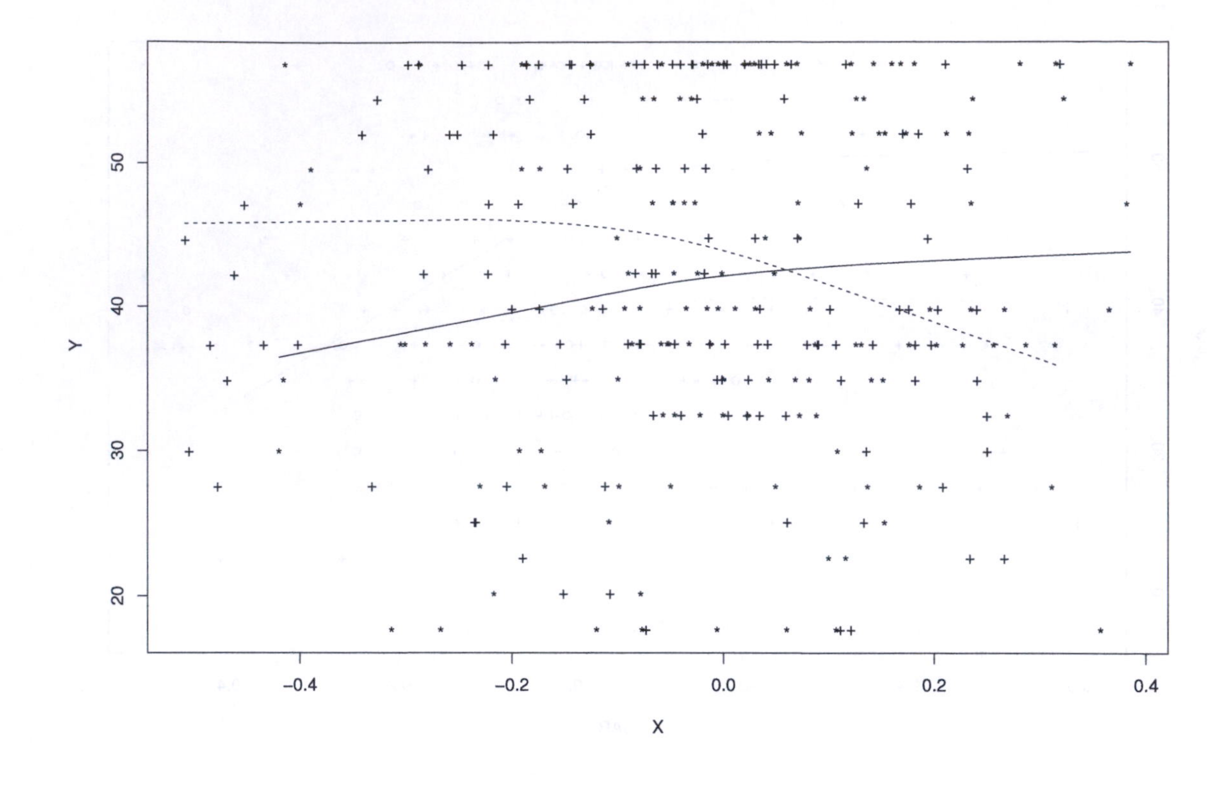

Figure 14.10 Regression lines for predicting perceived health based on the cortisol awakening response. The solid line is the estimated regression line prior to intervention and the points after intervention are indicated by a +.

3. Repeat the previous exercise, only use the data in column one as the independent variable.
4. A portion of the data from the Well Elderly 2 study is stored in the author's web page in the file A3_dat.txt. These are measures taken after intervention. Use the R function `qhdsm` to create a smooth for the 0.2, 0.5, and 0.8 quantiles using the data with the label MAPAGLOB as the independent variable (a measure of meaningful activities) and the data with the label LSIZ (life satisfaction) as the dependent variable. What does the smooth suggest about using the R function `khomreg` in Section 6.8.1 to test the hypothesis of homoscedasticity with leverage points removed?
5. For the data in the previous exercise, test the hypothesis that the 0.8 quantile regression line has a zero slope with leverage points removed.
6. Repeat Exercise 4, only use the data in the file B3_dat.txt. Comment on the results.
7. The author's web page contains the file skull_data.txt, which contains data reported by Thompson and Randall-Maciver (1905) dealing with skull measurements of male Egyptians from different time periods. The four measurements are contained in the first four columns and the time period is indicated by the values

in column five. Using the default values for the arguments, create a smooth using `lplot` and the data in columns one (the independent variable) and columns 4. What does the smooth suggest regarding curvature? Check this issue using `linchk`.

8. For the data in the previous exercise, what might be affecting the test that there is curvature. Use R to check on this possibility?

9. For the Well Elderly 2 data in Exercises 4 and 6, compare the regression lines using `ancJN` using the data labeled CESD (depressive symptoms) as the independent variable and LSIZ as the dependent variable. Use the data in the file A3 for the first group (the experimental group) and the file B1 as the second (control) group. Perform the analysis with leverage points removed.

10. Repeat the previous exercise, only now use the R function `ancova` to compare the regression lines. At which X values do you get a significant difference with $\alpha = 0.05$? Compute the adjusted p-values.

11. R has a built-in data set called leuk. The third column indicates survival times, in weeks, for patients diagnosed with acute myelogenous leukemia. The first column indicates the patient's white blood cell count at the time of diagnosis. Compute Pearson's correlation and compare the results to the strength of association based on Cleveland's smoother, which is returned by the R function `lplot`.

12. For the Well Elderly 2 data in the file B3_dat.txt, use the R function `ancova` to compare males and females based on the variable MAPAGLOB (meaningful activities) using PEOP (personal support) as a covariate.

CHAPTER 15

BASIC METHODS FOR ANALYZING CATEGORICAL DATA

CONTENTS

Categorical data refers to any variable where the responses can be classified into a set of categories, such as right versus wrong, ratings on a 10-point scale, or religious affiliation. Two types of categorical variables are distinguished: nominal and ordinal. The first refers to any situation where observations are classified into categories without any particular order. Brands of cars, marital status, race, and gender are examples of categorical variables. In contrast, ordinal random variables are variables having values that have order. For example, cars might be classified as small, medium, or large; incomes might be classified as low, medium, or high; or gymnasts might be rated by a judge on their ability to do floor exercises. Both nominal and ordinal random variables can be analyzed with the methods in this chapter. Special methods for analyzing ordinal data are available; they offer the potential of more power, but no details are given here. For a book dedicated to this issue, see Agresti (2010). Like other topics covered in this book, methods for analyzing categorical data have grown tremendously in recent years and an entire book is needed to cover the many techniques that have been derived. Books devoted to this topic include Agresti (2002, 2007, 2010), Powers and Xie (2008), Lloyd (1999) and Simonoff (2003).

The simplest case is where there are two categories only, which can be analyzed with the binomial distribution using methods covered in previous chapters. This chapter covers some basic extensions relevant to contingency tables where there are more than two categories. Another extension covered in this chapter deals with regression where the dependent (outcome) variable is binary. As will be indicated, regression methods described in previous chapters are not well suited for this special case.

15.1 GOODNESS OF FIT

This section focuses on a generalization of the binomial probability function to situations where observations are classified into one of K categories. For instance, n respondents might be asked which of four brands of ice cream they prefer. Here, the probabilities associated with these four categories are denoted by p_1, p_2, p_3, and p_4. For example, p_1 is the probability that a randomly sampled participant chooses the first brand and p_4 is the probability of choosing brand 4. To keep the illustration simple, it is assumed that all participants choose one and only one brand. A basic issue is whether there is any difference among the four probabilities. In more formal terms, a goal might be to test

$$H_0 : p_1 = p_2 = p_3 = p_4 = 1/4.$$

More generally, when dealing with J categories, the goal is to test

$$H_0 : p_1 = \cdots = p_J = \frac{1}{J}. \tag{15.1}$$

Generalizing even further, a goal might be to test

$$H_0 : p_1 = p_{01}, \ldots, p_J = p_{0J}, \tag{15.2}$$

where $p_{01}, \ldots, p_{0J}$ are specified constants.

When randomly sampling from an infinite population, or a finite population with replacement, Equation (15.2) can be tested with what is called the *chi-squared goodness-of-fit test.* When the null hypothesis is true, the expected number of successes in the jth cell is

$$m_j = np_{0j},$$

where

$$n = \sum n_j$$

is the total number of observations, and n_j is the number of observations corresponding to the jth group. Suggested by Karl Pearson in 1900, the most common test of the null hypothesis is based on the statistic

$$X^2 = \sum \frac{(n_j - m_j)^2}{m_j}. \tag{15.3}$$

For large sample sizes, X^2 has, approximately, a chi-squared distribution with $J-1$ degrees of freedom. The greater the difference between what is observed (n_j) and what is expected (m_j), the larger is X^2. If X^2 is large enough, the null hypothesis is rejected. More specifically, reject if

$$X^2 > \chi^2_{1-\alpha},$$

the $1-\alpha$ quantile of a chi-squared distribution with $J-1$ degrees of freedom.

There is a technical issue that should be stressed. Although not typically made explicit, the chi-squared test is based in part on normality. As noted in Section 3.10, the chi-squared distribution is defined in terms of standard normal distributions. Roughly, an extension of the central limit theorem, as described in Section 4.5, plays a role for the situation at hand. Put another way, any claim that a chi-squared test avoids the assumption of normality is incorrect.

15.1.1 R Functions chisq.test and pwr.chisq.test

By default, the built-in R function

```
chisq.test(x,p = rep(1/length(x), length(x)))
```

performs the chi-squared test of the hypothesis given by Equation (15.1), assuming that the argument `x` is a vector containing count data. The hypothesis given by Equation (15.2) is tested by specifying the hypothesized values via the argument `p`. Assuming that the R package `pwr` has been installed, the R command `library(pwr)` provides access to the R function

```
pwr.chisq.test(w = NULL, N = NULL, df = NULL, sig.level = 0.05, power =
NULL),
```

which performs a power analysis when dealing with the chi-squared test of independence. The argument `w` is a measure of effect size given by

$$w = \sqrt{\sum \frac{(p_j - p_{0j})^2}{p_{0j}}}.$$

What constitutes a large effect size will depend on the situation. Cohen (1988) has suggested that as a rough guide, $w = 0.1$, 0.3, and 0.5 constitute small, medium, and large effect sizes, respectively.

EXAMPLE

Continuing the example where participants are asked to choose their favorite brand of ice cream from among four brands, imagine that the observed counts are 20, 30, 15, 22. (For instance, 20 participants chose the first brand as their favorite.) The goal is to test the hypothesis that all four brands have the same probability of being chosen with the Type I error probability set to $\alpha = 0.05$. First we illustrate the computations without resorting to R. The total sample size is $n = 20 + 30 + 15 + 22 = 87$. So if the null hypothesis is true, the expected number of observations in each cell is $87/4 = 21.75$ and the test statistic is

$$X^2 = \frac{1}{21.75}((20 - 21.75)^2 + (30 - 21.75)^2 + (15 - 21.75)^2 + (22 - 21.75)^2) = 5.368.$$

The 0.05 critical value is 7.82, so fail to reject.

Here is how the same analysis can be done with R. First, use the R command

```
oc=c(20, 30, 15, 22)
```

to store the observed counts in the R variable `oc`. The R command

```
chisq.test(oc)
```

returns a p-value of 0.1468. So if it is desired that the probability of a Type I error be less than or equal to 0.05, fail to reject the hypothesis that all four probabilities are equal to 0.25. The command

```
pwr.chisq.test(w=.3,N=87,df=3)
```

Table 15.1 Hypothetical Results on Personality Versus Blood Pressure

	Personality		
Blood Pressure	A	B	Total
High	8	5	13
Not High	67	20	87
Total	75	25	100

reports that with a sample size of 87 and an effect size $w = 0.3$, power is 0.643 when testing at the 0.05 level. Consequently, to the extent it is deemed important to detect a difference among the probabilities when $w = 0.3$, this suggests not accepting the null hypothesis.

EXAMPLE

Imagine that adults are classified as belonging to one of five political organizations. Further imagine that the probabilities associated with these five categories are known to be 0.05, 0.10, 0.15, 0.30, 0.40 at a particular point in time. A year later, it is desired to determine whether the proportion of adults has changed based on a random sample of 100 adults. If the observed counts for the five categories are 3, 17, 10, 22, 48, the R command

```
chisq.test(x=c(3, 17, 10, 22, 48),p=c(0.05, 0.10, 0.15, 0.30, 0.40))
```

returns a p-value of 0.025 suggesting that the probabilities have changed.

Note that this leaves open the issue of which of the probabilities differ from the hypothesized values. Momentarily focus on whether an adult belongs to the first category or some other category. Then we are dealing with a binomial distribution with a hypothesized probability of 0.05 and the methods in Section 4.11 can be used to compute a confidence interval for the true probability. The same can be done for the other probabilities. For the second category, the observed count is 17 and the R command `binomci(17,100)` returns a 0.95 confidence interval equal to (0.104, 0.258). This interval does not contain the hypothesized value, 0.10, suggesting that the probability associated with the second category differs from 0.10.

15.2 A TEST OF INDEPENDENCE

One of the more basic goals is testing the hypothesis that two categorical variables are independent. Imagine, for example, that we classify individuals as having personality Type A or B. Simultaneously, each individual is also classified according to whether he or she has high blood pressure. Some hypothetical results are shown in Table 15.1. The issue of interest here is whether there is an association between personality type and the presence of high blood pressure. Said another way, if we ignore personality type, there is some probability that a participant will have high blood pressure. Of interest is whether this probability is altered when we are told that the individual has a Type A personality.

We begin with the simplest case: a 2-by-2 contingency table. Table 15.2 illustrates a commonly used notation for representing the unknown probabilities associated with a contingency table and Table 15.3 illustrates a common notation for the observed counts. For example, p_{11} is the probability that a randomly sampled person simultaneously has high blood pressure *and* a Type A personality, and $n_{11} = 8$ is the number of participants among the 100 sampled who have both of these characteristics. Similarly, $n_{21} = 67$, $n_{12} = 5$, and $n_{22} = 20$.

Table 15.2 Probabilities Associated with a Two-Way Table

	Personality		
Blood Pressure	A	B	Total
High	p_{11}	p_{12}	$p_{1+} = p_{11} + p_{12}$
Not High	p_{21}	p_{22}	$p_{2+} = p_{21} + p_{22}$
Sum	$p_{+1} = p_{11} + p_{21}$	$p_{+2} = p_{12} + p_{22}$	

Table 15.3 Notation for Observed Counts

	Personality		
Blood Pressure	A	B	Total
High	n_{11}	n_{12}	$n_{1+} = n_{11} + n_{12}$
Not High	n_{21}	n_{22}	$n_{2+} = n_{21} + n_{22}$
Total	$n_{+1} = n_{11} + n_{21}$	$n_{+2} = n_{12} + n_{22}$	n

In the illustration, the goal is to test the hypothesis that personality type and blood pressure are independent. If they are independent, then it follows from results in Chapter 3 that the cell probabilities must be equal to the product of the corresponding marginal probabilities. For example, if the probability of a randomly sampled participant having a Type A personality is $p_{+1} = 0.4$, and the probability of having high blood pressure is $p_{1+} = 0.2$, then independence implies that the probability of having a Type A personality *and* high blood pressure is

$$p_{11} = p_{1+}p_{+1} = 0.4 \times 0.2 = 0.08.$$

If, for example, $p_{11} = 0.0799999$, they are dependent although in some sense they are close to being independent. Similarly, independence implies that

$$p_{12} = p_{1+}p_{+2}$$

$$p_{21} = p_{2+}p_{+1}$$

$$p_{22} = p_{2+}p_{+2}.$$

The more general case is where there are R rows and C columns, in which case independence corresponds to a situation where for the ith row and jth column,

$$p_{ij} = p_{i+}p_{+j},$$

where

$$p_{i+} = \sum_{j=1}^{C} p_{ij},$$

$$p_{+j} = \sum_{i=1}^{R} p_{ij}$$

are the *marginal probabilities.*

For the special case of a 2-by-2 table, when the goal is to test the hypothesis of independence, the test statistic is

$$X^2 = \frac{n(n_{11}n_{22} - n_{12}n_{21})^2}{n_{1+}n_{2+}n_{+1}n_{+2}}.$$

When the null hypothesis of independence is true, X^2 has, approximately, a chi-squared distribution with 1 degree of freedom. For the more general case where there are R rows and C columns, the test statistic is

$$X^2 = \sum_{i=1}^{R}\sum_{j=1}^{C} \frac{n(n_{ij} - \frac{n_{i+}n_{+j}}{n})^2}{n_{i+}n_{+j}}, \tag{15.4}$$

and the degrees of freedom are

$$\nu = (R - 1)(C - 1).$$

If X^2 exceeds the $1 - \alpha$ quantile of a chi-squared distribution with ν degrees of freedom, which can be read from Table 3 in Appendix B, reject the hypothesis of independence. (As was the case in Section 15.1, an extension of the central limit theorem plays a role. That is, normality is being assumed.)

EXAMPLE

For the data in Table 15.1,

$$X^2 = \frac{100[8(20) - 67(5)]^2}{75(25)(13)(87)} = 1.4.$$

With $\nu = 1$ degree of freedom and $\alpha = 0.05$, the critical value is 3.84, and because $1.4 < 3.84$, you fail to reject. This means that you are unable to detect any dependence between personality type and blood pressure. Generally, the chi-squared test of independence performs reasonably well in terms of Type I errors (e.g., Hosmane, 1986), but difficulties can arise, particularly when the number of observations in any of the cells is relatively small. For instance, if any of the n_{ij} values is less than or equal to 5, problems might occur in terms of Type I errors. There are a variety of methods for improving upon the chi-squared test, but details are not given here. Interested readers can refer to Agresti (2002).

Cohen (1988, Section 7.2) discusses an approach to determining the sample size needed to achieve a specified amount of power when performing the test of independence covered in this section. Applying the method requires specifying an effect size, which is very similar to the effect size w introduced in Section 15.1. In particular, now the effect size is

$$w = \sqrt{\sum\sum \frac{(p_{ij} - p_{i+}p_{j+})^2}{p_{i+}p_{j+}}}.$$

As was the case in Section 15.1, Cohen (1988) suggests that as a rough guide, $w = 0.1$, 0.3, and 0.5 constitute small, medium, and large effect sizes, respectively. The R function `pwr.chisq.test`, introduced in Section 15.1, can be used to perform a power analysis.

15.2.1 R Function chi.test.ind

Assuming `x` is a matrix, the R function

```
chi.test.ind(x),
```

written for this book, performs the chi-squared test of independence. Based on the help files in R, it might seem that the test of independence described here can be performed with built-in R function `chisq.test`, which was described in Section 15.1. It should be noted, however, that the R command `chisq.test(x)` does not perform the test of independence just described.

EXAMPLE

Consider again the data in Table 15.1. The R command

Table 15.4 Performance of Political Leader

	Time 2		
Time 1	Approve	Disapprove	Total
Approve	794	150	944
Disapprove	86	570	656
Total	880	720	1600

```
x=matrix(c(8,5,67,20),ncol=2,byrow=T)
```

stores the data in a matrix, as shown in Table 15.1, and the command

```
chi.test.ind(x)
```

performs the chi-squared test of independence. The resulting p-value is 0.229. Because we failed to reject, this raises the issue of power. Imagine that, based on the sample size used here, which is $n = 100$, the goal is to have power 0.9 when $w = 0.1$ and $\alpha = 0.05$. The command

```
pwr.chisq.test(w=0.1,N=100,df=1,alpha=0.05)
```

indicates that power is only 0.17. The command

```
pwr.chisq.test(w=0.1,power=0.9,df=1)
```

returns $N = 1050.7$. That is, a sample size of 1051 is required to achieve power equal to 0.9 when $w = 0.1$.

15.3 DETECTING DIFFERENCES IN THE MARGINAL PROBABILITIES

Imagine that 1600 adults are asked whether they approve of some political leader resulting in the responses shown in Table 15.4. According to Table 15.4, 794 of the 1600 adults approved both times they were interviewed, and 150 approved at time 1 but not time 2. The last row and last column of Table 15.4 are called *marginal counts*. For example, 944 adults approved at time 1, regardless of whether they approved at time 2. Similarly, 880 adults approved at time 2, regardless of whether they approved at time 1.

A natural issue is whether the approval ratings have changed from time 1 to time 2. This issue can be addressed by computing a confidence interval for the difference between the probability of approving at time 1 minus the probability of approving at time 2. The problem is that we do not know the population values of the probabilities in Table 15.2, so they must be estimated based on a random sample of participants. The general notation for representing what you observe is shown in Table 15.3. In the illustration, $n_{11} = 794$ meaning that 794 adults approve at both times 1 and 2. The proportion of adults who approve at both times 1 and 2 is

$$\hat{p}_{11} = \frac{n_{11}}{n},$$

and in the illustration

$$\frac{n_{11}}{n} = \frac{794}{1600} = 0.496.$$

This means that the estimated probability of a randomly sampled adult approving at both times is 0.496. The other probabilities are estimated in a similar fashion. For example, the probability that a randomly sampled adult approves at time 1, regardless of whether the same person approves at time 2, is

$$\hat{p}_{1+} = \frac{944}{1600} = 0.59.$$

Now we can take up the issue of whether the approval rating has changed. In symbols, this issue can be addressed by computing a confidence interval for

$$\delta = p_{1+} - p_{+1}.$$

The usual estimate of δ is

$$\begin{aligned}\hat{\delta} &= \hat{p}_{1+} - \hat{p}_{+1} \\ &= \frac{n_{11}+n_{12}}{n} - \frac{n_{11}+n_{21}}{n} \\ &= \frac{n_{12}-n_{21}}{n}.\end{aligned}$$

In the illustration,

$$\hat{\delta} = \hat{p}_{1+} - \hat{p}_{+1} = 0.59 - 0.55 = 0.04$$

meaning that the change in approval rating is estimated to be 0.04. The squared standard error of $\hat{\delta}$ can be estimated with

$$\hat{\sigma}^2_{\delta} = \frac{1}{n}\{\hat{p}_{1+}(1-\hat{p}_{1+}) + \hat{p}_{+1}(1-\hat{p}_{+1}) - 2(\hat{p}_{11}\hat{p}_{22} - \hat{p}_{12}\hat{p}_{21})\};$$

so by the central limit theorem, an approximate $1-\alpha$ confidence interval for δ is

$$\hat{\delta} \pm z_{1-\alpha/2}\hat{\sigma}_{\delta}, \tag{15.5}$$

where, as usual, $z_{1-\alpha/2}$ is the $1-\alpha/2$ quantile of a standard normal distribution read from Table 1 in Appendix B. (For methods that might improve upon this confidence interval, see Lloyd, 1999.) In the illustration,

$$\begin{aligned}\hat{\sigma}^2_{\delta} &= \frac{1}{1600}\{0.59(1-0.59) + 0.55(1-0.55) - \\ &\quad 2(0.496(0.356) - 0.094(0.054))\} \\ &= 0.0000915,\end{aligned}$$

so

$$\hat{\sigma}_{\delta} = \sqrt{0.0000915} = 0.0096,$$

and a 0.95 confidence interval is

$$0.04 \pm 1.96(0.0096) = (0.021, 0.059).$$

That is, it is estimated that the probability of approval has dropped by at least 0.021 and as much as 0.059.

Alternatively, you can test the hypothesis

$$H_0 : p_{1+} = p_{+1},$$

which means that the approval rating is the same at both time 1 and time 2. Testing this hypothesis turns out to be equivalent to testing

$$H_0 : p_{12} = p_{21}.$$

An appropriate test statistic is

$$Z = \frac{n_{12} - n_{21}}{\sqrt{n_{12} + n_{21}}},$$

which, when the null hypothesis is true, is approximately distributed as a standard normal distribution. That is, reject if

$$|Z| > z_{1-\alpha/2},$$

where $z_{1-\alpha/2}$ is the $1 - \alpha/2$ quantile of a standard normal distribution. This turns out to be equivalent to rejecting if Z^2 exceeds the $1 - \alpha$ quantile of a chi-squared distribution with 1 degree of freedom, which is known as *McNemar's test.* In the illustration,

$$Z = \frac{150 - 86}{\sqrt{150 + 86}} = 4.2,$$

this exceeds $z_{0.975} = 1.96$, so you reject at the $\alpha = 0.05$ level and again conclude that the approval rating has dropped. A disadvantage of this approach is that it does not provide a confidence interval. (For a generalization of McNemar's test to situations where there are more than two rows and columns, commonly referred to as the generalized McNemar's or Stuart–Maxwell test, see Bhapkar, 1966; Stuart,1955 and Maxwell, 1970.)

15.3.1 R Functions contab and mcnemar.test

The R function

```
contab(dat,alpha=0.05)
```

computes a $1 - \alpha$ confidence interval for the difference between the marginal probabilities using Equation (15.5), where the argument `dat` is assumed to be a 2-by-2 matrix. The built-in R function

```
mcnemar.test(x,y=NULL)
```

performs McNemar's tests, where `x` is a 2-by-2 matrix or a factor object. The argument `y` is a factor object, which is ignored if `x` is a matrix.

EXAMPLE

R comes with a data set called `occupationalStatus`, which reports the occupational status of fathers and sons. There are eight occupational categories, but for illustrative purposes we focus on the first two. The data appear in R as

```
      destination
origin  1  2
     1 50 19
     2 16 40
```

where origin refers to the father's status and destination is the son's status. An issue might be whether the probability of a category 1 among fathers differs from the corresponding probability among sons. The R function `contab` returns

```
$delta
[1] 0.024

$CI
```

```
[1] -0.06866696  0.11666696

$p.value
[1] 0.6117234
```

So no difference in the two probabilities is detected. Or from the point of view of Tukey's three decision rule, make no decision about which probability is largest. It is left as an exercise to show that the R function `mcnemar.test` returns a p-value of 0.735.

15.4 MEASURES OF ASSOCIATION

This section takes up the common goal of measuring the association between two categorical variables. The p-value (or significance level) associated with the chi-squared test for independence described in Section 15.2 might seem reasonable, but it is known to be unsatisfactory (Goodman and Kruskal, 1954). Another approach is to use some function of the test statistic for independence, X^2, given by Equation (15.4). A common choice is

$$\phi = \frac{X}{\sqrt{n}}, \tag{15.6}$$

which is called the *phi coefficient*, but this measure, plus all other functions of X^2, has been found to have little value as measures of association (e.g., Fleiss, 1981). It is noted that the phi coefficient is tantamount to using Pearson's correlation. In the last example dealing with occupational status, if we take the values of possible outcomes to be 1 or 2 for fathers, and the same is done for sons, Pearson's correlation between fathers and sons is equal to the phi coefficient.

Some of the measures of association that have proven to be useful in applied work are based in part on conditional probabilities. First, however, some new notation is introduced. Momentarily consider the simplest situation where some outcome variable Y is binary, and some independent variable X is binary as well. For convenience, the values of both X and Y are assumed to be 0 or 1. Let

$$\pi(x) = P(Y = 1|X = x) \tag{15.7}$$

be the probability that the outcome variable is 1, given that the value of the $X = x$. So $\pi(1)$ is the probability that $Y = 1$ given that $X = 1$.

EXAMPLE

The notation is illustrated in Table 15.5, which in contrast to Table 15.2 indicates conditional probabilities rather than cell probabilities. Here, $Y = 1$ indicates high blood pressure and $X = 1$ indicates personality Type A. For instance, $\pi(1)$ is the probability of having high blood pressure given that an individual is a Type A personality and $\pi(0)$ is the probability of high blood pressure given that someone has a Type B personality. So $1 - \pi(0)$ is the probability of not having high blood pressure given that the individual has a Type B personality.

Each column in a contingency table has associated with it a quantity called the odds, which is the probability that a certain feature is present divided by the probability that it is not present. In the last example, the *odds* of having high blood pressure ($Y = 1$) among Type A personalities ($X = 1$) is

$$\frac{\pi(1)}{1 - \pi(1)}.$$

Table 15.5 Illustration of the Notation Based on Personality Versus Blood Pressure Example.

	Personality	
Blood Pressure	$X = 1$ (Type A)	$X = 0$ (Type B)
$Y = 1$ (High)	$\pi(1)$	$\pi(0)$
$Y = 0$ (Low)	$1 - \pi(1)$	$1 - \pi(0)$
Total	1	1

Note: Shown are the conditional probabilities associated with blood pressure, given an individual's personality type.

Table 15.6 Estimated Conditional Probabilities for Personality Versus Blood Pressure

	Personality	
Blood Pressure	A ($X = 1$)	B ($X = 0$)
High ($Y = 1$)	8/75	5/25
Not High ($Y = 0$)	67/75	20/25

Similarly, among Type B personalities ($X = 0$), the odds of having high blood pressure is

$$\frac{\pi(0)}{1 - \pi(0)}.$$

The log of the odds is called a *logit.* For $X = 1$ the logit is

$$g(1) = \ln\{\pi(1)/[1 - \pi(1)]\}$$

and for $X = 0$ it is

$$g(0) = \ln\{\pi(0)/[1 - \pi(0)]\}.$$

The odds ratio is the ratio of the odds for $X = 1$ to the odds of $X = 0$:

$$\Psi = \frac{(\pi(1))/[1 - \pi(1)]}{(\pi(0))/[1 - \pi(0)]}. \tag{15.8}$$

The term *log odds* refers to the log of the odds ratio, which is

$$\ln(\Psi) = g(1) - g(0).$$

The odds ratio has practical appeal because it reflects how much more likely it is for an outcome of interest among those with $X = 1$ than among those with $X = 0$.

EXAMPLE

Continuing the illustration dealing with blood pressure and personality types, Table 15.6 shows the estimated probabilities based on the data in Table 15.1. So, the odds of high blood pressure among Type A personalities is estimated to be 8/67= 0.1194. The odds of high blood pressure among Type B personalities is estimated to be 5/20 = 0.25. Consequently, the odds ratio is estimated to be $\hat{\Psi} = 0.1194/0.25 = 0.4776$. That is, among Type A personalities, the risk of hypertension is about half the risk among Type B personalities.

It is noted that in terms of the population probabilities in Table 15.2, the odds ratio can be written as

$$\Psi = \frac{p_{11}p_{22}}{p_{12}p_{21}},$$

Table 15.7 Mortality rates per 100,000 person-years from lung cancer and coronary artery disease for smokers and nonsmokers of cigarettes

	Smokers	Nonsmokers	Difference
Cancer of the lung	48.33	4.49	43.84
Coronary artery disease	294.67	169.54	125.13

and for this reason it is sometimes called the *cross-product ratio.* A simple way of writing the estimate of Ψ, based on the notation in Table 15.3, is

$$\hat{\Psi} = \frac{n_{11}n_{22}}{n_{12}n_{21}}.$$

Under independence, it can be shown that $\Psi = 1$. If $\Psi > 1$, then participants in column 1 (Type A personalities in the illustration) are more likely to belong to the first category of the second factor (high blood pressure) than are participants in column 2. If $\Psi < 1$, then the reverse is true.

Measures of association are generally open to the criticism that they reduce the data down to a point where important features can become obscured. This criticism applies to the odds ratio, as noted by Berkson (1958) and discussed by Fleiss (1981). Table 15.7 shows the data analyzed by Berkson on mortality and smoking. It can be seen that the estimated odds ratio is

$$\hat{\Psi} = \frac{10.8}{1.7} = 6.35,$$

and this might suggest that cigarette smoking has a stronger association with lung cancer than with coronary artery disease. Berkson argued that it is *only* the difference in mortality that permits a valid assessment of the effect of smoking on a cause of death. The difference for coronary artery disease is considerably larger than it is for smoking, as indicated in the last column of Table 15.7, indicating that smoking is more serious in terms of coronary artery disease. The problem with the odds ratio in this example is that it throws away all the information on the number of deaths due to either cause.

15.4.1 The Proportion of Agreement

Consider again Table 15.4 dealing with the approval of a political leader at two different times. The proportion of agreement is

$$p = p_{11} + p_{22}.$$

That is, p is the probability that, on both occasions, a randomly sampled individual approves or does not approve. The proportion of agreement is estimated with

$$\hat{p} = \frac{n_{11} + n_{22}}{n},$$

which in Table 15.4 is

$$\hat{p} = \frac{794 + 570}{1600} = 0.8525.$$

Notice that the probability of agreement can be viewed as the probability of success corresponding to a binomial distribution if we view the number of trials as being fixed and if random sampling is assumed. In the illustration, either the same response is given both times or it is not, and p represents the corresponding probability that the same response is given both times.

Table 15.8 Ratings of 100 Figure Skaters

	Rater A			
Rater B	1	2	3	Total
1	20	12	8	40
2	11	6	13	30
3	19	2	9	30
Total	50	20	30	100

The proportion of agreement can be generalized to a square contingency table with R rows and R columns. Imagine that two judges rate 100 women competing in a figure skating contest, each woman being rated on a three-point scale of how well she performs. If the ratings are as shown in Table 15.8, the proportion of agreement is

$$\hat{p} = \frac{n_{11} + n_{22} + n_{33}}{n} = \frac{20 + 6 + 9}{100} = 0.35.$$

This says that 35% of the time, the judges agree about how well a skater performs. More generally, the proportion of agreement for any square table having R rows and R columns is given by

$$\hat{p} = \frac{n_{11} + \cdots + n_{RR}}{n}.$$

That is, $\hat{p}$ is the proportion of observations lying along the diagonal of the table. A confidence interval for p can be computed using the methods in Section 4.11.

EXAMPLE

For the data in Table 15.8, the 0.95 confidence interval for the proportion of agreement, returned by the R command `binomci(35,100)`, is (0.262, 0.452).

15.4.2 Kappa

One objection to the proportion of agreement is that there will be some chance agreement even when two variables or factors are independent. In the rating of skaters, for example, even if the ratings by rater A are independent of the ratings by rater B, there will be instances where both raters agree. Introduced by Cohen (1960), the measure of agreement known as *kappa* is intended as a measure of association that adjusts for chance agreement. For the general case of a square table with R rows and R columns, the estimated chance of agreement under independence is

$$\hat{p}_c = \frac{n_{1.}n_{.1} + \cdots + n_{R.}n_{.R}}{n^2}.$$

The quantity $\hat{p} - \hat{p}_c$ is the amount of agreement beyond chance. Cohen's suggestion is to rescale this last quantity so that it has a value equal to 1 when there is perfect agreement. This yields

$$\hat{\kappa} = \frac{\hat{p} - \hat{p}_c}{1 - \hat{p}_c}.$$

In the illustration,

$$\hat{p}_c = \frac{40(50) + 30(20) + 30(30)}{10000} = 0.35,$$

so

$$\hat{\kappa} = \frac{0.35 - 0.35}{1 - 0.35} = 0.$$

That is, the estimate is that there is no agreement beyond chance. It is possible to have $\kappa < 0$ indicating less than chance agreement.

15.4.3 Weighted Kappa

One concern about kappa is that it is designed for nominal random variables. For ordinal data, such as in the rating illustration just given, the seriousness of a disagreement depends on the difference between the ratings. For example, the discrepancy between the two raters is more serious if one gives a rating of 1, but the other gives a rating of 3, as opposed to a situation where the second rater gives a rating of 2 instead. This section briefly notes that Fleiss (1981) describes a modification of kappa that uses weights to reflect the closeness of agreement. In particular, for a contingency table with R rows, set

$$w_{ij} = 1 - \frac{(i-j)^2}{(R-1)^2},$$

where the subscript i refers to the ith row of the contingency table and j is the jth column. Then a generalization of kappa is

$$\kappa_w = \frac{\sum\sum w_{ij}p_{ij} - \sum\sum w_{ij}p_{i+}p_{+j}}{1 - \sum\sum w_{ij}p_{i+}p_{+j}}. \tag{15.9}$$

Note that for $i = j$, $w_{ij} = 1$. This says that you give full weight to situations where there is agreement between the two judges, but for cell probabilities further away from the diagonal, the weights get smaller. That is, less weight is given to these probabilities in terms of agreement. Another set of weights that is used is

$$w_{ij} = 0.5^{|i-j|}.$$

(For results regarding how these weights compare to those given above, see Warren, 2012.)

For a 2×2 contingency table, yet another weighted version of kappa has been proposed that takes into account the relative importance of false negatives (e.g., deciding that a patient is not schizophrenic when in fact the reverse is true) to false positives (e.g., incorrectly deciding that a patient is schizophrenic). Imagine that some number ϱ can be specified that reflects the relative importance of false negatives to false positives, where $0 \le \varrho \le 1$. If each mistake is equally serious, then $\varrho = 0.5$. The weighted version of kappa is given by

$$\kappa_\varrho = \frac{p_{11}p_{22} - p_{12}p_{21}}{p_{11}p_{22} - p_{12}p_{21} + \varrho p_{12} + (1-\varrho)p_{21}}. \tag{15.10}$$

(See, for example, Spitzer, Cohen, Fleiss, and Endicott, 1967; Vandelle and Albert, 2009; and Kraemer, Periyakoil, and Noda, 2004, for more details.) Warren (2010) shows that this weighted version of kappa can be written in terms of the odds ratio. When $\varrho = 0.5$, $\kappa_\varrho = \kappa$.

15.4.4 R Function Ckappa

The R function

```
Ckappa(x,fleiss=FALSE, w=NULL)
```

computes kappa and weighted kappa, given by Equation (15.9), assuming that `x` is a square matrix. If `fleiss=TRUE`, the weights suggested by Fleiss are used. Otherwise, the function uses the weights $w_{ij} = 0.5^{|i-j|}$. Or some other set of weights can be specified via the argument `w`.

15.5 LOGISTIC REGRESSION

In many situations the goal is to predict the outcome of a binary random variable, Y, based on some predictor X. For example, the goal might be to predict whether someone will have a heart attack during the next five years given that this individual has a diastolic blood pressure of 110. A tempting approach is to apply least squares regression as described in Chapter 6. But this is unsatisfactory, roughly because the predicted value is never equal to the possible outcome, which is 1 for a success and 0 for a failure. An alternative approach is to predict the *probability* of a heart attack with an equation having the form

$$\hat{p} = \beta_0 + \beta_1 X,$$

where in the illustration, X is diastolic blood pressure. This approach is unsatisfactory as well. The basic problem is that probabilities have values between 0 and 1, whereas linear functions can have any value. What is needed is an equation where the predicted probability is guaranteed to be between 0 and 1. A commonly used approach for dealing with this problem is the *logistic regression* function

$$p(X) = \frac{\exp(\beta_0 + \beta_1 X)}{1 + \exp(\beta_0 + \beta_1 X)}. \tag{15.11}$$

(The notation $\exp(x)$ refers to the exponential function, which means that e is raised to the power x, where e, the base of the natural logarithm, is approximately 2.71828. For example, $\exp(2) = 2.71828^2 = 7.389$.) Here, β_0 and β_1 are unknown parameters that are estimated based on the data.[1] For the more general case where there are p predictors, $X_1, \dots, X_p$, the model is

$$p(X) = \frac{\exp(\beta_0 + \beta_1 X + \cdots + \beta_p X_p)}{1 + \exp(\beta_0 + \beta_1 X + \cdots + \beta_p X_p)}. \tag{15.12}$$

Focusing on the single predictor case, if the parameter $\beta_1 > 0$, then the logistic regression model assumes that the probability of success is a monotonic increasing function of X. (That is, the probability never decreases as X gets large.) If $\beta_1 < 0$, the reverse is true. The left panel of Figure 15.1 shows the regression line when $\beta_0 = 1$ and $\beta_1 = 0.5$. As is evident, curvature is allowed and predicted probabilities always have a value between 0 and 1. The right panel is the regression line when $\beta_0 = 0.5$ and $\beta_1 = -1$. So now the regression line is monotonic decreasing. (The predicted probability never increases.)

Situations might be encountered where the true regression line is increasing over some range of X values, but decreasing otherwise. A possible way of dealing with this problem is to consider a model having the form

$$p(X) = \frac{\exp(\beta_0 + \beta_1 X + \beta_2 X^2)}{1 + \exp(\beta_0 + \beta_1 X + \beta_2 X^2)}. \tag{15.13}$$

15.5.1 R Functions glm and logreg

R has a built-in function called

```
glm
```

[1] Typically, a maximum likelihood estimate is used. See, for example, Hosmer and Lemeshow (1989) for details.

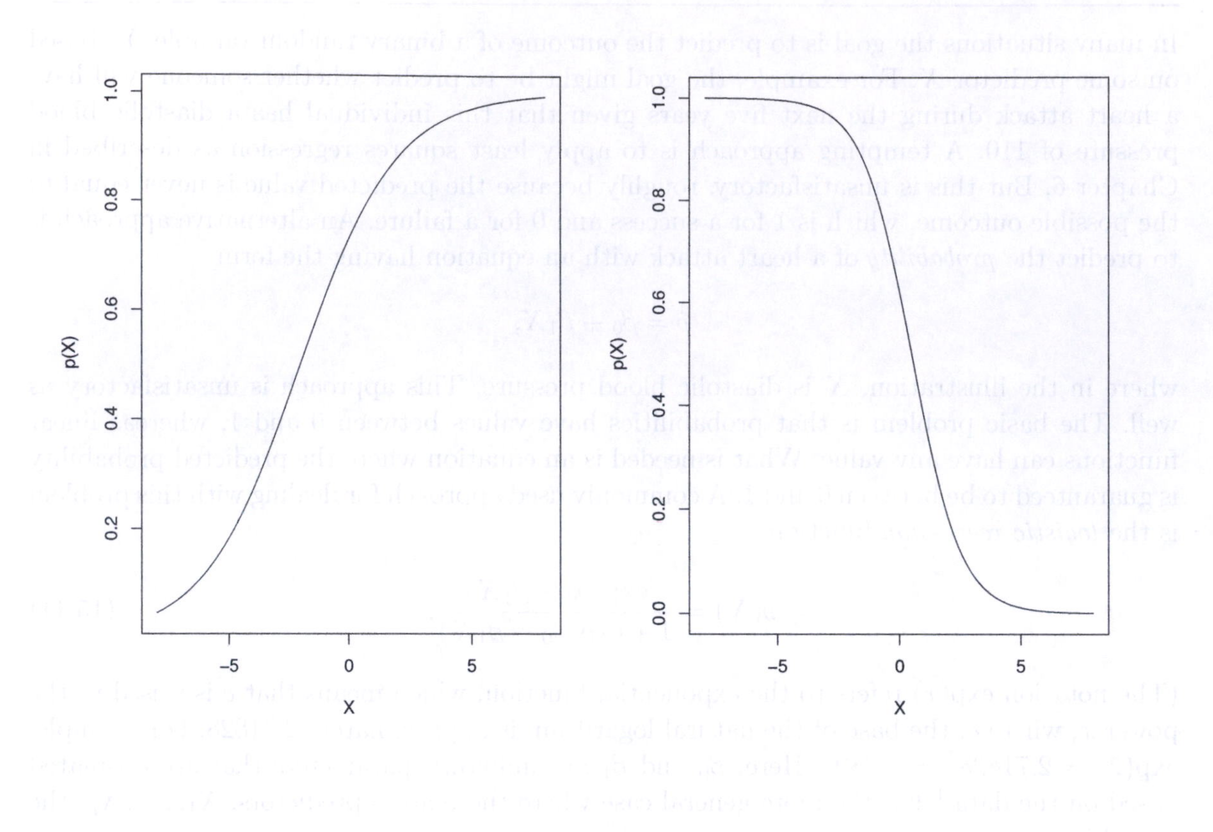

Figure 15.1 Examples of logistic regression lines based on Equation (15.11). In the left panel, $\beta_0 = 1$ and $\beta_1 = 0.5$. Because $\beta_1 > 0$, the predicted probability is monotonic increasing. In the right panel, $\beta_1 = -1$, and now the predicted probability is monotonic decreasing.

that can be used to deal with what are called generalized linear models. Briefly, *generalized linear models* refer to models where the outcome variable, Y, is assumed to have a distribution that belongs to what is called the exponential family of distributions. Included as special cases are the normal and binomial distributions. (One specifies which distribution is to be used via the argument `familiy`, which defaults to a normal distribution.) In addition, it is assumed that

$$\mu = \beta_0 + \beta_1 X.$$

The function relating $\beta_0 + \beta_1 X$ to the outcome of interest is called the link function, which is denoted here by g. The identity function $g(\mu) = \beta_0 + \beta_1 X$ corresponds to the regression methods covered in Chapter 6. Built-in R functions for testing hypotheses generally assume homoscedasticity. The link function, corresponding to the logistic regression model covered here, when there is only one predictor, is

$$g(\mu) = \frac{\exp(\beta_0 + \beta_1 X)}{1 + \exp(\beta_0 + \beta_1 X)}.$$

(Multiple predictors can be used as well.) The point is that several types of regression models are accommodated under a general framework. (See, for example, McCullagh and Nelder, 1992, for more details.)

When using R, the logistic regression model is specified via the command

```
glm(formula=y ~ x,family=binomial).
```

To obtain estimates of the parameters, use the command

```
summary(glm(formula=y ~ x,family=binomial)),
```

which also reports p-values when testing H_0: $\beta_0 = 0$ and H_0: $\beta_1 = 0$. The function expects the values in y to be 0's or 1's. With multiple predictors, x can be a matrix with p columns, or a formula can be used as illustrated, for example, in Section 6.2.3.

Leverage points are a concern when using logistic regression. To help address this problem, the function

```
logreg(x, y, xout = FALSE, outfun = outpro, ...)
```

is supplied, which eliminates leverage points when the argument xout=TRUE, and then it calls the R function glm. Now, the argument x must be a vector, or a matrix with p columns. By default, the projection method for detecting outliers (covered in Section 14.1.6) is used, but other methods can be specified via the argument outfun.

EXAMPLE

Hastie and Tibshirani (1990) report data on the presence or absence of kyphosis, a post-operative spinal deformity, in terms of the age of the patient, in months, the number of vertebrae involved in the spinal operation, and a variable called start, which is the beginning of the range of vertebrae involved. It is instructive to note that in the actual data set, the presence or absence of kyphosis was indicated by the value 'present' and 'absent,' respectively. Assume the data are stored in the R variable kypho, with column 2 containing the data on presence and absence. The command

```
flag=(kypho[,2]=="present")
```

converts these values to TRUE (present) and FALSE (absent) and stores the results in the R variable flag, which the R function logreg will take to be 1 and 0, respectively. Focusing on age, which is assumed to be stored in column 3 of the R variable kypho, the R command

```
logreg(kypho[,3],flag)
```

returns

```
                Estimate  Std. Error   z value     Pr(>|z|)
(Intercept) -1.809351277 0.530352625 -3.411600 0.0006458269
x            0.005441758 0.004821704  1.128596 0.2590681193
```

This indicates that the p-value, when testing H_0: $\beta_1 = 0$, is 0.259, so, in particular, fail to reject at the 0.05 level. Removing leverage points does not alter this result. (Compare this result to the example given in Section 15.5.)

15.5.2 A Confidence Interval for the Odds Ratio

The logistic regression model can be used even when the predictor X is binary. One practical advantage of this result is that it can be used to compute a confidence interval for the odds ratio, assuming that X has either the value 0 or 1. To illustrate how this is done, first note that the probabilities in Table 15.5 can be written as shown in Table 15.9. It can be shown that, as a result, the odds ratio is given by

$$\Psi = e^{\beta_1}.$$

Table 15.9 Table 15.5 Rewritten in Terms of the Logistic Regression Model

	Personality	
Blood Pressure	$X = 1$	$X = 0$
$Y = 1$	$\frac{e^{\beta_0+\beta_1}}{1+e^{\beta_0+\beta_1}}$	$\frac{e^{\beta_0}}{1+e^{\beta_0}}$
$Y = 0$	$\frac{1}{1+e^{\beta_0+\beta_1}}$	$\frac{1}{1+e^{\beta_0}}$
Total	1	1

Letting S_{β_1} denote the estimated standard error of $\hat{\beta}_1$, the estimate of β_1, a $1-\alpha$ confidence interval for the odds ratio is

$$\exp[\hat{\beta}_1 \pm z_{1-\alpha/2} \times S_{\beta_1}], \tag{15.14}$$

where $z_{1-\alpha/2}$ is the $1-\alpha/2$ quantile of a standard normal distribution.

15.5.3 R Function ODDSR.CI

The R function

```
ODDSR.CI(x, y = NULL, alpha = 0.05)
```

computes a $1-\alpha$ confidence interval for the odds ratio. If the argument y is not specified, it is assumed that the argument x is a 2-by-2 matrix that contains observed frequencies that are organized as shown in Table 15.9. Otherwise, x and y are assumed to be vectors containing 0's and 1's.

EXAMPLE

For the data in Table 15.6, a confidence interval for the odds ratio can be computed with the following commands:

```
x=matrix(c(8,67,5,20),ncol=2,nrow=2)
ODDSR.CI(x).
```

15.5.4 Smoothers for Logistic Regression

Special smoothers have been proposed for the case where the outcome variable Y is binary. One such estimator is described by Hosmer and Lemeshow (1989, p. 85, cf. Copas, 1983). Here, a slight modification of this estimator is used, where X has been standardized by replacing X with $(X - M_x)/\text{MADN}_x$, where M_x and MADN_x are the median and MADN based on the observed X values. The estimate of $P(x)$ is

$$\hat{P}(x) = \frac{\sum w_j Y_j}{\sum w_j}, \tag{15.15}$$

where

$$w_i = I_h e^{-(X_i - x)^2}$$

and $I_h = 1$ if $|x_i - x| < h$, otherwise $I_h = 0$. The choice $h = 1.2$ appears to perform relatively well. A slight modification and generalization of this estimator can be used with $p \geq 1$ predictors, which can be applied with the R function logSM described in the next section and appears to perform well compared to other estimators that have been proposed (Wilcox, 2010e, cf. Signorini and Jones, 2004).

15.5.5 R Functions logrsm, rplot.bin, and logSM

The R function

```
logrsm(x, y, fr = 1, plotit = T, pyhat = F, xlab = "X", ylab = "Y", xout =
                    F, outfun = outpro, ...)
```

computes the smoother based on Equation (15.15). An alternative approach is to use an appropriate version of the running-interval smoother, which includes the ability of dealing with more than one predictor. This can be accomplished with the R function

```
rplot.bin(x, y, est = mean, scat = TRUE, fr = 1.2, plotit = TRUE, pyhat =
FALSE, efr = 0.5, theta = 50, phi = 25, scale = FALSE, expand = 0.5, SEED =
TRUE, varfun = pbvar, nmin = 0, xout = FALSE, outfun = out, eout = F, xlab =
              "X", ylab = "Y", zlab = "", pr = TRUE).
```

(When there are two predictors, include the argument `zlab` to label the z-axis if desired.) The argument `fr` corresponds to the span and with `pyhat=T`, the function returns the predicted probability of $Y = 1$ for each value in `x`.

The R function

```
logSM(x,y, pyhat=FALSE, plotit=T, xlab="X", ylab="Y", zlab="Z", xout=FALSE,
outfun=outpro, pr=TRUE, theta=50, phi=25, expand=0.5, scale=FALSE, fr=2,...)
```

applies a slight modification and generalization of the smoother in Hosmer and Lemeshow (1989). It can be used with $p \geq 1$ predictors and appears to be a good choice for general use.

EXAMPLE

In the example in Section 15.1, a basic logistic regression model was used to investigate the probability of kyphosis given the age (in months) of the patient. Recall that the test of H_0: $\beta_1 = 0$ had a p-value of 0.259. However, look at Figure 15.2, which shows the estimate of the regression line based on the R function `logSM`, and note that the line is not monotonic. (The R function `logrsm` gives a similar result.) That is, initially the regression line increases with age, but at some point it appears to decrease. This is a concern because the basic logistic regression model that was fit to the data assumes that the regression line is monotonic. That is, if the regression line increases over some range of X values, the assumption is that it does not decrease for some other range of the predictor values. Figure 15.2 suggests considering instead the model given by Equation (15.13). If the predictor values are stored in the R variable `x`, this can be done with the commands

xsq=cbind(x, x^2)
logreg(xsq, y).

The p-values for H_0: $\beta_1 = 0$ and H_0: $\beta_2 = 0$ are 0.009 and 0.013, respectively. So in contrast to the results in the previous example, now there is an indication of an association.

15.6 EXERCISES

1. You randomly sample 200 adults, determine whether their income is high or low, and ask them whether they are optimistic about the future. The results are

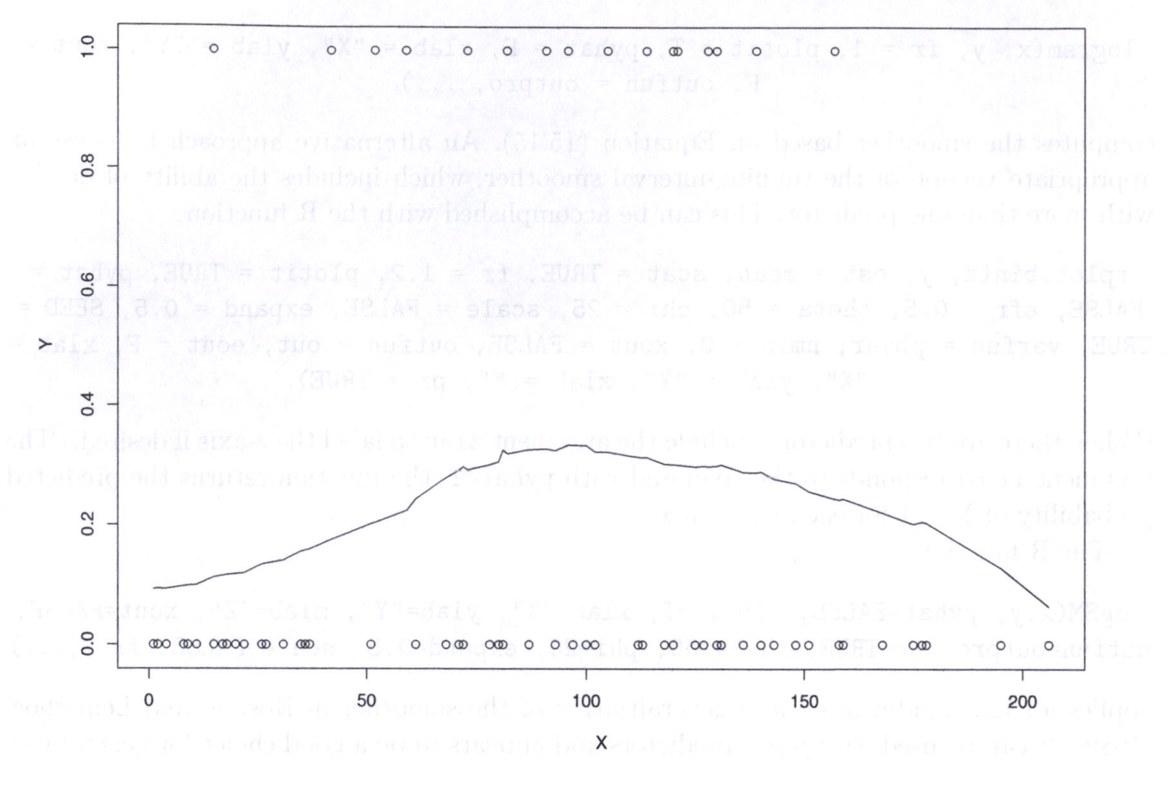

Figure 15.2 The estimated regression line suggests that the probability of kyphosis increases with age, up to a point, and then decreases, which is contrary to the assumption of the basic logistic regression model.

	Yes	No	Total
High	35	42	77
Low	80	43	123
Total	115	85	200

What is the estimated probability that a randomly sampled adult has a high income and is optimistic? Compute a 0.95 confidence interval for the true probability using the R function `binomci`. What is the estimated probability that someone is optimistic, given that the person has a low income?

2. Referring to the previous exercise, compute a 0.95 confidence interval for $\delta = p_{1+} - p_{+1}$. Test H_0: $p_{12} = p_{21}$.

3. In Exercise 1, would you reject the hypothesis that income and outlook are independent? Use $\alpha = 0.05$. Would you use the ϕ coefficient to measure the association between income and outlook? Why?

4. In Exercise 1, estimate the odds ratio and interpret the results.

5. You observe

Income (daughter)	Income (father)			Total
	High	Medium	Low	
High	30	50	20	100
Medium	50	70	30	150
Low	10	20	40	70
Total	90	140	90	320

Estimate the proportion of agreement, and compute a 0.95 confidence interval using the R function `binomci`.

6. Compute kappa for the data in Exercise 5. What does this estimate tell you?

7. Someone claims that the proportion of adults getting low, medium, and high amounts of exercise is 0.5, 0.3, and 0.2, respectively. To check this claim you sample 100 adults and find that 40 get low amounts of exercise, 50 get medium amounts, and 10 get high amounts. Test the claim with $\alpha = 0.05$.

8. For the data in Exercise 7, determine which if any of the claimed cell probabilities appear to be incorrect using the R function `binomci`. Use the Bonferroni method with the goal that the probability of one or more Type I errors will be at most 0.05.

9. For the data in Table 15.6, verify that odds ratio is 0.4776119.

10. Using the plasma retinol data in Section 1.3.3, and removing leverage points, ignore the former smokers and use the R function `logrsm` to estimate the likelihood that a participant was never a smoker given their plasma beta-carotene value. Does this suggest that a basic logistic model is reasonable? Test the hypothesis that there is no association using the basic logistic model. The data are stored on the author's web page in the file plasma_dat.txt. Note that the first 15 lines describe the variables and need to be skipped when reading the data into R.

APPENDIX A

ANSWERS TO SELECTED EXERCISES

CONTENTS

Detailed answers for all of the exercises can be downloaded from http://dornsife.usc.edu/labs/rwilcox/books.
(Or google the author's name and follow the links to his personal web page.) The answers are stored under books in the file CRC_answers.pdf.

Chapter 2

3. Two. **7**. $n = 88$. **10**. 51.36. **11**. Yes. **12**. One. **13**. 20%. **17**. 98, 350, 370, and 475. **18**. $\sum(x\text{-}\bar{X})^2(f_x/n)$. **25**. The lower and upper quartiles are approximately 125 and 50, respectively. So the boxplot rule would declare a value an outlier if it is less than 50−1.5(125−50) or greater than 125+1.5(125−50). **26**. 0.1.

Chapter 3

3. $\mu = 3$, $\sigma^2 = 1.6$. **4**. Smaller, $\sigma^2 = 1.3$. **5**. Larger. **7**. (a) 0.3, (b) 0.03/.3, (c) 0.09/.3 (d) 0.108/0.18. **8**. Yes. For example, probability of a high income given that they are under 30 is 0.03/0.3 = .1 which is equal to the probability of a high income. **9**. (a) 1253/3398, (b) 757/1828, (c) 757/1253, (d) no, (e) 1831/3398. **10**. Median = 2.5. The 0.1 quantile, say y, is given by $(y - 1) * (1/3) = 0.1$, so the 0.1 quantile is 1.3. The 0.9 quantile is 3.7. **11**. (a) $4 \times (1/5)$, (b) $1.5 \times (1/5)$, (c) $2 \times (1/5)$, (d) $2.2 \times (1/5)$, (e) 0. **12**. Median = −0.5. The 0.25 quantile, y, is given by $(y-(-3)) \times 1/5 = 0.25$, so $y = -1.75$. **13**. (a) $(c-(-1)) \times (1/2) = 0.9$, so $c = 0.8$, (b) $(c-(-1)) \times (1/2) = 0.95$, $c = 0.9$, (c) $(c-(-1)) \times (1/2) = 0.01$, $c = 0.98$. **14**. (a) 0.9, (b) $c = 0.95$, (c) $c = 0.99$. **15**. (a) 0, (b) 0.25, (c) 0.5. **16**. (a) 0, (b) 1/6, (c) 2/3, (d) 1/6. **17**. $y \times (1/60) = 0.8$, $y = 48$. **18**. (a) 0.0668, (b) 0.0062, (c) 0.0062, (d) 0.683. **19**. (a) 0.691, (b) 0.894, (c) 0.77. **20**. (a) 0.31, (b) 0.885, (c) 0.018, (d) 0.221. **21**. (a) −2.33, (b) 1.93, (c) −0.174, (d) 0.3. **22**. (a) 1.43, (b) −0.01, (c) 1.7, (d) 1.28. **23**. (a) 0.133, (b) 0.71, (c) 0.133, (d) 0.733. **24**. (a) 0.588, (b) 0.63, (c) 0.71, (d) 0.95. **26**. $c = 1.96$. **27**. 1.28. **28**. 0.16. **29**. 84.45. **30**. 1−0.91. **31**. 0.87. **32**. 0.001. **33**. 0.68. **34**. 0.95. **35**. 0.115. **36**. 0.043. **37**. Yes. **39**. No, for small departures from normality this probability can be close to one. **40**. No, for reasons similar to those in the previous exercise. **41**. Yes. **42**. Yes. **43**. $\mu = 2.3$, $\sigma = 0.9$, and $P(\mu - \sigma \leq X \leq \mu + \sigma) = 0.7$. **44**. (a) 0.75^5, (b) 0.25^5, (c) $1 - 0.25^5 - 5(0.75)(0.25)^4$. **45**. (a) 0.586, (b) 0.732, (c) 1−0.425, (d) 1−0.274. **46**. 0.4(25), 0.4(0.6)(25), 0.4, 0.4(0.6)/25.

Chapter 4

2. 1.28, 1.75, 2.326. **3**. 45 ± 1.96 **4**. 45 ± 2.58 **5**. Yes, the upper end of the confidence interval is 1158. **6**. (a) $65 \pm 1.96(22)/\sqrt{12}$, (b) $185 \pm 1.96(10)/\sqrt{22}$, (c) $19 \pm 1.96(30)/\sqrt{50}$. **8**. 9 and 8/10. **9**. 2.7 and 1.01/12. **10**. 2.7. **11**. 1.01. **12**. 94.3/8 and $\sqrt{94.3/8}$. **13**. 32. No. **14**. 93663.52/12. **15**. They inflate the standard error. **16**. The standard error is 6.78, the squared standard error is $6.78^2 = 45.97$. **17**. No. **18**. 10.9/25, Small departures from normality can inflate the standard error. **20**. (a) 0.0228, (b) 0.159. **21**. (a) 0.16, (b) 0.023, (c) 0.977 − 0.028. **22**. 0.023. **24**. 0.933−0.067. **25**. (a) 0.055, (b) 0.788, (c) 0.992, (d) 0.788−0.055. **26**. (a) 0.047, (b) 0.952, (c) 1 − 0.047, (d) 0.952 − 0.047. **27**. Sample from a heavy-tailed distribution. **28**. Sampling from a light-tailed distribution, the distribution of the sample mean will be well approximated by the central limit theorem. **29**. (a) $26\pm2.26(9)/\sqrt{10}$, (b) $132\pm2.09(20)/\sqrt{18}$, (c) $52\pm2.06(12)/\sqrt{25}$. **31**. (161.4, 734.7). **32**. (10.7, 23.5). **35**. (a) $52\pm2.13\sqrt{12}/(.6\sqrt{24})$ (b) $10\pm2.07\sqrt{30}/(.6\sqrt{36})$. **37**. (160.4, 404.99). **38**. Outliers. **39**. Outliers. **41**. No.

Chapter 5

1. $Z = -1.265$, Fail to reject. **2**. Fail to reject. **3**. (74.9, 81.1). **4**. 0.103. **5**. 0.206. **6**. $Z = -14$, reject. **7**. Reject **8**. (118.6, 121.4). **9**. Yes, because $\bar{X}$ is consistent with H_0. **10**. $Z = 10$, reject. **11**. $Z = 2.12$. Reject. **19**. Increase α. **20**. (a) $T = 1$, fail to reject. (b) $T = .5$, fail to reject. (c) $T = 2.5$, reject. **22**. (a) $T = 0.8$, fail to reject. (b) $T = 0.4$, fail to reject. (c) $T = 2$, reject. **24**. $T = 0.39$, fail to reject. **25**. $T = -2.61$. Reject. **26**. (a) $T_t = .596$, fail to reject. (b) $T_t = 0.298$, fail to reject. (c) $T_t = 0.894$, fail to reject. **28**. $T_t = -3.1$, reject. **29**. $T_t = 0.129$, fail to reject.

Chapter 6

4. 0.87. **6**. $b_1 = -0.0355$, $b_0 = 39.93$. **7**. $b_1 = .0039$, $b_0 = 0.485$. **9**. $b_1 = 0.0754$, $b_0 = -1.253$. **10**. One concern is that $X = 600$ lies outside the range of X values used to compute the least squares regression line. **11**. $r = -0.366$. Not significantly different from zero, so can't be reasonably certain about whether ρ is positive or negative. **14**. Health improves as vitamin intake increases, but health deteriorates with too much vitamin A. That is, there is a nonlinear association. **17**. Extrapolation can be misleading.

Chapter 7

3. $T = (45-36)/\sqrt{11.25(0.083333)} = 9.3$, $\nu = 48$, reject. **4**. $W = (45-36)/\sqrt{4/20+16/30} = 10.49$. Reject. **5**. Welch's test might have more power. **6**. $T = 3.79$, $\nu = 38$, reject. **7**. $W = 3.8$, $\nu = 38$, reject. **8**. With equal sample sizes and equal sample variances, T and W give the exact same result. This suggests that if the sample variances are approximately equal, it makes little difference which method is used. But if the sample variances differ enough, Welch's method is generally more accurate. **9**. $h_1 = 16$, $h_2 = 10$, $d_1 = 2.4$, $d_2 = 6$, $\nu = 16.1$, so $t = 2.12$. $T_y = 2.07$, fail to reject. **10**. 0.99 confidence interval is (−2.5, 14.5). **11**. $\nu = 29$, $t = 2.045$, CI is (1.38, 8.62), reject. **12**. CI is (1.39, 8.6), reject. **13**. $W = 1.2$, $\hat{\nu} = 11.1$, $t = 2.2$, fail to reject. **14**. $T_y = 1.28$, $\hat{\nu} = 9.6$, fail to reject. **15**. No, power might be low. **16**. 0.95 CI is (−7.4, 0.2). Fail to reject. **17**. Fail to reject. **20**. The data indicate that the distributions differ, so the confidence interval based on Student's T might be inaccurate. **24**. The distributions differ, so some would argue that by implication the means differ in particular. **26**. Power. **28**. One concern is that the second group appears to be more skewed than the first suggesting that probability coverage, when using means, might be poor. Another concern is that the first group has an outlier. **31**. If the tails of the distributions are sufficiently heavy, medians have smaller standard errors and hence can have more power. **33**. An improper estimate of the standard error is being used if extreme observations are discarded and methods for means are applied to the data that remain.

Chapter 8

4. The difference between the marginal trimmed means is, in general, not equal to the trimmed mean of the difference scores. **5**. This means that power, when using the difference between the marginal trimmed means, can be lower or higher than the power when using the trimmed mean of the difference scores.

Chapter 9

1. $(6.214+3.982+2.214)/3$. **4**. F = 4.210526 p-value = 0.04615848. **6**. Power of the test for equal variances might not be sufficiently high. And even with equal variances, little is gained using F in terms of controlling the probability of a Type I or achieving relatively high power.

Chapter 10

2. There are main effects for both Factors A and B, but no interaction. **3**. There is a disordinal interaction. **4**. Row 1: 10, 20, 30, Row 2: 20, 30, 40, and Row 3: 30, 40, 50. **7**. No, more needs to be done to establish that a disordinal interaction exists. It is necessary to reject $H_0 : \mu_1 = \mu_2$ as well as $H_0 : \mu_3 = \mu_4$ **9**. Disordinal interactions mean that interpreting main effects may not be straightforward. **10**. No, more needs to be done to establish that a disordinal interaction exists.

Chapter 11

4. Power depends on the measure of location that is used. For skewed distributions, changing measures of location results in testing different hypotheses, and, depending on how the groups differ, one approach could have more power than the other.

Chapter 12

1. Does not control the familywise error rate (the probability of at least one Type I error). **2**. MSWG=11.6. (a) $T = |15 - 10|/\sqrt{11.6(1/20 + 1/20)} = 4.64$ $\nu = 100 - 5 = 95$, reject. (b) $T = |15 - 10|/\sqrt{11.6(1/20 + 1/20)/2} = 6.565$, $q = 3.9$, reject. (c) $W = (15 - 10)/\sqrt{4/20 + 9/20} = 6.2$, $\hat{\nu} = 33$, $c = 2.99$, reject. (d) $(15 - 10)/\sqrt{.5(4/20 + 9/20)} = 8.77$, $q = 4.1$, reject. (e) $f = 2.47$, $S = \sqrt{4(2.47)(11.6)(1/20 + 1/20)} = 3.39$, reject. **3**. MSWG=8. (a) $T = |20\text{-}12|/\sqrt{8(1/10 + 1/10)} = 6.325$ $\nu = 50 - 5 = 45$, reject. (b) $T = |20 - 12|/\sqrt{8(1/10 + 1/10)/2} = 8.94$, $q = 4.01$, reject. (c) $W = (20 - 12)/\sqrt{5/10 + 6/10} = 7.63$, $\hat{\nu} = 37.7$, $c = 2.96$, reject. (d) $(20 - 12)/\sqrt{.5(5/10 + 6/10)} = 10.79$, $q = 4.06$, reject. (e) $f = 2.58$, $S = \sqrt{4(2.58)(8)(1/10 + 1/10)} = 4.06$, reject. (f) $f = 2.62$, $A = 11.3$, reject. **4**. Reject if the p-value is less than or equal to $0.05/6 = 0.0083$. So the fourth test is significant. **5**. Fourth test having significance level 0.001. **6**. None are rejected. **7**. All are rejected. **8**. MSWG goes up and eventually you will no longer reject when comparing groups 1 and 2. **9**. The same problem as in Exercise 8 occurs. **17**. One possibility is that the boxplot rule is missing outliers. Using the MAD-median rule, group 3 is found to have four outliers, but none are found based on the boxplot rules in Chapter 3. Comparing the standard errors of the means versus the standard errors when using a 20% trimmed mean, sometimes a 20% trimmed has a smaller standard error, but sometimes the reverse is true. Differences in skewness can affect power when comparing means even when there are no outliers.

Chapter 13

1. Other projections might find an outlier.

Chapter 14

12. Pearson's correlation is −0.33. The explanatory strength of the association returned by `lplot` is 0.61

Chapter 15

1. $\hat{p}_{11} = 35/200$, CI = (.126, .235), $\hat{p}_{1|2} = 80/123$. **2**. $\hat{\delta} = (42 - 80)/200 = -0.19$, $\hat{\sigma}_\delta = 0.0536$, 0.95 confidence interval is $(-0.29, -0.085)$, reject. **3**. $X^2 = 7.43$, reject. No, it is always an unsatisfactory measure of association. **4**. $\hat{\theta} = 0.448$. Participants with high incomes are about half as likely to be optimistic about the future. **5**. $\hat{p} = 0.4375$. CI = (0.385, 0.494). **6**. $\hat{\kappa} = 0.13$. Your estimate is that there is agreement beyond chance. **7**. $X^2 = 20.5$, reject. **8**. Using $\alpha = 0.05/3$, the confidence intervals are (0.29, 0.52), (0.385, 0.622), and (0.04, 0.194). First is not significant, but reject H_0: $p_2 = 0.3$ and H_0: $p_3 = 0.2$.

APPENDIX B

TABLES

CONTENTS

TABLE 1: Standard Normal Distribution

z	$P(Z \leq z)$	z	$P(Z \leq z)$	z	$P(Z \leq z)$	z	$P(Z \leq z)$
−3.00	0.0013	−2.99	0.0014	−2.98	0.0014	−2.97	0.0015
−2.96	0.0015	−2.95	0.0016	−2.94	0.0016	−2.93	0.0017
−2.92	0.0018	−2.91	0.0018	−2.90	0.0019	−2.89	0.0019
−2.88	0.0020	−2.87	0.0021	−2.86	0.0021	−2.85	0.0022
−2.84	0.0023	−2.83	0.0023	−2.82	0.0024	−2.81	0.0025
−2.80	0.0026	−2.79	0.0026	−2.78	0.0027	−2.77	0.0028
−2.76	0.0029	−2.75	0.0030	−2.74	0.0031	−2.73	0.0032
−2.72	0.0033	−2.71	0.0034	−2.70	0.0035	−2.69	0.0036
−2.68	0.0037	−2.67	0.0038	−2.66	0.0039	−2.65	0.0040
−2.64	0.0041	−2.63	0.0043	−2.62	0.0044	−2.61	0.0045
−2.60	0.0047	−2.59	0.0048	−2.58	0.0049	−2.57	0.0051
−2.56	0.0052	−2.55	0.0054	−2.54	0.0055	−2.53	0.0057
−2.52	0.0059	−2.51	0.0060	−2.50	0.0062	−2.49	0.0064
−2.48	0.0066	−2.47	0.0068	−2.46	0.0069	−2.45	0.0071
−2.44	0.0073	−2.43	0.0075	−2.42	0.0078	−2.41	0.0080
−2.40	0.0082	−2.39	0.0084	−2.38	0.0087	−2.37	0.0089
−2.36	0.0091	−2.35	0.0094	−2.34	0.0096	−2.33	0.0099
−2.32	0.0102	−2.31	0.0104	−2.30	0.0107	−2.29	0.0110
−2.28	0.0113	−2.27	0.0116	−2.26	0.0119	−2.25	0.0122
−2.24	0.0125	−2.23	0.0129	−2.22	0.0132	−2.21	0.0136
−2.20	0.0139	−2.19	0.0143	−2.18	0.0146	−2.17	0.0150
−2.16	0.0154	−2.15	0.0158	−2.14	0.0162	−2.13	0.0166
−2.12	0.0170	−2.11	0.0174	−2.10	0.0179	−2.09	0.0183
−2.08	0.0188	−2.07	0.0192	−2.06	0.0197	−2.05	0.0202
−2.04	0.0207	−2.03	0.0212	−2.02	0.0217	−2.01	0.0222
−2.00	0.0228	−1.99	0.0233	−1.98	0.0239	−1.97	0.0244
−1.96	0.0250	−1.95	0.0256	−1.94	0.0262	−1.93	0.0268
−1.92	0.0274	−1.91	0.0281	−1.90	0.0287	−1.89	0.0294
−1.88	0.0301	−1.87	0.0307	−1.86	0.0314	−1.85	0.0322
−1.84	0.0329	−1.83	0.0336	−1.82	0.0344	−1.81	0.0351
−1.80	0.0359	−1.79	0.0367	−1.78	0.0375	−1.77	0.0384
−1.76	0.0392	−1.75	0.0401	−1.74	0.0409	−1.73	0.0418
−1.72	0.0427	−1.71	0.0436	−1.70	0.0446	−1.69	0.0455
−1.68	0.0465	−1.67	0.0475	−1.66	0.0485	−1.65	0.0495
−1.64	0.0505	−1.63	0.0516	−1.62	0.0526	−1.61	0.0537
−1.60	0.0548	−1.59	0.0559	−1.58	0.0571	−1.57	0.0582
−1.56	0.0594	−1.55	0.0606	−1.54	0.0618	−1.53	0.0630

TABLE 1 continued

z	$P(Z \leq z)$	z	$P(Z \leq z)$	z	$P(Z \leq z)$	z	$P(Z \leq z)$
−1.52	0.0643	−1.51	0.0655	−1.50	0.0668	−1.49	0.0681
−1.48	0.0694	−1.47	0.0708	−1.46	0.0721	−1.45	0.0735
−1.44	0.0749	−1.43	0.0764	−1.42	0.0778	−1.41	0.0793
−1.40	0.0808	−1.39	0.0823	−1.38	0.0838	−1.37	0.0853
−1.36	0.0869	−1.35	0.0885	−1.34	0.0901	−1.33	0.0918
−1.32	0.0934	−1.31	0.0951	−1.30	0.0968	−1.29	0.0985
−1.28	0.1003	−1.27	0.1020	−1.26	0.1038	−1.25	0.1056
−1.24	0.1075	−1.23	0.1093	−1.22	0.1112	−1.21	0.1131
−1.20	0.1151	−1.19	0.1170	−1.18	0.1190	−1.17	0.1210
−1.16	0.1230	−1.15	0.1251	−1.14	0.1271	−1.13	0.1292
−1.12	0.1314	−1.11	0.1335	−1.10	0.1357	−1.09	0.1379
−1.08	0.1401	−1.07	0.1423	−1.06	0.1446	−1.05	0.1469
−1.04	0.1492	−1.03	0.1515	−1.02	0.1539	−1.01	0.1562
−1.00	0.1587	−0.99	0.1611	−0.98	0.1635	−0.97	0.1662
−0.96	0.1685	−0.95	0.1711	−0.94	0.1736	−0.93	0.1762
−0.92	0.1788	−0.91	0.1814	−0.90	0.1841	−0.89	0.1867
−0.88	0.1894	−0.87	0.1922	−0.86	0.1949	−0.85	0.1977
−0.84	0.2005	−0.83	0.2033	−0.82	0.2061	−0.81	0.2090
−0.80	0.2119	−0.79	0.2148	−0.78	0.2177	−0.77	0.2207
−0.76	0.2236	−0.75	0.2266	−0.74	0.2297	−0.73	0.2327
−0.72	0.2358	−0.71	0.2389	−0.70	0.2420	−0.69	0.2451
−0.68	0.2483	−0.67	0.2514	−0.66	0.2546	−0.65	0.2578
−0.64	0.2611	−0.63	0.2643	−0.62	0.2676	−0.61	0.2709
−0.60	0.2743	−0.59	0.2776	−0.58	0.2810	−0.57	0.2843
−0.56	0.2877	−0.55	0.2912	−0.54	0.2946	−0.53	0.2981
−0.52	0.3015	−0.51	0.3050	−0.50	0.3085	−0.49	0.3121
−0.48	0.3156	−0.47	0.3192	−0.46	0.3228	−0.45	0.3264
−0.44	0.3300	−0.43	0.3336	−0.42	0.3372	−0.41	0.3409
−0.40	0.3446	−0.39	0.3483	−0.38	0.3520	−0.37	0.3557
−0.36	0.3594	−0.35	0.3632	−0.34	0.3669	−0.33	0.3707
−0.32	0.3745	−0.31	0.3783	−0.30	0.3821	−0.29	0.3859
−0.28	0.3897	−0.27	0.3936	−0.26	0.3974	−0.25	0.4013
−0.24	0.4052	−0.23	0.4090	−0.22	0.4129	−0.21	0.4168
−0.20	0.4207	−0.19	0.4247	−0.18	0.4286	−0.17	0.4325
−0.16	0.4364	−0.15	0.4404	−0.14	0.4443	−0.13	0.4483
−0.12	0.4522	−0.11	0.4562	−0.10	0.4602	−0.09	0.4641
−0.08	0.4681	−0.07	0.4721	−0.06	0.4761	−0.05	0.4801
−0.04	0.4840	−0.03	0.4880	−0.02	0.4920	−0.01	0.4960

TABLE 1 continued

z	$P(Z \leq z)$	z	$P(Z \leq z)$	z	$P(Z \leq z)$	z	$P(Z \leq z)$
0.01	0.5040	0.02	0.5080	0.03	0.5120	0.04	0.5160
0.05	0.5199	0.06	0.5239	0.07	0.5279	0.08	0.5319
0.09	0.5359	0.10	0.5398	0.11	0.5438	0.12	0.5478
0.13	0.5517	0.14	0.5557	0.15	0.5596	0.16	0.5636
0.17	0.5675	0.18	0.5714	0.19	0.5753	0.20	0.5793
0.21	0.5832	0.22	0.5871	0.23	0.5910	0.24	0.5948
0.25	0.5987	0.26	0.6026	0.27	0.6064	0.28	0.6103
0.29	0.6141	0.30	0.6179	0.31	0.6217	0.32	0.6255
0.33	0.6293	0.34	0.6331	0.35	0.6368	0.36	0.6406
0.37	0.6443	0.38	0.6480	0.39	0.6517	0.40	0.6554
0.41	0.6591	0.42	0.6628	0.43	0.6664	0.44	0.6700
0.45	0.6736	0.46	0.6772	0.47	0.6808	0.48	0.6844
0.49	0.6879	0.50	0.6915	0.51	0.6950	0.52	0.6985
0.53	0.7019	0.54	0.7054	0.55	0.7088	0.56	0.7123
0.57	0.7157	0.58	0.7190	0.59	0.7224	0.60	0.7257
0.61	0.7291	0.62	0.7324	0.63	0.7357	0.64	0.7389
0.65	0.7422	0.66	0.7454	0.67	0.7486	0.68	0.7517
0.69	0.7549	0.70	0.7580	0.71	0.7611	0.72	0.7642
0.73	0.7673	0.74	0.7703	0.75	0.7734	0.76	0.7764
0.77	0.7793	0.78	0.7823	0.79	0.7852	0.80	0.7881
0.81	0.7910	0.82	0.7939	0.83	0.7967	0.84	0.7995
0.85	0.8023	0.86	0.8051	0.87	0.8078	0.88	0.8106
0.89	0.8133	0.90	0.8159	0.91	0.8186	0.92	0.8212
0.93	0.8238	0.94	0.8264	0.95	0.8289	0.96	0.8315
0.97	0.8340	0.98	0.8365	0.99	0.8389	1.00	0.8413
1.01	0.8438	1.02	0.8461	1.03	0.8485	1.04	0.8508
1.05	0.8531	1.06	0.8554	1.07	0.8577	1.08	0.8599
1.09	0.8621	1.10	0.8643	1.11	0.8665	1.12	0.8686
1.13	0.8708	1.14	0.8729	1.15	0.8749	1.16	0.8770
1.17	0.8790	1.18	0.8810	1.19	0.8830	1.20	0.8849
1.21	0.8869	1.22	0.8888	1.23	0.8907	1.24	0.8925
1.25	0.8944	1.26	0.8962	1.27	0.8980	1.28	0.8997
1.29	0.9015	1.30	0.9032	1.31	0.9049	1.32	0.9066
1.33	0.9082	1.34	0.9099	1.35	0.9115	1.36	0.9131
1.37	0.9147	1.38	0.9162	1.39	0.9177	1.40	0.9192
1.41	0.9207	1.42	0.9222	1.43	0.9236	1.44	0.9251
1.45	0.9265	1.46	0.9279	1.47	0.9292	1.48	0.9306
1.49	0.9319	1.50	0.9332	1.51	0.9345	1.52	0.9357

TABLE 1 continued

z	$P(Z \leq z)$	z	$P(Z \leq z)$	z	$P(Z \leq z)$	z	$P(Z \leq z)$
1.53	0.9370	1.54	0.9382	1.55	0.9394	1.56	0.9406
1.57	0.9418	1.58	0.9429	1.59	0.9441	1.60	0.9452
1.61	0.9463	1.62	0.9474	1.63	0.9484	1.64	0.9495
1.65	0.9505	1.66	0.9515	1.67	0.9525	1.68	0.9535
1.69	0.9545	1.70	0.9554	1.71	0.9564	1.72	0.9573
1.73	0.9582	1.74	0.9591	1.75	0.9599	1.76	0.9608
1.77	0.9616	1.78	0.9625	1.79	0.9633	1.80	0.9641
1.81	0.9649	1.82	0.9656	1.83	0.9664	1.84	0.9671
1.85	0.9678	1.86	0.9686	1.87	0.9693	1.88	0.9699
1.89	0.9706	1.90	0.9713	1.91	0.9719	1.92	0.9726
1.93	0.9732	1.94	0.9738	1.95	0.9744	1.96	0.9750
1.97	0.9756	1.98	0.9761	1.99	0.9767	2.00	0.9772
2.01	0.9778	2.02	0.9783	2.03	0.9788	2.04	0.9793
2.05	0.9798	2.06	0.9803	2.07	0.9808	2.08	0.9812
2.09	0.9817	2.10	0.9821	2.11	0.9826	2.12	0.9830
2.13	0.9834	2.14	0.9838	2.15	0.9842	2.16	0.9846
2.17	0.9850	2.18	0.9854	2.19	0.9857	2.20	0.9861
2.21	0.9864	2.22	0.9868	2.23	0.9871	2.24	0.9875
2.25	0.9878	2.26	0.9881	2.27	0.9884	2.28	0.9887
2.29	0.9890	2.30	0.9893	2.31	0.9896	2.32	0.9898
2.33	0.9901	2.34	0.9904	2.35	0.9906	2.36	0.9909
2.37	0.9911	2.38	0.9913	2.39	0.9916	2.40	0.9918
2.41	0.9920	2.42	0.9922	2.43	0.9925	2.44	0.9927
2.45	0.9929	2.46	0.9931	2.47	0.9932	2.48	0.9934
2.49	0.9936	2.50	0.9938	2.51	0.9940	2.52	0.9941
2.53	0.9943	2.54	0.9945	2.55	0.9946	2.56	0.9948
2.57	0.9949	2.58	0.9951	2.59	0.9952	2.60	0.9953
2.61	0.9955	2.62	0.9956	2.63	0.9957	2.64	0.9959
2.65	0.9960	2.66	0.9961	2.67	0.9962	2.68	0.9963
2.69	0.9964	2.70	0.9965	2.71	0.9966	2.72	0.9967
2.73	0.9968	2.74	0.9969	2.75	0.9970	2.76	0.9971
2.77	0.9972	2.78	0.9973	2.79	0.9974	2.80	0.9974
2.81	0.9975	2.82	0.9976	2.83	0.9977	2.84	0.9977
2.85	0.9978	2.86	0.9979	2.87	0.9979	2.88	0.9980
2.89	0.9981	2.90	0.9981	2.91	0.9982	2.92	0.9982
2.93	0.9983	2.94	0.9984	2.95	0.9984	2.96	0.9985
2.97	0.9985	2.98	0.9986	2.99	0.9986	3.00	0.9987

Note: This table was computed with IMSL subroutine ANORIN.

TABLE 2 Binomial Probability Function (values of entries are $P(X \leq k)$)

$n = 5$

k	p .05	.1	.2	.3	.4	.5	.6	.7	.8	.9	.95
0	0.774	0.590	0.328	0.168	0.078	0.031	0.010	0.002	0.000	0.000	0.000
1	0.977	0.919	0.737	0.528	0.337	0.188	0.087	0.031	0.007	0.000	0.000
2	0.999	0.991	0.942	0.837	0.683	0.500	0.317	0.163	0.058	0.009	0.001
3	1.000	1.000	0.993	0.969	0.913	0.813	0.663	0.472	0.263	0.081	0.023
4	1.000	1.000	1.000	0.998	0.990	0.969	0.922	0.832	0.672	0.410	0.226

$n = 6$

k	p .05	.1	.2	.3	.4	.5	.6	.7	.8	.9	.95
0	0.735	0.531	0.262	0.118	0.047	0.016	0.004	0.001	0.000	0.000	0.000
1	0.967	0.886	0.655	0.420	0.233	0.109	0.041	0.011	0.002	0.000	0.000
2	0.998	0.984	0.901	0.744	0.544	0.344	0.179	0.070	0.017	0.001	0.000
3	1.000	0.999	0.983	0.930	0.821	0.656	0.456	0.256	0.099	0.016	0.002
4	1.000	1.000	0.998	0.989	0.959	0.891	0.767	0.580	0.345	0.114	0.033
5	1.000	1.000	1.000	0.999	0.996	0.984	0.953	0.882	0.738	0.469	0.265

$n = 7$

k	p .05	.1	.2	.3	.4	.5	.6	.7	.8	.9	.95
0	0.698	0.478	0.210	0.082	0.028	0.008	0.002	0.000	0.000	0.000	0.000
1	0.956	0.850	0.577	0.329	0.159	0.062	0.019	0.004	0.000	0.000	0.000
2	0.996	0.974	0.852	0.647	0.420	0.227	0.096	0.029	0.005	0.000	0.000
3	1.000	0.997	0.967	0.874	0.710	0.500	0.290	0.126	0.033	0.003	0.000
4	1.000	1.000	0.995	0.971	0.904	0.773	0.580	0.353	0.148	0.026	0.004
5	1.000	1.000	1.000	0.996	0.981	0.938	0.841	0.671	0.423	0.150	0.044
6	1.000	1.000	1.000	1.000	0.998	0.992	0.972	0.918	0.790	0.522	0.302

$n = 8$

k	p .05	.1	.2	.3	.4	.5	.6	.7	.8	.9	.95
0	0.663	0.430	0.168	0.058	0.017	0.004	0.001	0.000	0.000	0.000	0.000
1	0.943	0.813	0.503	0.255	0.106	0.035	0.009	0.001	0.000	0.000	0.000
2	0.994	0.962	0.797	0.552	0.315	0.145	0.050	0.011	0.001	0.000	0.000
3	1.000	0.995	0.944	0.806	0.594	0.363	0.174	0.058	0.010	0.000	0.000
4	1.000	1.000	0.990	0.942	0.826	0.637	0.406	0.194	0.056	0.005	0.000
5	1.000	1.000	0.999	0.989	0.950	0.855	0.685	0.448	0.203	0.038	0.006
6	1.000	1.000	1.000	0.999	0.991	0.965	0.894	0.745	0.497	0.187	0.057
7	1.000	1.000	1.000	1.000	0.999	0.996	0.983	0.942	0.832	0.570	0.337

$n = 9$

k	p .05	.1	.2	.3	.4	.5	.6	.7	.8	.9	.95
0	0.630	0.387	0.134	0.040	0.010	0.002	0.000	0.000	0.000	0.000	0.000
1	0.929	0.775	0.436	0.196	0.071	0.020	0.004	0.000	0.000	0.000	0.000
2	0.992	0.947	0.738	0.463	0.232	0.090	0.025	0.004	0.000	0.000	0.000
3	0.999	0.992	0.914	0.730	0.483	0.254	0.099	0.025	0.003	0.000	0.000
4	1.000	0.999	0.980	0.901	0.733	0.500	0.267	0.099	0.020	0.001	0.000
5	1.000	1.000	0.997	0.975	0.901	0.746	0.517	0.270	0.086	0.008	0.001
6	1.000	1.000	1.000	0.996	0.975	0.910	0.768	0.537	0.262	0.053	0.008
7	1.000	1.000	1.000	1.000	0.996	0.980	0.929	0.804	0.564	0.225	0.071
8	1.000	1.000	1.000	1.000	1.000	0.998	0.990	0.960	0.866	0.613	0.370

TABLE 2 continued

$n = 10$

k	.05	.1	.2	.3	.4	p .5	.6	.7	.8	.9	.95
0	0.599	0.349	0.107	0.028	0.006	0.001	0.000	0.000	0.000	0.000	0.000
1	0.914	0.736	0.376	0.149	0.046	0.011	0.002	0.000	0.000	0.000	0.000
2	0.988	0.930	0.678	0.383	0.167	0.055	0.012	0.002	0.000	0.000	0.000
3	0.999	0.987	0.879	0.650	0.382	0.172	0.055	0.011	0.001	0.000	0.000
4	1.000	0.998	0.967	0.850	0.633	0.377	0.166	0.047	0.006	0.000	0.000
5	1.000	1.000	0.994	0.953	0.834	0.623	0.367	0.150	0.033	0.002	0.000
6	1.000	1.000	0.999	0.989	0.945	0.828	0.618	0.350	0.121	0.013	0.001
7	1.000	1.000	1.000	0.998	0.988	0.945	0.833	0.617	0.322	0.070	0.012
8	1.000	1.000	1.000	1.000	0.998	0.989	0.954	0.851	0.624	0.264	0.086
9	1.000	1.000	1.000	1.000	1.000	0.999	0.994	0.972	0.893	0.651	0.401

$n = 15$

k	.05	.1	.2	.3	.4	p .5	.6	.7	.8	.9	.95
0	0.463	0.206	0.035	0.005	0.000	0.000	0.000	0.000	0.000	0.000	0.000
1	0.829	0.549	0.167	0.035	0.005	0.000	0.000	0.000	0.000	0.000	0.000
2	0.964	0.816	0.398	0.127	0.027	0.004	0.000	0.000	0.000	0.000	0.000
3	0.995	0.944	0.648	0.297	0.091	0.018	0.002	0.000	0.000	0.000	0.000
4	0.999	0.987	0.836	0.515	0.217	0.059	0.009	0.001	0.000	0.000	0.000
5	1.000	0.998	0.939	0.722	0.403	0.151	0.034	0.004	0.000	0.000	0.000
6	1.000	1.000	0.982	0.869	0.610	0.304	0.095	0.015	0.001	0.000	0.000
7	1.000	1.000	0.996	0.950	0.787	0.500	0.213	0.050	0.004	0.000	0.000
8	1.000	1.000	0.999	0.985	0.905	0.696	0.390	0.131	0.018	0.000	0.000
9	1.000	1.000	1.000	0.996	0.966	0.849	0.597	0.278	0.061	0.002	0.000
10	1.000	1.000	1.000	0.999	0.991	0.941	0.783	0.485	0.164	0.013	0.001
11	1.000	1.000	1.000	1.000	0.998	0.982	0.909	0.703	0.352	0.056	0.005
12	1.000	1.000	1.000	1.000	1.000	0.996	0.973	0.873	0.602	0.184	0.036
13	1.000	1.000	1.000	1.000	1.000	1.000	0.995	0.965	0.833	0.451	0.171
14	1.000	1.000	1.000	1.000	1.000	1.000	1.000	0.995	0.965	0.794	0.537

$n = 20$

k	.05	.1	.2	.3	.4	p .5	.6	.7	.8	.9	.95
0	0.358	0.122	0.012	0.001	0.000	0.000	0.000	0.000	0.000	0.000	0.000
1	0.736	0.392	0.069	0.008	0.001	0.000	0.000	0.000	0.000	0.000	0.000
2	0.925	0.677	0.206	0.035	0.004	0.000	0.000	0.000	0.000	0.000	0.000
3	0.984	0.867	0.411	0.107	0.016	0.001	0.000	0.000	0.000	0.000	0.000
4	0.997	0.957	0.630	0.238	0.051	0.006	0.000	0.000	0.000	0.000	0.000
5	1.000	0.989	0.804	0.416	0.126	0.021	0.002	0.000	0.000	0.000	0.000
6	1.000	0.998	0.913	0.608	0.250	0.058	0.006	0.000	0.000	0.000	0.000
7	1.000	1.000	0.968	0.772	0.416	0.132	0.021	0.001	0.000	0.000	0.000
8	1.000	1.000	0.990	0.887	0.596	0.252	0.057	0.005	0.000	0.000	0.000
9	1.000	1.000	0.997	0.952	0.755	0.412	0.128	0.017	0.001	0.000	0.000
10	1.000	1.000	0.999	0.983	0.872	0.588	0.245	0.048	0.003	0.000	0.000
11	1.000	1.000	1.000	0.995	0.943	0.748	0.404	0.113	0.010	0.000	0.000
12	1.000	1.000	1.000	0.999	0.979	0.868	0.584	0.228	0.032	0.000	0.000
13	1.000	1.000	1.000	1.000	0.994	0.942	0.750	0.392	0.087	0.002	0.000
14	1.000	1.000	1.000	1.000	0.998	0.979	0.874	0.584	0.196	0.011	0.000
15	1.000	1.000	1.000	1.000	1.000	0.994	0.949	0.762	0.370	0.043	0.003
16	1.000	1.000	1.000	1.000	1.000	0.999	0.984	0.893	0.589	0.133	0.016
17	1.000	1.000	1.000	1.000	1.000	1.000	0.996	0.965	0.794	0.323	0.075
18	1.000	1.000	1.000	1.000	1.000	1.000	0.999	0.992	0.931	0.608	0.264
19	1.000	1.000	1.000	1.000	1.000	1.000	1.000	0.999	0.988	0.878	0.642

TABLE 2 continued
$n = 25$

					p						
k	.05	.1	.2	.3	.4	.5	.6	.7	.8	.9	.95
0	0.277	0.072	0.004	0.000	0.000	0.000	0.000	0.000	0.000	0.000	0.000
1	0.642	0.271	0.027	0.002	0.000	0.000	0.000	0.000	0.000	0.000	0.000
2	0.873	0.537	0.098	0.009	0.000	0.000	0.000	0.000	0.000	0.000	0.000
3	0.966	0.764	0.234	0.033	0.002	0.000	0.000	0.000	0.000	0.000	0.000
4	0.993	0.902	0.421	0.090	0.009	0.000	0.000	0.000	0.000	0.000	0.000
5	0.999	0.967	0.617	0.193	0.029	0.002	0.000	0.000	0.000	0.000	0.000
6	1.000	0.991	0.780	0.341	0.074	0.007	0.000	0.000	0.000	0.000	0.000
7	1.000	0.998	0.891	0.512	0.154	0.022	0.001	0.000	0.000	0.000	0.000
8	1.000	1.000	0.953	0.677	0.274	0.054	0.004	0.000	0.000	0.000	0.000
9	1.000	1.000	0.983	0.811	0.425	0.115	0.013	0.000	0.000	0.000	0.000
10	1.000	1.000	0.994	0.902	0.586	0.212	0.034	0.002	0.000	0.000	0.000
11	1.000	1.000	0.998	0.956	0.732	0.345	0.078	0.006	0.000	0.000	0.000
12	1.000	1.000	1.000	0.983	0.846	0.500	0.154	0.017	0.000	0.000	0.000
13	1.000	1.000	1.000	0.994	0.922	0.655	0.268	0.044	0.002	0.000	0.000
14	1.000	1.000	1.000	0.998	0.966	0.788	0.414	0.098	0.006	0.000	0.000
15	1.000	1.000	1.000	1.000	0.987	0.885	0.575	0.189	0.017	0.000	0.000
16	1.000	1.000	1.000	1.000	0.996	0.946	0.726	0.323	0.047	0.000	0.000
17	1.000	1.000	1.000	1.000	0.999	0.978	0.846	0.488	0.109	0.002	0.000
18	1.000	1.000	1.000	1.000	1.000	0.993	0.926	0.659	0.220	0.009	0.000
19	1.000	1.000	1.000	1.000	1.000	0.998	0.971	0.807	0.383	0.033	0.001
20	1.000	1.000	1.000	1.000	1.000	1.000	0.991	0.910	0.579	0.098	0.007
21	1.000	1.000	1.000	1.000	1.000	1.000	0.998	0.967	0.766	0.236	0.034
22	1.000	1.000	1.000	1.000	1.000	1.000	1.000	0.991	0.902	0.463	0.127
23	1.000	1.000	1.000	1.000	1.000	1.000	1.000	0.998	0.973	0.729	0.358
24	1.000	1.000	1.000	1.000	1.000	1.000	1.000	1.000	0.996	0.928	0.723

TABLE 3: Percentage Points of the Chi-Squared Distribution

ν	$\chi^2_{.005}$	$\chi^2_{.01}$	$\chi^2_{.025}$	$\chi^2_{.05}$	$\chi^2_{.10}$
1	0.0000393	0.0001571	0.0009821	0.0039321	0.0157908
2	0.0100251	0.0201007	0.0506357	0.1025866	0.2107213
3	0.0717217	0.1148317	0.2157952	0.3518462	0.5843744
4	0.2069889	0.2971095	0.4844186	0.7107224	1.0636234
5	0.4117419	0.5542979	0.8312111	1.1454763	1.6103077
6	0.6757274	0.8720903	1.2373447	1.6353836	2.2041321
7	0.9892554	1.2390423	1.6898699	2.1673594	2.8331099
8	1.3444128	1.6464968	2.1797333	2.7326374	3.4895401
9	1.7349329	2.0879011	2.7003908	3.3251143	4.1681604
10	2.1558590	2.5582132	3.2469759	3.9403019	4.8651857
11	2.6032248	3.0534868	3.8157606	4.5748196	5.5777788
12	3.0738316	3.5705872	4.4037895	5.2260313	6.3037949
13	3.5650368	4.1069279	5.0087538	5.8918715	7.0415068
14	4.0746784	4.6604300	5.6287327	6.5706167	7.7895403
15	4.6009169	5.2293501	6.2621403	7.2609539	8.5467529
16	5.1422071	5.8122101	6.9076681	7.9616566	9.3122330
17	5.6972256	6.4077673	7.5641880	8.6717682	10.0851974
18	6.2648115	7.0149183	8.2307510	9.3904572	10.8649368
19	6.8439512	7.6327391	8.9065247	10.1170273	11.6509628
20	7.4338474	8.2603989	9.5907822	10.8508148	12.4426041
21	8.0336685	8.8972015	10.2829285	11.5913391	13.2396393
22	8.6427155	9.5425110	10.9823456	12.3380432	14.0414886
23	9.2604370	10.1957169	11.6885223	13.0905151	14.8479385
24	9.8862610	10.8563690	12.4011765	13.8484344	15.6587067
25	10.5196533	11.5239716	13.1197433	14.6114349	16.4734497
26	11.1602631	12.1981506	13.8439331	15.3792038	17.2919159
27	11.8076019	12.8785095	14.5734024	16.1513977	18.1138763
28	12.4613495	13.5647125	15.3078613	16.9278717	18.9392395
29	13.1211624	14.2564697	16.0470886	17.7083893	19.7678223
30	13.7867584	14.9534760	16.7907562	18.4926147	20.5992126
40	20.7065582	22.1642761	24.4330750	26.5083008	29.0503540
50	27.9775238	29.7001038	32.3561096	34.7638702	37.6881561
60	35.5294037	37.4848328	40.4817810	43.1865082	46.4583282
70	43.2462311	45.4230499	48.7503967	51.7388763	55.3331146
80	51.1447754	53.5226593	57.1465912	60.3912201	64.2818604
90	59.1706543	61.7376862	65.6405029	69.1258850	73.2949219
100	67.3031921	70.0493622	74.2162018	77.9293976	82.3618469

TABLE 3 continued

ν	$\chi^2_{.900}$	$\chi^2_{.95}$	$\chi^2_{.975}$	$\chi^2_{.99}$	$\chi^2_{.995}$
1	2.7056	3.8415	5.0240	6.6353	7.8818
2	4.6052	5.9916	7.3779	9.2117	10.5987
3	6.2514	7.8148	9.3486	11.3465	12.8409
4	7.7795	9.4879	11.1435	13.2786	14.8643
5	9.2365	11.0707	12.8328	15.0870	16.7534
6	10.6448	12.5919	14.4499	16.8127	18.5490
7	12.0171	14.0676	16.0136	18.4765	20.2803
8	13.3617	15.5075	17.5355	20.0924	21.9579
9	14.6838	16.9191	19.0232	21.6686	23.5938
10	15.9874	18.3075	20.4837	23.2101	25.1898
11	17.2750	19.6754	21.9211	24.7265	26.7568
12	18.5494	21.0263	23.3370	26.2170	28.2995
13	19.8122	22.3627	24.7371	27.6882	29.8194
14	21.0646	23.6862	26.1189	29.1412	31.3193
15	22.3077	24.9970	27.4883	30.5779	32.8013
16	23.5421	26.2961	28.8453	31.9999	34.2672
17	24.7696	27.5871	30.1909	33.4087	35.7184
18	25.9903	28.8692	31.5264	34.8054	37.1564
19	27.2035	30.1434	32.8523	36.1909	38.5823
20	28.4120	31.4104	34.1696	37.5662	39.9968
21	29.6150	32.6705	35.4787	38.9323	41.4012
22	30.8133	33.9244	36.7806	40.2893	42.7958
23	32.0069	35.1725	38.0757	41.6384	44.1812
24	33.1962	36.4151	39.3639	42.9799	45.5587
25	34.3815	37.6525	40.6463	44.3142	46.9280
26	35.5631	38.8852	41.9229	45.6418	48.2899
27	36.7412	40.1134	43.1943	46.9629	49.6449
28	37.9159	41.3371	44.4608	48.2784	50.9933
29	39.0874	42.5571	45.7223	49.5879	52.3357
30	40.2561	43.7730	46.9792	50.8922	53.6721
40	51.8050	55.7586	59.3417	63.6909	66.7660
50	63.1670	67.5047	71.4201	76.1538	79.4899
60	74.3970	79.0820	83.2977	88.3794	91.9516
70	85.5211	90.5283	95.0263	100.4409	104.2434
80	96.5723	101.8770	106.6315	112.3434	116.3484
90	107.5600	113.1425	118.1392	124.1304	128.3245
100	118.4932	124.3395	129.5638	135.8203	140.1940

Note: This table was computed with IMSL subroutine CHIIN.

TABLE 4 Percentage Points of Student's T Distribution

ν	$t_{.9}$	$t_{.95}$	$t_{.975}$	$t_{.99}$	$t_{.995}$	$t_{.999}$
1	3.078	6.314	12.706	31.821	63.6567	318.313
2	1.886	2.920	4.303	6.965	9.925	22.327
3	1.638	2.353	3.183	4.541	5.841	10.215
4	1.533	2.132	2.776	3.747	4.604	7.173
5	1.476	2.015	2.571	3.365	4.032	5.893
6	1.440	1.943	2.447	3.143	3.707	5.208
7	1.415	1.895	2.365	2.998	3.499	4.785
8	1.397	1.856	2.306	2.897	3.355	4.501
9	1.383	1.833	2.262	2.821	3.245	4.297
10	1.372	1.812	2.228	2.764	3.169	4.144
12	1.356	1.782	2.179	2.681	3.055	3.930
15	1.341	1.753	2.131	2.603	2.947	3.733
20	1.325	1.725	2.086	2.528	2.845	3.552
24	1.318	1.711	2.064	2.492	2.797	3.467
30	1.310	1.697	2.042	2.457	2.750	3.385
40	1.303	1.684	2.021	2.423	2.704	3.307
60	1.296	1.671	2.000	2.390	2.660	3.232
120	1.289	1.658	1.980	2.358	2.617	3.160
∞	1.282	1.645	1.960	2.326	2.576	3.090

Note: Entries were computed with IMSL subroutine TIN.

TABLE 5: Percentage Points of the F Distribution, $\alpha = 0.10$

	ν_1								
ν_2	1	2	3	4	5	6	7	8	9
1	39.86	49.50	53.59	55.83	57.24	58.20	58.91	59.44	59.86
2	8.53	9.00	9.16	9.24	9.29	9.33	9.35	9.37	9.38
3	5.54	5.46	5.39	5.34	5.31	5.28	5.27	5.25	5.24
4	4.54	4.32	4.19	4.11	4.05	4.01	3.98	3.95	3.94
5	4.06	3.78	3.62	3.52	3.45	3.40	3.37	3.34	3.32
6	3.78	3.46	3.29	3.18	3.11	3.05	3.01	2.98	2.96
7	3.59	3.26	3.07	2.96	2.88	2.83	2.79	2.75	2.72
8	3.46	3.11	2.92	2.81	2.73	2.67	2.62	2.59	2.56
9	3.36	3.01	2.81	2.69	2.61	2.55	2.51	2.47	2.44
10	3.29	2.92	2.73	2.61	2.52	2.46	2.41	2.38	2.35
11	3.23	2.86	2.66	2.54	2.45	2.39	2.34	2.30	2.27
12	3.18	2.81	2.61	2.48	2.39	2.33	2.28	2.24	2.21
13	3.14	2.76	2.56	2.43	2.35	2.28	2.23	2.20	2.16
14	3.10	2.73	2.52	2.39	2.31	2.24	2.19	2.15	2.12
15	3.07	2.70	2.49	2.36	2.27	2.21	2.16	2.12	2.09
16	3.05	2.67	2.46	2.33	2.24	2.18	2.13	2.09	2.06
17	3.03	2.64	2.44	2.31	2.22	2.15	2.10	2.06	2.03
18	3.01	2.62	2.42	2.29	2.20	2.13	2.08	2.04	2.00
19	2.99	2.61	2.40	2.27	2.18	2.11	2.06	2.02	1.98
20	2.97	2.59	2.38	2.25	2.16	2.09	2.04	2.00	1.96
21	2.96	2.57	2.36	2.23	2.14	2.08	2.02	1.98	1.95
22	2.95	2.56	2.35	2.22	2.13	2.06	2.01	1.97	1.93
23	2.94	2.55	2.34	2.21	2.11	2.05	1.99	1.95	1.92
24	2.93	2.54	2.33	2.19	2.10	2.04	1.98	1.94	1.91
25	2.92	2.53	2.32	2.18	2.09	2.02	1.97	1.93	1.89
26	2.91	2.52	2.31	2.17	2.08	2.01	1.96	1.92	1.88
27	2.90	2.51	2.30	2.17	2.07	2.00	1.95	1.91	1.87
28	2.89	2.50	2.29	2.16	2.06	2.00	1.94	1.90	1.87
29	2.89	2.50	2.28	2.15	2.06	1.99	1.93	1.89	1.86
30	2.88	2.49	2.28	2.14	2.05	1.98	1.93	1.88	1.85
40	2.84	2.44	2.23	2.09	2.00	1.93	1.87	1.83	1.79
60	2.79	2.39	2.18	2.04	1.95	1.87	1.82	1.77	1.74
120	2.75	2.35	2.13	1.99	1.90	1.82	1.77	1.72	1.68
∞	2.71	2.30	2.08	1.94	1.85	1.77	1.72	.167	1.63

TABLE 5 continued

	ν_1									
ν_2	10	12	15	20	24	30	40	60	120	∞
1	60.19	60.70	61.22	61.74	62.00	62.26	62.53	62.79	63.06	63.33
2	9.39	9.41	9.42	9.44	9.45	9.46	9.47	9.47	9.48	9.49
3	5.23	5.22	5.20	5.19	5.18	5.17	5.16	5.15	5.14	5.13
4	3.92	3.90	3.87	3.84	3.83	3.82	3.80	3.79	3.78	3.76
5	3.30	3.27	3.24	3.21	3.19	3.17	3.16	3.14	3.12	3.10
6	2.94	2.90	2.87	2.84	2.82	2.80	2.78	2.76	2.74	2.72
7	2.70	2.67	2.63	2.59	2.58	2.56	2.54	2.51	2.49	2.47
8	2.54	2.50	2.46	2.42	2.40	2.38	2.36	2.34	2.32	2.29
9	2.42	2.38	2.34	2.30	2.28	2.25	2.23	2.21	2.18	2.16
10	2.32	2.28	2.24	2.20	2.18	2.16	2.13	2.11	2.08	2.06
11	2.25	2.21	2.17	2.12	2.10	2.08	2.05	2.03	2.00	1.97
12	2.19	2.15	2.10	2.06	2.04	2.01	1.99	1.96	1.93	1.90
13	2.14	2.10	2.05	2.01	1.98	1.96	1.93	1.90	1.88	1.85
14	2.10	2.05	2.01	1.96	1.94	1.91	1.89	1.86	1.83	1.80
15	2.06	2.02	1.97	1.92	1.90	1.87	1.85	1.82	1.79	1.76
16	2.03	1.99	1.94	1.89	1.87	1.84	1.81	1.78	1.75	1.72
17	2.00	1.96	1.91	1.86	1.84	1.81	1.78	1.75	1.72	1.69
18	1.98	1.93	1.89	1.84	1.81	1.78	1.75	1.72	1.69	1.66
19	1.96	1.91	1.86	1.81	1.79	1.76	1.73	1.70	1.67	1.63
20	1.94	1.89	1.84	1.79	1.77	1.74	1.71	1.68	1.64	1.61
21	1.92	1.87	1.83	1.78	1.75	1.72	1.69	1.66	1.62	1.59
22	1.90	1.86	1.81	1.76	1.73	1.70	1.67	1.64	1.60	1.57
23	1.89	1.84	1.80	1.74	1.72	1.69	1.66	1.62	1.59	1.55
24	1.88	1.83	1.78	1.73	1.70	1.67	1.64	1.61	1.57	1.53
25	1.87	1.82	1.77	1.72	1.69	1.66	1.63	1.59	1.56	1.52
26	1.86	1.81	1.76	1.71	1.68	1.65	1.61	1.58	1.54	1.50
27	1.85	1.80	1.75	1.70	1.67	1.64	1.60	1.57	1.53	1.49
28	1.84	1.79	1.74	1.69	1.66	1.63	1.59	1.56	1.52	1.48
29	1.83	1.78	1.73	1.68	1.65	1.62	1.58	1.55	1.51	1.47
30	1.82	1.77	1.72	1.67	1.64	1.61	1.57	1.54	1.50	1.46
40	1.76	1.71	1.66	1.61	1.57	1.54	1.51	1.47	1.42	1.38
60	1.71	1.66	1.60	1.54	1.51	1.48	1.44	1.40	1.35	1.29
120	1.65	1.60	1.55	1.48	1.45	1.41	1.37	1.32	1.26	1.19
∞	1.60	1.55	1.49	1.42	1.38	1.34	1.30	1.24	1.17	1.00

Note: Entries in this table were computed with IMSL subroutine FIN.

TABLE 6: Percentage Points of the F Distribution, $\alpha = 0.05$

	ν_1								
ν_2	1	2	3	4	5	6	7	8	9
1	161.45	199.50	215.71	224.58	230.16	233.99	236.77	238.88	240.54
2	18.51	19.00	19.16	19.25	19.30	19.33	19.35	19.37	19.38
3	10.13	9.55	9.28	9.12	9.01	8.94	8.89	8.85	8.81
4	7.71	6.94	6.59	6.39	6.26	6.16	6.09	6.04	6.00
5	6.61	5.79	5.41	5.19	5.05	4.95	4.88	4.82	4.77
6	5.99	5.14	4.76	4.53	4.39	4.28	4.21	4.15	4.10
7	5.59	4.74	4.35	4.12	3.97	3.87	3.79	3.73	3.68
8	5.32	4.46	4.07	3.84	3.69	3.58	3.50	3.44	3.39
9	5.12	4.26	3.86	3.63	3.48	3.37	3.29	3.23	3.18
10	4.96	4.10	3.71	3.48	3.33	3.22	3.14	3.07	3.02
11	4.84	3.98	3.59	3.36	3.20	3.09	3.01	2.95	2.90
12	4.75	3.89	3.49	3.26	3.11	3.00	2.91	2.85	2.80
13	4.67	3.81	3.41	3.18	3.03	2.92	2.83	2.77	2.71
14	4.60	3.74	3.34	3.11	2.96	2.85	2.76	2.70	2.65
15	4.54	3.68	3.29	3.06	2.90	2.79	2.71	2.64	2.59
16	4.49	3.63	3.24	3.01	2.85	2.74	2.66	2.59	2.54
17	4.45	3.59	3.20	2.96	2.81	2.70	2.61	2.55	2.49
18	4.41	3.55	3.16	2.93	2.77	2.66	2.58	2.51	2.46
19	4.38	3.52	3.13	2.90	2.74	2.63	2.54	2.48	2.42
20	4.35	3.49	3.10	2.87	2.71	2.60	2.51	2.45	2.39
21	4.32	3.47	3.07	2.84	2.68	2.57	2.49	2.42	2.37
22	4.30	3.44	3.05	2.82	2.66	2.55	2.46	2.40	2.34
23	4.28	3.42	3.03	2.80	2.64	2.53	2.44	2.37	2.32
24	4.26	3.40	3.01	2.78	2.62	2.51	2.42	2.36	2.30
25	4.24	3.39	2.99	2.76	2.60	2.49	2.40	2.34	2.28
26	4.23	3.37	2.98	2.74	2.59	2.47	2.39	2.32	2.27
27	4.21	3.35	2.96	2.73	2.57	2.46	2.37	2.31	2.25
28	4.20	3.34	2.95	2.71	2.56	2.45	2.36	2.29	2.24
29	4.18	3.33	2.93	2.70	2.55	2.43	2.35	2.28	2.22
30	4.17	3.32	2.92	2.69	2.53	2.42	2.33	2.27	2.21
40	4.08	3.23	2.84	2.61	2.45	2.34	2.25	2.18	2.12
60	4.00	3.15	2.76	2.53	2.37	2.25	2.17	2.10	2.04
120	3.92	3.07	2.68	2.45	2.29	2.17	2.09	2.02	1.96
∞	3.84	3.00	2.60	2.37	2.21	2.10	2.01	1.94	1.88

TABLE 6 continued

	ν_1									
ν_2	10	12	15	20	24	30	40	60	120	∞
1	241.88	243.91	245.96	248.00	249.04	250.08	251.14	252.19	253.24	254.3
2	19.40	19.41	19.43	19.45	19.45	19.46	19.47	19.48	19.49	19.50
3	8.79	8.74	8.70	8.66	8.64	8.62	8.59	8.57	8.55	8.53
4	5.97	5.91	5.86	5.80	5.77	5.74	5.72	5.69	5.66	5.63
5	4.73	4.68	4.62	4.56	4.53	4.50	4.46	4.43	4.40	4.36
6	4.06	4.00	3.94	3.87	3.84	3.81	3.77	3.74	3.70	3.67
7	3.64	3.57	3.51	3.44	3.41	3.38	3.34	3.30	3.27	3.23
8	3.35	3.28	3.22	3.15	3.12	3.08	3.04	3.00	2.97	2.93
9	3.14	3.07	3.01	2.94	2.90	2.86	2.83	2.79	2.75	2.71
10	2.98	2.91	2.85	2.77	2.74	2.70	2.66	2.62	2.58	2.54
11	2.85	2.79	2.72	2.65	2.61	2.57	2.53	2.49	2.45	2.40
12	2.75	2.69	2.62	2.54	2.51	2.47	2.43	2.38	2.34	2.30
13	2.67	2.60	2.53	2.46	2.42	2.38	2.34	2.30	2.25	2.21
14	2.60	2.53	2.46	2.39	2.35	2.31	2.27	2.22	2.18	2.13
15	2.54	2.48	2.40	2.33	2.29	2.25	2.20	2.16	2.11	2.07
16	2.49	2.42	2.35	2.28	2.24	2.19	2.15	2.11	2.06	2.01
17	2.45	2.38	2.31	2.23	2.19	2.15	2.10	2.06	2.01	1.96
18	2.41	2.34	2.27	2.19	2.15	2.11	2.06	2.02	1.97	1.92
19	2.38	2.31	2.23	2.16	2.11	2.07	2.03	1.98	1.93	1.88
20	2.35	2.28	2.20	2.12	2.08	2.04	1.99	1.95	1.90	1.84
21	2.32	2.25	2.18	2.10	2.05	2.01	1.96	1.92	1.87	1.81
22	2.30	2.23	2.15	2.07	2.03	1.98	1.94	1.89	1.84	1.78
23	2.27	2.20	2.13	2.05	2.00	1.96	1.91	1.86	1.81	1.76
24	2.25	2.18	2.11	2.03	1.98	1.94	1.89	1.84	1.79	1.73
25	2.24	2.16	2.09	2.01	1.96	1.92	1.87	1.82	1.77	1.71
26	2.22	2.15	2.07	1.99	1.95	1.90	1.85	1.80	1.75	1.69
27	2.20	2.13	2.06	1.97	1.93	1.88	1.84	1.79	1.73	1.67
28	2.19	2.12	2.04	1.96	1.91	1.87	1.82	1.77	1.71	1.65
29	2.18	2.10	2.03	1.94	1.90	1.85	1.81	1.75	1.70	1.64
30	2.16	2.09	2.01	1.93	1.89	1.84	1.79	1.74	1.68	1.62
40	2.08	2.00	1.92	1.84	1.79	1.74	1.69	1.64	1.58	1.51
60	1.99	1.92	1.84	1.75	1.70	1.65	1.59	1.53	1.47	1.39
120	1.91	1.83	1.75	1.66	1.61	1.55	1.50	1.43	1.35	1.25
∞	1.83	1.75	1.67	1.57	1.52	1.46	1.39	1.32	1.22	1.00

TABLE 7: Percentage Points of the F Distribution, $\alpha = 0.025$

	ν_1								
ν_2	1	2	3	4	5	6	7	8	9
1	647.79	799.50	864.16	899.59	921.85	937.11	948.22	956.66	963.28
2	38.51	39.00	39.17	39.25	39.30	39.33	39.36	39.37	39.39
3	17.44	16.04	15.44	15.10	14.88	14.74	14.63	14.54	14.47
4	12.22	10.65	9.98	9.61	9.36	9.20	9.07	8.98	8.90
5	10.01	8.43	7.76	7.39	7.15	6.98	6.85	6.76	6.68
6	8.81	7.26	6.60	6.23	5.99	5.82	5.70	5.60	5.52
7	8.07	6.54	5.89	5.52	5.29	5.12	5.00	4.90	4.82
8	7.57	6.06	5.42	5.05	4.82	4.65	4.53	4.43	4.36
9	7.21	5.71	5.08	4.72	4.48	4.32	4.20	4.10	4.03
10	6.94	5.46	4.83	4.47	4.24	4.07	3.95	3.85	3.78
11	6.72	5.26	4.63	4.28	4.04	3.88	3.76	3.66	3.59
12	6.55	5.10	4.47	4.12	3.89	3.73	3.61	3.51	3.44
13	6.41	4.97	4.35	4.00	3.77	3.60	3.48	3.39	3.31
14	6.30	4.86	4.24	3.89	3.66	3.50	3.38	3.29	3.21
15	6.20	4.77	4.15	3.80	3.58	3.41	3.29	3.20	3.12
16	6.12	4.69	4.08	3.73	3.50	3.34	3.22	3.12	3.05
17	6.04	4.62	4.01	3.66	3.44	3.28	3.16	3.06	2.98
18	5.98	4.56	3.95	3.61	3.38	3.22	3.10	3.01	2.93
19	5.92	4.51	3.90	3.56	3.33	3.17	3.05	2.96	2.88
20	5.87	4.46	3.86	3.51	3.29	3.13	3.01	2.91	2.84
21	5.83	4.42	3.82	3.48	3.25	3.09	2.97	2.87	2.80
22	5.79	4.38	3.78	3.44	3.22	3.05	2.93	2.84	2.76
23	5.75	4.35	3.75	3.41	3.18	3.02	2.90	2.81	2.73
24	5.72	4.32	3.72	3.38	3.15	2.99	2.87	2.78	2.70
25	5.69	4.29	3.69	3.35	3.13	2.97	2.85	2.75	2.68
26	5.66	4.27	3.67	3.33	3.10	2.94	2.82	2.73	2.65
27	5.63	4.24	3.65	3.31	3.08	2.92	2.80	2.71	2.63
28	5.61	4.22	3.63	3.29	3.06	2.90	2.78	2.69	2.61
29	5.59	4.20	3.61	3.27	3.04	2.88	2.76	2.67	2.59
30	5.57	4.18	3.59	3.25	3.03	2.87	2.75	2.65	2.57
40	5.42	4.05	3.46	3.13	2.90	2.74	2.62	2.53	2.45
60	5.29	3.93	3.34	3.01	2.79	2.63	2.51	2.41	2.33
120	5.15	3.80	3.23	2.89	2.67	2.52	2.39	2.30	2.22
∞	5.02	3.69	3.12	2.79	2.57	2.41	2.29	2.19	2.11

TABLE 7 continued

	ν_1									
ν_2	10	12	15	20	24	30	40	60	120	∞
1	968.62	976.71	984.89	993.04	997.20	1,001	1,006	1,010	1,014	1,018
2	39.40	39.41	39.43	39.45	39.46	39.46	39.47	39.48	39.49	39.50
3	14.42	14.33	14.26	14.17	14.13	14.08	14.04	13.99	13.95	13.90
4	8.85	8.75	8.66	8.56	8.51	8.46	8.41	8.36	8.31	8.26
5	6.62	6.53	6.43	6.33	6.28	6.23	6.17	6.12	6.07	6.02
6	5.46	5.37	5.27	5.17	5.12	5.06	5.01	4.96	4.90	4.85
7	4.76	4.67	4.57	4.47	4.41	4.36	4.31	4.25	4.20	4.14
8	4.30	4.20	4.10	4.00	3.95	3.89	3.84	3.78	3.73	3.67
9	3.96	3.87	3.77	3.67	3.61	3.56	3.51	3.45	3.39	3.33
10	3.72	3.62	3.52	3.42	3.37	3.31	3.26	3.20	3.14	3.08
11	3.53	3.43	3.33	3.23	3.17	3.12	3.06	3.00	2.94	2.88
12	3.37	3.28	3.18	3.07	3.02	2.96	2.91	2.85	2.79	2.72
13	3.25	3.15	3.05	2.95	2.89	2.84	2.78	2.72	2.66	2.60
14	3.15	3.05	2.95	2.84	2.79	2.73	2.67	2.61	2.55	2.49
15	3.06	2.96	2.86	2.76	2.70	2.64	2.59	2.52	2.46	2.40
16	2.99	2.89	2.79	2.68	2.63	2.57	2.51	2.45	2.38	2.32
17	2.92	2.82	2.72	2.62	2.56	2.50	2.44	2.38	2.32	2.25
18	2.87	2.77	2.67	2.56	2.50	2.44	2.38	2.32	2.26	2.19
19	2.82	2.72	2.62	2.51	2.45	2.39	2.33	2.27	2.20	2.13
20	2.77	2.68	2.57	2.46	2.41	2.35	2.29	2.22	2.16	2.09
21	2.73	2.64	2.53	2.42	2.37	2.31	2.25	2.18	2.11	2.04
22	2.70	2.60	2.50	2.39	2.33	2.27	2.21	2.14	2.08	2.00
23	2.67	2.57	2.47	2.36	2.30	2.24	2.18	2.11	2.04	1.97
24	2.64	2.54	2.44	2.33	2.27	2.21	2.15	2.08	2.01	1.94
25	2.61	2.51	2.41	2.30	2.24	2.18	2.12	2.05	1.98	1.91
26	2.59	2.49	2.39	2.28	2.22	2.16	2.09	2.03	1.95	1.88
27	2.57	2.47	2.36	2.25	2.19	2.13	2.07	2.00	1.93	1.85
28	2.55	2.45	2.34	2.23	2.17	2.11	2.05	1.98	1.91	1.83
29	2.53	2.43	2.32	2.21	2.15	2.09	2.03	1.96	1.89	1.81
30	2.51	2.41	2.31	2.20	2.14	2.07	2.01	1.94	1.87	1.79
40	2.39	2.29	2.18	2.07	2.01	1.94	1.88	1.80	1.72	1.64
60	2.27	2.17	2.06	1.94	1.88	1.82	1.74	1.67	1.58	1.48
120	2.16	2.05	1.95	1.82	1.76	1.69	1.61	1.53	1.43	1.31
∞	2.05	1.94	1.83	1.71	1.64	1.57	1.48	1.39	1.27	1.00

TABLE 8: Percentage Points of the F Distribution, $\alpha = 0.01$

	ν_1								
ν_2	1	2	3	4	5	6	7	8	9
1	4,052	4,999	5,403	5,625	5,764	5,859	5,928	5,982	6,022
2	98.50	99.00	99.17	99.25	99.30	99.33	99.36	99.37	99.39
3	34.12	30.82	29.46	28.71	28.24	27.91	27.67	27.50	27.34
4	21.20	18.00	16.69	15.98	15.52	15.21	14.98	14.80	14.66
5	16.26	13.27	12.06	11.39	10.97	10.67	10.46	10.29	10.16
6	13.75	10.92	9.78	9.15	8.75	8.47	8.26	8.10	7.98
7	12.25	9.55	8.45	7.85	7.46	7.19	6.99	6.84	6.72
8	11.26	8.65	7.59	7.01	6.63	6.37	6.18	6.03	5.91
9	10.56	8.02	6.99	6.42	6.06	5.80	5.61	5.47	5.35
10	10.04	7.56	6.55	5.99	5.64	5.39	5.20	5.06	4.94
11	9.65	7.21	6.22	5.67	5.32	5.07	4.89	4.74	4.63
12	9.33	6.93	5.95	5.41	5.06	4.82	4.64	4.50	4.39
13	9.07	6.70	5.74	5.21	4.86	4.62	4.44	4.30	4.19
14	8.86	6.51	5.56	5.04	4.69	4.46	4.28	4.14	4.03
15	8.68	6.36	5.42	4.89	4.56	4.32	4.14	4.00	3.89
16	8.53	6.23	5.29	4.77	4.44	4.20	4.03	3.89	3.78
17	8.40	6.11	5.18	4.67	4.34	4.10	3.93	3.79	3.68
18	8.29	6.01	5.09	4.58	4.25	4.01	3.84	3.71	3.60
19	8.18	5.93	5.01	4.50	4.17	3.94	3.77	3.63	3.52
20	8.10	5.85	4.94	4.43	4.10	3.87	3.70	3.56	3.46
21	8.02	5.78	4.87	4.37	4.04	3.81	3.64	3.51	3.40
22	7.95	5.72	4.82	4.31	3.99	3.76	3.59	3.45	3.35
23	7.88	5.66	4.76	4.26	3.94	3.71	3.54	3.41	3.30
24	7.82	5.61	4.72	4.22	3.90	3.67	3.50	3.36	3.26
25	7.77	5.57	4.68	4.18	3.85	3.63	3.46	3.32	3.22
26	7.72	5.53	4.64	4.14	3.82	3.59	3.42	3.29	3.18
27	7.68	5.49	4.60	4.11	3.78	3.56	3.39	3.26	3.15
28	7.64	5.45	4.57	4.07	3.75	3.53	3.36	3.23	3.12
29	7.60	5.42	4.54	4.04	3.73	3.50	3.33	3.20	3.09
30	7.56	5.39	4.51	4.02	3.70	3.47	3.30	3.17	3.07
40	7.31	5.18	4.31	3.83	3.51	3.29	3.12	2.99	2.89
60	7.08	4.98	4.13	3.65	3.34	3.12	2.95	2.82	2.72
120	6.85	4.79	3.95	3.48	3.17	2.96	2.79	2.66	2.56
∞	6.63	4.61	3.78	3.32	3.02	2.80	2.64	2.51	2.41

TABLE 8 continued

	ν_1									
ν_2	10	12	15	20	24	30	40	60	120	∞
1	6,056	6,106	6,157	6,209	6,235	6,261	6,287	6,313	6,339	6,366
2	99.40	99.42	99.43	99.45	99.46	99.46	99.47	99.48	99.49	99.50
3	27.22	27.03	26.85	26.67	26.60	26.50	26.41	26.32	26.22	26.13
4	14.55	14.37	14.19	14.02	13.94	13.84	13.75	13.65	13.56	13.46
5	10.05	9.89	9.72	9.55	9.46	9.38	9.30	9.20	9.11	9.02
6	7.87	7.72	7.56	7.40	7.31	7.23	7.15	7.06	6.97	6.88
7	6.62	6.47	6.31	6.16	6.07	5.99	5.91	5.82	5.74	5.65
8	5.81	5.67	5.52	5.36	5.28	5.20	5.12	5.03	4.95	4.86
9	5.26	5.11	4.96	4.81	4.73	4.65	4.57	4.48	4.40	4.31
10	4.85	4.71	4.56	4.41	4.33	4.25	4.17	4.08	4.00	3.91
11	4.54	4.40	4.25	4.10	4.02	3.94	3.86	3.78	3.69	3.60
12	4.30	4.16	4.01	3.86	3.78	3.70	3.62	3.54	3.45	3.36
13	4.10	3.96	3.82	3.66	3.59	3.51	3.43	3.34	3.25	3.17
14	3.94	3.80	3.66	3.51	3.43	3.35	3.27	3.18	3.09	3.00
15	3.80	3.67	3.52	3.37	3.29	3.21	3.13	3.05	2.96	2.87
16	3.69	3.55	3.41	3.26	3.18	3.10	3.02	2.93	2.84	2.75
17	3.59	3.46	3.31	3.16	3.08	3.00	2.92	2.83	2.75	2.65
18	3.51	3.37	3.23	3.08	3.00	2.92	2.84	2.75	2.66	2.57
19	3.43	3.30	3.15	3.00	2.92	2.84	2.76	2.67	2.58	2.49
20	3.37	3.23	3.09	2.94	2.86	2.78	2.69	2.61	2.52	2.42
21	3.31	3.17	3.03	2.88	2.80	2.72	2.64	2.55	2.46	2.36
22	3.26	3.12	2.98	2.83	2.75	2.67	2.58	2.50	2.40	2.31
23	3.21	3.07	2.93	2.78	2.70	2.62	2.54	2.45	2.35	2.26
24	3.17	3.03	2.89	2.74	2.66	2.58	2.49	2.40	2.31	2.21
25	3.13	2.99	2.85	2.70	2.62	2.54	2.45	2.36	2.27	2.17
26	3.09	2.96	2.81	2.66	2.58	2.50	2.42	2.33	2.23	2.13
27	3.06	2.93	2.78	2.63	2.55	2.47	2.38	2.29	2.20	2.10
28	3.03	2.90	2.75	2.60	2.52	2.44	2.35	2.26	2.17	2.06
29	3.00	2.87	2.73	2.57	2.49	2.41	2.33	2.23	2.14	2.03
30	2.98	2.84	2.70	2.55	2.47	2.39	2.30	2.21	2.11	2.01
40	2.80	2.66	2.52	2.37	2.29	2.20	2.11	2.02	1.92	1.80
60	2.63	2.50	2.35	2.20	2.12	2.03	1.94	1.84	1.73	1.60
120	2.47	2.34	2.19	2.03	1.95	1.86	1.76	1.66	1.53	1.38
∞	2.32	2.18	2.04	1.88	1.79	1.70	1.59	1.47	1.32	1.00

TABLE 9: Studentized Range Statistic, q, for $\alpha = 0.05$

	J (number of groups)									
ν	2	3	4	5	6	7	8	9	10	11
3	4.50	5.91	6.82	7.50	8.04	8.48	8.85	9.18	9.46	9.72
4	3.93	5.04	5.76	6.29	6.71	7.05	7.35	7.60	7.83	8.03
5	3.64	4.60	5.22	5.68	6.04	6.33	6.59	6.81	6.99	7.17
6	3.47	4.34	4.89	5.31	5.63	5.89	6.13	6.32	6.49	6.65
7	3.35	4.17	4.69	5.07	5.36	5.61	5.82	5.99	6.16	6.30
8	3.27	4.05	4.53	4.89	5.17	5.39	5.59	5.77	5.92	6.06
9	3.19	3.95	4.42	4.76	5.03	5.25	5.44	5.59	5.74	5.87
10	3.16	3.88	4.33	4.66	4.92	5.13	5.31	5.47	5.59	5.73
11	3.12	3.82	4.26	4.58	4.83	5.03	5.21	5.36	5.49	5.61
12	3.09	3.78	4.19	4.51	4.76	4.95	5.12	5.27	5.39	5.52
13	3.06	3.73	4.15	4.45	4.69	4.88	5.05	5.19	5.32	5.43
14	3.03	3.70	4.11	4.41	4.64	4.83	4.99	5.13	5.25	5.36
15	3.01	3.67	4.08	4.37	4.59	4.78	4.94	5.08	5.20	5.31
16	3.00	3.65	4.05	4.33	4.56	4.74	4.90	5.03	5.15	5.26
17	2.98	3.63	4.02	4.30	4.52	4.70	4.86	4.99	5.11	5.21
18	2.97	3.61	4.00	4.28	4.49	4.67	4.83	4.96	5.07	5.17
19	2.96	3.59	3.98	4.25	4.47	4.65	4.79	4.93	5.04	5.14
20	2.95	3.58	3.96	4.23	4.45	4.62	4.77	4.90	5.01	5.11
24	2.92	3.53	3.90	4.17	4.37	4.54	4.68	4.81	4.92	5.01
30	2.89	3.49	3.85	4.10	4.30	4.46	4.60	4.72	4.82	4.92
40	2.86	3.44	3.79	4.04	4.23	4.39	4.52	4.63	4.73	4.82
60	2.83	3.40	3.74	3.98	4.16	4.31	4.44	4.55	4.65	4.73
120	2.80	3.36	3.68	3.92	4.10	4.24	4.36	4.47	4.56	4.64
∞	2.77	3.31	3.63	3.86	4.03	4.17	4.29	4.39	4.47	4.55

TABLE 9 continued, $\alpha = 0.01$

	J (number of groups)									
ν	2	3	4	5	6	7	8	9	10	11
2	14.0	19.0	22.3	24.7	26.6	28.2	29.5	30.7	31.7	32.6
3	8.26	10.6	12.2	13.3	14.2	15.0	15.6	16.2	16.7	17.8
4	6.51	8.12	9.17	9.96	10.6	11.1	11.5	11.9	12.3	12.6
5	5.71	6.98	7.81	8.43	8.92	9.33	9.67	9.98	10.24	10.48
6	5.25	6.34	7.04	7.56	7.98	8.32	8.62	8.87	9.09	9.30
7	4.95	5.92	6.55	7.01	7.38	7.68	7.94	8.17	8.37	8.55
8	4.75	5.64	6.21	6.63	6.96	7.24	7.48	7.69	7.87	8.03
9	4.59	5.43	5.96	6.35	6.66	6.92	7.14	7.33	7.49	7.65
10	4.49	5.28	5.77	6.14	6.43	6.67	6.88	7.06	7.22	7.36
11	4.39	5.15	5.63	5.98	6.25	6.48	6.68	6.85	6.99	7.13
12	4.32	5.05	5.51	5.84	6.11	6.33	6.51	6.67	6.82	6.94
13	4.26	4.97	5.41	5.73	5.99	6.19	6.38	6.53	6.67	6.79
14	4.21	4.89	5.32	5.63	5.88	6.08	6.26	6.41	6.54	6.66
15	4.17	4.84	5.25	5.56	5.80	5.99	6.16	6.31	6.44	6.55
16	4.13	4.79	5.19	5.49	5.72	5.92	6.08	6.22	6.35	6.46
17	4.10	4.74	5.14	5.43	5.66	5.85	6.01	6.15	6.27	6.38
18	4.07	4.70	5.09	5.38	5.60	5.79	5.94	6.08	6.20	6.31
19	4.05	4.67	5.05	5.33	5.55	5.73	5.89	6.02	6.14	6.25
20	4.02	4.64	5.02	5.29	5.51	5.69	5.84	5.97	6.09	6.19
24	3.96	4.55	4.91	5.17	5.37	5.54	5.69	5.81	5.92	6.02
30	3.89	4.45	4.80	5.05	5.24	5.40	5.54	5.65	5.76	5.85
40	3.82	4.37	4.69	4.93	5.10	5.26	5.39	5.49	5.60	5.69
60	3.76	4.28	4.59	4.82	4.99	5.13	5.25	5.36	5.45	5.53
120	3.70	4.20	4.50	4.71	4.87	5.01	5.12	5.21	5.30	5.37
∞	3.64	4.12	4.40	4.60	4.76	4.88	4.99	5.08	5.16	5.23

Note: The values in this table were computed with the IBM SSP subroutines DQH32 and DQG32.

TABLE 10: Studentized Maximum Modulus Distribution

		C (the number of tests being performed)								
ν	α	2	3	4	5	6	7	8	9	10
2	.05	5.57	6.34	6.89	7.31	7.65	7.93	8.17	8.83	8.57
	.01	12.73	14.44	15.65	16.59	17.35	17.99	18.53	19.01	19.43
3	.05	3.96	4.43	4.76	5.02	5.23	5.41	5.56	5.69	5.81
	.01	7.13	7.91	8.48	8.92	9.28	9.58	9.84	10.06	10.27
4	.05	3.38	3.74	4.01	4.20	4.37	4.50	4.62	4.72	4.82
	.01	5.46	5.99	6.36	6.66	6.89	7.09	7.27	7.43	7.57
5	.05	3.09	3.39	3.62	3.79	3.93	4.04	4.14	4.23	4.31
	.01	4.70	5.11	5.39	5.63	5.81	5.97	6.11	6.23	6.33
6	.05	2.92	3.19	3.39	3.54	3.66	3.77	3.86	3.94	4.01
	.01	4.27	4.61	4.85	5.05	5.20	5.33	5.45	5.55	5.64
7	.05	2.80	3.06	3.24	3.38	3.49	3.59	3.67	3.74	3.80
	.01	3.99	4.29	4.51	4.68	4.81	4.93	5.03	5.12	5.19
8	.05	2.72	2.96	3.13	3.26	3.36	3.45	3.53	3.60	3.66
	.01	3.81	4.08	4.27	4.42	4.55	4.65	4.74	4.82	4.89
9	.05	2.66	2.89	3.05	3.17	3.27	3.36	3.43	3.49	3.55
	.01	3.67	3.92	4.10	4.24	4.35	4.45	4.53	4.61	4.67
10	.05	2.61	2.83	2.98	3.10	3.19	3.28	3.35	3.41	3.47
	.01	3.57	3.80	3.97	4.09	4.20	4.29	4.37	4.44	4.50
11	.05	2.57	2.78	2.93	3.05	3.14	3.22	3.29	3.35	3.40
	.01	3.48	3.71	3.87	3.99	4.09	4.17	4.25	4.31	4.37
12	.05	2.54	2.75	2.89	3.01	3.09	3.17	3.24	3.29	3.35
	.01	3.42	3.63	3.78	3.89	3.99	4.08	4.15	4.21	4.26
14	.05	2.49	2.69	2.83	2.94	3.02	3.09	3.16	3.21	3.26
	.01	3.32	3.52	3.66	3.77	3.85	3.93	3.99	4.05	4.10
16	.05	2.46	2.65	2.78	2.89	2.97	3.04	3.09	3.15	3.19
	.01	3.25	3.43	3.57	3.67	3.75	3.82	3.88	3.94	3.99
18	.05	2.43	2.62	2.75	2.85	2.93	2.99	3.05	3.11	3.15
	.01	3.19	3.37	3.49	3.59	3.68	3.74	3.80	3.85	3.89
20	.05	2.41	2.59	2.72	2.82	2.89	2.96	3.02	3.07	3.11
	.01	3.15	3.32	3.45	3.54	3.62	3.68	3.74	3.79	3.83
24	.05	2.38	2.56	2.68	2.77	2.85	2.91	2.97	3.02	3.06
	.01	3.09	3.25	3.37	3.46	3.53	3.59	3.64	3.69	3.73
30	.05	2.35	2.52	2.64	2.73	2.80	2.87	2.92	2.96	3.01
	.01	3.03	3.18	3.29	3.38	3.45	3.50	3.55	3.59	3.64
40	.05	2.32	2.49	2.60	2.69	2.76	2.82	2.87	2.91	2.95
	.01	2.97	3.12	3.22	3.30	3.37	3.42	3.47	3.51	3.55
60	.05	2.29	2.45	2.56	2.65	2.72	2.77	2.82	2.86	2.90
	.01	2.91	3.06	3.15	3.23	3.29	3.34	3.38	3.42	3.46
∞	.05	2.24	2.39	2.49	2.57	2.63	2.68	2.73	2.77	2.79
	.01	2.81	2.93	3.02	3.09	3.14	3.19	3.23	3.26	3.29

TABLE 10 continued

ν	α	11	12	13	14	15	16	17	18	19
2	.05	8.74	8.89	9.03	9.16	9.28	9.39	9.49	9.59	9.68
	.01	19.81	20.15	20.46	20.75	20.99	20.99	20.99	20.99	20.99
3	.05	5.92	6.01	6.10	6.18	6.26	6.33	6.39	6.45	6.51
	.01	10.45	10.61	10.76	10.90	11.03	11.15	11.26	11.37	11.47
4	.05	4.89	4.97	5.04	5.11	5.17	5.22	5.27	5.32	5.37
	.01	7.69	7.80	7.91	8.01	8.09	8.17	8.25	8.32	8.39
5	.05	4.38	4.45	4.51	4.56	4.61	4.66	4.70	4.74	4.78
	.01	6.43	6.52	6.59	6.67	6.74	6.81	6.87	6.93	6.98
6	.05	4.07	4.13	4.18	4.23	4.28	4.32	4.36	4.39	4.43
	.01	5.72	5.79	5.86	5.93	5.99	6.04	6.09	6.14	6.18
7	.05	3.86	3.92	3.96	4.01	4.05	4.09	4.13	4.16	4.19
	.01	5.27	5.33	5.39	5.45	5.50	5.55	5.59	5.64	5.68
8	.05	3.71	3.76	3.81	3.85	3.89	3.93	3.96	3.99	4.02
	.01	4.96	5.02	5.07	5.12	5.17	5.21	5.25	5.29	5.33
9	.05	3.60	3.65	3.69	3.73	3.77	3.80	3.84	3.87	3.89
	.01	4.73	4.79	4.84	4.88	4.92	4.96	5.01	5.04	5.07
10	.05	3.52	3.56	3.60	3.64	3.68	3.71	3.74	3.77	3.79
	.01	4.56	4.61	4.66	4.69	4.74	4.78	4.81	4.84	4.88
11	.05	3.45	3.49	3.53	3.57	3.60	3.63	3.66	3.69	3.72
	.01	4.42	4.47	4.51	4.55	4.59	4.63	4.66	4.69	4.72
12	.05	3.39	3.43	3.47	3.51	3.54	3.57	3.60	3.63	3.65
	.01	4.31	4.36	4.40	4.44	4.48	4.51	4.54	4.57	4.59
14	.05	3.30	3.34	3.38	3.41	3.45	3.48	3.50	3.53	3.55
	.01	4.15	4.19	4.23	4.26	4.29	4.33	4.36	4.39	4.41
16	.05	3.24	3.28	3.31	3.35	3.38	3.40	3.43	3.46	3.48
	.01	4.03	4.07	4.11	4.14	4.17	4.19	4.23	4.25	4.28
18	.05	3.19	3.23	3.26	3.29	3.32	3.35	3.38	3.40	3.42
	.01	3.94	3.98	4.01	4.04	4.07	4.10	4.13	4.15	4.18
20	.05	3.15	3.19	3.22	3.25	3.28	3.31	3.33	3.36	3.38
	.01	3.87	3.91	3.94	3.97	3.99	4.03	4.05	4.07	4.09
24	.05	3.09	3.13	3.16	3.19	3.22	3.25	3.27	3.29	3.31
	.01	3.77	3.80	3.83	3.86	3.89	3.91	3.94	3.96	3.98
30	.05	3.04	3.07	3.11	3.13	3.16	3.18	3.21	3.23	3.25
	.01	3.67	3.70	3.73	3.76	3.78	3.81	3.83	3.85	3.87
40	.05	2.99	3.02	3.05	3.08	3.09	3.12	3.14	3.17	3.18
	.01	3.58	3.61	3.64	3.66	3.68	3.71	3.73	3.75	3.76
60	.05	2.93	2.96	2.99	3.02	3.04	3.06	3.08	3.10	3.12
	.01	3.49	3.51	3.54	3.56	3.59	3.61	3.63	3.64	3.66
∞	.05	2.83	2.86	2.88	2.91	2.93	2.95	2.97	2.98	3.01
	.01	3.32	3.34	3.36	3.38	3.40	3.42	3.44	3.45	3.47

TABLE 10 continued

ν	α	20	21	22	23	24	25	26	27	28
2	.05	9.77	9.85	9.92	10.00	10.07	10.13	10.20	10.26	10.32
	.01	22.11	22.29	22.46	22.63	22.78	22.93	23.08	23.21	23.35
3	.05	6.57	6.62	6.67	6.71	6.76	6.80	6.84	6.88	6.92
	.01	11.56	11.65	11.74	11.82	11.89	11.97	12.07	12.11	12.17
4	.05	5.41	5.45	5.49	5.52	5.56	5.59	5.63	5.66	5.69
	.01	8.45	8.51	8.57	8.63	8.68	8.73	8.78	8.83	8.87
5	.05	4.82	4.85	4.89	4.92	4.95	4.98	5.00	5.03	5.06
	.01	7.03	7.08	7.13	7.17	7.21	7.25	7.29	7.33	7.36
6	.05	4.46	4.49	4.52	4.55	4.58	4.60	4.63	4.65	4.68
	.01	6.23	6.27	6.31	6.34	6.38	6.41	6.45	6.48	6.51
7	.05	4.22	4.25	4.28	4.31	4.33	4.35	4.38	4.39	4.42
	.01	5.72	5.75	5.79	5.82	5.85	5.88	5.91	5.94	5.96
8	.05	4.05	4.08	4.10	4.13	4.15	4.18	4.19	4.22	4.24
	.01	5.36	5.39	5.43	5.45	5.48	5.51	5.54	5.56	5.59
9	.05	3.92	3.95	3.97	3.99	4.02	4.04	4.06	4.08	4.09
	.01	5.10	5.13	5.16	5.19	5.21	5.24	5.26	5.29	5.31
10	.05	3.82	3.85	3.87	3.89	3.91	3.94	3.95	3.97	3.99
	.01	4.91	4.93	4.96	4.99	5.01	5.03	5.06	5.08	5.09
11	.05	3.74	3.77	3.79	3.81	3.83	3.85	3.87	3.89	3.91
	.01	4.75	4.78	4.80	4.83	4.85	4.87	4.89	4.91	4.93
12	.05	3.68	3.70	3.72	3.74	3.76	3.78	3.80	3.82	3.83
	.01	4.62	4.65	4.67	4.69	4.72	4.74	4.76	4.78	4.79
14	.05	3.58	3.59	3.62	3.64	3.66	3.68	3.69	3.71	3.73
	.01	4.44	4.46	4.48	4.50	4.52	4.54	4.56	4.58	4.59
16	.05	3.50	3.52	3.54	3.56	3.58	3.59	3.61	3.63	3.64
	.01	4.29	4.32	4.34	4.36	4.38	4.39	4.42	4.43	4.45
18	.05	3.44	3.46	3.48	3.50	3.52	3.54	3.55	3.57	3.58
	.01	4.19	4.22	4.24	4.26	4.28	4.29	4.31	4.33	4.34
20	.05	3.39	3.42	3.44	3.46	3.47	3.49	3.50	3.52	3.53
	.01	4.12	4.14	4.16	4.17	4.19	4.21	4.22	4.24	4.25
24	.05	3.33	3.35	3.37	3.39	3.40	3.42	3.43	3.45	3.46
	.01	4.00	4.02	4.04	4.05	4.07	4.09	4.10	4.12	4.13
30	.05	3.27	3.29	3.30	3.32	3.33	3.35	3.36	3.37	3.39
	.01	3.89	3.91	3.92	3.94	3.95	3.97	3.98	4.00	4.01
40	.05	3.20	3.22	3.24	3.25	3.27	3.28	3.29	3.31	3.32
	.01	3.78	3.80	3.81	3.83	3.84	3.85	3.87	3.88	3.89
60	.05	3.14	3.16	3.17	3.19	3.20	3.21	3.23	3.24	3.25
	.01	3.68	3.69	3.71	3.72	3.73	3.75	3.76	3.77	3.78
∞	.05	3.02	3.03	3.04	3.06	3.07	3.08	3.09	3.11	3.12
	.01	3.48	3.49	3.50	3.52	3.53	3.54	3.55	3.56	3.57

Note: This table was computed with the FORTRAN program in Wilcox (1986).

TABLE 11: Percentage Points of the Range of J Independent T Variates

α	$\nu = 5$	$\nu = 6$	$\nu = 7$	$\nu = 8$	$\nu = 9$	$\nu = 14$
			J=2 Groups			
.05	3.63	3.45	3.33	3.24	3.18	3.01
.01	5.37	4.96	4.73	4.51	4.38	4.11
			J=3 Groups			
.05	4.49	4.23	4.07	3.95	3.87	3.65
.01	6.32	5.84	5.48	5.23	5.07	4.69
			J=4 Groups			
.05	5.05	4.74	4.54	4.40	4.30	4.03
.01	7.06	6.40	6.01	5.73	5.56	5.05
			J=5 Groups			
.05	5.47	5.12	4.89	4.73	4.61	4.31
.01	7.58	6.76	6.35	6.05	5.87	5.33
			J=6 Groups			
.05	5.82	5.42	5.17	4.99	4.86	4.52
.01	8.00	7.14	6.70	6.39	6.09	5.53
			J=7 Groups			
.05	6.12	5.68	5.40	5.21	5.07	4.70
.01	8.27	7.50	6.92	6.60	6.30	5.72
			J=8 Groups			
.05	6.37	5.90	5.60	5.40	5.25	4.86
.01	8.52	7.73	7.14	6.81	6.49	5.89
			J=9 Groups			
.05	6.60	6.09	5.78	5.56	5.40	4.99
.01	8.92	7.96	7.35	6.95	6.68	6.01
			J=10 Groups			
.05	6.81	6.28	5.94	5.71	5.54	5.10
.01	9.13	8.14	7.51	7.11	6.83	6.10

α	$\nu = 19$	$\nu = 24$	$\nu = 29$	$\nu = 39$	$\nu = 59$
			J=2 Groups		
.05	2.94	2.91	2.89	2.85	2.82
.01	3.98	3.86	3.83	3.78	3.73
			J=3 Groups		
.05	3.55	3.50	3.46	3.42	3.39
.01	5.54	4.43	4.36	4.29	4.23
			J=4 Groups		
.05	3.92	3.85	3.81	3.76	3.72
.01	4.89	4.74	4.71	4.61	4.54
			J=5 Groups		
.05	4.18	4.11	4.06	4.01	3.95
.01	5.12	5.01	4.93	4.82	4.74
			J=6 Groups		
.05	4.38	4.30	4.25	4.19	4.14
.01	5.32	5.20	5.12	4.99	4.91
			J=7 Groups		
.05	4.55	4.46	4.41	4.34	4.28
.01	5.46	5.33	5.25	5.16	5.05
			J=8 Groups		
.05	4.69	4.60	4.54	4.47	4.41
.01	5.62	5.45	5.36	5.28	5.16
			J=9 Groups		
.05	4.81	4.72	4.66	4.58	4.51
.01	5.74	5.56	5.47	5.37	5.28
			J=10 Groups		
.05	4.92	4.82	4.76	4.68	4.61
.01	5.82	5.68	5.59	5.46	5.37

Reprinted, with permission, from Wilcox (1983).

TABLE 12: Critical Values for the One-Sided Wilcoxon Signed Rank Test

n	$\alpha = .005$	$\alpha = .01$	$\alpha = .025$	$\alpha = 0.05$
4	0	0	0	0
5	0	0	0	1
6	0	0	1	3
7	0	1	3	4
8	1	2	4	6
9	2	4	6	9
10	4	6	9	11
11	6	8	11	14
12	8	10	14	18
13	10	13	18	22
14	13	16	22	26
15	16	20	26	31
16	20	24	30	36
17	24	28	35	42
18	28	33	41	48
19	33	38	47	54
20	38	44	53	61
21	44	50	59	68
22	49	56	67	76
23	55	63	74	84
24	62	70	82	92
25	69	77	90	101
26	76	85	111	125
27	84	94	108	120
28	92	102	117	131
29	101	111	127	141
30	110	121	138	152
31	119	131	148	164
32	129	141	160	176
33	139	152	171	188
34	149	163	183	201
35	160	175	196	214
36	172	187	209	228
37	184	199	222	242
38	196	212	236	257
39	208	225	250	272
40	221	239	265	287

Note: Entries were computed as described in Hogg and Craig, 1970, p. 361.

TABLE 13 Critical Values, c_L, for the
One-Sided Wilcoxon–Mann–Whitney test, $\alpha = 0.025$

n_2	$n_1 = 3$	$n_1 = 4$	$n_1 = 5$	$n_1 = 6$	$n_1 = 7$	$n_1 = 8$	$n_1 = 9$	$n_1 = 10$
3	6	6	7	8	8	9	9	10
4	10	11	12	13	14	15	15	16
5	16	17	18	19	21	22	23	24
6	23	24	25	27	28	30	32	33
7	30	32	34	35	37	39	41	43
8	39	41	43	45	47	50	52	54
9	48	50	53	56	58	61	63	66
10	59	61	64	67	70	73	76	79

$\alpha = .005$

n_2	$n_1 = 3$	$n_1 = 4$	$n_1 = 5$	$n_1 = 6$	$n_1 = 7$	$n_1 = 8$	$n_1 = 9$	$n_1 = 10$
3	10	10	10	11	11	12	12	13
4	15	15	16	17	17	18	19	20
5	21	22	23	24	25	26	27	28
6	28	29	30	32	33	35	36	38
7	36	38	39	41	43	44	46	48
8	46	47	49	51	53	55	57	59
9	56	58	60	62	65	67	69	72
10	67	69	72	74	77	80	83	85

Entries were determined with the algorithm in Hogg and Craig, 1970, p. 373.

APPENDIX C

BASIC MATRIX ALGEBRA

CONTENTS

A matrix is a two-dimensional array of numbers or variables having r rows and c columns.

EXAMPLE

$$\begin{pmatrix} 32 & 19 & 67 \\ 11 & 21 & 99 \\ 25 & 56 & 10 \\ 76 & 39 & 43 \end{pmatrix}$$

is a matrix with four rows and three columns.

The matrix is said to be square if $r = c$ (the number of rows equals to the number of columns). A matrix with $r = 1$ ($c = 1$) is called a row (column) vector.

A common notation for a matrix is $\mathbf{X} = (x_{ij})$, meaning that $\mathbf{X}$ is a matrix where x_{ij} is the value in the ith row and jth column. For the matrix just shown, the value in the first row and first column is $x_{11} = 32$ and the value in the third row and second column is $x_{32} = 56$.

EXAMPLE

Within statistics, a commonly encountered square matrix is the correlation matrix. That is, for every individual, we have p measures with r_{ij} being Pearson's correlation between the ith and jth measures. Then the correlation matrix is $\mathbf{R} = (r_{ij})$. If $p = 3$, $r_{12} = .2$, $r_{13} = .4$ and $r_{23} = .3$, then

$$\mathbf{R} = \begin{pmatrix} 1 & .2 & .4 \\ .2 & 1 & .3 \\ .4 & .3 & 1 \end{pmatrix}.$$

(The correlation of a variable with itself is 1.)

The transpose of a matrix is just the matrix obtained when the rth row becomes the rth column. More formally, the transpose of the matrix $\mathbf{X} = (x_{ij})$ is

$$\mathbf{X}' = (x_{ji}),$$

which has c rows and r columns.

EXAMPLE

The transpose of the matrix

$$\mathbf{X} = \begin{pmatrix} 23 & 91 \\ 51 & 29 \\ 63 & 76 \\ 11 & 49 \end{pmatrix}$$

is

$$\mathbf{X}' = \begin{pmatrix} 23 & 51 & 63 & 11 \\ 91 & 29 & 76 & 49 \end{pmatrix}.$$

The matrix $\mathbf{X}$ is said to be symmetric if $\mathbf{X} = \mathbf{X}'$. That is, $x_{ij} = x_{ji}$. The built-in R function `t` computes the transpose of a matrix.

The diagonal of an r by r (square) matrix refers to x_{ii}, $i = 1, \ldots, r$. A diagonal matrix is an r by r matrix where the off-diagonal elements (the x_{ij}, $i \neq j$) are zero. An important special case is the identity matrix which has ones along the diagonal and zeros elsewhere. For example,

$$\begin{pmatrix} 1 & 0 & 0 \\ 0 & 1 & 0 \\ 0 & 0 & 1 \end{pmatrix}$$

is the identity matrix when $r = c = 3$. A common notation for the identity matrix is $\mathbf{I}$. An identity matrix with p rows and columns is created by the R command `diag(nrow=p)`.

Two $r \times c$ matrices, $\mathbf{X}$ and $\mathbf{Y}$, are said to be equal if for every i and j, $x_{ij} = y_{ij}$. That is, every element in $\mathbf{X}$ is equal to the corresponding element in $\mathbf{Y}$.

The sum of two matrices having the same number of rows and columns is

$$z_{ij} = x_{ij} + y_{ij}.$$

When using R, the R command X+Y adds to the two matrices, assuming both X and Y are R variables having matrix mode with the same number of rows and columns.

EXAMPLE

$$\begin{pmatrix} 1 & 3 \\ 4 & -1 \\ 9 & 2 \end{pmatrix} + \begin{pmatrix} 8 & 2 \\ 4 & 9 \\ 1 & 6 \end{pmatrix} = \begin{pmatrix} 9 & 5 \\ 8 & 8 \\ 10 & 8 \end{pmatrix}.$$

Multiplication of a matrix by a scalar, say a, is

$$a\mathbf{X} = (ax_{ij}).$$

That is, every element of the matrix $\mathbf{X}$ is multiplied by a. Using R, if the R variable `a=3`, and `X` is a matrix, the R command `a*X` will multiply every element in `X` by 3.

EXAMPLE

$$2\begin{pmatrix} 8 & 2 \\ 4 & 9 \\ 1 & 6 \end{pmatrix} = \begin{pmatrix} 16 & 4 \\ 8 & 18 \\ 2 & 12 \end{pmatrix}.$$

For an n by p matrix (meaning we have p measures for each of n individuals), the sample mean is

$$\bar{\mathbf{X}} = (\bar{X}_1, \ldots, \bar{X}_p),$$

the vector of the sample means corresponding to the p measures. That is,

$$\bar{X}_j = \frac{1}{n}\sum_{i=1}^{n} X_{ij},\ j = 1, \ldots, p.$$

If $\mathbf{X}$ is an r by c matrix and $\mathbf{Y}$ is a c by t matrix, so the number of columns for $\mathbf{X}$ is the same as the number of rows for $\mathbf{Y}$, the product of $\mathbf{X}$ and $\mathbf{Y}$ is the r by t matrix $\mathbf{Z} = \mathbf{XY}$, where

$$z_{ij} = \sum_{k=1}^{c} x_{ik}y_{kj}.$$

EXAMPLE

$$\begin{pmatrix} 8 & 2 \\ 4 & 9 \\ 1 & 6 \end{pmatrix} \begin{pmatrix} 5 & 3 \\ 2 & 1 \end{pmatrix} = \begin{pmatrix} 44 & 26 \\ 38 & 21 \\ 17 & 9 \end{pmatrix}.$$

When using R, the command

```
X%*%Y
```

will multiply the two matrices `X` and `Y`.

EXAMPLE

Consider a random sample of n observations, $X_1, \ldots, X_n$, and let $\mathbf{J}$ be a row matrix of ones. That is $\mathbf{J} = (1, 1, \ldots, 1)$. Letting $\mathbf{X}$ be a column matrix containing $X_1, \ldots, X_n$, then

$$\sum X_i = \mathbf{JX}.$$

The sample mean is

$$\bar{X} = \frac{1}{n}\mathbf{JX}.$$

The sum of the squared observations is

$$\sum X_i^2 = \mathbf{X}'\mathbf{X}.$$

Let $\mathbf{X}$ be an n by p matrix of p measures taken on n individuals. Then $\mathbf{X}_i$ is the ith row (vector) in the matrix $\mathbf{X}$ and $(\mathbf{X}_i - \bar{\mathbf{X}})'$ is a p by 1 matrix consisting of the ith row of $\mathbf{X}$ minus the sample mean. Moreover, $(\mathbf{X}_i - \bar{\mathbf{X}})'(\mathbf{X}_i - \bar{\mathbf{X}})$ is a p by p matrix. The (sample) covariance matrix is

$$\mathbf{S} = \frac{1}{n-1}\sum_{i=1}^{n}(\mathbf{X}_i - \bar{\mathbf{X}})'(\mathbf{X}_i - \bar{\mathbf{X}}).$$

That is, $\mathbf{S} = (s_{jk})$, where s_{jk} is the covariance between the jth and kth measures. When $j = k$, s_{jk} is the sample variance corresponding to the jth variable under study.

For any square matrix $\mathbf{X}$, the matrix $\mathbf{X}^{-1}$ is said to be the inverse of $\mathbf{X}$ if

$$\mathbf{X}\mathbf{X}^{-1} = \mathbf{I},$$

the identity matrix. If an inverse exists, $\mathbf{X}$ is said to be *nonsingular*; otherwise it is *singular*. The inverse of a nonsingular matrix can be computed with the R built-in function

`solve(m)`,

where `m` is any R variable having matrix mode with the number of rows equal to the number of columns.

EXAMPLE

Consider the matrix

$$\begin{pmatrix} 5 & 3 \\ 2 & 1 \end{pmatrix}.$$

Storing it in the R variable m, the command `solve(m)` returns

$$\begin{pmatrix} -1 & 3 \\ 2 & -5 \end{pmatrix}.$$

It is left as an exercise to verify that multiplying these two matrices together yields $\mathbf{I}$, the identity matrix.

EXAMPLE

It can be shown that the matrix

$$\begin{pmatrix} 2 & 5 \\ 2 & 5 \end{pmatrix}$$

does not have an inverse. The R function `solve`, applied to this matrix, reports that the matrix appears to be singular.

Consider any r by c matrix $\mathbf{X}$, and let k indicate any square submatrix. That is, consider the matrix consisting of any k rows and any k columns taken from $\mathbf{X}$. The *rank* of $\mathbf{X}$ is equal to the largest k for which a k by k submatrix is nonsingular.

The notation

$$\text{diag}\{x_1, \ldots, x_n\}$$

refers to a diagonal matrix with the values $x_1, \ldots, x_n$ along the diagonal. For example,

$$\text{diag}\{4, 5, 2\} = \begin{pmatrix} 4 & 0 & 0 \\ 0 & 5 & 0 \\ 0 & 0 & 2 \end{pmatrix}.$$

The R command `diag(X)` returns the diagonal values stored in the R variable `X`. If $r < c$, the r rows and the first r columns of the matrix `X` are used, with the remaining columns ignored. And if $c < r$, the c columns and the first r rows of the matrix `X` are used, with the remaining rows ignored.

The trace of a square matrix is just the sum of the diagonal elements and is often denoted by tr. For example, if

$$\mathbf{A} = \begin{pmatrix} 5 & 3 \\ 2 & 1 \end{pmatrix},$$

then

$$\text{tr}(\mathbf{A}) = 5 + 1 = 6.$$

The trace of a matrix can be computed with the R command

```
sum(diag(X)).
```

A block diagonal matrix refers to a matrix where the diagonal elements are themselves matrices.

EXAMPLE

If

$$\mathbf{V}_1 = \begin{pmatrix} 9 & 2 \\ 4 & 15 \end{pmatrix}$$

and

$$\mathbf{V}_2 = \begin{pmatrix} 11 & 32 \\ 14 & 29 \end{pmatrix},$$

then

$$\text{diag}(\mathbf{V}_1, \mathbf{V}_2) = \begin{pmatrix} 9 & 2 & 0 & 0 \\ 4 & 15 & 0 & 0 \\ 0 & 0 & 11 & 32 \\ 0 & 0 & 14 & 29 \end{pmatrix}.$$

Let $\mathbf{A}$ be an $m_1 \times n_1$ matrix, and let $\mathbf{B}$ be an $m_2 \times n_2$ matrix. The (right) Kronecker product of $\mathbf{A}$ and $\mathbf{B}$ is the $m_1 m_2 \times n_1 n_2$ matrix

$$\mathbf{A} \otimes \mathbf{B} = \begin{pmatrix} a_{11}\mathbf{B} & a_{12}\mathbf{B} & \ldots a_{1n_1}\mathbf{B} & \\ a_{21}\mathbf{B} & a_{22}\mathbf{B} & \ldots a_{2n_1}\mathbf{B} & \\ \vdots & \vdots & \vdots & \vdots \\ a_{m_1 1}\mathbf{B} & a_{m_1 2}\mathbf{B} & \ldots a_{m_1 n_1}\mathbf{B} & \end{pmatrix}.$$

Associated with every square matrix is a number called its determinant. The determinant of a 2-by-2 matrix is easily computed. For the matrix

$$\begin{pmatrix} a & b \\ c & d \end{pmatrix}$$

the determinant is $ad - cb$. For the more general case of a p-by-p matrix, algorithms for computing the determinant are available, but the details are not important here. (The R function `det` can be used.) If the determinant of a square matrix is 0, it has no inverse. That is, it is singular. If the determinant differs from 0, it has an inverse.

Eigenvalues (also called characteristic roots or characteristic values) and eigenvectors of a square matrix $\mathbf{X}$ are defined as follows. Let $\mathbf{Z}$ be a column vector having length p that differs from $\mathbf{0}$. If there is a choice for $\mathbf{Z}$ and a scalar λ that satisfies

$$\mathbf{XZ} = \lambda\mathbf{Z},$$

then $\mathbf{Z}$ is called an eigenvector of $\mathbf{X}$ and λ is called an eigenvalue of $\mathbf{X}$. Eigenvalues and eigenvectors of a matrix $\mathbf{X}$ can be computed with the R function `eigen`.

A matrix $\mathbf{X}^-$ is said to be a generalized inverse of the matrix $\mathbf{X}$ if

1. $\mathbf{XX}^-$ is symmetric.
2. $\mathbf{X}^-\mathbf{X}$ is symmetric.
3. $\mathbf{XX}^-\mathbf{X} = \mathbf{X}$.
4. $\mathbf{X}^-\mathbf{XX}^- = \mathbf{X}^-$.

The built-in R function `ginv` computes the generalized inverse of a matrix. (Computational details can be found, for example, in Graybill, 1983.)

REFERENCES

Acion, L., Peterson, J. J., Temple, S., & Arndt, S. (2006). Probabilistic index: An intuitive non-parametric approach to measuring the size of treatment effects. *Statistics in Medicine, 25*, 591–602.

Afshartous, D., & Preston, R. A. (2010). Confidence intervals for dependent data: Equating non-overlap with statistical significance. *Computational Statistics and Data Analysis, 54*, 2296–2305.

Agresti, A. (2002). *Categorical Data Analysis*, 2nd ed. Hoboken, NJ: Wiley.

Agresti, A. (2007). *An Introduction to Categorical Data Analysis*. New York: Wiley.

Agresti, A. (2010). *Analysis of Ordinal Categorical Data*, 2nd ed. New York: Wiley.

Agresti, A., & Coull, B. A. (1998). Approximate is better than "exact" for interval estimation of binomial proportions. *American Statistician 52*, 119–126.

Agresti, A., & Pendergast, J. (1986). Comparing mean ranks for repeated measures data. *Communications in Statistics—Theory and Methods, 15*, 1417–1433.

Ahmad, S., Ramli, N. M., & Midi, H. (2010). Robust estimators in logistic regression: A comparative simulation study. *Journal of Modern Applied Statistical Methods 9*: Issue 2, Article 18. http://digitalcommons.wayne.edu/jmasm/vol9/iss2/18.

Akritas, M. G. (1990). The rank transform method in some two-factor designs. *Journal of the American Statistical Association, 85*, 73–78.

Akritas, M. G. (1991), Limitations of the rank transform method procedure: A study of repeated measures designs, Part I. *Journal of the American Statistical Association, 86*, 457–460.

Akritas, M. G., & Arnold, S. F. (1994). Fully nonparametric hypotheses for factorial designs I: Multivariate repeated measures designs. *Journal of the American Statistical Association, 89*, 336–343.

Akritas, M. G., Arnold, S. F., & Brunner, E. (1997). Nonparametric hypotheses and rank statistics for unbalanced factorial designs. *Journal of the American Statistical Association, 92*, 258–265.

Akritas, M. G. Murphy, S. A., & LaValley, M. P. (1995). The Theil-Sen estimator with doubly censored data and applications to astronomy. *Journal of the American Statistical Association, 90*, 170–177.

Alexander, R. A., & Govern, D. M. (1994). A new and simpler approximation for ANOVA under variance heterogeneity. *Journal of Educational Statistics, 19*, 91–101.

Algina, J., Keselman, H. J., & Penfield, R. D. (2005). An alternative to Cohen's standardized mean difference effect size: A robust parameter and confidence interval in the two independent groups case. *Psychological Methods, 10*, 317–328.

Algina, J., Oshima, T. C., & Lin, W.-Y. (1994). Type I error rates for Welch's test and James's second-order test under nonnormality and inequality of variance when there are two groups. *Journal of Educational Statistics, 19*, 275–291.

Allison, P. D. (2001). *Missing Data*. Thousand Oaks, CA: Sage.

Andersen, E. B. (1997). *Introduction to the Statistical Analysis of Categorical Data*. New York: Springer.

Anderson, T. W. (2003). *An Introduction to Multivariate Statistical Analysis.* New York: Wiley.

Andrews, D. F., Bickel, P. J., Hampel, F. R., Huber, P. J., Rogers, W. H., & Tukey, J. W. (1972). *Robust Estimates of Location: Survey and Advances.* Princeton University Press, Princeton, NJ.

Asiribo, O., & Gurland, J. (1989). Some simple approximate solutions to the Behrens–Fisher problem. *Communications in Statistics–Theory and Methods, 18,* 1201–1216.

Atkinson, A. C., Riani, M., & Torti, F. (2016). Robust methods for heteroskedastic regression. *Computational Statistics and Data Analysis,104,* 209–222.

Bahadur, R., & Savage, L. (1956). The nonexistence of certain statistical procedures in nonparametric problems. *Annals of Statistics, 25,* 1115–1122.

Bai, Z.-D., Chen, X. R. Miao, B. Q., & Rao, C. R. (1990). Asymptotic theory of least distance estimate in multivariate linear model. *Statistics, 21,* 503–519.

Bakker, M., & Wicherts, J. M. (2014). Outlier removal, sum scores, and the inflation of the type I error rate in t tests. *Psychological Methods, 19,* 409–427

Banik, S., & Kibria, B. M. G. (2010). Comparison of some parametric and nonparametric Type one sample confidence intervals for estimating the mean of a positively skewed distribution. *Communications in Statistics–Simulation and Computation, 39,* 361–380.

Barnett, V., & Lewis, T. (1994). *Outliers in Statistical Data.* New York: Wiley.

Baron, R. M., & Kenny, D. A. (1986). The moderator-mediator variable distinction in social psychological research: Conceptual, strategic, and statistical considerations. *Journal of Personality and Social Psychology 51,* 1173–1182.

Barrett, J. P. (1974). The coefficient of determination—Some limitations. *Annals of Statistics, 28,* 19–20.

Barrett, B. E., & Ling, R. F. (1993). General classes of influence measures for multivariate regression. *Journal of the American Statistical Association, 87,* 184–191.

Barry, D. (1993). Testing for additivity of a regression function. *Annals of Statistics, 21,* 235–254.

Basu, S., & DasGupta, A. (1995). Robustness of standard confidence intervals for location parameters under departure from normality. *Annals of Statistics, 23,* 1433–1442.

Bathke, A. C., Solomon, W. H., & Madden, L. V. (2008). How to compare small multivariate samples using nonparametric tests. *Computational Statistics and Data Analysis, 52,* 4951–4965.

Beal, S. L. (1987). Asymptotic confidence intervals for the difference between two binomial parameters for use with small samples. *Biometrics, 43,* 941–950.

Beasley, T. M. (2000). Nonparametric tests for analyzing interactions among intra-block ranks in multiple group repeated measures designs. *Journal of Educational and Behavioral Statistics, 25,* 20–59.

Beasley, W. H., DeShea, L., Toothaker, L. E., Mendoza, J. L., Bard, D. E., & Rodgers, J. (2007). Bootstrapping to test for nonzero population correlation coefficients using univariate sampling. *Psychological Methods, 12,* 414–433. doi:10.1037/1082-989X.12.4.414

Beasley, W. H. & Zumbo, B. D. (2003). Comparison of aligned Friedman rank and parametric methods for testing interactions in split-plot designs. *Computational Statistics & Data Analysis, 42,* 569–593.

Bellman, R. E. (1961). *Adaptive Control Processes.* Princeton, NJ: Princeton University Press.

Belloni, A. & Chernozhukov, V. (2013). Least squares after model selection in high-dimensional sparse models. *Bernoulli, 19,* 521–547.

Belsley, D. A., Kuh, E., & Welsch, R. E. (1980). *Regression Diagnostics: Identifying Influential Data and Sources of Collinearity.* New York: Wiley.

Ben, M. G., Martínez, E., & Yohai, V. J. (2006). Robust estimation for the multivariate linear model based on a τ-scale. *Journal of Multivariate Analysis, 90*, 1600–1622.

Benjamini, Y., & Hochberg, Y. (1995). Controlling the false discovery rate: A practical and powerful approach to multiple testing. *Journal of the Royal Statistical Society, B, 57*, 289–300.

Benjamini, Y., Hochberg, Y., & Stark, P. B. (1998). Confidence intervals with more power to determine the sign: Two ends constrain the means. *Journal of the American Statistical Association, 93*, 309–317.

Benjamini, Y., & Hochberg, Y. (2000). On the adaptive control of the false discovery rate in multiple testing with independent statistics. *Journal of Educational and Behavioral Statistics, 25*, 60–83.

Benjamini Y., & Yekutieli D. (2001) The control of the false discovery rate in multiple testing under dependency. *Annals of Statistics, 29*, 1165–1188.

Berger, R. L. (1996). More powerful tests from confidence interval p values. *American Statistician, 50*, 314–318.

Berk, K. N., & Booth, D. E. (1995). Seeing a curve in multiple regression. *Technometrics, 37*, 385–398.

Berkson, J. (1958). Smoking and lung cancer: Some observations on two recent reports. *Journal of the American Statistical Association, 53*, 28–38.

Bernhardson, C. (1975). Type I error rates when multiple comparison procedures follow a significant F test of ANOVA. *Biometrics, 31*, 719–724.

Bernholdt, T., & Fischer, P. (2004). The complexity of computing the MCD-estimator. *Theoretical Computer Science, 326*, 383–393.

Berry, K. J., & Mielke, P. W. (2000). A Monte Carlo investigation of the Fisher z-transformation for normal and nonnormal distributions. *Psychological Reports 87*, 1101-1114.

Bhapkar V.P. (1966). A note on the equivalence of two test criteria for hypotheses in categorical data. *Journal of the American Statistical Association, 6*, 228–235.

Bickel, P. J., & Lehmann, E. L. (1975). Descriptive statistics for nonparametric models II. Location. *Annals of Statistics, 3*, 1045–1069.

Bishara, A., & Hittner, J. A. (2012). Testing the significance of a correlation With non-normal data: Comparison of Pearson, Spearman, transformation, and resampling approaches. *Psychological Methods, 17*, 399–417.

Bishop, T., & Dudewicz, E. J. (1978). Exact analysis of variance with unequal variances: Test procedures and tables. *Technometrics, 20*, 419–430.

Bjerve, S., & Doksum, K. (1993). Correlation curves: Measures of association as functions of covariate values. *Annals of Statistics, 21*, 890–902.

Blair, R. C., Sawilowsky, S. S., & Higgins, J. J. (1987). Limitations of the rank transform statistic in tests for interactions. *Communications in Statistics–Simulation and Computation, 16*, 1133–1145.

Bloch, D. A., & Moses, L. E. (1988). Nonoptimally weighted least squares. *American Statistician, 42*, 50–53.

Blyth, C. R. (1986). Approximate binomial confidence limits. *Journal of the American Statistical Association, 81*, 843–855.

Boik, R. J. (1981). A priori tests in repeated measures designs: Effects of nonsphericity. *Psychometrika, 46*, 241–255.

Boik, R. J. (1987). The Fisher-Pitman permutation test: A non-robust alternative to the normal theory F test when variances are heterogeneous. *British Journal of Mathematical and Statistical Psychology, 40*, 26–42.

Boos, D. B., & Hughes-Oliver, J. M. (2000). How large does n have to be for Z and t intervals? *American Statistician, 54*, 121–128.

Boos, D. B., & Zhang, J. (2000). Monte Carlo evaluation of resampling-based hypothesis tests. *Journal of the American Statistical Association, 95*, 486–492.

Booth, J. G., & Sarkar, S. (1998). Monte Carlo approximation of bootstrap variances. *American Statistician, 52*, 354–357.

Bourel, M., Fraiman, R., & Ghattas, B. (2014). Random average shifted histograms. *Computational Statistics and Data Analysis, 79*, 149–164.

Bowman, A.W., & Young, S.G. (1996). Graphical comparison of nonparametric curves. *Applied Statistics 45*, 83Ŭ–98.

Box, G. E. P. (1953). Non-normality and tests on variances. *Biometrika, 40*, 318–335.

Box, G. E. P. (1954). Some theorems on quadratic forms applied in the study of analysis of variance problems, I. Effect of inequality of variance in the one-way model. *Annals of Mathematical Statistics, 25*, 290–302.

Bradley, J. V. (1978) Robustness? *British Journal of Mathematical and Statistical Psychology, 31*, 144–152.

Breiman, L. (1995). Better subset regression using the nonnegative garrote. *Technomterics, 37*, 373–384.

Breiman, L., & Friedman, J. H. (1985). Estimating optimal transformations for multiple regression and correlation (with discussion). *Journal of the American Statistical Association, 80*, 580–619.

Breiman, L., & Friedman, J. H. (1997). Predicting multivariate responses in multiple linear regression. *Journal of the Royal Statistical Society Series B, 59*, 3–54.

Brown, C., & Mosteller, F. (1991). Components of variance. In D. Hoaglin, F. Mosteller & J. Tukey (Eds.) *Fundamentals of Exploratory Analysis of Variance*, pp. 193–251. New York: Wiley.

Brown, L. D., Cai, T. T., & DasGupta, A. (2002). Confidence intervals for a binomial proportion and asymptotic expansions. *Annals of Statistics 30*, 160–201.

Brown, M. B., & Forsythe, A. (1974a). The small sample behavior of some statistics which test the equality of several means. *Technometrics, 16*, 129–132.

Brown, M. B., & Forsythe, A. (1974b). Robust tests for the equality of variances. *Journal of the American Statistical Association, 69*, 364–367.

Browne, R. H. (2010). The t-test, p value and its relationship to the effect size and $P(X > Y)$. *American Statistician, 64*, 30–33. (Correction, 64, 195).

Brunner, E., & Dette, H. (1992). Rank procedures for the two-factor mixed model. *Journal of the American Statistical Association, 87*, 884–888.

Brunner, E., Dette, H., & Munk, A. (1997). Box-type approximations in non-parametric factorial designs. *Journal of the American Statistical Association, 92*, 1494–1502.

Brunner, E., & Denker, M. (1994). Rank statistics under dependent observations and applications to factorial designs. *Journal of Statistical Planning and Inference, 42*, 353–378.

Brunner, E., Munzel, U., & Puri, M. L. (1999). Rank-score tests in factorial designs with repeated measures. *Journal of Multivariate Analysis, 70*, 286–317.

Brunner, E., & Munzel, U. (2000). The nonparametric Behrens-Fisher problem: Asymptotic theory and small-sample approximation. *Biometrical Journal, 42*, 17–25.

Brunner, E., Domhof, S. & Langer, F. (2002). *Nonparametric Analysis of Longitudinal Data in Factorial Experiments*. New York: Wiley.

Brunner, E., Konietschke, F., Pauly, M. & Puri, M.L. (2016). Rank-Based Procedures in Factorial Designs: Hypotheses about Nonparametric Treatment Effects. arXiv:1606.03973.

Carling, K. (2000). Resistant outlier rules and the non-Gaussian case. *Computational Statistics and Data Analysis, 33*, 249–258.

Carroll, R. J., & Ruppert, D. (1982). Robust estimation in heteroscdastic linear models. *Annals of Statistics, 10*, 429–441.

Carroll, R. J., & Ruppert, D. (1988). *Transformation and Weighting in Regression.* New York: Chapman and Hall.

Carlson, M., Wilcox, R., Chou, C.-P., Chang, M., Yang, F., Blanchard, J., Marterella, A., Kuo, A., & Clark, F. (2009). Psychometric Properties of Reverse-Scored Items on the CES-D in a Sample of Ethnically Diverse Older Adults Unpublished Technical Report, Occupational Science and Occupational Therapy, University of Southern California.

Cerioli, A. (2010). Multivariate outlier detection with high-breakdown estimators. *Journal of the American Statistical Association, 105*, 147–156.

Chakraborty, B. (2001). On affine equivariant multivariate quantiles. *Annals of the Institute of Statistical Mathematics, 53*, 380–403.

Chang, C. (2013). *R Graphics Cookbook.* Sebastopol, CA: O'Reilly Media.

Chatterjee, S., & Hadi, A. S. (1988). *Sensitivity Analysis in Linear Regression Analysis.* New York: Wiley.

Chaudhuri, P. (1996). On a geometric notion of quantiles for multivariate data. *Journal of the American Association, 91*, 862–872.

Chen, L. (1995). Testing the mean of skewed distributions. *Journal of the American Statistical Association, 90*, 767–772.

Chen, S., & Chen, H. J. (1998). Single-stage analysis of variance under heteroscedasticity. *Communications in Statistics–Simulation and Computation, 27*, 641–666.

Chen, X & Luo, X. (2004). Some modifications on the application of the exact Wilcoxon–Mann–Whitney test. *Simulation and Computation, 33*, 1007–1020.

Chernick, M. R. (1999). *Bootstrap Methods: A Pracitioner's Guide.* New York: Wiley.

Choi, K., & Marden, J. (1997). An approach to multivariate rank tests in multivariate analysis of variance. *Journal of the American Statistical Association, 92*, 1581–1590.

Chow, G. C. (1960). Tests of equality between sets of coefficients in two linear regressions. *Econometrika, 28*, 591–606.

Chung, E., & Romano J. P. (2013). Exact and asymptotically robust permutation tests. *Annals of Statistics, 41*, 484–507.

Clark, F., Jackson, J., Carlson, M., Chou, C.-P., Cherry, B. J., Jordan-Marsh M., Knight, B. G., Mandel, D., Blanchard, J., Granger, D. A., Wilcox, R. R., Lai, M. Y., White, B., Hay, J., Lam, C., Marterella, A., & Azen, S. P. (2012). Effectiveness of a lifestyle intervention in promoting the well-being of independently living older people: Results of the Well Elderly 2 Randomised Controlled Trial. *Journal of Epidemiology and Community Health, 66*, 782–790. doi:10.1136/jech.2009.099754

Clark, S. (1999). *Towards the Edge of the Universe: A Review of Modern Cosmology.* New York: Springer.

Clemons, T. E., & Bradley Jr., E. L. (2000). A nonparametric measure of the overlapping coefficient. *Computational Statistics and Data Analysis, 34* 51–61.

Cleveland, W. S. (1979). Robust locally-weighted regression and smoothing scatterplots. *Journal of the American Statistical Association, 74*, 829–836.

Cleveland, W. S. (1985). *The Elements of Graphing Data.* Summit, NJ: Hobart Press.

Cliff, N. (1996). *Ordinal Methods for Behavioral Data Analysis.* Mahwah, NJ: Erlbaum.

Clopper, C. J., & Pearson, E. S. (1934). The use of confidence or fiducial limits illustrated in the case of the binomial. *Biometrika, 26*, 404–413.

Coakley, C. W., & Hettmansperger, T. P. (1993). A bounded influence, high breakdown, efficient regression estimator. *Journal of the American Statistical Association, 88*, 872–880.

Cochran, W. G., & Cox, G. M. (1950). *Experimental Design.* New York: Wiley.

Coe, P. R., & Tamhane, A. C. (1993). Small sample confidence intervals for the difference, ratio, and odds ratio of two success probabilities. *Communications in Statistics—Simulation and Computation, 22*, 925–938.

Cohen, J. (1988). *Statistical Power Analysis for the Behavioral Sciences*, 2nd ed. New York: Academic Press.

Cohen, J. (1994). The earth is round ($p < .05$). *American Psychologist, 49*, 997–1003.

Cohen, M., Dalal, S. R., & Tukey, J. W. (1993). Robust, smoothly heterogeneous variance regression. *Applied Statistics, 42*, 339–353.

Cohen, J., Cohen, P., West, S., & Aiken, L. S. (2003). *Applied Multiple Regression/Correlation Analysis for the Behavioral Sciences*, 3rd ed. Mahwah, NJ: Erlbaum.

Cojbasic, V. & Tomovic, A. (2007). Nonparametric confidence intervals for population variance of one sample and the difference of variances of two samples.

Coleman, J. S. (1964). *Introduction to Mathmematical Sociology.* New York: Free Press.

Comrey, A. L. (1985). A method for removing outliers to improve factor analytic results. *Multivariate Behavioral Research, 20*, 273–281.

Conerly, M. D., & Mansfield, E. R. (1988) An approximate test for comparing heteroscedastic regression models. *Journal of the American Statistical Association, 83*, 811–817.

Conover, W., Johnson, M., & Johnson, M. (1981). A comparative study of tests for homogeneity of variances, with applications to the outer continental shelf bidding data. *Technometrics, 23*, 351–361.

Cook, R. D. (1993). Exploring partial residuals plots. *Technometrics, 35*, 351–362.

Cook, R. D., & Weisberg, S. (1992). *Residuals and Influence in Regression.* New York: Chapman and Hall.

Cook, C., & Setodji, M. C. (2003). A model-free test for reduced rank in multivariate regression. *Journal of the American Statistical Association, 98*, 340–351.

Copas, J. B. (1983). Plotting p against x. *Applied Statistics, 32*, 25–31.

Crawley, M. J. (2007). *The R Book.* New York: Wiley.

Cressie, N. A. C., & Whitford, H. J. (1986). How to use the two sample t-test. *Biometrical Journal, 28*, 131–148.

Cribari-Neto, F. (2004). Asymptotic inference under heteroscedasticity of unknown form. *Computational Statistics and Data Analysis, 45*, 215–Ű233

Cribbie, R. A., Fiksenbaum, L., Keselman, H. J., & Wilcox, R. R. (2010). Effects of Nonnormality on Test Statistics for One-Way Independent Groups. Unpublished technical report, Department of Psychology, York University

Cronbach, L. J. (1987). Statistical tests for moderator variables: Flaws in analyses recently proposed. *Psychological Bulletin, 102*, 414–417.

Cronbach, L. J., Gleser, G. C., Nanda, H., & Rajaratnam, N. (1972). *The Dependability of Behavioral Measurements.* New York: Wiley.

Croux, C., & Haesbroeck, G. (2000). Principal component analysis based on robust estimators of the covariance or correlation matrix: Influence functions and efficiencies. *Biometrika, 87*, 603–618.

Croux, C., Rousseeuw, P. J., & Hössjer, O. (1994). Generalized S-estimators. *Journal of the American Statistical Association, 89*, 1271–1281.

Croux, C., Van Aelst, S., & Dehon, C. (2003). Bounded influence regression using high breakdown scatter matrices. *Annals of the Institute of Statistical Mathematics, 55*, 265–285.

Crowder, M. J., & Hand, D. J. (1990). *Analysis of Repeated Measures.* London: Chapman.

Dana, E. (1990). Salience of the self and salience of standards: Attempts to match self to standard. Unpublished Ph.D. dissertation, University of Southern California.

Daniels, M. J. and Hogan, J. W. (2008). *Missing Data in Longitudinal Studies: Strategies for Bayesian Modeling and Sensitivity Analysis.* Boca Raton, FL: Chapman & Hall/CRC.

Dantzig, G. (1940). On the non-existence of tests of "Student's" hypothesis having power functions independent of σ. *Annals of Mathematical Statistics, 11*, 186.

Davidson, R., & MacKinnon, J. G. (2000). Bootstrap tests: How many bootstraps?

Econometric Reviews, 19, 55–68.

Davies, L., & Gather, U. (1993). The indentification of multiple outliers (with discussion). *Journal of the American Statistical Association, 88*, 782–792.

Davison, A. C., & Hinkley, D. V. (1997). *Bootstrap Methods and Their Application.* Cambridge: Cambridge University Press.

Dawson, M. E., Schell, A. M., Hazlett, E. A., Nuechterlein, K. H., & Filion, D. L. (2000). On the clinical and cognitive meaning of impaired sensorimotor gating in schizophrenia. *Psychiatry Research, 96*, 187–197.

Derksen, S., & Keselman, H. J. (1992). Backward, forward and stepwise automated subset selection algorithms: Frequency of obtaining authentic and noise variables. *British Journal of Mathematical and Statistical Psychology, 45*, 265–282.

Dette, H. (1999). A consistent test for the functional form of a regression based on a difference of variances estimator. *Annals of Statistics, 27*, 1012–1040.

Devlin, S. J., Gnanadesikan, R., & Kettenring, J. R. (1981). Robust estimation of dispersion matrices and principal components. *Journal of the American Statistical Association, 76*, 354–362.

De Winter, J., Gosling, S. D., & Potter, J. (2016). Comparing the Pearson and Spearman Correlation Coefficients Across Distributions and Sample Sizes: A Tutorial Using Simulations and Empirical Data. *Psychological Methods, 21.* http://dx.doi.org/10.1037/met0000079.

Dielman, T., Lowry, C., & Pfaffenberger, R. (1994). A comparison of quantile estimators. *Communications in Statistics–Simulation and Computation, 23*, 355–371.

Dietz, E. J. (1989). Teaching regression in a nonparmatric statistics course. *American Statistician, 43*, 35–40.

Diggle, P. J.,Heagerty, P., Liang, K.-Y., & Zeger, S. L. (1994). *Analysis of Longitudinal Data.* Oxford: Oxford University Press.

Doksum, K. A. (1974). Empirical probability plots and statistical inference for nonlinear models in the two-sample case. *Annals of Statistics, 2*, 267–277.

Doksum, K. A. (1977). Some graphical methods in statistics. A review and some extensions. *Statistica Neerlandica, 31*, 53–68.

Doksum, K. A., Blyth, S., Bradlow, E., Meng, X., & Zhao, H. (1994). Correlation curves as local measures of variance explained by regression. *Journal of the American Statistical Association, 89*, 571–582.

Doksum, K. A., & Samarov, A. (1995). Nonparametric estimation of global functionals and a measure of the explanatory power of covariates in regression. *Annals of Statistics, 23*, 1443–1473.

Doksum, K. A., & Sievers, G. L. (1976). Plotting with confidence: graphical comparisons of two populations. *Biometrika, 63*, 421–434.

Doksum, K. A., & Wong, C.-W. (1983). Statistical tests based on transformed data. *Journal of the American Statistical Association, 78*, 411–417.

Donner, A., & Wells, G. (1986). A comparison of confidence interval methods for the intraclass correlation coefficient. *Biometrics, 42*, 401–412.

Donoho, D. L. (1988). One-sided inference about functionals of a density. *Annals of Statistics, 16*, 1390–1420.

Drouet, D., & Kotz, S. (2001). *Correlation and Dependence.* New Jersey: World Scientific Press.

Duncan, G. T., & Layard, M. W. (1973). A Monte-Carlo study of asymptotically robust tests for correlation. *Biometrika, 60*, 551–558.

Dunnett, C. W. (1980a). Pairwise multiple comparisons in the unequal variance case. *Journal of the American Statistical Association, 75*, 796–800.

Dunnett, C. W. (1980b). Pairwise multiple comparisons in the homogeneous variance, unequal sample size case. *Journal of the American Statistical Association, 75*, 796–800.

Dunnett, C. W., & Tamhane, A. C. (1992). A step-up multiple test procedure. *Journal of the American Statistical Association, 87*, 162–170.

Efromovich, S. (1999). *Nonparametric Curve Estimation: Methods, Theory and Applications.* New York: Springer-Verlag.

Efron, B., Hastie, T., Johnstone, I., & Tibshirani, R. (2004). Least angle regression (with discussion and rejoinder). *Annals of Statistics, 32*, 407–499.

Efron, B., & Hastie, T. (2016). Computer Age Statistical Inference: Algorithms, Evidence, and Data Science. New York: Cambridge University Press.

Efron, B., & Tibshirani, R. J. (1993). *An Introduction to the Bootstrap.* New York: Chapman and Hall.

Efron, B., & Tibshirani, R. J. (1997). Improvements on cross-validation: The .632+ bootstrap method. *Journal of the American Statistical Association, 92*, 548–560.

Elashoff, J. D., & Snow, R. E. (1970). A case study in statistical inference: Reconsideration of the Rosenthal-Jacobson data on teacher expectancy. Technical report no. 15, School of Education, Stanford University.

Emerson, J. D., & Hoaglin, D. C. (1983). Stem-and-leaf displays. In D. C. Hoaglin, F. Mosteller, & J. W. Tukey (Eds.) *Understanding Robust and Exploratory Data Analysis*, pp. 7–32. New York: Wiley.

Engelen, S. Hubert, M., & Vanden Branden, K . (2005). A comparison of three procedures for robust PCA in high dimensions. *Australian Journal of Statistics, 2*, 117–126.

Erceg-Hurn, D. M. & Steed, L. G. (2011), Does exposure to cigarette health warnings elicit psychological reactance in smokers? *Journal of Applied Social Psychology, 41*, 219-237. doi: 10.1111/j.1559-1816.2010.00710.x.

Eubank, R. L. (1999). *Nonparametric Regression and Spline Smoothing.* New York: Marcel Dekker.

Ezekiel, M. (1924). A method for handling curvilinear correlation for any number of variables. *Journal of the American Statistical Association, 19*, 431–453.

Fabian, V. (1991). On the problem of interactions in the analysis of variance. *Journal of the American Statistical Association, 86*, 362–367.

Fairley, D. (1986). Cherry trees with cones? *American Statistician, 40*, 138–139.

Fan, J. (1992). Design-adaptive nonparametric regression. *Journal of the American Statistical Association, 87*, 998–1004.

Fan, J. (1993). Local linear smoothers and their minimax efficiencies. *Annals of Statistics, 21*, 196–Ű216.

Fan, J., & Gijbels, I. (1996). *Local Polynomial Modeling and Its Applications.* Boca Raton, FL: CRC Press.

Fears, T. R., Benichou, J., & Gail, M. H. (1996). A reminder of the fallibiltiy of the Wald Statistic. *American Statistician, 50*, 226–227.

Feng, D., & Cliff, N. (2004). Monte Carlo evaluation of ordinal d with improved confidence interval. *Journal of Modern Applied Statistical Methodsm 3*: Issue 2, Article 6. http://digitalcommon s.wayne.edu/jmasm/vol3/iss2/6.

Ferretti, N., Kelmansky, D., Yohai, V. J., & Zamar, R. (1999). A class of locally and globally robust regression estimates. *Journal of the American Statistical Association, 94*, 174–188.

Field, A., Miles, J. & Field, Z. (2012). *Discovering Statistics Using R.* Thousand Oaks: Sage.

Filzmoser, P., Maronna, R., & Werner, M. (2008). Outlier identification in high dimensions. *Computational Statistics and Data Analysis, 52*, 1694–1711.

Finner, H., & Gontscharuk, V. (2009). Controlling the familywise error rate with plug-in estimator for the proportion of true null hypotheses. *Journal of the Royal Statistical Society, B, 71*, 1031–1048.

Finner, H., & Roters, M. 2002). Multiple hypotheses testing and expected number of type I errors. *Annals of Statistics, 30*, 220–238.

Fisher, R. A. (1932). *Statistical Methods for Research Workers.* Oliver and Boyd, Edinburgh

Fisher, R. A. (1935). The fiducial argument in statistical inference. *Annals of Eugenics, 6*, 391–398.

Fisher, R. A. (1941). The asymptotic approach to Behren's integral, with further tables for the d test of significance. *Annals of Eugenics, 11*, 141–172.

Fitzmaurice, G. M. , Davidian, M., Verbeke, G., & Molenberghs, G. (2009). *Longitudinal Data Analysis: A Handbook of Modern Statistical Methods.* Boca Raton, FL: Chapman & Hall.

Fitzmaurice, G.M., Laird, N. M., & Ware, J. H. (2004). *Applied Longitudinal Analysis.* New York: Wiley.

Fleiss, J. L. (1981). *Statisical Methods for Rates and Proportions*, 2nd ed. New York: Wiley.

Fligner, M. A., & Policello II, G. E. (1981). Robust rank procedures for the Behrens-Fisher problem. *Journal of the American Statistical Association, 76*, 162–168.

Fox, J. (2001). *Multiple and Generalized Nonparametric Regression.* Thousands Oaks, CA: Sage.

Freedman, D., & Diaconis, P. (1981). On the histogram as density estimator: L_2 theory. *Z. Wahrsche. verw. Ge., 57*, 453–476.

Freedman, D., & Diaconis, P. (1982). On inconsistent M-estimators. *Annals of Statistics, 10*, 454–461.

Friedrich, S., Brunner, E., & Pauly, M. (2017). Permuting longitudinal data despite all the dependencies. *Journal of Multivariate Analysis, 153*, 255–265.

Frigge, M., Hoaglin, D. C., & Iglewicz, B. (1989). Some implementations of the boxplot. *American Statistician, 43*, 50–54.

Fung, K. Y. (1980). Small sample behaviour of some nonparametric multi-sample location tests in the presence of dispersion differences. *Statistica Neerlandica, 34*, 189–196.

Fung, W.-K. (1993). Unmasking outliers and leverage points: A confirmation. *Journal of the American Statistical Association, 88*, 515-519.

Fúquene, J. A., Cook, J. D., & Pericchi, L. R. (2009). A case for robust Bayesian priors with applications to clinical trials. *Bayesian Analysis, 4*, 817–846.

Galton, F. (1888). Co-relations and their measurement, chiefly anthropometric data. *Proceedings of the Royal Society London, 45*, 135–145

Gamage, J., Mathew, T., & Weerahandi, S. (2004). Generalized p-values and generalized confidence regions for the multivariate Behrens–ŰFisher problem and MANOVA, *Journal of Multivariate Analysis, 88*, 177–Ű189.

Games, P. A., & Howell, J. (1976). Pairwise multiple comparison procedures with unequal n's and/or variances: A Monte Carlo study. *Journal of Educational Statistics, 1*, 113–125.

Gelman, A. (2008). Objections to Bayesian statistics. *Bayesian Analysis, 3*, 445–450.

Gelman, A., Carlin, J., Stern, H., Dunson, D., Vehtari, A., & Rubin, D. (2013). *Bayesian Data Analysis* 3rd ed. New York: Chapman & Hall/CRC.

Gelman, A., & Shalizi, C. R. (2010). Philosophy and the practice of Bayesian statistics. *British Journal of Mathematical and Statistical Psychology, 66*, 8–38. DOI: 10.1111/j.2044-8317.2011.02037.x.

Giltinan, D. M., Carroll, R. J., & Ruppert, D. (1986). Some new estimation methods for weighted regression when there are possible outliers. *Technometrics, 28*, 219–230.

Gleason, J. R. (1993). Understanding elongation: The scale contaminated normal family. *Journal of the American Statistical Association, 88*, 327–337.

Gleser, L. J. (1992). The importance of assessing measurement reliability in multivariate regression. *Journal of the American Statistical Association, 87* 696–707.

Godfrey, L. G. (2006). Tests for regression models with heteroskedasticity of unknown form. *Computational Statistics and Data Analysis, 50*, 2715–2733.

Goldberg, K. M., & Iglewicz, B. (1992). Bivariate extensions of the boxplot. *Technometrics, 34*, 307–320.

Goldstein, H., & Healey, M. J. R. (1995). The graphical presentation of a collection of means. *Journal of the Royal Statistical Society: Series A, 158*, 175–177.

Goodman, L. A., & Kruskal, W. H. (1954). Measures of association for cross- classifications. *Journal of the American Statistical Association, 49*, 732-736.

Graybill, F. A. (1983). *Matrices with Applications in Statistics.* Belmont, CA: Wadsworth.

Green, P. J., & Silverman, B. W. (1993). *Nonparametric Regression and Generalized Linear Models: A Roughness Penalty Approach.* Boca Raton, FL: CRC Press.

Grimm. G & Paul R. Yarnold, P. R. (1995). *Reading and Understanding Multivariate Statistics.* American Psychological Association.

Guo, J. H., & Luh, W. M. (2000). An invertible transformation two-sample trimmed t-statistic under heterogeneity and nonnormality. *Statistics & Probability Letters, 49*, 1–7.

Györfi, L., Kohler, M., Krzyzk, & A. Walk, H. (2002). *A Distribution-Free Theory of Nonparametric Regression.* New York: Springer-Verlag.

Hald, A. (1952). *Statistical Theory with Engineering Applications.* New York: Wiley.

Hald, A. (1998). *A History of Mathematical Statistics from 1750 to 1930.* New York: Wiley.

Hall, P. (1988a). On symmetric bootstrap confidence intervals. *Journal of the Royal Statistical Society, Series B, 50*, 35–45.

Hall, P. (1988b). Theoretical comparison of bootstrap confidence intervals. *Annals of Statistics, 16*, 927–953.

Hall, P. (1992). On the removal of skewness by transformation. *Journal of the Royal Statistical Society, Series B, 54*, 221–228.

Hall, P. G., & Hall, D. (1995). *The Bootstrap and Edgeworth Expansion.* New York: Springer Verlag.

Hall, P., & Sheather, S. J. (1988). On the distribution of a Studentized quantile. *Journal of the Royal Statistical Society, B, 50*, 380–391.

Hampel, F. R. (1975). Beyond location parameters: Robust concepts and methods (with discussion). *Bull. ISI, 46*, 375–391.

Hampel, F. R., Ronchetti, E. M., Rousseeuw, P. J., & Stahel, W. A. (1986). *Robust Statistics: The Approach Based on Influence Functions.* New York: Wiley.

Hand, D. J., & Crowder, M. J. (1996). *Practical Longitudinal Data Analysis.* London: Chapman and Hall.

Hand, D. J., &Taylor, C. C. (1987) *Multivariate Analysis of Variance and Repeated Measures.* New York: Chapman and Hall.

Härdle, W. (1990). *Applied Nonparametric Regression.* Econometric Society Monographs No. 19, Cambridge, UK: Cambridge University Press.

Harpole, J. K., Woods, C. M., Rodebaugh, T. L., Levinson, C. A. & Lenze, E. J. (2014). How bandwidth selection algorithms impact exploratory data analysis using kernel density estimation. *Psychological Methods, 19*, 428–443.

Harrell, F. E., & Davis, C. E. (1982). A new distribution-free quantile estimator. *Biometrika, 69*, 635–640.

Hastie, T. J., & Tibshirani, R. J. (1990). *Generalized Additive Models.* New York: Chapman and Hall.

Hawkins, D. M., & Olive, D. (1999). Applications and algorithms for least trimmed sum of absolute deviations regression. *Computational Statistics and Data Analysis, 28*, 119–134.

Hayes, A. F., & Cai, L. (2007). Further evaluating the conditional decision rule for comparing two independent groups. *British Journal of Mathematical and Statistical Psychology, 60*, 217–244.

Hayter, A. (1984). A proof of the conjecture that the Tukey-Kramer multiple comparison procedure is conservative. *Annals of Statistics, 12*, 61–75.

Hayter, A. (1986). The maximum familywise error rate of Fisher's least significant difference test. *Journal of the American Statistical Association, 81*, 1000–1004.

He, X., & Portnoy, S. (1992). Reweighted LS estimators converge at the same rate as the initial estimator. *Annals of Statistics, 20*, 2161–2167.

He, X., Simpson, D. G., & Portnoy, S. L. (1990). Breakdown robustness of tests. *Journal of the American Statistical Association, 85*, 446–452.

He, X., & Zhu, L. X. (2003). A lack-of-fit test for quantile regression. *Journal of the American Statistical Association, 98*, 1013–1022.

Headrick, T. C., & Rotou, O. (2001). An investigation of the rank transformation in multiple regression. *Computational Statistics and Data Analysis, 38*, 203–215.

Heiser, D. A. (2006). Statistical tests, tests of significance, and tests of a hypothesis (Excel). *Journal of Modern Applied Statistical Methods, 5*, 551–566.

Hettmansperger, T. P. (1984). *Statistical Inference Based on Ranks.* New York: Wiley.

Hettmansperger, T. P., & McKean, J. W. (1977). A robust alternative based on ranks to least squares in analyzing linear models. *Technometrics, 19*, 275–284.

Hettmansperger, T. P., & McKean, J. W. (1998). *Robust Nonparametric Statistical Methods.* London: Arnold.

Hettmansperger, T. P., Möttönen, J. & Oja, H. (1997). Affine-invariant one-sample signed-rank tests. *Journal of the American Statistical Association, 92*, 1591–1600.

Hettmansperger, T. P., & Sheather, S. J. (1986). Confidence intervals based on interpolated order statistics. *Statistics and Probability Letters 4*, 75–79.

Hewett, J. E., & Spurrier, J. D. (1983). A survey of two stage tests of hypotheses: Theory and application. *Communications in Statistics-Theory and Methods, 12*, 2307–2425.

Hittner, J. B., May, K., & Silver, N. C. (2003). A Monte Carlo evaluation of tests for comparing dependent correlations. *Journal of General Psychology, 130*, 149–168.

Hoaglin, D. C., Iglewicz, B., & Tukey, J. W. (1986). Performance of some resistant rules for outlier labeling. *Journal of the American Statistical Association, 81*, 991–999.

Hoaglin, D. C., & Iglewicz, B. (1987). Fine-tuning some resistant rules for outlier labeling. *Journal of the American Statistical Association, 82*, 1147–1149.

Hochberg, Y. (1975). Simultaneous inference under Behrens-Fisher conditions: A two sample approach. *Communications in Statistics, 4*, 1109–1119.

Hochberg, Y. (1988). A sharper Bonferroni procedure for multiple tests of significance. *Biometrika, 75*, 800–802.

Hodges, J. L., Ramsey, P. H., & Wechsler, S. (1990). Improved significance probabilities of the Wilcoxon test. *Journal of Educational Statistics, 15*, 249–265.

Hoenig, J. M., & Heisey, D. M. (2001). The abuse of power: The pervasive fallacy of power calculations for data analysis. *American Statistician 55*, 19–24.

Hogg, R. V., & Craig, A. T. (1970). *Introduction to Mathematical Statistics.* New York: Macmillan.

Hollander, M., & Sethuraman, J. (1978). Testing for agreement between two groups of judges. *Biometrika, 65*, 403–411.

Holmes, C. (2014). Robust (Bayesian) inference. http://mlss2014.hiit.fi/mlss_files/Holmes-MLSS.pdf.

Hommel, G. (1988). A stagewise rejective multiple test procedure based on a modified Bonferroni test. *Biometrika, 75*, 383–386.

Hosmane, B. S. (1986). Improved likelihood ratio tests and Pearson chi-square tests for independence in two dimensional tables. *Communications Statistics—Theory and Methods, 15*, 1875–1888.

Hosmer, D. W., & Lemeshow, S. L. (1989). *Applied Logistic Regression.* New York: Wiley.

Hössjer, O. (1994). Rank-based estiamtes in the linear model with high breakdown point. *Journal of the American Statistical Association, 89*, 149–158.

Huber, P. J. (1964). Robust estimation of location parameters. *Annals of Mathematical Statistics, 35*, 73–101.

Huber, P. J. (1981). *Robust Statistics.* New York: Wiley.

Huber, P. J., & Ronchetti, E. (2009). *Robust Statistics,* 2nd ed. New York: Wiley.

Huber, P. J. (1993). Projection pursuit and robustness. In S. Morgenthaler, E. Ronchetti, & W. Stahel (Eds.) *New Directions in Statistical Data Analysis and Robustness.* Boston: Birkhäuser Verlag.

Hubert, M., Rousseeuw, P. J., & Verboven, S. (2002). A fast method for robust principal components with applications to chemometrics. *Chemometrics and Intelligent Laboratory Systems, 60*, 101–111.

Hubert, M., Rousseeuw, P. J., & Vanden Branden, K. (2005). ROBPCA: A new approach to robust principal component analysis. *Technometrics, 47*, 64–79.

Hubert, M., & Vandervieren, E. (2008). An adjusted boxplot for skewed distributions. *Computational Statistics and Data Analysis, 52*, 5186–5201.

Huberty, C. J. (1989). Problems with stepwise methods—better alternatives. *Advances in Social Science Methodology, 1*, 43–70.

Huitema, B. E. (2011). *The Analysis of Covariance and Alternatives*, 2nd ed. New York: Wiley.

Hussain, S. S., & Sprent, P. (1983). Non-parametric regression. *Journal of the Royal Statistical Society, 146*, 182–191.

Huynh, H., & Feldt, L. S. (1976). Estimation of the Box correction for degrees of freedom from sample data in randomized block and split-plot designs. *Journal of Educational Statistics, 1*, 69–82.

Hyndman, R. B., & Fan, Y. (1996). Sample quantiles in social packages. *American Statistician, 50*, 361–365.

Iman, R. L. (1974). A power study of a rank transform for the two-way classification model when interaction may be present. *Canadian Journal of Statistics, 2*, 227–239.

Iman, R. L., & Davenport, J. M. (1980). Approximations of the critical region of the Friedman statistic. *Communications in Statistics, A9*, 571–595.

Iman, R. L., Quade, D., & Alexander, D. A. (1975). Exact probability levels for the Krusakl-Wallis test. *Selected Tables in Mathematical Statistics, 3*, 329–384.

Jaeckel, L. A. (1972) Estimating regression coefficients by minimizing the dispersion of residuals. *Annals of Mathematical Statistics, 43*, 1449–1458.

James, G. S. (1951). The comparison of several groups of observations when the ratios of the population variances are unknown. *Biometrika, 38*, 324–329.

Jeyaratnam, S., & Othman, A. R. (1985). Test of hypothesis in one-way random effects model with unequal error variances. *Journal of Statistical Computation and Simulation, 21*, 51–57.

Jhun, Myoungshic & Choi, Inkyung (2009). Bootstrapping least distance estimator in the multivariate regression model. *Computational Statistics and Data Analysis 53*, 4221–4227.

Johansen, S. (1980). The Welch–James approximation to the distribution of the residual sum of squares in a weighted linear regression. *Biometrika, 67*, 85–93.

Johansen, S. (1982). Amendments and corrections: the WelchŰJames approximation to the distribution of the residual sum of squares in a weighted linear regression. *Biometrika, 69*, 491.

Johnson, N. J. (1978). Modifed t tests and confidence intervals for asymmetrical populations. *Jouranl of the American Statistical Association, 73*, 536–576.

Johnson, P., & Neyman, J. (1936). Tests of certain linear hypotheses and their application to some educational problems. *Statistical Research Memoirs 1*, 57–93.

Jones, L. V., & Tukey, J. W. (2000). A sensible formulation of the significance test. *Psychological Methods, 5*, 411–414.

Jones, R. H. (1993). *Longitudinal Data with Serial Correlation: A State-Space Approach.* London: Chapman Hall.

Jorgensen, J., Gilles, R. B., Hunt, D. R., Caplehorn, J. R. M., & Lumley, T. (1995). A simple and effective way to reduce postoperative pain after laparocsoopic cholecystectomy. *Australian and New Zealand Journal of Surgery, 65*, 466–469.

Judd, C. M., Kenny, D. A., & McClelland, G. H. (2001), Estimating and testing mediation and moderation in within-subjects designs. *Psychological Methods, 6*, 115–134.

Kaiser, L., & Bowden, D. (1983). Simultaneous confidence intervals for all linear contrasts of means with heterogeneous variances. *Communications in Statistics—Theory and Methods, 12*, 73–88.

Kallenberg, W. C. M., & Ledwina, T. (1999). Data-driven rank tests for independence. *Journal of the American Statistical Association, 94*, 285–310.

Kendall, M. G., & Stuart, A. (1973). *The Advanced Theory of Statistics*, Vol. 2. New York: Hafner.

Kepner, J. L., & Robinson, D. H. (1988). Nonparametric methods for detecting treatment effects in repeated-measures designs. *Journal of the Statistical Association, 83*, 456–461.

Keselman, J. C., Cribbie, R., & Holland, B. (1999). The pairwise multiple comparison multiplicity problem: An alternative approach to familywise and comparisonwise Type I error control. *Psychological Methods, 4*, 58–69.

Keselman, J. C., Rogan, J. C., Mendoza, J. L., & Breen, L. J. (1980). Testing the validity conditions of repeated measures F tests. *Psychological Bulletin, 87*, 479–481.

Keselman, H. J., Algina, J., Boik, R. J., & Wilcox, R. R. (1999). New approaches to the analysis of repeated measurements. In B. Thompson (ed.). *Advances in Social Science Methodology, 5*, 251–268. Greenwich, CT: JAI Press.

Keselman, H. J., & Wilcox, R. R. (1999). The "improved" Brown and Forsyth test for mean equality: Some things can't be fixed. *Communications in Statistics—Simulation and Computation, 28*, 687–698.

Keselman, H. J., Wilcox, R. R., Taylor, J., & Kowalchuk, R. K. (2000). Tests for mean equality that do not require homogeneity of variances: Do they really work? *Communications in Statistics–Simulation and Computation, 29*, 875–895.

Keselman,H. J., Algina, J., Wilcox, R. R., & Kowalchuk, R. K. (2000). Testing repeated measures hypotheses when covariance matrices are heterogeneous: Revisiting the robustness of the Welch-James test again. *Educational and Psychological Measurement, 60*, 925–938.

Keselman, H. J., Othman, A. & Wilcox, R. R. (2016). Generalized linear model analyses for treatment group equality when data are non-normal. *Journal of Modern and Applied Statistical Methods,15*, 32–61.

Keselman, H. J., Othman, A. R., Wilcox, R. R., & Fradette, K. (2004). The new and improved two-sample t test. *Psychological Science, 15*, 47–51.

Kim, P. J., & Jennrich, R. I. (1973). Tables of the exact sampling distribution of the two-sample Kolmogorov-Smirnov criterion, D_{mn}, $m \leq n$. In H. L. Harter & D. B. Owen (Eds.) *Selected Tables in Mathematical Statistics*, Vol. I. Providence, RI: American Mathematical Society.

Kim, S.-J. (1992). A practical solution to the multivariate Behrens-Fisher problem. *Biometrika, 79*, 171–176.

Kirk, R. E. (1995). *Experimental Design.* Monterey, CA: Brooks/Cole.

Koenker, R. (1994). Confidence intervals for regression quantiles. In P. Mandl & M. Huskova (Eds.) Asymptotic Statistics. Proceedings of the Fifth Prague Symposium, 349–359.

Koenker, R., & Bassett, G. (1978). Regression quantiles. *Econometrika, 46* 33–50.

Koenker, R., Bassett, G. (1981). Robust tests for heteroscedasticity based on regression quantiles. *Econometrika, 50*, 43–61.

Koenker, R., & Ng, P. (2005). Inequality Constrained Quantile Regression. *Sankhya, The Indian Journal of Statistics, 67*, 418-440.

Koenker, R., Ng, P., & Portnoy, S. (1994). Quantile smoothing splines. *Biometrika, 81*, 673–680.

Koenker, R., & Portnoy, S. (1990). M estimation of multivariate regressions. *Journal of the American Statistical Association, 85*, 1060–1068.

Koenker, R., & Xiao, Z. J. (2002). Inference on the quantile regression process. *Econometrica, 70*, 1583–1612.

Koller, M., & Stahel, W. A. (2011). Sharpening Wald-type inference in robust regression for small samples. *Computational Statistics and Data Analysis, 55*, 2504–2515.

Kolmogorov, A, N. (1956). *Foundations of the Theory of Probability.* New York: Chelsea.

Konietschke, F., & Pauly, M. (2012). A studentized permutation test for the nonparametric Behrens-Fisher problem in paired data. *Electronic Journal of Statistics, 6*, 1358–1372. DOI: 10.1214/12-EJS714.

Konietschke, F., Bathke, A. C., Harrar, S. W., & Pauly, M. (2015). Parametric and nonparametric bootstrap methods for general MANOVA. *Journal of Multivariate Analysis, 140*, 291–301.

Kramer, C. (1956). Extension of multiple range test to group means with unequal number of replications. *Biometrics, 12*, 307–310.

Kraemer, H. C., Periyakoil, V. S., & Noda, A. (2004). Tutorial in biostatistics: Kappa coefficient in medical research. *Statistics in Medicine, 21*, 2109–2129.

Krause, A., & Olson, M. (2000). *The Basics of S and S-PLUS.* New York: Springer.

Krishnamoorthy, K., Lu, F., & Mathew, T. (2007). A parametric bootstrap approach for ANOVA with unequal variances: Fixed and random models. *Computational Statistics and Data Analysis, 51*, 5731–5742.

Krutchkoff, R. G. (1988). One-way fixed effects analysis of variance when the error variances may be unequal. *Journal of Statistical Computation and Simulation, 30*, 259–271.

Krzanowski, W. J. (1988) *Principles of Multivariate Analysis. A User's Perspective.* New York: Oxford University Press.

Kulinskaya, E., & Staudte, R. G. (2006). Interval estimates of weighted effect sizes in the one-way heteroscedastic ANOVA. *British Journal of Mathematical and Statistical Psychology, 59*, 97–111.

Kulinskaya, E., Morgenthaler, S. and Staudte, R. (2010). Variance Stabilizing the Difference of two Binomial Proportions. *American Statistician, 64*, 350–356.

Kuo, L., & Mallick, B. (1998). Variable selection for regression models. *Sankhya, Series B, 60*, 65–81.

Larsen, W. A., & McCleary, S. J. (1972). The use of partial residual plots in regression analysis. *Technometrics, 14*, 781–790.

Lax, D. A. (1985). Robust estimators of scale: Finite sample performance in long-tailed symmetric distributions. *Journal of the American Statistical Association, 80*, 736–741.

Le, C. T. (1994). Some tests of linear trend of variances. *Communications in Statistics—Theory and Methods, 23*, 2269–2282.

Le, C. T. (1998). *Applied Categorical Data Analysis* New York: Wiley.

Lee, H., & Fung, K. F. (1985). Behavior of trimmed F and sine-wave F statistics in one-way ANOVA. *Sankhya: The Indian Journal of Statistics, 47*, 186–201.

Lee, J., Sun, D., Sun, Y., & Taylor, J. (2016). Exact post-selection inference with the lasso. *Annals of Statistics, 44*, 907–927.

Lee, S., & Ahn, C. H. (2003). Modified ANOVA for unequal variances. *Communications in Statistics–Simulation and Computation, 32*, 987–1004.

Li, G. (1985). Robust regression. In D. Hoaglin, F. Mosteller & J. Tukey (Eds.), *Exploring Data Tables, Trends, and Shapes*, pp. 281–343. New York: Wiley.

Li, G., & Chen, Z. (1985). Projection-pursuit approach to robust dispersion and principal components: Primary theory and Monte Carlo. *Journal of the American Statistical Association, 80*, 759–766.

Liang, H., Su, H. and Zou, G. (2008). Confidence intervals for a common mean with missing data with applications in an AIDS study. *Computational Statistics and Data Analysis, 53*, 546–553.

Lin, P., & Stivers, L. (1974). On the difference of means with missing values. *Journal of the American Statistical Association, 61*, 634–636.

Little, R. J. A., & Rubin, D. (2002). *Statistical Analysis with Missing Data*, 2nd ed. New York: Wiley.

Liu, R. G., Parelius, J. M., & Singh, K. (1999). Multivariate analysis by data depth. *Annals of Statistics, 27*, 783–840.

Liu, R. G., & Singh, K. (1997). Notions of limiting P values based on data depth and bootstrap. *Journal of the American Statistical Association, 92*, 266–277.

Livacic-Rojas, P., Vallejo, G., & Fernández, P. (2010). Analysis of Type I error rates of univariate and multivariate procedures in repeated measures designs. *Communications in Statistics–Simulation and Computation, 39*, 624–640.

Lix, L. M, Keselman, H. J., & Hinds, A. (2005). Robust tests for the multivariate Behrens-Fisher problem. *Computer Methods .and Programs in Biomedicine, 77*, 129–139.

Lloyd, C. J. (1999). *Statistical Analysis of Categorical Data.* New York: Wiley.

Locantore, N., Marron, J. S., Simpson, D. G., Tripoli, N., & Zhang, J. T. (1999). Robust principal components for functional data. *Test, 8*, 1–28.

Loh, W.-Y. (1987). Does the correlation coefficient really measure the degree of clustering around a line? *Journal of Educational Statistics, 12*, 235–239.

Lombard, F. (2005). Nonparametric confidence bands for a quantile comparison function. Technometrics, 47, 364–369.

Long, J. D., & Cliff, N. (1997). Confidence intervals for Kendall's tau. *British Journal of Mathematical and Statistical Psychology, 50*, 31–42.

Long, J. S., & Ervin, L. H. (2000). Using heteroscedasticity consistent standard errors in the linear regression model. *American Statistician, 54*, 217–224.

Lord, F. M., & Novick, M. R. (1968). *Statistical Theories of Mental Test Scores.* Reading, MA: Addison-Wesley.

Ludbrook, J., & Dudley, H. (1998). Why permutation tests are superior to t and F tests in biomedical research. *American Statistician, 52*, 127–132.

Lumley, T. (1996). Generalized estimating equations for ordinal data: A note on working correlation structures. *Biometrics, 52*, 354–361.

Lunneborg, C. E. (2000). *Data Analysis by Resampling: Concepts and Applications.* Pacific Grove, CA: Duxbury.

Lyon, J. D., & Tsai, C.-L. (1996). A comparison of tests for homogeneity. *Statistician, 45*, 337–350.

Mair, P. & Wilcox, R. (2016). Robust statistical methods in R using the WRS2 package. Unpublished technical report.

Mallows, C. L. (1973). Some comments on C_p. *Technometrics, 15*, 661–675.

Mallows, C. L. (1986). Augmented partial residuals. *Technometrics, 28*, 313–319.

Manly, B. F. J. (2004). *Multivariate Statistical Methods: A Primer*, 3rd ed. New York: Chapman & Hall/CRC.

Mann, H. B., & Whitney, D. R. (1947). On a test of whether one of two random variables is stochastically larger than the other. *Annals of Mathematical Statistics, 18*, 50–60.

Marazzi, A. (1993). *Algorithms, Routines and S functions for Robust Statistics.* New York: Chapman and Hall.

Mardia, K. V., Kent, J. T., & Bibby, J. M. (1979). *Multivariate Analysis.* San Diego, CA: Academic Press.

Markowski, C. A., & Markowski, E. P. (1990). Conditions for the effectiveness of a preliminary test of variance. *American Statistician, 44*, 322–326.

Maronna, R. A. (2005). Principal components and orthogonal regression based on robust scales. *Technometrics, 47*, 264–273.

Maronna, R. A. (2011). Robust ridge regression for high-dimensional data. *Technometrics, 53*, 44–53.

Maronna, R. A., Martin, D. R., & Yohai, V. J. (2006). *Robust Statistics: Theory and Methods.* New York: Wiley.

Maronna, R., & Morgenthaler, S. (1986). Robust regression through robust covariances. *Communications in Statistics–Theory and Methods, 15*, 1347–1365.

Maronna, R., Yohai, V. J. (2010). Correcting MM estimates for "fat" data sets. *Computational Statistics and Data Analysis, 54*, 3168–3173.

Maronna, R., Yohai, V. J., & Zamar, R. (1993). Bias-robust regression estimation: A partial survey. In S. Morgenthaler, E. Ronchetti & W. A. Stahel (Eds.) *New Directions in Statistical Data Analysis and Robustness.* Boston: Verlag.

Maronna, R. A., & Zamar, R. H. (2002). Robust estimates of location and dispersion for high-dimensional datasets. *Technometrics, 44*, 307–317

Martin, M. A. (2007). Bootstrap hypothesis testing for some common statistical problems: A critical evaluation of size and power properties. *Computational Statistics and Data Analysis, 51*, 6321–6342.

Matuszewski, A., & Sotres, D. (1986). A simple test for the Behrens–Fisher problem. *Computational Statistics and Data Analysis, 3*, 241–249.

Maxwell, A.E. (1970). Comparing the classification of subjects by two independent judges. *British Journal of Psychiatry, 116*, 651– 655.

McCullagh, P., & Nelder, J. A. (1992). *Generalized Linear Models*, 2nd ed. London: Chapman and Hall.

McCullough, B. D. & Wilson, B. (2005). On the accuracy of statistical procedures in Microsoft Excel. *Computational Statistics and Data Analysis, 49*, 1244–1252.

McKean, J. W., & Schrader, R. M. (1984). A comparison of methods for studentizing the sample median. *Communications in Statistics— Simulation and Computation, 13*, 751–773.

McKean, J. W., & Vidmar, T. J. (1994). A comparison of two rank-based methods for the analysis of linear models. *American Statistician, 48*, 220–229.

McKnight, P. E., McKnight, K. M., Sidani, S. & Figueredo, A. J. (2007). *Missing Data: A Gentle Introduction.* New York: Guilford Press.

Mee, R. W. (1990). Confidence intervals for probabilities and tolerance regions based on a generalization of the Mann-Whitney statistic. *Journal of the American Statistical Association, 85*, 793–800.

Mehrotra, D. V. (1997). Improving the Brown-Forsythe solution to the generalized Behrens-Fisher problem. *Communications in Statistics–Simulation and Computation, 26*, 1139–1145.

Miller, A. J. (1990). *Subset Selection in Regression.* London: Chapman and Hall.

Miller, R. G. (1974). The jackknife-a review. *Biometrika, 61*,

Miller, R. G. (1976). Least squares regression with censored data. *Biometrika, 63*, 449–464.

Molenberghs, G., & Kenward, M. (2007). *Missing Data in Clinical Studies.* New York: Wiley.

Montgomery, D. C., & Peck, E. A. (1992). *Introduction to Linear Regression Analysis.* New York: Wiley.

Mooney, C. Z., & Duval, R. D. (1993). *Bootstrapping: A Nonparametric Approach to Statistical Inference.* Newbury Park, CA: Sage.

Morey, R. D., Hoekstra, R., Rouder, J. N., Lee, M. D., & Wagenmakers, E.-J. (2016). The fallacy of placing confidence in confidence intervals. *Psychonomic Bulletin Review, 23*, 103–123. DOI 10.3758/s13423-015-0947-8.

Morgenthaler, S., & Tukey, J. W. (1991). *Configural Polysampling.* New York: Wiley.

Moser, B. K., Stevens, G. R., & Watts, C. L. (1989). The two-sample t-test versus Satterthwaite's approximate F test. *Communications in Statistics—Theory and Methods, 18*, 3963–3975.

Möttönen, J., & Oja, H. (1995). Multivariate spatial sign and rank methods. *Nonparametric Statistics, 5*, 201–213.

Muirhead, R. J. (1982). *Aspects of Multivariate Statistical Theory.* New York: Wiley.

Müller, H.-G. (1988). *Nonparametric Regression Analysis.* New York: Springer-Verlag.

Munzel, U. (1999). Linear rank score statistics when ties are present. *Statistics and Probability Letters, 41*, 389–395.

Munzel, U., & Brunner, E. (2000a). Nonparametric test in the unbalanced multivariate one-way design. *Biometrical Journal, 42*, 837–854.

Munzel, U., & Brunner, E. (2000b). Nonparametric methods in multivariate factorial designs. *Journal of Statistical Planning and Inference, 88*, 117–132.

Naranjo, J. D., & Hettmansperger, T. P. (1994) Bounded influence rank regression. *Journal of the Royal Statistical Society, B, 56*, 209–220.

Nelder, J. A., & Mead, R. (1965). A simplex method for function minimization. *Computer Journal, 7*, 308–313.

Neuhäuser, M., Lösch, C., & Jöckel, K-H (2007). The Chen-Luo test in case of heteroscedasticity. *Computational Statistics and Data Analysis, 51*, 5055–5060.

Newcomb, S. (1882). Discussion and results of observations on transits of Mercury from 1677 to 1881. *Astronomical Papers, 1*, 363–487.

Newcombe, R. R. (2006). Confidence intervals for an effect size measure based on the Mann-Whitney statistics. Part 2: Asymptotic methods and evaluation. *Statistics in Medicine, 25*, 559–573.

Ng, T. H. M. (2009). *Significance Testing in Regression Analyses.* Unpublished doctoral dissertation, Dept. of Psychology, University of Southern California.

Ng, M., & Wilcox, R. R. (2009). Level robust methods based on the least squares regression estimator. *Journal of Modern and Applied Statistical Methods, 8*, 384–395.

Ng, M., & Wilcox, R. R. (2010). The small-sample efficiency of some recently proposed multivariate measures of location. *Journal of Modern and Applied Statistical Methods, 9*, 28–42.

Noguchi, K., Marmolejo-Ramos, F. (2016). *American Statistician, 70*, 325–334.

Norton, J. D. (2011). Challenges to Bayesian confirmation theory. In D. M. Gabbay, P. Thagard & J. Wood (Eds.), *Handbook of the Philosophy of Science*, Vol. 7. New York: Elsevier, pp. 391–440.

Oberhelman, D., & Kadiyala, R. (2007). A test for the equality of parameters of separate regression models in the presence of heteroskedasticity. *Communications in Statistics–Simulation and Computation, 36*, 99–121.

Olejnik, S., Li, J., Supattathum, S., & Huberty, C. J. (1997). Multiple testing and statistical power with modified Bonferroni procedures. *Journal of Educational and Behavioral Statistics, 22*, 389–406.

Olive, D. J. (2004). A resistant estimator of multivariate location and dispersion. *Computational Statistics & Data Analysis, 46*, 93–102.

Olson, C. L. (1974). Comparative robustness of six tests of multivariate analysis of variance. *Journal of the American Statistical Assocation, 69*, 894–908.

Olsson, D. M., & Nelson, L. S. (1975). The Nelder-Mead simplex procedure for function minimization. *Technometrics, 17*, 45–51.

Overall, J., & Tonidandel, S. (2010). The case for use of simple difference scores to test the significance of differences in mean rates of change in controlled repeated measurements designs. *Multivariate Behavioral Research, 45*, 806–827.

Owen, A. B. (2001). *Empirical Likelihood.* New York: Chapman and Hall.

Ozdemir, A. F., & Wilcox, R. R. (2010). New results on the small-sample efficiency of some robust estimators. Technical Report, Department of Statistics, Dokuz Eylul University, Izmir, Turkey.

Pagurova, V. I. (1968). On a comparison of means of two normal samples. *Theory of Probability and Its Applications, 13*, 527–534.

Parra-Frutos, I. (2014). Controlling the Type I error rate by using the nonparametric bootstrap when comparing means. *British Journal of Mathematical and Statistical Psychology, 67*, 117–132. DOI: 10.1111/bmsp.12011

Parra-Frutos, I. (2016). Preliminary tests when comparing means. *Computational Statistics, 31*, 1607–1631. DOI 10.1007/s00180-016-0656-4.

Parrish, R. S. (1990). Comparison of quantile estimators in normal sampling. *Biometrics, 46*, 247–257.

Patel, K. M., & Hoel, D. G. (1973). A nonparametric test for interactions in factorial experiments. *Journal of the American Statistical Association, 68*, 615–620.

Pedersen, W. C., Miller, L. C., Putcha-Bhagavatula, A. D., & Yang, Y. (2002). Evolved sex differences in sexual strategies: The long and the short of it. *Psychological Science, 13*, 157–161.

Peña, D., & Prieto, F. J. (2001). Multivariate outlier detection and robust covariance matrix estimation. *Technometrics, 43*, 286–299.

Piepho, H.-P. (1997). Tests for equality of dispersion in bivariate samples—review and empirical comparison. *Journal of Statistical Computation and Simulation, 56*, 353–372.

Poon, W. Y., Lew, S. F., & Poon, Y. S. (2000). A local influence approach to identifying multiple multivariate outliers. *British Journal of Mathematical and Statistical Psychology, 53*, 255–273.

Powers, D. A., & Xie, Y. (2008). *Statistical Methods for Categorical Data Analysis.* Bingley, UK: Emerald Group.

Pratt, J. W. (1964). Robustness of some procedures for the two-sample location problem. *Journal of the American Statistical Association, 59*, 665–680.

Pratt, J. W. (1968). A normal approximation for binomial, F, beta, and other common, related tail probabilities, I. *Journal of the American Statistical Association, 63*, 1457–1483.

Price, R. M., & Bonett, D. G. (2001). Estimating the variance of the median. *Journal of Statistical Computation and Simulation, 68*, 295–305.

Quade, D. (1979). Using weighted rankings in the analysis of complete blocks with additive block effects. *Journal of the American Statistical Association, 74*, 680–683.

Racine, J., & MacKinnon, J. G. (2007). Simulation-based tests that can use any number of simulations. *Communications in Statistics—Simulation and Computation, 36*, 357–365.

R Development Core Team (2010). R: A language and environment for statistical computing. R Foundation for Statistical Computing, Vienna, Austria. ISBN 3-900051-07-0, URL http://www.R-project.org.

Raine, A., Buchsbaum, M., & LaCasse, L. (1997). Brain abnormalities in murderers indicated by positron emission tomography. *Biological Psychiatry, 42*, 495–508.

Ramsey, P. H. (1980). Exact Type I error rates for robustness of Student's t test with unequal variances. *Journal of Educational Statistics, 5*, 337–349.

Ramsey, P. H., & Ramsey, P. P. (2007). Testing variability in the two-sample case. *Communications in Statistics–Simulation and Computation, 36*, 233–248.

Randal, J. A. (2008). A reinvestigation of robust scale estimation in finite samples. *Computational Statistics and Data Analysis, 52*, 5014–5021.

Rao, C. R. (1948). Tests of significance in multivariate analysis. *Biometrika, 35*, 58–79.

Rao, P. S. R. S., Kaplan, J., & Cochran, W. G. (1981). Estimators for the one-way random effects model with unequal error variances. *Journal of the American Statistical Association, 76*, 89–97.

Rasch, D., Teuscher, F., & Guiard, V. (2007). How robust are tests for two independent samples? *Journal of Statistical Planning and Inference, 137*, 2706 –2720.

Rasmussen, J. L. (1989). Data transformation, Type I error rate and power. *British Journal of Mathematical and Statistical Psychology, 42*, 203–211.

Reiczigel, J. , Abonyi-Tóth, Z., & Singer, J. (2008). *Computational Statistics and Data Analysis 52*, 5046–5053.

Reiczigel, J. Zakariás, I., & Rózsa, L. (2005). A bootstrap test of stochastic equality of two populations. *American Statistician, 59*, 156–161.

Renaud, O., & Victoria-Feser, M-P. (2010). A robust coefficient of determination for regression. *Journal of Statistical Planning and Inference, 140*, 1852–1862.

Rencher, A. C. (2002). *Methods of Multivariate Analysis.* New York: Wiley.

Rocke, D. M. (1996). Robustness properties of S-estimators of multivariate location and shape in high dimensions. *Annals of Statistics, 24*, 1327–1345.

Rocke, D. M., & Woodruff, D. L. (1996). Identification of outliers in multivariate data. *Journal of the American Statistical Association 91*, 1047–1061.

Rogan, J. C., Keselman, H. J., & Mendoza, J. L. (1979). Analysis of repeated measurements. *British Journal of Mathematical and Statistical Psychology, 32*, 269–286.

Rom, D. M. (1990). A sequentially rejective test procedure based on a modified Bonferroni inequality. *Biometrika, 77*, 663–666.

Romano, J. P. (1990). On the behavior of randomization tests without a group invariance assumption. *Journal of the American Statistical Association 85*, 686–692.

Rosenthal, R., & Jacobson, L. (1968). *Pygmalion in the Classroom: Teacher Expectations and Pupil's Intellectual Development.* New York: Holt, Rinehart and Winston.

Rosner, B., & Glynn, R. J. (2009). Power and sample size estimation for the Wilcoxon rank sum test with application to comparisons of C statistics from alternative prediction models. *Biometrics, 65*, 188–197.

Rousseeuw, P. J. (1984). Least median of squares regression. *Journal of the American Statistical Association, 79*, 871–880.

Rousseeuw, P. J., & Leroy, A. M. (1987). *Robust Regression & Outlier Detection.* New York: Wiley.

Rousseeuw, P. J., & van Zomeren, B. C. (1990). Unmasking multivariate outliers and leverage points (with discussion). *Journal of the American Statistical Association, 85*, 633–639.

Rousseeuw, P. J., & Ruts, I. (1996). AS 307: Bivariate location depth. *Applied Statistics, 45*, 516–526.

Rousseeuw, P. J., Ruts, I. & Tukey, J. W. (1999). The bagplot: A bivariate boxplot. *American Statistician, 53*, 382–387.

Rousseeuw, P. J., Hubert, M. (1999). Regression depth. *Journal of the American Statistical Association, 94*, 388–402.

Rousseeuw, P. J., van Driessen, K. (1999). A fast algorithm for the minimum covariance determinant estimator. *Technometrics, 41*, 212–223.

Rousseeuw, Van Aelst, Van Driessen & Agulló (2004). Robust multivariate regression. *Technometrics, 46*, 293–305.

Ruscio, J., & Mullen, T. (2012). Confidence intervals for the probability of superiority effect size measure and the area under a receiver operating characteristic curve. *Multivariate Behavioral Research, 47*, 201–223.

Rust, S. W., & Fligner, M. A. (1984). A modification of the Kruskal-Wallis statistic for the generalized Behrens-Fisher problem. *Communications in Statistics–Theory and Methods, 13*, 2013–2027.

Rutherford, A. (1992). Alternatives to traditional analysis of covariance. *British Journal of Mathematical and Statistical Psychology, 45*, 197–223.

Rutter, C. M., & Elashoff, R. M. (2006). Analysis of longitudinal data: Random coefficient regression modeling. *Statistics in Medicine, 13*, 1211–1231.

Ryan, G. W., & Leadbetter, S. D. (2002). On the misuse of confidence itnervals for two means in testing for the significance of the difference between the means. *Journal of Modern Applied Statistical Methods, 1*, 473–478.

Sackrowitz, H., & Samuel-Cahn, E. (1999). P values as random variables— expected P values. *American Statistician, 53*, 326–331.

Salk, L. (1973). The role of the heartbeat in the relations between mother and infant. *Scientific American, 235*, 26–29.

Samarov, A. M. (1993). Exploring regression structure using nonparametric functional estimation. *Journal of the American Statistical Association, 88*, 836–847.

Santner, T. J., Pradhan, V., Senchaudhuri, P., Mehta, C. R., & Tamhane, A. (2007). Small-sample comparisons of confidence intervals for the difference of two independent binomial porportions. *Computational Statistics and Data Analysis, 51*, 5791–5799.

Sarkar, S. K. (2002). Some results on false discovery rate in stepwise multiple testing procedures. *Annals of Statistics, 30*, 239–257.

Satterthwaite, F. E. (1946), An approximate distribution of estimates of variance components. *Biometrics Bulletin, 2*, 110–114.

Saunders, D. R. (1955). The "moderator variable" as a useful tool in prediction. In Proceedings of the 1954 Invitational Conference on Testing Problems (pp. 54–58). Princeton, NJ: Educational Testing Service.

Saunders, D. R. (1956). Moderator variables in prediction. *Educational and Psychological Measurement, 16*, 209–222.

Scariano, S. M., & Davenport, J. M. (1986). A four-moment approach and other practical solutions to the Behrens-Fisher problem. *Communications in Statistics–Theory and Methods, 15*, 1467–1501.

Scheffé, H. (1959). *The Analysis of Variance.* New York: Wiley.

Schenker, N., & Gentleman, J. F. (2001). On judging the significance of differences by examining the overlap between confidence intervals. *American Statistician, 55*, 182–186.

Schilling, M., & Doi, J. (2014). A coverage probability approach to finding an optimal binomial confidence procedure. *American Statistician, 68*, 133–145.

Schrader, R. M., & Hettmansperger, T. P. (1980). Robust analysis of variance. *Biometrika, 67*, 93–101.

Schroër, G., & Trenkler, D. (1995). Exact and randomization distributions of Kolmogorov-Smirnov tests two or three samples. *Computational Statistics and Data Analysis, 20*, 185–202.

Scott, D. W. (1979). On optimal and data-based histograms. *Biometrika, 66*, 605–610.

Scott, W. A. (1955). Reliability of content analysis: The case of nominal scale coding. *Public Opinion Quarterly, 19, 321–325.*

Searle, S. R. (1971). *Linear Models.* New York: Wiley.

Sen, P. K. (1968). Estimate of the regression coefficient based on Kendall's tau. *Journal of the American Statistical Association, 63*, 1379–1389.

Serfling, R. J. (1980). *Approximation Theorems of Mathematical Statistics.* New York: Wiley.

Shaffer, J. P. (1974). Bidirectional unbiased procedures. *Journal of the American Statistical Association, 69*, 437–439.

Shao, J., & Tu, D. (1995). *The jackknife and the bootstrap.* New York: Springer-Verlag.

Shao, J. (1996). Bootstrap model selection. *Journal of the American Statistical Association, 91*, 655–665.

She, Y., Li, S. & Wu, D. (2016). Robust orthogonal complement principal component analysis. *Journal of the American Statistical Association, 111*, 763–771.

Sheather, S. J., & McKean, J. W. (1987). A comparison of testing and confidence intervals for the median. *Statistical Probability Letters, 6*, 31–36.

Shevlyakov, G., & Smirnov, P. (2011). Robust estimation of the correlation coefficient: An attempt of survey. *Austrian Journal of Statistics, 40*, 147–156.

Signorini, D. F., & Jones, M. C. (2004). Kernel estimators for univariate binary regression. *Journal of the American Statistical Association, 99*, 119–126.

Silverman, B. W. (1986). *Density Estimation for Statistics and Data Analysis.* New York: Chapman and Hall.

Sim, C. H., Gan, F. F., & Chang, T. C. (2005). Outlier labeling with boxplot procedures. *Journal of the American Statistical Association, 100*, 642–652.

Simonoff, J. S. (2003). *Analyzing Categorical Data.* New York: Springer.

Singer, J. D., & Willett, J. B. (2003). *Applied Longitudinal Data Analysis: Modeling Change and Event Occurrence.* New York: Oxford University Press.

Singh, K. (1998). Breakdown theory for bootstrap quantiles. *Annals of Statistics, 26*, 1719–1732.

Small, C. G. (1990). A survey of multidimensional medians. *International Statistical Review, 58*, 263–277.

Smith, C. A. B. (1956). Estimating genetic correlations. *Annals of Human Genetics, 44*, 265–284.

Snedecor, G. W., & Cochran, W. (1967). *Statistical Methods.* 6th Edition Ames, IA: University Press.

Snow, R. E. (1995). Pygmalion and Intelligence? *Current Directions in Psychological Science*, 4, 169–172.

Sockett, E. B., Daneman, D. Clarson, & C. Ehrich, R. M. (1987). Factors affecting and patterns of residual insulin secretion during the first year of type I (insulin dependent) diabetes mellitus in children. *Diabetes, 30*, 453–459.

Spitzer, R. L., Cohen, J., Fleiss, J. L., & Endicott, J. (1967). Quantification of agreement in psychiatric diagnosis. *Archives of General Psychiatry 17*, 83–87.

Srihera, R., & Stute, W. (2010). Nonparametric comparison of regression functions. *Journal of Multivariate Analysis, 101*, 2039-2059.

Staudte, R. G., & Sheather, S. J. (1990). *Robust Estimation and Testing.* New York: Wiley.

Stein, C. (1945). A two-sample test for a linear hypothesis whose power is independent of the variance. *Annals of Statistics, 16*, 243–258.

Sterne, T.E. (1954). Some remarks on confidence or fiducial limits. *Biometrika 41*, 275–278.

Stevens, S. S. (1951). Mathematics, measurement, and psychophysics. In S. Stevens (ed.) *Handbook of Experimental Psychology*, pp. 1–49. New York: Wiley.

Stigler, S. M. (1986). *The History of Statistics: The Measurement of Uncertainty before 1900.* Cambridge, MA: Belknap Press.

Storer, B. E., & Kim, C. (1990). Exact properties of some exact test statistics for comparing two binomial proportions. *Journal of the American Statistical Association, 85*, 146–155.

Stromberg, A. J. (1993). Computation of high breakdown nonlinear regression parameters. *Journal of the American Statistical Association, 88*, 237–244.

Strube, M. (1988). Bootstrap Type I error rates for the correlation coefficient: An examination of alternate procedures. *Psychological Bulletin, 104*, 290–292.

Struyf, A., & Rousseeuw, P. J. (2000). High-dimensional computation of the deepest location. *Computational Statistics and Data Analysis, 34*, 415–426.

Stuart A. (1955). A test for homogeneity of the marginal distributions in a two-way classification. *Biometrika, 42*, 412–416.

Stuart, V. M. (2009). Exploring robust alternatives to Pearson's r through secondary analysis of published behavioral science data. Unpublished PhD dissertation, Dept of Psychology, University of Southern California.

Stute, W., Manteiga, W. G., & Quindimil, M. P. (1998). Bootstrap approximations in model checks for regression. *Journal of the American Statistical Association, 93*, 141–149.

Sutton, C. D. (1993). Computer-intensive methods for tests about the mean of an asymmetrical distribution. *Journal of the American Statistical Association, 88*, 802–810.

Tabachnick, B. G., & Fidell, L. S. (2006). *Using Multivariate Statistics*, 5th ed. New York: Allyn & Bacon.

Tamhane, A. C. (1977). Multiple comparisons in model I one-way ANOVA with unequal variances. *Communications in Statistics–Theory and Methods*, *6*, 15–32.

Theil, H. (1950). A rank-invariant method of linear and polynomial regression analysis. *Indagationes Mathematicae,12*, 85–91.

Thompson, A., & Randall-Maciver, R. (1905). *Ancient Races of the Thebaid.* Oxford: Oxford University Press.

Thompson, G. L. (1991). A unified approach to rank tests for multivariate and repeated measures designs. *Journal of the American Statistical Association, 86*, 410–419.

Thompson, G. L., & Ammann, L. P. (1990). Efficiencies of interblock rank statistics for repeated measures designs. *Journal of the American Statistical Association, 85*, 519–528.

Tibshirani, R. (1988). Estimating transformations for regression via additivity and variance stabilization. *Journal of the American Statistical Association, 83*, 394–405.

Tibshirani, R. (1996). Regression shrinkage and selection via the lasso. *Journal of the Royal Statistical Society, B, 58*, 267–288.

Tibshirani, R. J., Taylor, J., Lockhart, R. & Tibshirani, R. (2016). Exact post-selection inference for sequential regression procedures. *Journal of the American Statistical Association, 111*, 600–620. DOI: 10.1080/01621459.2015.1108848.

Todorov, V., & Filzmoser, P. (2010). Robust statistic for the one-way MANOVA. *Computational Statistics and Data Analysis, 54*, 37–48.

Tryon, W. W. (2001). Evaluating statistical difference, equivalence, and indeterminacy using inferential confidence intervals: An integrated alternative method of conducting null hypothesis statistical tests. *Psychological Methods, 6*, 371–386.

Tukey, J. W. (1960a). A survey of sampling from contaminated normal distributions. In I. Olkin et al. (Eds.) *Contributions to Probability and Statistics* (pp. 448-485). Stanford, CA: Stanford University Press.

Tukey, J. W. (1960b). Conclusions vs. descisions. *Technometrics, 2*, 423–433.

Tukey, J. W. (1977). *Exploratory Data Analysis.* Reading, MA: Addison-Wesley.

Tukey, J. W. (1991). The philosophy of multiple comparisons. *Statistical Science, 6*, 100–116.

Tukey, J. W., & McLaughlin D. H. (1963). Less vulnerable confidence and significance procedures for location based on a single sample: Trimming/Winsorization 1. *Sankhya A, 25*, 331–352.

Van Aelst, S., & Willems, G., (2005) Multivariate regression S-estimators for robust estimation and inference. *Statistica Sinica, 15*, 981–1001.

Vandelle, S., & Albert, A. (2009). Agreement between two independent groups of raters. *Psychometrika, 74*, 477–491.

Velina, M., Valeinis, J., Luca Greco, L. & George Luta, G. (2016). Empirical likelihood-based ANOVA for trimmed means. *International Journal of Environmental Research and Public Health, 13*, 953. doi:10.3390/ijerph13100953.

Velleman, P. F., & Wilkinson, L. (1993). Nominal, ordinal, interval and ratio typologies are misleading. *American Statistician, 47*, 65–72.

Venables, W. N., & Smith, D. M. (2002). *An Introduction to R.* Bristol, UK: Network Theory Ltd.

Verzani, J. (2004). *Using R for Introductory Statistics.* CRC Press: Boca Raton, FL.

Vexler, A., Liu, S. Kang, L., & Hutson, A. D. (2009). Modifications of the empirical likelihood interval estimation with improved coverage probabilities. *Communications in Statistics–Simulation and Computation, 38*, 2171–2183.

Victoroff, J., Quota, S., Adelman, J. R., Celinska, M. A., Stern, N. Wilcox, R., & Sapolsky, R. M. (2010). Support for religio-political aggression among teenaged boys in Gaza: Part I: Psychological Findings. *Aggressive Behavior, 36*, 219–231.

Vonesh, E. (1983). Efficiency of repeated measures designs versus completely randomized designs based on multiple comparisons. *Communications in Statistics–Theory and Methods, 12*, 289–302.

Wald, A. (1955). Testing the difference between the means of two normal populations with unknown standard deviations. In T. W. Anderson et al. (Eds.), *Selected Papers in Statistics and Probability by Abraham Wald.* New York: McGraw-Hill.

Wang, L., & Qu, A. (2007). Robust tests in regression models with omnibus alternatives and bounded influence. *Journal of the American Statistical Association, 102*, 347–358.

Wang, Q.H. & Rao, J. N. K. (2002). Empirical likelihood-based inference in linear models with missing data. *Scandanavian Journal of Statistics, 29*, 563–576.

Warren, M. J. (2010). A Kraemer-type rescaling that transforms the odds ratio into the weighted kappa coefficient. *Psychometrika, 75*, 328–330.

Warren, M. J. (2012). Cohen's quadratically weighted kappa is higher than linearly weighted kappa for tridiagonal agreement tables. *Statistical Methodology, 9*, 440–444.

Wechsler, D. (1958). *The Measurement and Appraisal of Adult Intelligence.* Baltimore: Williams and Wilkins.

Weerahandi, S. (1995). ANOVA under unequal error variances. *Biometrics, 51*, 589–599.

Welch, B. L. (1938). The significance of the difference between two means when the population variances are unequal. *Biometrika, 29*, 350–362.

Welch, B. L. (1951). On the comparison of several mean values: An alternative approach. *Biometrika, 38*, 330–336.

Westfall, P. (1988). Robustness and power of tests for a null variance ratio. *Biometrika, 75*, 207–214.

Westfall, P. H., & Young, S. S. (1993). *Resampling Based Multiple Testing.* New York: Wiley.

Wickham, H. (2009). *ggplot2: Elegant Graphics for Data Analysis.* New York: Springer.

Wilcox, R. R. (1983). A table of percentage points of the range of independent t variables. *Technometrics, 25*, 201–204.

Wilcox, R. R. (1986). Improved simultaneous confidence intervals for linear contrasts and regression parameters. *Communications in Statistics– Simulation and Computation, 15*, 917–932.

Wilcox, R. R. (1992). An improved method for comparing variances when distributions have non-identical shapes. *Computational Statistics and Data Analysis, 13*, 163–172.

Wilcox, R. R. (1993a). Some results on the Tukey-McLaughlin and Yuen methods for trimmed means when distributions are skewed. *Biometrical Journal, 36*, 259–273.

Wilcox, R. R. (1993b). Comparing one-step M-estimators of location when there are more than two groups. *Psychometrika, 58*, 71–78.

Wilcox, R. R. (1994a). A one-way random effects model for trimmed means. *Psychometrika, 59*, 289–306.

Wilcox, R. R. (1994b). The percentage bend correlation coefficient. *Psychometrika, 59*, 601–616.

Wilcox, R. R. (1995). Simulation results on solutions to the multivariate Behrens-Fisher problem via trimmed means. *Statistician, 44*, 213–225.

Wilcox, R. R. (1996a). Estimation in the simple linear regression model when there is heteroscedasticity of unknown form. *Communications in Statistics— Theory and Methods, 25*, 1305–1324.

Wilcox, R. R. (1996b). Confidence intervals for the slope of a regression line when the error term has non-constant variance. *Computational Statistics and Data Analysis, 22*, 89–98.

Wilcox, R. R. (1996c). *Statistics for the Social Sciences.* San Diego, CA: Academic Press.

Wilcox, R. R. (1997). ANCOVA based on comparing a robust measure of location at empirically determined design points. *British Journal of Mathematical and Statistical Psychology, 50*, 93–103.

Wilcox, R. R. (1998). Simulation results on extensions of the Theil-Sen regression estimator. *Communications in Statistics–Simulation and Computation, 27*, 1117–1126.

Wilcox, R. R. (2000a). Some exploratory methods for studying curvature in robust regression. *Biometrical Journal, 42*, 335–347.

Wilcox, R. R. (2000b). Rank-based tests for interactions in a two-way design when there are ties. *British Journal of Mathematical and Statistical Psychology, 53*, 145–153.

Wilcox, R. R. (2001a). *Fundamentals of Modern Statistical Methods: Substantially Increasing Power and Accuracy.* New York: Springer.

Wilcox, R. R. (2001b). Robust regression estimators that reduce contamination bias and have high efficiency when there is heteroscedasticity. Unpublished technical report, Dept. of Psychology, University of Southern California.

Wilcox, R. R. (2001c). Rank-based multiple comparisons for interactions in a split-plot design. *British Journal of Mathematical and Statistical, 53*, 145–153.

Wilcox, R. R. (2001d). Pairwise comparisons of trimmed means for two or more more groups. *Psychometrika, 66*, 343–256.

Wilcox, R. R. (2002). Comparing the variances of independent groups. *British Journal of Mathematical and Statistical Psychology, 55*, 169–176.

Wilcox, R. R. (2003). Inferences based on a skipped correlation coefficient. *Computational Statistics and Data Analysis, 44*, 223–236.

Wilcox, R. R. (2004a). Inferences based on a skipped correlation coefficient. *Journal of Applied Statistics, 31*, 131–144.

Wilcox, R. R. (2004b). An extension of Stein's two-stage method to pairwise comparisons among dependent groups based on trimmed means. *Sequential Analysis, 23*, 63–74.

Wilcox, R. R. (2005). A comparison of six smoothers when there are multiple predictors. *Statistical Methodology, 2*, 49–57.

Wilcox, R. R. (2006a). Comparing medians. *Computational Statistics and Data Analysis, 51*, 1934–1943.

Wilcox, R. R. (2006b). A note on inferences about the median of difference scores. *Educational and Psychological Measurement, 66*, 624–630.

Wilcox, R. R. (2006c). Pairwise comparisons of dependent groups based on medians. *Computational Statistics and Data Analysis, 50* 2933–2941.

Wilcox, R. R. (2006d). Some results on comparing the quantiles of dependent groups. *Communications in Statistics–Simulation and Computation, 35*, 893–900.

Wilcox, R. R. (2006e). Testing the hypothesis of a homoscedastic error term in simple, nonparametric regression. *Educational and Psychological Measurement, 66*, 85–92.

Wilcox, R. R. (2007). An omnibus test when using a quantile regression estimator with multiple predictors. *Journal of Modern and Applied Statistical Methods, 6*, 361–366.

Wilcox, R. R. (2008a). Some small-sample properties of some recently proposed multivariate outlier detection techniques. *Journal of Statistical Computation and Simulation, 78*, 701–712.

Wilcox, R. R. (2008b). Robust principal components: A generalized variance perspective. *Behavioral Research Methods, 40*, 102–108.

Wilcox, R. R. (2008c). A test of independence via quantiles that is sensitive to curvature. *Journal of Modern and Applied Statistics, 7*, 11–20.

Wilcox, R. R. (2009a). Robust multivariate regression when there is heteroscedasticity. *Communications in Statistics–Simulation and Computation, 38*,1–13.

Wilcox, R. R. (2009b). Robust ANCOVA using a smoother with bootstrap bagging. *British Journal of Mathematical and Statistical Psychology 62*, 427–437.

Wilcox, R. R. (2009c). Comparing Pearson correlations: Dealing with heteroscedasticity and non-normality. *Communications in Statistics–Simulation and Computation, 38*, 2220–2234.

Wilcox, R. R. (2010a). A note on principal components via a robust generalized variance Unpublished Technical Report, Dept. of Psychology, University of Southern California.

Wilcox, R. R. (2010b). Comparing robust nonparametric regression lines via regression depth. *Journal of Statistical Computation and Simulation, 80*, 379–387.

Wilcox, R. R. (2010c). Measuring and detecting associations: Methods based on robust regression estimators that allow curvature. *British Journal of Mathematical and Statistical Psychology, 63*, 379–393.

Wilcox, R. R. (2010d). Regression: Comparing predictors and groups of predictors based on robust measures of association *Journal of Data Science, 8*, 429–441.

Wilcox, R. R. (2011). Comparing two dependent groups: Dealing with missing values. *Journal of Data Science, 9*, 1–13.

Wilcox, R. R. (2012). Nonparametric regression when estimating the probability of success. *Journal of Statistical Theory and Practice, 6*, 1–9.

Wilcox, R. R. (2013). A heteroscedastic method for comparing regression lines at specified design points when using a robust regression estimator. *Journal of Data Science, 11*, 281–291.

Wilcox, R. R. (2015a). Comparing the variances of two dependent variables. *Journal of Statistical Distributions and Applications, 2:7*. DOI: 10.1186/s40488-015-0030-z. URL: http://www.jsdajournal.com/content/2/1/7.

Wilcox, R. R. (2015b). Global comparisons of medians and other quantiles in a one-way design when there are tied values. *Communications Statistics–Simulation and Computation* DOI:10.1080/03610918.2015.1071388.

Wilcox, R. (2015c). Within groups ANCOVA: Multiple comparisons at specified design points using a robust measure of location when there is curvature. *Journal of Statistical Computation and Simulation, 86*, 3236–3246. http://dx.doi.org/10.1080/00949655.2014.962536.

Wilcox, R. R. (2016a). Comparisons of two quantile regression smoothers. *Journal of Modern and Applied Statistical Methods, 15*, 62–77. http://digitalcommons.wayne.edu/jmasm/vol15/iss1/5.

Wilcox, R. (2016b). The running interval smoother: A confidence band having some specified simultaneous probability coverage. Unpublished Technical Report, Dept. of Psychology, University of Southern California.

Wilcox, R. R. (2017). *Introduction to Robust Estimation and Hypothesis Testing*, 4th ed. San Diego, CA: Academic Press.

Wilcox, R. R. (in press a). Linear regression: Heteroscedastic confidence bands for the typical value of Y, given X, having some specified simultaneous probability coverage. *Journal of Applied Statistics.*

Wilcox, R. (in press b). Regression: an inferential method for determining which independent variables are most important. *Journal of Applied Statistics.*

Wilcox, R. (in press c). The regression smoother lowess: A confidence band that allows heteroscedasticity and has some specified simultaneous probability coverage. *Journal of Modern and Applied Statistical Methods.*

Wilcox, R. R., & Charlin, V. (1986). Comparing medians: A Monte Carlo study. *Journal of Educational Statistics, 11*, 263–274.

Wilcox, R. R., Charlin, V., & Thompson, K. L. (1986). New Monte Carlo results on the robustness of the ANOVA F, W, and F^* statistics. *Communications in Statistics–Simulation and Computation, 15*, 933–944.

Wilcox, R. R., & Clark, F. (2013). Robust regression estimators when there are tied values. *Journal of Modern and Applied Statistical Methods, 12*, 20–34.

Wilcox, R. R., & Clark, F. (2015). Robust multiple comparisons based on combined probabilities from independent tests. *Journal of Data Science, 13*, 43–52.

Wilcox, R. R., & Costa, K. (2009). Quantile regression: On inferences about the slopes corresponding to one, two or three quantiles. *Journal of Modern and Applied Statistical Methods, 8*, 368–375.

Wilcox, R. R., & Erceg-Hurn, D. (2012). Comparing two dependent groups via quantiles. *Journal of Applied Statistics, 39*, 2655–2664.

Wilcox, R. R., Erceg-Hurn, D., Clark, F., & Carlson, M. (2014). Comparing two independent groups via the lower and upper quantiles. *Journal of Statistical Computation and Simulation, 84*, 1543–1551. DOI: 10.1080/00949655.2012.754026.

Wilcox, R. R., & Keselman, H. J. (2002a). Power analyses when comparing trimmed means. *Journal of Modern Statistical Methods, 1*, 24–31.

Wilcox, R. R., & Keselman, H. J. (2002b). Within groups multiple comparisons based on robust measures of location. *Journal of Modern Applied Statistical Methods, 1*, 281–287.

Wilcox, R. R., & Keselman, H. J. (2006a). Detecting heteroscedasticity in a simple regression model via quantile regression slopes. *Journal of Statistical Computation and Simulation, 76*, 705–712.

Wilcox, R. R., & Keselman, H. J. (2006b). A skipped multivariate measure of location: One- and two-sample hypothesis testing. In S. Sawilowsky (ed.) *Real Data Analysis* (pp. 125–138). Charlotte, NC: IAP.

Wilcox, R. R., Keselman, H. J., Muska, J., & Cribbie. R. (2000). Repeated measures ANOVA: Some new results on comparing trimmed means and means. *British Journal of Mathematical and Statistical Psychology, 53*, 69–82.

Wilcox, R. R., & Muska, J. (1999). Measuring effect size: A non-parametric analogue of ω^2. *British Journal of Mathematical and Statistical Psychology, 52*, 93–110.

Wilcox, R. R., & Muska, J. (2001). Inferences about correlations when there is heteroscedasticity. *British Journal of Mathematical and Statistical Psychology, 54*, 39–47.

Wilcox, R. R., & Muska, J. (2002). Comparing correlation coefficients. *Communications in Statistics—Simulation and Computation, 31*, 49–59.

Wilcox, R. R., & Keselman, H. J. (2006). Detecting heteroscedasticity in a simple regression model via quantile regression slopes. *Journal of Statistical Computation and Simulation, 76*, 705-712.

Wilcox, R.R., Schönbrodt, F.D. (2016). The WRS package for robust statistics in R (version 0.30). Retrieved from https://github.com/nicebread/WRS.

Wilcox, R. R., & Tian, T. (2011). Measuring effect size: A robust heteroscedastic approach for two or more groups. *Journal of Applied Statistics, 38*, 1359–1368.

Wilcoxon, F. (1945). Individual comparisons by ranking methods. *Biometrics, 1*, 80–83.

Williams, N., Stanchina, J., Bezdjian, S., Skrok, E., Raine A., & Baker, L. (2005). Porteus' mazes and executive function in children: Standardize administration and scoring, and relationships to childhood aggression and delinquency. Unpublished manuscript, Dept. of Psychology, University of Southern California.

Williams, V. S. L., Jones, L. V., & Tukey, J. W. (1999). Controlling error in multiple comparisons, with examples from state-to-state differences in educational achievement. *Journal of Educational and Behavioral Statistics, 24*, 42–69.

Wisnowski, J. W., Montgomery, D. C., & Simpson, J. R. (2001). A comparative analysis of multiple outlier detection procedures in the linear regression model. *Computational Statistics and Data Analysis, 36*, 351–382.

Woodruff, D. L., & Rocke, D. M. (1994). Computable robust estimation of multivariate location and shape in high dimension using compound estimators. *Journal of the American Statistical Association, 89*, 888–896.

Wu, C. F. J. (1986). Jackknife, bootstrap, and other resampling methods in regression analysis. *The Annals of Statistics, 14*, 1261–1295.

Wu, P. C. (2002). Central limit theorem and comparing means, trimmed means one-step M-estimators and modified one-step M-estimators under non-normality. Unpublished doctoral dissertation, Dept. of Education, University of Southern California.

Yanagihara, H., & Yuan, K. H. (2005). Three approximate solutions to the multivariate Behrens-Fisher problem. *Communications in Statistics– Simulation and Computation, 34*, 975–988.

Yohai, V. J. (1987). High breakdown point and high efficiency robust estimates for regression. *The Annals of Statistics, 15*, 642–656.

Yohai, V. J., & Zamar, R. H. (1988). High breakdown point estimates of regression by means of the minimization of an efficient scale. *Journal of the American Statistical Association, 83*, 406–414.

Young, S.G., & Bowman, A.W. (1995). Nonparametric analysis of covariance. *Biometrics 51*, 920Ű–931.

Yu, M. C., & Dunn, O. J. (1982). Robust test for the equality of two correlations: A Monte Carlo study. *Educational and Psychological Measurement, 42*, 987–1004.

Yuan, K.-H., & Chan, W. (2011). Biases and standard errors of standardized regression coefficients. *Psychometrika, 76*, 670–691.

Yuen, K. K. (1974). The two sample trimmed t for unequal population variances. *Biometrika, 61*, 165–170.

Zani, S., Riani, M., & Corbellini, A. (1998). Robust bivariate boxplots and multiple outlier detection. *Computational Statistics and Data Analysis, 28*, 257–270.

Zhu, X., Chen, F., Guo, X., & Zhu, L. (2016). Heteroscedasticity testing for regression models: A dimension reduction-based model adaptive approach. *Computational Statistics and Data Analysis, 103*, 263–283.

Zimmerman, D. W. (2004). A note on preliminary tests of equality of variances. *British Journal of Mathematical and Statistical Psychology, 57*, 173–182.

Zou, H. (2006). The adaptive lasso and its oracle properties. *Journal of the American Statistical Association, 101*, 1418–1429.

Zou, H., & Hastie, T. (2005). Regularization and variable selection via the elastic net. *Journal of the Royal Statistical Society, Series B 67*, 301–320.

Zuo, Y. (2000a). A note on finite sample breakdown points of projection based multivariate location and scatter statistics. *Metrika, 51*, 259–265.

Zuo, Y (2000b). Finite sample tail behavior of the multivariate trimmed mean based on Tukey-Donoho halfspace depth. *Metrika, 52*, 69–75.

Zuo, Y., & Serfling, R. (2000a). General notions of statistical depth function. *Annals of Statistics, 28*, 461–482.

Zuo, Y., & Serfling, R. (2000b). Structural properties and convergence results for contours of sample statistical depth functions, *Annals of Statistics, 28*, 483–499.

Zuur, A. F., Ieno, E. N., & Meesters, E. (2009). *A Beginner's Guide to R*. New York: Springer.

Index

* 9 7 8 1 4 9 8 7 9 6 7 8 1 *
T - #0097 - 020919 - C0 - 279/216/39 [41] - CB - 9781498796781